Synthetic Organic Photochemistry

MOLECULAR AND SUPRAMOLECULAR PHOTOCHEMISTRY

Series Editors

V. Ramamurthy
Professor
Department of Chemistry
Tulane University
New Orleans, Louisiana

Kirk S. Schanze
Professor
Department of Chemistry
University of Florida
Gainesville, Florida

1. Organic Photochemistry, *edited by V. Ramamurthy and Kirk S. Schanze*
2. Organic and Inorganic Photochemistry, *edited by V. Ramamurthy and Kirk S. Schanze*
3. Organic Molecular Photochemistry, *edited by V. Ramamurthy and Kirk S. Schanze*
4. Multimetallic and Macromolecular Inorganic Photochemistry, *edited by V. Ramamurthy and Kirk S. Schanze*
5. Solid State and Surface Photochemistry, *edited by V. Ramamurthy and Kirk S. Schanze*
6. Organic, Physical, and Materials Photochemistry, *edited by V. Ramamurthy and Kirk S. Schanze*
7. Optical Sensors and Switches, *edited by V. Ramamurthy and Kirk S. Schanze*
8. Understanding and Manipulating Excited-State Processes, *edited by V. Ramamurthy and Kirk S. Schanze*
9. Photochemistry of Organic Molecules in Isotropic and Anisotropic Media, *edited by V. Ramamurthy and Kirk S. Schanze*
10. Semiconductor Photochemistry and Photophysics, edited by *V. Ramamurthy and Kirk S. Schanze*
11. Chiral Photochemistry, *edited by Yoshihisa Inoue and V. Ramamurthy*
12. Synthetic Organic Photochemistry, *edited by Axel G. Griesbeck and Jochen Mattay*

Synthetic Organic Photochemistry

edited by

Axel G. Griesbeck
Universität zu Köln,
Köln, Germany

Jochen Mattay
Universität Bielefeld,
Bielefeld, Germany

CRC Press
Taylor & Francis Group
Boca Raton London New York

CRC Press is an imprint of the
Taylor & Francis Group, an **informa** business

CRC Press
Taylor & Francis Group
6000 Broken Sound Parkway NW, Suite 300
Boca Raton, FL 33487-2742

First issued in paperback 2019

ISBN-13: 978-0-8247-5736-6 (hbk)
ISBN-13: 978-0-367-39351-9 (pbk)

Library of Congress Cataloging-in-Publication Data

A catalog record for this book is available from the Library of Congress.

Preface

Photochemistry is a highly valuable tool for modern organic synthesis. It has had a very strong first period of prosperity in the early fifties and sixties of the last century, when numerous light-induced reactions were discovered, modified and applied to synthetic problems. A milestone of development is related to the famous Woodward-Hoffmann rules and the photochemical reactions which served as the experimental basis. Subsequently, expectations from the "synthetic community" were high and the actual output in fact remarkable, as chemists became aware of the reactivity potential of electronic excited states of a molecule. Indeed, photochemistry means in many cases a multiplication of reactivity options, i.e. in addition to the ground state, the electronically excited singlets and triplets show different chemical behavior and often differ so remarkably that they behave as completely new molecules. If you don't want to use them, however, they simply disappear again by radiative or non-radiative pathways (a key principle also of *green chemistry*). The techniques of time-resolved spectroscopy approaching new fascinating time domains ("femtochemistry") in the last decades, enable the photochemistry community to describe a photophysical process in full detail and also to predict photochemical reactivity.

In spite of these exceptional advantages, light-induced reactions were only little by little accepted by the synthetic organic community. Up to 1995, only about 1% of the procedures and reactions in *Organic Syntheses* and in *Organic Reactions* dealt with photochemistry! In our opinion this results from two major problems:

First, photochemistry is always linked to photophysical aspects—the nature of the excited state, its lifetime, its radiative and non-radiative deactivation paths have to be considered (and to be optimized) in order to design a productive light-induced process—which makes things less easy. Second, photochemical processes require appropriate (and sometimes also expensive) equipment, a fact which often discouraged the interested user.

In the meanwhile, interdisciplinary research became more and more essential and photochemistry became a perfect example as a powerful bridge between chemistry, physics, and biology, between material science, life science, and synthesis.

Until recently organic photochemistry has only partially focused on stereoselective synthesis, one of the major challenges and research areas in modern organic synthesis. This situation has dramatically changed in the last decade and highly chemo-, regio-, diastereo- as well as enantioselective reactions have been developed. Chemists all over the world became aware of the fascinating synthetic opportunities of electronically excited molecules and definitely this will lead to a new period of prosperity. Photochemical reactions can be performed at low temperatures, in the solid or liquid state or under gas-phase conditions, with spin-selective direct excitation or sensitization, and even multi-photon processes start to enter the synthetic scenery.

With contributions from 24 international subject authorities, *Synthetic Organic Photochemistry* comprises a leading-edge presentation of the most recent and in-demand applications of photochemical methodologies. Outlining a wide assortment of reaction types, entailing cycloadditions, cyclizations, isomerizations, rearrangements, and other organic syntheses, this reference also ties in critical considerations that overlap in modern photochemistry and organic chemistry, such as stereoselectivity. Select experimental procedures demonstrate the industrial and academic value of reactions presented in the text.

Containing a remarkable 2113 references, this volume

- reviews [2+2], [4+2], and ene photooxygenation reactions,
- illustrates photocycloadditions of alkenes to excited alkenes or carbonyls,
- clarifies abstractions of γ- and ($\gamma \pm n$) hydrogens by excited carbonyls,
- describes di-π-methane and oxa-di-π-methane rearrangements,
- explores photoinduced electron transfer cyclizations using radical ions,
- studies photogenerated nitrene additions to π-bonds,
- surveys reactions with photoinduced aromatic nucleophilic substitutions,

- covers several other photochemical methods for specific syntheses,
- inspects medium effects on photoinduced processes.

Synthetic Organic Photochemistry is an ideal resource for photochemists, organic and physical chemists, and graduate students in these disciplines.

Axel G. Griesbeck
Jochen Mattay

Contents

Contributors

Hans-Werner Abraham Institut für Chemie der Humboldt-Universität zu Berlin, Berlin, Germany

Antonia R. Agarrabeitia Departamento de Quimica Organica I, Facultad de Ciencias Quimicas, Universidad Complutense, Madrid, Spain

Angelo Albini Dipartimento Chimica Organica, Università di Pavia, Pavia, Italy

Diego Armesto Departamento de Quimica Organica I, Facultad de Ciencias Quimicas, Universidad Complutense, Madrid, Spain

Jens Otto Bunte Organische Chemie I, Fakultät für Chemie, Universität Bielefeld, Bielefeld, Germany

Edward L. Clennan Department of Chemistry, University of Wyoming, Laramie, Wyoming, U.S.A.

Maurizio Fagnoni Dipartimento Chimica Organica, Università di Pavia, Pavia, Italy

Steven A. Fleming Department of Chemistry and Biochemistry, Brigham Young University, Provo, Utah, U.S.A.

Axel G. Griesbeck Institut für Organische Chemie, Universität zu Köln, Köln, Germany

Norbert Hoffmann UMR CNRS et Université de Reims, Champagne-Ardenne, Reims, France

Maria Rosaria Iesce   Università di Napoli Federico II, Napoli, Italy

Yoshihisa Inoue   Department of Molecular Chemistry, Osaka University and ICORP/JST, Suita, Japan

V. Jayathirtha Rao   Indian Institute of Chemical Technology, Hyderabad, India

Lakshmi S. Kaanumalle   Department of Chemistry, Tulane University, New Orleans, Louisiana, U.S.A.

Paul Margaretha   University of Hamburg, Hamburg, Germany

Jochen Mattay   Organische Chemie I, Fakultät für Chemie, Universität Bielefeld, Bielefeld, Germany

Tadashi Mori   Osaka University, Suita, Japan

Olaf Mühling   Institut für Organische Chemie der Humboldt-Universität zu Berlin, Berlin, Germany

Arunkumar Natarajan   Department of Chemistry, Tulane University, New Orleans, Louisiana, U.S.A.

Michael Oelgemöller   Bayer CropScience K.K., Yuki City, Ibaraki, Japan

Maria J. Ortiz   Departamento de Quimica Organica I, Facultad de Ciencias Quimicas, Universidad Complutense, Madrid, Spain

V. Ramamurthy   Department of Chemistry, Tulane University, New Orleans, Louisiana, U.S.A.

Roberto A. Rossi   Universidad Nacional de Córdoba, Ciudad Universitaria, Córdoba, Argentina

Scott McN. Sieburth   Department of Chemistry, Temple University, Philadelphia, Pennsylvania, U.S.A.

Peter J. Wagner   Department of Chemistry, Michigan State University, East Lansing, Michigan, U.S.A.

Pablo Wessig   Institut für Organische Chemie der Humboldt-Universität zu Berlin, Berlin, Germany

1
Synthetic Organic Photochemistry

Axel G. Griesbeck
Institut für Organische Chemie, Universität zu Köln, Köln, Germany

Jochen Mattay
Universität Bielefeld, Bielefeld, Germany

1.1. INTRODUCTION

This book is on synthetic organic photochemistry. With emphasis on "synthetic."

Considering only the electronically excited states of organic molecules relevant for photochemistry, essentially the first excited singlet and triplet states (in rare cases, contrary to the expectations of the Kasha rule, also the second excited states), the reactivity options are *triplicated* when compared to ground-state reactivity. Furthermore, photoinduced electron transfer (PET) is capable of activating organic substrates by one-electron oxidation and/or reduction and thus, in optimal cases, singlet excited states are available to direct photon excitation, triplet states by triplet sensitization (or via rapid and efficient ISC from the corresponding excited singlet), radical anions by PET with an appropriate donor (e.g., an amine or thioether), and radical cations by PET with an appropriate acceptor (e.g., cyanoaromatics or pyrylium salts). Reactivity options are thus *quintupled* when compared to ground-state reactivity.

The photochemists toolbox has several selective instruments in store to check this multiplication in reactivity:

(a) Selective *direct excitation* is possible using monchromatic light sources or appropriate filter systems (either liquid filter solutions of glass filters) when applying polychromatic light sources.

1

Additonally, triplet quenchers can be added in order to selectivity pattern the reactivity of the first excited singlet state;

(b)   Selective generation of the *triplet excited state* is possible by applying appropriate triplet sensitizers, most convenient (whenever possible) is the use of sensitizing solvents such as acetone. Less frequently used, but important for mechanistic analyses is the use of chemoluminescent precursors which thermally decompose to triplet excited states (as often applied for triplet carbonyl photochemistry and the correspoding 1,2-dioxetanes as triplet precursors);

(c)   Using the Rehm-Weller equation(s), an estimate of the optimal redox properties of PET sensitizers is possible and thus the selective and quantum-efficient generation of either radical anion or radical cation of the photoactive substrate. In *photoinduced electron transfer processes* either the oxidant or the reducant can be excited electronically, thus this parameter is an additional bonus added to the PET process.

Yield is an evergreen in synthetic chemistry; but, in addition to chemical yields, also quantum yields have to be considered when trying to design an efficient photochemical process. Quantum yields for photochemical reactions Φ_r vary from zero (very bad) to 10^5 (very good, but a speciality for some light-initiated radical chain reactions). Reaction quantum yields of 2–10 look attractive and indicate quantum-chain processes often to be found in PET processes like PET-oxygenations, quantum yields of 0.1 to 1.0 are typical for carbonyl photochemistry like Norrish processes or Paternò-Büchi reactions. Even lower quantum yields can still come along with good chemical yields, but they are a waste of energy and quite often are highly sensitive against competing reactions and require defined reaction conditions.

Reports concerning the mechanisms of photochemical processes have been published in overwhelming quantity in the last decades. These investigations often originate from the interface of organic, physical, and theoretical chemistry as well as laser spectroscopy (cf. femtochemistry). Though often highly simplifying, the results prove to be useful for problems in synthetic organic chemistry. Especially all aspects of stereoselectivity have become more and more important in the last decades and also found their way into modern organic photochemistry. This will become clear from the following chapters, where stereochemistry plays a central role in nearly all processes described and nearly all target molecules choosen.

A glance at Fig. 1 already illustrates the broadness of photochemical reactions, ranging from carbonyl reactions and photooxygenations to

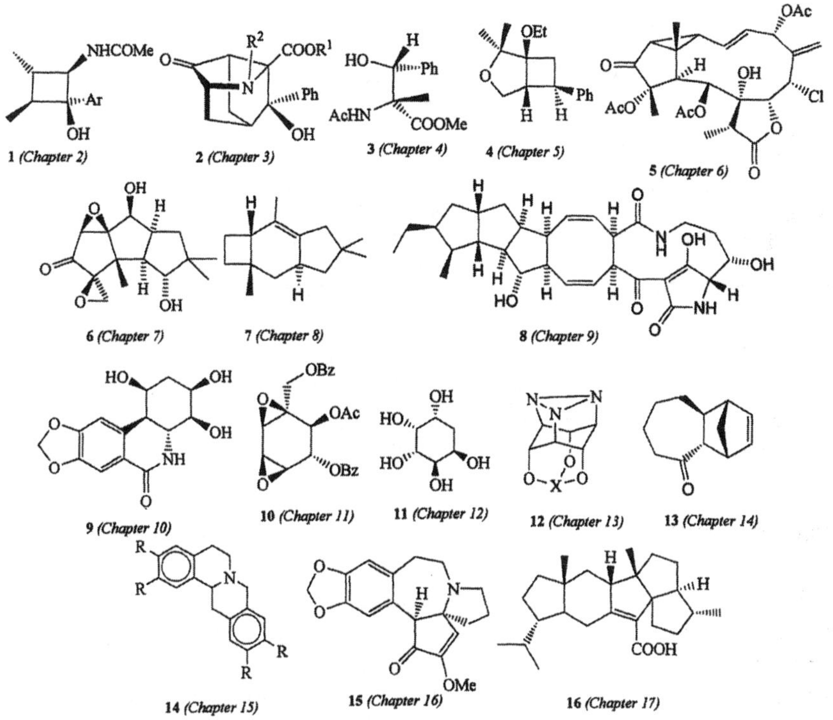

Figure 1 Illustrative target molecules chosen from 16 chapters.

electron transfer cyclizations and nitrene additions. The chapters in this book are ordered in the way of group transformations, i.e. hydrogen transfer and cycloadditions to carbonyl groups, alkene cycloadditions, transformations of 1,4-dienes, β,γ-unsaturated carbonyl compounds, α,β-unsaturated carbonyl compounds, 1,3-diene-photocycloadditions, photocyclizations induced by electron transfer, oxygenation, amination by nitrene addition, alkene photoisomerization, activation of benzylic functions, aromatic substitution, ortho-, meta-, and para-photocycloaddition to arenes. As an outlook into the possible future of photochemistry, the last chapter describes photochemistry in constrained media. Seventeen senior authors and further coauthors have compiled this collection of reactions which includes more than 2000 references.

The first three chapters by Wagner, Wessig, and Griesbeck deal with typical carbonyl chemistry: Norrish type II reactions followed by Yang-cyclization, homologous Norrish type II reactions (i.e. hydrogen abstractions from non γ-positions), and Paternò-Büchi [2+2]-photocycloadditions. The enantiomerically pure β-amido-cyclobutanol **1** is formed from a chiral

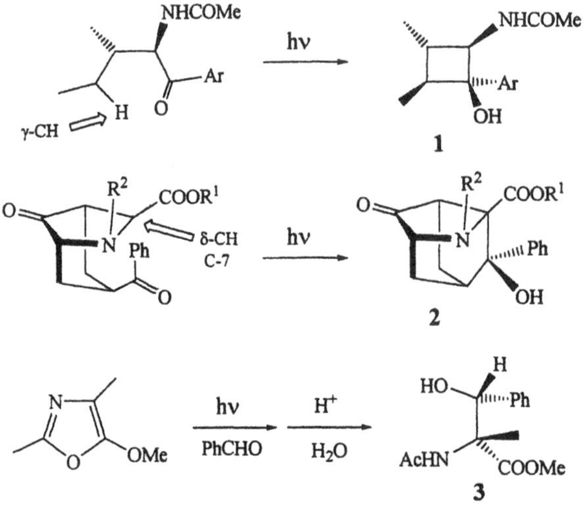

Figure 2   Photochemical hydrogen transfer and carbonyl cycloaddition.

valerophenone derivative (from isoleucine—a readily available compound from the pool of chiral amino acids) [1]. Two new stereogenic centers are formed in a highly controlled fashion via an interemediate 1,4-biradical which is formed by a γ-hydrogen transfer to the excited carbonyl triplet. When conformationally possible, other than γ-hydrogens can be active in homolytic transfer as outlined for the synthesis of the enantiomerically pure tricyclic α-amino acid **2**. On irradiation of the bicyclic precursor diketone, prepared in a few steps from cheap 4-hydroxyproline, a fully stereoselective cyclization to the tricyclic amino acids **2** is observed. In the photochemically initiated CH-transfer step, the stereogenic center at C-7 is destroyed. The (triplet) biradical formed can, however, only be attacked from one side, and thus, no epimerization is observed at C-7 [2].

A completely different product class, α-amido, β-hydroxy carboxylic acids, can be obtained from the [2+2]-photocycloaddition of aldehydes to 5-methoxyoxazoles and subsequent hydrolysis of the bicyclic oxetanes [3]. Compound **3** is available from the triplet benzaldehyde addition to dimethyl 5-methoxyoxazole in diastereomerically pure (*erythro*-selective) form.

The next three chapters by Fleming, Armesto, and Rao deal with different aspects of alkene photochemistry: alkene [2+2]-photocycloadditions to other alkenes, di-π-methane (DPM) rearrangements of 1,4-dienes and oxa-di-π-methane (ODPM) rearrangements of β,γ-unsaturated carbonyl compounds. Photocycloaddition of an ether-tethered 1,6-diene by Cu(I)-catalysis leads to the *exo*-selective formation of the bicyclic tetrahydrofuran derivative **4** [4]. By direct electronic excitation of a

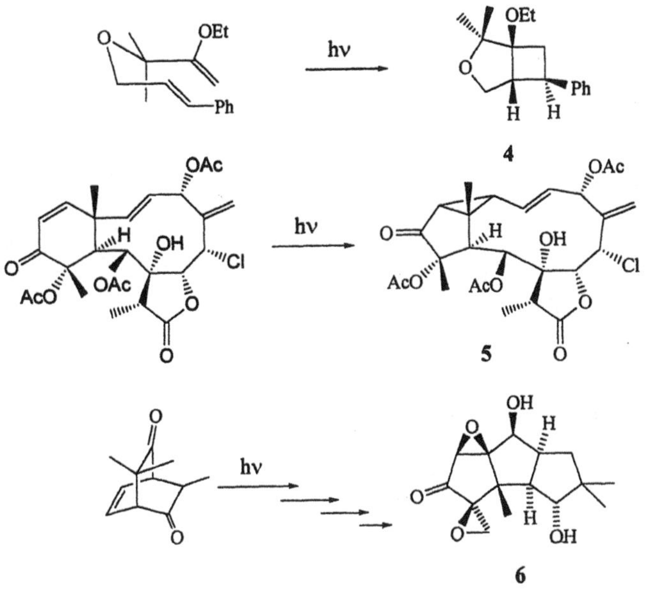

Figure 3 Alkene photocycloaddition and DPM/ODPM rearrangements.

deconjugated enone/ene system, the DPM rearrangement product *erythrolide A* **5** was obtained in excellent yields [5]. This is probably the first observation of the involvement of this specific photoreaction in the production of natural products. Solvent triplet-sensitized (photolysis in acetone) ODPM rearragement of a bifunctional enone results in the formation of a diquinane, that could be converted, by subsequent transformations involving the annulation of the third five-membered ring as well as epoxidation and hydroxylation steps to the natural product *corioline* (**6**) [6].

The next three chapters are by Margaretha, Sieburth, and Mattay, and describe enone photochemistry, [4+4]-photocycloaddition reactions, and photoinduced electron transfer processes for the synthesis of ring systems. The fungal metabolite *sterpurene* (**7**) can be obtained in a multistep sequence starting with a [2+2] enone-ene photocycloaddition resulting in a bicyclooctanone from 3-methylcyclohex-2-enone and ethylene [7]. A spectacular [4+4]-photocycloaddition reaction transforms the macrocyclic polyunsaturated ring system of *alteramide A* in quantitative yield into an annulated 1,5-cyclooctadiene simply by irradiation with solar light [8]. A ring/substituent pattern which is present in isoquinolines such as *2,7-dideoxypancratistatin* is generated by photoinduced electron transfer cyclization of an electron-rich benzalamine derivative and results in the formation of the tetrahydroisoquinoline **9** [9].

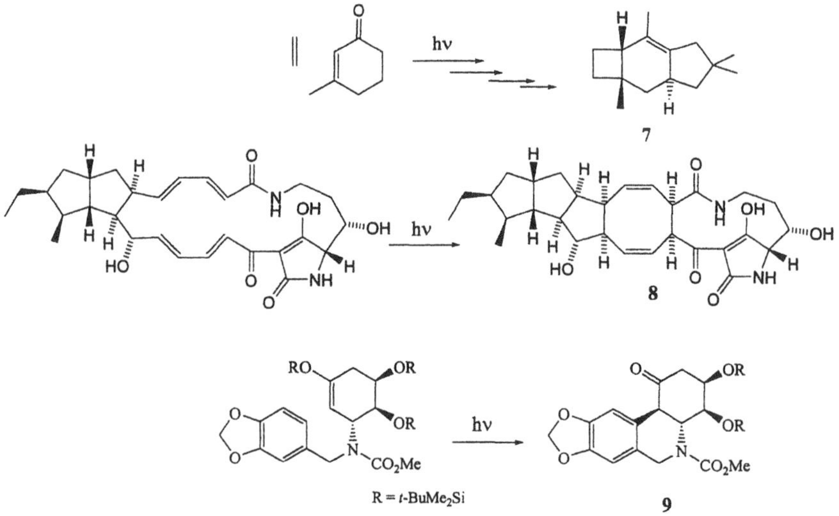

Figure 4   Enone-, 4 + 4-cycloadditions and PET reactions.

The next three chapters deal with the generation of highly reactive intermediates, singlet oxygen and nitrenes, respectively. In contrast to the chapters before, these approaches involve heteroatom transfer and not the formation of C–C bonds. Three "type II processes", i.e. reactions with singlet molecular oxygen, are described in detail by Iesce and Clennan: [2+2] and [4+2]-cycloaddition reactions as well as ene reactions. The naturally occuring cyclohexane diepoxides are a favorable group of target molecules for singlet oxygen [4+2]-cycloaddition: the enantiomerically pure *boesenoxide* (**10**) is produced from a cyclohexa-1,3-diene precursor and two deprotection/ protection steps [10]. The synthesis of the pentol *talo-quercitol* (**11**) starts with cyclohexa-1,4-diene which is thermally dioxygenated and acetal-protected. Singlet oxygen ene-reaction delivers an allylic hydroperoxide which is transformed into **11** [11]. In a third chapter, nitrene generation and addition to C–C double bonds are described by Abraham. These reactions can give rise to unusual structures and, as shown for the *cis,cis*-trialkyltriaziridine **12**, also addition to N–N double bonds is possible [12].

The next three chapters are by Inoue and Mori, Albini, and Rossi, and deal with alkene photoisomerization reactions, the modification of benzylic positions and photochemical aromatic substitution reactions. (*E*)-2-cyclo-heptenone is produced upon irradiation of the Z-isomer at −50 °C and can be trapped by cyclopentadiene to afford the adduct **13** [13]. Benzyl-substituted dihydroisoquinolinium derivatives can be used for the photochemical synthesis of tetrahydroisoquinolines. The corresponding

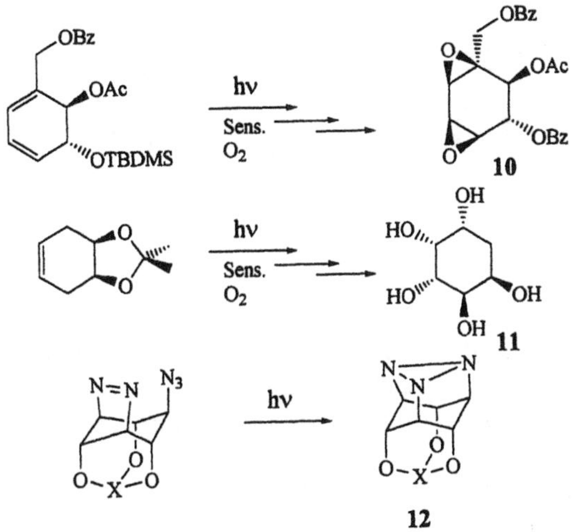

Figure 5 Singlet oxygen and nitrene reactions.

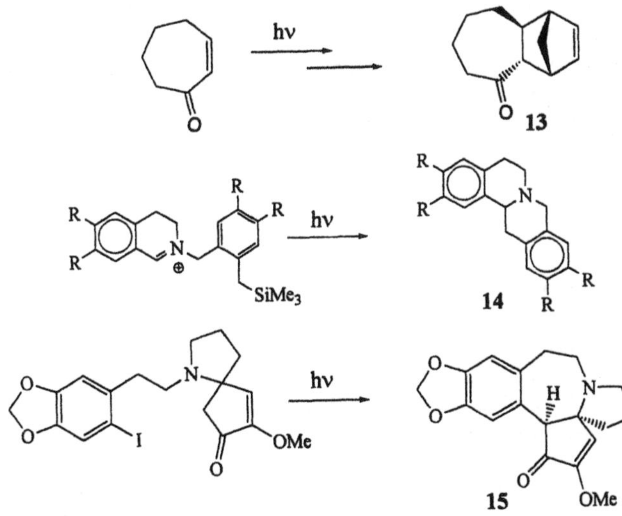

Figure 6 Photoisomerization and photosubstitutions.

perchlorates have been successfully cyclized in the synthesis of the
protoberbine alkaloids *xylopinine* and *stylopine*. The reaction proceeds via
SET from the xylyl donor to the iminium moiety, fragmentation of the
benzylsilane radical cation and carbon–carbon bond formation to give **14**

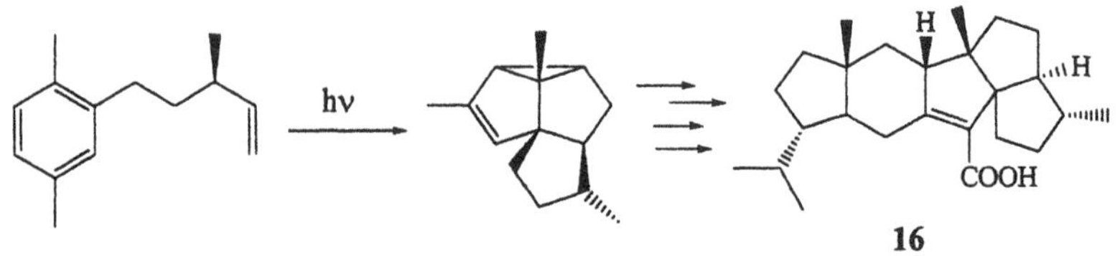

Figure 7   Arene photochemistry.

[14]. Photochemical aromatic substitution initiated by a reductive step as in S$_{RN}$1 reactions can be used for the synthesis of *cephalotaxinone* (**15**). The corresponding iodoketone precursor cyclizes in liquid ammonia under photolysis [15].

The last two chapters by Hoffmann and Ramamurthy deal with a collection of photochemical reactions with arenes, the *ortho-*, *meta-* and *para* photocycloadditions and with a conceptually exciting concept in organic photochemistry, the use of contrained media. *Retigeranic acid* (**16**, by formal asymmetric synthesis) was synthesized via a fabulous reaction sequence involving an intramolecular *meta* photocycloaddition as key step [16].

These examples and many more can be found in the following chapters. Additional examples including experimental details have also been collected by us in an experimental course book [17].

Cologne and Bielefeld, August 2003.

REFERENCES

1. Griesbeck AG, Heckroth H. J Am Chem Soc 2002; 124:396.
2. Wessig P. Synlett 1999; 1465.
3. Griesbeck AG, Bondock S. Can J Chem 2003; 81:555.
4. Avasthi K, Raychaudhuri SR, Salomon RG. J Org Chem 1984; 49:4322.
5. Look SA, Fenical W, Van Engen D, Clardy J. J Am Chem Soc 1984; 106:5026.
6. Demuth M, Ritterskamp P, Weigt E, Schaffner K. J Am Chem Soc 1986; 108:4149.
7. Ishii S, Zhao S, Mehta G, Knors CJ, Helquist P. J Org Chem 2001; 66:3449.
8. Shigemori H, Bae M-A, Yazawa K, Sasaki T, Kobayashi J. J Org Chem 1992; 57:4317.
9. Pandey G, Murugan A, Balakrishnan M. Chem Commun 2002; 624.
10. Hathaway SJ, Paquette LA. Tetrahedron 1985; 41:2037.
11. Maras A, Secen H, Sütbeyaz Y, Balci M. J Org Chem 1998; 63:2039.
12. Klingler O, Prinzbach H. Angew Chem Int Ed 1987; 26:566.
13. Corey EJ, Tada M, LaMahieu R, Libit L. J Am Chem Soc 1965; 87:2051.
14. Ho DG, Mariano PS. J Org Chem 1988; 53:5113.

15. Semmelhack MF, Chong BP, Stauffer RD, Rogerson TD, Chong A, Jones LD. J Am Chem Soc 1975; 97:2507.
16. Wender PA, Singh SK. Tetrahedron Lett 1990; 31:2517.
17. Mattay J, Griesbeck A. Photochemical Key Steps in Organic Synthesis. Weinheim, New York, Basel, Cambridge, Tokyo: VCH, 1994.

2

Abstraction of γ-Hydrogens by Excited Carbonyls

Peter J. Wagner
Michigan State University, East Lansing, Michigan, U.S.A.

2.1. HISTORICAL BACKGROUND

2.1.1. General Summary

Although the occurrence of intermolecular hydrogen atom abstraction from solvent by excited ketones was known one hundred years ago, understanding of its intramolecular counterpart was achieved only in the past 40 years. A wide variety of carbonyl compounds undergo photoinduced intramolecular hydrogen atom abstraction to form biradicals, which then undergo two common competing reactions: coupling to produce cyclic alcohols, and disproportionation back to ketone or to various enols. The next chapter of this book is devoted specifically to the chemistry initiated by β-, δ-, and more remote intramolecular hydrogen abstractions. The 1,4-biradicals formed by γ-hydrogen abstraction undergo unique reactions that other biradicals do not, and they do not disproportionate to enols as other biradicals can. Thus this special chapter.

There are four distinct processes initiated by γ-hydrogen abstraction in excited carbonyl compounds: Norrish type II photoelimination, Yang photocyclization (cyclobutanol formation), Yang photoenolization (*o*-xylylenol formation), and β-cleavage of radicals from carbons adjacent to the radical sites of the 1,4-biradicals. Some of these require unique structures and generate distinct products.

Since hydrogen atom abstraction by an excited carbonyl closely resembles hydrogen atom abstraction by alkoxy radicals, it is mainly compounds with n,π^* excited states, namely ketones and aldehydes, that

11

perform this reaction. There are a few reports of simple esters that undergo photo-induced reactions, but carboxylic acid derivatives in general are not very reactive, since they do not have low-energy n,π^* excited states [1]. An exception occurs when electron transfer from a γ-substituent to the excited carbonyl induces a 1,5-proton transfer [2].

2.1.2. 1,4-Biradical Formation from Simple Ketones

The first and still best known example of such intramolecular hydrogen abstraction is the "type II" photoelimination discovered by Norrish [3], who found that dialkyl ketones with γ C–H bonds cleave to methyl ketones and alkenes rather than to acyl and alkyl radicals, the earlier discovered "Type I" cleavage. Later workers found that both cleavage processes compete in certain ketones and that overall quantum yields are particularly low whenever the type II process occurs. For years the "type II" reaction was considered to be concerted; a 1,5-hydrogen transfer together with C–C bond cleavage in a six-atom cyclic transition state could lead to the alkene and the enol tautomer of the product ketone. Such a process would now be called a retro-ene reaction. In a key experiment, Calvert and Pitts verified by IR that the enol is indeed formed first and then is rapidly converted to ketone [4]. However, the Yangs had already discovered competing cyclobutanol formation and suggested that cleavage and cyclization both arise from a 1,4-biradical intermediate formed by γ-hydrogen abstraction by the excited carbonyl group [5]. Nonetheless, suggestions were still made that the cyclization could occur concertedly.

Before the concerted vs. two-step question was further elucidated, another basic mechanistic puzzle was raised. One research group found that type II cleavage of 2-pentanone was quenched by biacetyl [6], which was known to quench excited triplets rapidly. Another group found that the reaction of 2-hexanone was not quenched under the same conditions [7]. The two groups obviously differed as to which excited state undergoes the reaction. The apparent conflict was neatly solved by the revelation that each of the two ketones reacts from both states, with 2-hexanone undergoing more unquenchable singlet reaction than 2-pentanone [8,9].

2.1.3. Normal Behavior of 1,4-Biradicals

Before the mid-1960s, most studies were performed on aliphatic ketones until Wan and Pitts reported that phenyl alkyl ketones also undergo the reaction [10]. Wagner and Hammond then showed that the type II reaction of phenyl ketones is completely triplet derived and suggested that its notoriously low quantum yields are caused by disproportionation of Yang's

1,4-biradical intermediate back to ketone [11]. Wagner soon discovered that added Lewis bases markedly increase the quantum yields of triplet type II reactions, often to 100% [12,13]. This behavior was attributed to suppression of biradical reversion to ketone by hydrogen bonding of the biradical's hydroxy group to the Lewis base: the H-bond is broken by disproportionation, but not by cleavage or cyclization. In 1972 Wagner and Zepp succeeded in trapping the biradicals with mercaptans [14]. Yang and Elliot had already found that the triplet, but not singlet, component of 5-methyl-2-hexanone's photoreactivity also produces extensive racemization at the γ-carbon, which could be caused only by disproportionation of a 1,4-biradical [15]. Wagner and Kelso found that (+)-4-methyl-1-phenyl-1-hexanone also undergoes photoinduced racemization, the quantum yield for racemization equaling 1 minus the quantum yield for type II products [16]. This fact equated racemization at the γ-carbon with what had been called "radiationless decay" and thus proved that what had been thought to be physical decay of the excited state instead was chemical reversion of a biradical intermediate to ground state ketone [16]. All of this work firmly established 1,4-biradicals as intermediates in the triplet reaction but left open how much they are involved in the singlet reaction. Some unquenchable cyclization reactions indicate that singlet 1,4-biradicals are indeed formed, even if they may not account for all cleavage reaction [17]. Although cleavage and cyclization of 1,4-biradicals are linked mechanistically, the cleavage process is so easy to measure that Norrish type II elimination has been widely studied to gain basic mechanistic information about biradicals and about hydrogen abstraction reactions of excited ketones.

Phenylglyoxylate esters undergo type II elimination in a unique fashion in that the alcohol portion is oxidized to a ketone or aldehyde while the benzoylcarboxy portion forms a hydroxyketene [18]. The intermediate 1,4-biradicals do not cyclize, presumably because of the strain in a beta-lactone.

In contrast, Urry and Trecker had shown in 1962 that α-diketones undergo photocyclization to 2-hydroxycyclobutanones, the intermediate biradical cyclizing but not cleaving (which would form a hydroxyketene) [19].

2.1.4. β-Cleavage of Radicals from 1,4-Biradicals

Various δ-substituted valerophenones [20] and 2-halo- or 2-thiylethanol esters of phenylglyoxylic acid [21] form 1,4-biradicals that undergo radical β-cleavage of the substituents to form 4-benzoyl-1-butene or vinyl phenylglyoxylate, respectively, together with HX or RSH. This β-cleavage of halogen atoms and thiyl radicals from biradicals was an important discovery: that biradicals can behave like independent monoradicals at two different sites demonstrated that they are aptly named. Moreover, in the valerophenone study prior knowledge of biradical lifetimes allowed the determination of relative rate constants for many common radical β-cleavage reactions. This knowledge then allowed estimation of lifetimes for the glyoxylate 1,4-biradicals.

2.1.5. Photoenolization of *o*-Alkylphenyl Ketones

In 1961 Yang and Rivas reported that irradiation of some *o*-alkyl benzophenones in the presence of a good dienophile yields products of Diels–Alder addition of the dienophile to an *o*-xylylene. They surmised that γ-hydrogen abstraction occurs from the *o*-alkyl group, producing a conjugated 1,4-biradical that relaxes to an enolic *o*-xylylene [22]. That methanol-*O*-d resulted in benzylic deuteration of the starting ketone also suggested the presence of an enolic intermediate. There have been several forms of photo-induced enolization reported since Yang's discovery of this example, which can properly be called Yang photoenolization.

For some time the synthetic potential of this reaction as a source of Diels-Alder adducts underwent considerable study. One outcome of these studies was the realization that of the four possible *o*-xylylenol isomers, only ones with the enolic OH group pointed out (the "*E*-photoenol") reacted with dienophiles [23]. Mechanistic studies picked up in the 1970s, after Matsuura and Kitaura reported that, in the absence of dienophiles, benzocyclobutenols are formed from 2,6-dialkylphenyl ketones but not from simple *o*-alkylphenyl ketones [24]. Previously the absence of cyclobutenol products had been quite puzzling and led to suggestions that they were formed from the initial biradical but underwent rapid electrocyclic opening to the *o*-xylylenols.

In the mid-1970s it was discovered that there are two kinetically distinct triplets of *o*-alkylphenyl ketones, one with the carbonyl oxygen syn to the *o*-alkyl group and thus highly reactive, and the anti rotamer which eventually abstracts a hydrogen from the *o*-alkyl group after irreversible rate-determining rotation to the syn isomer [25]. This finding prompted Wirz and coworkers to see if there were two kinetically distinct rotamers of the *o*-xylylenols; and flash kinetics showed that there is a very short-lived one (~ 100 ns) and a quite long-lived one (ms) [26]. The former was identified as the *Z*-isomer and its rapid decay was ascribed to a highly favorable 1,5-sigmatropic rearrangement in which the enolic proton essentially protonates the other end of a dienol to regenerate the starting ketone. The realization that both syn and anti *o*-xylylenols are formed and that the former rapidly disappears explained why only the anti isomer is trapped by dienophiles. This was just one of many studies of this reaction by flash kinetics, and over the decades structural assignments have had to be altered as the presence of additional intermediates was realized. For example Scaiano and coworkers found that the initial 1,4-biradical, which is the triplet state of the o-xylylenol, decays relatively slowly ($\sim$ 1 µs) to the ground state o-xylylenol [27].

In 1991 Wagner and coworkers discovered that simple o-alkylphenyl ketones do indeed form stable benzocyclobutenols, and that their formation is quenched by the presence of acid [28]. This finding established that the benzocyclobutenols are formed from the o-xylylenols, which of course undergo very rapid acid catalyzed reketonization. Moreover, the benzo-cyclobutenols are formed as single diastereomers in most cases, which indicates that only one of the four possible xylylenol isomers undergoes electrocyclization. The reason for their late discovery is that at temperatures above 60° they undergo thermal opening to o-xylylenols which reketonize.

2.2. STATE OF THE ART MECHANISTIC MODEL

2.2.1. Acyclic Ketones

Scheme 1 shows the chronology of reactions involving γ-hydrogen abstraction in simple ketones [29]. Light absorption produces an n,π^* singlet that undergoes competing hydrogen abstraction and intersystem crossing to a triplet state. The hydrogen abstraction process produces an avoided crossing or conical intersection wherein a fraction α of the developing biradical forms a metastable biradical and the rest reverts to ground state reactant. The singlet biradical undergoes mainly cleavage to enol and olefin.

The triplet ketone forms a biradical that can cleave, cyclize, and revert to starting ketone by disproportionation. The presence of Lewis base

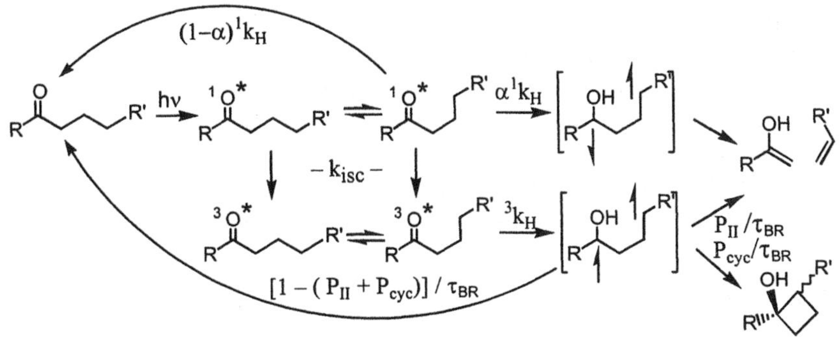

Scheme 1

solvents or additives suppresses this disproportionation and enhances quantum yields of products, the percent suppression increasing with concentration and basicity of the additive. Thus n,π^* states of both multiplicities undergo hydrogen abstraction in ways that cause large scale reversion to ground state reactant.

The scheme contains a tacit message, namely that hydrogen abstraction requires specific orientations of the carbonyl oxygen and the C–H bond with respect to each other. Flexible acyclic ketones exist in several conformations, some or most of which are unreactive for geometric reasons. For some time it was thought that the C–H bond needed to be positioned such that the H lies on the long axis of the oxygen's half-occupied n orbital, that being the verdict from several theoretical analyses of the reaction. That may indeed be the most favorable intrinsic transition state orientation for just the C=O and the H, but it clearly is not a necessity when the rest of the molecule does not cooperate. In situations that provide rigidity to the molecule, such as in keto-steroids and in crystalline ketones, hydrogen abstraction occurs even when the C=O–H alignment is far from ideal. Scheffer has reported many examples of efficient cyclization and cleavage of crystalline ketones, where molecular rigidity is enforced by the crystal lattice. His results indicate that hydrogen abstraction occurs over a fairly wide range of C=O–H–C dihedral angles and that the distance between O and H is the major determinant of relative rates of γ-hydrogen abstraction [30].

Another puzzle is provided by the fact that singlet 1,4-biradicals undergo little, if any, cyclization and disproportionation back to ketone, whereas triplet 1,4-biradicals undergo both, often in large measure. In that regard, the scheme does not include separate steps for intersystem crossing of the triplet biradical to the singlet state before it forms products. Current thinking is split between such a two-step process and a one step process whereby movement along a reaction coordinate induces sufficient singlet character in the biradical to allow formation of products. Their relatively long lifetimes allow triplet biradicals to react along three different reaction coordinates. It is commonly thought that singlet biradicals react so rapidly that they form only the product closest in geometry to the biradical geometry. However, the initial biradical formed by singlet state hydrogen abstraction has a geometry more suited for cyclization than cleavage, but it tends to undergo mostly the latter, probably because of entropic differences.

2.2.2. α-Ketoesters

The photocleavage of phenylglyoxylate esters is a simple Norrish type II process with a secondary mechanistic twist, namely that continued irradiation causes the hydroxyketene to decompose to benzaldehyde and carbon monoxide rather than tautomerize to phenylglyoxal. A similar oxidation of the alcohol portion occurs in α-alkoxyacetophenones [31].

2.2.3. α-Diketones

Photocyclization of α-diketones is unique in that hydrogen abstraction apparently occurs only from the inner lobe of one carbonyl's n orbital, such that acylcyclobutanols are not formed. The behavior was once thought to reflect mainly a conformational preference but through bond coupling of the two oxygen n orbitals is also responsible, as the zwitterionic biradical structure of the n,π^* state depicts. Some early reports that photoinduced cleavage to two acyl radicals occurs appear to involve hydrates, diketones being particularly susceptible to addition of water.

neither observed

2.2.4. β-Cleavage of Radicals from 1,4-Biradicals

The β-cleavage of halogen atoms and thiyl radicals from biradicals appears to follow a simple enough mechanism, as shown in Sch. 2. After being formed, the 5-substituted 1,4-biradical can undergo radical cleavage in competition with its normal decay reactions. That leaves an

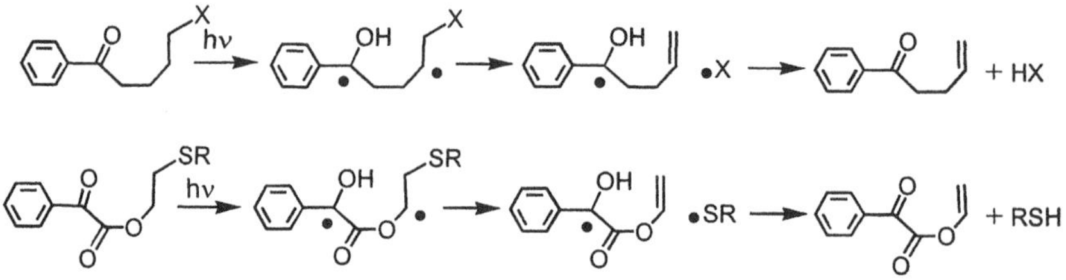

Scheme 2

electron-deficient atom or radical next to an electron-rich α-hydroxy radical. Consequently the hydroxy hydrogen gets transferred either as an atom by a charge transfer induced disproportionation or as a proton following complete electron transfer. Coupling of the two initial radicals would form an unstable pseudo-hemiketal that also would revert to ketone.

2.2.5. *o*-Alkylphenyl Ketones

As described in the historical section, it has taken some 40 years for the development of a relatively complete mechanism for the behavior of the 1,4-biradicals formed by excited *o*-alkylphenyl ketones and aldehydes. Although there is evidence that some acetophenones react from their n,π^* singlets, most enolization occurs from triplet n,π^* states [26]. The excited state hydrogen abstraction occurs from the *syn* isomer and thus the biradical initially generated has the OH in a *syn* geometry. Given the conjugation of the excited benzoyl group, the hydroxy radical site initially is conjugated to the benzene ring, while the other site is twisted orthogonally to obey Hund's rule about two same-spin electrons not occupying the same orbital. Since *o*-xylylenols are formed with their OH groups both *syn* and *anti*, the initial biradical must be able to rotate its two radical sites. Since this twisted biradical is the excited triplet of the *o*-xylylenol product [27], it converts to ground state xylylenol by a 90° twist of the nonconjugated radical site, just as triplet alkenes and dienes do. It is this twist that produces two xylylenols.

Since the stereochemistry of the benzocyclobutenol and Diels–Alder products is determined by the geometry of the xylylenols and both reactions occur stereoselectively, it became important to understand why specific *o*-xylylenols are formed from a given carbonyl compound. Since the initial triplet biradical can undergo conformational conversion, it likely relaxes to ground state *o*-xylylenols from its lowest energy rotamer, the geometry of which determines which two xylylenols are formed. Comparison of the selectivities displayed by several *ortho*-substituted acylbenzenes indicated that the *o*-xylylenols are formed from a biradical in which the radical site with

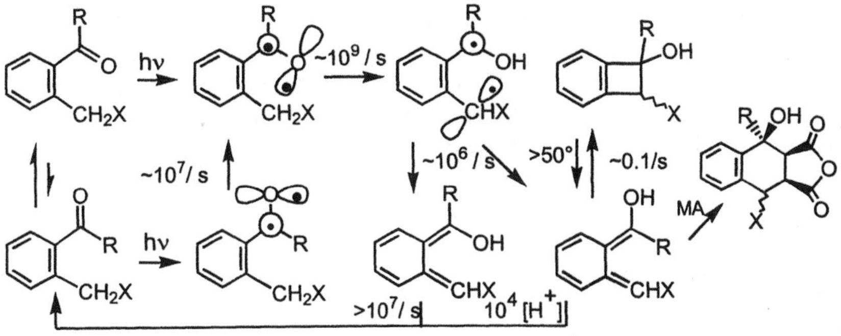

Scheme 3

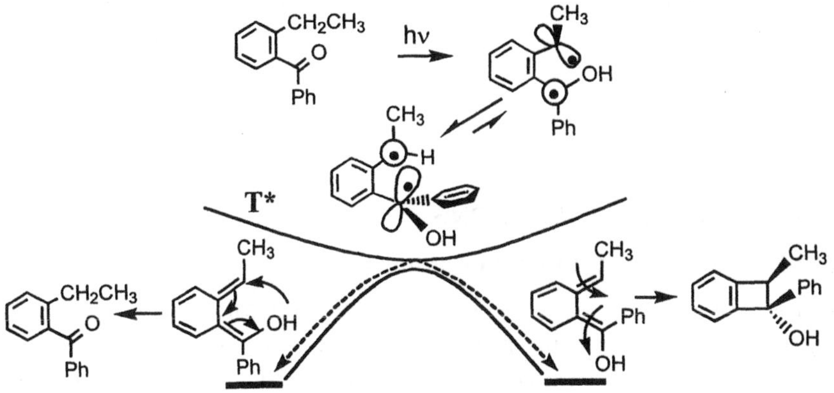

Scheme 4

the worst radical-stabilizing substituents is conjugated with the central benzene ring and the site with the best radical-stabilizing substituents is twisted 90° out of conjugation with the central benzene ring [32]. The *syn/anti* orientation of the two substituents in the conjugated site, determined by steric factors, is maintained in both *o*-xylylenols, whereas the other site twists 90° in both directions to provide two ground state *o*-xylylenols with each substituent on one end being *anti* in one isomer and *syn* in the other. Sch. 4 displays one example of this triplet biradical relaxation.

2.2.6. Benzoate Esters: Electron Transfer

Although esters do not have low-lying n,π^* states, their carbonyl groups make them excellent electron acceptors. Thus with a good electron acceptor on the benzoate phenyl ring, irradiation forms a very electrophilic π,π^* excited state. With a good electron donor on a γ-carbon, electron transfer to the benzoate group allows a 1,5-proton transfer to the carbonyl, resulting in the same kind of 1,4-biradical formed by simple hydrogen atom abstraction and typical Norrish type II cleavage products [2].

2.3. SCOPE AND LIMITATIONS

2.3.1. Factors of General Importance

Since carbonyls can react from both singlet and triplet n,π^* states, it is important to recognize the general features that determine reaction rates and product yields. Equation (1) describes the actual rate of any photochemical reaction, where I_a is the rate of light absorption in photons/L-s and Φ_{pr} is the overall quantum yield of product formation, which depends on how many competitive reactions *each* intermediate undergoes. Equation (2) defines Φ_{pr}, specifically for γ-hydrogen abstraction, as the sum of the probabilities that the various competing reactions in the singlet and triplet manifolds lead to product, where α is the probability that singlet H-abstraction produces a biradical, ^1k_H and ^3k_H are the rate constants for hydrogen abstraction, τ_S and τ_T are the excited state lifetimes (which normally are determined primarily by rates of chemical reaction), k_{isc} is the rate of intersystem crossing, and the ^1P's and ^3P's are the percentage of biradicals that cyclize or cleave. Equation (3) fully defines the quantum yield for intersystem crossing of excited singlet ketone, where kfl is the rate of fluorescence, which normally competes only slightly. Equations (4) and (5) define the quantum efficiencies for cyclobutanol formation and type II cleavage. It must be remembered that low overall quantum yields do not necessarily mean low chemical yields, simply slow reactions; but $\Phi_{\text{cyc}}/\Phi_{\text{II}}$ does equal the cyclobutanol/cleavage ratio.

$$- d[K]/dt = d[\text{products}]/dt = I_a \times \Phi_{\text{pr}} \tag{1}$$

$$\Phi_{\text{pr}} = \alpha^1 k_H \tau_S + \Phi_{\text{isc}}{}^3 k_H \tau_T ({}^3 P_{\text{cyc}} + {}^3 P_{\text{II}}) \tag{2}$$

$$\Phi_{\text{isc}} = k_{\text{isc}} \tau_S = k_{\text{isc}}/(k_{\text{isc}} + {}^1 k_H + {}^1 k_{fl}) \tag{3}$$

$$\Phi_{\text{cyc}} = \alpha^1 k_H \tau_S {}^1 P_{\text{cyc}} + \Phi_{\text{isc}}{}^3 k_H \tau_T {}^3 P_{\text{cyc}} \tag{4}$$

$$\Phi_{\text{II}} = \alpha^1 k_H \tau_S {}^1 P_{\text{II}} + \Phi_{\text{isc}}{}^3 k_H \tau_T {}^3 P_{\text{II}} \tag{5}$$

There is only one limitation of general scope, namely the possibility for more remote hydrogen abstraction, especially from carbons next to strong electron donors, and for competing intramolecular Paterno-Buchi reaction with appropriately located double bonds. Chapters 3 and 4 cover these processes and may provide the information necessary to determine the likelihood of such reactions competing with γ-hydrogen abstraction in a particular compound.

2.3.2. Aliphatic Ketones

As Secs. 2.1 and 2.2 point out, nonconjugated ketones and aldehydes undergo γ-hydrogen abstraction from both singlet and triplet n,π^* states. They all undergo S → T intersystem crossing with rate constants $\sim 2 \times 10^8 \, \text{s}^{-1}$, so that the singlet/triplet reaction ratio depends mainly on $^1 k_H$ values. Both $^1 k_H$ and $^3 k_H$ values vary with the type of C–H bond, simple acyclic ketones showing tertiary/secondary/primary $^1 k_H$ ratios of 20/9/1 and $^3 k_H$ ratios of 4/1/0.1, all values in units of $10^8 \, \text{s}^{-1}$. Thus 2-pentanone reacts mainly from its triplet state while 5-methyl-2-hexanone reacts mainly from its singlet state [8] and 4-octanone forms ~ 15 times as much 2-pentanone as 2-hexanone [33]. The size of α typically ranges from 0.8 to 0.16 [29]; and the $^3 P_{\text{cyc}}$ and $^3 P_{\text{II}}$ values vary significantly in different compounds, with $^1 P_{\text{cyc}}$ being very small for singlet states [34], for reasons not well understood. A quite comprehensive early review of aliphatic ketone photochemistry includes data for a large number of compounds [35].

There is an important reason to distinguish between singlet and triplet reaction, since biradicals are longer-lived in their triplet states than in their singlet states, giving the triplets more time to undergo conformational interconversions. This fact has a long history and often decides both the stereo- and chemoselectivity of biradical reactions, in the latter case the competition between cleavage and cyclization.

There is a well established way to enforce only excited singlet reaction, namely the inclusion of a triplet quencher at a concentration sufficient to quench all triplets before they can react. For this to work the lifetime of the excited triplet must be known as well as the bimolecular rate constant for its quenching by the chosen quencher. Conjugated dienes have often been used, sometimes as solvent, since they quench triplet ketones with near-diffusion controlled rate constants. Naphthalene and 1-methylnaphthalene are better in that they absorb enough light below 330 nm to sensitize formation of excited singlet ketone [36], although disposing of them afterwards is not as easy as distilling off pentadiene.

The naphthalene effect has been employed very effectively by Giese in maximizing the stereoselectivity for cyclization of the 1,5-biradicals formed by irradiation of *N*-(2-acylethyl)amino acids to prolines [37]. Of course 1,5-biradicals cannot cleave, so they cyclize. Since the Norrish type II 1,4-biradicals formed from excited singlets undergo very little cyclization, the only stereochemistry they can produce is the *cis/trans* ratio of the olefin cleavage product. Thus γ-hydrogen abstraction by singlet excited ketones is limited synthetically because most of the excited states do not form biradicals and the biradicals mainly cleave. However, if cleavage is what is desired, then a singlet state reaction is preferred, even if the quantum efficiency is low.

Norrish type I α-cleavage occurs with rate constants close to those for γ-hydrogen abstraction and therefore often competes, especially in ketones with α-substituents [35]. This problem is especially serious for cyclic ketones; for example 2-(*n*-propyl)cyclopentanone rearranges to 7-octenal with a quantum yield of 37%, while its type II cleavage yield is only 1% [35]. In contrast, 2-(*n*-propyl)cyclohexanone undergoes both α-cleavage and γ-hydrogen abstraction in only 3–4% quantum efficiency, since triplet cyclohexanone cleaves only 1/10 as rapidly as triplet cyclopentanone. The 2-alkyl group must be equatorial for any γ-hydrogens to be close enough to the carbonyl for H-abstraction, since *trans*-4-*t*-butyl-2-*n*-propylcyclo-hexanone, with the propyl group axial, undergoes no type II reaction [38].

2.3.3. Aryl Alkyl Ketones

With intersystem crossing rates $\geq 10^{10} \, s^{-1}$, aryl ketones react almost exclusively from their n,π^* triplet states, although a few examples of $^1n,\pi^*$ reactivity have been reported. Rate constants for γ-hydrogen abstraction [39] in butyrophenone, valerophenone, and γ-methylvalerophenone are 0.7, 16, and $50 \times 10^7 \, s^{-1}$, showing the usual oxy-radical 1°/2°/3° C–H selectivity and being high enough to compete with bimolecular hydrogen abstraction from alkane solvents and with Norrish type I α-cleavage in α-alkylated ketones.

Substituents on the benzene ring and at the δ-position can lower rate constants in two separate ways. Electron-donating (o-, m-, or p-) and conjugatively electron-withdrawing (para) ring substituents invert the relative energies of $^3n,\pi^*$, and $^3\pi,\pi^*$ states so as to lower the population of the reactive $^3n,\pi^*$ states. This change affects all triplet oxy-radical-like reactions, so that competing product ratios undergo little if any change, the slower overall rate of the photoreactions being the only significant effect. Electron-withdrawing substituents on the δ-carbon lower k_H by a simple inductive effect, showing a Hammett ρ_I value of −1.85. However, rates of competing bimolecular or α-cleavage reactions are not affected, such that lower chemical yields of type II elimination and cyclobutanol formation are possible.

2.3.4. α-Diketones

These compounds, which absorb violet and blue light, undergo their special kind of γ-hydrogen abstraction to form mainly hydroxycyclobutanone products. Rate constants for the triplet abstraction are only 1% as large as in monoketones, slow enough that phosphorescence competes. The 1,4-biradicals only cyclize and do not cleave. In some 1-phenyl-1,2-diones enolization of the 2-carbonyl occurs, formally by a β-hydrogen transfer [40]. There also are cases in which hydroxycyclopentanones are formed [41]. These were originally ascribed to γ-hydrogen abstraction followed by transfer of the hydroxy hydrogen to the remaining carbonyl, but current knowledge makes it more likely that δ-hydrogen abstraction occurs in such cases.

As mentioned in Sec. 2.2, α-diketones form hydrates readily, so care must be taken to minimize this to prevent cleavage of the two acyl groups.

2.3.5. o-Alkylphenyl Ketones

A very wide variety of such compounds have been studied in the past four decades, primarily for synthesis of six-membered rings by Diels–Alder cycloaddition of the initial o-xylylenol photoproducts to dienophiles. The benzocyclobutenol products also are of synthetic interest, in that those formed from 2-alkylphenyl ketones revert to their *o*-xylylene isomers at relatively low temperatures, making them isolable precursors for Diels–Alder reactions. They can be modified first, as follows: strong acids cause their slow isomerization to an equilibrium mixture of diastereomers, presumably by S_N1 substitution at the protonated hydroxy group. Nucleophiles other than water can of course replace the hydroxy group, which also can be esterified. Treatment with strong base causes them to revert to their ketone photo-precursor. The benzocyclobutenols formed from 2,6-dialkylphenyl ketones are much more thermally stable, surviving very hot GC injectors, and have not drawn as much synthetic attention.

Interestingly, there is one class of *o*-substituted phenyl ketones that do not follow the usual photochemical reaction path. Whereas ketones with alkoxymethyl, benzyl, ethyl, and other alkyl substituents form benzocyclobutenols, (*o*-acylphenyl)acetic acid derivatives do not. The nitrile gets transformed to the amide [42], the parent acid undergoes decarboxylation in benzene solvent [43], and esters and amides seem to be photoinert [43]. This behavior all seems rooted in the lowest energy rotamer of the initial triplet biradical having a geometry that forms only *o*-xylylenols with their OH group *syn*, which rapidly revert to starting ketone.

2.3.6. Pyruvate Esters

These have been extensively studied by Neckers and coworkers [21]. Those which undergo γ-hydrogen abstraction cleave cleanly to aldehyde or ketone, and thus provide a possible methodology for environmentally friendly oxidation of alcohols. The secondary photolysis of the hydroxyketene forms benzaldehyde and carbon monoxide, both of which could be considered nuisances, although it might be possible to trap the carbene intermediate in order to make a 1-phenylcyclopropanol.

Due to the low rate constants for γ-hydrogen abstraction (~1% of those for analogous phenyl ketone triplets), more remote hydrogen abstractions and cycloadditions to remote double bonds commonly compete and sometimes are the only reactions of pyruvates.

2.4. SYNTHETIC POTENTIAL: REACTIVITY AND SELECTIVITY PATTERNS

Intramolecular reactions display chemo- and regioselectivity that are largely dependent on the proximity and orientation of reactive functional groups. Thus double bonds, aryl groups, amino groups, and thioalkoxy groups all react rapidly with n,π^* states by charge or electron transfer; ketones with any of them close to the carbonyl typically undergo γ-hydrogen transfer in low yields. Competition from type I α-cleavage is serious only in when an α-carbon is tertiary or substituted with strong radical stabilizing groups.

2.4.1. Selectivity of Hydrogen Abstraction

In the absence of competitive reactions, what determines which C–H bonds are abstracted? Generally, hydrogen atoms are abstracted much faster from γ-carbons than from δ- or more distant carbons, due to a mixture of enthalpic and entropic factors. Hexanophenone and longer phenones undergo mainly type II elimination, photocyclization to cyclopentanols being barely detectable. The γ/δ rate constant ratio is ~20/1 both experimentally [44] and theoretically [45]. The behavior of δ-methoxyvalerophenone exemplifies how the inductive effects of a nearby substituent (the methoxy group) can modify C–H reactivity so that δ- and γ-hydrogen abstraction occur equally [46]. Some compounds have more than one hydrogen-bearing γ-carbon. The behavior of 4-octanone [33] mentioned in Sec. 2.3.2 provides an example of C–H bond strength (1° vs. 2°) determining the regioselectivity of γ-hydrogen abstraction, while the low type II yield from o-methylvalerophenone [25] reflects the closer proximity to the carbonyl oxygen of the o-tolyl methyl group relative to that of the γ-carbon on the butyl group.

What if a compound has two separate keto groups? Since intramolecular energy transfer is quite fast, the carbonyl with the lower excitation energy normally is excited more than the other carbonyl and therefore might be expected to be the more reactive. The behavior of 1-benzoyl-4-(p-anisoyl)butane provides a good example of how different factors affect product ratios. At 25° the acetophenone/p-methoxyacetophenone product ratio is 20/1, indicating that 95% of the γ-hydrogen abstraction is done by the triplet benzoyl group, even though population of the lower energy triplet anisoyl groups is 10 times greater [47]. However, hydrogen abstraction by a triplet anisoyl group is known to be only 1/200 as fast as it is by a triplet benzoyl group. Thus the benzoyl triplet's 200-fold greater reactivity counteracts its 10-fold lower concentration.

2.4.2. Chemoselectivity of 1,4-Biradical Reactions

Once γ-hydrogen abstraction occurs, all selectivity is determined by the behavior of the 1,4-biradical, including the type, distribution, and stereochemistry of products. The 1,4-biradicals formed by different kinds of compounds behave differently. How the biradicals partition among cyclization, type II cleavage, and reversion to ground state ketone varies wildly. Since each process is very exothermic, selectivity is kinetically controlled, with the kinetics strongly dependent on geometric factors and multiplicity. Since disproportionation simply regenerates ground state reactants that recycle to biradicals while light is still available, it does not affect the cyclization/cleavage ratio, except in rare cases when it is the only biradical reaction.

The singlet biradicals formed from $^1n,\pi^*$ states appear to mainly cleave, even though their initial geometry might be expected to favor both cyclization and disproportionation, neither of which appears to occur much at all. Such expectation would be based on the belief that singlet biradicals react so rapidly that they undergo little or no geometric change before reacting. Experiments and basic theory both indicate that cleavage requires the two SOMOs of the biradical to be aligned as parallel as possible to the α–β bond between the two radical sites. Once this was thought to require an anti staggered biradical conformation, but it is now realized that the SOMOs can be parallel in gauche conformations as well. Since cleavage is preferred over cyclization both enthalpically and entropically, the observed predominance of cleavage indicates that singlet biradicals do not undergo barrier-free product formation, their partitioning reflecting differences in free energy barriers along different reaction coordinates.

The behavior of triplet biradicals has received a great amount of study, most of it aimed at understanding how T → S intersystem crossing (isc) affects their lifetimes and product formation. Inasmuch as their lifetimes are determined by the rates at which they form different products, their selectivity has been suggested to reflect different amounts of spin-orbit coupling (soc) and thus different isc rates for different conformers [48]. This view of events fits both current theories on the timing of isc and product formation: either irreversible T → S isc precedes product formation; [48,49]; or isc occurs during product formation as the biradical travels along the proper reaction coordinates [50]. The first, two-step mechanism originally assumed that once the biradical has become a singlet it instantaneously forms whichever product its geometry favors. However, the actual behavior of singlet type II 1,4-biradicals does not fit that picture. Likewise product distributions from triplet 1,4-biradicals clearly reflect likely energy barriers.

Acyclic ketones

Various computations indicate that triplet type II 1,4-biradicals exist as a mixture of several conformers with geometries disposed toward product formation, as shown in Sch. 5. The initial biradical rotates such that the p orbital on the hydroxy-substituted site is aligned nearly parallel to the 2,3 C–C bond as required for cyclization and cleavage. The computed percent population of four conformers are shown for valerophenone (R = CH$_3$), but they do not correlate with the product distribution as listed in Table 1.

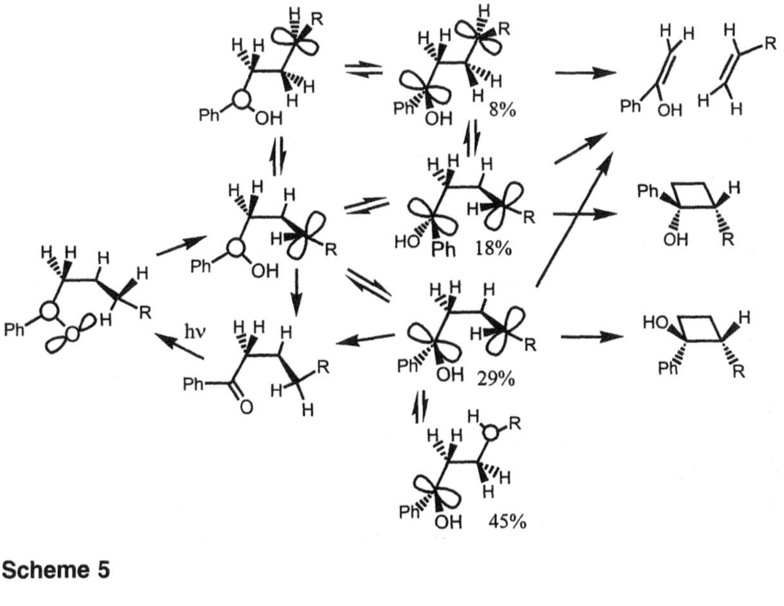

Scheme 5

Table 1 Some Representative 1,4-Biradical Partitioning Ratios.

Ketone		Benzene ϕ_{elim}	Benzene ϕ_{cyc} $(t/c)^a$	Alcohol ϕ_{elim}	Alcohol ϕ_{cyc} $(t/c)^a$
$PhCOCH_2CH_2CH_3$	1	0.36	0.04	~0.90	~0.10
$PhCOCHMeCH_2CH_3$	2	0.28	0.12 (>20)		
$PhCOCMe_2CH_2CH_3{}^b$	3	0.007	0.057		
$PhCOCH_2CHMeCH_3$	4	0.26	0.045		
$PhCOCH_2CMe_2CH_3$	5	0.19	0.01		
$PhCOCH_2CH_2CH_2CH_3$	6	0.30	0.06 (3:1)	0.88	0.12 (2:1)
$PhCOCH_2CH_2CH_2\text{-}C2H_5$	7	0.24	0.06 (3:1)	0.82	0.12
$PhCOCH_2CH_2CH_2CMe_3$	8	0.22	0.086 (2:1)	0.83	0.17 (1:1)
$PhCOCHMeCH_2CH_2CH_3$	9	0.17	0.12 (3:1)		
$PhCOCMe_2CH_2CH_2CH_3{}^b$	10	0.04	0.08 (4:3)	~0.10	~0.20 (4:5)
$PhCOCH_2CH_2CHMe_2$	11	0.26	0.04		
$PhCOCHFCH_2CH_2CH_3$	12	0.34	0.35		
$PhCOCF_2CH_2CH_2CH_3$	13	<0.01	0.60		
$PhCOCH_2CH_2CH_2Ph$	14	0.47	0.05		
$PhCOCH_2CH_2CH_2OMe$	15	0.21	0.11 (7:1)		
$PhCO(CH_2)_3CH_2OMe^c$	16	0.33	0.10 (>20)	0.47	0.8 (4:1)
$PhCO(CH_2)_3CH_2CO_2Me$	17	0.58	0.17		
$PhCOCH_2\text{-}O\text{-}CH_2CH_3$	18	0.35	0.62 (11:1)		
$PhCOCH_2CH_2CH=O$	19	<0.05	0.30	<0.05	0.40
$PhCOc\text{-}C_4H_7$	20	0.01	0.03		
$PhCOc\text{-}C_5H_9$	21	0.40	0		
$PhCOc\text{-}C_6H_{11}$	22	0.008	0.005		

^{a}2-Substituent/1-phenyl; ^{b}competing α-cleavage; ^{c}competing δ-H-abstraction.

The behavior of **2** and **3** compared to **1** and of **9** and **10** compared to **6** shows how α-alkylation increases the normally low cyclization/cleavage ratio [51], while **12** and **13** show how α-fluorination does likewise [52]. In both cases altered biradical geometries are thought to be responsible, although the relatively high energy of a difluoroenol has been suggested to hinder cleavage from **13**. The high cyclization yields for **18** and **19** clearly cannot be due to steric hindrance. Cleavage of **19** would form a ketene and cyclization may be aided by internal H-bonding; cleavage of **18** forms an aldehyde, while cyclization of the biradical is eased by the oxygen atom.

Internal hydrogen-bonding in the biradicals from **17** and **18** increases overall quantum yields for product formation by impeding disproportionation. Likewise most polar solvents enhance quantum yields but lower the cyclization/cleavage ratio, presumably because H-bonding of solvent to the hydroxyl group increases steric hindrance to bond rotations.

Cyclic Compounds

Acylcycloalkanes undergo γ-hydrogen abstraction with different efficiencies and the 1,4-biradicals show widely varying cleavage/cyclization selectivity, as the table indicates. These variations apparently reflect the different abilities of the half-occupied *p* orbitals in biradicals of different ring-size to align for cleavage. A second substituent on the acyl-containing ring-carbon tends to significantly enhance cyclization efficiency, just as in the acyclic ketones. The high cleavage efficiency of benzoylcyclopentane produces a γ-allyl product that undergoes further γ-hydrogen abstraction to produce a biradical that cyclizes to both 4- and 6-membered rings.

The same behavior holds for some *endo*-benzoyl-substituted bicycloalkanes [53].

2.4.3. Stereoselectivity of 1,4-Biradical Reactions

Type II cleavage generates double bonds; but there has not been much study of *cis/trans* selectivity in this process. Photolysis of β,γ-diphenylbutyro-phenone forms exclusively *trans*-stilbene, the thermodynamically favored product [54]. This experiment was designed to determine whether cleavage of the triplet 1,4-biradical might form triplet stilbene, which would have decayed to a 55:45 *trans/cis* mixture of stilbenes. Instead the preferred conformation of the biradical determines product geometry; and that principle probably holds for type II cleavage of most compounds.

There has been considerable interest in the factors that control the stereoselectivity of cyclobutanol formation. Three main factors were identified quite early: pre-existing conformational preferences due to steric effects or to internal hydrogen bonding; solvation of the OH group; and variable rotational barriers for cyclization. More recently Griesbeck has proposed that orbital orientation favoring soc produces another form of conformational preference in triplet biradicals [55]. These factors have different importance depending upon the molecule.

Acyclic Ketones

Valerophenone **6** and α-methylbutyrophenone **2** both form 2-methyl-1-phenylcyclobutanol; but **2** forms only the *trans* product, whereas the former cyclizes with a small preference for the thermodynamically favorable *trans* isomer. Ketone **2** provides an example of stereoselectivity determined by pre-existing conformational preference in the biradical, while **6** demonstrates selectivity driven by nonbonded interactions created during the bond rotations required for cyclization as well as a small energy difference between the two biradical conformers most disposed to cyclize.

Perusal of Table 1 reveals that the less substituted ketones cyclize with a 3- or 4-fold preference for the *trans* cyclobutanol, whereas the more sterically congested ketones cyclize with very little selectivity. In contrast, ketones containing ether oxygen atoms show very high selectivity, presumably due to internal H-bonding of the OH to the ether oxygen of the biradical. When the solvent is a Lewis base, stereoselectivity is decreased significantly, again reflecting the increased bulk of the solvated OH group. *In all cases basic solvents decrease the stereoselectivity of cyclization.*

Cyclic Compounds

Acylcyclohexanes have received considerable study and, especially as solids, undergo highly stereoselective cyclization, since H-abstraction occurs with the phenyl twisted away from the α-methyl, where it stays until the biradical cyclizes [56]. Addition of a chiral ammonium salt to the benzene ring fosters enantioselective cyclization, presumably by inducing a tilt of the benzoyl group that leads to preferred abstraction from one of the two γ carbons [57].

Scheffer also has provided examples of phase dependent cyclization diastereoselectivity [58].

2.4.4. α-Diketones

There does not seem to have been much study of the stereoselectivity of cyclization of α-diketones. Scheffer and coworkers reported that 1,2-cyclodecanedione forms only the *cis*-1-hydroxybicyclo-[6,2,0]decan-2-one when irradiated as crystals, but a 7:1 *cis/trans* ratio in benzene [59]. In solution the products proceed to cleave to a β-ketoaldehyde, probably by Norrish type I cleavage. This problem can affect any α-hydroxyketone photoproduct if light below 330 nm is used to irradiate the α-diketone. Since α-diketones absorb above 400 nm, use of such wavelengths can avoid the problem.

2.4.5. *o*-Alkylphenyl Ketones

As described in Sec. 2.2, the *o*-xylylenols formed by hydrogen abstraction cyclize thermally to benzocyclobutenols, which at high temperatures reopen to the same *o*-xylylenols, which can undergo rapid [4+2]-cycloaddition to dienophiles or revert to starting ketone. In early studies of Diels–Alder products formed by this process it was suggested that product stereo-selectivity could be due to steric effects during cycloaddition. However, it now appears that the stereochemistry of both the Diels–Alder adducts and the cyclobutenols are determined by the geometry of the *o*-xylylenols, which is established by decay of the twisted biradical triplet to ground state *o*-xylylenol, the geometry of the biradical first being determined by the relative ability of different substituents to stabilize the two radical sites. Prediction of this geometry can be intuitive but requires computational methods in some cases. It has not been established how the geometry of the biradical formed by excited singlet hydrogen abstraction may affect the *o*-xylylenol geometry.

2.4.6. Biradical Rearrangements

Since each end of a 1,4-biradical is able to undergo normal radical behavior in competition with biradical decay [60], appropriate substituents can effect rearrangements leading to longer biradicals that cyclize to larger rings [61], 60 two examples of which are shown below, another being that of γ-vinylvalerophenone shown above.

2.4.7. Summary of Synthetic Potential

Norrish type II elimination itself provides a unique method for inserting a double bond into a structure in a way that need not disturb other functional groups. It also produces enols in a neutral environment, which already has allowed detailed study of the kinetics of acid/base catalysis [62]. As regards cyclization, prediction of its yield is difficult; its stereochemistry is more predictable and can be fine-tuned by choice of solvent as well as by functional groups to which the OH group can hydrogen-bond. Type II cleavage of glyoxylate esters provides a clean nonionic method to oxidize alcohols. Photoenolization of *o*-allkylphenyl ketones to *o*-xylylenols provides one-step access to both bimolecular and intramolecular Diels–Alder formation of various tetralin derivatives. Prediction of stereochemistry requires computation of the lowest energy conformer of the initially formed biradical. The monoradical reactions of 1,4-biradicals also have synthetic potential: β-cleavage of halogen atoms yields acids, making the overall reaction possibly useful for photoimaging; rearrangements can lead to larger cyclic alcohols.

2.5. REPRESENTATIVE PROCEDURES

Excited state γ-hydrogen abstraction being an intramolecular reaction and thus kinetically unimolecular results in relatively simple experimental procedures. As in any photochemical reaction, only wavelengths absorbed by the reacting chromophore produce the desired products, care to avoid exciting other chromophores in the molecule also being necessary. Thus the major requirement is finding the right light source, sometimes together with proper filtering to narrow the range of wavelengths emitted. Since monoketones undergo n,π^* absorption in the 310–370 nm range, simple medium pressure mercury lamps usually suffice. The simplest method generally is to place a de-oxygenated glass tube containing the reactant close to the light source. Vacuum-walled immersion wells with the lamp in the center allow temperature-control of a reactant solution in the outer chamber. Several lasers produce linear beams in the required wavelength range and allow irradiation of distant samples in any sort of vessel that has a transparent window.

In many cases irradiation of neat compound, as liquid or solid, is quite adequate. Irradiation of solids works best with small crystals packed in sealed glass tubes. Since product ratios are often solvent-dependent, reactions often need to be performed in solution. Beer's Law must be remembered, since the amount of light absorbed and thus the rate of reaction decreases as reactant concentration decreases. A general potential problem is possible light absorption by products. Fortunately the cyclobutanols and alkenes formed by simple type II chemistry generally do not present this problem, although the type II ketone product absorbs the same as the reactant. Fortunately this is not much of a problem: at high concentrations of reactant ketone in the proper solvents, excited product ketone only transfers energy to ground state reactant ketone [63].

The aldehyde products formed from glyoxylate esters not only absorb light but are highly reactive substrates for attack by excited ketones and can be depleted by over-irradiation.

The proclivity for the 2-hydroxycyclobutanones formed from α-diketones to cleave to γ-ketoaldehydes is mentioned in Sec. 2.4.4, but this occurs only at near-uv wavelengths and not above 400 nm.

Finally, the *o*-xylylenols formed from *o*-alkylphenyl ketones may undergo photo-induced electrocyclic reactions, one example having been discovered recently [64].

A general potential problem is possible light absorption by products. Fortunately the cyclobutanols and alkenes formed by simple type II chemistry generally do not present this problem, although the type II ketone product absorbs the same as the reactant. Fortunately this is not much of a problem: at high concentrations of reactant ketone in the proper solvents, excited product ketone only transfers energy to ground state reactant ketone [63]. In contrast, the aldehyde products formed from glyoxylate esters not only absorb light but are highly reactive substrates for attack by excited ketones and can be depleted by over-irradiation. Likewise, the proclivity for the 2-hydroxycyclobutanones formed from α-diketones to cleave to γ-ketoaldehydes is mentioned in Sec. 2.4.4, but this occurs only at near-uv wavelengths and not above 400 nm. Finally, the *o*-xylylenols formed from *o*-alkylphenyl ketones may undergo photo-induced electrocyclic reactions, one example having been discovered recently [65]. In all cases where products absorb with results detrimental to those desired, one must simply stop the reaction at partial conversion, separate products from reactant, and start over. Since both cleavage and cyclization products are usually formed even without further complications, the Norrish type II reaction requires significant product separation efforts.

Another potential problem is the presence of other functional groups on the molecule that may react with the excited carbonyl, such as double bonds and good electron-donors. Amino groups in particular react rapidly with excited ketones; and their ability to quench γ-hydrogen abstraction by electron transfer was demonstrated 30 years ago [66]. However, recent synthetic efforts have emphasized protecting amino groups by acylating them, thus minimizing their electron-donor potential. One example is the photocyclization of a variety of α-amino-ketones to α-aminocyclobutanols [67]. The ketones were synthesized by Friedel-Crafts acylation of toluene with appropriate *N*-acetylated α-amino acid chlorides. Solutions containing 0.2 mmol of ketone in 50 mL of benzene were irradiated under nitrogen at 15°C for 5 h with phosphor-coated low pressure mercury lamps emitting in the 330–370 nm range. Reaction progress was followed by TLC and GC analysis. Products were isolated and purified by normal procedures, with chemical yields of 75–80%. Unlike the case for simple alkanophenones, cyclobutanol formation occurred in proportions comparable to or greater than type II elimination; and type I cleavage competed in proportions ≤20%.

The type I α-cleavage was most likely promoted by the radical-stabilizing ability of the α-acylamino group. The high yields of cyclization apparently were promoted by H-bonding of the OH to the *N*-acetyl group in the 1,4-biradical intermediates. Thus acylation provided two benefits, one preventing internal quenching and one inducing regio- and stereoselectivity.

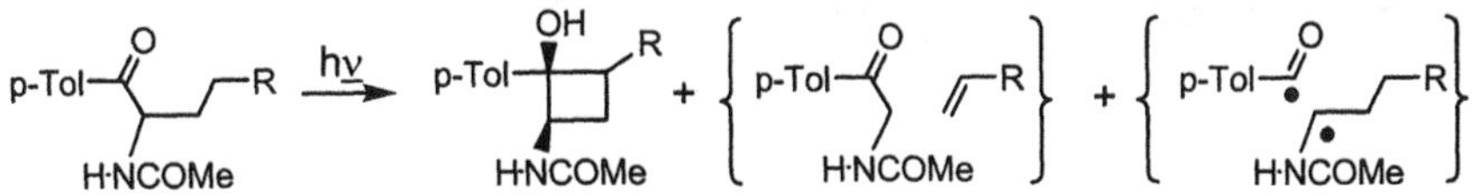

2.6. TARGET MOLECULES: NATURAL AND NONNATURAL PRODUCT STRUCTURES

The Norrish type II reaction is occasionally used in synthetic processes, mostly when other procedures leading to a desired product are difficult or costly. Simple type II elimination has been put to use in several ways:

Glycosides made from γ-hydroxy ketones undergo photoelimination to give *O*-vinyl glycosides, which can undergo a Claisen rearrangement in certain deoxy sugars [68].

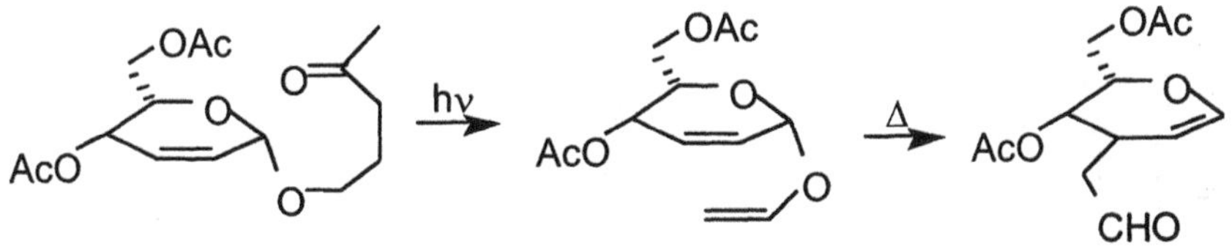

Irradiation of a phenylglyoxylamides forms β-lactams [69]. Irradiation of a *para*-substituted phenylglyoxylamide as a suspension of crystals in hexane formed a β-lactam in > 91% yield and 99% ee, the selectivity induced by a chiral ammonium cation paired with the carboxylate anion [70].

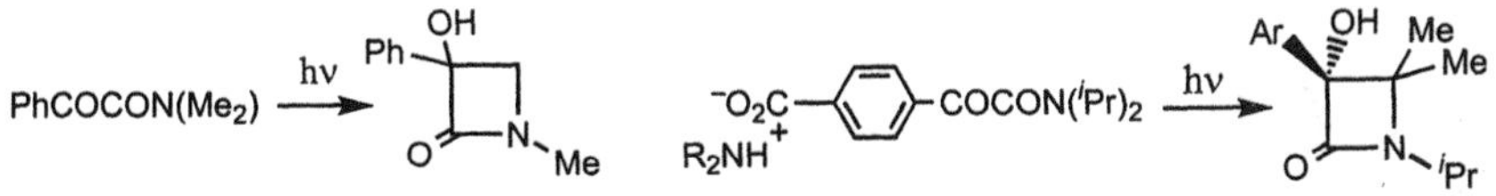

Possibly the most unusual use of type II elimination was Goodman and Berson's preparation of *meta*-xylylene [71].

A somewhat similar approach led to ene-diyne syntheses, in 1:1 *E/Z* ratio [72].

Properly alkylated bicyclo[4.2.0]-octan-2-ones formed by 2+2 photocycloaddition of alkenes to cyclohexenones cleave to δ,ε-unsaturated ketones, one example leading to a sesquiterpene synthesis [73].

The enol formed by irradiation of α-disubstituted indanones and tetralones bearing at least one hydrogen in the γ-position undergoes enantioselective tautomerization to ketone in the presence of catalytic amounts of optically active aminoalcohols [74].

There have been many succesful syntheses of tricyclic structures by Diels–Alder reactions of the *o*-xylylenols formed from *o*-alkylphenyl ketones. Below is an intramolecular example aimed at the synthesis of podophyllotoxin [75].

REFERENCES

1. Coyle JD. Chem. Rev 1978; 78:97.
2. Decosta DP, Bennett AK, Pincock JA. J Am Chem Soc 1999; 121:3785.
3. Norrish RGW, Appleyard MES. J Chem Soc 1934; 874.
4. McMillan GR, Calvert JG, Pitts JN Jr. J Am Chem Soc 1964; 86:3602.
5. Yang NC, Yang D-H. J Am Chem Soc 1958; 80:2913.
6. Ausloos P, Rebbert RE. J Am Chem Soc 1964; 86:4512.
7. Michael JL, Noyes WA Jr. J Am Chem Soc 1963; 85:1027.
8. Wagner PJ, Hammond GS. J Am Chem Soc 1965; 87:4009.
9. Dougherty TJ. J Am Chem Soc 1965; 87:4011.
10. Baum EJ, Wan JKS, Pitts, JN Jr. J Am Chem Soc 1966; 88:2652.
11. Wagner PJ, Hammond GS. J Am Chem Soc 1966; 88:1245.
12. Wagner PJ. J Am Chem Soc 1967; 89:5898.
13. Wagner PJ, Kochevar IE, Kemppainen AE. J Am Chem Soc 1972; 94:7489.
14. Wagner PJ, Zepp RG. J Am Chem Soc 1972; 94:287.
15. Yang NC, Elliot SP. J Am Chem Soc 1969; 91:7550.
16. Wagner PJ, Kelso PA, Zepp RG. J Am Chem Soc 1972; 94:7480.
17. Yang NC, Morduchowitz A, Yang D-H. J Am Chem Soc 1963; 85:1017.
18. Hu S, Neckers DC. J Org Chem 1996; 61:6407.
19. Urry WH, Trecker DJ. J Am Chem Soc 1962; 84:118.
20. Wagner PJ, Sedon JH, Lindstrom MJ. J Am Chem Soc 1978; 100:2579.
21. Hu SK, Neckers DC. J Org Chem 1997; 62:7827.
22. Yang NC, Rivas C. J Am Chem Soc 1961; 83:2213.
23. Sammes PG. Tetrahedron 1976; 32:405.
24. Matsuura T, Kitaura Y. Tetrahedron Lett 1967; 3309; Kitaura Y, Matsuura T. Tetrahedron 1971; 27:1597.
25. Wagner PJ, Chen C-P. J Am Chem Soc 1976; 98:239.

26. Haag R, Wirz J, Wagner PJ. Helv Chim Acta 1977; 60:2595.
27. Das PK, Encinas MV, Small RD Jr. Scaiano JC. J Am Chem Soc 1979; 101:6965.
28. Wagner PJ, Subrahmanyam D, Park B-S. J Am Chem Soc 1991; 113:709.
29. Wagner PJ. Acc Chem Res 1971; 4:168; Wagner PJ. Topics Curr Chem 1976; 66:1.
30. Scheffer JR. Org Photochem 1987; 8:249.
31. Turro NJ, Lewis FD. J Am Chem Soc 1970; 92:311.
32. Wagner PJ, Sobczak M, Park B-S. J Am Chem Soc 1998; 120:2488.
33. Wagner PJ. Tetrahedron Lett 1968; 5385.
34. Coulson DR, Yang NC. J Am Chem Soc 1966; 88:4511; Barltrop JA, Coyle JC. Tetrahedron Lett 1968; 3235.
35. Dalton JC, Turro NJ. Ann Rev Phys Chem 1970; 21:499–560.
36. Wagner PJ. Mol Photochem 1971; 3:169.
37. Giese B, Barbosa F, Stahelin C, Sauer S, Wettstein PH, Wyss C. Pure Appl Chem 2000; 72:1623.
38. Turro NJ, Weiss DS. J Am Chem Soc 1968; 90:2185.
39. Wagner P, Park B-S. Org Photochem 1991; 11:227.
40. Wagner PJ, Zepp RG, Liu K-C, Thomas M, Lee T-J, Turro NJ. J Am Chem Soc 1976; 98:8125.
41. (a) Burkoth T, Ullman E. Tetrahedron Lett 1970; 145; (b) Bishop R, Hamer NK. J Chem Soc C 1193; 1970; (c) Hamer NK, Samuel CJ. J Chem Soc Perkin 1973; 2:1316.
42. Lu AL, Bovonsombat P, Agosta WC. J Org Chem 1996; 61:3729.
43. Sobczak M, Wagner PJ. Org Lett 2002; 4:379.
44. Wagner PJ, Kelso PA, Kemppainen AE, Zepp RG. J Am Chem Soc 1972; 94:7500.
45. Dorigo AE, McCarrick MA, Loncharich RJ, Houk KN. J Am Chem Soc 1990; 112:7508.
46. Wagner PJ, Zepp RG. J Am Chem Soc 1971; 93:4958.
47. Wagner PJ, Nakahira T. J Am Chem Soc 1973; 95:8474 .
48. Scaiano JC. Tetrahedron 1982; 38:819.
49. Caldwell RA. Pure Appl Chem 1984; 56:1167.
50. (a) Closs G, Redwine OD. J Am Chem Soc 1985; 107:4543; (b) Wagner PJ. Acc Chem Res 1989; 22:83; (c) Michl J, Bonacic-Koutecky V. Electronic Aspects of Organic Photochemistry. New York: Wiley, 1990.
51. Lewis FD, Hilliard TA. J Am Chem Soc 1972; 94:3852.
52. Wagner PJ, Thomas MJ. J Am Chem Soc 1976; 98:241.
53. Lewis FD, Johnson RW, Ruden RA. J Am Chem Soc 1972; 94:4292.
54. Caldwell RA, Fink PM. Tetrahedron Lett 1969; 2987; Wagner PJ, Kelso PA. Tetrahedron Lett 1969; 4153.
55. Griesbeck AG, Mauder H, Stadtmüller S. Acc Chem Res 1994; 27:70.
56. Vishnumurthy K, Cheung E, Scheffer JR, Scott C, Org Lett 2002; 4:1071.
57. Leibovich M, Olovsson G, Sundarababu G, Ramamurthy V, Scheffer JR, Trotter J. J Am Chem Soc 1996; 118:1219.

58. Evans SV, Omkaram N, Scheffer JR, Trotter J. Tetrahedron Lett 1986; 27:1419.
59. Olovsson G, Scheffer JR, Trotter J, Wu C-H. Tetrahedron Lett 1997; 38:6549.
60. Wagner PJ, Liu K-C. J Am Chem Soc 1974; 96:5952; Encinas MV, Wagner PJ, Scaiano JC. J Am Chem Soc 1980; 102:1357.
61. Wagner PJ, Liu K-C, Noguchi Y. J Am Chem Soc 1981; 103:3837.
62. Haspra P, Sutteer A, Wirz J. Angew Chem Int Ed. 1979; 18:617; Chiang Y, Kresge AJ, Tang YS, Wirz J. J Am Chem Soc 1984; 106:460.
63. Wagner PJ, Kochevar IE, Kemppainen AE. J Am Chem Soc 1972; 94:7489.
64. Sobczak M, Wagner PJ. Tetrahedron Lett 1998; 39:2523.
65. Wagner PJ, Kemppainen AE, Jellinek T. J Am Chem Soc 1972; 94:7512.
66. Griesbeck AG, Heckroth H. J Am Chem Soc 2002; 124:396.
67. Cottier L, Remy G, Descotes G. Synthesis 1979; 711.
68. Toda F, Miyamoto H, Kanemoto K. J Chem Soc Chem Commun. 1995; 1719.
69. Natarajan A, Wang K, Ramamurthy V, Scheffer JR, Patrick B. Org Lett 2002; 4:1443.
70. Goodman JL, Berson JA. J Am Chem Soc 1984; 106:1867.
71. Nuss JM, Murphy MM, Tetrahedron Lett 1994; 35:37.
72. Manh DDK, Ecoto J, Fetizon M, Colin H, Diez-Masa J-C. J Chem Soc Chem Commun 1981; 953.
73. Henin F, Muzart J, Pete JP, Mboungoumpassi A, Rau H, Angew Chem Int Edit Engl 1991; 30:416.
74. Kraus GA, Wu Y. J Org Chem 1993; 57:2922.

3
Abstraction of ($\gamma \pm n$)-Hydrogen by Excited Carbonyls

Pablo Wessig and Olaf Mühling
*Institut für Organische Chemie der Humbold-Universität zu Berlin,
Berlin, Germany*

3.1. INTRODUCTION AND HISTORICAL BACKGROUND

The photochemistry of ketones is mainly dominated by two types of reactions, both of them are connected with the partial radical character of the oxygen atom in the $n-\pi^*$-excited state. The first one includes the attack of this oxygen atom onto a C–C double bond, is called *Paternò-Büchi* reaction and will be discussed in the following chapter. The second one, comprising the predominantly intramolecular abstraction of a hydrogen atom, was already introduced in the preceding chapter, which dealt with the abstraction of γ-hydrogen atoms. This reaction is above all connected with the names Norrish and Yang. It was Norrish and his coworker Appleyard nearly seventy years ago [1], who discovered that ketones undergo a cleavage reaction upon irradiation. Nowadays, this process is called the *Norrish Type II* cleavage. Even though they assumed a concerted mechanism, they certainly didn't suspect anything about the second kind of products, which are much more interesting for preparative chemists. In 1958 only Yang and Yang recognized the formation of cyclobutanes besides the *Norrish Type II* cleavage products [2]. This discovery was the basis of a very versatile and valuable ring closure reaction that should therefore be named the *"Norrish–Yang reaction"*.

For reasons discussed in detail in the preceding chapter, the γ-position with respect to the excited carbonyl group is preferably attacked by the oxygen atom and 1,4-biradicals are formed. This chapter deals with reactions that are initiated by hydrogen abstraction from a non-γ-position as well as with methods to shift the spin density of one of the two radical centers to the adjacent atom and thus the formation of $(1,4 \pm 1)$-biradicals from 1,4-biradicals and 1,6-biradicals from 1,5-biradicals. Due to the great resemblance of the initial steps of both γ- and $(\gamma \pm n)$-hydrogen abstraction, most of the historical background and the elucidation of the mechanism, provided in the preceding chapter are also applied to this chapter and, therefore, should not be duplicated.

The historical roots of cyclization reactions that are based on $(\gamma \pm n)$-hydrogen abstraction are not clearly discernible. The first examples of such reactions were published in the first half of the 1970s. Wagner [3], Lappin [4a], and Pappas [4b] succeeded in the preparation of five-membered rings (cyclopentanes [3a], indanes [3b], and benzofuranes [4]) through δ-hydrogen abstraction by blocking the γ-position, and Roth [5] discovered the formation of cyclopropanes by β-hydrogen abstraction. The latter reaction is accompanied by a photoelectron transfer (PET), which is also important for the hydrogen abstraction from remote positions. More than 10 years later, the first syntheses of six-membered rings by the Norrish–Yang reaction appeared. Thus, Sauers investigated the transannular cyclization of cyclodecanone to decaline-4a-ol [6] and pointed out the importance of hydrogen back transfer for the regioselectivity of the Norrish–Yang reaction. The "blocking strategy," already successful in the preparation of five-membered rings, allowed a photochemical access to naphthalines [7] and piperidine-2-ones [8].

The preparation of macrocyclic compounds, one of the great challenges for organic chemists, was also achieved using the Norrish–Yang reaction. The obvious difficulties in these cyclizations caused by entropic factors were surmounted by two different approaches. The first one is characterized by the presence of a rigid molecular scaffold or a suited molecular constitution that prevents hydrogen abstraction from near positions with respect to the carbonyl group. Thus, Breslow [9] and Baldwin [10] reported the photochemical formation of macrocyclic compounds by irradiation of benzophenones bearing ester linked alkyl groups in the *para*-position. Kraus and Wu [11] have developed a route to eight-membered rings by irradiation of various phenylglyoxylates. As a rigid building block served cyclopropanes, benzene rings, cyclohexanes, and *cis*-configured alkenes. The second approach is based on an intramolecular photoinduced electron transfer and has proven to be very efficient. While only few examples are known in which the chromophoric group was an aromatic ketone [12], a variety of macrocyclic rings were prepared by the irradiation of phthalimide derivatives [13].

An entirely different approach was previously reported by Wessig [14]. The initial step an intramolecular γ- or δ-hydrogen abstraction providing

1,4- or 1,5-biradicals in the usual manner. A leaving group is tethered at the carbon atom, adjacent to the hydroxy-substituted radical center, and causes a very rapid elimination reaction. In the resulting new biradical the spin density is shifted by one atom and, consequently, the ring in the final products consists of $(1,4 \pm 1)$ or $(1,5 \pm 1)$ atoms. This approach is called *spin center shift*. In this way, cyclopropanes [14a,b], 1,3-oxazines [14c], and indanones [14d] were prepared.

3.2. STATE OF THE ART MECHANISTIC MODELS

The basic mechanistic details of γ-hydrogen abstraction discussed in the preceding chapter apply to the ($\gamma \pm n$)-hydrogen abstraction to the greatest possible extent and, therefore, the reader is referred to this chapter. Here, especially regio- and stereoselectivity phenomena that are connected with the preparation of cyclic systems consisting of $(4 \pm n)$ atoms, i.e., three-, five- and six-membered rings, macrocycles as well as bicyclic compounds will be discussed.

The following mechanistic remarks concerning the regioselectivity are subdivided in three parts. The first part deals with concepts that are within the framework of the "classical" Norrish–Yang reaction, i.e., the reaction underlies a homolytic hydrogen transfer. In the second part the cyclization reactions that are based on photoinduced electron transfer (PET) will be discussed and in the third part the *spin center shift* approach is elucidated.

As pointed out above, the γ-hydrogen abstraction by an excited carbonyl group is strongly favored compared to other positions. While the homolytic β-hydrogen abstraction is unfavorable due to the ring strain of the corresponding cyclic transition state, the ($\gamma + n$)-hydrogen abstractions suffer from increasing entropic disadvantage. The six-membered transition state of the γ-hydrogen abstraction represents an optimal compromise between these two contrary factors. This situation can be partially altered if the C–H bond energy of the δ-C–H bonds (or more remote C–H bonds) is remarkably decreased compared with those of the γ-C–H bonds. A classic example is the photochemical behavior of δ-methoxyvalerophenone **1**. On irradiation both cyclobutane **2** and cyclopentane **3** are formed (Sch. 1) [3a].

As a rule, such a decrease of C–H bond energy cannot completely overcome the preference of the γ-hydrogen abstraction. A more reliable concept is based on the structural blocking of all positions between the desired ($\gamma + n$)-position and carbonyl group with the exception of the β-position, which doesn't need to be blocked. Blocking is the introduction of

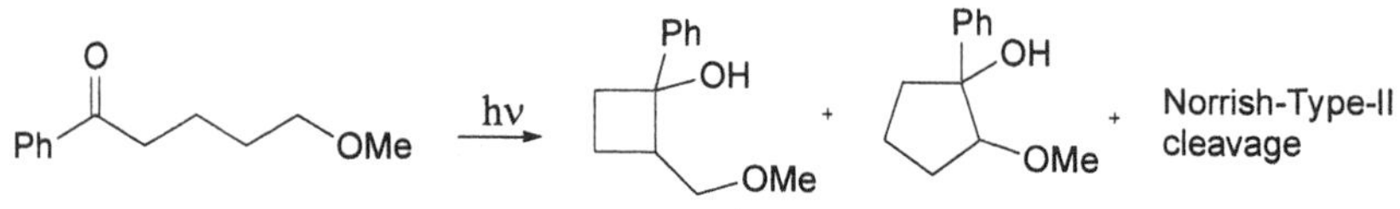

1 **2** **3**

solvent:			
benzene	10%	3.4%	33%
t-BuOH	9%	10%	17%

(Quantum yields)

Scheme 1

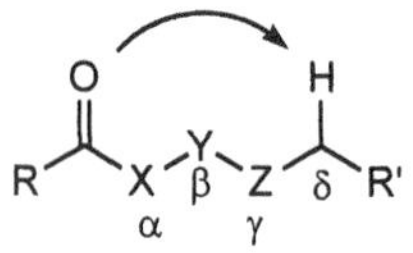

- **a** X,Y = C(sp^3), Z = O
- **b** X,Y = C(sp^3), Z = NR''
- **c** X,Y = C(sp^3), Z = CR''R'''
- **d** X-Y = *o*-aryl, Z = O
- **e** X-Y = *o*-aryl, Z = NR''
- **f** X-Y = *o*-aryl, Z = CR''R'''
- **g** X = C(sp^3), Y-Z = *o*-aryl
- **h** X = C(sp^3), Y-Z = CO-NR''

I

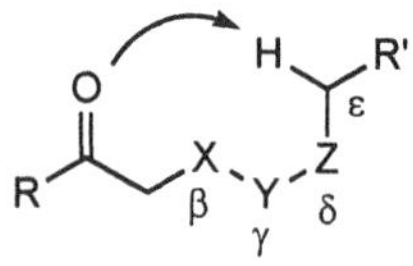

- **a** X = C(sp^3), Y-Z = *o*-aryl
- **b** X = CO, Y-Z = *o*-aryl
- **c** X = C(sp^3), Y-Z = CO-NR''
- **d** X-Y = *o*-aryl, Z = O

II

Scheme 2

atoms or atom groups that do not bear any hydrogen atoms that are susceptible to the excited carbonyl group. The atoms or atom groups that fulfill this blocking function can be heteroatoms (O, N), quaternary carbon atoms, amide groups, or aromatic rings. It should be noted that the hydrogen atoms of aromatic rings are generally not attacked by excited carbonyl groups because of the relatively high C–H bond energy. In Sch. 2 the different "blocking methods" are summarized. The two types **I** and **II**, which represent the reactants for the preparation of five- and six-membered rings respectively, should be classified. The main feature of type **I** is that the γ-position is blocked by an atom group Z and thus a hydrogen abstraction from the δ-position is forced, whereas in type **II** both the γ- and the δ-position are blocked by groups Y and Z and, consequently,

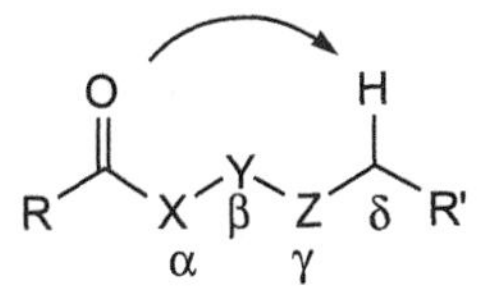

Figure 1

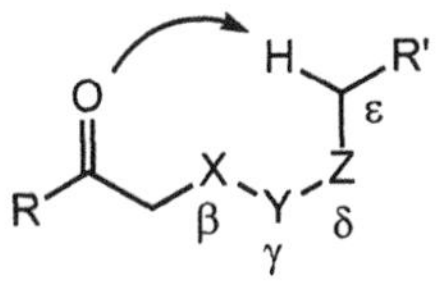

Figure 2

the hydrogen abstraction can proceed from the ε-position. The remaining groups X and Y in type **I** and X in type **II** may also be involved depending on their structure. Various combinations of the atoms and atom groups, mentioned above, provide 12 reactant systems **Ia–h** (Fig. 1) and **IIa–d** (Fig. 2). Their photochemical behavior will be discussed in the following sections.

Though the approach to prevent a hydrogen abstraction from a normally preferred position by introducing blocking groups was quite successful, it comes up against limiting factors, even if larger than six-membered rings have to be prepared. To solve this problem one needs a process in which the rate is still faster than the homolytic hydrogen transfer. This applies to the photoinduced electron transfer (PET) in many cases. The prerequisite of such a PET process is the presence of a functional group with a relatively low oxidation potential, such as amines, thioethers, alkenes, or arenes. The general mechanism is briefly depicted in Sch. 3. Initially, an electron is transferred from the group Y to the excited carbonyl group that provides the radical anion–radical cation pair **IV**. It is an important characteristic feature of this step that it can take place even in case of relatively large distances between the donor group Y and the excited carbonyl group. The strong electron-withdrawing effect of the radical cation, formed from Y induces an increased C–H-acidity at the neighboring carbon atoms and consequently a proton transfer results that gives the biradical **V**. This proton transfer occurs mostly (but not always) from the more remote C–H-position. In the last step, the biradical cyclizes to the products **VI** (Sch. 3).

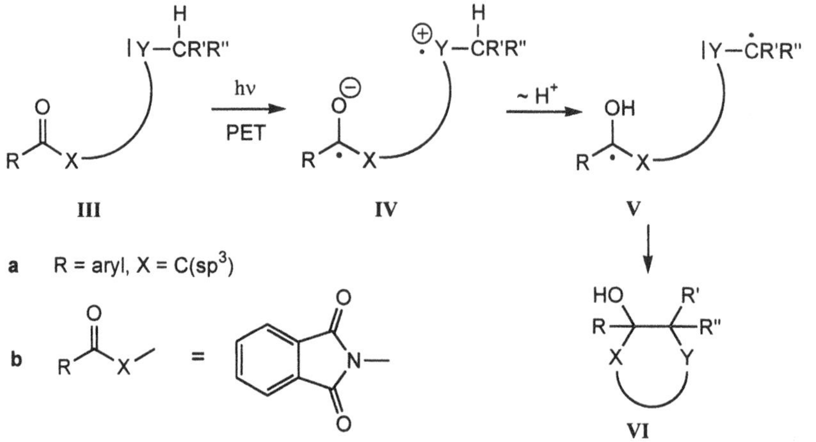

Scheme 3

There were relatively few examples reported, in which the chromophoric group was an aryl ketone (**IIIa**) whereas much more PET induced cyclizations with phthalimides (**IIIb**) are known, sometimes other imides, such as succinimides and maleimides were also used. Phthalimides differ from aryl ketones in some respects despite their apparently similar photophysical properties. On one hand, their reduction potential is remarkably lower in comparison with aryl ketones [13]. On the other hand, the radical anion derived from phthalimides is clearly more stable than the corresponding species from aryl ketones [15]. Both facts increase the thermodynamic driving force of a PET and facilitates applications that are unknown from aryl ketones. In this context, one of the most successful approaches is a PET-induced decarboxylation–cyclization route, developed by Griesbeck [13]. The details of this interesting method will be discussed in Sec. 3.4.6.

All of the so far described concepts have in common that the size of the initially formed biradical, i.e., the number of atoms between the radical centers, directly determines the ring size of the product. The recently developed *spin center shift* [14] considerably extends the scope of the Norrish–Yang reaction. The basic idea of this approach is to tether a leaving group X at the carbon atom adjacent to the carbonyl group. After the intramolecular hydrogen abstraction the hydroxy-substituted biradical undergoes a very rapid elimination of the acid HX that provides a new, keto-substituted biradical. The spin density of the formerly hydroxy-substituted radical center has been shifted to one of the adjacent atoms. Depending on the position of the leaving group, two reactant types **VII** and

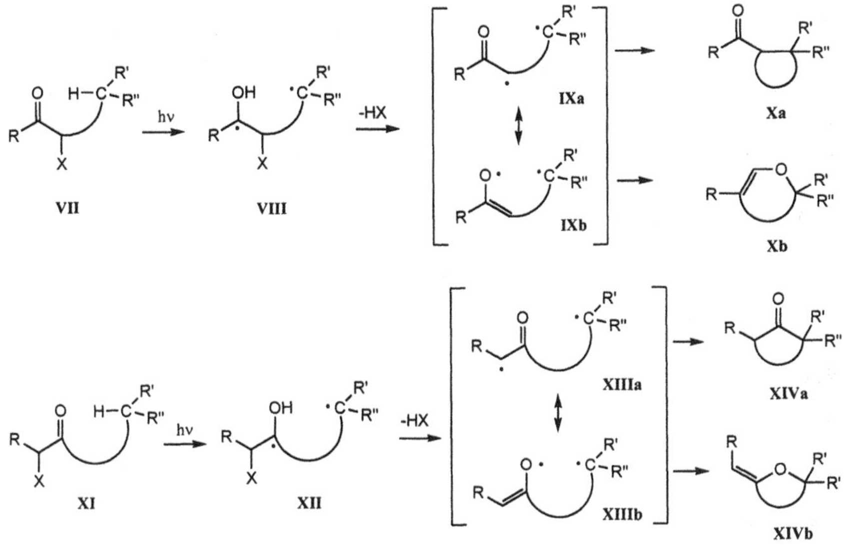

Scheme 4

XI must be distinguished. The photochemical behavior of **VII** and **XI** is summarized in Sch. 4. In type **VII**, the leaving group is placed at one atom of the chain between the reaction centers whereas in type **XI** this leaving group is tethered to an atom behind this chain. The biradicals **IX** and **XIII**, formed after the elimination step, can be considered both as C–C-biradicals **IXa** and **XIIIa** respectively and as O–C-biradicals **IXb** and **XIIIb** respectively. It is noteworthy that all of the carbocyclic (**Xa, XIVa**) and heterocyclic (**Xb, XIVb**) products may be selectively prepared depending on the structure of the reactants.

Until now three basic concepts for the preparation of rings differing from four-membered ones by means of the Norrish–Yang reaction have been discussed, yet two important mechanistic factors must be mentioned. Until now the question of the spin state, of the excited carbonyl compound remained unmentioned. Generally, it depends on the structure of the chromophoric group whether the photochemical reaction course is dominated by a singlet state, a triplet state, or both. The most important difference between these states is the extremely short-lived singlet state in contrast to the relatively long-lived triplet state of the biradicals formed. The consequences of this difference will be discussed in detail in Sec. 3.3.1.

Another important phenomenon, which must not be underestimated, is the tendency of hydroxy-substituted biradicals to undergo the reverse reaction to the reactants by back-transfer of the hydrogen atom. Several examples are known, in which the intrinsic regioselectivity of the hydrogen abstraction by the excited carbonyl group is entirely covered up by the distinct inclination of the preferred biradical to undergo this back-transfer. Therefore, the quantum yields of the product formation and consequently the maximal possible extend of the reverse reaction should always be taken into account in connection with mechanistic considerations.

3.3. SCOPE AND LIMITATIONS

3.3.1. Suitable Reactants

Not every carbonyl compound is a suitable reactant for the Norrish–Yang reaction. In most carbonyl compounds with one or two heteroatoms adjacent to the carbonyl group (carboxylic acids, esters, carbonates, amides, urethanes, and ureas) the electron-donating character of the heteroatom causes an increase of the n,π^* gap. Consequently, these carbonyl compounds have larger n,π^* excitation energies than ketones or aldehydes and are less important for photochemical reactions. Aldehydes were rarely used as reactants in the Norrish–Yang reaction as well most probably owing to the low C–H bond energy in the formyl group. In the following section the most common chromophores and their photophysical properties will be presented.

3.3.1.1. Ketones

3.3.1.1.1. Aliphatic Ketones. Only few applications of the synthesis of carbo- or heterocycles via Norrish–Yang reaction with $(\gamma \pm n)$-hydrogen abstraction starting from dialkyl ketones are known. From a practical chemist point of view, aliphatic ketones are less suitable, because the n,π^* absorption band of aliphatic ketones is located in an unfavorable wavelength region (270–300 nm). Most common glasses are transparent for wavelengths >300 nm (pyrex, solidex, normal lab glass) or >190 nm (e.g., quartz glass). In addition to that, the photophysical properties are undesirable. The n,π^*-lowest singlet excited state of aliphatic ketones undergoes an intersystem crossing to the lowest triplet excited state much slower ($k_{ISC} = 10^8/s$) than the n,π^*-lowest singlet state of aromatic ketones ($k_{ISC} > 10^{10}/s$). As aliphatic ketones have a longer singlet lifetime, products are produced from both the excited singlet and triplet state. From the singlet state the yields of cyclization products are near-zero, because the efficiency of the hydrogen back-transfer is much higher. Contrary to aromatic ketones, the irradiation of aliphatic ketones without γ-hydrogen atom preferentially leads to *Norrish Type I* cleavage products due to their large excitation energies.

3.3.1.1.2.   Aromatic Ketones.   Aromatic ketones are characterized by a strong spin-orbit coupling, because the S_1 and the T_2 state are nearly degenerated. Therefore a rapid intersystem crossing ($k_{ISC} > 10^{10}/s$) of most aromatic ketones usually leads to exclusive triplet reactivity, so that they are much more appropriate for Norrish–Yang reactions [16]. From a synthetic point of view, the absorption band for the n,π^*-transition (300–330 nm) is much more favorable in comparison to that of aliphatic ketones (see Sec. 3.3.1.1.1). However, one must observe that ring substituents have an substantial effect on the efficiency of the triplet hydrogen abstraction [17]. (For details see Sec. 3.3.2.2.)

3.3.1.1.3.   1,2-Dicarbonyl Compounds.

3.3.1.1.3.1.   1,2-DIKETONES.   A large variety of both alkyl- and aryl-1,2-diketones are used as reactants for the Norrish–Yang reaction [18], but no synthetic application with ($\gamma \pm n$)-hydrogen abstraction has been established.

3.3.1.1.3.2.   2-KETO ESTERS.   The Norrish–Yang reaction of 2-keto esters has been the subject of only a limited number of studies [19,20, 21d,22]. Encinas and his coworkers investigated the photochemistry of some alkyl esters of benzoylformic acid [23a]. These aromatic 2-keto esters have a n,π^* absorption maxima >330 nm. Similar to aromatic monocarbonyl compounds all observed products originate from an excited triplet state.

In contrast, aliphatic 2-keto esters show a n,π^* absorption of a low extinction coefficient in the 300–350 nm region [23b]. Excited aliphatic 2-keto esters are not completely converted into the triplet state. Upon addition of either a triplet sensitizer or a triplet quencher, each of the two spin states can be promoted [19].

3.3.1.2.   Imides

Surprisingly, imides possess a photochemical behavior that is very similar to that of ketones. On regarding the imides as *N*-acyl amides it becomes clear that the electron-withdrawing character of the acyl group compensates the electron-donating character of the N-atom adjacent to the carbonyl group. Therefore, imides have n,π^*-excitation energies comparable to those of ketones.

3.3.1.2.1.   Alicyclic Imides.   In contrast to their aliphatic ketone counterparts, *N*-alkyl alicyclic imides (succinimides and glutarimides derivatives) could obtain a certain importance in the Norrish–Yang reaction. With a n,π^* absorption band between 230 and 270 nm, the 254 nm emission of a low-pressure mercury lamp is convenient for exciting

alicyclic imides. Contrary to aliphatic ketones, the *Norrish type I* cleavage doesn't play an important role due to the large C–N bond energy in the imide group.

3.3.1.2.2. Phthalimides. Like their aromatic ketone counterparts, *N*-alkyl phthalimides participate in hydrogen atom abstraction reactions to form a large variety of heterocyclic compounds. However, despite many similarities between phthalimides and aromatic ketones, there are some important differences. In contrast to aromatic ketones, electronically excited phthalimides are not transformed quantitatively into the triplet state, and thus they may react from both the singlet and the triplet state. In addition, phtalimides are much more inclined to photoinduced single-electron-transfer (PET) reactions [24,25]. (For details see Sec. 3.3.2.1.)

3.3.2. Interaction with Functional Groups

In the following section, the effect of different functional groups on the Norrish–Yang reaction will be discussed.

3.3.2.1. Photoinduced Single-Electron-Transfer (PET)

In the presence of an electron-rich functional group electronically excited carbonyl compounds can undergo photoinduced single-electron-transfer (PET) instead of hydrogen abstraction and the electronically excited carbonyl group acts as the electron-acceptor. The equation which describes the change of the free energy associated with the electron transfer is the Rehm–Weller relationship [26]. This equation can easily be used to determine (if all photophysical and electrochemical data are available) or to estimate the thermochemistry of such a reaction. A simplified version is given in Eq. (1).

$$\Delta G_{PET} = E^0_{ox.}(D) - E^0_{red.}(A) + E_{coul} - E^*_{00} \qquad (1)$$

in which $E^0_{ox.}(D)$ is the standard oxidation potential of the donor, $E^0_{red.}(A)$ the standard reduction potential of the acceptor, E^*_{00} the excitation energy, and E_{coul} a coulombic interaction term in a given solvent. Therefore, it is necessary that the excitation energy E^*_{00} is higher than the difference between the redox potentials of donor and acceptor $E^0_{ox.}(D) - E^0_{red.}(A)$ [ΔG_{PET} have to be smaller than 0]. In their original work Rehm and Weller showed that an empirical relationship exists between the rate of electron transfer (k_q) and ΔG_{PET} [26]. If ΔG_{PET} is less than -21 kJ/mol the electron transfer is diffusion-controlled for bimolecular reactions and, consequently, rotation-controlled in the intramolecular case. An intramolecular electron transfer of an excited carbonyl compound primarily leads to a zwitterionic biradical instead of a nonionic biradical. Mostly a mesolytic cleavage [27] of a CH-bond adjacent to the radical cation and therefore a proton transfer to the radical anion or rarely an addition of a solvent molecule (e.g., ROH) follows the electron transfer. The resulting $1,n$-biradical combines to give the corresponding carbo- or heterocyclic compound. Typical electron-donating partners for electronically excited carbonyl compounds are arenes, alkenes, amines, and thioethers.

As mentioned in Sec. 3.3.1.2.2 PET-promoted cyclization reactions are above all a domain of phthalimides in contrast to aromatic ketones. It is well-known that the PET process of phthalimides occurs via the S_1 (π,π^*) [28] excited state which has an excitation energy of about 335 kJ/mol [29] and, therefore, is unsignificantly higher than the triplet energy of aromatic ketones (e.g., valerophenone: $E_{00}^* = 300$–311 kJ/mol, depending on the solvent used) [30]. However, the reduction potential of pthalimides is remarkably lower compared with aromatic ketones (e.g., N-methylphtalimide in DMF: $E_{red.}^0 = -1.37$ V [25a], valerophenone: $E_{red.}^0 = -2.07$ V [30]). The radicals **A** and the radical anions **B** formed by hydrogen or electron transfer onto the triplet excited phthalimide are more stable than the corresponding species derived from aryl ketones (**C** and **D**), even though the triplet energies are very close (**A** is more stable than **C** by about 29 kJ/mol, Sch. 5) [31].

This stabilization of the radical intermediates, arising from a better mesomeric stabilization of radicals in the phthalimide moiety, consequently increase the exoenergicity of reactions and, according to the Bell–Evans–Polanyi principle, lowers the activation barrier and thus enables processes that are unknown from ketones. The unique photochemical reactivity of phthalimides will be demonstrated with some examples.

With regard to the simplified version of the Rehm–Weller relationship (1) follows that in the presence of an electron donor with an oxidation potential less than ca. 2.1 V an exergonic electron transfer can occur [25b].

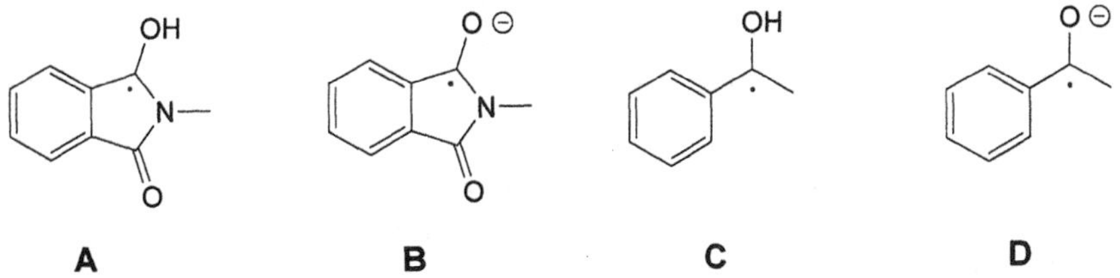

    A   **B**   **C**   **D**

Scheme 5

3.3.2.2. CT-Quenching

It is well-known that the carbonyl triplets of aromatic ketones are efficiently deactivated by β-aryl [7a,32] or β-vinyl [33] groups. This quenching process requires a n,π^* triplet state and typically leads to triplet lifetimes of about 1 ns at room temperature and is rather insensitive to substituent effects. It is suggested that this quenching process involves charge transfer interactions [32a].

3.3.2.3. Substituent Effects

3.3.2.3.1. Ring Substituents. One has to bear in mind that two main effects of ring substituents on the Norrish–Yang reaction of aromatic ketones exist.

First of all, ring substituents have a substantial effect on the efficiency of the triplet hydrogen abstraction. Aromatic ketones can have either an n,π^* or a π,π^* lowest triplet state, of which the n,π^* triplet state (with a distinct spin localization at the oxygen atom) is much more reactive than the π,π^* triplet state (with a delocalized spin density). In ring-unsubstituted phenyl ketones the n,π^* triplet is below the lowest π,π^* triplet. Comprehensive investigations of Wagner on ring-substituted phenyl ketones [17a] have illustrated that the substituent effects on triplet reactivity involve several factors and don't follow a simple rule. The essential message for a practical chemist is that in most cases an electron-withdrawing group can increase and an electron-donating substituent can reduce the rate constant for triplet hydrogen abstraction. However, only very strong electron donors (-OR, -NR$_2$ with R = H or alkyl) are able to produce large enough gaps between the n,π^*- and π,π^*-lowest triplet ($\pi,\pi^* \ll n,\pi^*$), so that reactivity becomes undetectable. In polynuclear aromatic ketones and biphenylyl ketones the difference between the n,π^*- and π,π^*-lowest triplet state is large enough that only $\pi-\pi^*$ reactivity can be observed.

Second of all, on certain conditions an appropriate *ortho*-substituent in electronically excited aryl ketones (-R, -OR, -NR$_2$ with R = alkyl) can serve as alternative hydrogen donor, i.e., an hydrogen atom may be abstracted from this *ortho*-substituent rather than from the side chain. In this way, carbo- and heterocycles with an annulated aromatic ring are accessible via Norrish–Yang reaction. (For synthetic applications see Secs. 3.4.2 and 3.4.3.)

3.3.2.3.2. Substituents in α-Position. To rationalize the ratio between *Norrish type II* cleavage and cyclization (see preceding chapter) the influence of α-substituents on cyclization were studied exclusively on straight-chain aryl ketones with preferential γ-hydrogen abstraction.

However, the *Norrish Type II* cleavage doesn't matter in carbonyl compounds with ($\gamma \pm n$)-hydrogen abstraction. In this case, one has to bear in mind that electron-donating α-substituents can promote the *Norrish Type I* cleavage by stabilizing the resulting radical. For that reason, α-cleavage can compete with the desired cyclization.

A completely different situation arises if an adequate leaving group X is introduced at the carbon atom adjacent to the carbonyl group. In this case an elimination of the acid HX occurs at the stage of the $1,n$-biradical (after hydrogen abstraction) to form a ($1,n \pm 1$)-biradical. (For details see Secs. 3.4.1.2, 3.4.2.2, and 3.4.3.2.)

3.4. SYNTHETIC POTENTIAL: REACTIVITY AND SELECTIVITY PATTERN

3.4.1. Cyclopropanes

3.4.1.1. By β-Hydrogen Abstraction

The precondition for the synthesis of cyclopropanes as products of a Norrish–Yang reaction is the formation of 1,3-biradicals and consequently a hydrogen abstraction from the β-position with respect to the carbonyl group. The geometrical parameters of the corresponding five-membered transition state, especially the O–H–C bond angle, differ substantially from the ideal parameters reported by Scheffer [34]. Therefore, it is not surprising that only few syntheses of cyclopropanes via the Norrish–Yang reaction are known [35,36]. In most of these cases, it was either proven or considered to be very likely that the initial step is a photoinduced electron transfer (PET). In addition, the hydrogen atom is transferred not homolytically, but a proton shift occurs after PET. These conditions limit the preparative scope because the electron rich functional groups responsible for the PET cause a considerable sensitivity of the cyclopropanes to oxidative ring opening.

The best investigated compounds that react in this manner are β-dialkylamino propiophenones **4** [35]. Upon irradiation, they are converted into triplet-betains **5** by intersystem crossing (ISC), PET and afterwards a proton migration takes place to give the triplet 1,3-biradicals **6**. After ISC back to the singlet state, these biradicals cyclize to cyclopropanes **7**, which are sensitive against oxygen and therefore readily converted into α,β-unsaturated ketones **8** (Sch. 6).

The cyclization to **7** proceeds with high diastereoselectivity. Furthermore, from a mechanistic point of view it is noteworthy that, if enantiomerically pure aminoketones **4** are used, only one enantiomer of the cyclopropanes is obtained [35e]. Obviously, the biradicals **6** are so

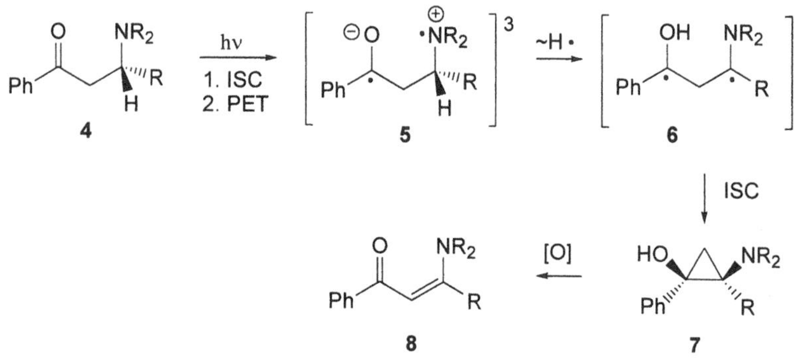

Scheme 6

short-lived (<1 ns) that no rotations can take place around the C–C-single bonds and, consequently, the chirality information of the reactant **4** is conserved in the course of the photochemical cyclization. This observation is a very early and hardly noticed example of the memory effect of chirality in photochemical reactions.

3.4.1.2. By γ-Hydrogen Abstraction and Subsequent Spin Center Shift

The task to generate a 1,3-biradical in order to prepare a cyclopropane, may also be achieved by an elegant reaction cascade called *spin center shift*. The method utilizes a well-known behavior of monoradicals for the chemistry of biradicals that are produced in the Norrish–Yang reaction. Thus, if a leaving group X is introduced at the carbon atom adjacent to the carbonyl group (**9**), an elimination of the acid HX occurs at the stage of the biradicals **10**, which proceeds fast enough to compete with "classic" Norrish–Yang processes (cleavage and cyclization). As a result, the hydroxy-substituted site of biradicals **10** is converted into an enolate radical (oxoallyl radical) and the spin density is shifted to the adjacent atom. In fact, a 1,4-biradical (**10**) is transformed into a 1,3-biradical (**11**) [14a]. After ISC the biradicals **11** cyclize to the benzoylcyclopropanes **12** (Sch. 7, Table 1).

The method tolerates a wide range of functional groups. Though the lifetimes of biradicals **11** are not yet measured, the experimental results strongly suggest that they are very short (<25 ns). This can be concluded especially from the irradiation of the cyclopropyl substituted compound **9f**, which cyclized in good yields to the *bis*-cyclopropyl ketone **12f** without any detectable opening of the cyclopropyl ring. In general, cyclopropylcarbinyl

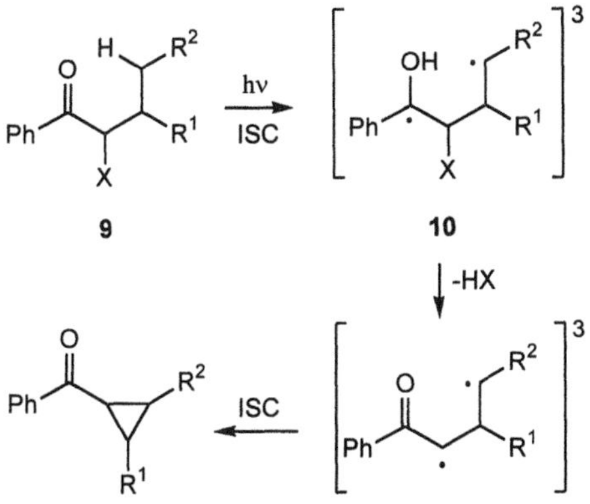

Scheme 7

Table 1 Yields of Cyclopropanes **12**.

12	R^1	R^2	X	Yield (%)
a	H	H	OMs	87
b	Me	H	OMs	90
c	COOMe	H	ONO_2	59
d	H	OBn	OMs	46
e	H	Ph	OMs	78
f	H	$C_3H_5{}^a$	OTs	68
g	H	$C_2H_3{}^b$	OTs	80

Ms = Methanesulfonyl, Ts = 4-Toluenesulfonyl.
[a]Cyclopropyl.
[b]Vinyl.

radicals undergo a very fast ring opening to butenyl radicals ($k = 4.0 \times 10^7/s$) and this reaction is often used to determine radical lifetimes ("radical clock") [37].

The cyclopropane synthesis is also suitable for the preparation of highly strained bicyclic hydrocarbons such as [2.1.0]bicyclopentanes (**14**) and spiropentanes (**16**) [14a,b]. The formation of the spiropentane **16** is particularly remarkable as it is the result of a homolytic hydrogen abstraction from a cyclopropane ring. Those processes are very rarely observed due to the relatively high C–H-bond energies of cyclopropanes (Sch. 8).

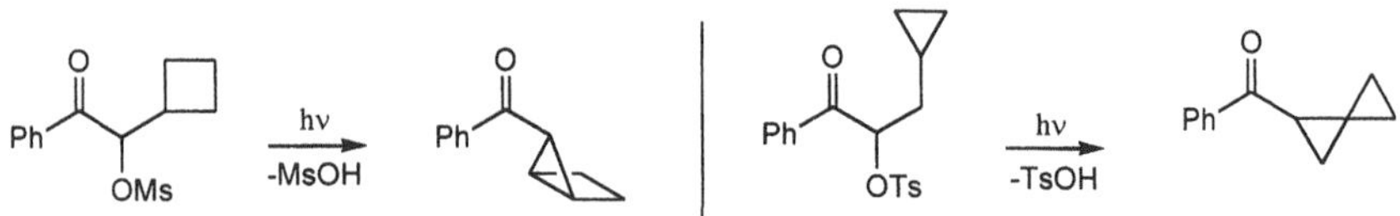

Scheme 8

3.4.2. Five-Membered Rings

3.4.2.1. By δ-Hydrogen Abstraction

As outlined in Sec. 3.2, the most convenient way to five-membered rings is to block the γ-position in order to avoid the hydrogen abstraction from this position. In the following section, each of the formal blocking types **Ia–h**, summarized in Sch. 2, will be verified with concrete examples.

Ia: $X, Y = C(sp^3)$, $Z = O \rightarrow$ Tetrahydrofurans

Taking into account the wide-spread occurrence of molecules containing the tetrahydrofuran substructure, it is astonishing that there are only few examples of synthesis of these heterocycles by means of the Norrish–Yang reaction [38]. Interestingly, both aliphatic and aromatic ketones were used as exemplified by the photocyclization of ketones **17** [38b], **19** [38c], and **21** [3a] (Sch. 9).

Ib: $X, Y = C(sp^3)$, $Z = NR'' \rightarrow$ Pyrrolidines

In 1981 Henning and his coworkers developed a synthesis of pyrrolidines from β-aminopropiophenones [39a]. It should be noted that the introduction of an electron withdrawing group at the nitrogen atom was decisive for circumventing the cyclopropane formation dicussed in the preceding section. In due course, Henning thoroughly investigated the influence of various substituents and developed a diastereoselective route to substituted prolines [39b,c]. Based on this work, enantiomerically pure prolines were prepared. Thus, on irradiation the *N*-substituted glycine amide **23** bearing a C_2-symmetric chiral auxiliary gave the proline amide **24** in high yields and in a fully diastereoselective manner [40]. Besides the auxiliary controlled asymmetric synthesis, the utilization of the chiral pool of natural products is wide-spread and was also applied on the photochemical synthesis of prolines. Starting from aspartic acid **25**, the glycine esters **26** were prepared and stereoselectively cyclized to pyrrolidines **27**, which may be regarded both as L- and D-prolines. Remarkably, the three protective groups in **27b** are orthogonal to each other and can be removed independently, which makes this pyrrolidine derivative become a very

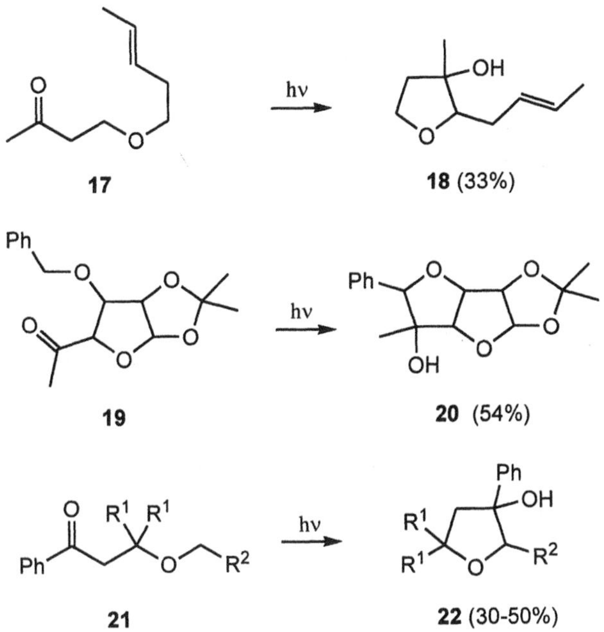

Scheme 9

attractive building block for peptide synthesis [41]. The stereoselectivity of the formation of pyrrolidines **24** and **27** demands that the intermediate biradicals have sufficient lifetimes to allow conformational equilibration. This applies only to triplet biradicals which are formed exclusively from phenyl ketones. A completely different situation is present when the photoreaction commences with singlet-excited ketones and singlet biradicals are formed. Due to the extremely short lifetimes of these singlet biradicals, conformational changes are almost impossible at this stage and, consequently, the stereochemical information of the reactant can be conserved even if a chirality center is formally destroyed by hydrogen abstraction. Such a behavior was recently observed on the irradiation of alanine derivative **28** bearing an α-ketoester side chain at the nitrogen atom. If the triplet state of the ketone is quenched by addition of naphthalene, a considerable memory effect of chirality with respect to the alanine moiety is observed, though the diastereoselectivity was as low as expected (Sch. 10) [19].

Ic: $X, Y = C(sp^3)$, $Z = C''R''' \rightarrow$ Cyclopentanes

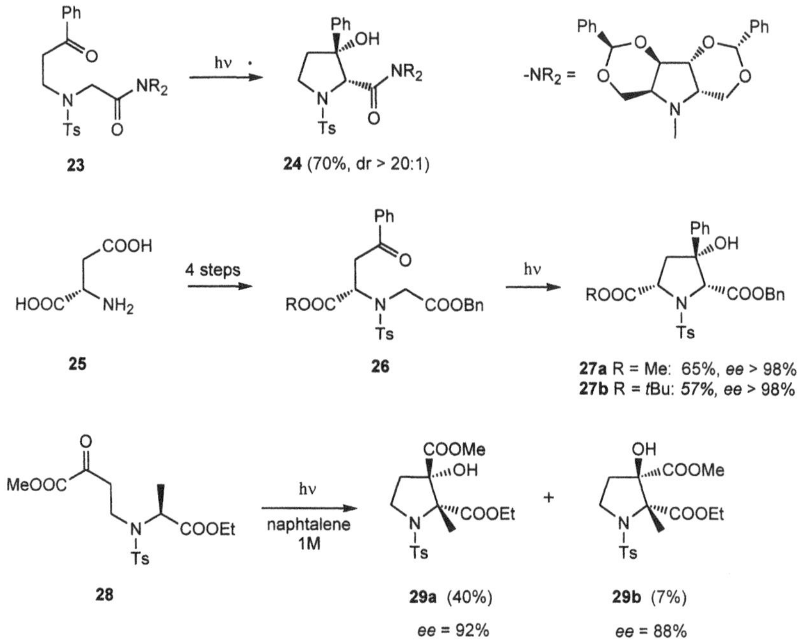

23   **24** (70%, dr > 20:1)

25   **26**   **27a** R = Me: 65%, ee > 98%
                                               27b R = tBu: 57%, ee > 98%

28   **29a** (40%)   +   **29b** (7%)

                    ee = 92%   ee = 88%

Scheme 10

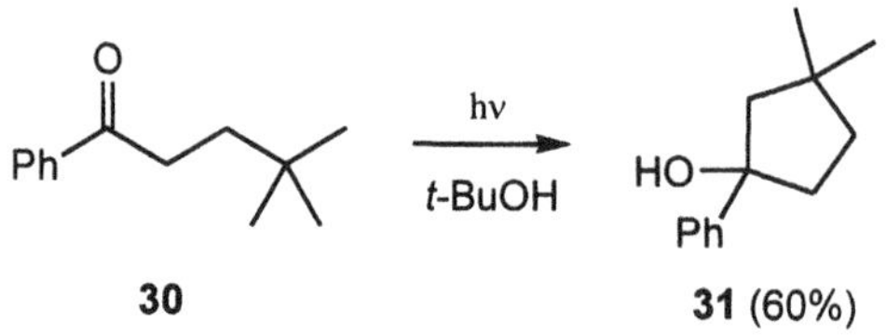

30   **31** (60%)

Scheme 11

There is only one example of this approach, which was elaborated by Wagner in 1972 (Sch. 11) [3a]. The formation of cyclopentane **3** from δ-methoxyvalerophenone **1** has already been mentioned in Sec. 3.2 (cf. Sch. 1).

Id: $X, Y = o$-Aryl, $Z = O \rightarrow$ Benzofuranes

The replacement of the two sp^3-carbon atoms between the oxygen atom and the carbonyl group in type **Ia** by an aromatic ring furnishes the *o*-alkoxy-alkyl-aryl ketones **32**, the photochemical behavior of which has been

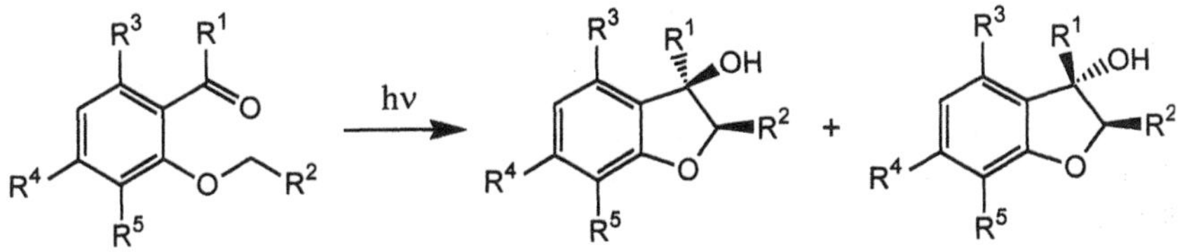

Scheme 12

Table 2 Yields of Benzofuranes **33**.

R^1	R^2	R^3	R^4	R^5	Yield (%)[a]	Ref.
Ph	H	H	H	CO-Ph	77	42a
Ph	Ph	H	H	CO-Ph	100	42a
Me	Ph	H	H	CO-Me	17	42a
Me	OMe	O-MOM	OMe	H	47	42b
Me	OMe	O-MOM	H	H	77	42b
Me	H	OMe	H	H	55	42b
(CH₂)₂OH	OMe	O-MOM	H	H	30	42b
Ph	Ph	H	O(C₁₂H₂₅)	H	67	4a
H	COOEt	H	H	H	46	42c
Me	COOEt	H	H	H	86	42c
COOMe	Ph	H	H	H	89	42d

MOM = MeOCH₂.
[a]Yield of **33a** + **33b**.

intensively studied. The variability of substituents at the two side-chains as well as at the aromatic core, tolerated by the photocyclization, is remarkable. Furthermore, in many cases only one diastereomer of the benzofuranes **33** is obtained upon irradiation of **32** (Sch. 12, Table 2) [4,42].

Ie: X, Y = o-Aryl, Z = NR″ → Indoles

A special case exists if the two carbon atoms between the carbonyl group and the nitrogen atom (cf. type **Ib**) are members of an aromatic ring. Due to the strong electron donating character of the nitrogen atom, the lowest excited state of these compounds reveals π–π*-character, and consequently there is hardly any Norrish–Yang reactivity to be observed. But the introduction of electron acceptor groups holds pitfalls, too. Taking into account that the nitrogen atom is a part of the chromophore, it is

not very surprising that the photochemical behavior of compounds
34 (R² = Tosyl) is mainly characterized by a homolytic cleavage of the
N–S-bond. Reactions like that were described previously for N-tosyl-β-
amino-α,β-unsaturated ketones [43]. Only the trifluoromethane sulfonyl
group (Tf) facilitates the formation of 2,3-dihydroindoles **35** [20]. Another
interesting approach to suppress the electron donating character of the
nitrogen atom is to let it become a part of quaternary ammonium salts such
as **36**, which cyclizes upon irradiation giving the indoline **37** (Sch. 13) [44].

If: $Y = o$-Aryl, $Y, Z = CR''R''' \rightarrow$ Indanes
Ig: $Y = C(sp^3)$, $Y, Z = o$-Aryl $\rightarrow$ Indanes

These two concepts were seldom used for preparative purposes, but
proved to be useful for mechanistic considerations (Sch. 14) [3b,45,46].

Ih: $X = C(sp^3)$, $Y, Z = CO-NR'' \rightarrow$ Pyrrolidin-2-ones (γ-Lactams)

The necessity of protecting the nitrogen atom by an electron
withdrawing group to avoid PET, mentioned in connection with type **Ib**,
can also be carried out by the introduction of a carbonyl group in the
β-position, resulting in β-ketoamides **42**. On irradiation they provided an

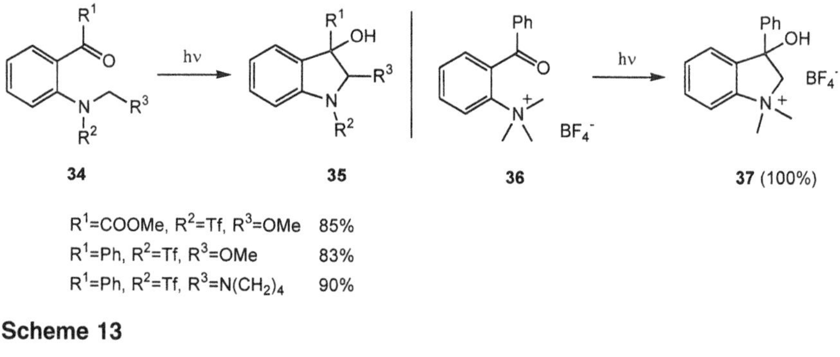

34   35   36   37 (100%)

R¹=COOMe, R²=Tf, R³=OMe 85%
R¹=Ph, R²=Tf, R³=OMe   83%
R¹=Ph, R²=Tf, R³=N(CH₂)₄ 90%

Scheme 13

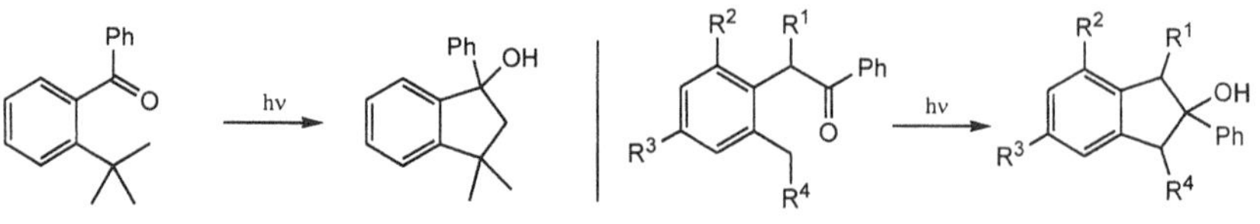

38   39   40   41

Scheme 14

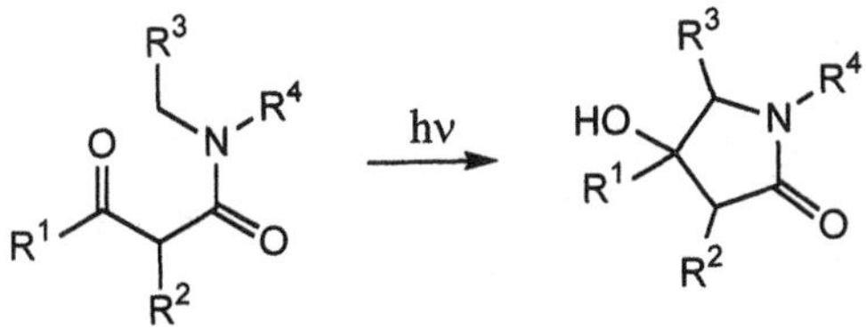

Scheme 15

Table 3 Yields of γ-Lactams **43**.

Entry	R^1	R^2	R^3	R^4	Yield (%)	Ref.
a	COOMe	H	-(CH$_2$)$_3$-		44	21a
b	Ph	H	Ph	CH$_2$Ph	83	21c
c	Ph	Et	Me	Et	66	21d
d	Ph	iBu	Ph	CH$_2$Ph	86	21e

access to γ-lactams **43** (Sch. 15) [21]. The reaction possesses a considerable variability with respect to the residues R^1–R^4, as shown with the examples summarized in Table 3. Notably, δ-hydrogen transfer proceeds even in cases in which γ-hydrogen atoms in R^2 are present (**42c,d**).

3.4.2.2. By γ-Hydrogen Abstraction and Subsequent Spin Center Shift

The idea of the *spin center shift*, which was successfully applied to the preparation of cyclopropanes (cf. Sec. 3.4.1.2) can be extended to five-membered rings. Accordingly, *o*-alkyl substituted aryl ketones **44** that bear a leaving group in the α-position undergo γ-hydrogen abstraction giving 1,4-biradicals **45**, which entirely eliminate the acid HX, providing 1,5-biradicals **46**. After ISC the 1,5-biradicals cyclize to indanones **47** in good yields (Sch. 16, Table 4) [14d]. This reaction was originally developed for photochemically removable protective groups [47], but not used to prepare such substances so far.

3.4.3. Six-Membered Rings

3.4.3.1. By ε-Hydrogen Abstraction

In a similar manner as described for five-membered rings, the synthesis of six-membered rings may be attained by blocking positions bearing available

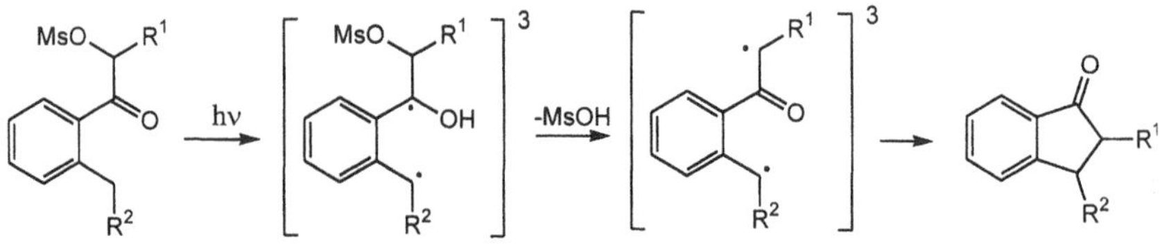

Scheme 16

Table 4 Yields of Indanones **47**.

Entry	R^1	R^2	Yield (%)	Solvent
a	H	H	42	CH_2Cl_2
b	Me	H	73	CH_2Cl_2
c	Et	H	69[a]	MeOH
d	iPr	H	46[b]	MeOH
e	Ph	H	52	MeOH
f	COOEt	H	82	tBuOH
g	$CON(CH_2)_4$	H	70	tBuOH
h	H	Me	70	MeOH

[a]+3% **47a**.
[b]+27% **47a**.

C–H-bonds. Though, both the γ- and the δ-position have to be occupied by such blocking moieties. The four formal blocking types **II**, depicted in Sch. 2, will be discussed here in detail.

IIa: $X = C(sp^3)$, $Y–Z = o$-Aryl → Naphthalines
IIb: $X = C(O)$, $Y–Z = o$-Aryl → Naphthalines

This approach is taken by compounds **48**. The preparative applicability of compounds bearing two sp^3-carbon atoms between the carbonyl group and the aromatic ring (**48a**) is limited, because the quantum yields of product formation are very low ($\Phi = 0.0012$ in MeOH for **48a**). The low reactivity is attributed to a rapid internal CT quenching as a result of an efficient interaction between the β-aryl and the carbonyl group in the excited state [7a]. For a moment, the behavior of diketones **48b–d** is surprising, because they bear two carbonyl groups. Based both on enthalpic and entropic considerations one would expect, that preferably the carbonyl group adjacent to the aromatic ring reacts, affording benzocyclobutenes.

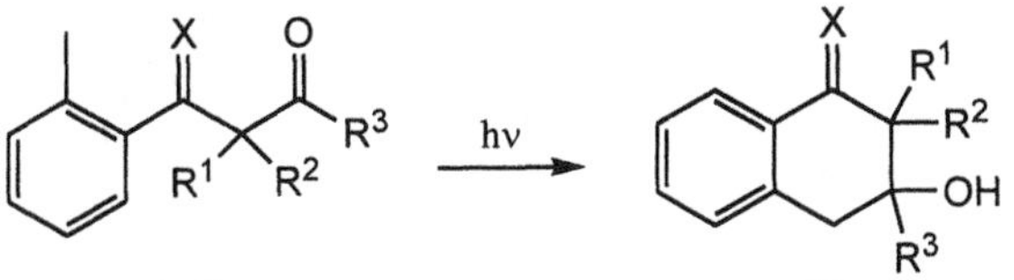

Scheme 17

Table 5 Yields of Naphthalines **49**.

Entry	R^1	R^2	R^3	X	Yield (%)	Ref.
a	H	Me	Ph	CH_2	60–90	7a
b	H	H	Me	O	35	7b
c	Me	Me	Prop	O	85	7c
d	Me	Me	Me	O	88	7c

Indeed, these products are isolated if irradiation is performed in hexane whereas the utilization of methanol as a solvent furnishes benzocyclobutenes only in trace amounts. The unusual outcome of the reaction was explained either by a hydrogen shift from a primarily formed hydroxy benzyl radical to the aliphatic carbonyl group [7b] or by a radical attack of the o-benzyl radical onto the aliphatic carbonyl group forming an alkoxy radical [7c] (Sch. 17, Table 5).

IIc: $X = C(sp^3)$, $Y-Z = CO-NR'' \rightarrow \delta$-Lactams

From a synthetic point of view, the extension of the concept, described in Sec. 3.4.2.1 (type **Ih**), on γ-ketoamides, does not only provide an attractive method but also a molecular system that is very valuable for the improvement of the mechanistic understanding of the Norrish–Yang reaction. Thus, on irradiation γ-ketoamides **50** cyclize to δ-lactams **51** in good yields [48]. It should be noted that the reaction outcome strongly depends on the solvent used. Whereas the irradiation in dichloromethane gives the δ-lactams **51**, the only products formed in diethylether were pinacols [49]. The formation of cyclopropanes from **50** [50] described earlier is probably based on a misinterpretation of the spectral data [49].

The δ-lactams bear two newly formed stereogenic centers and thus two diastereomers may be formed. The ratio of these diastereomers depends on the ring size of the cyclic secondary amine moiety in **50**, on the substituents and on the solvent. It has been shown that small rings and substituents at

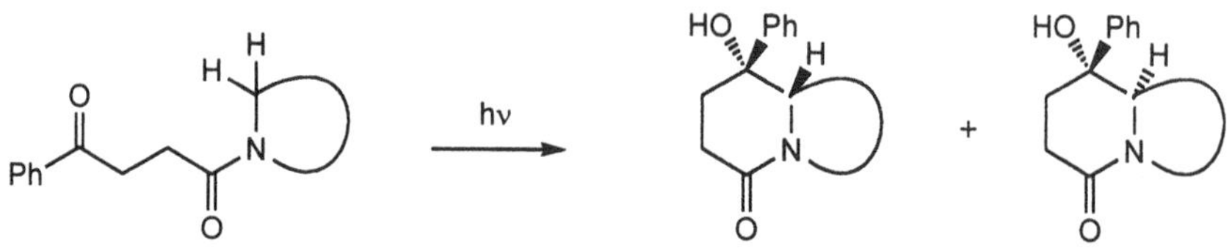

Scheme 18

Table 6   Yields of δ-Lactams **51**.

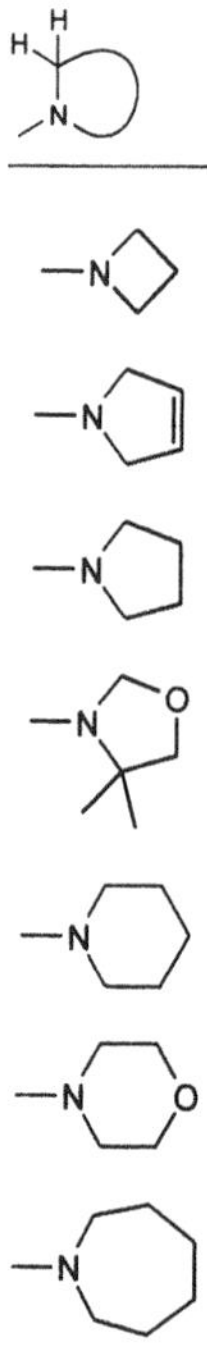

	51a [%]	51b [%]	*de* (51a)
—N (azetidine)	67	20	>99
	54	9	71
—N (2,5-dihydropyrrole)	55	0	>99
	33	8	61
—N (pyrrolidine)	34	8	61
	48	6	78
—N (oxazolidine, dimethyl)			
—N (piperidine)			
—N (morpholine)			
—N (azepane)			

the carbon atom adjacent to the nitrogen atom increase the diastereoselectivity in favor of **51a** (Sch. 18, Table 6). This could be substantiated by consideration of different triplet biradical conformers [48].

The stereochemical situation is more complicated if a chirality center is already present in the reactant. In this case, two phenomena must be

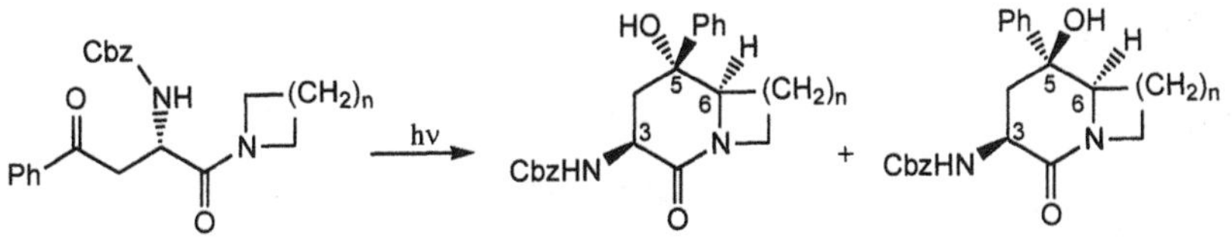

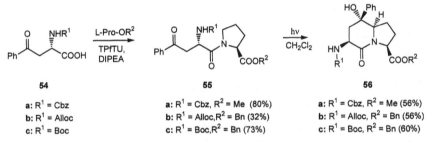

Scheme 19

n=1 67%   29% (de=40%)
n=2 54%   16% (de=54%)
n=3 41%   9% (de=64%)
n=4 68%   26% (de=45%)

54

a: R^1 = Cbz
b: R^1 = Alloc
c: R^1 = Boc

55

a: R^1 = Cbz, R^2 = Me (80%)
b: R^1 = Alloc, R^2 = Bn (32%)
c: R^1 = Boc, R^2 = Bn (73%)

56

a: R^1 = Cbz, R^2 = Me (56%)
b: R^1 = Alloc, R^2 = Bn (56%)
c: R^1 = Boc, R^2 = Bn (60%)

TPfTU=O-Pentafluorphenyl-N,N,N,N-tetramethyluronium tetrafluoroborate

Scheme 20

considered: the simple diastereoselectivity and the asymmetric induction. Commencing with aspartic acid, the enantiomerically pure benzoylalanine amides **52** were prepared and irradiated to give the ornithine lactams **53** (Sch. 19) [51]. Later results with other substituents were also obtained by Griesbeck [52].

Despite the fact that the asymmetric induction from the chirality center C3 on C6 proceeds with high efficiency (>99%), the simple diastereoselectivity with respect to the bond C5–C6 is only moderate. That is why it was not very promising to complicate the system by introducing of additional chirality centers. Therefore, the photochemical behavior of dipeptides **55** is surprising. Prepared from N-protected benzoylalanines **54** by peptide coupling with proline esters, they cyclize to "bicyclic β-turn dipeptides" **56**, which are potential peptide mimetics, in good yields. The high stereoselectivity was explained by a synergistic effect of the two chirality centers at the stage of triplet biradicals (Sch. 20) [53].

IId: $X, Y = o\text{-Aryl}, Z = O \rightarrow$ Chromanes

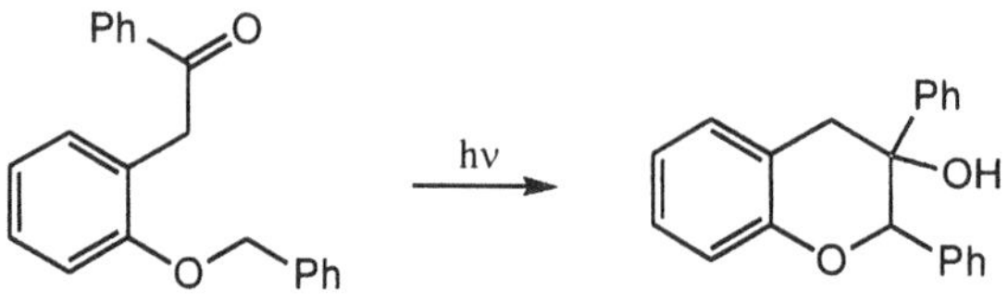

57   **58** (80%)

Scheme 21

Little is known about the synthesis of six-membered rings that bear an oxygen atom. In 1985, Meador and Wagner described the photocyclization of α-(o-benzyloxyphenyl)acetophenone **57**, which may formally be derived from ketones **32** (cf. Sec. 3.4.2.1, type **Id**) by insertion of a methylene group between the keto group and the aromatic ring. The desired 3-hydroxychromane **58** was obtained in 80% yield as a 1.6/1 mixture of diastereomers (Sch. 21) [54].

3.4.3.2. By δ-Hydrogen Abstraction and Subsequent Spin Center Shift

The idea of a *spin center shift* as an extension of the synthetic scope of the Norrish–Yang reaction was already discussed twice in this chapter. A third example applies to the preparation of six-membered rings. Thus, by treating β-ketoamides **42** (Sec. 3.4.2.1, **Ih**) with hypervalent iodine(III) reagents the α-methanesulfonyloxy-β-ketoamides **59** are obtained in excellent yields. On irradiation, the 1,5-biradicals **60** are formed, which are converted into biradicals **61** by elimination of methanesulfonic acid.

In contrast to the applications of *spin center shift* mentioned above, the enolate radical moiety of biradicals **61** reacts solely as oxygen radicals **61-A** to give the oxazinones **62**. This result was explained considering the relative energy of different biradical conformers (Sch. 22, Table 7) [55].

3.4.4. Bi- and Tricyclic Compounds

The formation of bi- and tricyclic compounds by transannular photochemical cyclization of monocyclic reactants mostly follows special rules that are caused by the molecular rigidity, equilibria of ring conformers and distinct stereoelectronic effects and, therefore, will be discussed in a separate section. Naturally, it is no longer possible to strictly structure the reactions according to the size and the type of the ring and, furthermore, vastly different examples are known. Therefore, only some representative examples will be discussed in this section and the selection does not claim to be complete.

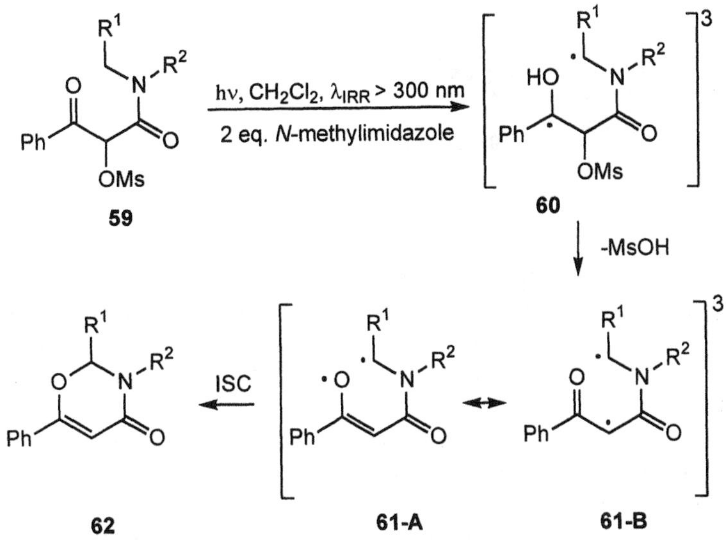

Scheme 22

Table 7 Yields of Oxazinones **62**.

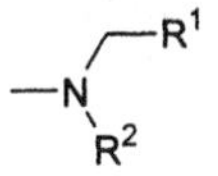

R^1, R^2 on N	Yield (%)
-NMe$_2$	~~48~~
azetidinyl	42
pyrrolidinyl	57
piperidinyl	64
azepanyl	42
morpholinyl	

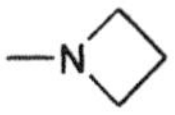

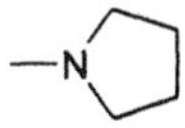

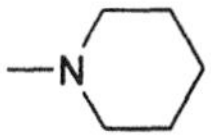

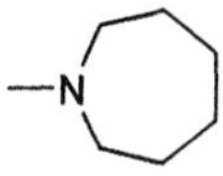

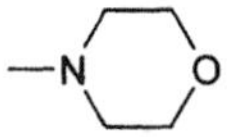

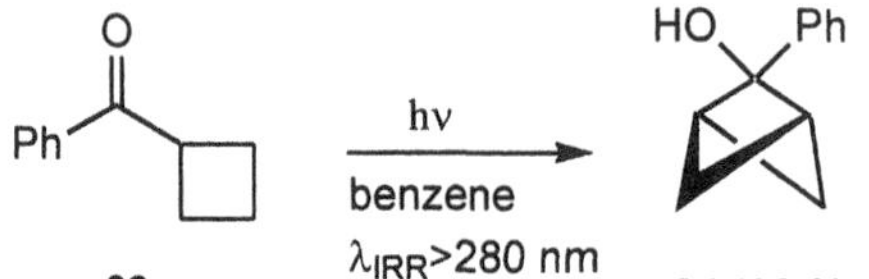

Scheme 23

One of the first examples of a transannular hydrogen abstraction is Padwa's bicyclo[1.1.1]pentane synthesis published in 1967 [56]. By virtue of the considerable ring strain of bicyclo[1.1.1]pentanes the formation of **64** from **63** impressively demonstrates the synthetic scope of the Norrish-Yang reaction (Sch. 23).

Two other examples of the synthesis of bicyclic compounds whose skeleton consists only of carbon atoms are shown in Sch. 24, each of them is instructive in its own way. Upon irradiation of the cyclohexane derivative **65**, not the expected product from a hydrogen abstraction from the allylic position, but the bicyclo[3.3.1]heptane **67** was obtained by cyclization of the less stable biradical **66-B** [57]. The reason for this is the exclusive hydrogen back-transfer of the more stable biradical and it underlines the importance of this process in regioselectivity phenomena of the Norrish–Yang reaction.

The bicyclo[2.2.1]heptane **71** is not directly formed in the course of a Norrish–Yang reaction but the primarily formed 1,4-biradical is trapped intramolecularly by an allyl group in a kind of 1,5-*exo*-cyclization of a hexenyl radical well-known from monoradical chemistry [58]. Disappointingly, in most cases the cyclization of alkenyl radicals proceeds too slowly [59] to compete with the "normal" reactions of the Norrish–Yang biradicals and thus this concept seems not to be commonly applicable. Furthermore, the decay quantum yield of **68** is very low, which is caused by a charge transfer quenching of the excited carbonyl group by the vinyl group (Sch. 24).

Different C–H-bond energies as well as different ring conformers of the reactant may be responsible for the photochemical reaction course of 1,4-ditosylpiperazines **72**. From the chair-like conformer **72-A**, the 3,8-diazabicyclo[3.2.1]octanes **73** are formed by a hydrogen abstraction from one of the alkyl groups affixed to the 6-position of the piperazine ring in yields between 3% and 6%, whereas 2,5-diazabicyclo[2.2.1]heptanes **74** are formed from conformer **72-B** in yields between 15% and 29% (Sch. 25) [60].

The regioselective attack of nonactivated C–H-bonds by an excited carbonyl group is demonstrated with the transannular cyclization of bicyclic diketones **76**, which can easily be prepared from cyclic ketones **75** according to Stetter et al. [61]. Upon irradiation, the selectively excited benzoyl

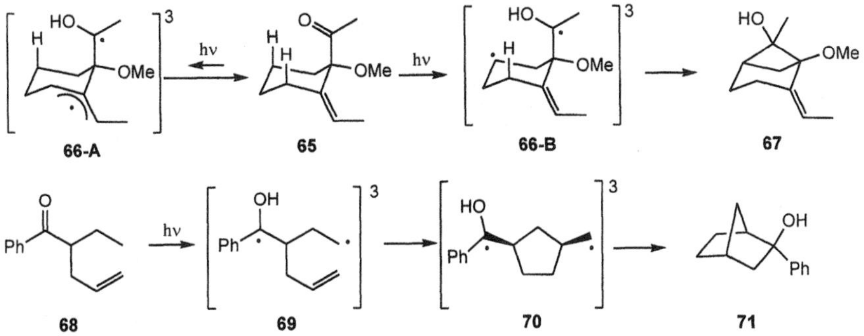

Scheme 24

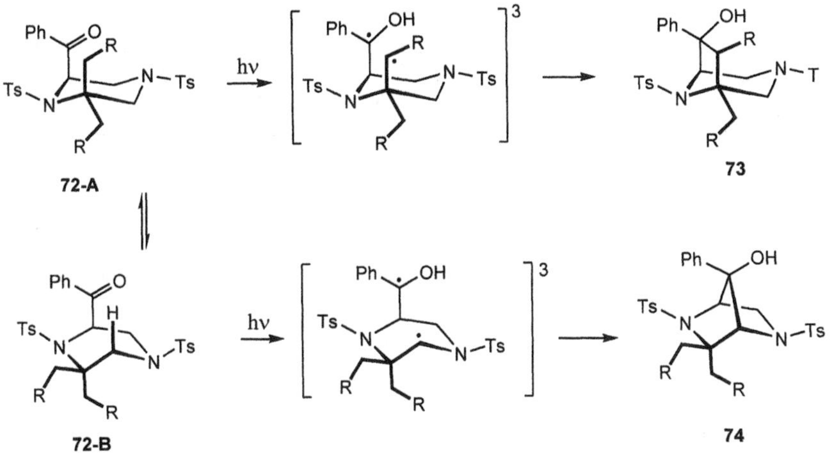

Scheme 25

group abstracts a hydrogen atom from the formal δ-position, because the γ-position is unattainable due to the rigid molecular framework. Whereas the noradamantane **77a** was obtained as a mixture of diastereomers, the heterocyclic compounds **77b,c** were formed fully diastereoselectively with respect to the newly formed C–C-bond despite lower yields. This result was substantiated by the directing effect of intramolecular hydrogen bonds at the stage of triplet biradicals (Sch. 26) [62].

The strategy outlined in Sch. 26 was extended to enantiomerically pure tricyclic α-amino acids. On irradiation the bicyclic diketones **78** prepared in a few steps from cheap 4-hydroxyproline undergo a fully

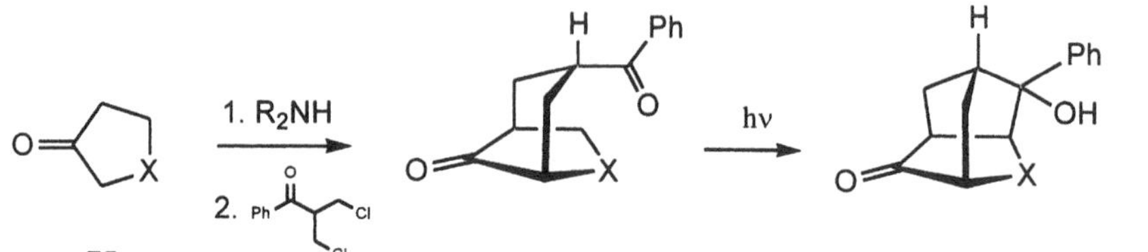

a X = CH$_2$ 71%
b X = NCbz   55%
c X = CH$_2$NCbz 37%

Scheme 26

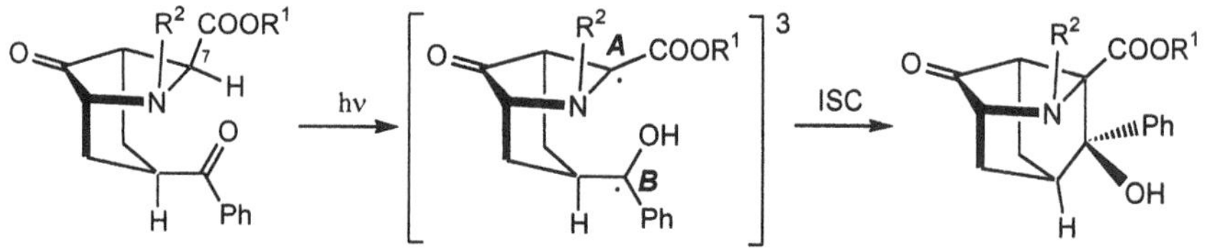

78

a R^1=Bn, R^2=Boc
b R^1=Bn, R^2=Cbz
c R^1=tBu, R^2=Cbz

79

80

a 60%
b 57%
c 67%

Scheme 27

stereoselective cyclization to the tricyclic amino acids **80**. Interestingly, in the photochemical key step, the chirality center at C-7 is destroyed by the hydrogen abstraction. Naturally, the formed radical center **A** of biradical **79** can only be attacked by the radical center **B** from the underside of the pyrrolidine ring and, therefore, no racemization can take place (Sch. 27) [63].

3.4.5. Macrocycles

In this section it will be shown, how special molecular constitution and conformation may force the formation of macrocycles. Furthermore, examples will be discussed in which the photoinduced electron transfer (PET) from electron rich functional groups to the excited keto carbonyl group initiates a macrocyclization. The exceptionally successful PET assisted macrocyclizations of phthalimides will be treated separately in Sec. 3.4.6.

The *spin center shift* approach is based on the very rapid acid elimination in hydroxysubstituted radicals, bearing a leaving group at the adjacent carbon atom. Another very rapid process is the ring opening of cyclopropyl carbinyl radicals, which proceeds fast enough to be able to compete with the "normal" biradical cyclization. This idea was pursued by Kraus and Wu [22]. Upon irradiation, the α-ketoesters **81** undergo 1,5-hydrogen migration that affords the 1,4-biradicals **82**. Besides the cyclopropylcarbonyl compounds **83**, which are products of the Norrish Type II cleavage, the unsaturated seven-membered-ring lactons **85** are obtained in the course of an opening of the cyclopropane ring. The presence of the phenyl group is necessary for the formation of lactons **85**, because it accelerates the opening of the cyclopropane ring dramatically (Sch. 28, Table 8) [64].

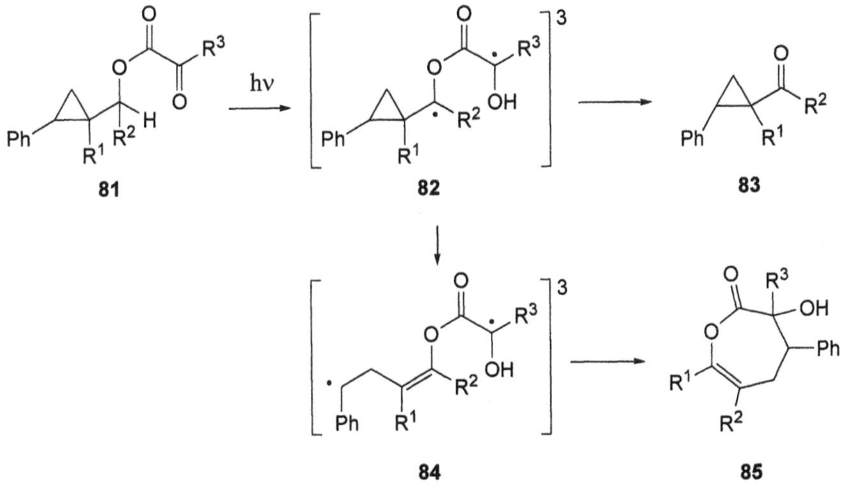

Scheme 28

Table 8 Yields of **83** and **85**.

R^1	R^2	R^3	Yield of **83**	Yield of **85**
H	H	Ph	traces	25%
	(CH$_2$)$_4$	Ph	38%	30%
	(CH$_2$)$_4$	CH$_3$	9%	36%

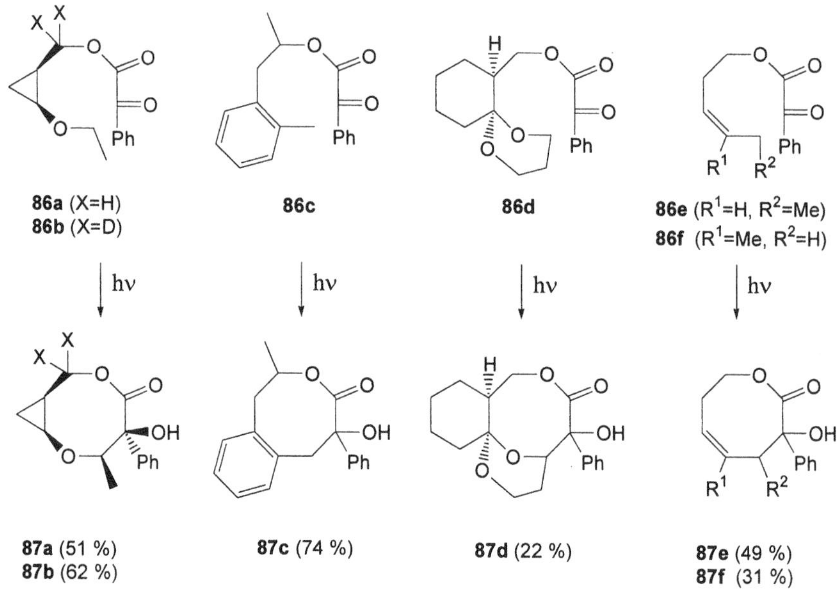

86a (X=H)   **86c**   **86d**   **86e** (R^1=H, R^2=Me)
86b (X=D)   **86f** (R^1=Me, R^2=H)

87a (51 %)   **87c** (74 %)   **87d** (22 %)   **87e** (49 %)
87b (62 %)   **87f** (31 %)

Scheme 29

In the same work, Kraus and Wu described the preparation of eight-membered-ring lactons **87** in (partly) remarkable high yields from phenylglyoxylates **86**. The formation of the deuterated lacton **87b** unambiguously revealed that the reaction really proceeds via a 1,9-hydrogen migration and not by a rearrangement of initially formed 1,4-biradicals. It seems that, compared with C–H-bonds in γ-position, the remote hydrogen transfer in **86** is facilitated by both a restriction of conformational degrees of freedom and a decrease of the bond energy of the remote C–H-bond (Sch. 29).

A concept called "remote functionalization", which based on the photochemical preparation of macrocyclic *para*-cyclophanes and subsequent oxidative cleavage of the macrocyclus, was developed by Breslow [65] and Baldwin [66]. Commencing with alkyl esters of benzophenone-4-carboxylic acid **88**, a remote hydrogen abstraction took place on irradiation and, after ISC, macrocyclic lactons **90** were formed. The regioselectivity of the hydrogen abstraction is low and consequently each carbon atom between C-10 and C-19 of the carbon chain was attacked. Therefore, the preparative value of this method appears to be limited. A dramatic improvement of the regioselectivity was achieved when the conformationally rigid steroid skeleton was attached in place of the flexible alkyl

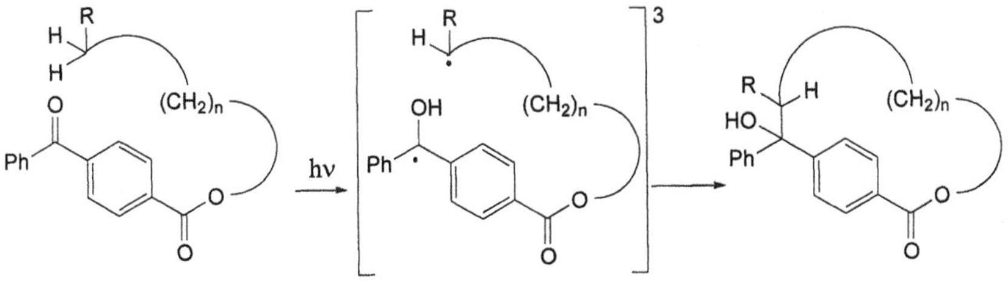

88 **89** **90**

Scheme 30

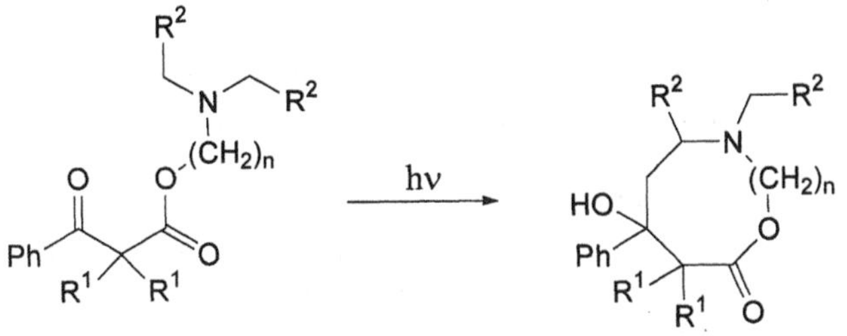

91 **92**

Scheme 31

chain. Though, cyclization was only observed to a small extend and the main product (65%) was the result of a radical disproportionation (Sch. 30).

The photoinduced electron transfer (PET) was used for the preparation of eight- to 10-membered-ring lactons by Hasegawa [67]. Thus, the aminoalkyl esters of benzoylacetic acids **91** were irradiated and, after an electron transfer from the nitrogen atom to the excited carbonyl group, a proton transfer took place, mainly from a remote methylene group. The best yields of macrocycles **92** were attained when two benzyl groups were tethered to the nitrogen atom and the hydrogen atoms of the methylene group of the benzoylacetic acid moiety were replaced by methyl groups (Sch. 31, Table 9).

3.4.6. Photochemistry of Imides

As discussed in Sec. 3.3.1.4, besides ketones also imides can be used as reactants in the Norrish–Yang reaction. For clarity, photochemical reactions of imides, which are initiated by a hydrogen abstraction both

Table 9 Yields of Lactons **92**.

R^1	R^2	n	Yield (%)
H	H	2	33
H	Ph	2	52
Me	Ph	2	63
H	H	3	34
H	Ph	4	25

from the γ-position and other positions with respect to the excited carbonyl group will be outlined in this section.

Phthalimides represent the most frequently used type of imides in photochemical syntheses. They were originally investigated by Kanaoka and later many valuable preparative applications were developed mainly by Griesbeck.

A remarkable reaction of phtalimides is the photodecarboxylation of *N*-phthaloyl amino acids that might be initiated by a hydrogen abstraction from the carboxyl group [68]. A very versatile cyclization method, which is based on the irradiation of the potassium salts of *N*-phthaloyl amino acids in an acetone/water mixture is owed to the Griesbeck group. Under these special conditions a simple decarboxylation does not occur if the carboxyl group is in the δ-position (or in more remote positions with respect to the excited carbonyl group), but the resulting diradicals preferably cyclize. In this way both small-sized rings such as pyrrolizidine **94** [69] as well as macrocycles **96** can be obtained [70] (Sch. 32).

Another fascinating photochemical modification of phthaloyl amino acids utilizes the two following reactions of the primarily formed diradicals that have rarely been observed in the ketone photochemistry, too: the radical disproportionation and the photoelimination. As a result, β,γ-unsaturated amino acid derivatives are accessible, which are very interesting as peptide building blocks as well as versatile synthons. As shown in Sch. 33, the yields of the unsaturated compounds **98** strongly depend on whether the β-carbon atom bears one or two substituents. Whereas the valine and isoleucine derivatives **97b,c** afforded the desired products in good yields, the method is ineffective from a synthetic point of view if amino butyric acid or norvaline is used (**98a,d**) [71]. The photoelimination seems to be more generally applicable. Thus, compounds **99** that are prepared from L-methionine or L-homoserine and bear a leaving group in the γ-position, provided the vinyl glycine derivative **100** in good yields [72].

Besides these applications, which have in common that an analogous behavior of aryl ketones has rarely been observed, now cyclization reactions

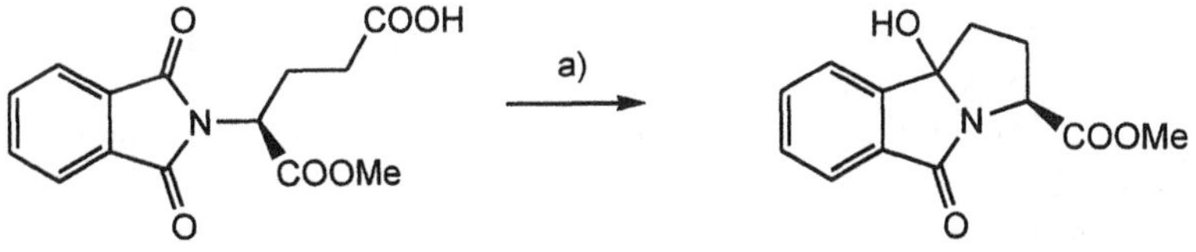

93 → **94**

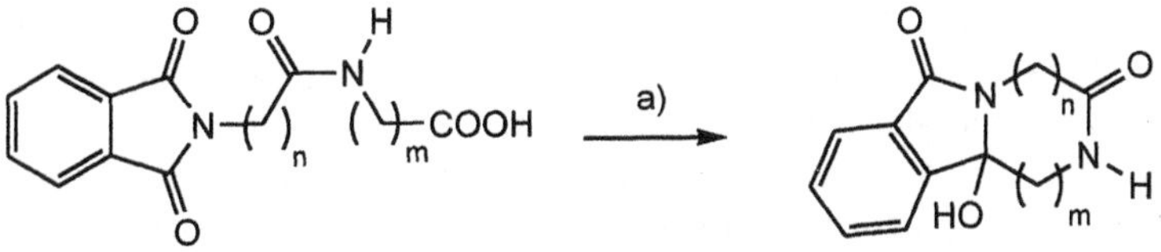

95 → **96**

n = 1-3, m = 11, 26-68 %
n = 10, m = 1, 10, 71-80%

a) hν / K₂CO₃ / acetone / H₂O

Scheme 32

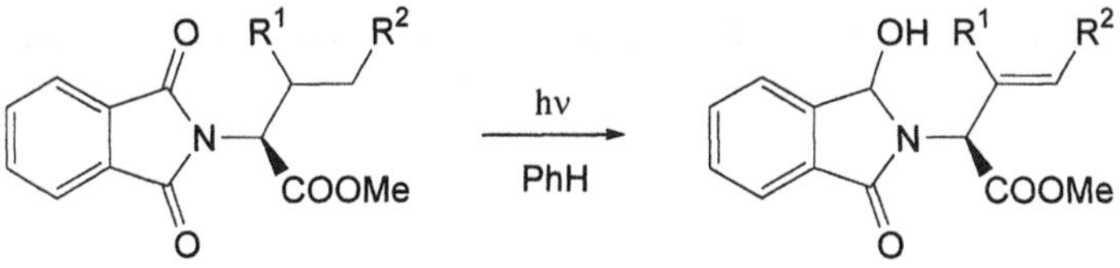

97a R¹=R²=H
97b R¹=Me, R²=H
97c R¹=Et, R²=H
97d R¹=H, R²=Me

98a 20%
98b 85%
98c 75%
98d 15%

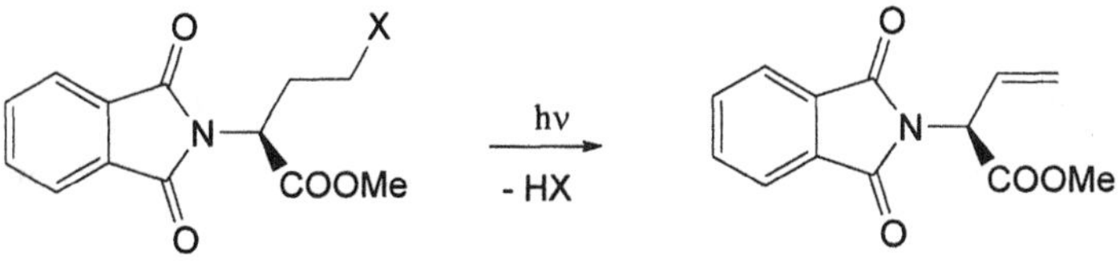

99 X=SOMe, Cl, Br

100 (69-85%)

Scheme 33

have to be presented, which are also known from aryl ketones, though PET processes often play an important role.

The γ-hydrogen abstraction affords bicyclic azetidines **102** which are *O,N*-acetals that undergo an immediate ring expansion yielding the benzazepinediones **103** [71,73]. If the residue R^1 bears hydrogen atoms in the β-position, then once more a photochemical hydrogen transfer takes place and the formed 1,4-diradicals cleave in most cases giving the 6-unsubstituted benzazepinedione [71,74], though a cyclization was observed in one case [75]. It should be noted that the reaction from the *S*-substituted cystein derivatives **101f** only proceeds in the presence of the triplet quencher piperylene, whereas otherwise the annulated products **104** are formed. The reason seems to be the different behavior of the radical anion/radical cation pair formed by an initial PET from the sulfur atom. In the singlet case, a rapid reverse electron transfer occurs and products are only formed from the homolytic γ-hydrogen transfer. In contrast to that, the triplet ion pair persists some time and a proton migration from the kinetically more acidic remote CH-position may take place [74]. If only δ-CH's are available as in the phthaloyl *tert*-leucine ester **105** the expected pyrrolizidine **106** is formed in excellent yields (Sch. 34, Table 10) [76].

Less intensively investigated than phthalimides was the photochemistry of maleimides, succinimides, and glutarimides.

In 1983 Bryant and Coyle described the cyclization of *N*-(ω-morpholino-alkyl)-maleimides **107**, which is initiated by a photo-induced electron transfer (Sch. 35) [77]. To the best of our knowledge,

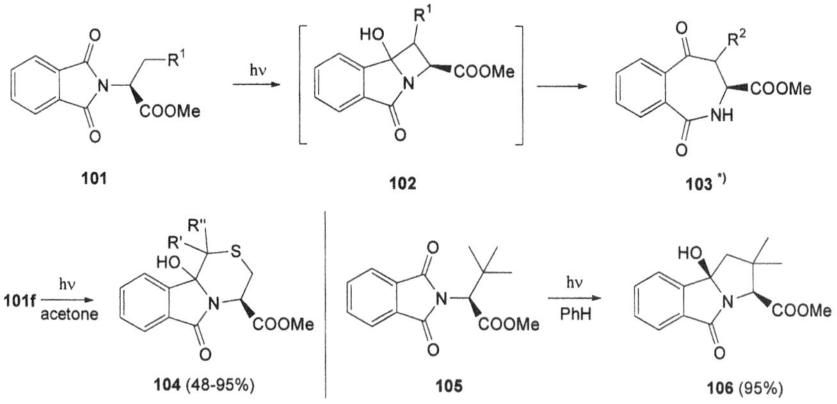

*) For R^2 see Table 10.

Scheme 34

Table 10   Yields of Benzoacepines **103**.

101	R^1	Yields of 103	R^2
a	H	60	$=R^1$
b	Me	60	$=R^1$
c	3,4-(OCMe$_2$O)Ph	100	$=R^1$
d	Et	60	H
e	CHMe$_2$	67	H
f	SCHR'R''	95[a]	H

[a]Irradiation in the presence of piperylene.

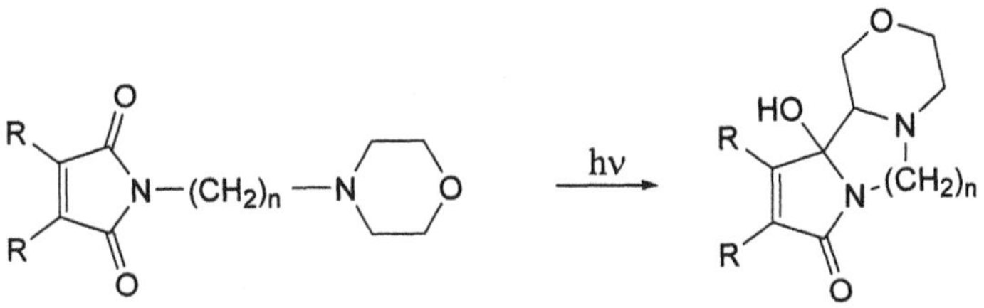

107

a   R = H, n = 2

b   R'-R' = (CH$_2$)$_4$, n = 2

c   R'-R' = (CH$_2$)$_4$, n = 3

108

33%

37%

17%

Scheme 35

no applications of maleimides are known, which are based on a homolytic hydrogen transfer.

The photochemistry of succinimides was also investigated by Kanaoka for the first time nearly 25 years ago [24]. The most frequently observed pathway of *N*-substituted succinimides **109** is the γ-hydrogen abstraction, followed by the formation of azetidinols **110**. These highly strained compounds could not be isolated, but immediately undergo a ring-opening of the *O,N*-acetal moiety to give the ε-lactams **111** as final products in a similar manner as described above for phthalimides. Interestingly, in contrast to the analog maleimide **107a** the irradiation of morpholinoethyl succinimide **112** afforded the ε-lactam **113** (Sch. 36, cf. Sch. 33) [77b].

The ring enlargement approach depicted in Sch. 35 was recently applied by Thiem et al. on a series of sugar-derived succinimides [78].

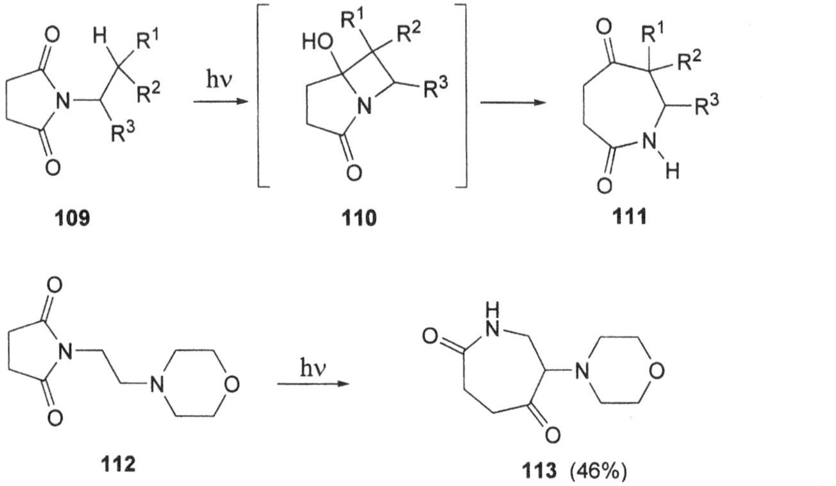

109   **110**   **111**

112   **113** (46%)

Scheme 36

The photoreactive succinimide moiety was introduced in various positions of the pyrane ring of suitable protected monosaccharides. Interestingly, the regioselectivity of the hydrogen abstraction by the excited succinimide depends sensitively on the ring conformation of the pyrane ring, which is in turn influenced by the substituents and protective groups used.

In N-glycosides **114**, the pyrane ring can exist in two different chair-like ring conformations. If the residue R^3 is a hydrogen atom (**114a**), the conformation **114-A** is preferred and, consequently, the bicyclic lactams **115** are obtained as products of a γ-hydrogen abstraction (cf. Sch. 35). If R^3 is replaced by a more bulky CH_2OSiR_3-group, the conformation **114-B**, in which the succinimide moiety adopts an axial position, predominates. Now, a weak C–H-bond in 5-position becomes attainable. The ratio of γ-hydrogen abstraction from 2-position and δ-hydrogen abstraction from 5-position sensitively depends on the protective groups of 3- and 6-hydroxy groups. Whereas the TBS-protected compound **114b** afforded a mixture of products **115** and **116**, the TMS-protected compound **114c** produced the bicyclic lactam **116** exclusively (Sch. 37, Table 11) [78a,b,d].

Spirocyclic derivatives **118** of pyranoses were prepared in moderate yields by tethering the succinimide on C-6 of the protected monosaccharide skeleton [24c]. The photochemical behavior of 1,6-anhydro sugars **119** [78e], **121** [78f), and **123** [78e] that bear the succinimide moiety in 2-, 3-, and

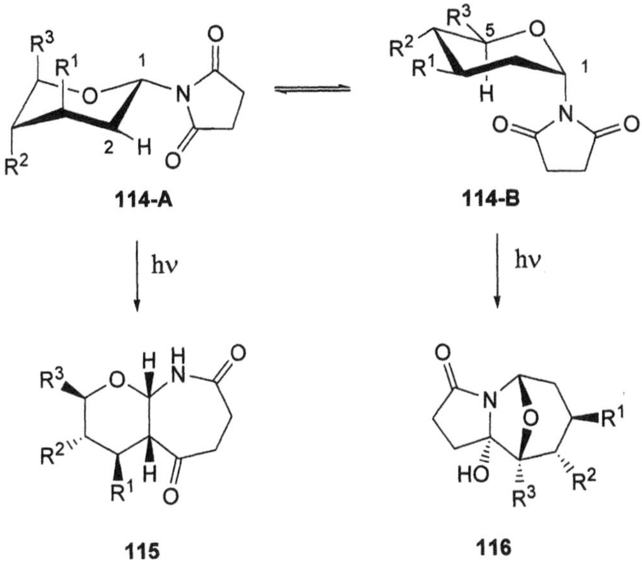

Scheme 37

Table 11 Irradiation of Glycosylimides **114**.

Entry	R^1	R^2	R^3	115	116
a	OTBS	OTBS	H	52%	—
b	OTBS	OTBS	CH_2OTBS	21%	62%
c	OTMS	OTMS	CH_2OTMS	—	83%

4-position respectively, is also summarized in Sch. 38. Finally, it should be noted that also glutarimides were used occasionally [78e,f].

3.5. REPRESENTATIVE EXPERIMENTAL PROCEDURES

3.5.1. *trans*-1-Benzoyl-2-methyl-cyclopropane (12b) [14b]

Irradiation of the ketone **9b** was performed in CH_2Cl_2 (400 mL) at a concentration of 0.01 mmol/mL in the presence of N-methylimidazole (2 equiv.), using a high pressure mercury arc lamp (150 W). Light of wavelength below 300 nm was absorbed using a Pyrex™ glass jacket between the lamp and the reaction vessel. The reaction was monitored by TLC and

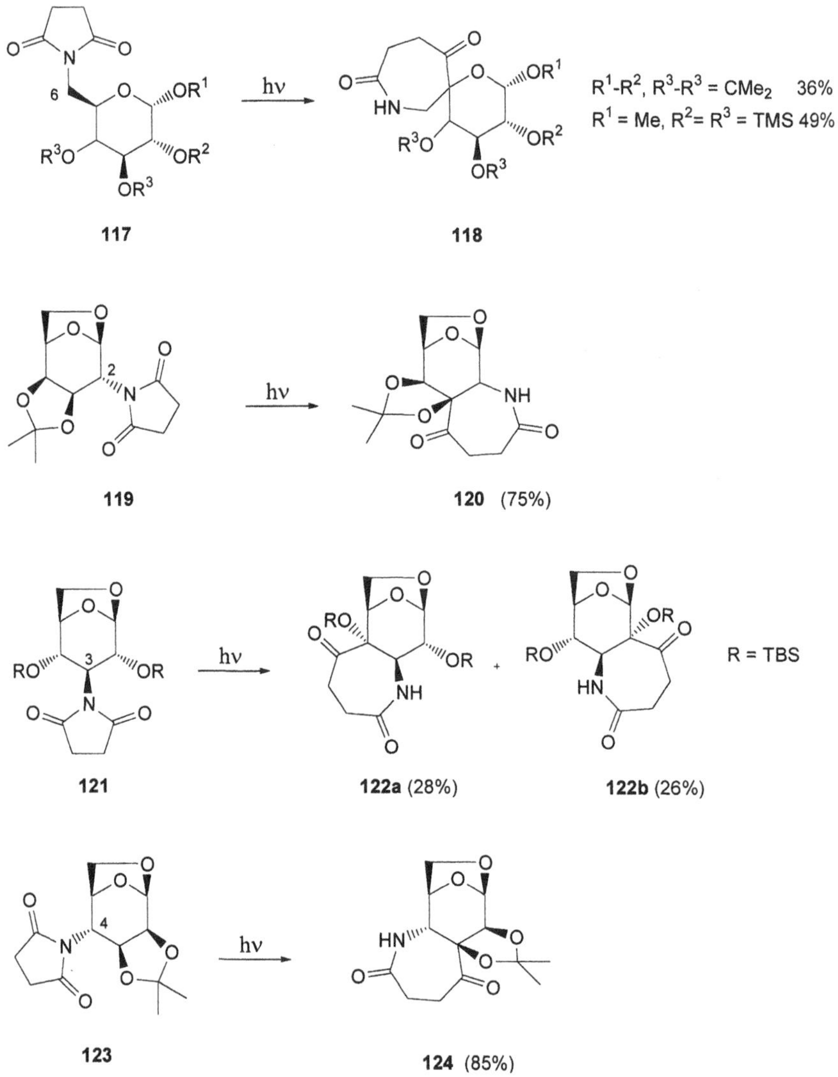

117   118

R^1-R^2, R^3-R^3 = CMe_2 36%
R^1 = Me, R^2= R^3 = TMS 49%

119   120 (75%)

121   122a (28%)   122b (26%)   R = TBS

123   124 (85%)

Scheme 38

aborted when the reactant had completely disappeared (45 min). The solution was washed with H_2O (2 × 100 mL), dried over anhydrous $MgSO_4$ and concentrated in vacuo to 10% of the original volume. To the solution was added silica gel, and the remaining solvent was removed under reduced pressure. The residue was purified immediately by flash

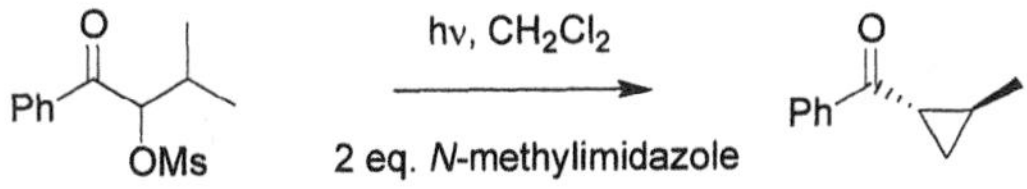

9b   **12b (90%)**

Figure 3

column chromatography (petroleum ether/ethyl acetate mixtures: 100:5) to give *trans*-1-Benzoyl-2-methyl-cyclopropane **12b** as colorless oil (90% yield) (Fig. 3). IR (film, cm^{-1}): 1665, 1448, 1402, 1222, 699. ^{1}H-NMR (300 MHz, CDCl$_3$, ppm): 8.03–7.99 (*m*, 2H); 7.57–7.45 (*m*, 3H); 2.45–2.38 (*m*, 1H); 1.67–1.56 (*m*, 1H); 1.54–1.47 (*m*, 1H); 1.23 (*d*, $J = 5.9$, 3H); 0.94–0.87 (*m*, 1H). ^{13}C-NMR (75.5 MHz, CDCl$_3$, ppm): 200.0 (C=O); 138.0 (aromat. Cq); 132.5, 128.4, 127.9 (aromat. CH); 26.3 (CH); 21.2 (CH); 20.0 (CH$_2$; 18.2 (CH$_3$). EI-MS: m/z (%): 160 (15, M$^+$), 105 (100), 77 (49), 57 (19), 55 (19), 51 (19), 43 (21). Anal. calcd. for C$_{11}$H$_{12}$O (160.21): C, 82.46; H, 7.55. Found: C, 81.78; H, 7.54.

3.5.2. (2*R*,4a*S*,5a*S*,8*R*,9a*R*,9b*R*)-Perhydro-1-[(2*R*,3*R*)-3-hydroxy-3-phenyl-*N*-tosylprolyl]-2,8-diphenyl-2H,5H,8H-*bis*[1,3]dioxino[5,4-b:4′,5′-d]pyrroline (24) [40]

A solution of the ketone **23** in cyclohexane/benzene 4:1 (ca. 10^{-2} mol/L) was rinsed with dry, O$_2$-free Argon for 30 min. The solution was irradiated until no reactant was detectable by TLC (ca. 1 h), using a high pressure mercury arc lamp (150 W). Light of wavelength below 300 nm was absorbed using a Pyrex™ glass jacket between the lamp and the reaction vessel. After evaporation, the crude photoproducts were separated by flash column chromatography and purified by medium-pressure liquid chromatography to give **24** as a colorless solid (70% yield) (Fig 4.). M.p.: 222–225 °C. IR (KBr, cm^{-1}): 3436, 2924, 1676, 1349, 1138. ^{1}H-NMR (300 MHz, CDCl$_3$, ppm): $\delta = 7.85$–7.01 (*m*, 19 aromat. H); 5.57, 5.41 (2*s*, H–C(2), H–C(8)); 5.55, 5.03 (2*d*, $J(4A, 4B) = 13.46$, $J(6A, 6B) = 16.21$, H$_A$–C(4), H$_A$–C(6)); 5.06 (*s*, H-C(2′)); 4.39, 4.12 (2*s*, H–C(9a), H–C(9b)); 4.28 (*d*, $J(4a, 6B) = 2.32$, H–C(4a)); 4.00–3.94 (*m*, H$_B$–C(4), H$_B$–C(6)); 3.85–3.63 (*m*, 2H–C(5′)); 3.66 (*s*, OH); 3.22 (*s*, H-C(5a)); 2.40 (*s*, *Me*C$_6$H$_4$); 2.28–2.16 (*m*, 2H–C(4′)). ^{13}C-NMR (75.5 MHz, CDCl$_3$, ppm): $\delta = 168$ (C=O); 144.2, 143.3, 137.5, 137.0, 129.5, 129.2, 129.1, 128.5, 128.3, 128.2, 127.6, 127.5, 126.1, 126.0, 124.5 (aromat. C); 100.0, 99.2 (C(2), C(8)); 83.6 (C(3′)); 78.4, 77.2 (C(9a), C(9b)); 71.0 (C(2′)); 67.3, 64.3 (C(4), C(6)); 57.2, 54.3 (C(4a), C(5a)); 47.6, 43.6 (C(4′), C(5′)); 21.6 (*Me*C$_6$H$_4$). FAB-MS: m/z (%): 683

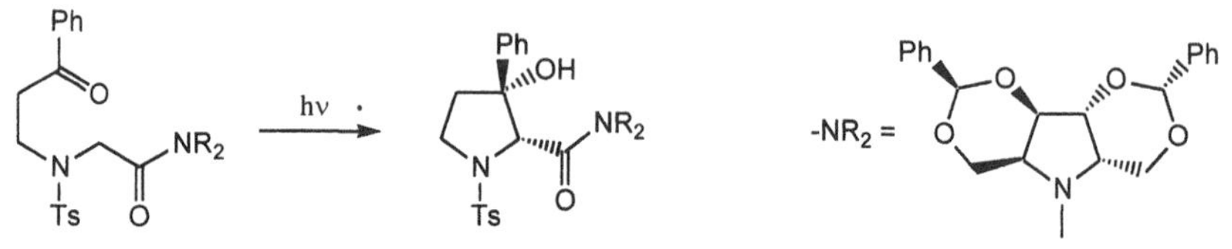

23   **24** (70%, dr > 20:1)

Figure 4

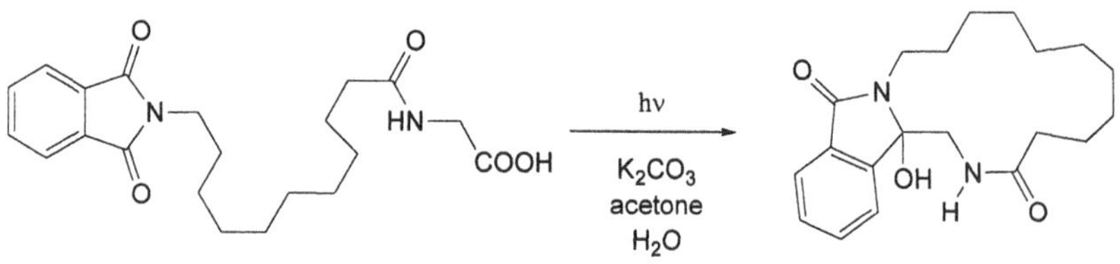

95b   **96b** (71%)

Figure 5

(81, $[M + H]^+$), 577 (20), 527 (20), 316 (100), 144 (33), 105 (62), 91 (96), 77 (22).

3.5.3. 1,2,3,6,7,8,9,10,11,12,13,14,15,21b-Tetradecahydro-21b-hydroxy-17H-[1,5]diazacyclo-heptadecino[2,1-a] isoindole-4,17(5H)-dione (96b) [70]

A suspension of K_2CO_3 (10 mmol) and **95b** (2 mmol) in water (3 mL) and acetone (200 mL) was irradiated ($\lambda = 300$ nm ± 5 nm) in a Pyrex tube for 12–24 h while purging with a slow stream of nitrogen and cooling to ca. 15 °C. After decantation, the solvent was evaporated at 40 °C/10 Torr and the residue was washed with cold Et_2O. The resulting light-yellow precipitate was crystallized from acetone to give **96b** in 71% yield (Fig. 5). M.p.: 161–162 °C (from acetone/H_2O). IR (KBr, cm^{-1}): 3292, 3078, 2930, 2856s, 1678, 1640s, 1552, 1407, 1049, 706. ^{1}H-NMR (250 MHz, CDCl$_3$, ppm): δ = 1.19–1.42 (*m*, 14H); 1.45–1.78 (*m*, 4H); 1.89 (*m*, 1H); 2.14 (*m*, 2H); 2.40 (*m*, 1H); 3.05–3.29 (*m*, 4H); 4.56 (*s*, OH); 5.98 (*t*, *J* = 5.4, 1H); 7.39 (*ddd*, *J* = 3.2, 5.3, 7.4, 1H); 7.49–7.56 (*m*, 2H); 7.60 (*d*, *J* = 7.4, 1H). ^{13}C-NMR (63 MHz, CDCl$_3$, ppm): δ = 25.4 (CH$_2$); 25.5 (CH$_2$); 26.0 (CH$_2$); 26.2 (CH$_2$); 26.3 (CH$_2$); 26.5 (CH$_2$); 26.8 (CH$_2$); 26.9 (CH$_2$); 35.3 (CH$_2$); 36.2 (CH$_2$); 37.2 (CH$_2$); 38.6 (CH$_2$); 81.1 (C); 122.1 (CH); 123.4 (CH); 129.6 (CH); 131.7 (C); 132.1 (CH); 146.2 (C); 167.2 (C); 173.3 (C).

3.6. TARGET MOLECULES: NATURAL AND NONNATURAL PRODUCT STRUCTURES

See target.cdx.

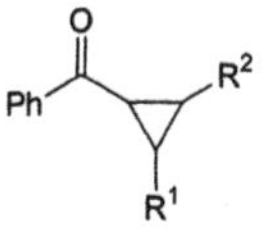

12

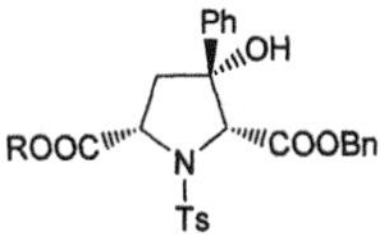

27 R = Me, _t_Bu

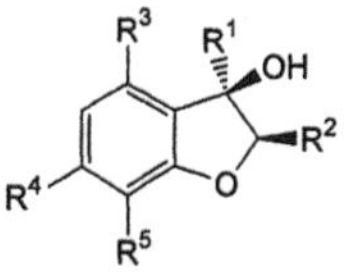

33

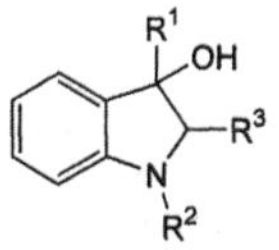

35

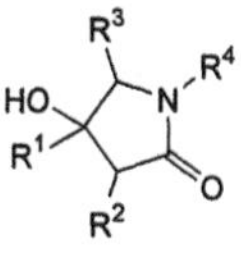

43

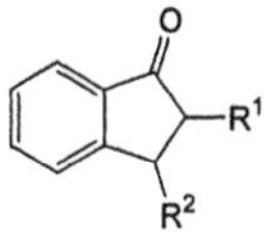

47

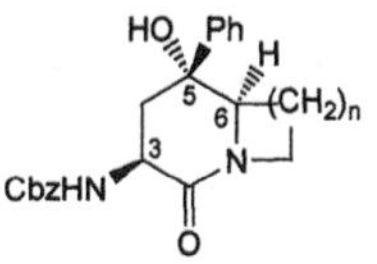

53

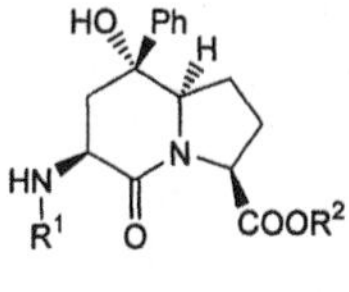

56

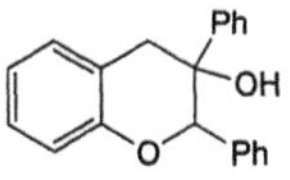

58

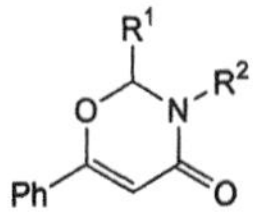

62

64

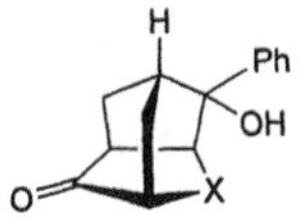

77

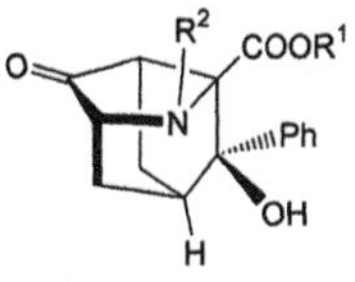

80

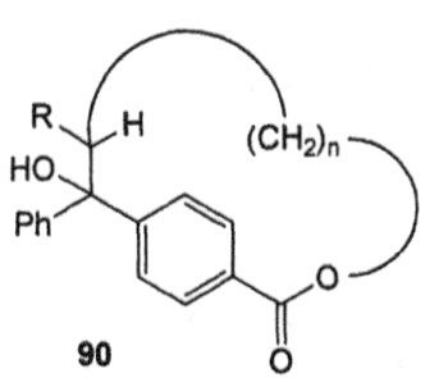

90

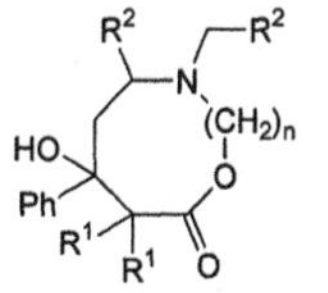

92

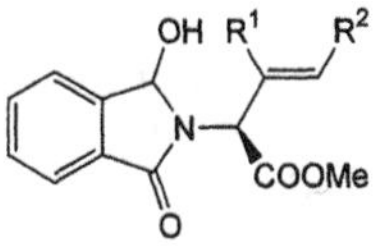

98

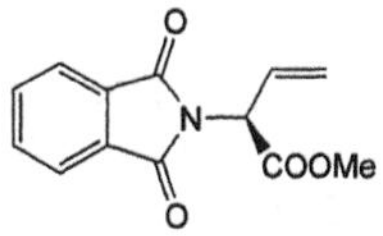

100

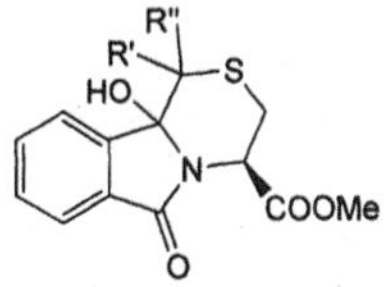

104

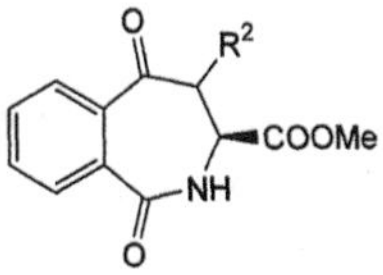

103

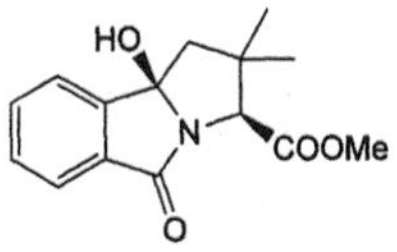

106

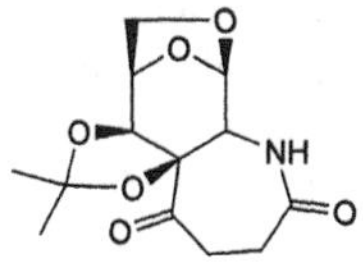

120

REFERENCES

1. Norrish RGW, Appleyard MES. J Chem Soc 1934; 874.
2. Yang NC, Yang Ding-Djung H. J Am Chem Soc 1958; 80:2913.
3. a) Wagner PJ, Kelso PA, Kemppainen AE, Zepp RG. J Amer Chem Soc 1972; 94:7500; b) Wagner PJ, Hasegawa T, Zhou B, Ward DL. J Amer Chem Soc 1991; 113:9640.
4. a) Lappin GR, Zanucci JS. J Org Chem 1971; 36:1808; b) Pappas SP, Zehr RD Jr. J Am Chem Soc 1971; 93:7112.
5. a) Roth HJ, George H. Arch Pharm 1970; 303:725; b) Roth HJ, ElRaie MH, Schrauth T. Arch Pharm 1974; 307:584.
6. Sauers RR, Huang S-Y. Tetrahedron Lett 1990; 31:5709.
7. a) Zhou B, Wagner PJ. J Am Chem Soc 1989; 111:6796; b) Hornback JM, Poundstone ML, Vadlamani B, Graham SM, Gabay J, Patton ST. J Org Chem 1988; 53:5597; c) Yoshioka M, Nishizawa K, Arai M, Hasegawa T. J Chem Soc Perkin Trans 1 1991; 541.
8. Azzouzi A, Dufour M, Gramain J-C, Remuson R. Heterocycles 1988; 27:133.
9. a) Breslow R, Winnik MA. J Am Chem Soc 1969; 91:3083; b) Breslow R, Baldwin SW. J Am Chem Soc 1970; 92:732.
10. Baldwin JE, Bhantnagar AK, Harper RW. J Chem Soc, Chem Commun 1970; 659.
11. Kraus GA, Wu Y. J Am Chem Soc 1992; 114:8705.
12. Hasegawa T, Miyata K, Ogawa T, Yoshihara N, Yoshioka M. J Chem Soc, Chem Commun 1985, 363.
13. Griesbeck AG, Henz A, Hirt J. Synthesis 1996; 1261 and references cited therein.
14. a) Wessig P, Mühling O. Angew Chemie Int Ed Eng 2001; 40:1064; b) Wessig P, Mühling O. Helv Chim Acta 2003. In press; c) Wessig P, Schwarz JU, Lindemann U, Holthausen MC. Synthesis 2001; 1258; d) Wessig P, Müller G. Unpublished results.
15. Wessig P. Synthesis and modifications of amino acids and peptides via diradicals. In: Renaud P, Sibi M, eds. Radicals in Organic Synthesis. Wiley, 2001.
16. Wagner PJ. In: Horspool WM, Song PS, eds. CRC Handbook of Organic Photochemistry and Photobiology. New York: CRC, 1995:449.
17. a) Wagner PJ, Kemppainen AE, Schott HN. J Am Chem Soc 1971; 93:5604; b) Wagner PJ, Truman RJ, Scaiano JC. J Am Chem Soc 1985; 107:7093.
18. Rubin MB. In: Horspool WM, Song PS, eds. CRC Handbook of Organic Photochemistry and Photobiology. New York: CRC, 1995:436.
19. Giese B, Wettstein P, Staehelin C, Barbosa F, Neuburger M, Zehnder M, Wessig P. Angew Chem Int Ed 1999; 38:2586.
20. Seiler M, Schumacher A, Lindemann U, Barbosa F, Giese B. Synlett 1999; 1588.
21. a) Gramain JC, Remuson R, Vallee-Goyet D, Guilhem J, Lavaud C. J Nat Prod 1991; 54:1062; b) Gramain JC, Remuson R, Vallee D. J Org Chem 1985; 50:710; c) Hasegawa T, Moribe J, Yoshioka M. Bull Chem Soc Jpn 1988; 61:1437; d) Hasegawa T, Arata Y, Mizuno K, Masuda K, Yoshihara N. J

Chem Soc, Perkin Trans 1 1986; 541; e) Hasegawa T, Arata Y, Mizuno K. J Chem Soc, Chem Commun 1983; 395.

22. Kraus GA, Wu Y. J Am Chem Soc 1992; 114:8705.

23. a) Encinas MV, Lissi EA, Zanocco A, Stewart LC, Scaiano JC. Can J Chem 1984; 62:386; b) Withesell JK, Younathan JN, Hurst JR, Fox MA. J Org Chem 1985; 50:5499.

24. a) Kanaoka Y, Hatanaka Y. J Org Chem 1976; 41:400; b) Kanaoka Y. Acc Chem Res 1978; 11:407; c) Kanaoka Y, Okajima H, Hatanaka Y, Terashima M. Heterocycles 1978; 11:455.

25. a) Griesbeck AG, Mauder H. In: Horspool WM, Song PS, eds. CRC Handbook of Organic Photochemistry and Photobiology. New York: CRC, 1995:513; b) Yoon UC, Mariano PS. Acc Chem Res 2001; 34:523.

26. Rehm D, Weller A. Isr J Chem 1970; 8:259.

27. Maslak P, Navaraez JN. Angew Chem Int Ed 1990; 29:302.

28. Maruyama K, Kubo Y. J Org Chem 1981; 46:3612.

29. Coyle JD, Newport GL, Harriman A. J Chem Soc, Perkin Trans 2 1978; 133.

30. Wagner PJ, Siebert EJ. J Am Chem Soc 1981; 103:7329.

31. Wessig P. Unpublished results.

32. a) Netto-Ferreira JC, Leigh WJ, Scaiano JC. J Am Chem Soc 1985; 107:2617; b) Carlson GLB, Quina FH, Zarnegar BM, Whitten DG. J Am Chem Soc 1975; 97:347; c) Moorthy JN, Patterson WS, Bohne C. J Am Chem Soc 1997; 119:11094; d) Moorthy JN, Monahan SL, Sunoj RB, Chandrasekhar J, Bohne C. J Am Chem Soc 1999; 121:3093.

33. a) Morrison H, Tisdale V, Wagner PJ, Liu K-C. J Am Chem Soc 1975; 97:7189; b) Wagner PJ, Liu K-C, Noguchi Y. J Am Chem Soc 1981; 103:3837.

34. a) Ariel S, Ramamurthy V, Scheffer JR, Trotter J. J Am Chem Soc 1983; 105:6959; b) Scheffer JR, Garcia-Garibay M, Nalamasu O. Org Photochem 1987; 8:249.

35. a) Roth HJ, ElRaie MH, Schrauth T. Arch Pharm 1974; 307:584; b) Abdul-Baki A, Rotter F, Schrauth T, Roth HJ. Arch Pharm 1978; 311:341; c) Roth HJ, George H. Arch Pharm 1970; 303:725; d) Weigel W, Schiller S, Henning H-G. Tetrahedron 1997; 53:7855; e) Weigel W, Schiller S, Reck G, Henning H-G. Tetrahedron Lett 1993; 42:6737; f) Weigel W, Wagner PJ. J Am Chem Soc 1996; 118:12858.

36. a) Zimmerman H, Nuss J, Tantillo A. J Org Chem 1988; 53:3792; b) Yoshioka M, Miyazoe S, Hasegawa T. J Chem Soc Perkin Trans 1 1993; 22:2781.

37. Newcomb M, Choi SY, Horner JH. J Org Chem 1999; 64:1225.

38. a) Roy SC, Adhikari S. Tetrahedron 1993; 49:8415; b) Carless HAJ, Swan DI, Haywood DJ. Tetrahedron 1993; 49:1665; c) Araki Y, Arai Y, Endo T, Ishido Y. Chem Lett 1989; 1; d) Cottier L, Descotes G. Tetrahedron 1985; 41:409; e) Cottier L, Descotes G, Grenier MF, Metras F. Tetrahedron 1981; 37:2515.

39. a) Henning H-G, Dietzsch T, Fuhrmann J. J Prakt Chem 1981; 323:435; b) Haber H, Buchholz H, Sukale R, Henning H-G. J Prakt Chem 1985; 327:51; c) Walther K, Kranz U, Henning H-G. J Prakt Chem 1987; 329:859.

40. Wessig P, Wettstein P, Giese B, Neuburger M, Zehnder M. Helv Chim Acta 1994; 77:829.
41. Steiner A, Wessig P, Polborn K. Helv Chim Acta 1996; 79:1843.
42. a) Wagner PJ, Meador MA, Park B-S. J Am Chem Soc 1990; 112:5199; b) Kraus GA, Thomas PJ, Schwinden MD. Tetrahedron Lett 1990; 31:1819; c) Horaguchi T, Tsukada C, Hasegawa E, Shimizu T, Suzuki T, Tanemura K. J Heterocyclic Chem 1991; 28:1273; d) Pappas SP, Pappas BC, Backwell JE Jr. Journal of Organic Chemistry 1967; 32:3066.
43. a) Henning H-G, Amin M, Wessig P. J Prakt Chem 1993; 42:335; b) Wessig P, Amin M, Henning H-G, Schulz B, Reck G. J Prakt Chem 1994; 336:169.
44. Wagner PJ, Cao Q. Tetrahedron Lett 1991; 32:3915.
45. a) Wagner PJ, Meador MA, Giri BP, Scaiano JC. J Am Chem Soc 1985; 107:1087; b) Wagner PJ, Giri BP, Scaiano JC, Ward DL, Gabe E, Lee FL. J Am Chem Soc 1985; 107:5483.
46. a) Zand A, Park B-S, Wagner PJ. J Org Chem 1997; 62:2326; b) Wagner PJ, Rabon P, Park BS, Zand AR, Ward DL. J Am Chem Soc 1994; 116:589; c) Wagner PJ, Meador MA, Zhou B, Park BS. J Am Chem Soc 1991; 113:9630.
47. a) Bergmark WR, Barnes C, Clark J, Paparian S, Marynowski S. J Org Chem 1985; 50:5612; b) Klán P, Zabadal M, Heger D. Org Lett 2000; 2:1569.
48. Lindemann U, Reck G, Wulff-Molder D, Wessig P. Tetrahedron 1998; 54:2529.
49. Lindemann U, Neuburger M, Neuburger-Zehnder M, Wulff-Molder D, Wessig P. J Chem Soc Perkin Trans 2 1999; 2029.
50. Henning H-G, Berlinghoff R, Mahlow A, Köppel H, Schleinitz K-D. J Prakt Chem 1981; 323:914.
51. Lindemann U, Wulff-Molder D, Wessig P. Tetrahedron: Asymmetry 1998; 9:4459.
52. Griesbeck AG, Heckroth H, Schmickler H. Tetrahedron Lett 1999; 40:3137.
53. Wessig P. Tetrahedron Lett 1999; 40:5987.
54. Meador MA, Wagner PJ. J Org Chem 1985; 50:419.
55. Wessig P, Schwarz J, Lindemann U, Holthausen MC. Synthesis 2001; 1258.
56. a) Padwa A, Alexander E. Izw Akad Nauk SSSR Ser Khim 1967; 89:6376; b) Padwa A, Shefter E, Alexander E. J Am Chem Soc 1968; 90:3717; c) Padwa A, Alexander E, Niemcyzk M. J Am Chem Soc 1969; 91:456.
57. Blondeel G, De Bruyn A, De Keukeleire. Tetrahedron Lett 1984; 25:2055.
58. Wagner PJ, Liu K, Noguchi Y. J Am Chem Soc 1981; 103:3837.
59. Newcomb M, Renaud P, Sibi MP, eds. Radicals in Organic Synthesis. Vol. 1. Weinheim: Wiley-VCH, 2001:329.
60. Wessig P, Legart F, Hoffmann B, Henning H-G. Liebigs Ann Chem 1991; 979.
61. Stetter H, Rämsch K-D, Elfert K. Liebigs Ann Chem 1974; 1322.
62. Wessig P, Schwarz J, Wulff-Molder D, Reck G. Monatshefte Chem 1997; 128:849.
63. Wessig P. Synlett 1999; 1465.
64. Newcomb M, Manek MB. J Am Chem Soc 1990; 112:9662.
65. a) Breslow R, Winnik MA. J Am Chem Soc 1969; 91:3083; b) Breslow R, Baldwin SW. J Am Chem Soc 1970; 92:732.

66. Baldwin JE, Bhantnagar AK, Harper RW. J Chem Soc Chem Commun 1970; 659.
67. a) Hasegawa T, Miyata K, Ogawa T, Yoshihara N, Yoshioka M. J Chem Soc, Chem Commun 1985; 363; b) Hasegawa T, Ogawa T, Miyata K, Karakizawa A, Komiyama M, Nishizawa Yoshioka M. J Chem Soc Perkin Trans I 1990; 901.
68. Sato Y, Nakai H, Mizoguchi T, Kawanishi M, Hatanaka Y, Kanaoka Y. Chem Pharm Bull 1982; 30:1263.
69. Griesbeck AG, Henz A, Peters K, Peters E-M, v. Schnering H-G. Angew Chem Int Ed 1995; 34:474.
70. Griesbeck AG, Henz A, Kramer W, Lex J, Nerowski F, Olgemöller M, Peters K, Peters E-M. Helv Chim Acta 1997; 80:912.
71. Griesbeck AG, Mauder H, Müller I. Chem Ber 1992; 125:2467.
72. Griesbeck AG, Hirt J. Liebigs Ann 1995; 1957.
73. Griesbeck AG, Henz A, Hirt J, Ptatschek V, Engel T, Löffler D, Schneider FW. Tetrahedron 1994; 50:701.
74. Griesbeck AG, Hirt J, Peters K, Peters E, v. Schnering H-G. Chem Eur J 1996; 2:1388.
75. Griesbeck AG. EPA Newsletters 1998; 62:3.
76. Griesbeck AG, Mauder H. Angew Chem 1992; 104:97.
77. a) Bryant LRB, Coyle JD. Tetrahedron Lett 1983; 24:1841; b) Coyle JD, Bryant LRB. J Chem Soc Perkin Trans 1 1983; 12:2857.
78. a) Sowa CE, Thiem J. Angew Chem Int Ed 1994; 33:1979; b) Sowa CE, Kopf J, Thiem J. J Chem Soc, Chem Commun 1995; 211; c) Sowa CE, Stark M, Heidelberg TH, Thiem J. Synlett 1996, 227; d) Thiering S, Sowa CE, Thiem J. J Chem Soc Perkin Trans 1 2001; 801; e) Thiering S, Sund C, Thiem J, Giesler A, Kopf J. Carbohydrate Res 2001; 336:271; f) Sund C, Thiering S, Thiem J, Kopf J, Stark M. Monatshefte Chem 2002; 133:485.

4
Photocycloadditions of Alkenes to Excited Carbonyls

Axel G. Griesbeck
Institut für Organische Chemie, Universität zu Köln, Köln, Germany

4.1. HISTORICAL BACKGROUND

The first photocycloaddition of an aromatic carbonyl compound (benzaldehyde) to an alkene (2-methyl-2-butene) was published by Paternò and Chieffi. They described this long-term experiment (104 days) with the words: "In una prima esperienza abbiamo esposto un miscuglio equimolecolare di amilene (gr. 43) e di aldeide benzoica (gr. 67) in un tubo chiuso, dal 5 dicembre 1907 al 20 marzo 1908, cioè per circa tre mesi e mezzo della stagione invernale" [1]. This and other experiments were repeated by Büchi and his coworkers in the mid-1950s and the oxetane *constitution* of the photogenerated product was confirmed [2]. The *regioselectivity* of this specific transformation, however was not correctly established until a publication by Yang et al. in 1964 [3]. In the following time, two features of this reaction were investigated in detail: the influence of the state properties of the electronically excited species and of the alkene properties on rate, efficiency, as well as selectivity of the oxetane formation. Furthermore, a detailed discussion of diastereo- and enantioselective modifications of intramolecular variants is presented in the following chapters. After discussion of the features of the carbonyl addends, the olefinic reaction partners are discussed. A number of extensive reviews about special features of this reaction type appeared in the last two decades which are recommended for further reading [4].

4.2. STATE OF THE ART MECHANISTIC MODELS

Considering the orbital interactions between alkenes and $n\pi^*$ excited carbonyls, Turro has classified two possible primary trajectories: (i) the nucleophilic attack of the alkene towards the cabonyl half-filled n orbital, characterized as "perpendicular approach" and (ii) the nucleophilic attack of the carbonyl by its half-filled π^* orbital towards the alkenes empty π^* orbital, characterized as "parallel approach" [5]. First-order orbital correlation diagrams are in line with this model and predict the formation of a carbon–carbon bonded 1,4-biradical for the parallel and a carbon-oxygen bonded biradical for the perpendicular approach [6]. What approach dominates, is controlled by the relative positions of the alkene HOMO and LUMO. For the interaction between electron-rich alkenes with carbonyl compounds the perpendicular approach is favored, for electron-deficient alkenes the parallel approach. Several ab initio calculations have been published [7]. A recent MC-SCF-study resulted in the prediction that ISC from triplet to singlet leads to the same biradical ground-state pathways that are entered via singlet photochemistry [8]. Following this line of argumentation, the product-determining molecular geometry is expected to be similar either from the first excited singlet or triplet state of the carbonyl reactant. The biradical model has been used for decades in order to describe chemo- and regioselectivity phenomena (Fig. 1).

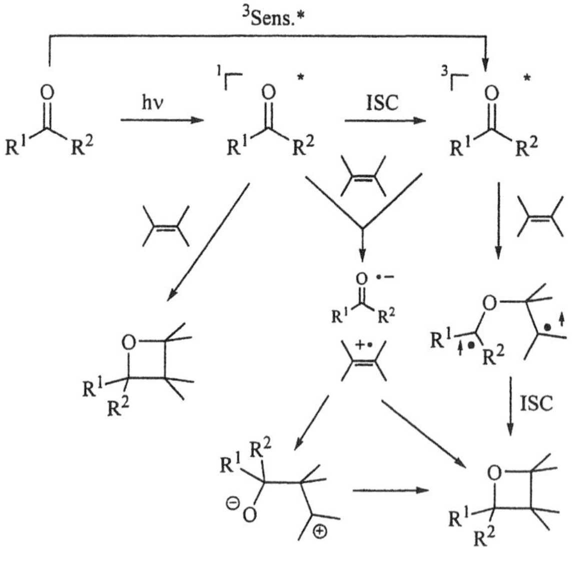

Figure 1

At least for triplet 1,4-biradicals (2-oxatetramethylenes or *preoxetanes*) this assumption has been confirmed by several experimental facts. Trapping experiments (e.g., with triplet oxygen and sulfur dioxide) [9] as well as the application of radical clocks [10] revealed, that short-lived (1–10 ns) triplet 1,4-biradicals are formed when triplet excited carbonyl compounds interact with alkenes. Spectroscopic evidences for these species came from laser flash photolysis experiments of electron-rich olefines with benzophenone [11]. In this case also the radical anion of the ketone was detected, demonstrating that electron transfer processes can sometimes interfere with the formation of oxetanes. The existence of an exciplex as precursor of the 1,4-biradicals as well as of the radical ion pair was deduced from a correlation of oxetane formation with the fluorescence quenching of singlet excited ketones by electron-rich and electron-deficient alkenes [12]. A series of experiments on substrate diastereoselectivity have been performed in order to differentiate between reactive states ($n\pi^*$ vs. $\pi\pi^*$) and multiplicities (singlet vs. triplet). Simple rules have been deduced which could be used as first approximations: triplet $n\pi^*$ carbonyls lead to a lower degree in stereoselectivity than singlet $n\pi^*$ carbonyls [13].

4.3. SCOPE AND LIMITATIONS

4.3.1. Aromatic Aldehydes and Ketones

There is an impressive number of Paternò-Büchi reactions with substituted benzophenones or benzaldehydes. A special group of aromatic carbonyl reagents are heteroarenes such as acyl derivatives of furan and thiophene which could in principle undergo [2+2] cycloaddition to the C=O group as well as to one of the ring C=C groups. This reaction periselectivity is controlled by the heteroatom and the substitution pattern which influences the nature of the lowest excited triplet state [14]. Cantrell has published a series of experiments with 2- and 3-substituted benzoyl-, acetyl-, and formylthiophenes and furans. The aldehydes **1** and **3** ($X = S, O$) gave with high selectivity the oxetanes **2,4** with 2,3-dimethyl-2-butene as alkene addend (Sch. 1) [15]. Similar results were reported for 2-benzoylfuran and thiophene **5** ($X = S, O$) whereas the corresponding 2-acetyl substrates **6** gave mixtures of [2+2] and [4+2] photoproducts **7–9** [16].

Vargas and Rivas reported the photocycloaddition of acetylseleno-phene (**10**) to tetrasubstituted alkenes resulting in two [2+2]-photoadducts, one of them (**11**) involves the C=C bond of the monoalkene and the β,γ-double bond of the selenophene. The second product **12** is a Paternò-Büchi photoadduct involving the carbonyl group of the selenophene and the C=C

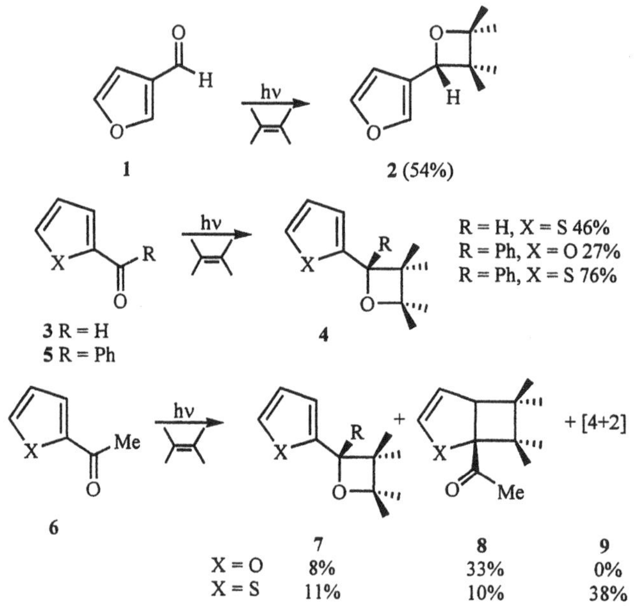

1   **2 (54%)**

R = H, X = S 46%
R = Ph, X = O 27%
R = Ph, X = S 76%

3 R = H
5 R = Ph
                          4

                                    + [4+2]

6

	7	**8**	**9**
X = O	8%	33%	0%
X = S	11%	10%	38%

Scheme 1

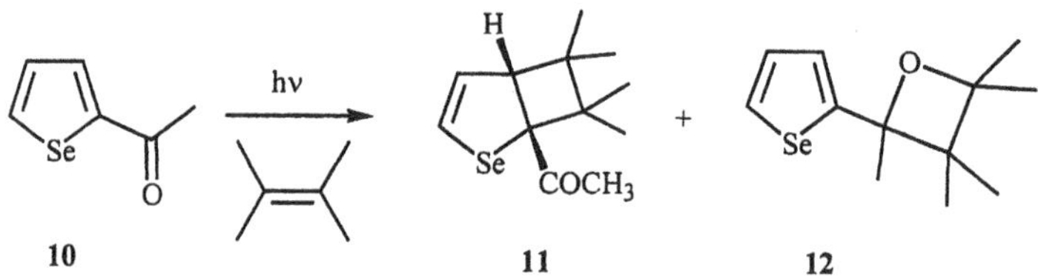

10   **11**   **12**

Scheme 2

bond of the monoalkene (Sch. 2) [17]. In contrast to 2-acetylthiophenone which also yields a [4+2]-cycloaddition product, no such adducts were found with 2-acetyl selenophene. This was rationalized by the difference in size of the heteroatom, which might disfavor a Diels-Alder approach for the selenium compound in contrast to the corresponding sulfur- and oxygen heterocycles.

Photoinduced [2+2] cycloadditions of 1H-1-acetylindole-2,3-dione (**13**) with alkenes gave spiro-oxetanes **14** in moderate to high yields, displaying the typical triplet n-π^* reactivity of acetylisatin (Sch. 3) [18]. The regioselectivity and diastereoselectivity of these reactions depends on the

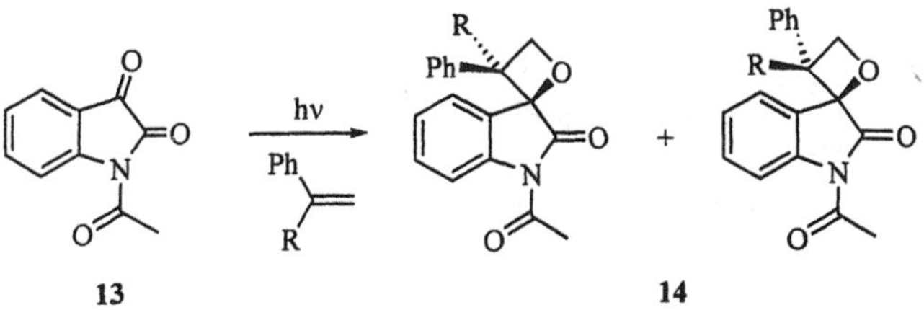

13   **14**

Scheme 3

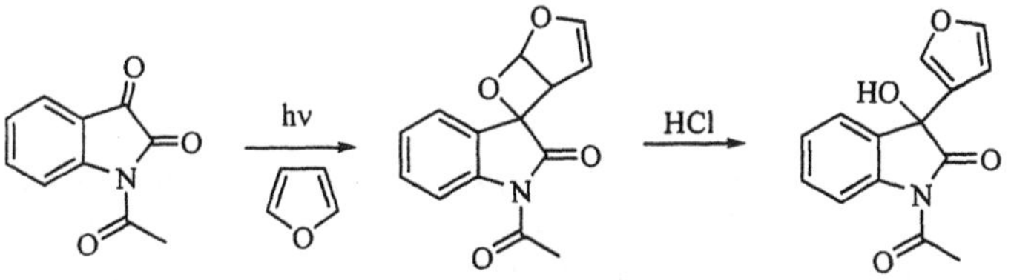

13   **15**   **16**

Scheme 4

reaction mechanism: reactions involving alkenes with high oxidation potential exclude single-electron transfer (SET) processes, and thus the regioselectivity can be rationalized by considering frontier molecular orbital interactions of the two addends and the diastereoselectivity by applying the Salem–Rowland rules for triplet to singlet biradical intersystem crossing (vide infra).

As an extension of this work, photoinduced [2+2]-cycloadditions of 1-acetylisatin (**13**) with cyclic enolethers (furan, benzofuran, 2-phenylfuran, 8-methoxypsoralen), and acyclic enolethers (*n*-butyl vinyl ether and vinyl acetate) were investigated which afforded the spiro-oxetanes in high yields (82–96%) and with high regio- and diastereoselectivity (Sch. 4) [19]. Treatment of the furan-derived oxetane **15** with acid resulted in oxetane ring opening and yielded the 3-(furan-3-yl)indole derivative **16**.

4.3.2. Carboxylic Acid Derivatives and Nitriles

The photochemistry of carboxylic acid derivatives has been summarized by Coyle [20]. For arene carboxylic acid esters it has been shown that [2+2] cycloaddition competes with hydrogen abstraction by the excited ester from an allylic position of the alkene. The addition of methyl benzoate **17** to

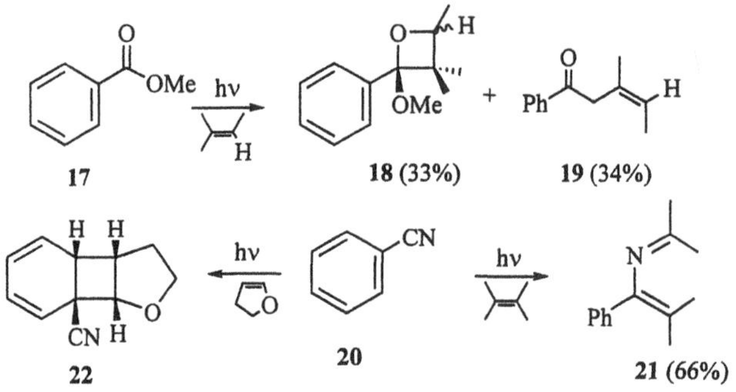

Scheme 5

2-methyl-2-butene gave a 1:1 mixture of the Paternò-Büchi adduct **18** and the coupling product **19** [21]. Less electron rich alkenes, e.g., cyclopentene did preferentially add towards the benzene ring of **17** in a *ortho-* and *meta-* cycloaddition manner. Furans could also be added photochemically to methyl benzoate and other arenecarboxylic acid esters. The resulting bicyclic oxetanes could be transformed into a series of synthetically valuable products [22]. [2+2]-Cycloadducts and/or their cleavage- or rearrangement products have also been described for photoreactions of alkenes with diethyloxalate [23], benzoic acid [24], and carbamates [25]. The site selectivity of alkene addition to benzonitrile **20** has been studied with a series of cyclic and acyclic olefins (Sch. 5). Again, two reaction modes could be observed: [2+2]-cycloaddition to the nitrile group leading to 2-azabutadienes (deriving from the primarily formed azetines, e.g., **21** from 2,3-dimethyl-2-butene) [26] and *ortho*-cycloaddition to the benzene ring (e.g., **22** from 2,3-dihydrofuran) [27].

Recently, Hu and Neckers reported that triplet excited states of alkyl phenyl glyoxylates react rapidly and with high chemical yields with electron-rich alkenes forming oxetanes **24** with high regio and stereoselectivity (Sch. 6) [28]. The intramolecular γ-hydrogen abstraction (Norrish type II) cannot compete with intermolecular reactions in most cases. When less electron-rich alkenes were used, the Norrish type II reaction became competitive.

Griesbeck and Mattay described photocycloaddition of methyl and ethyl trimethyl pyruvates (**25**) with di-isopropyl-1,3-dioxol. In contrast to the reaction with ethyl pyruvate, the bicyclic oxetane **26** was formed with very high (>98%) diastereoisomeric excess (Sch. 7) [29]. An X-ray analysis revealed the unusual *endo-tert*-butyl configuration. Semiempirical calcula-tion indicated that this clearly is the kinetic product formed by a biradical

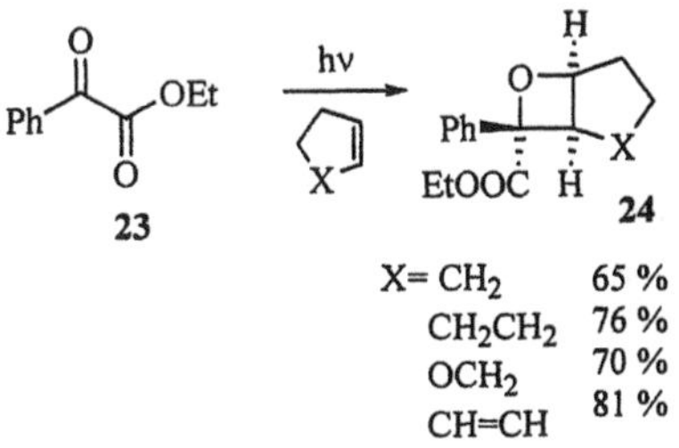

X= CH$_2$ 65 %
CH$_2$CH$_2$ 76 %
OCH$_2$ 70 %
CH=CH   81 %

Scheme 6

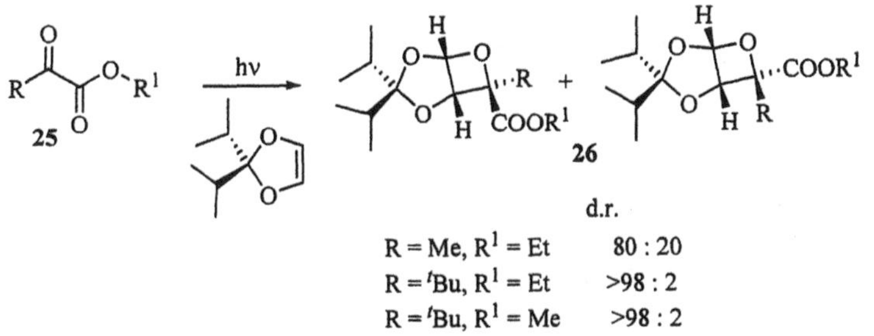

d.r.

R = Me, R^1 = Et   80 : 20
R = ^{t}Bu, R^1 = Et   >98 : 2
R = ^{t}Bu, R^1 = Me   >98 : 2

Scheme 7

combination reaction which might be controled by favorable spin-orbit coupling geometries.

4.3.3. α-Ketocarbonyl Compounds, Acyl Cyanides

α-Diketones undergo primary photochemical addition to olefins to form [4+2] and [2+2] cycloadducts in competition to hydrogen abstraction, α-cleavage, and enol formation. The product ratio **28:29:30:31** of the biacetyl (**27**)/2,3-dimethyl-2-butene photoreaction strongly depends on the solvent polarity and reaction temperature, indicating an exciplex intermediate with pronounced charge separation and possibly free radical ions from a photo-electron-transfer (PET) step (Sch. 8) [30].

Similar product ratios were reported for the methyl pyruvate/2,3-dimethyl-2-butene photoreaction. In this case, however, a state selectivity effect is responsible for the formation of the different ether and alcohol products [31]. Obviously the existence of allylic hydrogens favors the formation of unsaturated acyclic products via hydrogen migration steps at

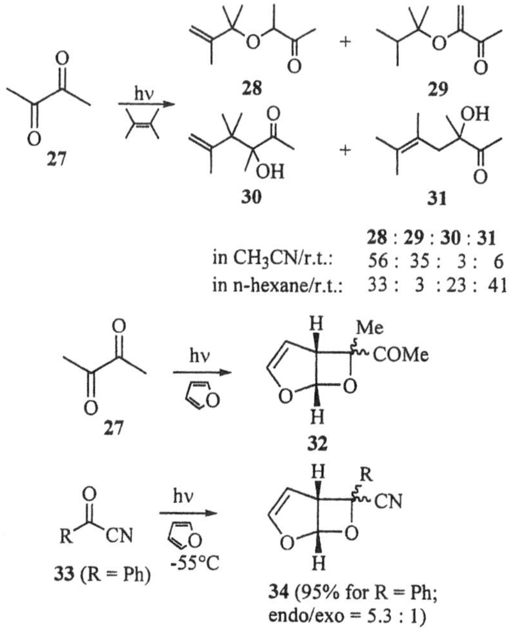

in CH₃CN/r.t.:
in n-hexane/r.t.:

$$28 : 29 : 30 : 31$$
in CH$_3$CN/r.t.: 56 : 35 : 3 : 6
in n-hexane/r.t.: 33 : 3 : 23 : 41

34 (95% for R = Ph;
endo/exo = 5.3 : 1)

Scheme 8

the triplet biradical level. More electron rich alkenes without (or with unfavorable) allylic hydrogens do give oxetanes with excited α-dicarbonyl compounds. Furan, indene [32] as well as isopropenyl ethyl ether [33] were converted to the corresponding [2+2] cycloadducts (e.g., **32**) with biacetyl **27** and methyl pyruvate, respectively. Chiral phenyl pyruvates have been investigated intensively as carbonyl addends with medium to remarkably high diastereoselectivities (vide infra). Benzoyl cyanie **33** gives mixtures of cycloaddition and coupling products with 2,3-dimethyl-2-butene [34], whereas the addition of **33** and a series of other acyl cyanides to furan is chemoselective and leads to bicyclic oxetanes **34** with variable *endo/exo* ratios [35].

Symmetric $^3n\pi^*$-excited 1,2-diarylethanediones (**35**) undergo highly regio- and stereoselective head to head additions to various captodative-substituted alkenes (2-aminopropenenitriles **36**) forming oxetanes **37** in moderate to good yield (Sch. 9) [36].

Also unsymmetric 1,2-diarylethanediones **38** result in the formation of cycloadducts **39** [37]. Only one regioisomer of the *a priori* conceivable regioisomers has been detected for each case (Sch. 10). The connectivity and the preferred configuration of the products is rationalized in terms of the

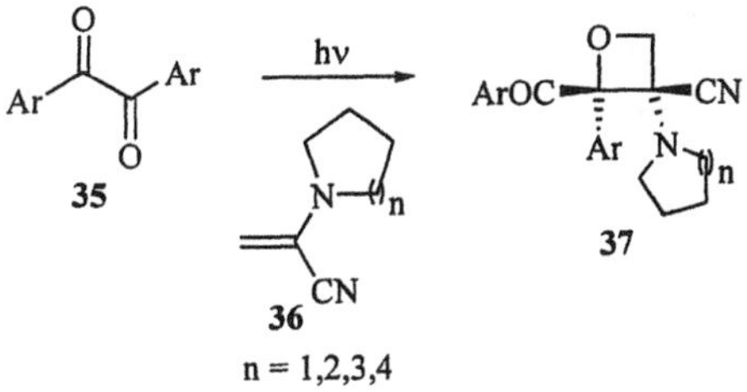

Scheme 9

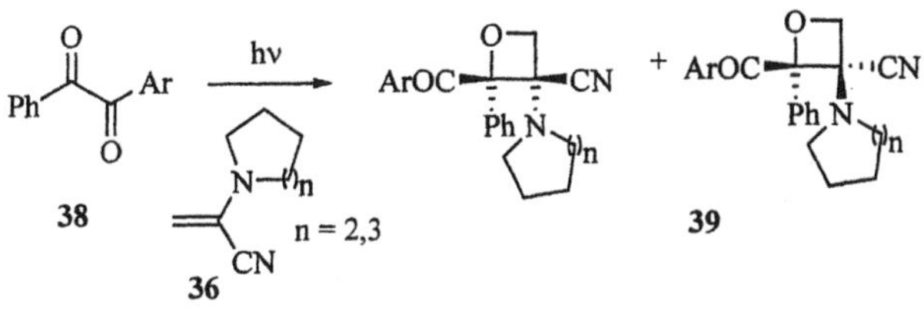

Scheme 10

geometry of the more stable (including captodative stabilization) 1,4-biradical intermediate.

In all cases, the cyclic amino substituent in the major diastereoisomer is oriented *cis* to the aryl moiety of the participating aroyl group whereas the nonparticipating aroyl group is *cis* to the nitrile function. Whereas 2-naphthyl-substituted diones from oxetanes, 1,1′-naphthyl- and 1-(1-naphthyl)-2-phenyl ethanedione are unreactive. The latter effect probably reflects the $\pi\pi^*$ nature of the lowest excited triplet state of these diones.

4.3.4. Enones and Ynones

There is a striking difference between the photochemical reactivity of α,β-unsaturated enones and the corresponding ynones. Whereas many cyclic enones undergo [2+2] cycloaddition to alkenes at the C=C double bond of the enone (probably from the triplet $\pi\pi^*$ state) to yield cyclobutanes, acyclic enones easily deactivate radiationless by rotation about the central C–C single bond. Ynones on the other hand behave much more like alkyl-substituted carbonyl compounds and add to (sterically less encumberd) alkenes to yield oxetanes (Sch. 11) [38,39]. The *regioselectivity* of the Paternò-Büchi reaction is similar to that of aliphatic or aromatic carbonyl compounds with a preference for primary attack at the less substituted carbon atom (e.g., **41** and **42** from the reaction of but-3-in-2-one **40** with

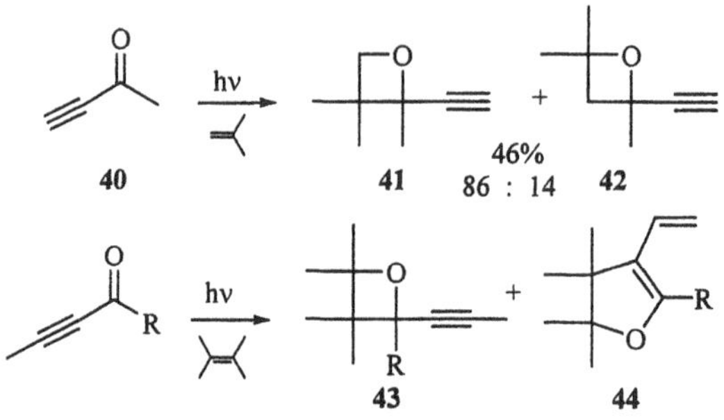

Scheme 11

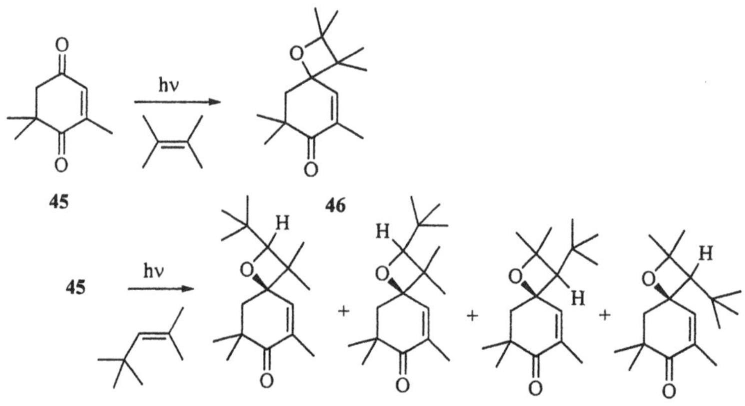

47 (9.8 : 5.1 : 3.6 : 81.5)

Scheme 12

isobutylene). A serious drawback is the low *chemoselectivity*. For most substrates a (formal) [3 + 2] cycloaddition is the major reaction path which constitutes a possible rearrangement at the preoxetane biradical level [40]. A detailed kinetic and spectroscopic investigation of the reaction showed that the [2+2] adducts **43** are formed from the singlet biradical precursors whereas the [3 + 2] adducts **44** derive from the corresponding triplet biradicals [41].

The first examples of exclusive oxetane formation upon olefin photoaddition to the cyclohexen-1,4-dione 4-oxoisophorone **45**, leading to two novel 2-substituted 1-oxaspiro[3,5]non-5-en-7-ones **46** and **47**, respectively, was reported by Catalani and coworkers (Sch. 12) [42]. They demonstrated that the chemoselectivity of the olefin-enone photochemistry can be directed to exclusive oxetane formation when sufficient steric hindrance prevails cyclobutane formation.

4.3.5. Quinones

The photoaddition of 1,4-benzoquinone **48** to electron-donor substituted alkenes is an efficient process which leads to spiro-oxetanes (e.g., **49a**) in high yield [43]. The use of quinines as carbonyl compounds is advantageous because of their long-wavelenght shifted $n\pi^*$ transitions (430–480 nm). Most reactive are strained alkenes such as norbornene or norbornadiene were also rearrangement products are formed (Sch. 13) [44]. Due to their convenient absorption behavior, benzoquinones have been used to study trapping reactions of intermediates. Wilson and Musser have reported the first oxygen trapping experiments using the 1,4-benzoquinone **48**/*tert*-butyl-ethene system [45].

Whereas a triplet 1,4-biradical has been assigned as the most probable intermediate at that time, later work on intramolecular trapping reaction favored the assumption of radical ion pairs [46]. Efficient lactonization reaction to form **51** during irradiation of pent-4-enoic acid **50** and **48** accounts for an olefin radical cation which undergoes electrophilic addition towards the carboxyl group. Another type of rearrangement has been detected in the photoreaction of tetramethylallene and **48** [47]. 5-Hydroxy-indan-2-ones **52** are formed in high yields probably via instable spiro-oxetanes **49b** as intermediates.

Kochi and coworkers reported the photochemical addition of various stilbenes and chloroanil (**53**) which is controlled by the charge-transfer (CT)

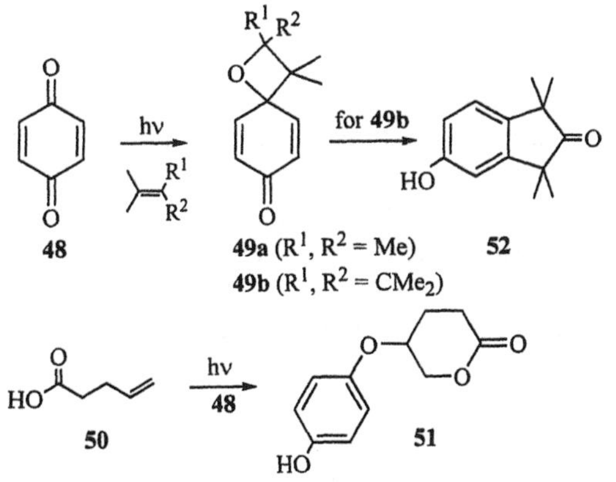

Scheme 13

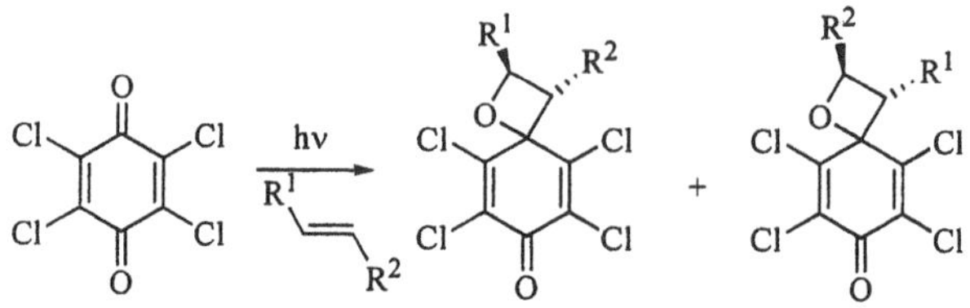

	53	R^1 = Ph, R^2 = Ph
		R^1 = 4-Cl-C$_6$H$_4$, R^2 = Ph
		R^1 = 4-Me-C$_6$H$_4$, R^2 = Ph

Scheme 14

activation of the precursor electron donor–acceptor (EDA) complex (Sch. 14) [48]. The [2+2] cycloaddition products **54** were established by an X-ray structure of the *trans*-oxetane formed selectively in high yields.

Time-resolved (fs/ps) spectroscopy revealed that the (singlet) ion-radical pair is the primary reaction intermediate and established the electron-transfer pathway for this Paternò-Büchi transformation. The alternative pathway via direct electronic activation of the carbonyl component led to the same oxetane regioisomers in identical ratios. Thus, a common electron-transfer mechanism applies involving quenching of the excited quinone acceptor by the stilbene donor to afford a triplet ion-radical intermediate which appear on the ns/µs time scale. The spin multiplicities of the critical ion-pair intermediates in the two photoactivation paths determine the time scale of the reaction sequences and also the efficiency of the relatively slow ion-pair collapse ($k_c \cong 10^8$/s) to the 1,4-biradical that ultimately leads to the oxetane product **54**.

4.3.6. Alkenes, Alkyl-, and Aryl-Substituted

Arnold and coworkers reported a series of reactions between benzophenones and monoalkenes. In most cases the oxetanes were the major products and could be isolated in good yields [49]. Benzaldehyde **55**, which has a triplet $n\pi^*$ as the reactive state [50], is less chemoselective and gave with cyclohexene besides oxetane **56** also several hydrogen abstraction and radical coupling products **57–58** (Sch. 15) [51].

A clear evidence for a long-lived intermediate came from investigations of the stereoselectivity of the Paternò-Büchi reaction with *cis* and *trans* 2-butene as substrates. when acetone [52] or benzaldehyde [53] was used as carbonyl addends, complete stereo-randomization was observed. Acetelde-hyde and 2-naphthaldehyde showed stereoselective addition reactions which accounts for the *singlet* $n\pi^*$ as the reactive state [43]. Fleming and Gao

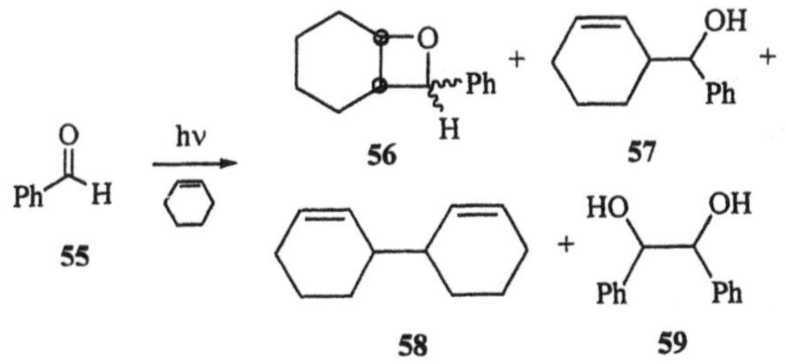

Scheme 15

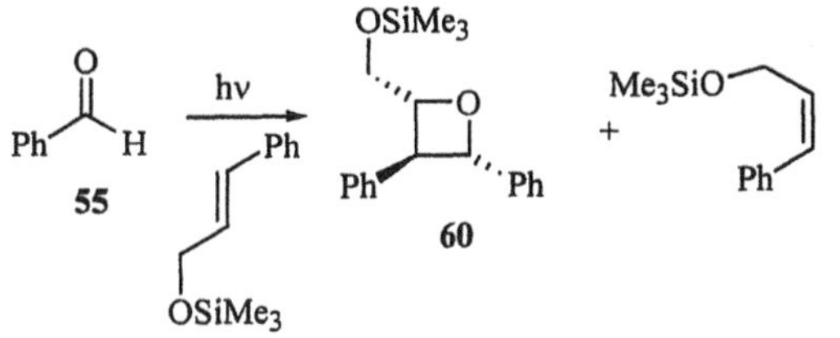

Scheme 16

reported the photocycloaddition of the trimethylsilyl ether of cinnamic with benzaldehyde to proceed with high stereoselectivity to give the *trans*-oxetane **60** in 20% yield (Sch. 16) [54]. In competition to the Paternò-Büchi reaction, *cis–trans*-isomerization leads to *cis* and *trans* isomers of the substrate in a 1.6:1 ratio.

4.3.7. Alkenes, Electron Donor-Substituted

The regioselectivity of the Paternò-Büchi reaction with acyclic enol ethers is substantially higher than with the corresponding unsymmetrically alkyl-substituted olefins. This effect was used for the synthesis of a variety of 3-alkoxyoxetanes and a series of derivatives [55]. The diastereoisomeric *cis*- and *trans*-1-methoxy-1-butenes were used as substrates for the investigation of the spin state influence on reactivity, regio- and stereoselectivity [56]. The use of trimethylsilyloxyethene **62** as electron rich alkene is advantageous and several 1,3-anhydroapiitol derivatives such as **63** could be synthesized via photocycloaddition with 1,3-diacetoxy-2-propanone **61** (Sch. 17) [57].

Branched-erythrono-1,4-lactones are accessible from the oxetane **66** which was derived thermally from diethyl mesoxalate **64** and 2,2-di-isopropyl-1,3-dioxole **65** [58]. An impressive improvement in the regio-selectivity of oxetane formation was discovered with 2,3-dihydrofuran **67** as

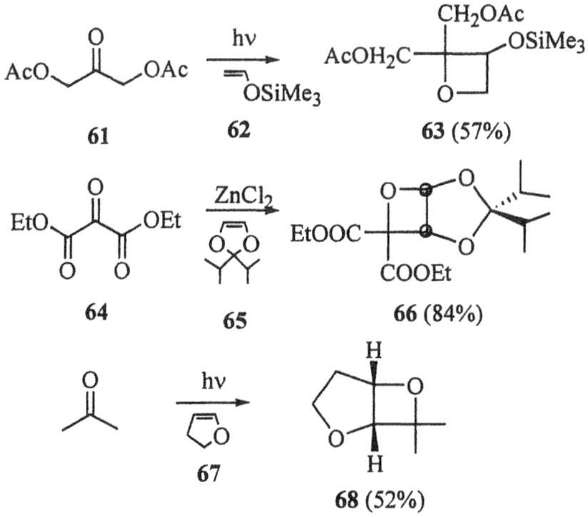

Scheme 17

alkene addend. For the acetone/2,3-dihydrofuran cycloadduct **68** a >200:1 ratio of the two possible regioisomers was determined [59]. Acyclic thioenol ethers also have been investigated in their photocycloaddition behavior with benzophenone. In contrast to acyclic enol ethers these substrates exhibit high regioselectivity with almost exclusive formation of the 3-alkylthio oxetanes [60].

Photocycloaddition reactions of aromatic aldehydes with cyclic ketene silylacetates have been investigated by Abe and coworkers [61]. Regio- and diastereoselective formation of the bicyclic 2-alkoxyoxetanes **69** was observed in high yields. Hydrolysis of these acid-labile cycloadducts with neutral water efficiently gave aldol-type adducts **70** with high *threo*-selectivity (Sch. 18).

The Paternò-Büchi photocycloaddition of silyl O,X-ketene acetals (with X = O, S, Se) and aromatic aldehydes was intensively investigated by Abe and coworkers in the last decade [62]. The regioselectivity of the reaction (**71** vs. **72**) is highly affected by the heteroatom (Sch. 19) [63,64]. The regioselectivity is rationalized by (a) the relative stability of the 1,4-biradicals and (b) the relative nucleophilicity of sp²-carbons in the respective O,X-ketene acetal.

Recently, the photocycloaddition of L-ascorbic acid derivatives **73** with 4-chlorobenzaldehyde and benzylmethyl ketone was described which led to preferential attack on the less hindered α-face of the enone to give the oxetanes **74** and **75** (Sch. 20), respectively, with approximately 2:1 regioselectivity (33% de both) [65]. When the substrate is changed to

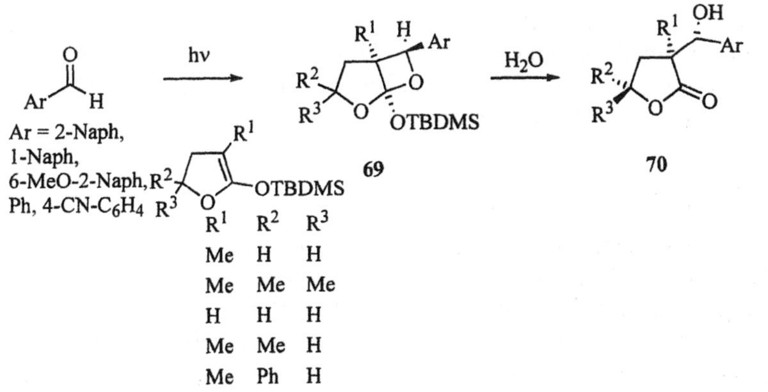
hν
H_2O
Ar = 2-Naph,
1-Naph,
6-MeO-2-Naph,
Ph, 4-CN-C_6H_4
OTBDMS
69
70
R1 R2 R3
Me H H
Me Me Me
H H H
Me Me H
Me Ph H

Scheme 18

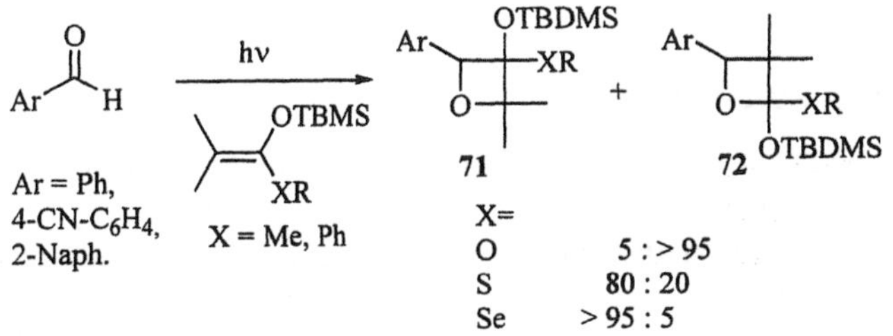
hν
OTBMS
XR
Ar = Ph,
4-CN-C_6H_4,
2-Naph.
X = Me, Ph
OTBDMS
71
72
X=
O 5 : > 95
S 80 : 20
Se > 95 : 5

Scheme 19

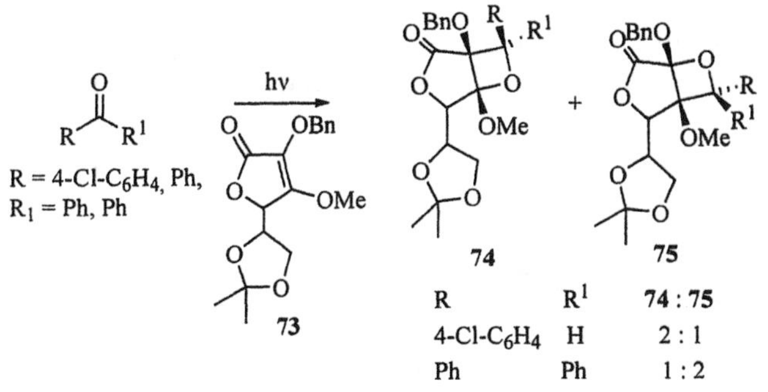
hν
R = 4-Cl-C_6H_4, Ph,
R_1 = Ph, Ph
73
74
75
R R1 74 : 75
4-Cl-C_6H_4 H 2 : 1
Ph Ph 1 : 2

Scheme 20

benzophenone, the regioselectivity was reversed, even though the facial selectivity remained the same (35% de). This was proposed to be the result of a mechanistic switchover, from a 1,4-biradical process for benzophenone to a photoinduced electron transfer process for the other substrates.

4.3.8. Alkynes

Oxetenes (oxets) have been postulated as primary photoadducts between carbonyl compounds to alkyl- and aryl-substituted alkynes and alkylthio-acetylens [66]. The first evidence for an unstable intermediate with a lifetime of several hours at −35 °C was reported for the benzaldehyde **55**/2-butyne-photoproduct **76** [67]. On further irradiation in the presence of excess benzaldehyde, a *bis*-oxetane **77** was formed (Sch. 21).

At elevated temperatures rapid ring-opening to α,β-unsaturated ketones occurs, which are the major products for photocycloaddition of alkynes with carbonyl addends at room temperature. The parent oxetene has been prepared from 3-hydroxyoxetane [68].

4.3.9. Allenes and Ketenimines

The photocycloaddition of a variety of carbonyl compounds with methyl-substituted allenes has been reported to proceed with high quantum yields (0.59 for acetophenone/tetramethylallene) [69] to give 1:1 and 1:2 adducts [70]. The 2-alkylideneoxetanes are useful precursors for cyclobutanones, e.g., **79a,b** from the benzophenone/tetramethylallene-cycloadduct **78** (Sch. 22) [71].

Upon prolonged irradiation in the presence of an excess ketone, the monoadducts are converted into 1,5- and 1,6-dioxaspiro[3.3]heptanes (e.g., **80a,b**) [53]. The regioisomeric 2- and 3-imino-oxetanes could be prepared by photolysis of ketenimines in the presence of aliphatic or aromatic ketones [72].

The photocycloaddition of aliphatic aldehydes to 1,1-dimethylallene was investigated by Howell et al. [73]. The major products, the 2-alkylidene

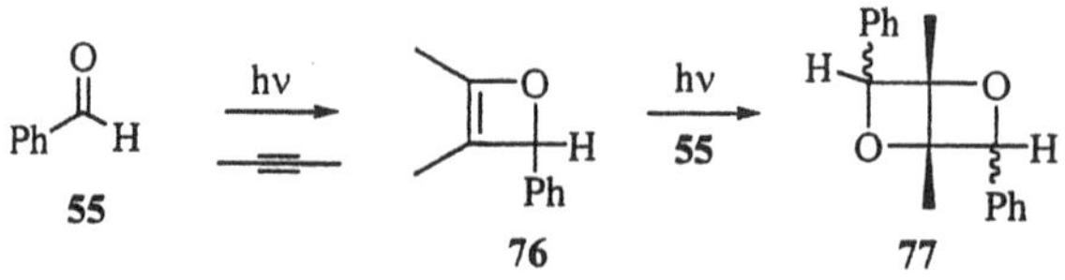

Scheme 21

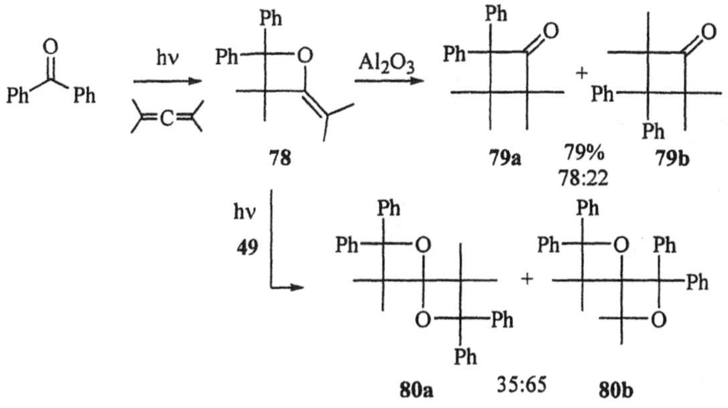

Scheme 22

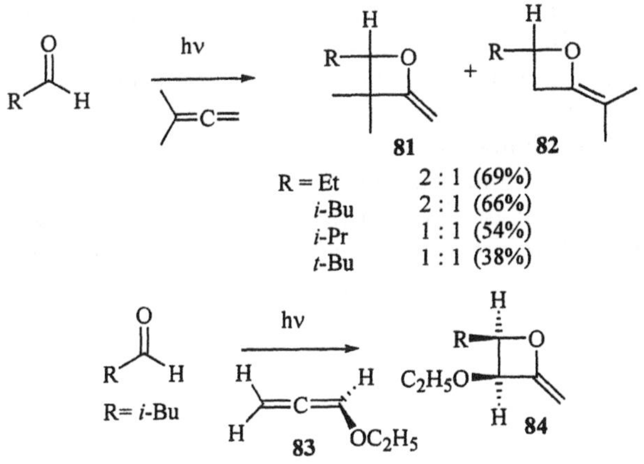

Scheme 23

oxetanes **81** and **82** were obtained in a 2:1 ratio (Sch. 23). This low degree in regioselectivity was rationalized by both steric and electronic factors. On the other hand, the allene **83** with enolether structure reacted with isovaleraldehyde with a high degree of regioselectivity to give the 2-methylene oxetane **84**.

4.3.10. Dienes and Enynes

Because of their low triplet energies (55–60 kcal/mol), 1,3-dienes are often used as quenchers for excited triplet states of carbonyl compounds. Besides

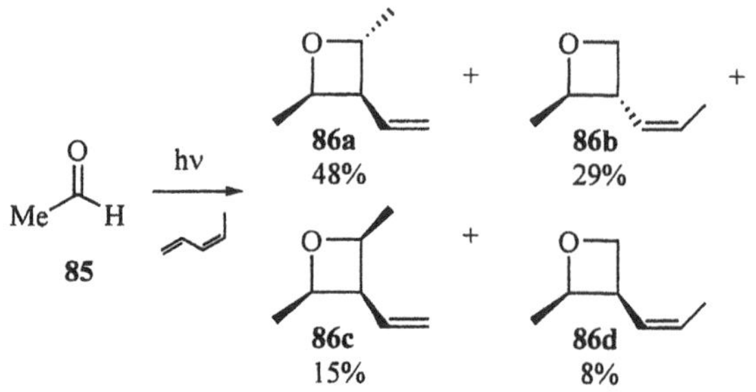

Scheme 24

physical quenching, however, also cycloaddition leading to oxetanes can occur as side reaction, as demonstrated for benzophenone [74]. Chemical yields are low because of competing diene dimerization and hydrogen abstraction reactions. The corresponding photoreactions with aliphatic ketones [75] or aldehydes [76] are much more effective in the sense that oxetanes are formed with high quantum yields and good chemical yields by a mechanism involving the singlet excited carboxyl added. Carless and Maitra have shown that the photocycloaddition of acetaldehyde **85** to the diastereomeric *E*- and *Z*-penta-1,3-dienes is highly regio- and stereoselective (e.g., oxetanes **86a–d** from the *Z*-isomer) which also accounts for a singlet mechanism (Sch. 24) [77]. A *locoselective* reaction has been reported for the benzophenone cycloaddition to an 1,3-enyne with exclusive addition to the C=C double bond [78].

4.3.11. Furans

The photocycloaddition of benzophenone to furan **87a** was originally described by Schenck et al. [79]. Additionally to the 1:1 adduct **88** also two regioisomeric 2:1 adducts **89a,b** were isolated [80], the structure of **89a** was revised by Toki and Evanega [81]. All prostereogenic carbonyl addends when photochemically added to furan showed regioselectivities >99:1 in favor of the bicyclic acetal product (Sch. 25).

This is also the case for 2-substituted furans, however, mixtures of acetal- and ketal-type oxetanes (e.g., for **87b**) were obtained [82]. The use of furans with steric demanding substituents (e.g., **87c**) [66a,83], or a acetyl substituent (e.g., **87d**) [66b] at the 2-position largely improves the regio-selectivity. A huge number of carbonyl compounds have been investigated in the last 10 years by Zamojski [84] and especially by Schreiber [85] who used furan-carbonyl adducts as intermediates in total synthesis of natural

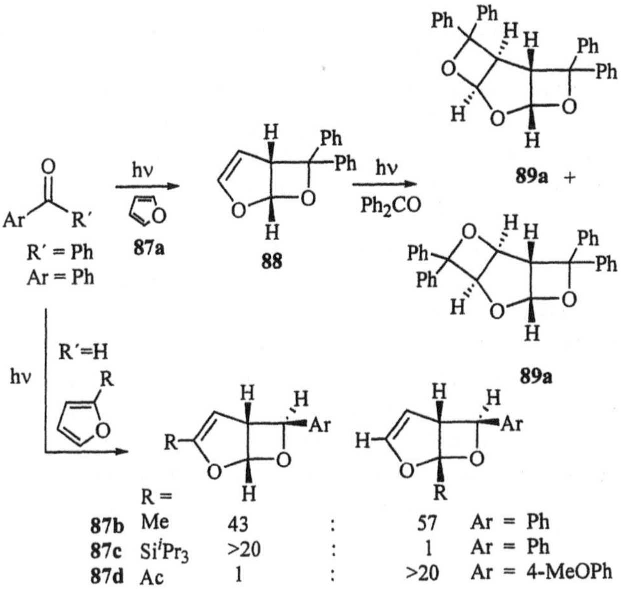

Scheme 25

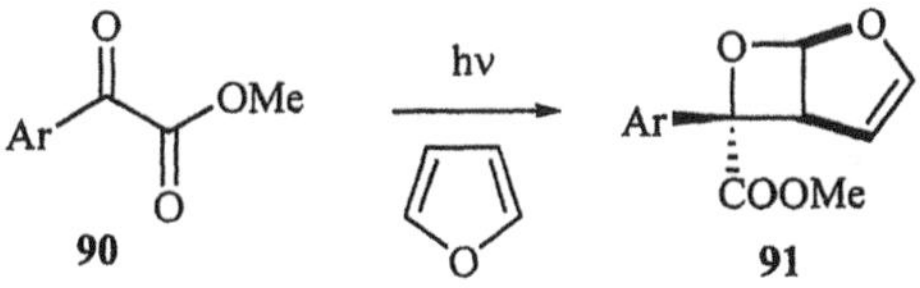

Scheme 26

products. Acid-catalyzed rearrangement of these adducts is a useful method for the synthesis of 3-substituted furans [86].

The photocycloaddition of methyl arylglyoxylates **90** with cyclo-1,3-dienes was investigated by Hu and Neckers (Sch. 26) [87]. These reactions proceed with high regioselectivity whereas the diastereoselectivity strongly depends on the nature of the aryl substituent.

Oxetanes **91** (shows with furan as the diene component) were formed with high *endo*-aryl selectivity with bulky aryl groups while insignificant stereoselectivity was observed with glyoxylates containing sterically less demanding aryl groups. This observations was rationalized by the stability of the intermediate triplet 1,4-biradical geometries during the ISC process.

The rates of the competing *ortho*-hydrogen abstraction (for *ortho*-methylated aryl groups) varied significantly among the substrates.

D'Auria and coworkers investigated the photocycloaddition of 5-methyl-2-furyl-phenylmethanol **92** with benzophenone resulting in two adducts **93** and **94** in a 1:1 ratio while the addition to 4,4′-dimethoxy-benzophenone, benzaldehyde or 4-methoxybenzaldehyde, respectively, gave merely the adducts **93** (Sch. 27) [88].

The photocycloaddition of carbonyl compounds with 2-siloxyfurans **95** has been investigated in detail by Abe and coworkers [89]. The stereoselective formations of *exo*-oxetanes **96** and **97** were observed in high yields (Sch. 28). The regioselectivity was found to be largely dependent upon the nature of the carbonyl component, the substitutents at the furan ring, and the excited state of the carbonyls (singlet vs. triplet). Aldehydes resulted in bicyclic oxetanes **96** and **97**, respectively, with low regioselectivity independent of the nature of the excited states and the substituents at the furan. Triplet excited ketones gave regioselectively the *exo*-oxetanes, except for 4-methyl-2-siloxyfuran.

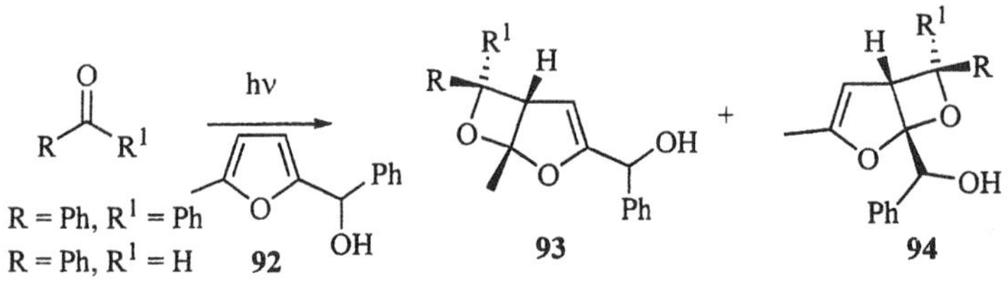

Scheme 27

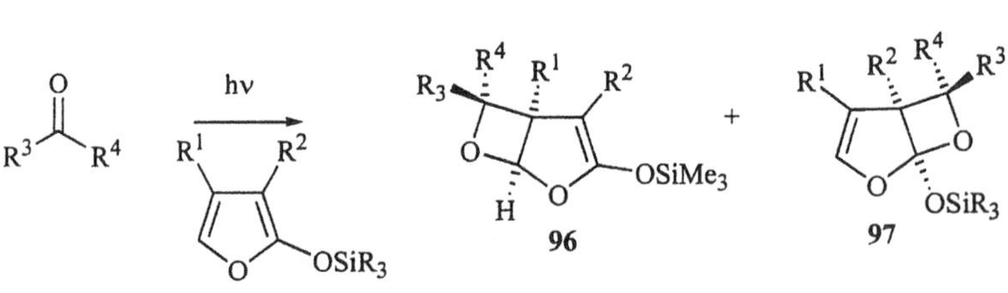

d.r.	SiR_3	R^1	R^2	R^3	R^4
61 : 39	TIPS	H	H	Me	H
54 : 46	TIPS	H	H	Ph	H
40 : 60	TBDMS	Me	H	Ph	H

Scheme 28

4.3.12. Other Heteroaromatic Substrates

Methylsubstituted thiophenes afford oxetanes with high regioselectivity when reacted with excited benzophenone [90]. Pyrroles, imidazoles, and indoles behave similarly when substituted at the nitrogen atom with electron-acceptor groups [91]. Pyrroles, when alkyl-substituted at the nitrogen atom, however, gave rearranged pyrroles, probably via an oxetane intermediate [92]. The photocycloaddition of 2,4,5-trimethyloxazole **98** to carbonyl compounds afforded the bicyclic oxetanes **99** with high regio and excellent (*exo*) diastereoselectivity [93]. Hydrolytic ring opening of bicyclic oxetane yielded *erythro* α-actamido-β-hydroxy ketones **100** (Sch. 29).

4.3.13. Strained Hydrocarbons

The photocycloaddition of triplet benzophenone to norbornene has been originally reported by Scharf and Korte (Sch. 30) [94]. The photoproduct **101** which is formed in high *exo*-selectivity could be thermally cleaved to the δ,ε-unsaturated ketone **102**, an application of the "carbonyl-olefin-metathesis" (COM) concept [95].

The 1,4-biradical formed in the interaction of norbornene with *o*-dibenzoyl-benzene was trapped in an intramolecular fashion by the

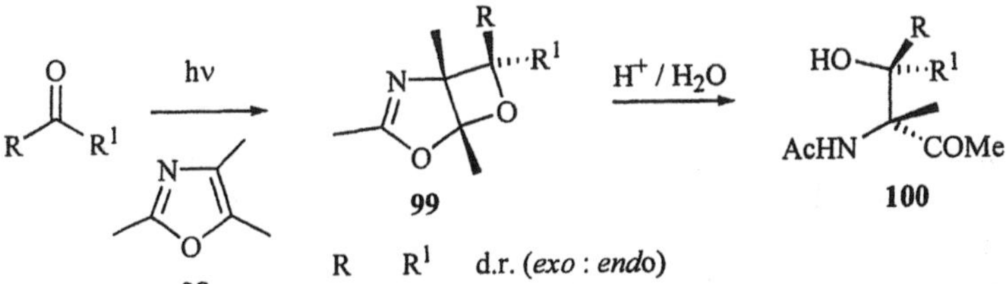

R	R¹	d.r. (*exo* : *endo*)
Ph	H	>99 : 1
Et	H	>99 : 1
t-Bu	H	>99 : 1
Ph	COOMe	74 : 26

Scheme 29

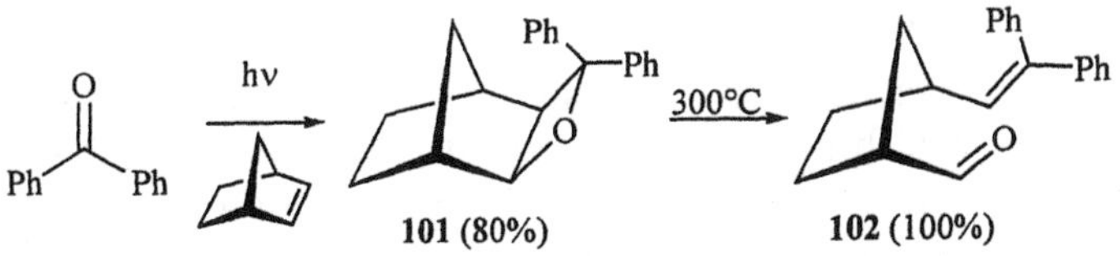

Scheme 30

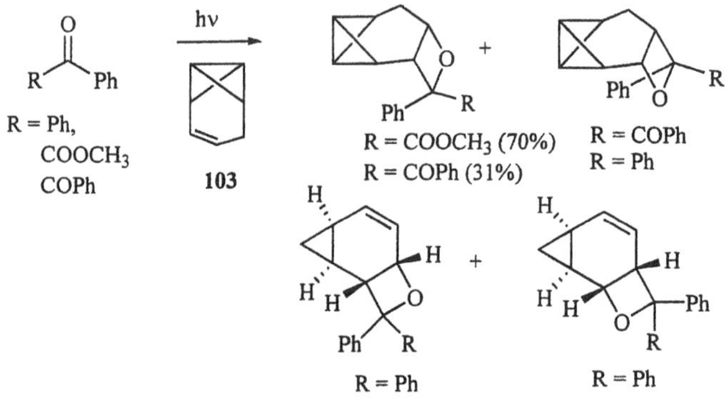

Scheme 31

second carbonyl moiety [96]. A highly regioselective reaction of triplet benzophenone was reported with 5-methylenenorborn-2-ene with preferential attack towards the *exo* C=C-double bond [97]. A number of publications have been appeared which discuss photocycloaddition reactions of triplet carbonyl compounds to norbornadiene and quadricyclane and the competition between Paternò-Büchi reaction and the sensitized norbornadiene/quadricyclane interconversion [98]. Oxetane formation has also been reported for the photoreaction of biacetyl as well as *para*-quinones with benzvalene [99].

The irradiation of methyl phenylglyoxylate, benzil, benzophenone as well as 1,4-benzoquinone in the presence of homobenzvalene **103** gave, as products of the Paternò-Büchi reaction, oxetane derivatives which contain the tricyclo[4.1.0.0^{2,7}] heptane subunit as well as ring-opened products (Sch. 31) [100].

In the case of benzophenone, the cycloaddition competes with the isomerization of **103** to cycloheptatriene. Exclusive isomerization was observed with acetophenone and acetone. Carbonyl compounds with triplet energies lower than 69 kcal/mol prefer the cycloaddition path. Cyclopent-2-en-1-one is an exception to this rule: in spite of its triplet energy of 74 kcal/mol, 2 + 2 cycloadducts were formed rather efficiently.

4.3.14. Alkenes, Electron Acceptor-Substituted

In contrast to photocycloaddition reactions of carbonyl compounds to electron-rich alkenes (which proceed with a low degree of stereoselectivity in the case of triplet excited carbonyls), reactions with electron-deficient alkenes, such as cyanoalkenes, are, although rather inefficient, but highly stereoselective [101]. Kinetic analysis showed that these reactions involve the

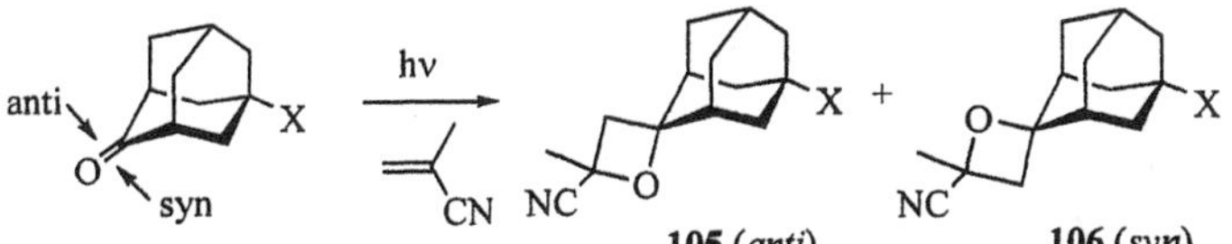

Scheme 32

interaction with the singlet excited carbonyl via a parallel approach [102]. An important side reaction is the photosensitized geometrical isomerization of alkene C=C double bond, e.g., cis-1,2-dicyanoethylene and acetone gives the cis-oxetane and trans-1,2-dicyanoethylene. 2-Norbornanone was used as a model reagent for investigation of the influence of steric hindrance on the face-selectivity of oxetane formation with electron-donor and electron-acceptor substituted alkenes [103]. Chung and coworkers have reported, that the photocycloaddition of methacrylonitrile to 5-substituted adamant-2-ones **104** produces two geometrically isomeric oxetanes **105** and **106**, respectively, with the oxetane-ring oxygen and the substitutent at C-5 in *anti* or *syn* positions (Sch. 32) [104]. This selectivity is discussed for the *syn*-face attack in terms of transition state hyperconjugation.

4.3.15. Exocyclic Olefins

The photochemistry of ketones in the presence of exocyclic olefins has not yet been systematically studied. Chung and Ho reported the photochemistry of acetone in the presence of several exocyclic olefins. Surprisingly, homoalkylation occurred resulting in a series of 4-cycloalkylbutan-2-ones (with quantum yields of 0.14 ± 0.01) rather than the expected Paternò-Büchi reaction (Sch. 33). With perdeuterated acetone, the photocycloaddition path increased due to the primary kinetic isotope effect (as shown for products **108** and **109**, respectively, from 2-methyleneadamantane **107**) [105].

Similar effects were obtained for methylenecycloalkanes **108** with preferential formation of the photo-Conia products **111**. Increasing ring size in **110** as well as H/D exchange favored the formation of the Paternò-Büchi products **112**.

4.4. SYNTHETIC POTENTIAL: REACTIVITY AND SELECTIVITY PATTERN

The combination of two prostereogenic substrate molecules, the carbonyl and the alkene component, in the course of a Paternò-Büchi reaction, leads

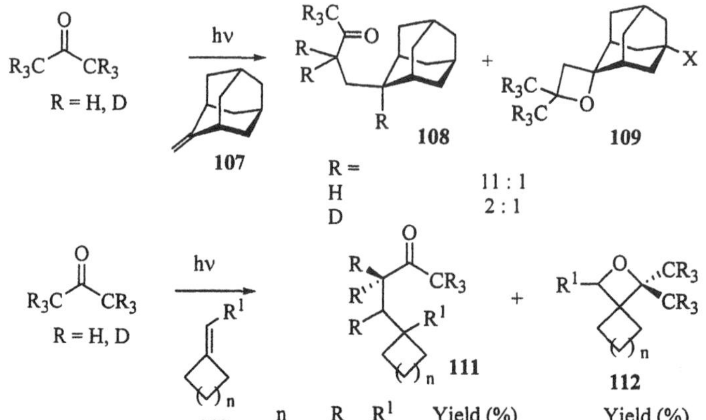

n	R	R^1	Yield (%) 111	Yield (%) 112
1	H	H	14	trace
2	H	H	12	trace
3	H	H	60	trace
3	H	CH_3	17	17
3	D	CH_3	5	32

Scheme 33

to a photoadduct with three new stereogenic centres. Control of the relative and absolute configuration of these stereogenic centers is a challenge for synthetic chemistry in that many interesting products could in principle be derived from oxetane precursors. A detailed knowledge of the photophysical properties of the electronically excited compound (which is in most cases the carbonyl addend) is necessary to understand (and predict) the stereochemical result of such a [2+2] cycloaddition reaction. Therefore the configuration of the excited state, its lifetime and IC as well as ISC properties should be known. For clean transformations the carbonyl group should be the only absorbing chromophore in the reaction mixture (i.e., the product should not absorb at the wavelength used), the solvent should not interfere with the cycloaddition step by competing reactions (e.g., hydrogen abstraction) and the polarity influence of the reaction medium on biradical or photoinduced electron transfer (PET) steps ought to be carefully investigated. Considering the simple model for the spatial distribution of electron density in the $n\pi^*$ state, a nonprostereogenic $n\pi^*$ excited carbonyl compounds has two pairs of enantiotopic faces for interaction with an alkene addend. Analogously to ground state nucleophilic addition reaction, considering the *Umpolung* effect, electron-deficient alkenes should preferentially interact with the nucleophilic π^* orbital. Such as orientation has been named the *parallel* approach. On the other hand, electron-rich alkenes should preferentially interact with the electrophilic n-orbital perpendicular to the π plane. Consequently such an orientation has been named the *perpendicular* approach. Analyses of the product stereochemistry cannot uncover these primary orientation phenomena. A study of Stern-Volmer kinetics of several 2-norbornanones with the electron-rich *cis*-diethoxyethylene (*c*-DEE) and the electron-poor *trans*-dicyanoethylene (*t*-DCE) as quenchers did corroborate this model [103]. In these cases the carbonyl compounds are chiral and exhibit two pairs of diastereotopic faces for interaction with the alkene addends. In an additional paper it was shown by Turro and Farrington that the *t*-DCE addition is a highly stereoselective process indicating an exclusive interaction with the first excited singlet state of the carbonyl compound [103]. An increase in steric hindrance toward the approach from the *exo*-side reduced the rate of fluorescence quenching, however, the efficiency of oxetane formation was increased. Therefore it was interpreted that for sterically more hindered ketones a possible exciplex intermediate is much more effectively transformed into the photoproduct. This may be due to a puckering effect of the $n\pi^*$-excited carbonyl group which leads to enhanced efficiency of photoproduct formation from the exciplex intermediate. Ketones with homotopic faces have been widely used to determine the stereoselectivity in Paternò-Büchi reactions with E- and Z-isomers of electron-rich and electron-poor olefins.

The photocycloaddition of acetone to *cis*- or *trans*-2-butene leads to a mixture of diastereomeric oxetanes in a constant ratio of 64:36 ($\pm 2\%$) independent from the substrate configuration [106]. No isomerization of the alkene substrates could be detected at low to medium conversion. The photocycloaddition of acetone to *cis*- and *trans*-1-methoxy-1-butene has been studied in order to show the divergent stereoselectivity of singlet and triplet excited carbonyl states [107]. By use of piperylene as triplet quencher and investigation of the concentration (quencher as well as substrate) dependency of the stereoselectivity of the Paternò-Büchi reaction a consistant mechanism was established. At high concentrations of the alkene or the triplet quencher maximum d.r. (diastereomeric ratio) values of 82:18 and 27:73 were found for the diastereoisomeric oxetanes, respectively. Therefore it was concluded that the stereochemistry of the initial butene is retained in the oxetanes when singlet excited acetone attacks the alkene, whereas a scrambling effect is obtained for the triplet case indicating a long lived triplet biradical intermediate.

In some cases, however, a high degree of stereoselectivity could be obtained even with "pure" triplet excited carbonyl compounds. In these cases, e.g., the photocycloaddition of benzophenone to several methyl vinyl sulphides **113**, the intermediary triplet 1,4-biradical preferentially undergoes one of two possible cyclization modes after intersystem crossing (Sch. 34) [60].

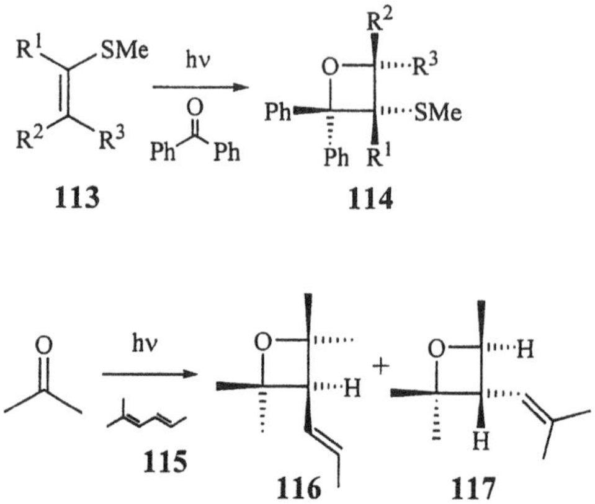

113   114

115

116   117

Scheme 34

The nonstereospecific (using Zimmerman's definition) [108] nature of this reaction has been demonstrate by the use of stereoisomeric substrates which lead to oxetanes **114** in identical diastereomeric ratios. Another possibility for obtaining stereoisomerically pure oxetanes is the use of alkenes which simultaneously serve as quenchers for the carbonyl triplets, e.g., dienes. The photocycloaddition of acetone and 2-methyl-2,4-hexadiene (**115**) represents such a process leading to two regioisomeric oxetanes **116** and **117** where the substrate configuration is retained in **117** [75].

4.4.1. Prostereogenic Carbonyl Groups: Product Stereoselectivity

Simple (noninduced) diastereoselectivity in general describes a selection process where two stereogenic elements (or more) are generated in a chemical process without stereogenic elements present already in the starting materials, whereas induced diastereoselectivity describes a selection process where stereogenic elements are generated in a chemical process from substrates with at least one stereogenic elements already present. Thus, in case of Paternò-Büchi reactions, the combination of two prostereogenic substrate molecules leads to a photoadduct with maximum three new stereogenic centers with a characteristic simple diastereoselectivity.

4.4.1.1. Simple Diastereoselectivity

In contrast to the results with acetone (vide infra) the photoaddition of acetaldehyde with *trans* (and analogous with the *cis*-isomer) 2-butene is highly stereoselective (or stereospecific [108]) [53]. Aromatic carbonyl compounds such as benzaldehyde show low stereoselectivites independent from their concentrations (in contrast to aliphatic ketones) due to fast intersystem crossing into the triplet manifold of the carbonyl substrate. If, however, the configuration of the triplet excited state switches form $n\pi^*$ to $\pi\pi^*$ the product stereoselectivity rises again. In these cases (e.g., 2-naphthaldehyde) the carbonyl singlet is the reactive state in photocycloaddition reactions involving the carbonyl group. Obviously the quantum yields for such a process are low (about 5% of triplet reactivity) because of efficient deactivation via the triplet $\pi\pi^*$ states. Another concentration study has been published by Jones II et al. where they describe the photocycloaddition of aliphatic aldehydes to medium ring cycloalkenes. The variation in photoadduct distribution is due to stereospecific addition of the aldehyde singlet (dominant at high alkene concentrations) accompanied by a less stereoselective triplet pathway at lower concentrations of the cycloalkenes [109].

In case of Paternò-Büchi reaction of cycloalkenes with prostereogenic aromatic carbonyl compounds (which show rapid ISC to the triplet excited carbonyl) less clear results were obtained. The reaction of cyclohexene with benzaldehyde was reported in the literature [51] and the spectral data of the main product **56** (35%) described as consistent with the assignment of *exo*-stereochemistry (Sch. 35) [110].

A detailed study of the photocycloaddition of (triplet) benzaldehyde with several cycloalkenes revealed that *endo*-stereoselectivity is an inherent property for these processes [111]. Especially the addition of benzaldehyde to 2,3-dihydrofuran was significant: only one regioisomer **118** (analogous to the reaction with acetone) [59] is formed in a diastereomeric ratio of 88:12. Increasing steric demand of the carbonyl addend leads to an increase in diastereoselectivity, e.g., for mesityl aldehyde only the *endo*-diastereomer **118b** could be detected [112]. Substituent effects have also been described for methyl-substituted cycloalkenes which were in accord with the postulated principle for control of stereoselectivity in triplet reactions [113]. This selectivity effects could be explained by the assumption of certain 1,4-diradical conformers which fulfill the prerequisites (Salem rules) [114] for rapid intersystem crossing to form the 1,4-singlet biradicals. These can interconvert without spin barrier into the products and thus exhibit a "conformational memory" [111,113]. In contrast to the triplet pathway, singlet excited carbonyl compounds such as the 1- or 2-naphthaldehydes do add with high *exo* stereoselectivity to 2,3-dihydrofuran [115]. Both singlet *and* triplet excited carbonyl compounds underwent photocycloaddition to *furan* with high *exo* selectivity to give adducts **119** (Sch. 36).

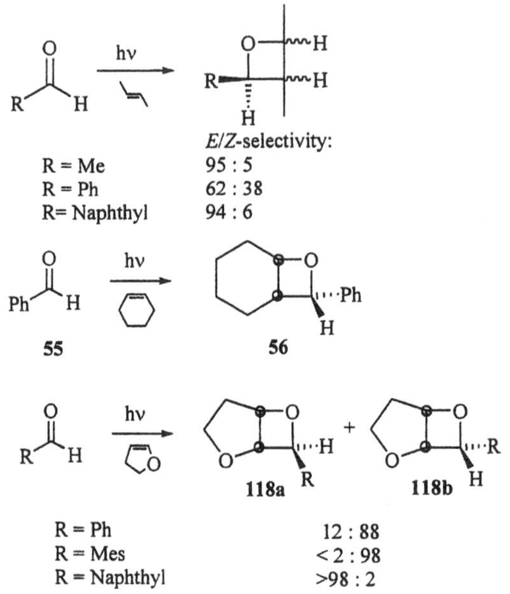
hν
E/Z-selectivity:
R = Me 95 : 5
R = Ph 62 : 38
R= Naphthyl 94 : 6
55
56
118a
118b
R = Ph 12 : 88
R = Mes <2 : 98
R = Naphthyl >98 : 2

Scheme 35

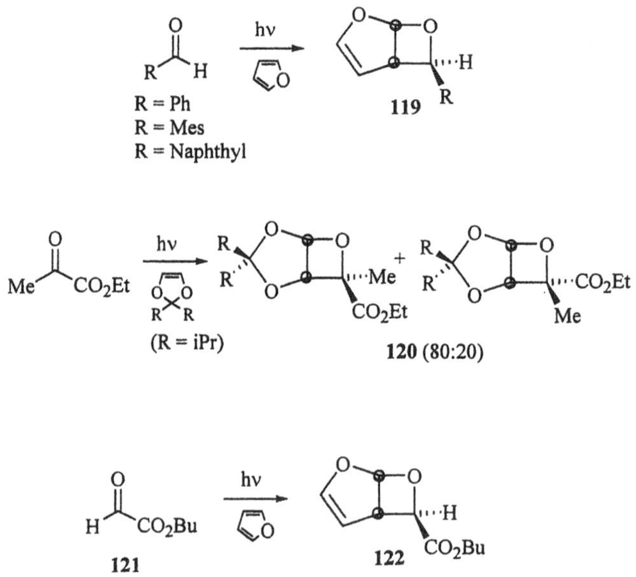
hν
R = Ph
R = Mes
R = Naphthyl
119
(R = iPr)
120 (80:20)
121
122

Scheme 36

Additionally only one regioisomer is formed in these reactions which show high efficiency and proceed with excellent yields [79,82,116]. The oxetanes formed in these reactions are acid-labile compounds and can be converted into the corresponding aldol products. Schreiber et al. used this property for the synthesis of several key compounds as substrates for the synthesis of natural products by "photo-aldol reaction" [85,117]. Besides the addition of aldehydes to furan, also ketones and esters as well as substituted furans had been used as staring materials [22,118]. In most cases the locoselectivity (for furans with unsymmetric substitution pattern) is moderate [119], but the regioselectivity (exclusive formation of the acetal product) and the stereoselectivity stays high (>500:1 in some cases) [84]. The same degree of regio- and steroselectivity has been observed for other heterocyclic substrates such as thiophenes, oxazoles, pyrazoles, and many more [90,91]. A somewhat lower degree in stereoselectivity was reported for the photocycloaddition of carbonyl compounds to 1,3-cyclopentadiene (with propanal: *exo/endo* = 80:20) [76] and 1,3-cyclohexadiene (with acetaldehyde: *exo/endo* = 88:12) [120]. Another group of highly efficient carbonyl addends are α-keto esters such as alkyl pyruvates, alkyl glyoxylates, and esters of phenylk glyoxylic acid. Photocycloadditon with electron-rich cycloalkenes such as dioxoles and furans have been reported to proceed highly regio- and diastereoselectively. The Paternò-Büchi reaction of 2,2-di-isopropyl-1,3-dioxole with ethyl pyruvate leads to a 80:20 mixture of diastereomeric oxetanes **120** [121]. Zamojski has report a highly selective reaction of triplet excited *n*-butyl glyoxylate **121** with furan [122]. Similar to the reactions with aromatic aldehydes [116] only the *exo*-photoproduct **122** is formed.

The simple diastereoselectivity of the photocycloaddition of electronically excited carbonyl compounds with electron rich olefins was studied as a function of the substituent size—at identical starting conditions ignoring the electronic state involved in the reaction mechanism [123]. The [2+2] photocycloaddition of 2,3-dihydrofuran with different aldehydes in the nonpolar solvent benzene resulted in oxetanes **118** with high regioselectivity and suprising simple diastereoselectivites: the addition to acetaldehyde resulted in 45:55 mixture of *endo* and *exo* diastereoisomer, with increasing the size of the α-carbonyl substituent (Me, Et, *i*-Bu, *t*-Bu), the simple diastereoselectivity increased with preferential formation of the *endo* stereoisomer (Sch. 37).

The benzaldehyde addition which was most intensively investigated gave a 88:12 mixture of *endo* and *exo* diastereoisomers **118**. Thus, the thermodynamically less stable stereoisomers (>1.5 kcal/mol, from ab initio calculation) were formed preferentially. To further enlarge the phenyl substituent, *ortho*-tolyl and mesitaldehyde as well as

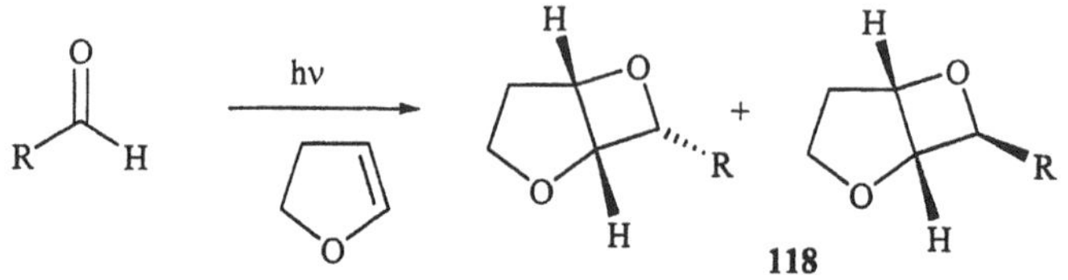

Scheme 37

R =	d.r. (endo : exo)
Me	45 : 55
Et	58 : 42
i-Bu	67 : 33
t-Bu	91 : 9

Scheme 38

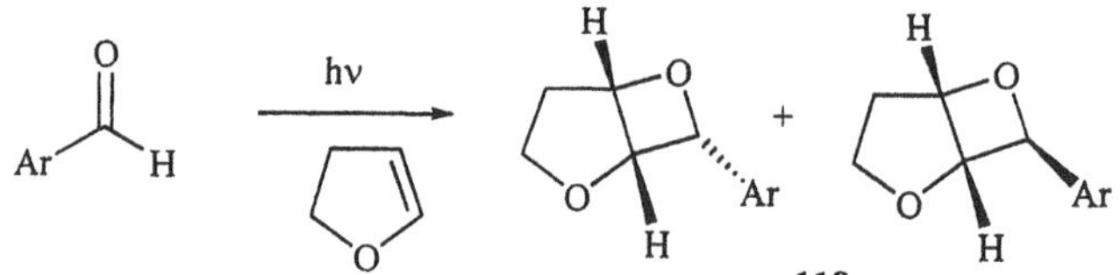

Ar =	d.r. (endo : exo)
Ph	88 : 12
o-Tol.	92 : 8
Mes.	>98 : 2
2,4-di-tert-Bu-Ph	>98 : 2

2,4-di-*tert*-butyl-6-methylbenzaldehyde were applied and actually the diastereoselectivity did further increase (Sch. 38).

In contrast to the diastereoselectivity of the 2,3-dihydrofuran photocycloaddition, the photocycloaddition of furan with aromatic as well as aliphatic aldehydes proceeded with unusually high *exo*-diastereoselectivity to give the bicyclic oxetanes **119** in good yield (Sch. 39) [123]. The diastereoselectivity of reaction of furan with acetaldehyde and benzaldehyde (*exo/endo*) were 19:1 and 212:1, respectively.

The exchange of the hydrogen in benzaldehyde by a methoxy group completely inverted the diastereoselectivity in the photocycloaddition with furan. Further modification of the α-substituent in the benzoyl substrates disclosed a distinct dependence of the *exo/endo* ratio on the size of this substituent. The photocycloaddition of acetophenone with furan gave only one product, whereas a 77:23 mixture of diastereoisomers resulted from the addition of benzoyl cyanide. Increasing the size of the aryl group from

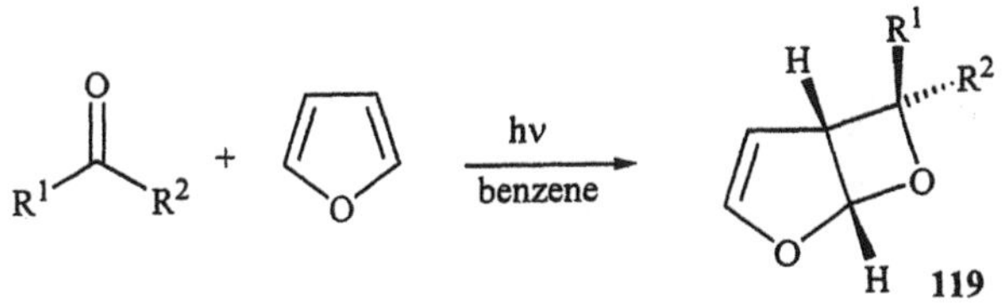

$R^1 =$	$R^2 =$	d.r. (exo-R^1 : $endo$-R^1)
Ph	H	212 : 1
Ph	Me	>49 : 1
Ph	CN	3.7 : 1
Ph	CO_2Me	1 : 9
Ph	OMe	1 : 19
Ph	CO_2R	<1 : 49
Mes	CN	16 : 1
^tBu	CO_2Me	<1 : 49

Scheme 39

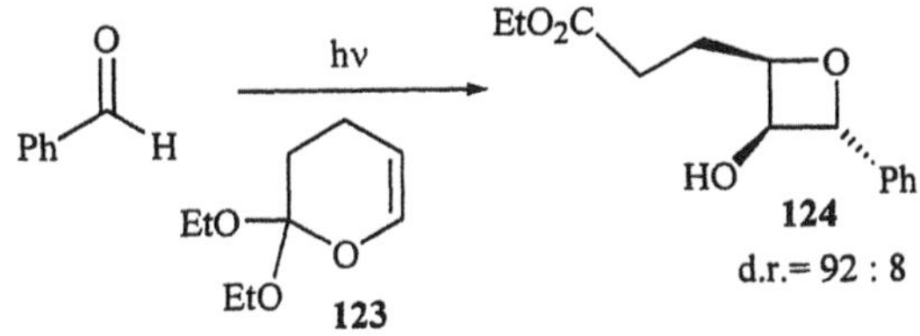

Scheme 40

phenyl to mesityl in aroyl cyanides led to an increase in exo-diastereoselectivity from 3.7:1 up to 16:1.

Steric hindrance can also reach a critical value during bond formation and might favor the formation of thermodynamically more stable product. Park et al. described the photocycloaddition of benzaldehyde to 2,2-diethoxy-3,4-dihydro-2H-pyran (**123**) as preferentially resulting in the exo-phenyl product **124** (Sch. 40) [124].

Bach and coworkers investigated the photocycloaddition of N-acyl, N-alkyl enamines **125** with benzaldehyde [125]. The 3-amido oxetanes **126** were formed with excellent regioselectivity (analogous to reactions with enolethers—vide supra) and good diastereoselectivity (Sch. 41). Enamines, not deactivated by acylation at the nitrogen atom are poor substrates for Paternò-Büchi reactions due to preferred electron transfer reactivity (formation of the corresponding enamine radical cation and subsequent reactions).

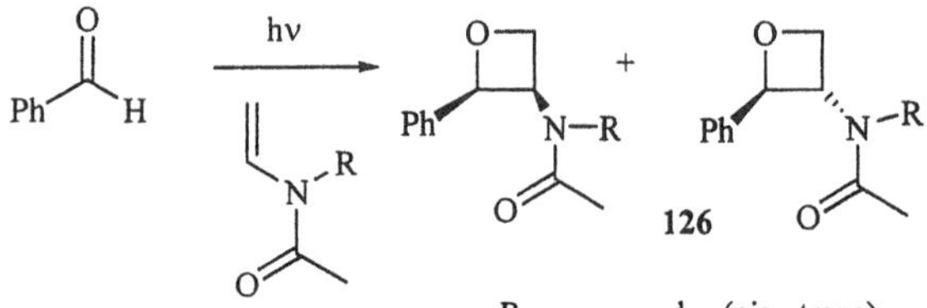

125 126

R=	d.r. (cis : trans)
H	79 : 21
Bn	89 : 11
Pr	90 : 10

Scheme 41

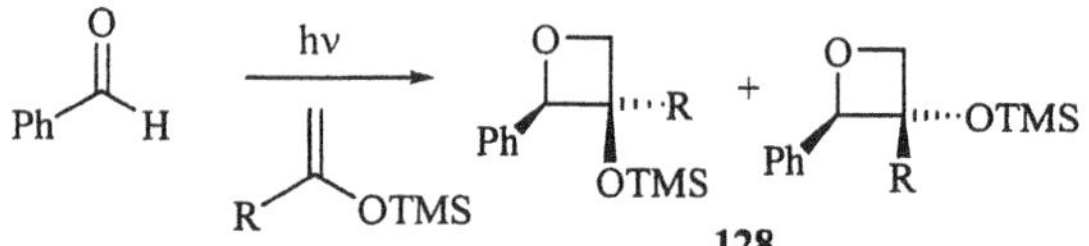

127 128

R =	d.r. (Z: E)
Me	70 : 30
Et	83 : 17
i-Pr	88 : 12
t-Bu	91 : 9
Ph	92 : 8

Scheme 42

Silylenolethers **127** were also investigated by Bach et al. in the last decade [126]. A series of photocycloaddition reactions of benzaldehyde with these substrates showed a trend in stereoselectivity which at first sight was in contradiciton to the rules described above, i.e., the thermodynamically more stable *Z* diastereoisomers **128** (with respect to the substituent at C-2 and C-3) were formed preferentially and the *Z*/*E* ratio increases with increasing size of the C-3 substituent (Sch. 42).

This peculiar stereoselectivity might be attributed to a memory effect from the approach geometry between the triplet excited benzaldehyde and the alkene. Abe and coworkers have also observed a comparable stereochemical effect in the Paternò-Büchi reaction of 4-cycanobenzaldehyde with *O*-silylated thioketene acetals **129** (Sch. 43) resulting in the highly functionalized oxetanes **130** [64].

Recently, Bach and coworkers reported the photocycloaddition of α-alkyl-substituted enecarbamates **131** to benzaldehyde affording 3-amino oxetanes **132** (Sch. 44) in moderate to good yields (46–71%) [127]. The α-phenyl substituted enecarbamate did not lead to a photocycloaddition

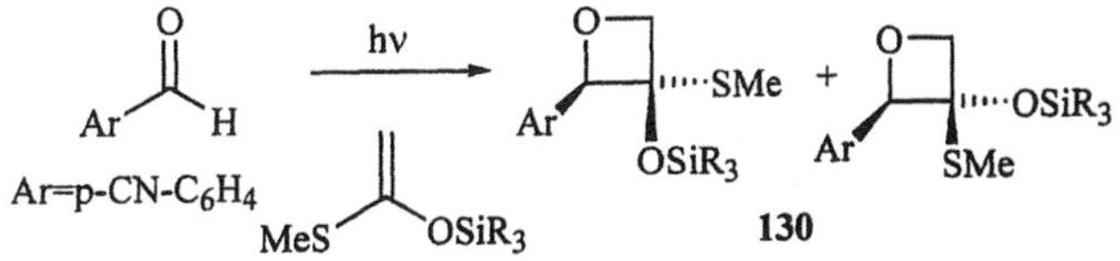

129 130

SiR_3 =	d.r. (E: Z)
Si^tBuMe_2	20 : 80
$SiMe_3$	18 : 82

Scheme 43

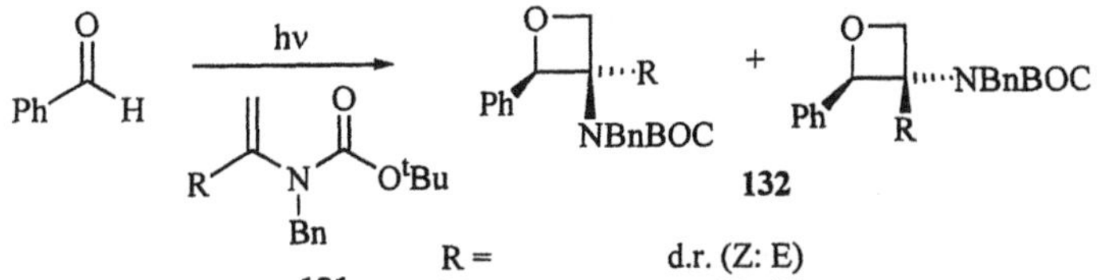

131 132

R =	d.r. (Z: E)
Me	29 : 71
Et	34 : 66
i-Pr	54 : 46
Cyclohex	57 : 43
H	87 : 13

Scheme 44

product presumably due to rapid energy transfer (tripler sensitization) from the elctronically excited aldehyde. An increase in steric bulk of the alkyl substituent *R* shifted the diastereomeric ratio *Z*/*E* in direction of the thermodynamically more stable *Z* product.

4.4.1.2. Induced Diastereoselectivity

The first report concerning an "asymmetric" Paternò-Büchi reaction with a chiral carbonyl component was reported in 1979 by Gotthardt and Lenz [128]. The photocycloaddition of the enantiomerically pure menthyl ester of phenylglyoxylic acid **133** with 2,3-dimethyl-2-butene gave the oxetane **134** with a diastereomeric excess of only 37% (Sch. 45).

The corresponding chiral glyoxylates gave even lower diastereomeric excesses of 5 ± 2% when reacted with furan [129]. The unique behavior of chiral phenyl glyoxylates has later been demonstrated by Scharf and coworkers. Despite the fact that in all cases the stereogenic centers are localized in the alcohol part of the α-keto ester and therefore remarkably far

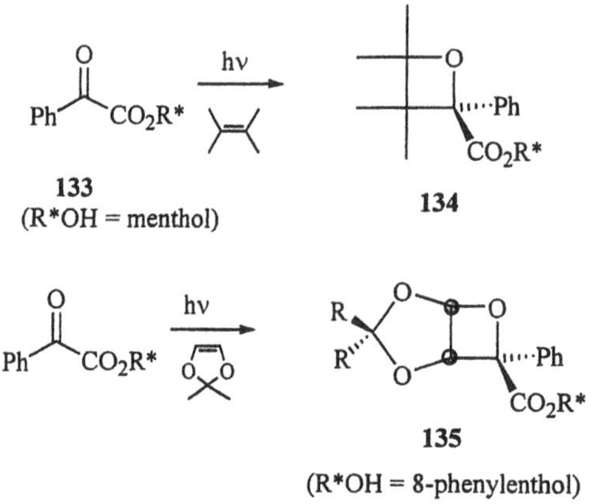

133
(R*OH = menthol)

134

135
(R*OH = 8-phenylenthol)

Scheme 45

away from the reactive (triplet excited) carbonyl group, the (induced) diastereoselectivities were exceedingly high (>96% in many cases) with 8-phenylmenthol as chiral auxiliary [130]. The temperature dependence of the auxiliary-induced diastereoselectivity has been studied by the same group [131]. As substrates for the photoaddition of phenyl glyoxylates several electron-rich cycloalkenes were used which had already been studied in their reaction behavior with α-keto esters [132]. Besides normal isoselectivity effects in most cases also an *inversion* of the induced diastereoselectivity was found at certain temperatures. Consequently this effect was named *isoinversion principle* and in the meanwhile demonstrated to be a general phenomena for many two-step reactions [133]. The inherent (noninduced) diastereoselectivity for these photoadditions (e.g., oxetane **135** from a dioxole derivative and the 8-phenylmenthyl phenyl glyoxylate) was >96% with the phenyl group being directed into the *endo*-position. The applications of Cram-like model compounds such as isopropylidene glyceraldehyde [134] or acyclic α-chiral ketones [84] did, however, not lead to high stereoselectivities.

4.4.2. Effect of Temperature on Diastereoselectivity

Recently, Adam and coworkers reinvestigated the Paternò-Büchi reaction of *cis*- as well as *trans*-cyclo-octene (**136**) with electronically excited benzophenone (Sch. 46) [135]. They reported an unprecedented temperature-dependent diastereoselectivity, where the more stable substrate diastereoisomer (*cis*-cyclo-octene) leads, with increasing reaction temperature, to

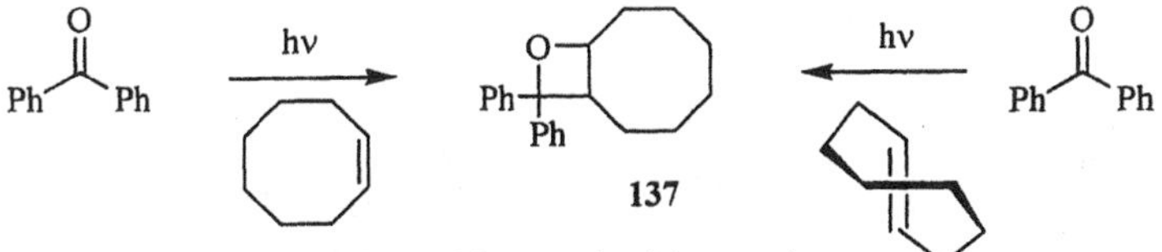

Scheme 46

increasing amounts of the less stable product diastereoisomer **137** (*trans*-oxetane).

Adam rationalized the unprecedented experimental facts for the [2+2]-photocycloaddition of the diastereoisomeric cyclo-octenes with benzophenone in terms of a consistent mechanism [136]: (i) The *cis*-**136** displays a remarkable temperature dependence in that the *trans*-2 oxetane is favored with increasing temperature. (ii) For *trans*-**136** the *trans* geometry is preserved in *trans*-cycloadduct over a broad temperature range of 180 °C. (iii) The extent of *trans* to *cis* isomerization in the cycloaddition with the *trans*-cyclo-octene increase with temperature.

A huge number of acyclic and cyclic alkenes and alkadienes have been used as addends for Paternò-Büchi reaction with carbonyl compounds and many of them are mentioned in the former chapters. Dioxoles [137] and 2,3-dihydro-oxazoles [138] have been found to be remarkably effective addends. These alkenes are highly electron-rich compounds with low-lying ionization potentials and also serve as potent electron-donor substrates. This property make the application of electron-rich alkenes sometimes critical because PET (photo-electron-transfer) reactions can interfere with the "normal" photocycloadditons via triplet 1,4-biradicals. The use of unpolar solvents is therefore recommended, however, an analysis of the energetics (Rehm–Weller relationship) should always be included [139]. Intensive investigations of the diastereofacial selectivity of ketone photocycloaddition to norbornene and norbornadiene have been reported. The biacetyl addition to norbornene (**138a**) is highly (>24:1) *exo*-selective whereas the *syn*-7-*tert*-butyl derivative **138b** showed inverted (<1:30) *exo*-selectivity (Sch. 47) [140].

Introduction of a hydroxy group at the 7-*syn*-position of norbornene (**138c**) re-inverts the diastereoselectivity: in this case the *exo*-adduct **139c** is formed with d.e. >97%. The later effect could be due to hydrogen bonding which precomplexes the excited carbonyl species. A similar photoreaction has been reported for norbornadiene [141]. As was shown by Gorman et al., ketone triplets do not add to but interconvert norbornadiene it to

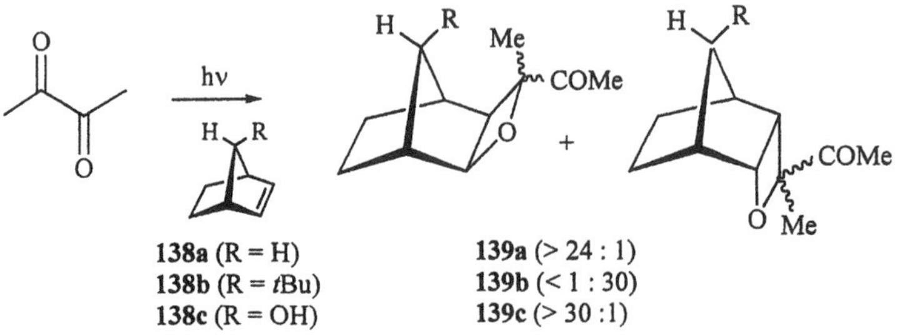

O
‖
⟍ ⟋ ⟍ ⟋
‖
O

hv
⟶

H ⟍ R
⟋
↗ ↘

138a (R = H)
138b (R = *t*Bu)
138c (R = OH)

H ⟍ R
Me
COMe
O

+

H ⟍ R
COMe
O Me

139a (> 24 : 1)
139b (< 1 : 30)
139c (> 30 :1)

Scheme 47

quadricyclane isomer which serves a substrate for the subsequent Paternò-Büchi reaction [98,142].

4.4.3. Effect of Hydrogen Bonding on Diastereoselectivity of Paternò-Büchi Reaction

The Paternò-Büchi reaction of 2-furylmethanol derivatives **140** with benzophenone was recently investigated by D'Auria and coworkers [143]. The regio- and stereoselecitivity of the reaction was exceedingly high and rationalized by assuming a decisive hydroxy bonding interaction favoring the approach of the electronically excited carbonyl group toward one of the diastereotopic faces of the furan ring and formation of only one diastereoisomeric oxetane **141** (Sch. 48). When the OH group was masked by means of a methyl ether, the reactivity dropped remarkably.

Hydrogen bonding is a possible reason also for the diastereoselectivity of the [2+2] photocycloaddition of benzophenone also in case of acyclic chiral allylic alcohols **142** (Sch. 49). These substrates afforded only one regioisomer of the diastereomeric *threo, erythro* oxetanes **143** and preferentially the *threo* isomer [144]. The diastereoselectivity was remarkably reduced in the presence of a protic solvent (methanol as a competitive intermolecular hydrogen bonding substrate) and totally disappeared in case of the silylated substrate [145].

Bach and coworkers investigated the effect of *ground-state* hydrogen bonding interactions in a complex between a rigid chiral host **144** which assists the control of enantiofacial- and diastereofacial selectivity (Sch. 50) [146]. Irradiation of this chiral host with 3,4-dihydro-1H-pyridino-2-one **145** in benzene gave the oxetanes **146** with high diastereoselectivity, where the attack of the electronically excited aldehyde occurs exclusively on one of the two enantiotopic faces of the alkene. The *N*-methylated chiral host, which is no longer hydrogen-bond active, gave no significant diastereomeric excess.

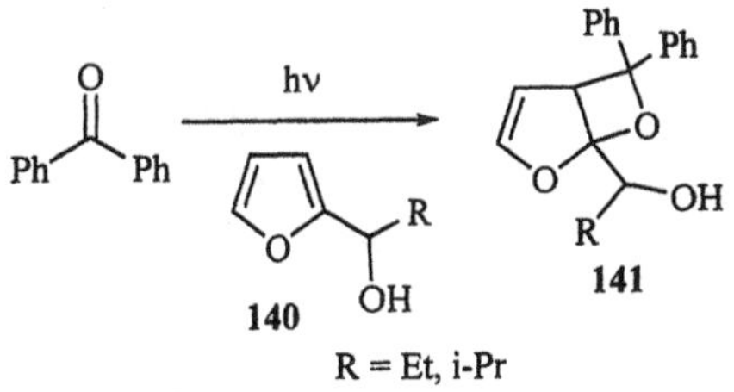

Scheme 48

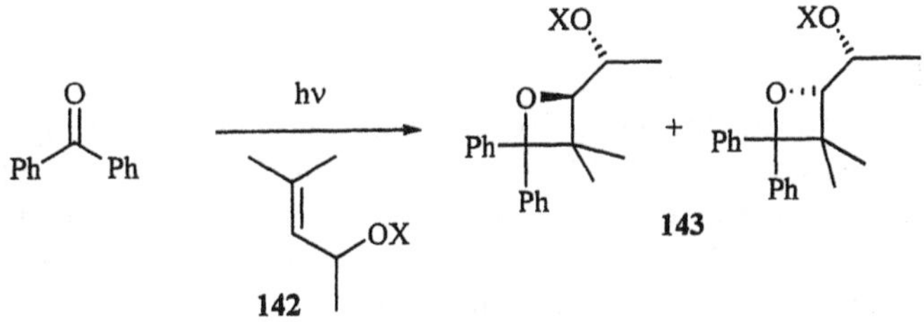

X =	solvent	d.r. (threo : erythro)
H	C_6D_6	90 : 10
H	C_6D_6/CD_3COD (1:1)	69 : 31
$SiMe_2{}^tBu$	C_6D_6	52 : 48

Scheme 49

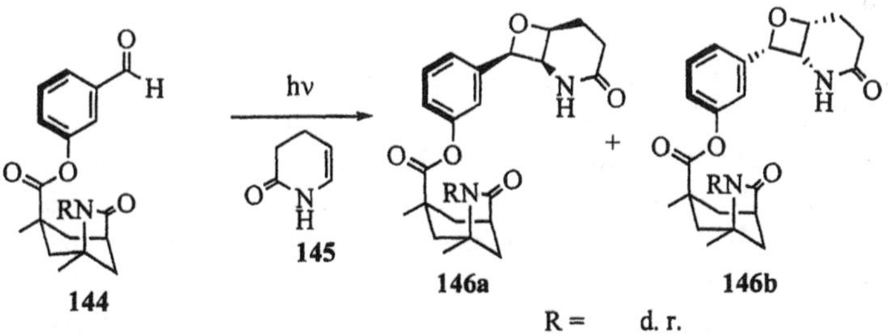

R =	d. r.
H	83 : 17
Me	50 : 50

Scheme 50

Recently, the effect of hydrogen bonding in the first excited singlet vs. the first excited triplet state of aliphatic aldehydes in the photocycloaddition to allylic alcohols and acetates (**147**) was compared (Sch. 51) [147]. The simple diastereoselectivity was nearly the same, but the presence of

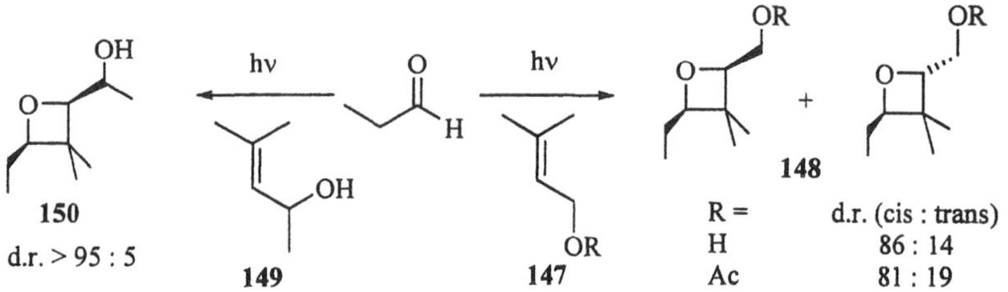

Scheme 51

hydrogen bonding interactions increases the rate of the Paternò-Büchi reaction. Moreover, the effect of hydrogen bonding for simple diastereoselectivity was completely different than that for the induced diastereoselectivity, which was confirmed by comparison with the (highly diastereoselective) mesitylol (**149**) photocycloaddition to aliphatic aldehydes resulting in the formation of the *trans* oxetane **150**.

4.4.4. Asymmetric Induction via Chiral Carbonyl Compounds

Numerous applications were developed by the Scharf group using chiral α-keto esters in [2+2] photocycloadditions to electron-rich cycloalkenes which have already been mentioned in the text [130–133,137,138]. An optically active acetal was used as substrate in the synthesis of prostaglandins. The induced (*anti*) diastereoselectivity was high (>90%), the inherent diastereoselectivity was not reported [148]. An interesting solvent dependency of the face-selecitivty was reported for 3,4,6-tri-*O*-acetyl-D-glucal as carbohydrate substrate [149]. Oppenländer and Schönholzer studied the diastereoface differentiation in the Paternò-Büchi reaction using the 4-(*S*)isopropyl-2-benzoyl-2-oxazoline **151** (prepared from condensation of phenylglyoxylic acid with (*S*)-valinol) as a chiral auxiliary (Sch. 52) [150]. The [2+2] photocycloaddition gave a mixture of diastereoisomeric *l*- and *u*- (*like* and *unlike*) oxetanes **152** in equal amounts.

4.4.5. Asymmetric Induction via Chiral Alkenes

The induced diastereoselectivity in a Paternò-Büchi reaction resulting from a stereogenic center in the alkene part was recently described by Bach and coworkers in the photocycloaddition of chiral silylenol ethers **153** with benzaldehyde (Sch. 53) [151]. The substituents *R* at the stereogenic center

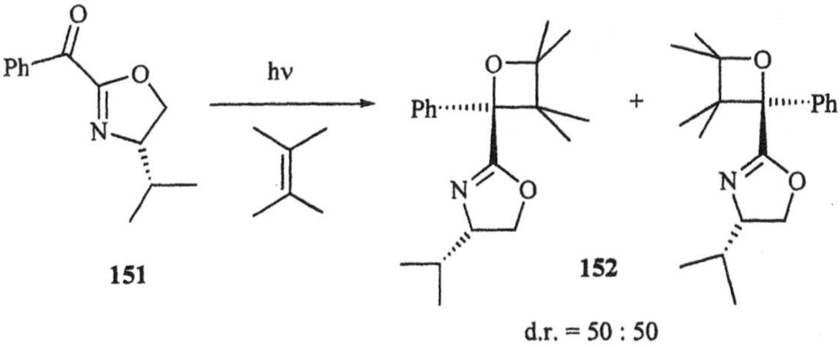

151   **152**

d.r. = 50 : 50

Scheme 52

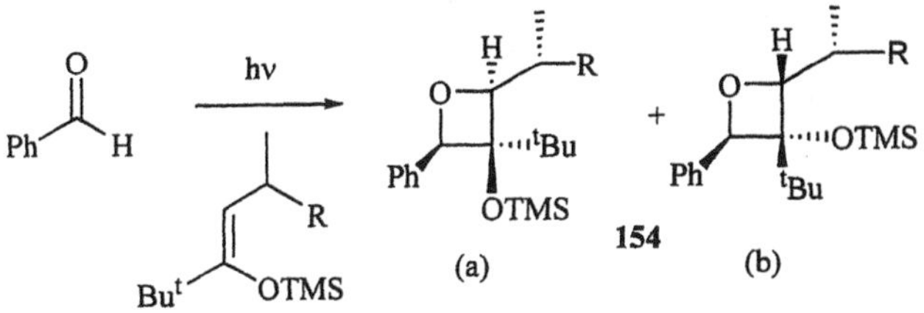

R =	d. r.
Et	61 : 39
i-Pr	70 : 30
Ph	71 : 29
t-Bu	95 : 5

Scheme 53

were varied in order to evaluate the influence of steric bulk and possible electronic effects. In accord with the 1,3-allylic strain model, the facial diastereoselectivity was maximum with large ($R = t$-Bu, SiMe$_2$Ph) and polar ($R =$ OMe) substituents at the γ-position of silyl enol ether (d.r. of oxetanes **154** is >95/5).

Several efforts were also made by the same group to achieve chiral auxiliary control in *N*-acyl enamines **155**, but only moderate diastereoselectivities for the 3-amido oxetanes **156** were reported (Sch. 54) [152]. The photocycloaddition of an axial chiral *N*-acyl enamine **157** with benzaldehyde resulted in oxetane (**158**) formation with moderate diastereomeric excess [153].

The stereogenic center in a dihydropyrrol derivative was used as the key directing element in the synthesis of natural product (+)-preussin via a [2+2] photocycloaddition with benzaldehyde [154].

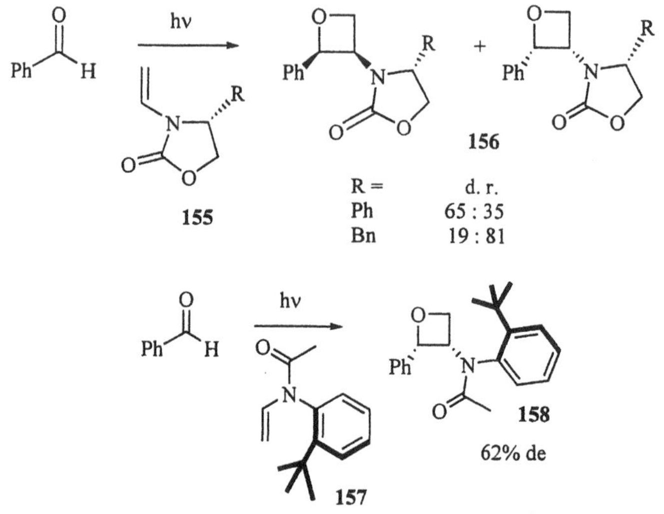

R =   d. r.
Ph   65 : 35
Bn   19 : 81

Scheme 54

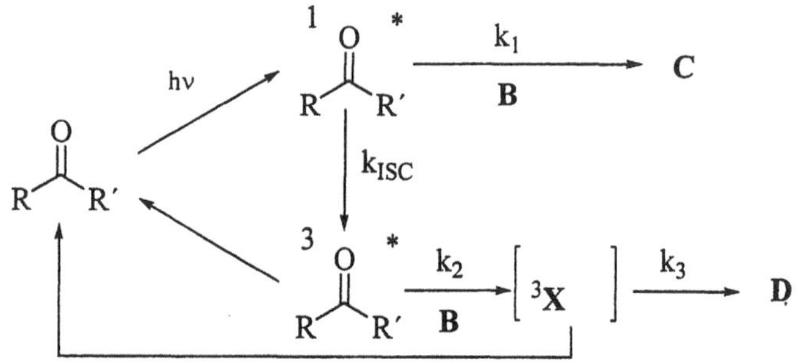

Scheme 55

The difference between the simple diastereoselectivities in photo-cycloaddition reactions following the singlet vs. the triplet route were studied by determination of the concentration dependence of the Paternò-Büchi reaction [155]. Carbonyl substrates which have both reactive singlet and triplet states, exhibit one characteristic substrate concentration where a 1:1 ratio of single and triplet reactivity, i.e., *spinselectivity*, could be detected. The shape of these concentration/diastereoselectivity corrrelations reflects the different kinetic contributions to this complex reaction scenario.

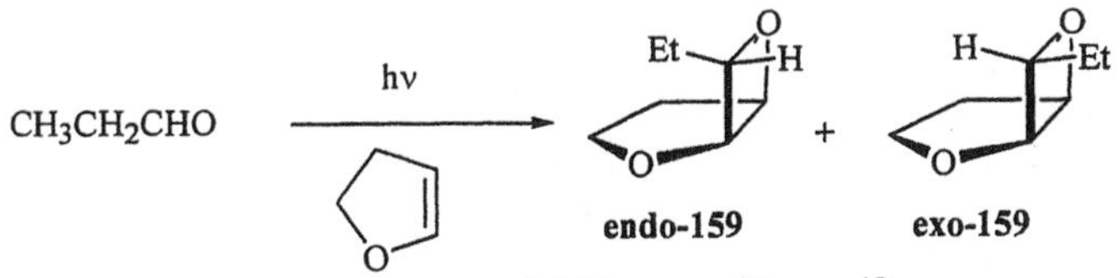

CH_3CH_2CHO $\xrightarrow{h\nu}$

endo-**159**   exo-**159**

0.5 M	52	:	48
0.05 M	78	:	22
0.005 M	85	:	15

Scheme 56

The mixtures of diastereoisomers formed in the singlet and triplet pathway, respectively, are termed C and D (Sch. 55).

The selectivity C/D is controlled by the geometry of the conical intersection for the singlet reaction and by the optimal ISC-geometry of the 2-oxatetramethylene biradical (^3X) for the triplet reaction. The situation is described for the propionaldehyde/2,3-dihydrofuran photocycloaddition reaction. At low concentrations (0.005 M, triplet conditions), the diastereoselectivity (endo-**159**/exo-**159**) approaches a maximum endo/exo value of 85:15 (Sch. 56). At high concentrations (0.5 M, singlet conditions), the diastereoselectivity decreased to 52:48.

The concentration dependence of the photocycloaddition of cis- and trans-cyclo-cocetene **136** with aliphatic aldehydes was studied in order to test the spin-selectivity with respect to the cis/trans oxetane ratio and/or the endo/exo-selectivity relevant for the cis-photoadduct [156]. The results indicate a moderate but still significant spin correlation effect in the Paternò-Büchi reaction of cyclo-octene with aliphatic aldehydes. The exo-diastereoisomers were formed with similar probability as the endo-diastereoisomers in the singlet carbonyl manifold, whereas the triplet excited aldehydes preferred the formation of the endo-diastereoisomer and trans fused products.

4.4.6. Intramolecular Paternò-Büchi Reactions

The reactivity and selectivity trends described for the intermolecular version of the Paternò-Büchi reaction are also valid for the intramolecular version. Special aspects of intramolecular photocycloadditions have been reviewed [157]. A crucial role for the reactivity mode of unsaturated carbonyl compounds is the tether length, i.e., the distance between reactive carbonyl and C–C double bond. Conjugated systems, i.e., α,β-unsaturated carbonyls, prefer photodeconjugation or photocycloaddition invovling the C–C double

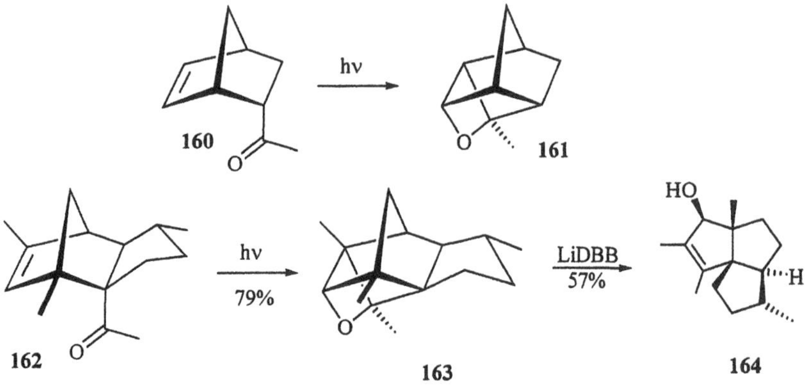

Scheme 57

bond of the Michael system. Homoconjugated system, i.e., β,γ-unsaturated carbonyls, represent substrates for the oxa-di-π-methane rearrangement (see Chap. 7), but, in special cases, can also result in bicyclic ("housane"-type) oxetanes. The major side reaction, however, is Norrish I cleavage resulting in an allylic radical. Increasing tether lengths lead to highly photoactive substrates for the intramolecular Paternò-Büchi photocycloaddition. Most promising substrates are γ,δ-unsaturated ketones and aldehydes, especially when connected by a rigid hydrocarbon system. The first examples in the norbornane series has been reported by Sauers et al. for acetylnorbornene (**160**) [158], where exclusively the tetracyclic oxetane **161** is formed, nicely demonstrating the *rule of five* approach [159] with primary bond formation between the carbonyl oxygen and the proximate alkene carbon atom (Sch. 57).

An elegant application of this photocycloaddition in the field of triquinane synthesis has been reported by Reddy and Rawal [160]. The initially formed oxetane **163**, formed from the Diels-Alder adduct **162**, is easily cleaved reductively (by use of LiDBB = lithium di-*tert*-butyl-biphenylide) to give **164**. This approach was also used for the construction of structurally diverse di- and (propellane-type as well as linear or angular anellated) triquinanes [161]. The classic linear triquinane *hirsutene* is also available via this route [162]. Star-like molecules like the tiene **167** are available via a sequence of intramolecular photocycloaddition (from the 1,3-cyclohexadiene/acylallene adduct **165**) and oxetane (**166**) ring-opening (Sch. 58) [163]. Further examples of intramolecular Paternò-Büchi reactions with acyclic substrates are described in two reviews [157].

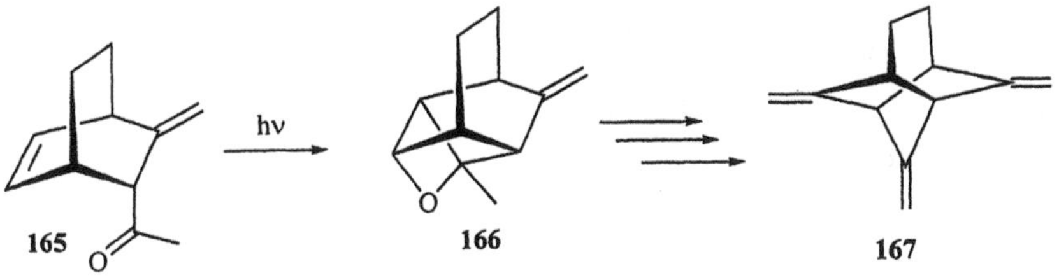

Scheme 58

4.5. REPRESENTATIVE EXPERIMENTAL PROCEDURES

4.5.1. Photocycloaddition of Benzaldehyde to 2,3-Dihydrofuran [308 nm XeCl Excimer Photolysis on the 1 mol Scale]

The Paternò-Büchi [2+2]-photocycloaddition of 2,3-dihydrofuran (70.0 g, 1 mol) and benzaldehyde (74.0 g, 0.7 mol) in ethanol was performed in a XeCl excimer preparative radiation system. The excimer reactor which has been constructed for scaleable preparative organic photoreactions consists of a commercially available 60 cm 3 kW XeCl excimer lamp which is operated in a vertical position and surrounded by a falling-film setup. The latter consists of a cooled solvent reservoir (1.5–5 L solvent volume), a magnetic stirrer and a solvent supply tube which transports the solution to the edge of the falling film glass barrel. Two apertures are used for inert gas supply and temperature and pH-determination. After 14 h of irradiation, the solvent was evaporated and 130 g of the oxetane **168** was isolated (75%, purified by distillation, b.p. 127–129 °C/10 Torr) [164].

4.5.2. Photocycloaddition of Benzaldehyde to 2,4-Dimethyl-5-methoxyoxazole with Subsequent Ring Opening

A mixture of the 2,4-dimethyl-5-methoxyoxazole (5 mmol) and the benzaldehyde (5 mmol) was dissolved in 50 mL benzene, the solution transferred to a vacuum-jacket quartz tube and degassed with a steady stream of N_2 gas (Sch. 59). The reaction mixture was irradiated at 10 °C in a Rayonet photoreactor (RPR-208, $\lambda = 300 \pm 10$ nm) for 24 h. The solvent was evaporated (40 °C, 20 Torr) and the crude product was purified by preparative thick-layer chromatography (EA:H = 1:4) to give 0.59 g (72%) of the oxetane **169** as a colorless oil. To a solution this photoproduct (2 mmol) in 20 mL methylene chloride, 0.5 mL of conc. HCl was added. The mixture is stirred in an open flask at room temperature for 2 h and the

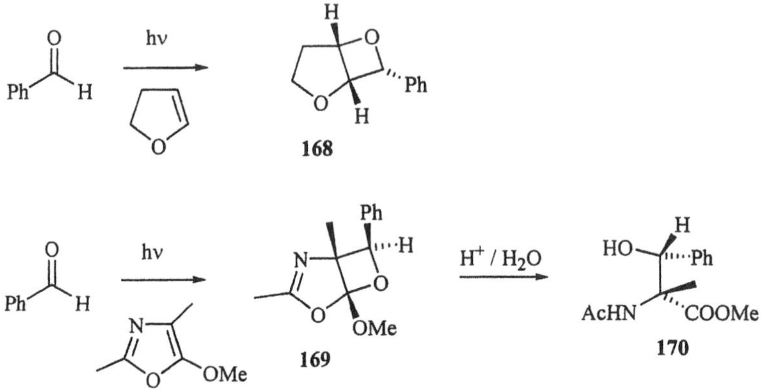

Scheme 59

reaction was controlled by TLC. The reaction mixture was quenched by pouring into water and extracted with methylene chloride (3 × 20 mL). The organic layer was washed with 5% $NaHCO_3$, brine, dried over anhydrous $MgSO_4$. The solvent was removed in vacuo and the residual oil was purified by preparative thick layer chromatography to yield 0.32 g (65%) of the **170** as a colorless oil [165].

4.6. TARGET MOLECULES: NATURAL AND NONNATURAL PRODUCT STRUCTURES

Several naturally occuring terpenoids contain the oxetane ring structure and are in the focus of modern organic synthesis (Fig. 2). The anti-tumor active taxol and taxane derivatives (taxoids, e.g., from *Taxus yunnanensis* [166] and oxetane ring-opened species [167], also from other *Taxus* species) [168] are among the most attractive target molecules, however, no successful photochemical approach has been reported in the literature [169]. Another terpenoid compounds with neurotrophic activity are merrilactone, a complex pentacyclic bislactone [170], partho-oxetine [171], and a guaiane dimer [172]. A novel signaling molecule, bradyoxetin, involved in symbiotic gene regulation has a surprisingly symmetric *bis*-oxetane structure [173]. Oxetanocin is a microbial nucleoside with promising biological and pharmaceutical activities and the enantioselective preparation of this oxetane [174] as well as its carbocylic analogue, carba-oxetanocin, is an actual synthetic problem. Among the longest-known compounds with an

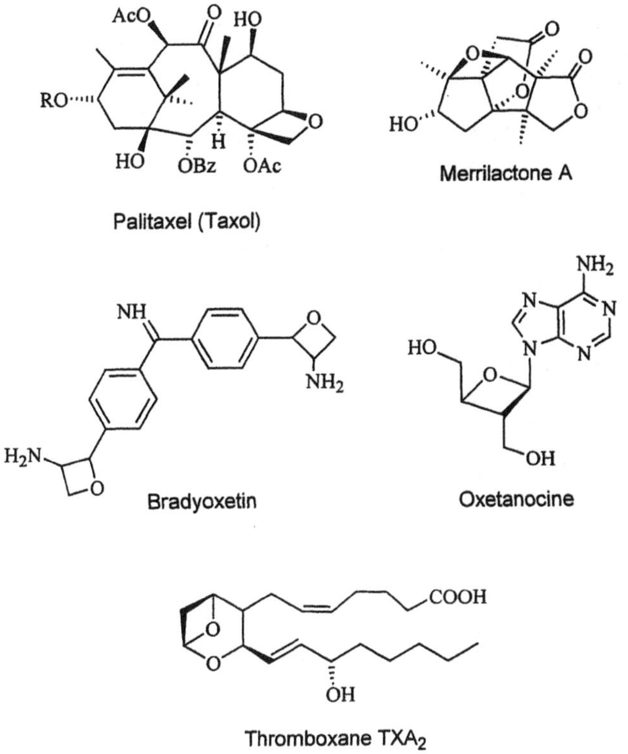

Palitaxel (Taxol)

Merrilactone A

Bradyoxetin

Oxetanocine

Thromboxane TXA_2

Figure 2

oxetane ring as central active element is the eicosanoide thromboxane TXA_2, the active substance inducing blood platelet aggregation. Photochemical attempts to this compound, ring-expanded derivatives [175], and fluoro-analogous have been reported [176].

REFERENCES

1. Paternò E, Chieffi G. Gazz Chim Ital 1909; 341.
2. Büchi G, Inman CG, Lipinsky ES. J Am Chem Soc 1954; 76:4327.
3. Yang NC, Nussim M, Jorgenson MJ, Murov S. Tetrahedron Lett 1964; 3657.
4. (a) Jones G II. Org Photochem 1981; 5:1; (b) Carless, HAJ. In: Horspool WM, ed. Synthetic Organic Photochemistry. New York: Plenum Press, 1984:425; (c) Carless, HAJ. In: Coyle JD, ed. Photochemistry in Organic Chemistry. The Royal Society of Chemistry, Special Publication, 1986; 57:95; (d) Demuth M,

Mikhail G. Synthesis 1989; 145; (e) Porco JA Jr, Schreiber SL. In: Trost BM, Fleming I, Paquette LA, eds. Comprehensive Organic Synthesis 1991; Pergamon Press; 5:151; (f) Inoue Y. Chem Rev 1992; 92:741; (g) Griesbeck AG. In: Horspool WA, Song PS, eds. CRC Handbook of Organic Photochemistry and Photobiology. Vol. 522. CRC Press, 1995:536.

5. Turro NJ, Dalton JC, Dawes K, Farrington G, Hautala R, Morton D, Niemczyk M, Schore N. Acc Chem Res 1972; 5:92.
6. Dauben WG, Salem L, Turro NJ. Acc Chem Res 1975; 8:41.
7. (a) Salem L. J Am Chem Soc 1974; 96:3486; (b) Bigot B, Devaquet A, Turro NJ. J Am Chem Soc 1981; 103:6.
8. Palmer IJ, Ragazos IN, Bernardi F, Olivucci M, Robb MA. J Am Chem Soc 1994; 116:2121.
9. Wilson RM. Org Photochem 1985; 7:339.
10. Shimizu N, Ishikawa M, Ishikura K, Nishida S. J Am Chem Soc 1974; 96:6456.
11. Freilich SC, Peters KS. J Am Chem Soc 1981; 103:6255.
12. (a) Barltrop JA, Carless HAJ. J Am Chem Soc 1972; 94:1951; (b) Dalton JC, Wriede PA, Turro NJ. J Am Chem Soc 1970; 92:1318.
13. Turro NJ, Wriede PA. J Am Chem Soc 1970; 92:320.
14. (a) Arnold DR, Clarke BM. Can J Chem 1975; 53:1; (b) Arnold DR, Hadjiantoniou. Can J Chem 1978; 56:1970.
15. Cantrell TS. J Org Chem 1977; 42:3774.
16. Cantrell TS. J Org Chem 1974; 39:2242.
17. Vargas F, Rivas C. J Photochem Photobiol A: Chem 2001; 138:1.
18. Xue J, Zhang Y, Wu T, Kunfun H, Hua XJ. J Chem Soc, Perkin Trans 1 2001; 183.
19. Zhang Y, Xue J, Gao Y, Fun HK, Xu JH. J Chem Soc, Perkin Trans 1 2002; 345.
20. Coyle JD. Chem Rev 1978; 78:97.
21. Cantrell TS, Allen AC. J Org Chem 1989; 54:135.
22. Cantrell TS, Allen AC, Ziffer H. J Org Chem 1989; 54:140.
23. Tominaga T, Odaira Y, Tsutsumi S. Bull Chem Soc Jpn 1967; 40:2451.
24. Cantrell TS. J Am Chem Soc 1973; 95:2714.
25. Tominaga T, Tsutsumi S. Tetrahedron Lett 1969; 3175.
26. Cantrell TS. J Org Chem 1977; 42:4238.
27. Mattay J, Runsink J, Heckendorn R, Winkler T. Tetrahedron 1987; 43:5781.
28. Hu S, Neckers DC. J Org Chem 1997; 62:564.
29. Buhr S, Griesbeck AG, Lex J, Mattay J, Schroer J. Tetrahedron Lett 1996; 37:1195.
30. Turro NJ, Shima K, Chung CJ, Tanielian C, Kanfer S. Tetrahedron Lett 1980; 2775.
31. Shima K, Sawada T, Yoshinaga H. Bull Chem Soc Jpn 1978; 51:608.
32. Ryang HS, Shima K, Sakurai H. J Org Chem 1973; 38:2860.
33. Shima K, Kawamura T, Tanabe K. Bull Chem Soc Jpn 1974; 47:2347.
34. Cantrell TS. J Chem Soc Chem Commun 1975; 637.
35. Zagar C, Scharf HD. Chem Ber 1991; 124:967.

36. Döpp D, Memarian HR, Fischer MA, Van Eij AMO, Varma CAGO. Chem Ber 1992; 125:983.
37. Döpp D, Fischer MA. Rec Trav Chim Pays-Bas 1995; 114:498.
38. Jorgenson MJ. Tetrahedron Lett 1966; 5811.
39. Kwiatkowski GT, Selley DB. Tetrahedron Lett 1968; 3471.
40. Hussain S, Agosta WC. Tetrahedron 1981; 37:3301.
41. Saba S, Wolff S, Schröder C, Margaretha P, Agosta WC. J Am Chem Soc 1983; 105:6902.
42. Catalani LH, Rezende DB, Campos IA. J Chem Res 2000; 111.
43. Bryce-Smith D, Gilbert A, Johnson MG. J Chem Soc (C) 1967; 383.
44. Bunce NJ, Hadley M. Can J Chem 1975; 53:3240.
45. Wilson RM, Musser AK. J Am Chem Soc 1980; 102:1720.
46. Fehnel EA, Brokaw FC. J Org Chem 1980; 45:578.
47. Ishibe N, Hashimoto K, Yamaguchi Y. J Chem Soc, Perkin Trans 1 1975; 318.
48. Sun D, Hubig SM, Kochi JK. J Org Chem 1999; 64:2250.
49. Arnold DR, Hinman RL, Glick AH. Tetrahedron Lett 1964; 1425.
50. Yang NC, Loeschen R, Mitchell DJ. Am Chem Soc 1967; 89:5465.
51. Bradshaw JS. J Org Chem 1966; 31:237.
52. Carless HAJ. Tetrahedron Lett 1973; 3173.
53. Yang NC, Kimura M, Eisenhardt W. J Am Chem Soc 1973; 95:5058.
54. Fleming SA, Gao JJ. Tetrahedron Lett 1997; 38:5407.
55. (a) Schroeter SH, Orlando CM. J Org Chem 1969; 34:1181; (b) Schroeter SH. J Org Chem 1969; 34:1188.
56. Turro NJ, Wriede PA. J Am Chem Soc 1968; 90:6863.
57. Araki Y, Nagasawa JI, Ishido Y. Carbohydr Res 1981; 91:77.
58. Mattay J, Buchkremer K. Helv Chim Acta 1988; 71:981.
59. Carless HAJ, Haywood DJ. J Chem Soc Chem Commun 1980; 1067.
60. Morris TH, Smith EH, Walsh R. J Chem Soc Chem Commun 1987; 964.
61. Abe M, Ikeda M, Nojima M. J Chem Soc, Perkin Trans I 1998; 3261.
62. Abe M, Tachibana K, Fujimoto K, Nojima M. Synthesis 2001; 1243.
63. Abe M, Shirodai Y, Nojima M. J Chem Soc, Perkin Trans I 1998; 3253.
64. Abe M, Fujimoto K, Nojima M. J Am Chem Soc 2000; 122:4005.
65. Thopate SR, Kulkarni MG, Puranik VG. Angew Chem Int Ed Engl 1998; 37:1110.
66. (a) Büchi G, Kofron JT, Koller E, Rosenthal D. J Am Chem Soc 1956; 78:876; (b) Miyamoto T, Shigemitsu Y, Odaira Y. J Chem Soc Chem Commun 1969; 1410; (c) Bradshaw JS, Knudsen RD, Parish WW. J Org Chem 1975; 40:529; (d) Mostered A, Matser HJ, Bos HJT. Tetrahedron Lett 1974; 4179.
67. Friedrich LE, Bower JD. J Am Chem Soc 1973; 95:6869.
68. Friedrich LE, Lam PY-S. J Org Chem 1981; 46:306.
69. Gotthardt H, Steinmetz R, Hammond GS. J Org Chem 1968; 33:2774.
70. Arnold DR, Glick AH. J Chem Soc Chem Commun 1966; 813.
71. Gotthardt H, Hammond GS. Chem Ber 1974; 107:3922.
72. (a) Singer LA, Davis GA, Knutsen RL. J Am Chem Soc 1972; 94:1188; (b) Singer LA, Davis GA. J Am Chem Soc 1973; 95:8638.

73. Howell AR, Fan R, Truong A. Tetrahedron Lett 1996; 37:8651.
74. Barltrop JA, Carless HAJ. J Am Chem Soc 1971; 93:4794.
75. (a) Hautala RR, Dawes K, Turro NJ. Tetrahedron Lett 1972; 1229; (b) Barltrop JA, Carless HAJ. J Am Chem Soc 1972; 94:8761.
76. (a) Kubota T, Shima K, Toki S, Sakurai H. J Chem Soc Chem Commun 1969; 1462; (b) Shima K, Kubota T, Sakurai H. Bull Chem Soc Jpn 1976; 49:2567.
77. Carless HAJ, Maitra AK. Tetrahedron Lett 1977; 1411.
78. Carless HAJ. Tetrahedron Lett 1972; 2265.
79. Schenck GO, Hartmann W, Steinmetz R. Chem Ber 1963; 96:498.
80. Ogata M, Watanable H, Kano H. Tetrahedron Lett 1967; 533.
81. (a) Toki S, Sakurai H. Tetrahedron Lett 1967; 4119; (b) Evanega GR, Whipple EB. Tetrahedron Lett 1967; 2163.
82. (a) Toki S, Shima K, Sakurai H. Bull Chem Soc Jpn 1965; 38:760; (b) Sekretár S, Rudá J, Stibrányi L. Coll Czech Chem Commun 1984; 49:71.
83. (a) Schreiber SL, Desmaele D, Porco JA Jr. Tetrahedron Lett 1988; 29:6689; (b) Carless HAJ, Halfhide AFE. J Chem Soc, Perkin Trans I 1992; 1081.
84. (a) Jarosz S, Zamojski A. Tetrahedron 1982; 38:1453; (b) Kozluk T, Zamojski A. Tetrahedron 1983; 39:805.
85. (a) Schreiber SL, Satake K. J Am Chem Soc 1983; 105:6723; (b) Schreiber SL, Satake K. J Am Chem Soc 1984; 106:4186.
86. (a) Kitamura T, Kawakami Y, Imegawa T, Kawanishi M. Synth Commun 1977; 7:521; (b) Jarosz S, Zamojski A. J Org Chem 1979; 44:3720.
87. Hu S, Neckers DC. J Chem Soc, Perkin Trans 2 1999; 1771.
88. D'Auria M, Emanuele L, Poggi G, Racioppi R, Romanielli G. Tetrahedron 2002; 58:5045.
89. Abe M, Torii E, Nojima M. J Org Chem 2000; 65:3426.
90. Rivas C, Pacheco D, Vargas F, Ascanio J. J Heterocyclic Chem 1981; 18:1065.
91. (a) Rivas C, Bolivar RA. J Heterocyclic Chem 1976; 13:1037; (b) Nakano T, Rodriquez W, de Roche SZ, Larrauri JM, Rivas C, Pérez C. J Heterocyclic Chem 1980; 17:1777; (c) Julian DR, Tringham GD. J Chem Soc Chem Commun 1973; 13.
92. (a) Matsuura T, Banba A, Ogura K. Tetrahedron 1971;27:1211; (b) Jones G II, Gilow HM, Low J. J Org Chem 1979; 44:2949.
93. Griesbeck AG, Fiege M, Lex J. Chem Commun 2000; 589.
94. Scharf HD, Korte F. Tetrahedron Lett 1963; 821.
95. Jones G II, Schwartz SB, Marton MT. J Chem Soc Chem Commun 1973; 374.
96. Shigemitsu Y, Yamamoto S, Miyamoto T, Odaira Y. Tetrahedron Lett 1975; 2819.
97. Gorman AA, Leyl RL, Parekh CT, Rodgers MAJ. Tetrahedron Lett 1976; 1391.
98. (a) Barwise AJG, Gorman AA, Leyl RL, Parekh CT, Smith PG. Tetrahedron 1980; 36:397; (b) Gorman AA, Leyl RL, Rodgers MAJ, Smith PG. Tetrahedron Lett 1973; 5085.
99. Christl M, Braun M. Angew Chem Int Ed Engl 1989; 28:601.
100. Christl M, Braun M. Liebigs Ann 1997; 1135.

101. (a) Beereboom JJ, von Wittenau MS. J Org Chem 1965; 30:1231; (b) Turro NJ, Wriede P, Dalton JC, Arnold D, Glick A. J Am Chem Soc 1967; 89:3950.
102. (a) Turro NJ, Wriede PA, Dalton JC. J Am Chem Soc 1968; 90:3274; (b) Dalton JC, Wriede PA, Turro NJ. J Am Chem Soc 1970; 92:1318.
103. (a) Turro NJ, Farrington GL. J Am Chem Soc 1980; 102:6051; (b) Turro NJ, Farrington GL. J Am Chem Soc 1980; 102:6056.
104. Chung WS, Liu YD, Wang NJ. J Chem Soc, Perkin Trans 2 1995; 581.
105. Chung WS, Ho CC. Chem Commun 1997; 317.
106. Carless HAJ. Tetrahedron Lett 1973; 3173.
107. Turro NJ, Wriede PA. J Am Chem Soc 1970; 92:320.
108. Zimmerman HE, Singer L, Thyagarajan BS. J Am Chem Soc 1959; 81:108.
109. Jones II G, Khalil ZH, Phan XT, Chen TJ, Welankiwar S. Tetrahedron Lett 1981; 22:3823.
110. Jones II G. In: Padwa A, ed. Organic Photochemistry, Marcel Dekker: New York, Basel, Vol. 5. 1981:1.
111. Griesbeck AG, Stadtmüller S. J Am Chem Soc 1990; 112:1281.
112. Griesbeck AG, Stadtmüller S. Chem Ber 1990; 123:357.
113. Griesbeck AG, Stadtmüller S. J Am Chem Soc 1991; 113:6923.
114. Salem L, Rowland C. Angew Chem Int Ed Engl 1972; 11:92.
115. Griesbeck AG, Mauder H, Peters K, Peters EM, von Schnering HG. Chem Ber 1991; 124:407.
116. Shima K, Sakurai H. Bull Chem Soc Jpn 1966; 39:1806.
117. (a) Schreiber SL, Hoveyda AH, Wu HJ. J Am Chem Soc 1983; 105:660; (b) Schreiber SL. Science 1985; 227:857.
118. Feigenbaum A, Pete JP, Poquet-Dhimane AL. Heterocycles 1988; 27:125.
119. Exceptions due to steric or stereoelectronic reasons are known: (a) Carless HAJ, Halfhide AFE. J Chem Soc, Perkin Trans I 1992; 1081; (b) Schreiber SL, Desmaele D, Porco JA Jr. Tetrahedron Lett 1988; 29:6689.
120. Hoye TR, Richardson WS. J Org Chem 1989; 54:688.
121. Mattay J, Buchkremer K. Heterocycles 1988; 27:2153.
122. Zamojski A, Kozluk T. J Org Chem 1977; 42:1089.
123. Griesbeck AG, Buhr S, Fiege M, Schmickler H, Lex J. J Org Chem 1998; 63:3847.
124. Park SK, Lee SJ, Baek K, Yu CM. Bull Korean Chem Soc 1998; 19:35.
125. Bach T, Schröder J. J Org Chem 1999; 64:1265.
126. Bach T, Jödicke K. Chem Ber 1993; 126:2457.
127. Bach T, Schröder J. Synthesis 2001; 1117.
128. Gotthardt H, Lenz W. Angew Chem Int Ed Engl 1979; 18:868.
129. Jarosz S, Zamojski A. Tetrahedron 1982; 38:1447.
130. Nehrings A, Scharf HD, Runsink J. Angew Chem Int Ed Engl 1985; 25:877.
131. Buschmann H, Scharf HD, Hoffman N, Plath M, Rusink J. J Am Chem Soc 1989; 111:5367.
132. (a) Koch H, Runsink J, Scharf HD. Tetrahedron Lett 1983; 24:3217; (b) Koch H, Scharf HD, Runsink J, Leismann H. Chem Ber 1985; 118:1485; (c) Pelzer R, Jütten P, Scharf HD. Chem Ber 1989; 122:487.

133. Buschmann H, Scharf HD, Hoffman N, Esser P. Angew Chem Int Ed Engl 1991; 30:477.
134. Schreiber SL, Satake K. Tetrahedron Lett 1986; 27:2575.
135. Adam W, Stegmann VR. J Am Chem Soc 2002; 124:3600.
136. Adam W, Stegmann VR, Weinkoetz S. J Am Chem Soc 2001; 123:2452.
137. Meier L, Scharf HD. Synthesis 1987; 517.
138. Weuthen M, Scharf HD, Runsink J. Chem Ber 1987; 120:1023.
139. Mattay J. Angew Chem 1987; 99:849.
140. Sauers RR, Valenti PC, Tavss E. Tetrahedron Lett 1975; 3129.
141. Kubota T, Shima K, Sakurai H. Chem Lett 1972; 343.
142. Gorman AA, Leyl RL. Tetrahedron Lett 1972; 5345.
143. D'Auria M, Racioppi R, Romaniello G. Eur J Org Chem 2000; 3265.
144. Adam W, Peters K, Peters EM, Stegmann VR. J Am Chem Soc 2000; 122:2958.
145. Adam W, Stegmann VR. Synthesis 2001; 1203.
146. Bach T, Bergmann H, Harms K. J Am Chem Soc 1999; 121:10650.
147. Griesbeck AG, Bondock S. J Am Chem Soc 2001; 123:6191.
148. Morton DR, Morge RA. J Org Chem 1978; 43:2093.
149. Araki Y, Senna K, Matsuura K, Ishido Y. Carbohydr Res 1978; 60:389.
150. Oppenländer T, Schönholzer P. Helv Chim Acta 1989; 72:1792.
151. Bach T, Jödicke K, Wibbeling B. Tetrahedron 1996; 52:10861.
152. Bach T, Schröder J, Brandl T, Hecht J, Harms K. Tetrahedron 1998; 54:4507.
153. Bach T, Schröder J, Harms K. Tetrahedron Lett 1999; 40:9003.
154. Bach T, Brummerhop H, Harms K. Chem Eur J 2000; 6:3838.
155. Griesbeck AG, Fiege M, Bondock S, Gudipati MS. Org Let 2000; 2:3623.
156. Griesbeck AG, Bondock S. Photochem Photobiol Sci 2002; 1:81.
157. (a) Porco JA Jr, Schreiber SL. In: Trost BM, Fleming I, Paquette LA, eds. Comprehensive Organic Synthesis. Vol. 5. Pergamon Press; 1991:178; (b) Carless HAJ. In: Horspool WA, Song PS, eds. CRC Handbook of Organic Photochemistry and Photobiology. CRC Press, 1995:560.
158. Sauers RR, Rousseau AD, Byrne B. J Am Chem Soc 1975; 97:4947.
159. Liu RS, Hammond GS. J Am Chem Soc 1967; 89:4936.
160. Reddy TJ, Rawal VH. Org Lett 2000; 2:2711.
161. Dvorak CA, Dufour C, Iwasa S, Rawal VH. J Org Chem 1998; 63:5302.
162. Rawal VH, Fabré A, Iwasa S. Tetrahedron Lett 1995; 38:6851.
163. Gleiter R, Herb T, Borzyk O, Hyla-Kryspin I. Liebigs Ann 1995; 357.
164. Griesbeck AG, Maptue N, Bondock S, Oelgemöller M. Photochem Photobiol Sci 2003; 2:450.
165. Griesbeck AG, Bondock S. Can J Chem 2003; 81, in print.
166. Zhang H, Tadeda Y, Sun H. Phytochemistry 1995; 39:1147.
167. Li SH, Zhang HJ, Niu XM, Yao P, Sun HD, Fong HS. Planta Med 2002; 68:253.
168. Shen YC, Lo KL, Chen CY, Kuo YH, Hung MC. J Nat Prod 2000; 63:720.
169. Gan CY, Gable RW, Lambert JN. Aust J Chem 1999; 52:629.

170. Huang JM, Yokoyama R, Yang CS, Fukuyama Y. Tetrahedron Lett 2000; 41:6111.
171. Ortega A, Maldonado E. Phytochemistry 1986; 25:699.
172. Martins D, Osshiro E, Roque NF, Marks V, Gottlieb HE. Phytochemistry 1998; 48:677.
173. Loh J, Carlson RW, York WS, Stacey G. Proc Nat Acad Sci 2002; 99:14446.
174. (a) Saksena AK, Ganguly AK, Girijavallabhan VM, Pike RE, CHen YT, Puar MS. Tetrahedron Lett 1992; 33:7721; (b) Hambalek R, Just G. Tetrahedron Lett 1990; 31:5445; (c) Wilson FX, Fleet GWJ, Vogt K, Wang Y, Witty DR, Choi S, Storer R, Myers PL, Wallis CJ. Tetrahedron Lett 1990; 31:6931.
175. Carless HAJ, Fekarurhobo GK. J Chem Soc Chem Commun 1984; 667.
176. Fried J, Kittisopikul S, Hallinan EA. Tetrahedron Lett 1984; 25:4329.

5
Photocycloaddition of Alkenes to Excited Alkenes

Steven A. Fleming
Brigham Young University, Provo, Utah, U.S.A.

5.1. INTRODUCTION

A number of reviews have been published concerning the synthetic utility of photochemical 2+2 cycloadditions [1]. We have previously summarized the observed regio- and stereocontrol of photocycloaddition between alkenes and excited π systems [2]. Control of solution photochemistry has been demonstrated with use of templates, tethering of reagents, and substituent stabilization to affect the regioselectivity and/or stereoselectivity. The types of photoreactions that have been investigated using these techniques are numerous.

There are several reasons for the interest in controlled photocyclo-additions. First, a cycloaddition (the 2+2, for example) allows access to the four-member ring system. Second, investigation of the regio- and stereochemical outcome of the cyclization process allows for a better understanding of the mechanistic pathway the reaction takes. The reaction is studied not only for synthetic exploitation, but for basic understanding of the photochemical process.

A third reason for the interest in the 2+2, in particular, is that it allows for a logical analysis of orbital symmetry arguments. Few photochemical reactions are as predictable as the Diels-Alder reaction, but the 2+2 photocycloaddition is a good starting point for comparison.

Synthetic utility for the 2+2 photocycloaddition has been a driving force for understanding the mechanism of cyclization. Stereocontrol of the 2+2 photocycloaddition allows an efficient synthesis of small ring

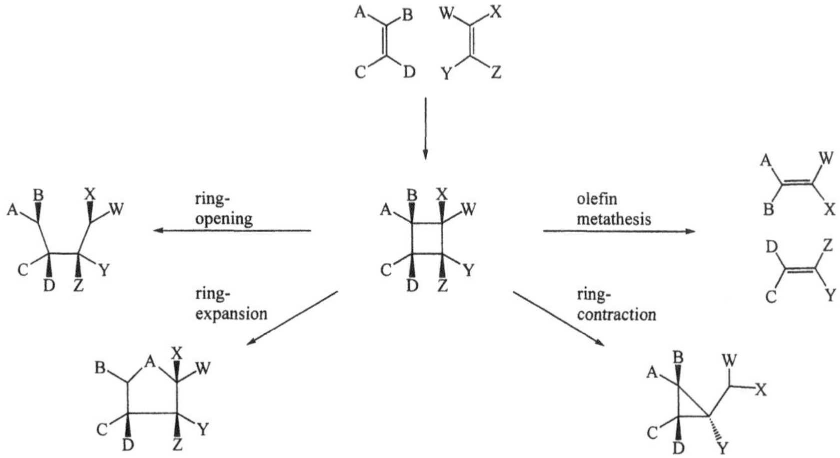

Scheme 1 Cyclobutane ring manipulation for synthetic applications.

compounds with a potential of four adjacent stereocenters. The cyclobutane ring itself can be further manipulated by ring expansion [3], ring contraction [4], or ring opening [5] (see Sch. 1). Thus, stereocontrol of this carbon–carbon bond forming process could be a valuable tool for synthetic chemists.

5.2. MECHANISTIC DETAILS

There are only a few examples of singlet photoreaction between alkenes that are not tethered or constrained in close proximity. In these cases, calculations have suggested the presence of exciplex [6] and/or diradical intermediates [7]. The short lifetime of the typical alkene singlet excited state (on the order of 10–20 ns) [8] limits the chance for productive collisions. On the other hand, photochemistry between electron rich and electron poor alkenes such as tetracyanoethylene and methoxy substituted alkenes, provides evidence for an electron transfer process [9]. These matched pairs benefit from ground state attraction and a resulting preorientation that enhances the alkene orbital overlap. Alternatively, electron transfer pathways have been accessed by employing electron transfer sensitizers (see Sch. 2), DCA is dicyanoanthracene) [10].

Semiempirical calculations on the 2+2 photocycloaddition between two propene molecules has suggested that there is a preference for a head-to-tail approach and a concerted pathway [7]. The authors suggest that intermolecular cycloaddition between alkenes with electron withdrawing groups should favor formation of the *trans*-1,3-disubstituted cyclobutane.

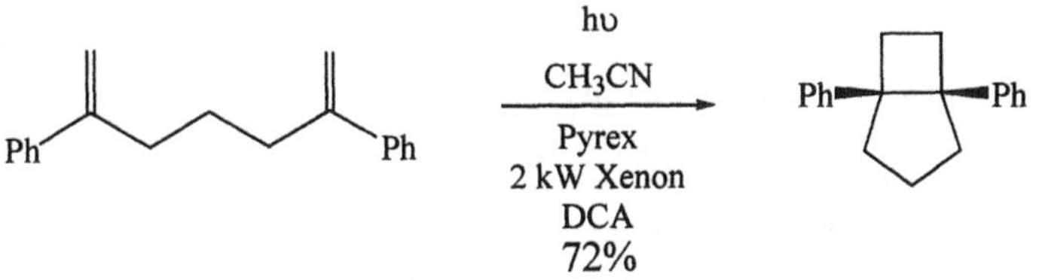

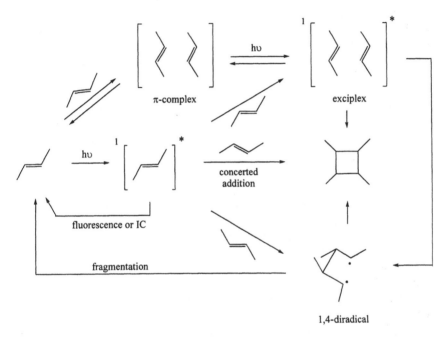

Scheme 2 Photocycloaddition via radical/cation and radical/anion pathway.

Scheme 3 Stepwise or concerted 2+2 cycloaddition in the singlet manifold.

The photochemistry of tethered alkenes is more predictable than the nontethered situation. There is evidence of π-stacking for closely held moieties which presumably improves the orbital interactions between the alkenes. Exciplex formation is likely involved when the reacting groups are within 5 angstroms. Cornil et al. have shown that π systems can couple within this distance [11]. This exciplex could lead to a concerted cycloaddition from the excited state which would be consistent with the observed products. Although stepwise addition (see Sch. 3) cannot be ruled out even in these tethered singlet reactions. Ring closure must be very rapid if diradicals are involved, since no radical-trapped species have been found.

Penn and coworkers studied the 2+2 photocycloaddition between 1,2-diphenylcycloalkenes and various alkenes [12]. They provided evidence of exciplex formation for the diphenylcyclobutene chromophore paired with dimethylhexadiene. This exciplex led to 2+2 cycloaddition. The more sterically hindered methylated 1,2-diphenylcyclobutene did not form an exciplex with alkenes and, as a result, it did not undergo photocycloaddition.

Alkene systems that use copper as a tethering agent should probably be included in this discussion of excited alkene photocycloaddition to alkenes. Although it can be argued that the copper complex rather than the alkene is the chromophore in these examples. Mechanistic details of this singlet photochemistry are found in Salomon's work. He invokes an alkene–copper–alkene complex excited state that has two absorption bands. The shorter wavelength band (236 nm) is a metal to alkene charge transfer (MLCT) excitation and the more accessible band (272 nm) is probably an alkene ligand to metal charge transfer (LMCT) process. He points out that it is not clear which excitation process leads to cycloaddition. The excited complex can undergo a concerted cycloaddition or a sequential cationic or radical cyclization to yield the cyclobutane photoadduct and reform the copper I species (see Sch. 4) [13]. The cationic cyclization would follow from the LMCT excited state and the ring forming step would be an electrophilic abstraction or a reductive elimination. The radical cyclization would originate from a MLCT excitation where the final step of the catalytic cycle would be a radical reductive elimination.

There are a number of examples in the literature of tripet alkene excited states adding to a second alkene in a 2+2 fashion. Stereochemical analysis of the photoproducts is the best analysis of the mechanism for these examples. Complete stereoselectivity in the triplet reaction is not typical. The triplet intermediate formed after the first bond of the 2+2 is made, has a lifetime that allows for bond rotation prior the final cyclization. As a result, the stereoselectivity should be significantly in favor of the thermodynamically more stable cyclobutane product. If there is reversibility in the first bond forming step of this stepwise reaction, then the regioselectivity (head-to-head vs. head-to-tail addition) can also be thermodynamically controlled (see Sch. 5). If there is no reversibility, then the observed regioselectivity may be explained by a Frontier Molecular Orbital argument based on a polarized excited state or perhaps a π-stacking preorientation aligns the aryl substituted alkenes in a head-to-head fashion.

Caldwell has published mechanistic details of the cycloaddition between an excited state 1-phenylcyclohexene in the triplet manifold and a ground state phenylcyclohexene [14]. His analysis is consistent with the mechanistic pathway indicated in Sch. 5. A diradical is the proposed excited state for the styryl group with its p-orbitals perpendicular to each

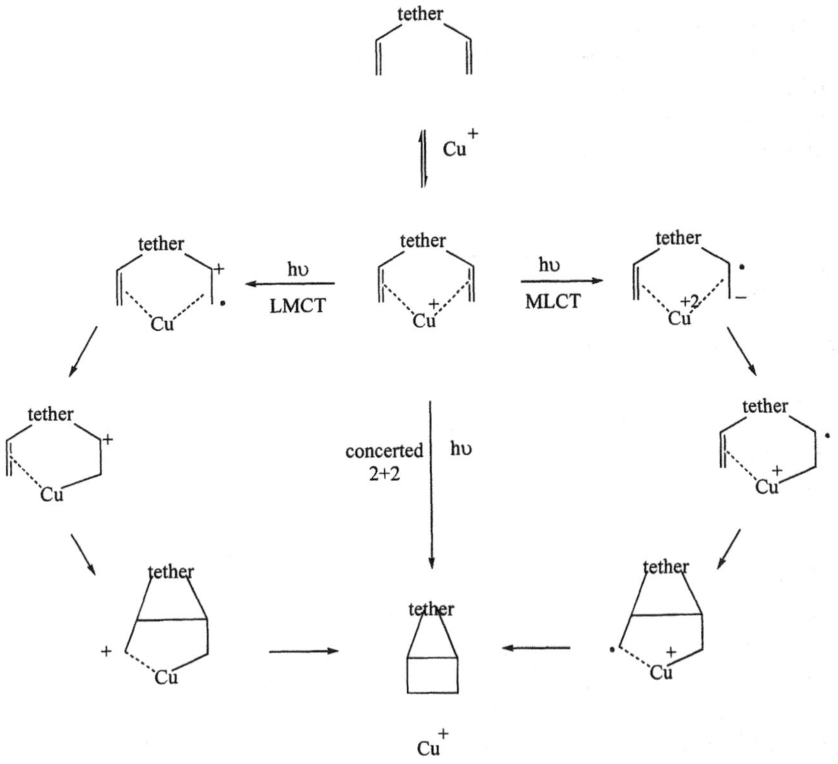

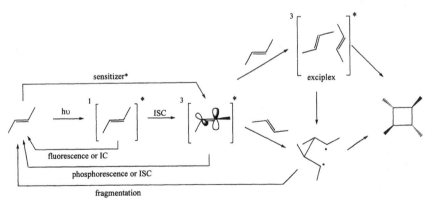

Scheme 4 Mechanistic details of the copper coordinated 2+2.

Scheme 5 Mechanistic pathway for triplet alkene–alkene photocycloaddition.

other. When the initial diradical adds to another styrenyl group, a triplet 1,4-diradical intermediate is formed and it has time to assume the more stable conformation before closing. The initial 1,2-diradical intermediate is supported by rate studies and quenching data. Caldwell points out that an exciplex formed prior to cycloaddition cannot be discounted.

5.3. EXAMPLES OF ALKENE + ALKENE PHOTOCYCLOADDITION

A. Nontethered Examples

One of the first examples of efficient intermolecular alkene photocyclo-additions was published by Lewis in 1976 [15]. He reported that irradiation of diphenylvinylene carbonate with various alkenes led to the formation of two cyclobutane products with very good selectivity. Furthermore, benzil triplet sensitization led to different product ratios than did the direct irradiation, which presumably went through a singlet mechanism (see Sch. 6). The rationalization for the observed selectivity stems from steric and electronic considerations. The triplet reactions favored the products with more thermodynamic stability, while the singlet reactions resembled concerted reactions, where the lowest-energy transition states had the greatest frontier orbital overlap. The product distribution indicates that π-stacking plays an important role in the reaction by stabilizing the transition states. The quantum yield data also suggest formation of a loose exciplex in both the singlet and triplet reactions.

The nontethered styrene cycloadditions initially reported by Dauben [16] have been further studied by Caldwell [14b]. The 1-phenylcyclohexene

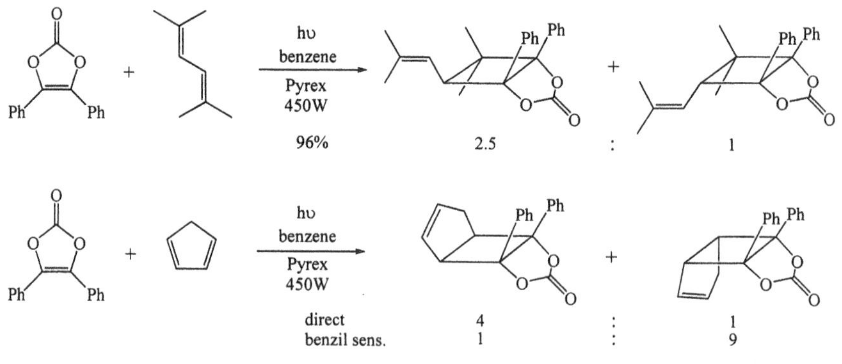

Scheme 6 Nontethered intermolecular 2+2 photocycloaddition.

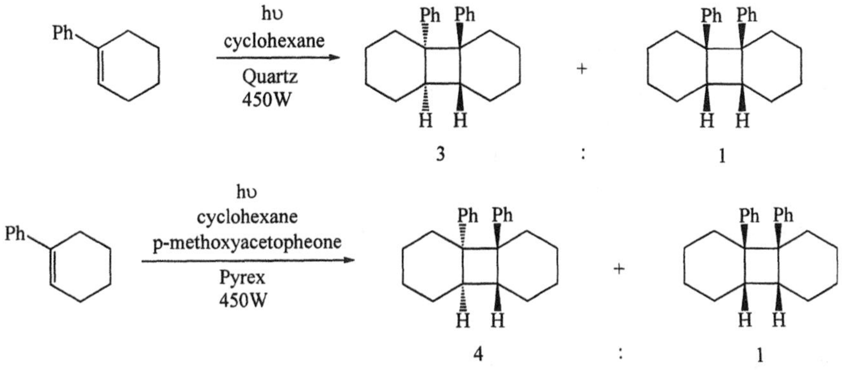

Scheme 7 Phenylcyclohexene photodimerization.

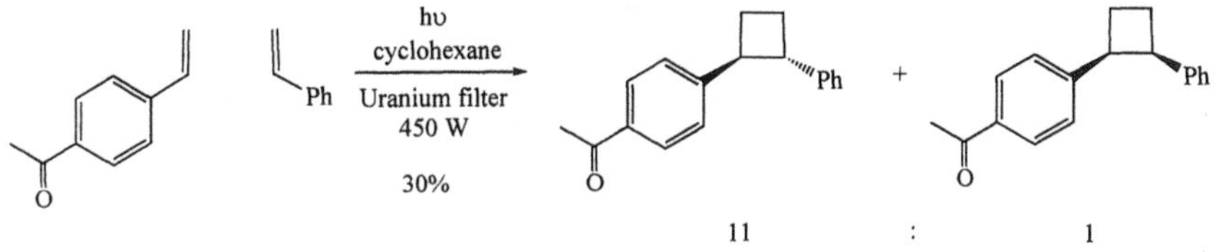

Scheme 8 Triplet manifold photocycloaddition of styrenyl groups.

photodimer is formed in a 3:1 ratio (*anti:syn*) of the head-to-head isomers (see Sch. 7). The cyclobutane photoproducts from the sensitized irradiation are the same and the ratio is only slightly altered (4:1). The triplet photo-cycloaddition of *p*-acylstyrene gives the products shown in Sch. 8 in a 11:1 ratio when short irradiation times are used. Prolonged irradiation leads to secondary photochemistry of the cyclobutanes.

Photoreactions of polyacetylenes and alkenes have also been investigated [17].

Shim and coworkers have published several examples of the 2+2 between aryl-substituted alkynes and alkenes. Their data suggests that the singlet and triplet pathways give rise to the observed cyclobutenes. Exciplex emission was detected for the alkyne–alkene pairs.

B. Tethered Photocycloaddition

Copper (I) coordination has been used as a tether to preorganize alkene reactants so that they will undergo [2+2] photocyclization. Both diallyl ethers and 1,6-heptadienes will cyclize to form cyclobutane products when

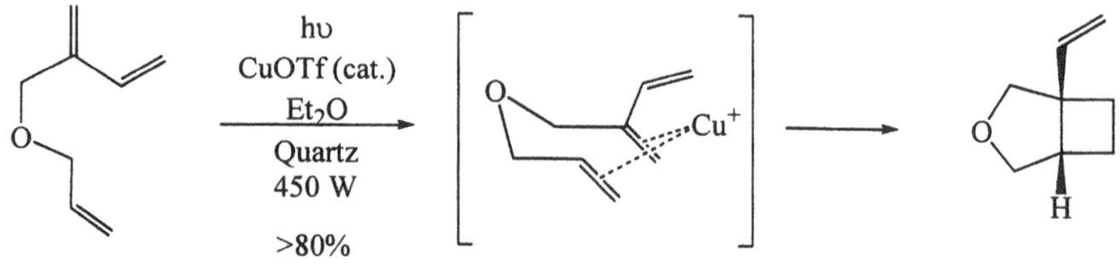

Scheme 9   Early example of copper catalyzed intramolecular photocycloaddition.

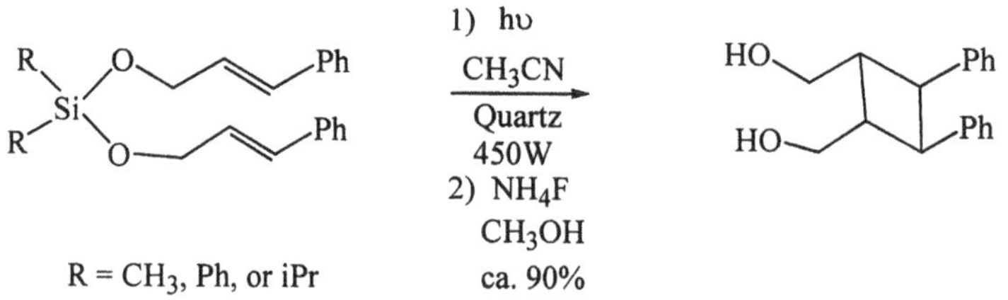

R = CH₃, Ph, or iPr

Scheme 10   Silicon tethered 2+2 photocycloaddition between styrenyl groups.

irradiated in the presence of copper triflate (see Sch. 9 [18]. Hydroxyl substitution on the carbon framework appears to assist with the copper coordination.

Various silyloxy tethered cycloadditions, where one of the alkenyloxy groups in each case is the chromophore (the cinnamyloxy group), have been reported [19]. Irradiation of compounds that have radical-stabilizing π systems result in cyclobutane formation in good to excellent yields (see Sch. 10]. When the tethered group is a simple alkene, then only *cis-trans* isomerization or hydrogen-abstraction products are obtained. The observed cyclobutane products all resulted from conformations in the excited state in the maximum possible π-overlap between the two ligands. This suggests a π-stacking exciplex is involved in the reaction pathway. A weak ground state complex may provide preorientation of the alkenes.

Alkyl tethered alcohols such as D-mannitol and L-erythritol [20], have been used to bring cinnamyl units together for photocycloadditions. However, this probably involves cycloaddition between an excited enone moiety and the tethered alkene. Other examples of carbon linked alkenes include perhaps the earliest example of alkene+alkene photocycloaddition. Liu and Hammond [21] reported the formation of cyclobutane from the triplet irradiation of myrcene (see Sch. 11).

Akabori et al. employed an interesting tethering system by linking cinnamyl groups onto the amine of a diazacrown ether [22]. Since the

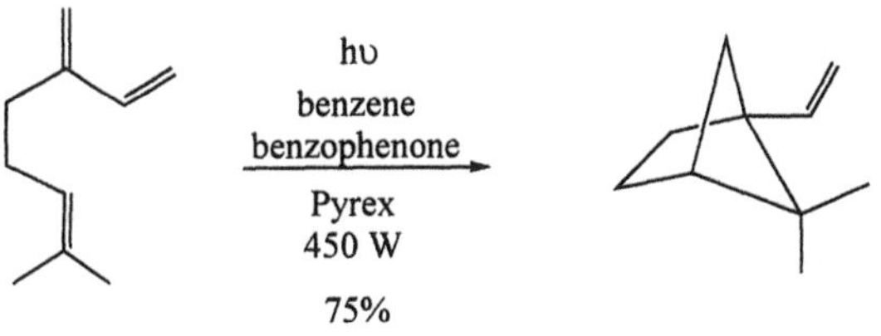

Scheme 11 Myrcene intramolecular photocycloaddition.

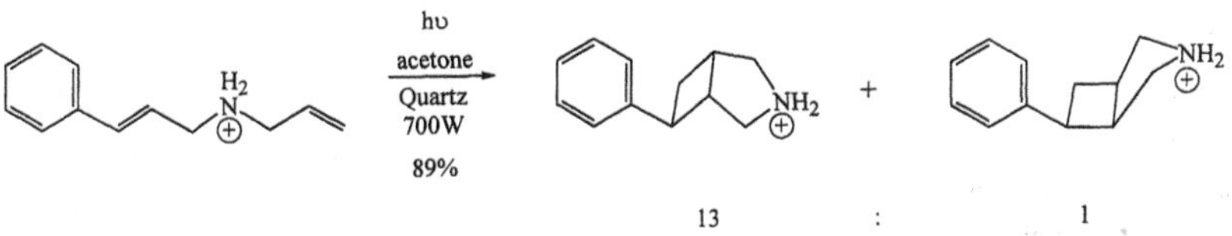

Scheme 12 Photocycloaddition of amine-tethered alkenes.

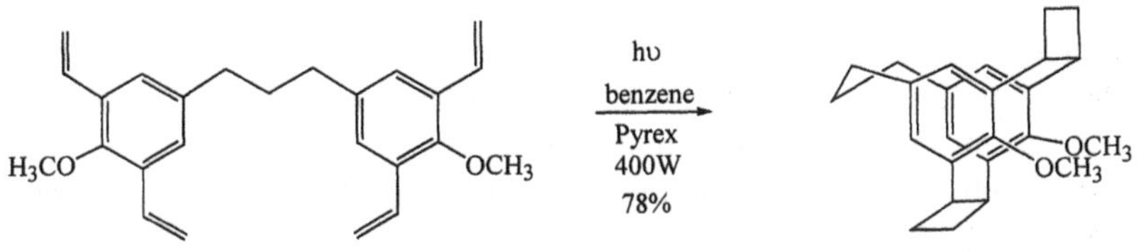

Scheme 13 Alkyl tethered styrenes.

chromophore for this example is the cinnamyl amide group, one might argue that it involves excited state alkene photochemistry rather than enone photochemistry. In a similar study, alkenes tethered by an amine group (see Sch. 12) have been shown to undergo 2+2 cycloaddition with high stereoselectivity. The selectivity is attributed to steric interactions in the triplet diradical intermediate [23]. The high degree of selectivity is quite impressive for this case, in light of the triplet pathway.

Perhaps the most common tethering is the use of alkyl linked styrenes that result in cyclophanes upon irradiation [24]. Okada et al. have reported several systems that undergo efficient alkene+alkene photocycloaddition (see Sch. 13) [25]. A common explanation for the effectiveness of the cycloaddition between these tethered styryl groups is that π-stacking orients the alkenes in close proximity so that the excited state alkene can readily interact with the adjacent ground-state alkene.

5.4. REGIO- AND STEREOSELECTIVITY

A. Nontethered Alkenes

Phenylcyclohexene has been shown to undergo a regio- and stereoselective intermolecular [2+2] photocycloaddition (see Sch. 7) [14a,16]. Both the singlet and the triplet pathways give cycloadducts where the major product is the more thermodynamically stable ring system. These examples favor the head-to-head cyclobutane product with the two aryl groups *trans* to each other, but it is not the exclusive product. The *trans:cis* ratio is 3:1 for the direct irradiation and 4:1 for the sensitized pathway. The *trans* stereo-selectivity may easily be rationalized as a result of sterics for the triplet cycloaddition since there should be sufficient time for bond rotation before the putative diradical intermediate closes to give the four-member ring. The singlet reaction presumably does not involve a diradical intermediate of significant lifetime so its selectivity is more difficult to explain. The preference for *trans* isomer may be inherent to the approach of the alkenes or, perhaps, it is a function of selective reversibility of diradical intermediates (see Sch. 3).

In Caldwell's styrene study (see Sch. 8) [14b], the regioselectivity is completely head-to-head. Only 1,2-diarylcyclobutane was observed upon irradiation at 355 nm. The stereoselectivity is approximately 11:1 in favor of the *trans* isomer in the *p*-acylstyrene to styrene cycloaddition.

A recent example of intermolecular photocycloaddition between acephenanthrylene demonstrates the role of π-stacking [26]. The singlet excited state gives only the *cis* dimer that the authors suggest arises from an excimer state. The triplet excited state results in formation of both *cis* and *trans* cyclobutanes. Interestingly, in the presence of oxygen the *cis* dimer is the only product due to quenching of the triplet excited state.

B. Tethered Alkenes

Irradiation of hydroxy-substituted 1,6-heptadienes in the presence of CuOTf results in formation of bicyclo[3.2.0]heptanes with complete regioselectivity and impressive stereoselectivity [27]. For example, when the *S* isomer of 6-methyl-1,6-heptadien-3-ol was irradiated in the presence of CuOTf the *endo*-bicyclo photoadduct shown in Sch. 14 was obtained in 98% e.e. Interestingly, the endoselectivity of this reaction increases in polar solvents. The authors argue that the copper intramolecular coordination between the hydroxyl group and the alkenes is favored in increasingly polar media.

Copper catalyzed 2+2 of ether tethered alkenes has also shown regio- and stereoselectivity [18a]. Irradiation of the substituted diallyl ether shown in Sch. 15 yields predominantly the *exo* product via a coordinated

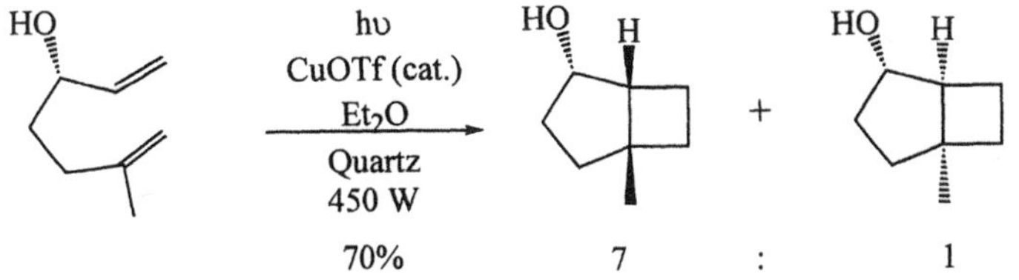

Scheme 14   Copper assisted photocycloaddition.

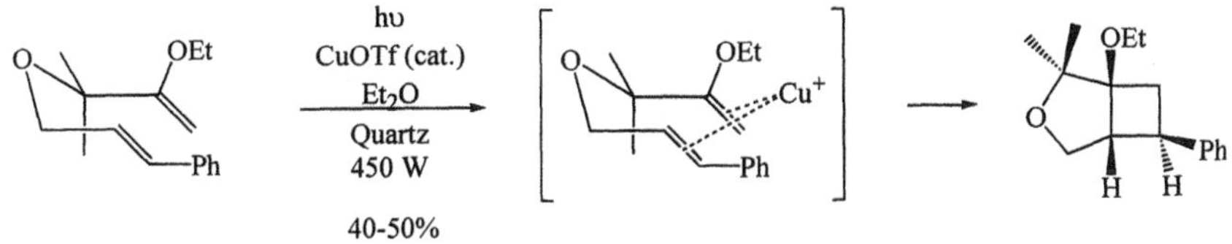

Scheme 15   Copper catalyzed photocycloaddition of diallyl ether.

species that avoids unfavorable steric interactions between the substituent groups. The fact that any *endo* product was formed suggests that there is some stepwise nature to this reaction (see Sch. 4). Although the stereoselectivity for the cycloaddition of the *cis* alkene was not reported, one might expect it to give a similar preference for the *exo* product.

Consistent with most tethered photocycloadditions, the photo-products obtained from silicon tethered irradiations are formed with complete regiocontrol [19]. Several examples of silyl tethered groups have been reported. Scheme 16 lists the variants that have been successful. As mentioned previously, the mechanistic pathway that is most consistent with the results involves interaction between the alkenes prior to cycloaddition. The experimental observation is that cycloadducts are formed only when the alkenes have extended π systems attached.

One silicon tethered example that is unique in its selectivity is the cinnamyl tethered silyl enol ether shown in Sch. 17. Unlike all of the other silyl tethered examples, this compound gives a photoadduct that is the result of a cross 2+2. However, it is the product expected if the cycloaddition is a stepwise process involving radical intermediates. It is also the product expected if the reaction pathway is controlled by π-stacking.

Another example of excellent photoproduct control is found in the cyclophane studies mentioned previously. The head-to-tail isomers that would result from a cross 2+2 in these tethered alkenes are significantly strained, which presumably accounts for the complete regioselectivity for the head-to-head products. Stereocontrol of the cycloaddition is usually

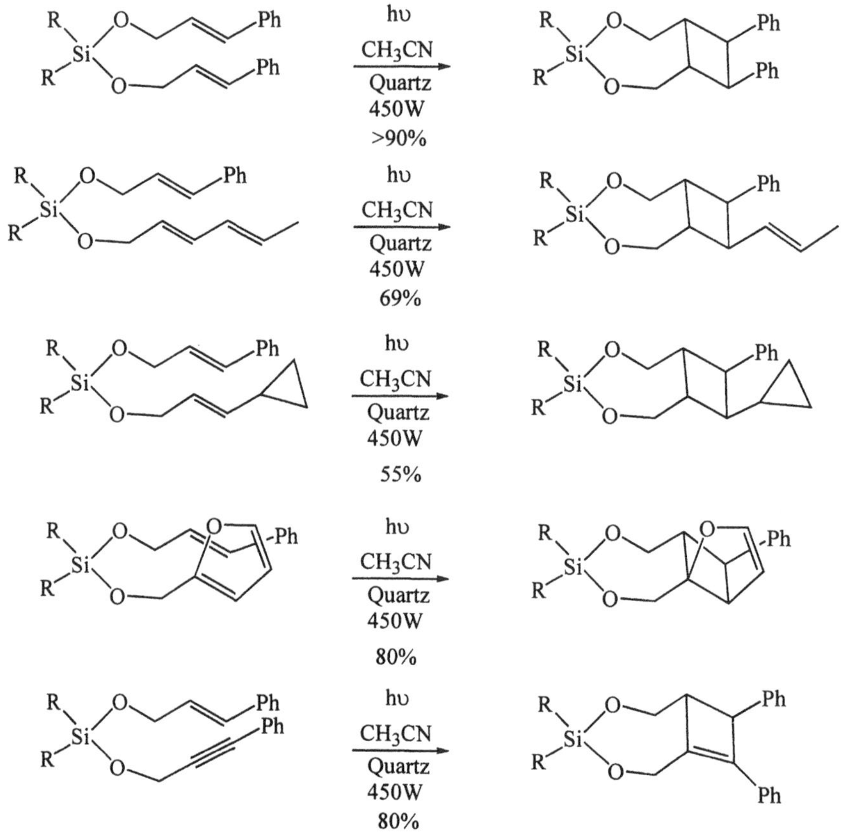

Scheme 16 Silyl tethered groups that undergo cycloaddition.

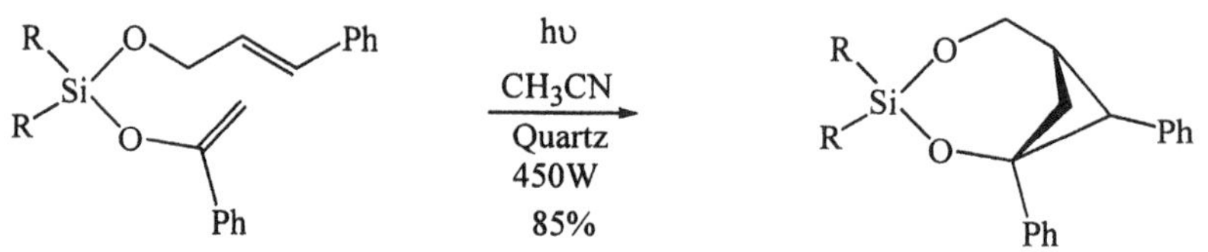

Scheme 17 Cross 2+2 from silyl enol ether.

based on steric arguments. For example, Okada et al. [28] reported that irradiation of the styrenyl groups tethered at the meta position (see Sch. 18 yielded exclusively the adduct which had the cyclobutane ring in the exo orientation. They suggested that this selectivity results from the steric

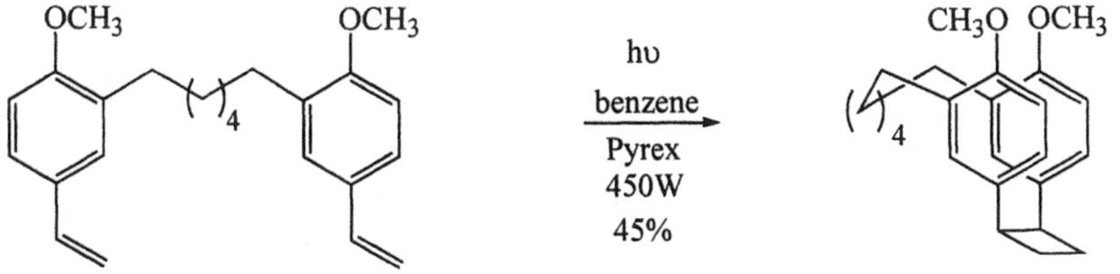

Scheme 18 Alkyl tethered styrenes that yield cyclophanes.

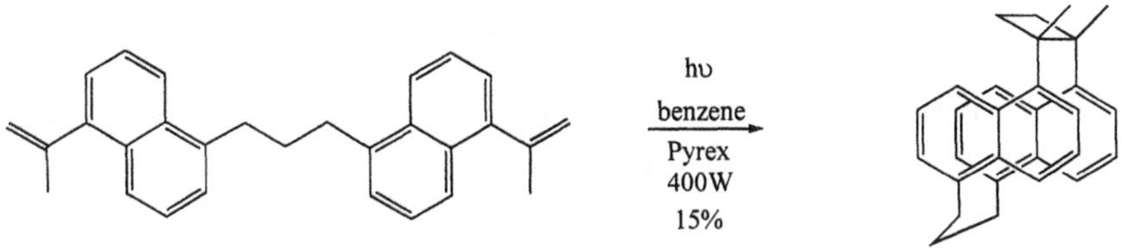

Scheme 19 Photocycloaddition of tethered vinylnaphthalenes.

interactions that would result between the tether and the cyclobutane in the endo product.

In the divinyl substituted tethered aryl systems shown previously in Sch. 13 the steric interaction between the cyclobutane and the *ortho* methoxy groups is obvious. The authors indicate that the observed product, the *exo–exo* isomer, has 6 kcal/mol less strain energy than the *exo–endo* compound. The suggestion that the major photoadduct is formed because it has less steric strain implicates reversibility in a reaction pathway which eventually leads to the most thermodynamically stable product. However, it does not preclude a reversibly formed exciplex or ground state complex.

In a similar study, the alkyl tethered vinyl naphthalenes shown in Sch. 19 were found to produce cyclobutanes with complete regio- and stereocontrol [29]. Like the phenyl analogues, sterics play an important role in the vinyl naphthalene cycloaddition. The tethered α-methyl substituted alkenylnaphthalene resulted in the endo adduct, whereas the β-methyl alkenyl reagent yielded only the *exo* adduct. This reversal in selectivity is due to the steric interaction between the methyl hydrogens and the ring hydrogens that would occur in the *exo* product of the α-methyl compound. It was also found that if the alkenes were trisubstituted, no cycloaddition occurred. The authors suggested that the lack of reactivity was due to alkene being too crowded to react efficiently. Otherwise, π-stacking, as well as the

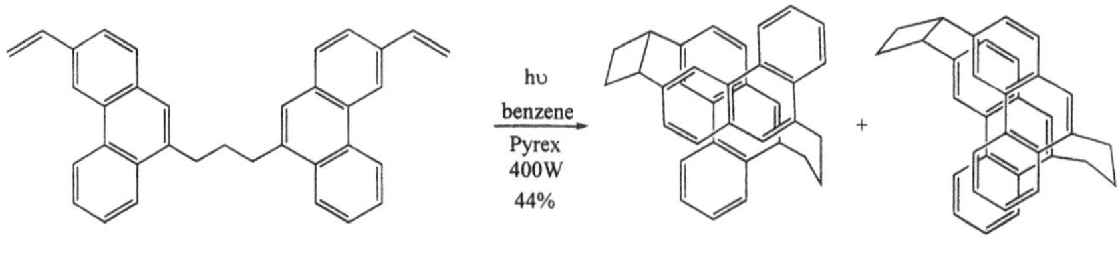

Scheme 20   Cycloaddition of tethered vinylphenanthrenes.

tether, helps the two alkene substituents come close enough together to cyclize upon irradiation.

A similar effect is observed in the reaction of tethered alkenyl-substituted phenanthrenes [30]. As expected, this example proceeds with total regiocontrol (see Sch. 20). Only the head-to-head isomer is formed. The stereocontrol is also impressive, the *exo* is the sole product observed. The authors suggest π-stacking plays an important role in the preorganization of the reactants.

In terms of synthetic methodology, the 2+2 photocycloaddition reaction can be an effective method for predictably forming cyclobutanes. Tethering the two reacting species clearly helps facilitate the reaction, provides control of the regiochemistry, and maximizes the stereocontrol as well. Tethered groups that hold the alkenes in close proximity give a more reliable result. Triplet sensitization is often necessary for alkenes that are not conjugated by aryl substitution. The triplet pathway usually yields the thermodynamically more stable product due to its stepwise nature. This selectivity is often excellent. But there is still considerable room for the study of stereoselective cyclobutane formation, especially in the area of asymmetric synthesis.

5.5.   EXPERIMENTAL PROCEDURES

A.   Photocycloaddition of Silyl-Tethered Alkenes

In a typical experiment: A solution of 0.01 M silane in distilled CH_3CN was deoxygenated by bubbling nitrogen for 30 min prior to irradiation. Nitrogen outgassing was continued as the solution was irradiated with a water-cooled Hanovia 450 W mercury medium pressure lamp through quartz for 2 h. The solution was concentrated in vacuo and the resulting oil subjected to chromatographic separation. Depending on the ease of hydrolysis of the silane (dialkoxy silanes are more prone to hydrolyze than mono-alkoxy silanes) the crude photomixture can be immediately subjected to desilation using NH_4F. This treatment involves refluxing the concentrated photomixture in a 0.5 M NH_4F in methanol solution using approximately 10 eq. of fluoride to silane [31]. Chromatographic purification of the resulting diol is much easier than the dialkoxysilane. Yields are typically 50–90%.

B. Photocycloaddition of Copper-Tethered Alkenes

A solution of 0.05 M diene in freshly distilled ether (sodium benzophenone ketyl) was treated with 0.1 molar equivalent of cupric trifluoromethane-sulfonate [32]. The solution was agitated by bubbling nitrogen until it was homogenous. Irradiation using a water-jacketed Hanovia 450 W mercury medium pressure lamp was followed by an ice cooled conc. aq. NH_4OH quench. The organic layer was dried over Na_2SO_4 then concentrated in vacuo. The crude photomixture was purified by chromatography. Yields are usually 70–80%.

5.6. TARGET MOLECULES

A. Cubane

Cubane-type compounds have been synthesized by a variety of routes. Most take advantage of the tethered 2+2 photocycloaddition. The first synthesis was reported by Eaton [33], although his synthetic route employed an enone+alkene photocycloaddition. An alkene+alkene cyclo-addition was used to synthesize the propellacubane shown in Sch. 21 [34]. A more recent report decribed the synthesis of permethylated cubane using the alkene + alkene photochemical approach [35]. That photocycloaddition is shown in Sch. 22. Although this particular reaction is inefficient, the product was made in sufficient quantity to allow for its complete characterization.

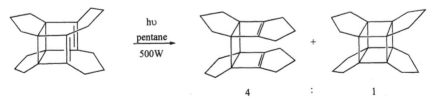

Scheme 21 Propellacubane synthesis.

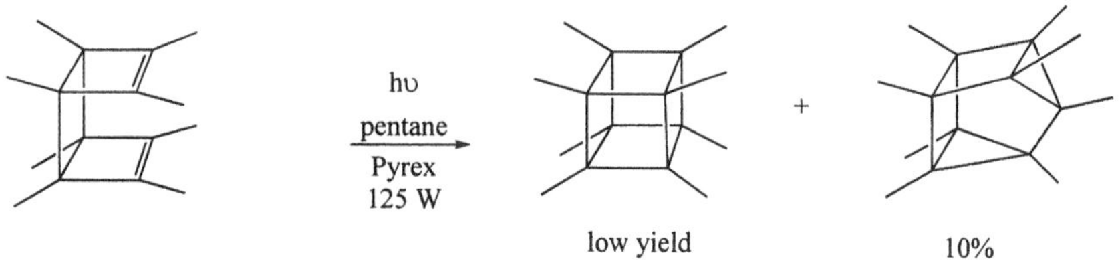

Scheme 22 Octamethylcubane formation.

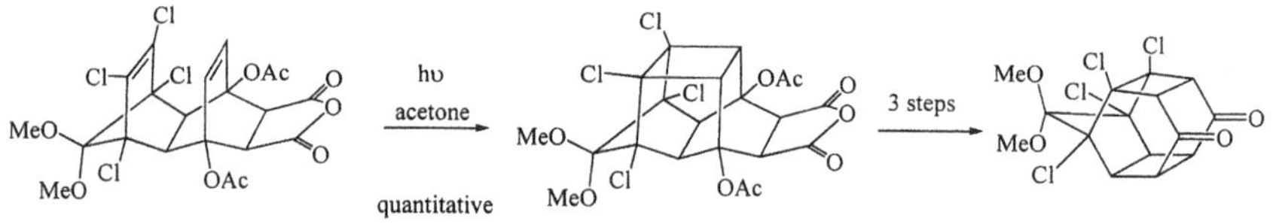

Scheme 23 Synthesis of homosecohexaprismane structure using 2+2 photo-cycloaddition.

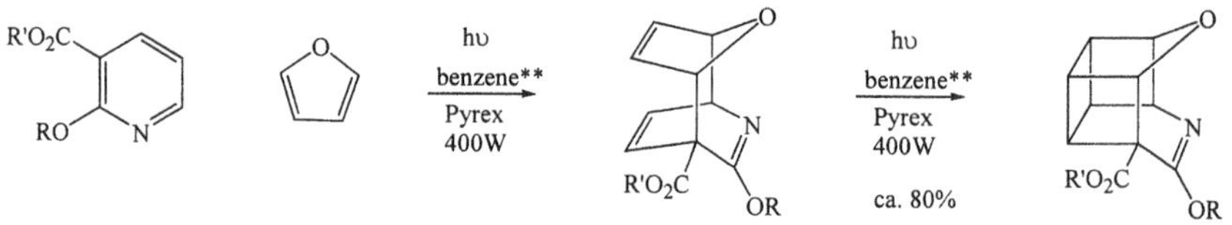

Scheme 24 Pyridine based cubane-like synthesis.

B. Cubane-like Structures

Other approaches to the novel multicyclic strained systems have been reported [36]. Chou et al. have demonstrated that novel cage structures can be obtained using the alkene + alkene methodology (see Sch. 23). More recently an aza analogue has been reported. This example is shown in Sch. 24.

C. High Energy Compounds

The imidazolinone shown in Sch. 25 has been dimerized by irradiating in acetone [37]. The *bis*-urea substituted cyclobutane that was obtained has been further modified to produce several high energy explosives. The most energetic product is the nitrourea shown. It is hydrolytically and thermally stable, but sensitive to impact.

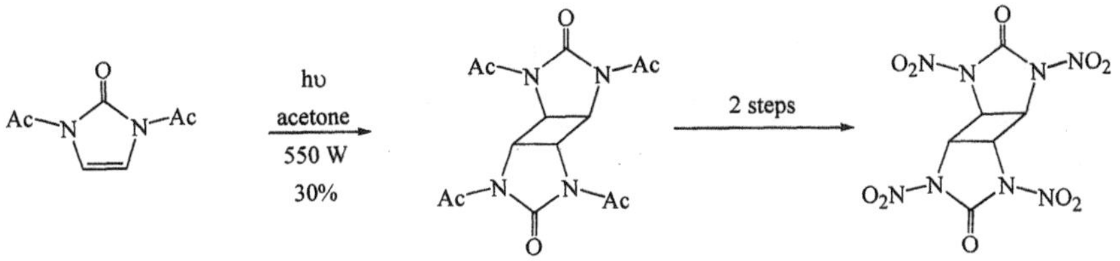

Scheme 25   Intermolecular photocycloaddition of imidazolinone.

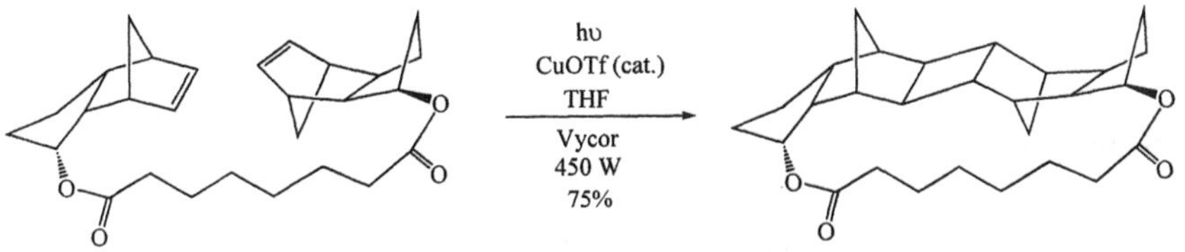

Scheme 26   Tethered norbornene photocycloaddition.

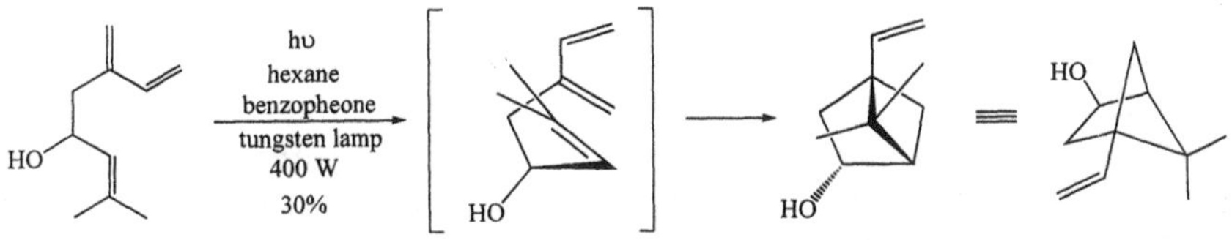

Scheme 27   Ipsdienol photocycloaddition.

D.   Cage Structures

Norbornene structures can be linked by long chain diesters. The copper catalyzed photocycloaddition of these alkenes gives rise to the cyclobutane shown in Sch. 26 [38]. The intramolecular reaction is stereoselective.

E.   Terpenoids

A recent publication described the carbon tethered 2+2 photocycloaddition of the terpene ipsdienol [39]. The pseudoterpene photoproduct (see Sch. 27) is formed as a single stereoisomer. The authors suggest that the hydroxyl group avoids the developing gem-dimethyl group due to steric interactions. They propose the transition state shown although the reaction is triplet sensitized and presumably proceeds in a step-wise fashion. The monoterpene,

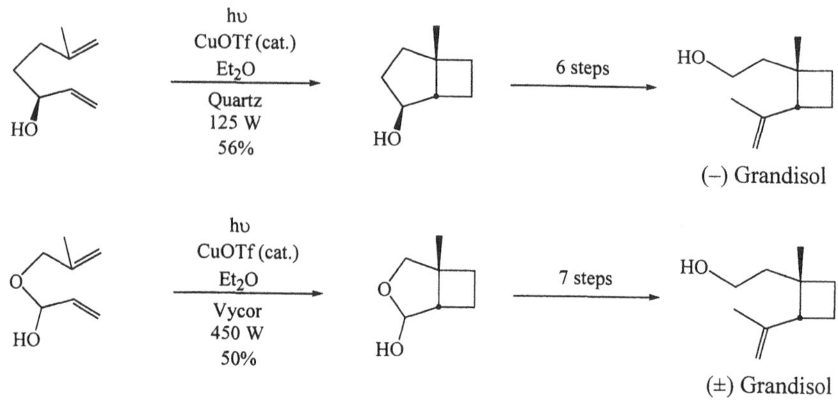

Scheme 28   Grandisol synthesis.

β-necrodol, has also been synthesized using an alkene+alkene photocy-cloaddition [40]. In this case the authors use a copper catalyzed approach.

F. Insect Pheromone

Grandisol has previously been synthesized using a number of photochemical procedures. Two recent reports demonstrate the utility of the copper catalyzed photocycloaddition (see Sch. 28) in the synthesis of the four member ring observed in the boll weevil pheromone. The first example [41] is a chiral synthesis and the second pathway [42] provides the racemic product.

REFERENCES

1. (a) Schreiber SL. Science 1985; 227:857; (b) Oppolzer W. Acc Chem Res 1982; 15:135; (c) Weedon A. Synthesis 1992; 95; For specific examples of recent applications see: (a) Bach T. Kather K. J Org Chem, 1996; 61:3900; (b) Somekawa K, Hara R, Kinnami K, Muraoka F, Suishu T, Shimo T. Chem Lett 1995; 407; (c) Meyers AI, Fleming SA. J Am Chem Soc. 1986; 108:306; (d) Crimmins MT, Gould LD. J Am Chem Soc. 1987; 109:6199; (e) Pattenden G, Robertson GM. Tetrahedron Lett 1986; 27:399; (f) Kitano Y, Fukuda J, Chiba K, Tada M. J Chem Soc, Perkin I 1996; 829.
2. Fleming SA, Bradford CL, Gao J. J Organic Photochemistry; Molecular and Supramolecular Photochemistry 1997; 1:187.
3. (a) Hegedus LS, Brown B. J. Org Chem Soc. 2000; 65:1865; (b) Kokubo K, Koizumi T, Yamaguchi H, Oshima T. Tetrahedron Lett 2001; 42:5025.

4. (a) Estieu K, Ollivier J, Salaün J. Tetrahedron Lett 1996; 37:623; (b) Chen B-C, Ngu K, Guo P, Liu W, Sundeen JE, Weinstein DS, Atwal KS, Ahmad S. Tetrahedron Lett 2001; 42:6227.

5. (a) Lange GL, Merica A, Chimanikire M. Tetrahedron Lett 1997; 38:6371; (b) Haque A, Ghatak A, Ghosh S, Ghoshal N. J Org Chem 1997; 62:5211.

6. (a) Caldwell RA. J Am Chem Soc 1980; 102:4004; (b) Michl J. J Photochem Photobiol 1977; 25:141; (c) Gilbert A, Baggott J. Essentials of Molecular Photochemistry, Blackwell Scientific Publications, 1991:270.

7. Bentzien J, Klessinger M. J Org Chem 1994; 59:4887.

8. Zimmerman HE, Kamm KS, Werthemann DP. J Am Chem Soc 1975; 97:3718.

9. (a) Takahashi Y, Okitsu O, Ando M, Miyashi T. Tetrahedron Lett 1994; 35:3953; (b) Kim T, Mirafzal GA, Bauld NL. Tetrahedron Lett 1993; 34:7201; (c) Sadlek O, Gollnick K, Polborn K, Griesbeck AG. Angew Chem Int Ed Engl 1994; 33:2300.

10. Alkene+alkene photocycloaddition in the presence of electron transfer sensitizers involves a similar path. For example, see: (a) Asaoka S, Ooi M, Jiang P, Wada T, Inoue Y. J Chem Soc, Perkin 2 2000; 73; (b) Goez M, Frisch I. J Am Chem Soc. 1995; 117:10486.

11. Cornil J, dos Santos DA, Crispin X, Silbey R, Brédas JL. J Am Chem Soc 1998; 120:1289.

12. Penn JH, Gan L-X, Chan EY, Loesel PD, Hohlneicher G. J Org Chem 1989; 54:601.

13. Salomon RG. Tetrahedron 1983; 39:485.

14. (a) Caldwell RA, Diaz JF, Hrncir DG, Unett DJ. J Am Chem Soc. 1994; 116:8138; (b) Unett DJ, Caldwell RA, Hrncir DC. J Am Chem Soc. 1996; 118:1682.

15. Lewis FD, Hirsch RH. J Am Chem Soc 1976; 98:5914.

16. Dauben WG, Riel HCHA, Robbins JD, Wagner GJ. J Am Chem Soc 1979; 101:6383.

17. (a) Lee TS, Lee SJ, Shim SC. J Org Chem 1990; 55:4544; (b) Chung CB, Kwon JH, Shim SC. Tetrahedron Lett 1993; 34:2143.

18. (a) Avasthi K, Raychaudhuri SR, Salomon RG. J Org Chem 1984; 49:4322; (b) Ghosh S, Patra D, Samajdar S. Tetrahedron Lett 1996; 37:2073; (c) Langer K, Mattay K. J Org Chem 1995; 60:7256; (d) Bach T, Spiegel A. Eur J Org Chem 2002; 645.

19. (a) Ward SC, Fleming SA. J Org Chem 1994; 59:6476; (b) Bradford CL, Fleming SA, Ward SC. Tetrahedron Lett 1995; 36:4189.

20. (a) Green BS, Rabinsohn Y, Rejtö M. Carbohydrate Res. 1975; 45:115; (b) Green BS, Hagler AT, Rabinsohn Y, Rejtö M. Israel J Chem 1976/1977; 15:124.

21. (a) Liu RSH, Hammond GS. J Am Chem Soc 1964; 86:1892; (b) Dauben WG, Cargill RL, Coates RM, Saltiel J. J Am Chem Soc 1966; 88:2742.

22. Akabori S, Kumagai T, Habata Y, Sato S. J Chem Soc, Perkin 1 1989; 1497.

23. Steiner G, Munschauer R, Klebe G, Siggel L. Heterocycles 1995; 40:319.

24. Greiving H, Hopf H, Jones PG, Bubenitschek P, Desvergne P, Bouas-Laurent H. J Chem Soc, Chem Commun 1994; 1075.

25. Okada Y, Ishii F, Kasai Y, Nishimura J. Tetrahedron 1994; 50:12159.
26. Plummer BF, Moore MJB, Wright J. J Org Chem 2000; 65:450.
27. Langer K, Mattay J, Heidbreder A, Möller M. Leibigs Ann Chem 1992; 257.
28. Okada Y, Ishii F, Nishimura J. Bull Chem Soc Jpn. 1993; 66:3828.
29. Nishimura J, Takeuchi M, Takahashi H, Sato M. Tetrahedron Lett. 1990; 31:2911.
30. (a) Takeuchi M, Nishimura J. Tetrahedron Lett 1992; 33:5563; (b) Nakamura Y, Fujii T, Nishimura J. Tetrahedron Lett 2000; 41:5563. See also: Nakamura Y, Kaneko M, Yamanaka N, Tani K, Nishimura J. Tetrahedron Lett. 1999; 40:4693.
31. Zhang W, Robins MJ. Tetrahedron Lett 1992; 33:1177.
32. Salomon RG, Kochi J. J Am Chem Soc 1973; 95:1889. Cupric trifluoro-methanesulfonate can be simply replaced by the commercially available copper (II) trifluoromethanesulfonate as well. See Ref. 27.
33. Eaton PE, Cole TW. J Am Chem Soc 1964; 86:962, 3157.
34. Gleiter R, Karcher M. Angew Chem 1988; 100:851.
35. Gleiter R, Brand S. Tetrahedron Lett 1994; 35:4969.
36. (a) Chou T-C, Yeh Y-L, Lin G-H. Tetrahedron Lett 1996; 37:8779; (b) Sakamoto M, Yagi T, Fujita S, Ando M, Mino T, Yamaguchi K, Fujita T. Tetrahedron Lett 2002; 43:6103; (c) Bojkova NV, Glass RS. Tetrahedron Lett 1998; 39:9125.
37. Fischer JW, Hollins RA, Lowe-Ma CK, Nissan RA, Chapman RD. J Org Chem 1996; 61:9340.
38. Chebolu R, Zhang W, Galoppini E, Gilardi R. Tetrahedron Lett. 2000; 41:2831. See also: Dave PR, Duddu R, Li J, Surapaneni R, Gilardi R. Tetrahedron Lett 1998; 39:5481.
39. Barbero A, Garcia C, Pulido FJ. Synlett 2001; 824.
40. Samajdar S, Ghatak A, Ghosh S. Tetrahedron Lett 1999; 40:4401.
41. Langer K, Mattay J. J Org Chem 1995; 60:7256.
42. Panda J, Ghosh S. Tetrahedron Lett 1999; 40:6693.

6

Di-π-Methane Rearrangement

Diego Armesto, Maria J. Ortiz, and Antonia R. Agarrabeitia
Universidad Complutense, Madrid, Spain

6.1. HISTORICAL BACKGROUND

The chemistry of alkenes is one of the pillars of organic chemistry. These substrates undergo many synthetically useful transformations such as electrophilic and nucleophilic additions, oxidations, reductions, [1+2] and [2+2]-cycloadditions, metathesis, polymerizations [1]. This chemistry also applies to dienes and higher polyenes. Conjugated dienes undergo specific reactions, such as conjugated additions and [4+2]-cycloadditions (Diels-Alder reactions) that have found important synthetic applications [1]. However, the double bonds of nonconjugated dienes behave as two separate alkenes in most reactions, an exception being 1,5-dienes that undergo Cope and Claisen rearrangements [1]. In particular, C–C double bonds in 1,4-dienes tend to react as two separated entities in most of the reactions mentioned above, although a few specific reactions of compounds containing the 1,4-diene have been observed in the chemistry of some natural products such as polyunsaturated fatty acids and prostaglandins.

The photochemical reactivity of alkenes is also of great interest [1,2]. Studies in this area have led to an expansion of the synthetic utility of these substances. Typical photochemical reactions include *cis-trans* isomerizations, inter- and intramolecular cycloadditions, photooxidations, and electrocyclic ring opening and closing of conjugated dienes and polyenes. Many of these photoreactions have thermal counterparts. In contrast, 1,4-dienes can undergo a unique photochemical transformation affording vinylcyclopropanes, as shown schematically in Sch. 1.

This process is known as the di-π-methane (DPM) rearrangement [3]. One of the first examples of this reaction was reported in 1966 by

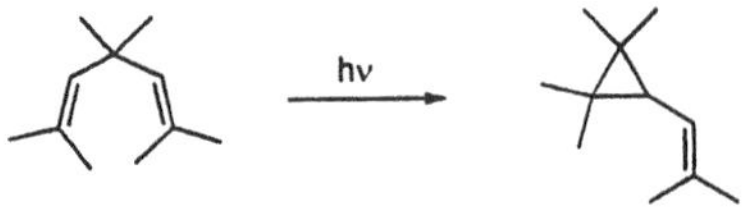

Scheme 1

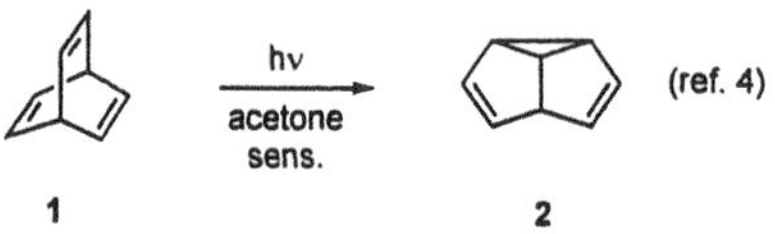

Scheme 2

Zimmerman et al. in the triplet-sensitized irradiation of barrelene **1** that yields semibullvalene **2** (Sch. 2) [4]. Zimmerman also recognized the generality of the process and proposed a mechanistic interpretation for the reaction that it is still commonly accepted [5]. Since then, a large number of studies by Zimmerman and his coworkers, and by other research groups worldwide, have demonstrated that the rearrangement is very general and that it usually occurs with a high degree of regioselectivity and stereochemical control [3].

In many instances, high chemical and quantum yields have been observed for this process. The reaction also takes place when an alkene unit is replaced by an aryl ring [3]. A representative example of the aryl-di-π-methane version of the rearrangement is found in the conversion of **3** to the cyclopropane **4**, in 93% yield (Sch. 3) [6]. The large number of different cyclopropane derivatives that can be obtained by using the DPM rearrangement reaction is remarkable, making the reaction particularly useful from a synthetic point of view [3].

The di-π-methane rearrangement is not restricted to 1,4-dienes. Other 1,4-unsaturated systems also undergo similar photoreactions. The extension of the rearrangement to β,γ-unsaturated ketones occurred almost simultaneously with the discovery of the DPM rearrangement [7]. These compounds react from their triplet exited state to give the corresponding cyclopropyl-ketones regioselectively, in what is known as the oxa-di-π-methane (ODPM) version of the rearrangement. The large number of studies which have been carried out on the photochemistry of β,γ-unsaturated ketones show that the ODPM reaction usually occurs with a high degree of diastereoselectivity and, under special conditions, even enantioselectivity [3a,3e,8]. Therefore, it is not surprising that the ODPM rearrangement has been applied as the key step in the synthesis of natural products and other highly complex molecules that

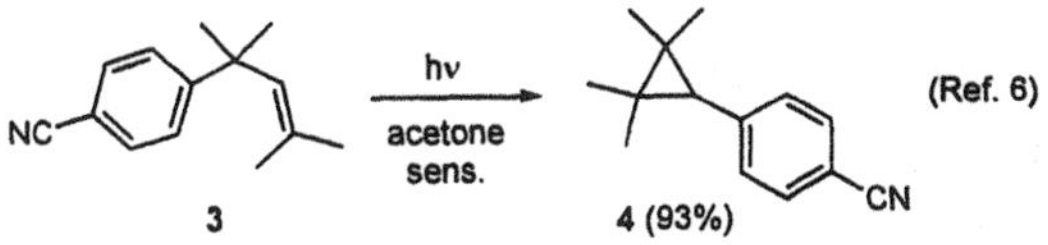

Scheme 3

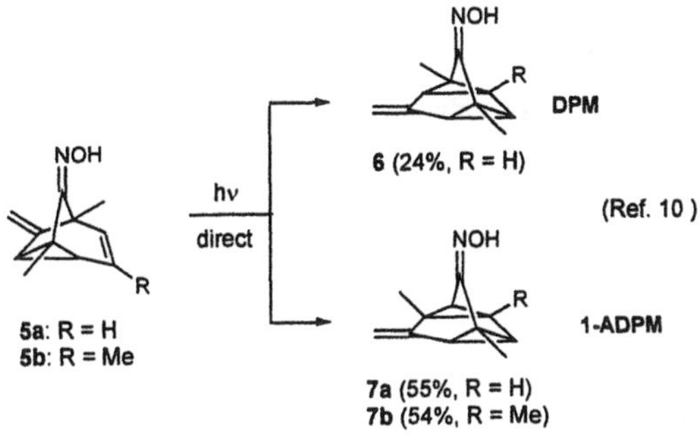

Scheme 4

are difficult to obtain by alternative preparative routes [8e]. The reaction has been extended recently to β,γ-unsaturated aldehydes that have been considered for many years unreactive in this mode [9]. Because of its important synthetic applications, the ODPM rearrangement is discussed in a separate Chapter.

In spite of the large number of studies carried out on the DPM and the ODPM rearrangements since 1966, ten years elapsed before the reaction was extended to other 1,4-unsaturated systems, particularly to C–N double bond derivatives. The first example of a 1-aza-di-π-methane (1-ADPM) rearrangement was reported by Nitta et al. in a study on the photoreactivity of the tricyclic oximes 5 [10]. Direct irradiation of compound 5a brought about the formation of the DPM product 6 and the 1-ADPM derivative 7a, in the first example of competition between these two processes. However, the methyl substituted derivative 5b yielded the 1-ADPM photoproduct 7b, exclusively (Sch. 4) [10a].

The first 1-ADPM rearrangement of an acyclic derivative, reported five years later, was observed in studies of the sensitized irradiation of the β,γ-unsaturated imine 8 that yielded the corresponding cyclopropylimine 9 exclusively (Sch. 5) [11]. This photoproduct undergoes hydrolysis during isolation to generate the corresponding cyclopropane carbaldehyde 10.

A series of studies in this area since then have demonstrated that the 1-ADPM rearrangement is as general as the DPM and ODPM counterparts

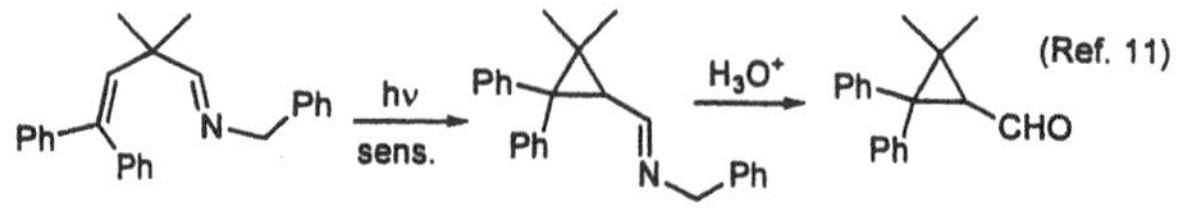

Scheme 5

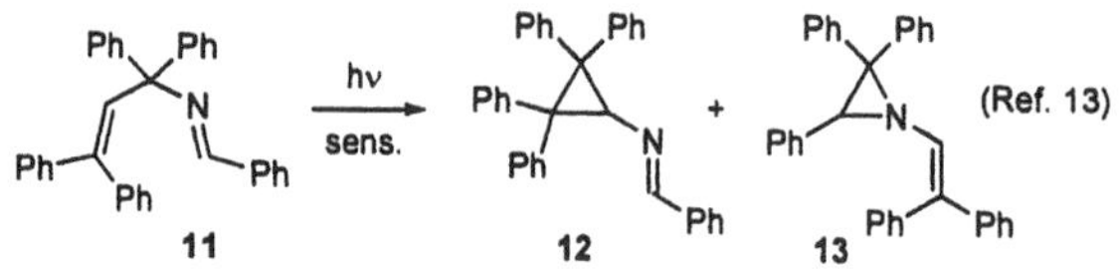

Scheme 6

and that it can be used to regioselectively form hydrolytically stable C–N double bond containing cyclopropylimine derivatives, such as oxime esters, acylhydrazones, and semicarbazones [3e,12].

For more than 30 years, di-π-methane processes were limited to the three different types of β,γ-unsaturated systems mentioned above. However, recent studies have shown that the rearrangement is applicable to 2-aza-1,4-dienes [13]. Thus, triplet-sensitized irradiation of compound **11** affords the cyclopropylimine **12** and the *N*-vinylaziridine **13** (Sch. 6). The photoreaction of **11** represents the first example of a 2-aza-di-π-methane rearrangement (2-ADPM) that brings about the formation of a heterocyclic product. The reaction has been extended to other 2-azadienes that yield the corresponding cyclopropylimines regioselectively [13b].

A few examples of DPM rearrangements in β,γ-unsaturated compounds in which the methane carbon is replaced by a boron or silicon atom have been described. However, the di-π-borate [14] and di-π-silane [15] alternatives of the rearrangement do not have the scope and synthetic applications of the other alternatives mentioned before.

Those interested in seeking detailed information about covering different versions of the rearrangement reaction should consult the many reviews which have been published over the years [3,8,12].

6.2. MECHANISTIC MODELS

The classical di-π-methane rearrangement can be considered as a 1,2 shift of one of the π units to the other followed or accompanied by ring closure to

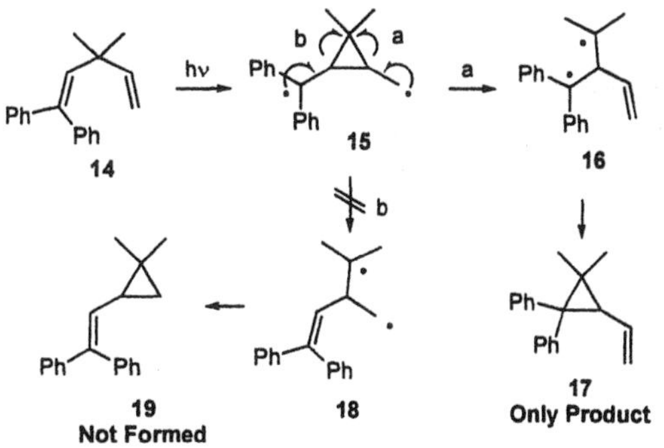

Scheme 7

form a cyclopropane ring. Concerted and biradical mechanisms have been proposed to account for the rearrangement. Studies on the reaction stereochemistry have shown that the configuration at C-1 and C-5 of the 1,4-diene is retained [16] while the configuration at C-3 (i.e., the sp^3-hybridized carbon) is inverted during rearrangement to the vinylcyclopropane [17,18]. These stereochemical features can be explained considering that the rearrangement takes place via a [$_\pi 2_a + _\pi 2_a + _\sigma 2_a$] concerted transition state. However, a biradical mechanism as shown in Sch. 7 for compound **14** has been normally used to justify the rearrangement because it provides a convenient way to understand the regioselectivity and the influence of substitution observed for the reaction [19]. Thus, compound **14** affords the vinylcyclopropane **17** exclusively.

The alternative regioisomer **19** is not formed in this reaction. This regioselectivity can be easily explained considering that irradiation of **14** generates the cyclopropyldicarbinyl biradical **15**. Ring opening in **15** by breaking of bond a affords the more stable 1,3-biradical **16**, the precursor of **17**. The alternative ring opening in **15** by breaking of bond b does not occur because biradical **18** is less stable than **16**. This interpretation permits a prediction of the regioselectivity of the reaction. However, it should be pointed out that the biradical species shown in Sch. 7 should not be considered necessarily as true intermediates but they might merely represent points along a reaction surface. The stereoselectivity observed for the reaction can also be explained by using the biradical mechanism if the rates of bond rotation in these species are lower that those for bond formation and cleavage. However, the issue of whether the DPM reaction occurs via a concerted or a biradical mechanism is still under discussion. Thus, while theoretical calculations support the concerted 1,2-vinyl migration [20], kinetic studies on the aryl-di-π-methane rearrangement favor the

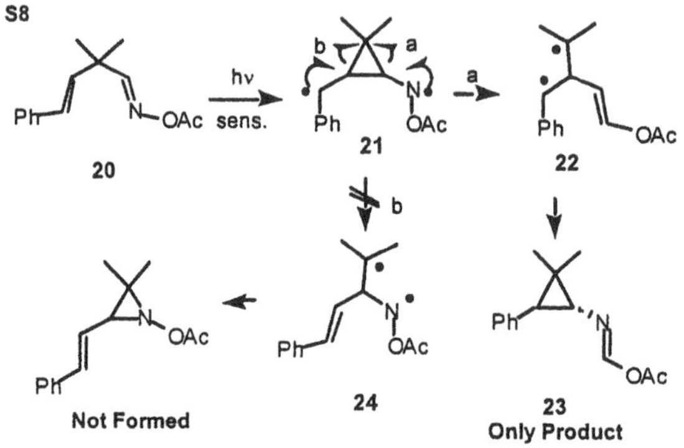

S8

Scheme 8

conventional biradical mechanism [21]. Nevertheless, the latter mechanistic approach is the most widely used since it depicts the reaction course in a very simple way.

Biradical mechanisms, similar to that shown in Sch. 7, have been used to account for the regioselectivity and the influence of substituents in the ODPM and 1-ADPM versions of the rearrangement [8,12]. In these instances the ring opening of the 1,4-cyclopropyl biradical intermediates always restores the C–O and C–N double bonds, affording the corresponding cyclopropylketones and cyclopropylimines, exclusively. This regioselectivity can be explained based on the stability of the intermediates as shown in Sch. 8 for the 1-ADPM rearrangement of oxime acetate **20** [22].

Thus, triplet-sensitized irradiation of **20** affords the cyclopropyl biradical **21**. Ring opening of bond a in **21** gives the 1,3-biradical **22**, the precursor of the observed photoproduct **23**. The alternative ring opening in **21**, by rupture of bond b, does not occur because the 1,3-biradical **24** is less stable than **22**. In this instance the reaction is also diastereoselective affording the *trans*-cyclopropane **23**, exclusively. Similar stereoselectivity affording predominantly or exclusively the most stable diastereoisomer has been observed in other instances [23].

However, the regioselectivity observed in the triplet-sensitized 2-ADPM reaction (Sch. 6) is more difficult to justify based on the differences in stability of the 1,3-biradical intermediates and further studies are necessary to understand the factors that control this novel rearrangement [13].

For more than 30 years di-π-methane rearrangements have stood as the paradigm reactions that occur in the excited state manifold exclusively. However, recent studies have shown that rearrangements of the di-π-methane type can also occur in ground state of radical-cation and radical-anion

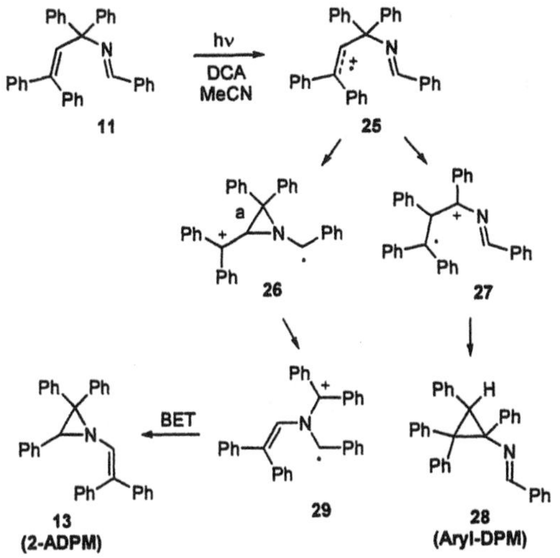

Scheme 9

intermediates. The first example of a di-π-methane rearrangement via radical-cation intermediates was reported in the irradiation of 2-azadiene **11**, using 9,10-dicyanoanthracene (DCA) as electron-acceptor sensitizer, that affords the *N*-vinylaziridine **13**, arising from a 2-ADPM rearrangement, and the cyclopropylimine **28**, resulting from an aryl-di-π-methane reaction (Sch. 9) [13].

Other 2-azadienes also react under these conditions to generate *N*-vinylaziridines regioselectively, via radical-cation intermediates [13]. In some instances, imine and olefin centered cation-radical intermediates undergo alternative reactions to produce isoquinoline and benzoazepine products. The formation of **13** and **28** can be justified by a mechanistic pathway involving the generation of an olefin-localized cation-radical **25**, which reacts by C–N bond formation to afford the aziridinyldicarbinyl radical-cation **26**. Ring opening in **26**, by breaking of bond a, generates the radical-cation **29** that undergoes back electron transfer and cyclization to yield **13**. A competitive route involving phenyl migration in **25** gives the radical-cation **27**, the precursor of **28** (Sch. 9) [13].

1-Aza-1,4-dienes also undergo 1-ADPM reactions via radical-cation intermediates, generated under SET-sensitized irradiation conditions, yielding the corresponding cyclopropane derivatives regioselectively, as shown in Sch. 10 for the 1-ADPM rearrangement of oxime acetate **30** that affords the cyclopropane **31** [24]. Interestingly, in this instance, the yield of products increases considerably when dicyanodurene (DCD) is used as electron acceptor sensitizer instead of DCA (Sch. 10) [24b].

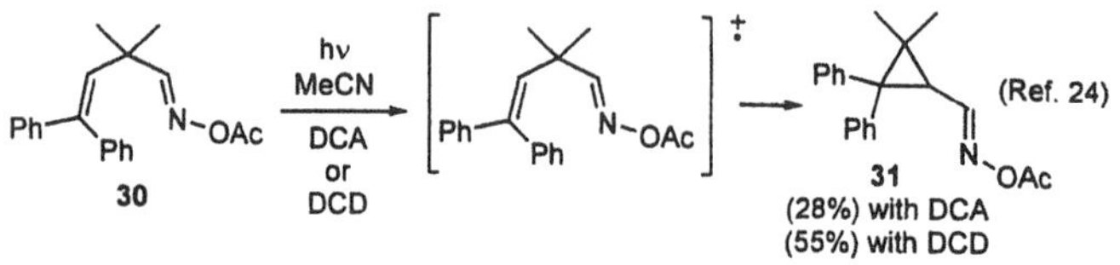

Scheme 10

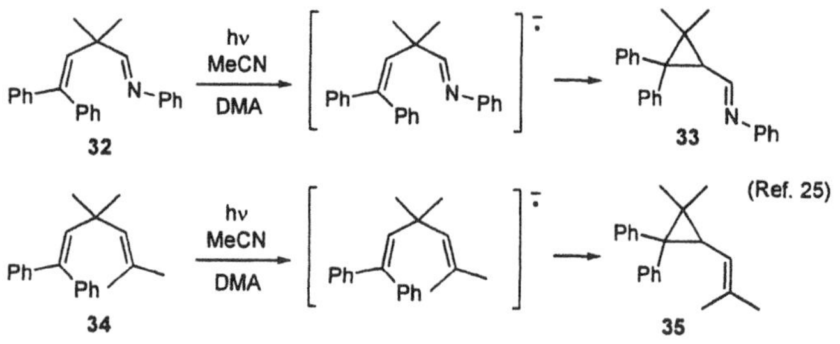

Scheme 11

Within the area of SET-promoted di-π-methane reactions, recent studies have shown that irradiation of 1-aza-1,4-dienes, such as **32**, and the 1,4-diene **34**, using *N,N*-dimethylaniline (DMA) as electron-donor sensitizer, leads to production of the corresponding cyclopropane derivatives **33** and **35** resulting from 1-ADPM and DPM rearrangements, respectively, in reactions that take place via radical-anion intermediates (Sch. 11) [25].

The results obtained in these studies on the photoreactivity of different β,γ-unsaturated systems under SET-sensitization have opened new lines of research in the area of di-π-methane photochemistry. However, further studies will be necessary to determine the scope and synthetic potential of these new reactions of the di-π-methane type via radical-ion intermediates.

6.3. SCOPE AND LIMITATIONS

The DPM reaction is an extremely general process. A large variety of bicyclic and polycyclic 1,4-dienes undergo the rearrangement in the triplet excited state to form, in many cases, highly constrained molecules in high yields. However, the reaction of acyclic 1,4-dienes has some limitations. Thus, disubstitution by alkyl, aryl, or different functional groups at the methane carbon seems to be essential for the success of the rearrangement reactions in these systems [26]. On the other hand, the DPM reaction in

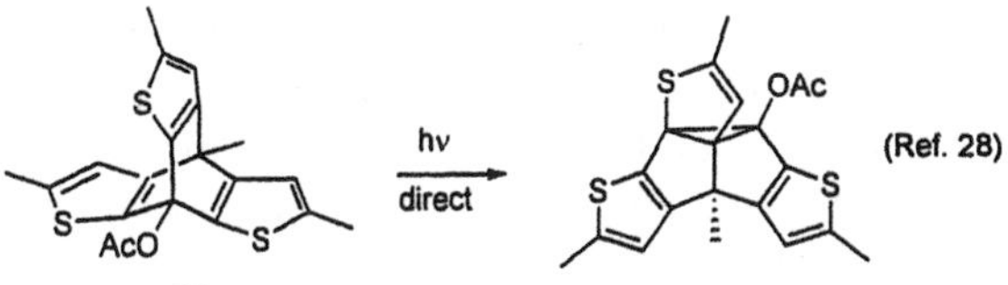

Scheme 12

36 → 37 (90%) (Ref. 28)

Scheme 13

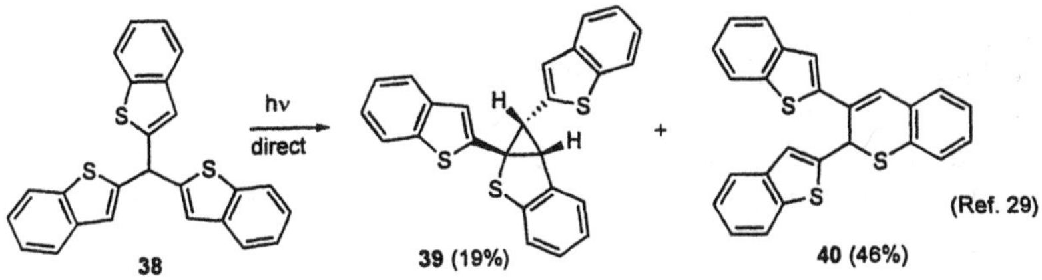

38 → 39 (19%) + 40 (46%) (Ref. 29)

most acyclic dienes usually occurs in the singlet excited state, therefore, the reaction is restricted to 1,4-dienes that absorb at approx. 250 nm or longer wavelengths.

The aryl-di-π-methane version of the rearrangement is also very general. The need for disubstitution at the methane carbon does not apply in this instance and examples of allylbenzenes, unsubstituted [27a,27b] and monosubstituted [27c–e] at C-3, undergoing the rearrangement have been reported. The reaction has been extended to compounds in which the two π-units participating in the rearrangement are heteroaromatic rings, broadening the scope of the aryl-di-π-methane process. Thus, irradiation of the thiophenetriptycene derivative **36** affords the diaryl-di-π-methane product **37** in 90% yield (Sch. 12) [28].

In another example, direct irradiation of compound **38** yields cyclopropane **39** and the thiopyrane derivative **40**, arising from secondary photolysis of **39** (Sch. 13). The formation of **39** is rationalized based on a di-π-methane rearrangement involving the participation of two benzo[*b*]thiophene rings present in **38** [29].

In spite of the limitations mentioned above, the number of 1,4-dienes that undergo the DPM reaction is very high and this rearrangement is among the most general of all types of photoreactions yet discovered.

The triplet 1-ADPM rearrangement has proved to be very general for a series of acyclic and bicyclic 1-aza-1,4-dienes affording the corresponding cyclopropanes in high yield and after remarkable short irradiation times [3e,12]. However, in the absence of substituents at position 4 of the 1-azadiene system, such as aryl rings or C–C double bonds, that would stabilize the biradical intermediates, the efficiency of the reaction and the yield of products diminish considerably [30–32].

Regarding the influence of substitution on nitrogen in the 1-ADPM process, the reaction has proved to be general for a series of imines, oxime esters, benzoylhydrazones, and semicarbazones [3e,12]. However, oxime acetates are usually the most convenient substrates for this reaction because they are easily accessible, hydrolytically stable and the chemical and quantum efficiencies of the corresponding reactions are usually higher than with the corresponding imines [31,33]. The 1-ADPM rearrangement of acyclic β,γ-unsaturated oxime acetates is limited to aldoxime acetates. The corresponding ketoxime derivatives usually undergo the rearrangement very inefficiently [31]. The possible influence of substitution at C-3 has not been explored yet in this reaction. Finally, owing to the fact that the 2-ADPM reaction and the rearrangements of the di-π-methane type via radical-cation and radical-anion intermediates have only recently been uncovered, their synthetic potential has not yet been established.

6.4. SYNTHETIC POTENTIAL: REACTIVITY AND SELECTIVITY PATTERN

6.4.1. The Reaction Regioselectivity

As mentioned above, the regioselectivity of the DPM reaction can be predicted in most cases by considering that competitive ring opening of the cyclopropyldicarbinyl biradical intermediate will occur to afford the most stable 1,3-biradical intermediate (Sch. 7). The aryl-di-π-methane reaction is also regioselective. In this instance the aromaticity of the aryl ring is always restored yielding arylcyclopropanes, exclusively. However, other situations must be considered. Studies of the influence of electron-donor and electron-acceptor functional groups on the regioselectivity of the rearrangement show that acyclic 1,4-dienes bearing electron donating groups at C-1 afford regioselectively the cyclopropane with the electron donating functional groups attached to the vinyl unit, while dienes with electron-withdrawing groups at C-1 yield the corresponding photoproduct in which the electron-withdrawing substituents are attached to the cyclopropane unit (Sch. 14) [19]. This regioselectivity is also observed in cases where the electron donating and electron withdrawing groups are situated at the *para*-position

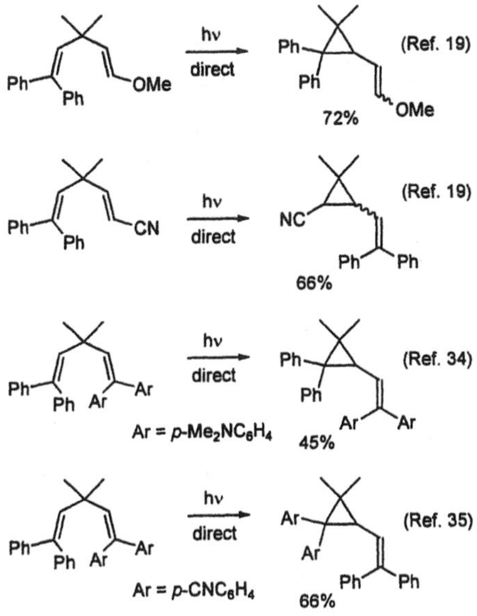

Scheme 14

of phenyl rings attached at C-1 and C-5 of the 1,4-diene skeleton (Sch. 14) [34,35].

Apart from the stability of the biradical intermediates, other factors should be taken into account in order to predict the result of the reaction of barrelene derivatives. Theses systems must be considered as tri-π-methane systems. Therefore, in order to predict the regioselectivity of the reaction, it is necessary to know which two of the three π units would be involved preferentially in the di-π-methane process. Some general rules have been established based on studies of a series of benzo-, naphtho-, and anthrabarrelenes. Thus, benzobarrelenes and anthrabarrelenes usually yield di-π-methane photoproducts resulting from vinyl–vinyl bridging and not from aryl–vinyl bridging. In the case of naphthobarrelenes the following pattern of preferred bridging reactivity is observed: α-naphtho-vinyl > vinyl–vinyl > β-naphtho–vinyl ≥ benzo–vinyl [36]. On the other hand, the regioselectivity in substituted benzobarrelene derivatives can be predicted based on the relative stability of the possible biradical intermediates. Thus, for example, triplet-sensitized irradiation of benzo-barrelene **41** yields semibullvalene **43**, exclusively. The alternative regioisomer **45** is not formed [37]. This result can be interpreted considering that vinyl–vinyl bridging in **41** occurs between positions 2 and 6 (arbitrary numbering) to afford biradical **42**, stabilized by conjugation with the cyano

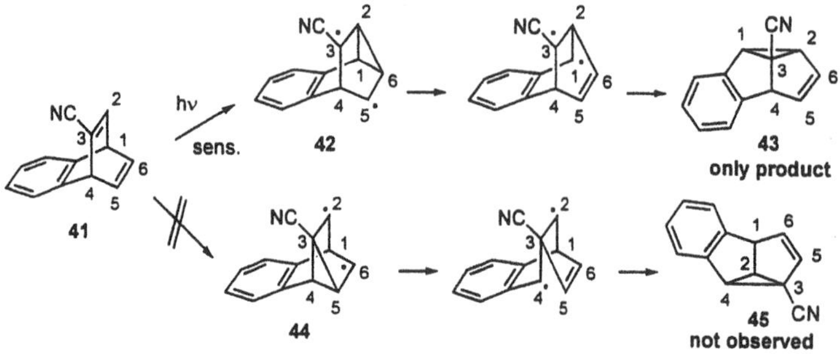

Scheme 15

group, the precursor of **43**. The alternative bridging in **41** between positions 3 and 5 does not occur because biradical **44** is less stable than **42** (Sch. 15). These general rules also apply to other bicyclic systems [38].

The 1-ADPM rearrangement of 1-aza-1,4-dienes affords the corresponding cyclopropylimines regioselectively. However, in order to predict the outcome of the reaction in pyrazinobarrelenes, quinoxalinobarrelenes, and other polycyclic systems, two different factors should be taken into account. One is the possible competition between the DPM and the 1-ADPM rearrangements, and the other is the influence of substitution on the stability of the reaction intermediates. However, in most of the cases studied, the 1-ADPM process takes place preferentially over the DPM reaction [3e]. An example of competition between these two reactions is shown in Sch. 4 for compound **5a** [10a].

6.4.2. The Excited State Involved in the Rearrangement

For practical proposes, it is important to know the nature of the excited state involved in any photoreaction. Some reactions occur in the singlet excited state while others take place in the triplet state manifold. The former are normally carried out by using direct irradiation and, therefore, the starting compound must absorb light in accessible UV-spectral regions. The latter reaction types usually require the use of a triplet sensitizer that absorbs the incident light and, after ISC, transfers its triplet energy to the diene reactant.

Di-π-methane rearrangements of acyclic and monocyclic 1,4-dienes usually occur in the singlet excited state and, therefore, the reaction takes place on direct irradiation. This is an important limitation since nonconjugated alkenes absorb below 250 nm. Consequently, the DPM reaction in acyclic and monocyclic dienes is restricted to compounds in

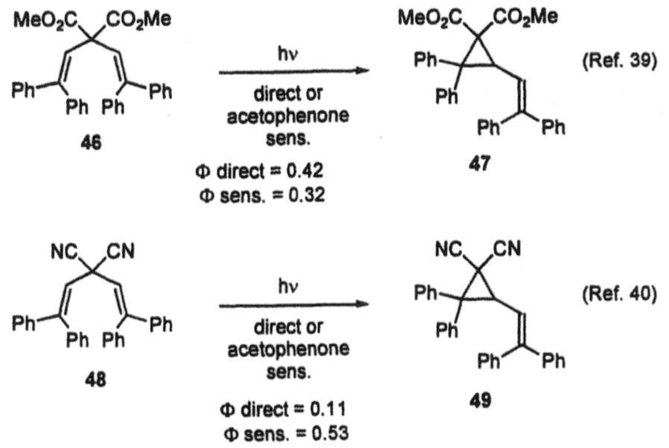

MeO$_2$C CO$_2$Me

Ph—Ph Ph—Ph
46

hv
direct or
acetophenone
sens.

Φ direct = 0.42
Φ sens. = 0.32

MeO$_2$C CO$_2$Me

Ph—
Ph

Ph Ph
47

(Ref. 39)

NC CN

Ph—Ph Ph—Ph
48

hv
direct or
acetophenone
sens.

Φ direct = 0.11
Φ sens. = 0.53

NC CN

Ph—
Ph

Ph Ph
49

(Ref. 40)

Scheme 16

which at least one of the double bonds is conjugated with functional groups that would shift the absorption of the alkene to longer wavelengths. However, acyclic 1,4-dienes with electron withdrawing groups (such as cyano and ester) at position 3 of the 1,4-diene system rearrange both on direct and triplet sensitized conditions. Thus, diene 46 affords cyclopropane 47 with comparable quantum yields in the triplet and singlet excited states [39], while the triplet-sensitized irradiation of diene 48 gives cyclopropane 49 five times more efficiently than on direct irradiation (Sch. 16) [40].

Dienes with electron withdrawing groups at C-1 also undergo the DPM rearrangement in the singlet and triplet excited states. In these instances the regioselectivity of the reaction depends on the excited state involved in the rearrangement [39,40]. Thus, for example, direct irradiation of diene 50 affords cyclopropane 47, in which the methoxycarbonyl groups appears in the cyclopropane ring, whereas in the triplet reaction the only product obtained is 51, with the electron withdrawing groups attached to the residual double bond (Sch. 17) [39]. In this instance, the quantum yield for the triplet reaction ($\Phi = 0.92$) is almost three times higher than for the direct run ($\Phi = 0.39$).

Bicyclic and polycyclic dienes usually rearrange in the triplet excited state, and therefore, the reaction must be carried out using a triplet sensitizer. The other versions of the rearrangement (ODPM, 1-ADPM, and 2-ADPM) occur in the triplet excited state, also. This is normally an advantage since nonconjugated double bonds can accept triplet energy from sensitizers with high triplet energies, such as acetone. Aromatic ketones, acetone, and polycyclic aromatic hydrocarbons are frequently used as triplet sensitizers. In considering which sensitizer should be used, one should take

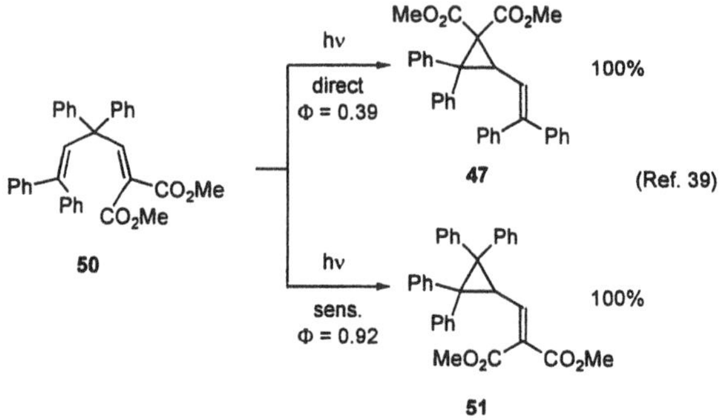

Scheme 17

into account the fact that the sensitizer should absorb light at longer wavelength than does the diene. Furthermore, the sensitizer's triplet energy should be higher than that of the alkene to ensure efficient energy transfer. Other factors that should be considered in selecting an appropriate sensitizer for a specific reaction are solubility, separation problems, and concentration. However, in most cases the choice is relatively simple since the absorption properties and triplet-energy values for a large number of compounds have been determined [41].

6.5. REPRESENTATIVE EXPERIMENTAL PROCEDURES

The following examples have been selected with the intention of illustrating the detailed experimental conditions used for direct and triplet-sensitized DPM rearrangements and for triplet-sensitized and SET-sensitized 1-ADPM reactions.

6.5.1. DPM Rearrangements

6.5.1.1.   6,6-Diphenyl-2-diphenylmethylenebicyclo[3.1.0]hex-3-ene (**53**) (Sch. 18) [42].

Direct irradiation of 1-diphenylmethylene-4,4-diphenylcyclohexa-2,5-diene (**52**): A solution of **52** (252 mg, 0.64 mmol) in t-BuOH (252 mL) was purged for 1 h with purified nitrogen and subsequently irradiated under a positive pressure of nitrogen for 1 h through a Pyrex filter using a Hanovia 450-W medium pressure Hg lamp. After completion of the irradiation, the solution was concentrated, leaving 261 mg of light yellow oil whose NMR showed quantitative conversion to **53**. The product was dissolved in Et_2O, filtered

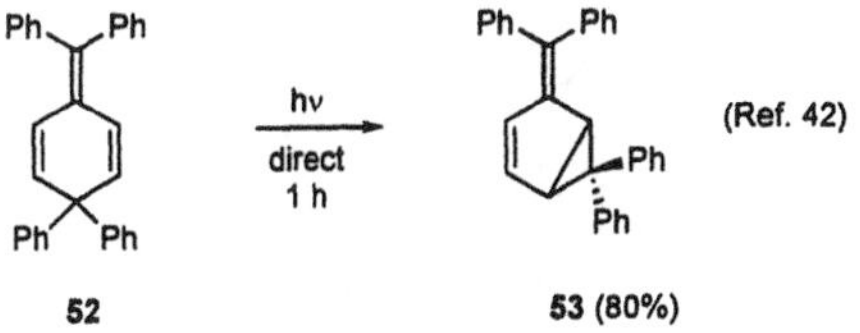

Scheme 18

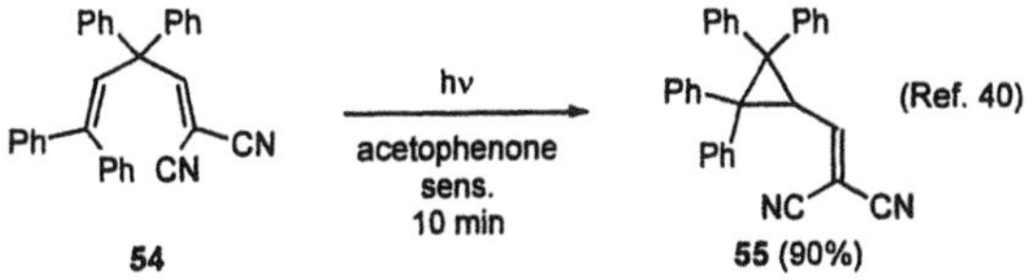

Scheme 19

through silica gel, and concentrated to give a colorless oil (237 mg, 93%). Crystallization from 95% EtOH gave **53** (202 mg, 80%).

6.5.1.2. 3-(2,2-Dicyanovinyl)-1,1,2,2-tetraphenylcyclopropane (55) (Sch. 19) [40]

Acetophenone-sensitized photolysis of 1,1-dicyano-3,3,5,5-tetraphenyl-penta-1,4-diene (**54**): A solution of **54** (500 mg, 1.18 mmol) and acetophenone (20.0 g, 167 mmol—this large amount is necessary in order to exclude direct excitation of the substrate) in benzene (525 mL) was purged for 1 h with purified nitrogen and subsequently irradiated under a positive pressure of nitrogen for 10 min through a Pyrex filter using a Hanovia 450-W medium pressure Hg lamp. After completion of the irradiation, the solvent and the sensitizer were removed under reduced pressure, affording 544 mg of an off-white fine crystalline solid. Recrystallization from EtOH gave **55** (448 mg, 90%), as colorless crystals.

6.5.2. 1-ADPM Rearrangements

6.5.2.1. *endo*-5-Methyl-3-phenylbicyclo[3.1.0]hexane-6-carbonitrile (58) (Sch. 20) [43]

Acetophenone-sensitized photolysis of 1-methyl-3-phenyl-2-cyclohexene-carbaldehyde oxime acetate (**56**): A solution of **56** (298 mg, 1.16 mmol)

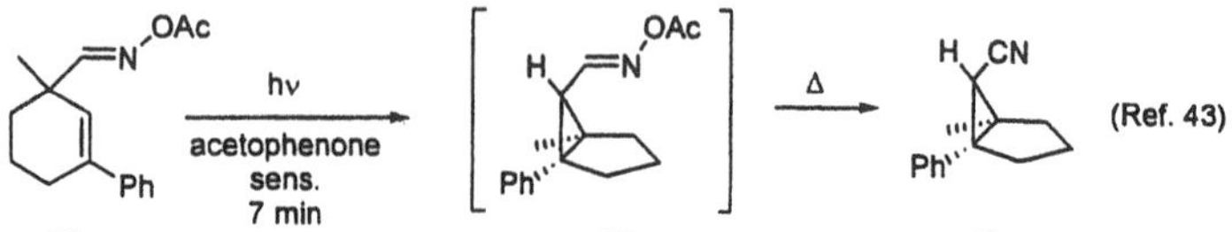

Scheme 20

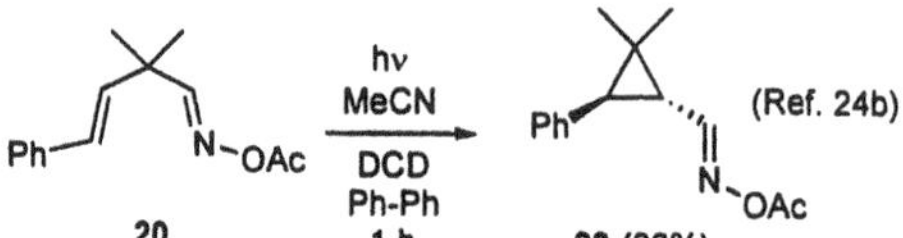

Scheme 21

and acetophenone (5.6 g, 46.6 mmol) in dry CH_2Cl_2 (450 mL) was purged for 1 h with argon and subsequently irradiated under a positive pressure of argon for 7 min through a Pyrex filter using a 400-W medium pressure Hg arc lamp. After completion of the irradiation, the solvent and the sensitizer were removed under reduced pressure and the products were separated by flash chromatography on silica gel using hexane/Et$_2$O (98:2) as eluent to afford starting oxime acetate **56** (12 mg, 4%) and the nitrile **58** (195 mg, 91%), as a colorless oil, resulting from elimination of acetic acid in the 1-ADPM product **57** during workup.

6.5.2.2. (E_{cyclo}, E_{C-N})-2,2-Dimethyl-3-phenylcyclopropanecarbaldehyde oxime acetate (**23**) (Sch. 21) [24b]

DCD-sensitized photolysis of (E_{C-N}, E_{C-C})-2,2-dimethyl-4-phenyl-3-butenal oxime acetate (**20**): A solution of **20** (263 mg, 1.14 mmol), biphenyl (20 mg, 0.13 mmol), and DCD (25 mg, 0.11 mmol) in dry acetonitrile (450 mL) was purged for 1 h with argon and subsequently irradiated under a positive pressure of argon for 1 h through a Pyrex filter using a 400-W medium pressure Hg arc lamp. After completion of the irradiation, the solvent was removed under reduced pressure and the sensitizer, the cosensitizer and the products were separated by flash chromatography on silica gel, using hexane/Et$_2$O (95:5) as eluent, to afford the starting oxime acetate **20** (87 mg, 33%) as a 3:1 mixture of (E_{C-N}, E_{C-C})/(E_{C-N}, Z_{C-C}) isomers and the (E_{cyclo}, E_{C-N})-cyclopropane **23** (131 mg, 60%) as an oil.

6.6. TARGET MOLECULES: NATURAL AND NONNATURAL PRODUCTS

Among all of the versions of the di-π-methane rearrangement, ODPM reactions have been widely used as key steps in the synthesis of natural products or highly strained molecules [8d,e]. However, in spite of the large number of studies carried out on the DPM reaction, there are only a few examples in which this rearrangement has been used in the synthesis of natural products. The first example, reported almost simultaneously by Baeckström [44] and Pattenden [45], involves the direct irradiation of dienoate **59** to afford methylchrysanthemate **60** in low yield. Interestingly, Baeckström also observed that triplet-sensitized irradiation of **59** afforded the other possible regioisomer **61**, showing for the first time that the regioselectivity of the rearrangement, in 1,4-dienes with electron withdrawing groups at C-1, depends on the excited state involved in the reaction (Sch. 22) [44].

The DPM rearrangement has provided a simple route for the construction of the carbon skeleton present in the sesquiterpenoid taylorione, isolated from *Mylia taylori* [46]. Thus, direct irradiation of the *E*-cyclopentenone **62** results in efficient, regiospecific DPM reaction affording the *trans*-cyclopropane **63** in 70% yield. Compound **63** was easily transformed into *trans*-desoxytaylorione **64** by using a Wittig reaction (Sch. 23).

In another study, erytrolide A (**66**), isolated from the Caribbean octocoral *Erythropodium caribaeorum*, is considered to originate in nature from terpenoid **65** by a DPM rearrangement, probably the first observation of the involvement of this photoreaction in the production of natural products [47]. In order to test this possibility, compound **65** was irradiated in benzene for 3 h, using a medium-pressure Hg arc lamp, affording **66** in 87%

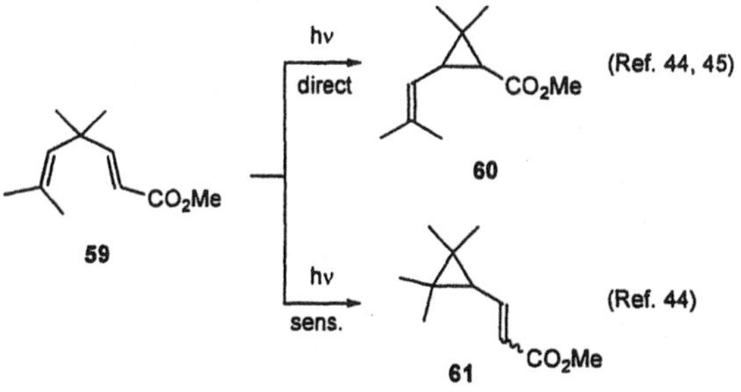

Scheme 22

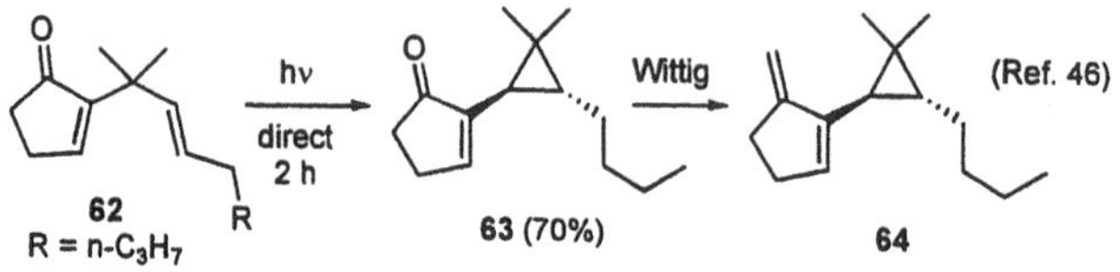

Scheme 23

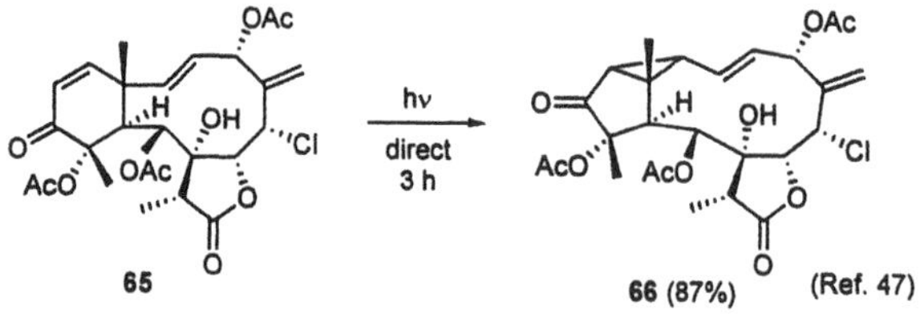

Scheme 24

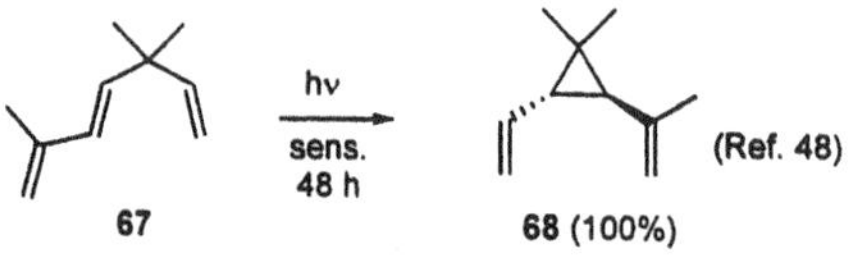

Scheme 25

yield. Irradiation of **65** in 5% methanolic seawater with sunlight for 8 days also afforded **66** in 37% yield (Sch. 24) [47].

Another example of a natural product that undergoes the DPM rearrangement is the artemisia triene **67**, which on direct irradiation results the cyclopropane **68** in quantitative yield (Sch. 25) [48].

Finally, the neoflavonoid latifolin, a major constituent of *Dalbergia latifolia*, and its dimethyl and diethyl derivatives, undergo aryl-di-π-methane rearrangements on both direct and triplet-sensitized irradiation [27d,e] Thus, Pyrex filtered irradiation of the dimethyl derivative **69** affords the *trans*-cyclopropane **70** in 90% yield (Sch. 26).

Sunlight was also effective in promoting this reaction which yielded **70** quantitatively after 5 or 6 day exposure times [27d]. Representative examples of cyclopropane derivatives that can be obtained by the DPM rearrangement, including isolated yields and references, are shown in Schs. 27 [18,49,50], 28 [51–53], 29 [53–56], 30 [57–59], and 31 [60–62].

The 1-ADPM rearrangement was considered for many years to be the main synthetic alternative overcoming the reported lack of ODPM reactivity

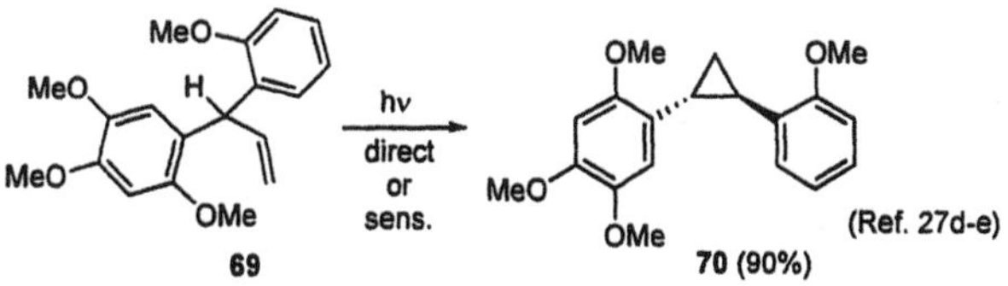

Scheme 26

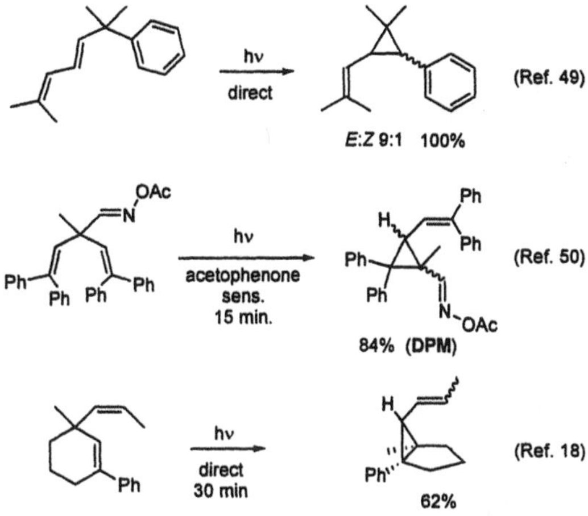

Scheme 27

of β,γ-unsaturated aldehydes [8e]. Thus, aldehydes, reported to be unreactive in the ODPM mode, could be transformed into the corresponding cyclopropyl aldehydes by a simple route consisting of (1) transformation into the corresponding imine; (2) triplet sensitized irradiation; and (3) hydrolysis of the cyclopropylimine to the corresponding aldehyde. However, recent studies have shown that, contrary to the common belief, β,γ-unsaturated aldehydes do undergo the ODPM reaction very efficiently in the triplet excited state manifold [9]. This observation reduces the synthetic importance of the 1-ADPM reaction. Nevertheless, in some instances alternative reactions, such as decarbonylation to the corresponding alkene, compete with the ODPM rearrangement [9c]. In these cases the 1-ADPM reaction of the corresponding oxime acetates could still be considered as an alternative for the synthesis of cyclopropanecarbaldehydes.

The 1-ADPM reaction has been used as the key step in the synthesis of a series of cyclopropanecarboxylic acids present in insecticidally-active pyrethrins and pyrethroids [63]. Thus, acetophenone-sensitized irradiation

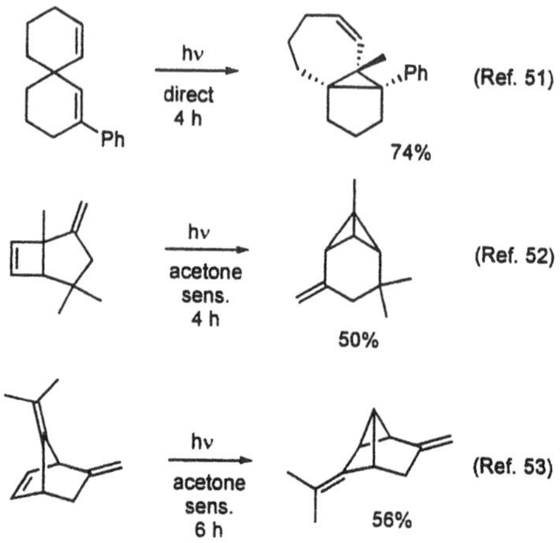

(Ref. 51)

74%

(Ref. 52)

50%

(Ref. 53)

56%

Scheme 28

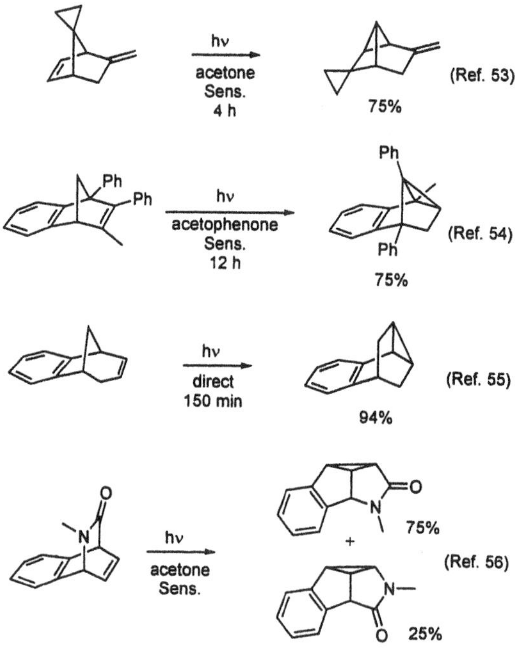

(Ref. 53)

75%

(Ref. 54)

75%

(Ref. 55)

94%

(Ref. 56)

75%

+

25%

Scheme 29

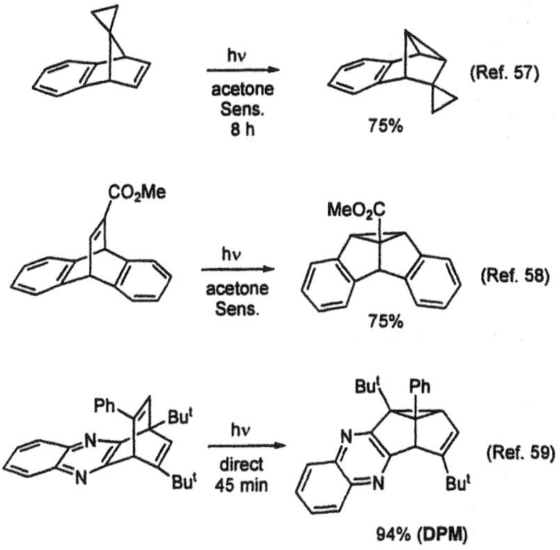

(Ref. 57)

75%

(Ref. 58)

75%

(Ref. 59)

94% (DPM)

Scheme 30

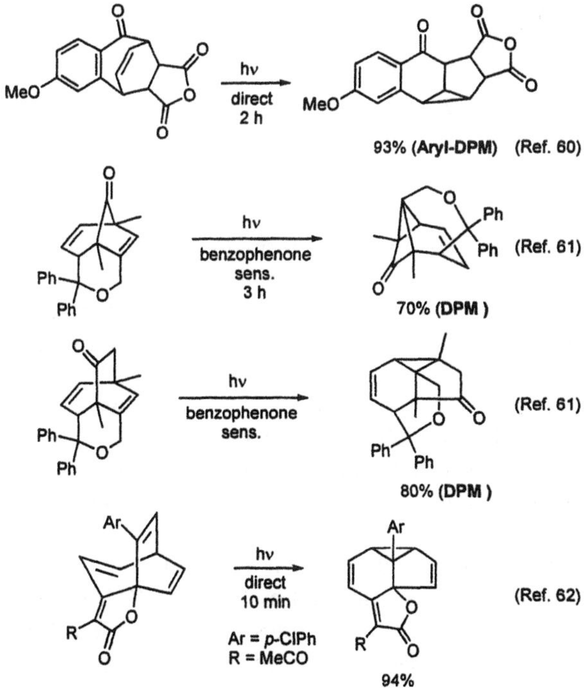

93% (Aryl-DPM) (Ref. 60)

(Ref. 61)

70% (DPM)

(Ref. 61)

80% (DPM)

(Ref. 62)

Ar = p-ClPh
R = MeCO

94%

Scheme 31

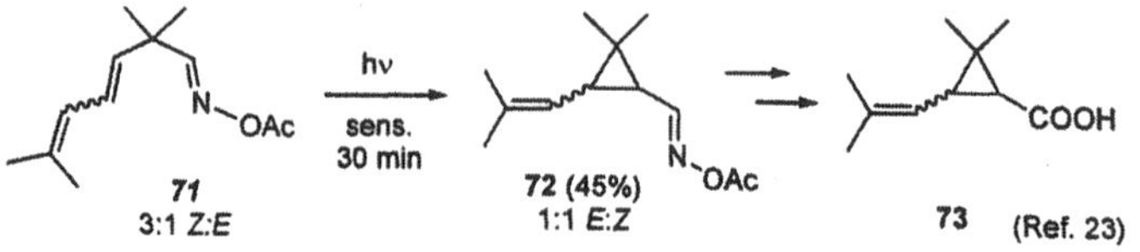

Scheme 32

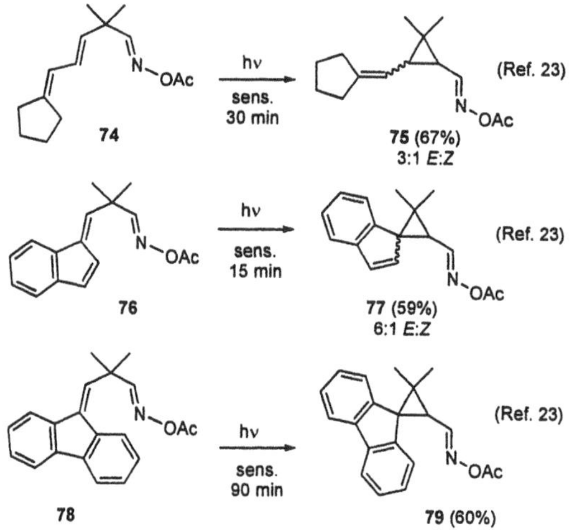

Scheme 33

of 1-azatriene **71**, for 30 min, brings about the formation of cyclopropane **72**, in 45% yield, as a 1:1 mixture of *cis:trans* diastereoisomers [23]. Compound **72** was converted into chrysanthemic acid **73** by a pathway involving transformation of the oxime acetate to the nitrile, followed by reduction to the aldehyde and oxidation to the acid (Sch. 32) [23].

As mentioned earlier the DPM rearrangement has been used in the synthesis of methyl chrysanthemate **60** [44,45]. However, the 1-ADPM reaction should be considered as a better alternative since the DPM reaction of **59** afforded compound **60** in 12–15% only, after 30 h of irradiation (Sch. 22).

Other cyclopropanecarboxylic acids, present in pyrethroids and of known insecticidal activity, are easily synthesized by the 1-ADPM reaction. Thus, triplet-sensitized irradiation of oxime acetates **74**, **76**, and **78** affords cyclopropanes **75**, **77**, and **79**, respectively, in good yields. These photoproducts can be transformed to the corresponding carboxylic acids by using conventional routes (Sch. 33) [23].

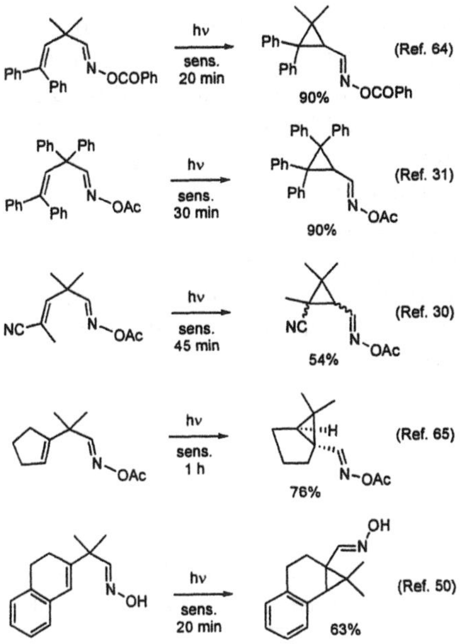

Scheme 34

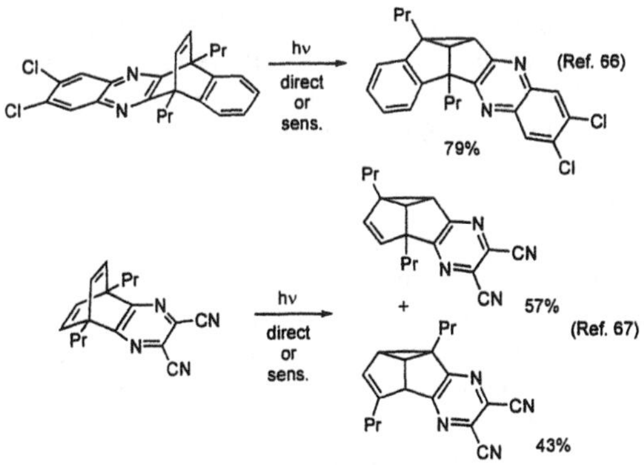

Scheme 35

Representative examples of cyclopropane derivatives that can be obtained by the 1-ADPM rearrangement, including isolated yields and references, are shown in Schs. 34 [30,31,50,64,65] and 35 [66,67].

In summary, di-π-methane rearrangements have proved to be very general for a large number of 1,4-unsaturated systems affording the corresponding cyclopropane derivatives in good isolated yields in most of the cases studied. These reactions have been used in the synthesis of molecules of high structural complexity that, very often, are difficult to obtain by alternative reaction routes. In addition, the rearrangements usually occur with high regio- and stereochemical control. The number of cyclopropane derivatives that can be obtained using the different versions of the rearrangement is remarkable making these reactions particularly useful from a synthetic point of view. The novel versions of the rearrangement, via radical-cation and radical-anion intermediates, have opened new lines of research in the area of di-π-methane photochemistry although their synthetic potential has not yet been established.

REFERENCES

1. (a) Patai S. ed. The Chemistry of Functional Groups, The Chemistry of Alkenes. New York: John Wiley & Sons, 1964; (b) Patai S. ed. The Chemistry of Double-Bonded Functional Groups, Supplement A3. New York: John Wiley & Sons, 1997.
2. For example see: (a) Calvert GJ, Pitts JN. Photochemistry. New York: John Wiley, 1966; (b) Turro NJ. Modern Molecular Photochemistry. Menlo Park, CA: Benjamin Cummings Publishing Company, 1978; (c) Gilbert A, Baggott J. Essential of Molecular Photochemistry. Oxford: Blackwell Scientific Publications, 1991; (d) Horspool WM, Armesto D. Organic Photochemistry. Chichester, England: Ellis Horwood and PTR Prentice Hall, 1992; (e) Horspool WM, Song P, eds. CRC Handbook of Organic Photochemistry and Photobiology. New York: CRC Press, 1995.
3. For reviews see: (a) Hixson SS, Mariano PS, Zimmerman HE. The di-π-methane and oxa-di-π-methane rearrangements. Chem Rev 1973; 73:531–551; (b) Zimmerman HE. The di-π-methane (Zimmerman) rearrangement. In: de Mayo P, ed. Rearrangements in Ground and Excited States. Vol. 3. New York: Academic Press, 1980:131–166; (c) De Lucchi O, Adam W. Di-π-methane photoisomerizations. In: Trost BM, Fleming I, Paquette LA, eds. Comprehensive Organic Synthesis. Vol. 5, Oxford: Pergamon Press, 1991:193–214; (d) Zimmerman HE. The di-π-methane rearrangement. In: Padwa A, ed. Organic Photochemistry. Vol. 11. New York: Marcel Dekker, Inc., 1991:1–36; (e) Zimmerman HE, Armesto D. Synthesis aspects of the di-π-methane rearrangement. Chem Rev 1996; 96:3065–3112.
4. Zimmerman HE, Grunewald GL. J Am Chem Soc 1966; 88:183–184.
5. Zimmerman HE, Binkley RW, Givens RS, Sherwin MA, J Am Chem Soc 1967; 89:3932–3933.
6. Zimmerman HE, Heydinger JA. J Org Chem 1991; 56:1747–1758.
7. Tenney LP, Boykin DW Jr, Lutz RE. J Am Chem Soc 1966; 88:1835–1836.

8. For reviews see: (a) Dauben GW, Lodder G, Ipaktschi J. Photochemistry of β,γ-unsaturated ketones. Top Curr Chem 1975; 54:73–114; (b) Houk KN. The photochemistry and spectroscopy of β,γ-unsaturated carbonyl compounds. Chem Rev 1976; 76:1–74; (c) Schuster DI. Photochemical rearrangements of β,γ-enones. In: de Mayo P, ed. Rearrangements in Ground and Excited States. Vol. 3. New York: Academic Press, 1980:167–279; (d) Demuth M. Oxa-di-π-methane photoisomerizations. In: Trost BM, Fleming I, Paquette LA, eds. Comprehensive Organic Synthesis. Vol. 5. Oxford: Pergamon Press, 1991:215–237; (e) Demuth M. Synthetic aspects of the oxadi-π-methane rearrangement. In: Padwa A, ed. Organic Photochemistry. Vol. 11. New York: Marcel Dekker, Inc., 1991:37–109.

9. (a) Armesto D, Ortiz MJ, Romano S. Tetrahedron Lett 1995; 36:965–968; (b) Armesto D, Ortiz MJ, Romano S, Agarrabeitia AR, Gallego MG, Ramos A. J Org Chem 1996; 61:1459–1466; (c) Armesto D, Ortiz MJ, Agarrabeitia AR, Aparicio-Lara S. Synthesis 2001; 1149–1158.

10. (a) Nitta M, Inoue O, Tada M. Chem Lett 1977; 1065–1068; (b) Nitta M, Kasahara I, Kobayashi T. Bull Chem Soc Jpn 1981; 54:1275–1276.

11. (a) Armesto D, Martin JF, Perez-Ossorio R, Horspool WM. Tetrahedron Lett 1982; 23:2149–2152; (b) Armesto D, Horspool WM, Martin J-AF, Perez-Ossorio R. J Chem Res 1986; (S):46–47; (M):631–648.

12. (a) Armesto D. The azadi-π-methane rearrangement. In: Horspool WM, Soon P-S, eds. CRC Handbook of Organic Photochemistry and Photobiology. New York: CRC Press, 1995:915–930; (b) Armesto D. Influence of nitrogen incorporation on the photoreactivity of some unsaturated systems. EPA Newsletter 1995; 53:6–21; (c) Armesto D, Ortiz MJ, Agarrabeitia AR. Novel photoreactions of azadienes and related compounds. Synthetic applications and mechanistic studies. In: Nalwa HS, ed. Advanced Functional Molecules and Polymers. Vol. 1. Amsterdam: Gordon and Breach, 2001:351–379; (d) Armesto D, Ortiz MJ, Agarrabeitia AR. Recent advances in di-π-methane photochemistry: A new look at the Classical Reaction. In: Ramamurthy V, Schanze KS, eds. Photochemistry of Organic Molecules in Isotropic and Anisotropic Media, Molecular and Supramolecular Photochemistry. Vol. 9. New York: Marcel Dekker, Inc., 2003:1–41.

13. (a) Armesto D, Caballero O, Amador U. J Am Chem Soc 1997; 119:12659–12660; (b) Armesto D, Caballero O, Ortiz MJ, Agarrabeitia AR, Martin-Fontecha M, Torres MR. J Org Chem 2003; 68:6661–6671.

14. (a) Park KP, Schuster GB. J Org Chem 1992; 57:2502–2504; (b) Eisch J, Shaffi B, Boleslawski MP. Pure & Appl Chem 1991; 63:365–368.

15. Vuper M, Barton TJ. J Chem Soc, Chem Commun 1982; 1211–1212.

16. (a) Zimmerman HE, Pratt AC. J Am Chem Soc 1970; 92:6267–6271; (b) Zimmerman HE, Baeckström P, Johnson T, Kurtz DW. J Am Chem Soc 1974; 96:1459–1465.

17. Zimmerman HE, Robbins JD, McKelvey RD, Samuel CJ, Sousa LR. J Am Chem Soc 1974; 96:4630–4643.

18. Mariano PS, Ko JK. J Am Chem Soc 1973; 95:8670–8678.

19. Zimmerman HE, Klun RT. Tetrahedron 1978; 43:1775–1803.
20. Reguero M, Bernardi F, Jones H, Olivucci M, Ragazos IN, Robb MA. J Am Chem Soc 1993; 115:2073–2074.
21. Lewis FD, Zuo X, Kalgutkar RS, Wagner-Brennan JM, Miranda MA, Font-Sanchis E, Perez-Prieto J. J Am Chem Soc 2001; 123:11883–11889.
22. Armesto D, Agarrabeitia AR, Horspool WM, Gallego MG. J Chem Soc, Chem Commun 1990; 934–936.
23. Armesto D, Gallego MG, Horspool WM, Agarrabeitia AR. Tetrahedron 1995; 51:9223–9240.
24. (a) Ortiz MJ, Agarrabeitia AR, Aparicio-Lara S, Armesto D. Tetrahedron Lett 1999; 40:1759–1762; (b) Armesto D, Ortiz MJ, Agarrabeitia AR, Aparicio-Lara S, Martin-Fontecha M, Liras M, Martinez-Alcazar MP. J Org Chem 2002; 67:9397–9405.
25. Armesto D, Ortiz MJ, Agarrabeitia AR, Martin-Fontecha M. J Am Chem Soc 2001; 123:9920–9921.
26. Zimmerman HE, Pincock JA. J Am Chem Soc 1973; 95:2957–2963.
27. (a) Hixson S. Tetrahedron Lett 1972; 13:1155–1158; (b) Hixson S. J Am Chem Soc 1972; 94:2507–2508; (c) Zimmerman HE, Steinmetz MG, Kreil CL. J Am Chem Soc 1978; 100:4146–4162; (d) Kumari D, Mukerjee SK. Tetrahedron Lett 1967; 8:4169–4172; (e) Walia S, Kulshrestha SK, Mukerjee SK. Tetrahedron 1986; 42:4817–4826.
28. Ishii A, Maeda K, Nakayama J, Hoshino M. Chem Lett 1993; 681–684.
29. Tanifuji N, Huang H, Shinagawa Y, Kobayashi K. Tetrahedron Lett 2003; 44:751–754.
30. Armesto D, Gallego MG, Horspool WM. Tetrahedron 1990; 46:6185–6192.
31. Armesto D, Horspool WM, Langa F, Ramos A. J Chem Soc, Perkin Trans 1 1991; 223–228.
32. Armesto D, Horspool WM, Gallego MG, Agarrabeitia AR. J Chem Soc, Perkin Trans 1 1992; 163–169.
33. Armesto D, Horspool WM, Langa F. J Chem Soc, Chem Commun 1987; 1874–1875.
34. Zimmerman HE, Gruenbaum WT. J Org Chem 1978; 43:1997–2005.
35. Zimmerman HE, Cotter BR. J Am Chem Soc 1974; 96:7445–7453.
36. Zimmerman HE, Virot-Villaume M-L. J Am Chem Soc 1973; 95:1274–1280.
37. Bender CO, Brooks DW, Cheng W, Dolman D, O'Shea SF, Shugarman S. Can J Chem 1979; 56:3027–3037.
38. (a) Paquette LA, Cottrell DM, Snow RA, Gifkins KB, Clardy J. J Am Chem Soc 1975; 97:3275–3276; (b) Paquette LA, Cottrell DM, Snow RA. J Am Chem Soc 1977; 99:3723–3733.
39. Zimmerman HE, Factor RE. Tetrahedron 1981; 37(suppl 1):125–141.
40. Zimmerman HE, Armesto D, Amezua MG, Gannett TP, Johnson RP, J Am Chem Soc 1979; 101:6367–6383.
41. Murov SL, Carmichael I, Hug GL. Handbook of Photochemistry. 2nd ed. New York: Marcel Dekker, Inc., 1993.
42. Zimmerman HE, Diehl DR. J Am Chem Soc 1979; 101:1841–1857.

43. Aparicio-Lara S, Tesina de Licenciatura. Universidad Complutense de Madrid, Madrid, Spain, 1995.
44. (a) Baeckström P. J Chem Soc, Chem Commun 1976; 476; (b) Baeckström P. Tetrahedron 1978; 34:3331–3335.
45. Bullivan MJ, Pattenden G. J Chem Soc, Perkin Trans 1 1976; 256–258.
46. Pattenden G, Whybrow D. J Chem Soc, Perkin Trans 1 1981; 1046–1051.
47. Look SA, Fenical W, Van Engen D, Clardy J. J Am Chem Soc 1984; 106:5026–5027.
48. Sasaki T, Eguchi S, Ohno M, Umemura T. Tetrahedron Lett 1970; 11:3895–3896.
49. Kristinsson H, Hammond GS. J Am Chem Soc 1967; 89:5968–5970.
50. Armesto D, Ortiz MJ, Ramos A, Horspool WM, Mayoral EP. J Org Chem 1994; 59:8115–8124.
51. Mariano PS, Steitle RB, Watson DG, Peters MJ, Bay E. J Am Chem Soc 1976; 98:5899–5906.
52. Ipaktschi J. Chem Ber 1972; 105:1996–2003.
53. Ipaktschi J. Chem Ber 1972; 105:1840–1853.
54. Padwa A, Rieker WF, Rosenthal RJ. J Am Chem Soc 1983; 105:4446–4456.
55. Hahn RC, Rothman LJ. J Am Chem Soc 1969; 91:2409–2410.
56. Paquette LA, Meisinger RH. Tetrahedron Lett 1970; 11:1479–1482.
57. Ipaktschi J. Tetrahedron Lett 1970; 11:3183–3184.
58. Ciganek E. J Am Chem Soc 1966; 88:2882–2883.
59. Nair V, Anilkumar G, Prabhakaran J, Maliakal D, Eigendorf GK, Williard PG. J Photochem Photobiol A: Chem 1997; 111:57–59.
60. Hassner A, Middlemiss D, Murray-Rust J, Murray-Rust P. Tetrahedron 1982; 38:2539–2546.
61. Nitta M, Sugiyama H. Bull Chem Soc Jpn 1982; 55:1127–1132.
62. Nair V, Nandakumar MV, Anilkumar G, Maliakal D, Vairamani M, Prabhakar S, Rath NP. J Chem Soc, Perkin Trans 1 2000; 3795–3798.
63. Naumann K. Synthetic Pyrethroid Insecticides. Vols. 4 and 5. Berlin: Springer-Verlag, 1990.
64. Armesto D, Horspool WM, Mancheño MJ, Ortiz MJ. J Chem Soc, Perkin Trans 1 1992; 2325–2329.
65. Armesto D, Ramos A. Tetrahedron 1993; 49:7159–7168.
66. Liao C-C, Yang P-H. Tetrahedron Lett 1992; 33:5521–5524.
67. Liao C-C, Hsieh H-P, Lin S-Y. J Chem Soc, Chem Commun 1990; 545–547.

7

Oxa-Di-π-Methane Rearrangements

V. Jayathirtha Rao
Indian Institute of Chemical Technology, Hyderabad, India

Axel G. Griesbeck
Institut für Organische Chemie, Universität zu Köln, Köln, Germany

7.1. HISTORICAL BACKGROUND

The photochemical rearrangement of β,γ-unsaturated carbonyl compounds involving 1,2-acyl migration coupled with cyclization leading to the formation of three membered rings is termed *oxa-di-π-methane* (ODPM) rearrangement. The photochemical oxa-di-π-methane rearrangement is analogous to the *di-π-methane* rearrangement (see the previous chapter). The first example of an ODPM rearrangement was reported [1] in 1966 for the β,γ-unsaturated ketone **A** and it was largely unnoticed (Sch. 1).

Later Swenton [2] suggested the first mechanistic pathway and Givens [3] subsequently described the mechanism for the photochemistry of the butyrolactone derivative **B** as analogous to the di-π-methane rearrangement. These developments were utilized by Dauben [4] and led to the description of the reaction as *oxa-di-π-methane* (ODPM) rearrangement. Since then numerous investigators were involved in this subject and a large number of publications appeared. It is essential at this point to mention that the photochemistry of β,γ-unsaturated ketones is dominated by ODPM rearrangement and acyl migration routes, although there are several other pathways like ketene formation, elimination of carbon monoxide, [2+2]-cycloaddition, or cis-trans isomerization available. The preference for ODPM rearrangement and acyl migration in these β,γ-unsaturated ketone photochemistry is directly reflected in the higher chemical yields and higher photochemical reaction quantum yields. This observation promoted many

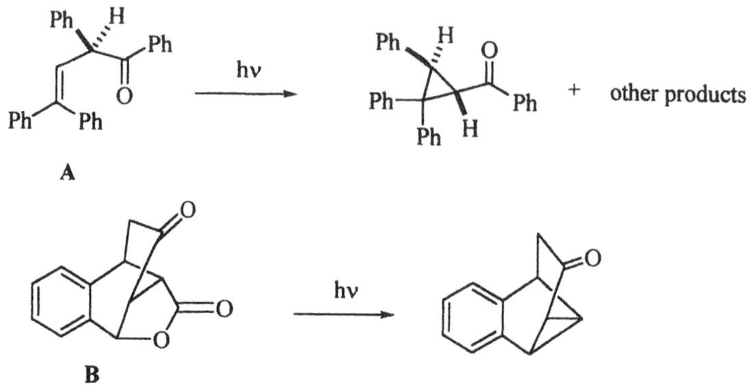

A

B

Scheme 1

workers to apply ODPM rearrangement reaction for the synthesis of a variety of complex natural products and intermediates, which are otherwise difficult to obtain by thermal routes. Specifically, ODPM rearrangement originates from the triplet manifold and this turned out to be advantageous for the high synthetic potential of the reaction. Triplet sensitization can be carried out on various substrates inducing selective ODPM rearrangements and leaving other functionalities of the substrate untouched. The large number of studies on this ODPM rearrangement has been reviewed several times [5–12] during the last three decades. In this chapter we have provided a brief introduction, state-of-the-art mechanism and mainly highlighted the synthetic potential of this rearrangement associated with β,γ-unsaturated ketones. Scope, limitations, and especially the stereochemical aspects of the ODPM rearrangement are discussed where appropriate. More details can be obtained by consulting the detailed reviews and the particular references cited therein.

7.2. STATE OF THE ART OF MECHANISTIC MODELS

Nature of the reactive excited state: The excited state involved in the ODPM rearrangement reactions is found to be a triplet excited state in most cases. In initial studies carried out on a series of model compounds (Sch. 2) revealed [13–15] that ODPM rearrangement can be achieved by direct excitation and also by triplet sensitization and this led to the proposal that the ODPM rearrangement can originate from either a singlet or a triplet excited state.

Detailed studies on a set of more elaborated model compounds (Sch. 3) [14,16,17] indicated, that indeed the ODPM rearrangement originates from

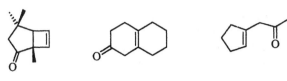

Scheme 2

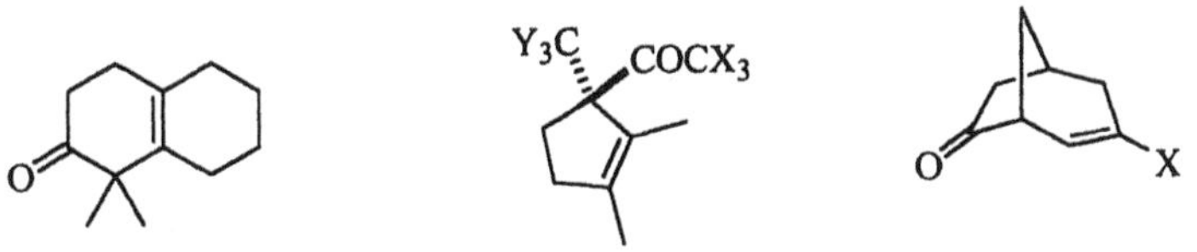

Scheme 3

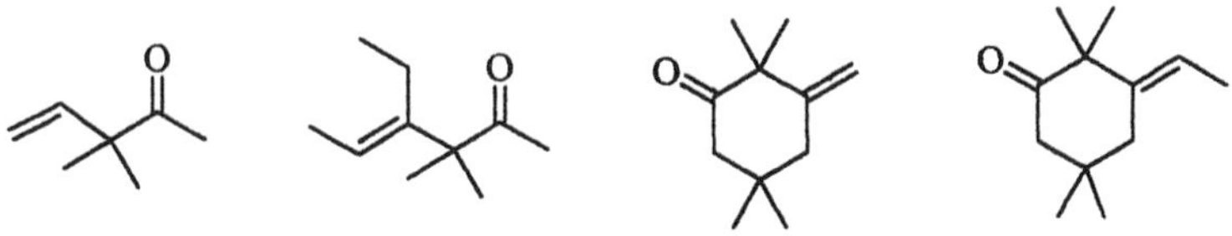

Scheme 4

the triplet excited state only. Theoretical studies suggested [18] that a $^3\pi\pi^*$ excited state is involved in these ODPM rearrangement reactions and this $^3\pi\pi^*$ excited state is well below the energy level of the corresponding $^3n\pi^*$ excited state.

Remarkably, one study [19] reported that ODPM rearrangement is also possible from the singlet excited state, but this is a matter of debate. For the triplet excited state of these β,γ-unsaturated ketones, it was clearly pointed out that it is $\pi\pi^*$ in origin based on low temperature phosphorescence measurements, theoretical and mechanistic studies [20–23]. Furtheron, these phosphorescene and mechanistic studies indicated that the $^3\pi\pi^*$ excited state is approx. 8–12 kcal/mol lower to the corresponding $^3n\pi^*$ excited state [25].

The acyclic and cycloalkenyl-derived β,γ-unsaturated ketones shown in Sch. 4 did not exhibit the ODPM rearrangement [24–26] neither upon direct excitation nor by triplet sensitization (Sch. 4).

A plausbile explanation reason is that these ketones have relatively less strained double bonds and undergo and efficient $E \rightarrow Z$ isomerization, thereby dissipating the excited state energy, termed as the "free rotor" effect. In another report [27] it was indicated that the 1,2-acyl shift can

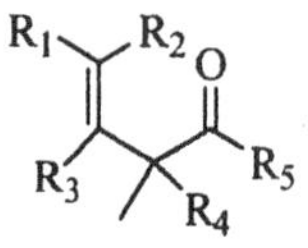

$R_1 = Ph, R_2 = H, R_3 = H, R_4 = Me, R_5 = i\text{-}Pr$
$R_1 = Ph, R_2 = H, R_3 = H, R_4 = Et, R_5 = Me$
$R_1 = H, R_2 = H, R_3 = Ph, R_4 = Me, R_5 = Me$
$R_1 = Ph, R_2 = Ph, R_3 = H, R_4 = Me, R_5 = Me$

Scheme 5

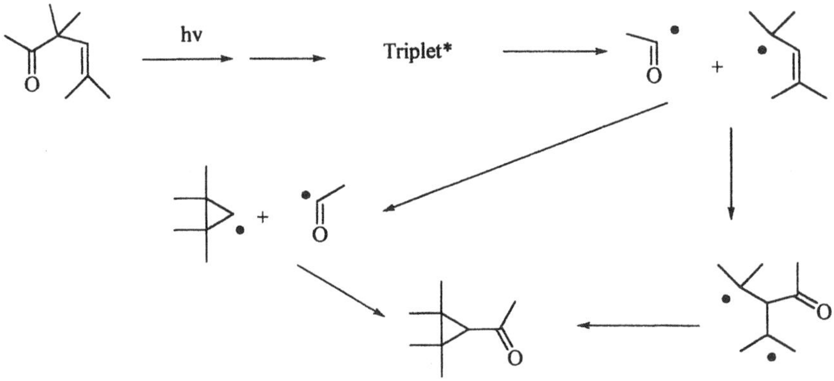

Scheme 6

compete with the "free rotor" effect to induce effectively the ODPM rearrangement in certain substrates (Sch. 5). It is found that the 1,3-acyl shift occurs in these β,γ-unsaturated ketones from the singlet (S_1) and the higher excited triplet (T_2) excited states of $n\pi^*$ origin [28–31].

Considering the data given above, indicates that the selection of a suitable sensitizer with appropriate triplet energy is one of the key factors in successfully carrying out photochemical ODPM rearrangements.

There are at least three types of mechanisms discussed in order to explain the ODPM rearrangement. The first mechanism (Sch. 6) is radicaloid in nature: in involves a Norrish type I cleavage leading to the formation of acyl and allyl radical and recombination of these radicaloid species to the ODPM rearrangement product in two ways.

The second mechanism is a concerted (Sch. 7) rearrangement involving symmetry-allowed processes. The third mechanism is a mixed mechanism, involving one concerted process leading to radicals and followed by radical combination to give the ODPM rearrangement products.

These different mechanistic routes were applied in order to explain selectivity phenomena within the ODPM rearrangement. The results shown in Sch. 9 support the assumption [32–34] of radical-type intermediates in the ODPM reaction, i.e., the stereorandom formation of products.

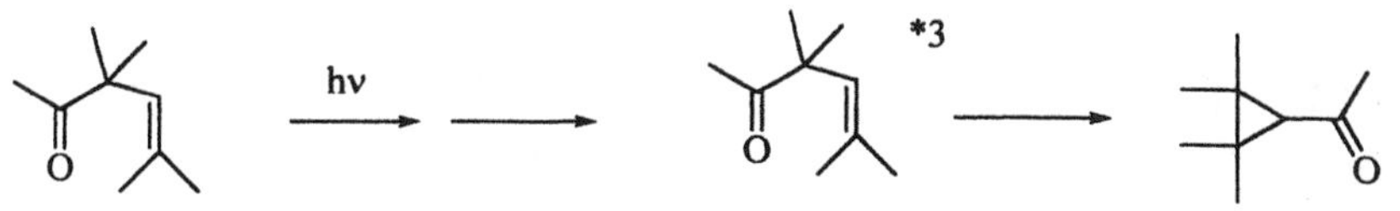
hν
*3

Scheme 7

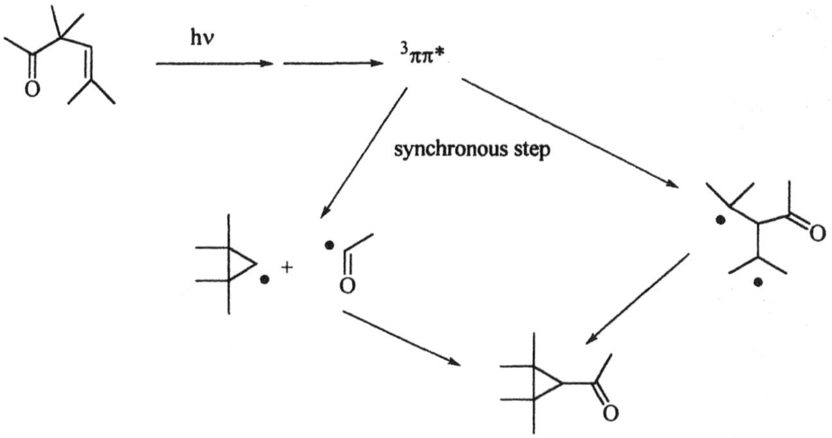
hν
³ππ*
synchronous step

Scheme 8

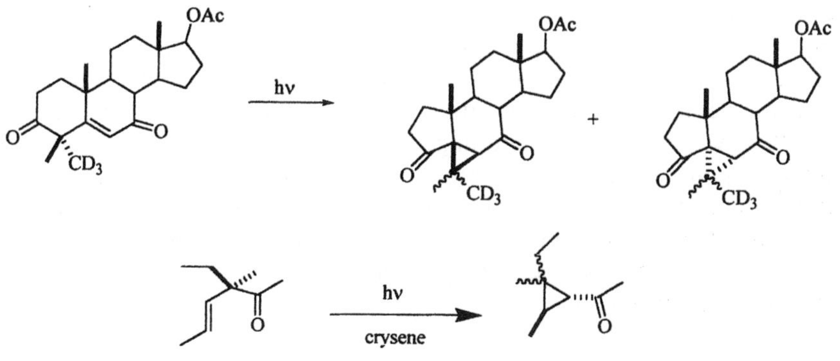
OAc
hν
CD₃
hν
crysene

Scheme 9

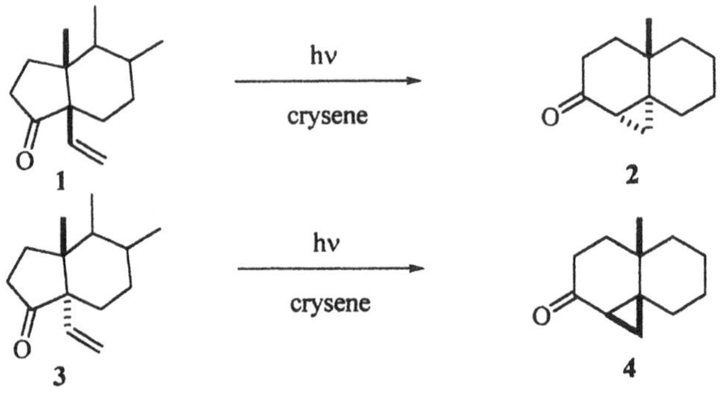

Scheme 10

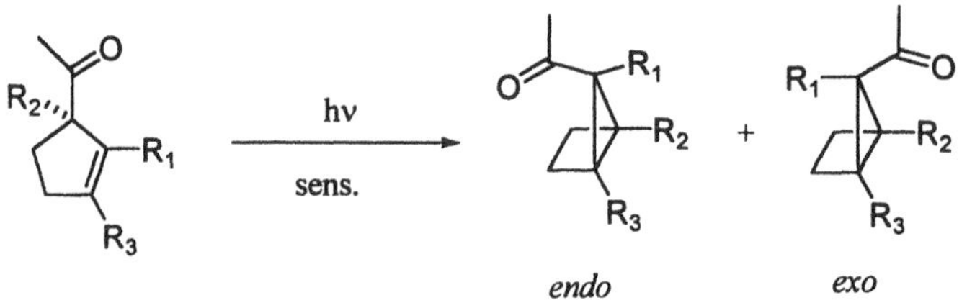

Scheme 11

Scheme 10 suggested a concerted reaction mechanism due to the high stereospecificity fo the respective transformations of the bicyclic substrates **1** and **3**, resulting in the corresponding diastereoisomeric cyclopropanes **2** and **4**, respectively [35]. Support for the mixed mechanism came for the example shown [16,23,36] in Sch. 11, were mixtures of diastereoisomers were formed. These mechanisms are still a matter of debate and create enough of curiosity to continue creative and innovative research in order to understand the ODPM rearrangement. Interesting questions are, why β,γ-unsaturated ketones upon excitation largely prefer the ODPM rearrangement pathway, even though several other pathways are available which in general dominate the photochemistry of ketones, also how the excitation energy is localized with in the molecule and utilized to cleave (with high mode selectivity) a particular chemical bond.

7.3. SCOPE AND LIMITATIONS

Cyclohexadienones were described to undergo ODPM rearrangement upon direct excitation [37] in highly polar solvents like trifluoroethanol due to a

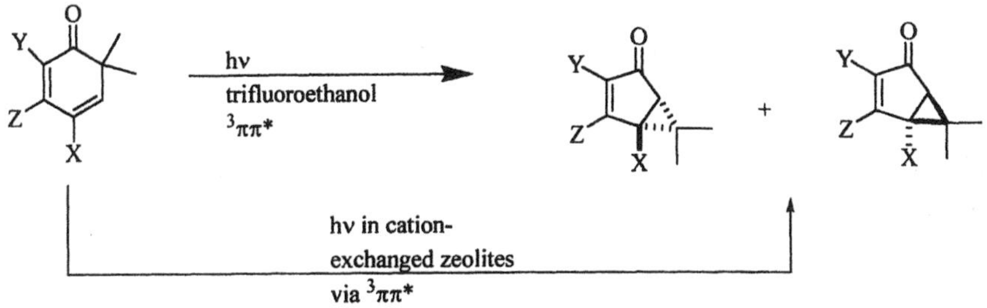

Scheme 12

change in order of the excited states leading to a ππ states as the lowest excited singlet states (Sch. 12) which, after intersystem crossing (ISC), results in the formation of the ^{3}ππ state.

   When the ODPM rearrangement of 2,4-cyclohexadienones is performed in constrained media [38], not only is the diastereoselectivity of the cyclopropane formation influenced but also the asymmetric induction exhibited by an chiral auxiliary covalently connected to the chromophore. The photochemistry of 2,4-cyclohexadienones is known to be strongly solvent-dependent with the α-cleavage preferred in unpolar solvents (presumably via the nπ*-state) and ODPM preferred in unpolar solvents (presumably via the ππ*-state). State switching can also be induced in alkali exchanged Y zeolites [39]. The diastereoselectivity of the ODPM rearrangement of **5** leading to the diastereoisomers **6** and **7**, respectively, strongly dependent on the alkali metal used in the MY-zeolite host. Also the aspect of enantioselectivity has been studied in the presence of chiral dopants like ephedrine (Sch. 13).

   Cyclopentenylmethylketones undergo ODPM rearrangement and as well as 1,3-acyl shift (Sch. 14) reactions in solution phase [40]. Schaffner et al. carried out the gas phase photolysis of cyclopentenylmethyl ketones (**8**) in the presence of scavengers [41]. Their studies enlightened the mixed type of mechanism responsible for ODPM rearrangement (to give **9**) and also the 1,3-acyl shift reaction (to give **10**). These studies indicate that one can conduct the ODPM rearrangement even in gas phase and this enhances the general scope of the reaction.

   van der Anton and Cerfontain [42] prepared several acyclic β,γ-unsaturated ketones for the purpose of studying ODPM rearrangement. All these compounds (Sch. 15) underwent efficient $E \rightarrow Z$ isomerization (free rotor effect), but in competition to that the compounds did also undergo a slow ODPM rearrangement.

   The results were explained by making use of substituent effect and the mechanism proposed is in favour of radicaloid type. In another study,

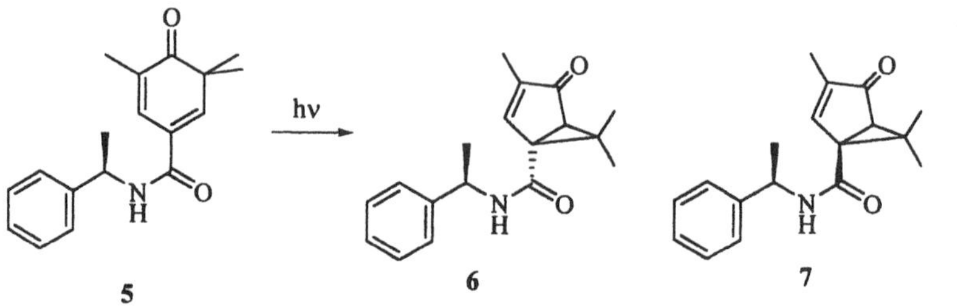

solution: d.e. = 5% (in trifluoroethanol)
LiY zeolite: d.e. = 17%
NaY zeolite: d.e. = 59%
KY zeolite: d.e. = 53%
RbY zeolite: d.e. = 39%
CsY zeolite: d.e. = 9%

Scheme 13

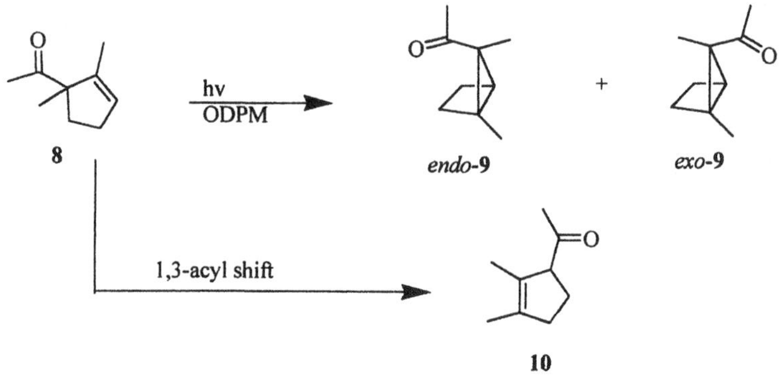

Scheme 14

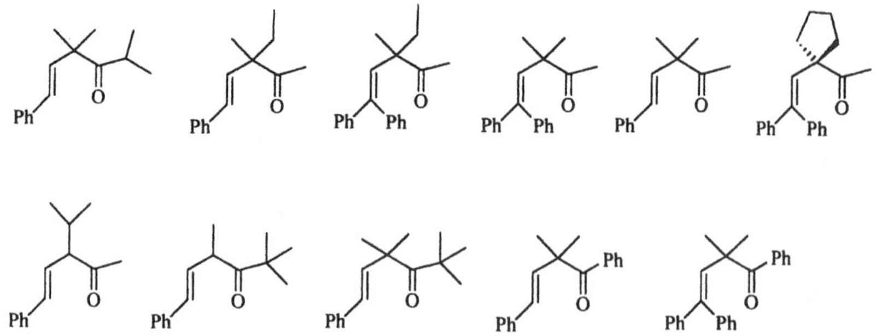

Scheme 15

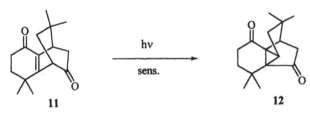

Scheme 16

Scheme 17

Koppes and Cerfontain [43] synthesized several substrates including (Sch. 16) β,γ- and γ,δ-dienones and studied their photochemistry.

Selective sensitization gave ODPM rearrangement products, whereas direct photolysis resulted in efficient $E \rightarrow Z$ isomerization and inefficient 1,3-acyl shift processes. The results indicated that γ,δ-double bonds or benzoyl groups present in these systems did not act as intramolecular triplet quencher. The new argument placed by the authors was the charge transfer interaction within the molecule enhances the internal conversion process there by a reduction in the ODPM rearrangement reactivity.

Margaretha et al. [44] described a nice example of ODPM rearrangement with the starting material **11** containing two carbonyl groups, where the double bond is situated α,β to one of the carbonyl groups and at the same time it becomes β,γ to another carbonyl group. Nevertheless, the product **12** (Sch. 17) was derived from ODPM rearrangement with high quantum yield ($\Phi = 0.67$) and high selectivity under a variety of photolytic conditions.

Highly strained substrates can be transformed into even more strained isomers by ODPM rearrangement. This has been shown by Murata et al. for the synthesis of a valence isomer of azulene [45]. Albeit the photochemical reaction yielded only 20–25% of the bicyclo[1.1.0]butane derivative **15**, the synthesis of the precursor cyclobutene (**14**) is straight-forward from the bicyclo[3.3.0]octenone **13** (Sch. 18). This substrate has obviously a diverse reactivity pattern when directly excited, however, triplet sensitization reduces these competitive pathways because alkene excitation is excluded. Also benzo-annulated azulene valence isomers were generated by this approach [46].

The photolyses of conformationally fixed β,γ-unsaturated ketones can also be performed in constrained media. This was described for the ODPM rearrangement of bicyclo[2.2.1]heptenone and bicyclo[2.2.2]octenone in MY zeolites [47]. It is remarkable that direct irradiation of the guest-incorporated thallium-Y (TlY) zeolites at 254 nm resulted in higher yields of ODPM products than the triplet-sensitized solution photolyses (Sch. 19). This indicates that the triplet carbonyl species are generated more efficiently in the supercages of the zeolites presumably because of the presence of heavy

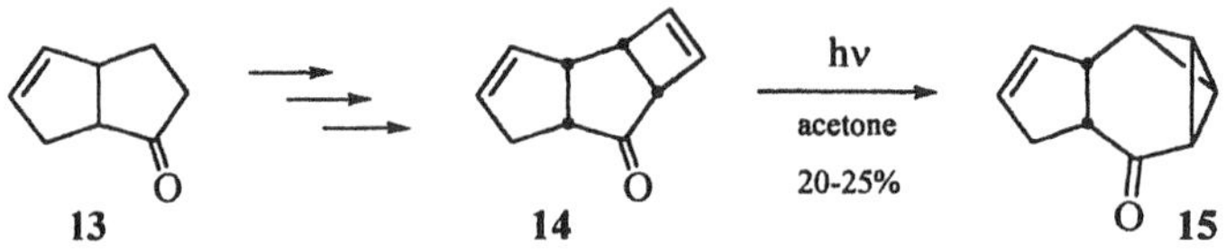

Scheme 18

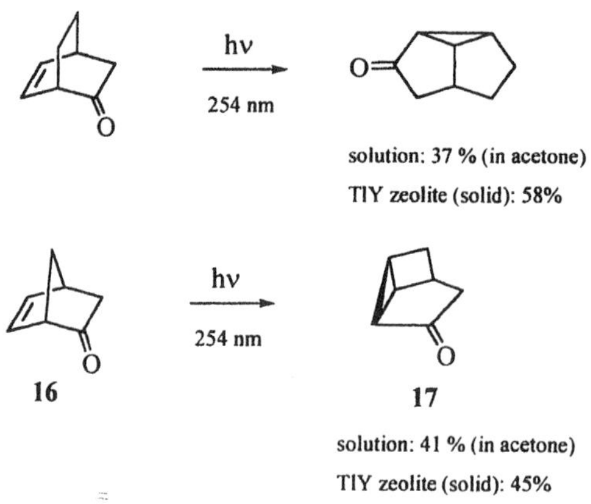

solution: 37 % (in acetone)
TlY zeolite (solid): 58%

solution: 41 % (in acetone)
TlY zeolite (solid): 45%

Scheme 19

atoms. The ODPM rearrangement of bicyclo[2.2.1] heptenone **16** leads to the highly strained housane derivative **17**.

Under normal circumstances, direct excitation of the carbonyl chromphore in bicyclo[2.2.2]octenones induces cleavage of the α-carbonyl C–C bond (Norrish I process) and the biradical generated can undergo a variety of subsequent reactions. This dichotomy of photochemical reactions depending on the spin state of the excited substrate is nicely corroborated in a report by Liao et al. on the photochemistry of 3,3-dialkoxybicyclo[2.2.2] oct-5-en-2-ones **18** [48]. The ether-bridged substrates, when irradiated in acetone solutions, results in the formation of a highly functionalized tetracyclic products **19**, whereas direct excitation (using the same wavelength as for sensitization—the sensitizer serving therefore also as an internal filter) in benzene solution resulted in decarbonylation and formation of functionalized cyclohexenes **20** (Sch. 20).

If rapid bond formation at the stage of the Norrish I fragments competes with decarbonylation, 1,2-acyl shift products are formed as the major components from direct photolysis. In case of the cyclobuteno-annulated bicyclo[2.2.2]octenone **21**, 1,2-acyl shift results in the cyclo-butanone **22**, whereas ODPM rearrangement (photolysis in acetone) results in the cyclobuteno-diquinane **23** in 44% yield (Sch. 21) [49].

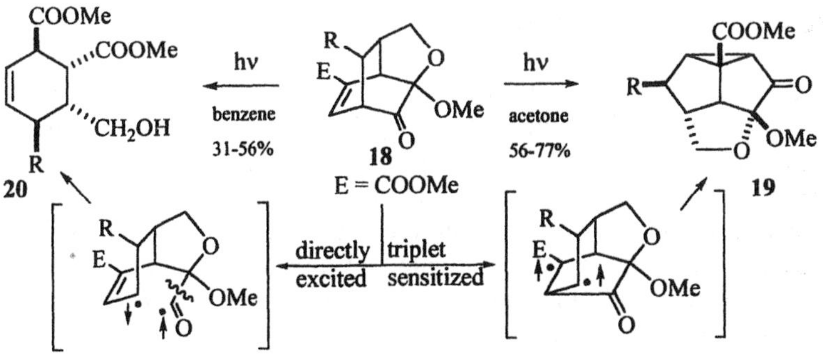

Scheme 20

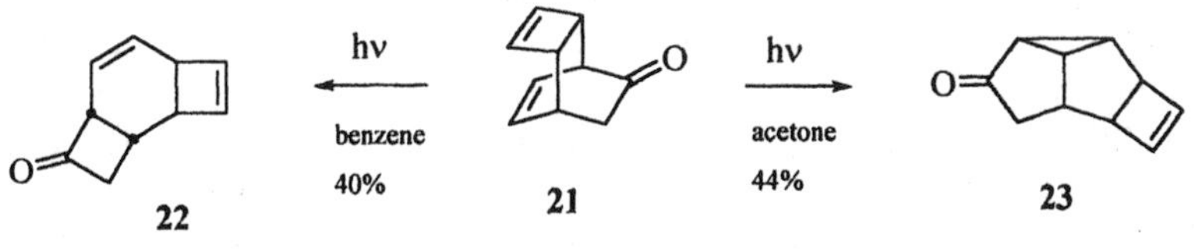

Scheme 21

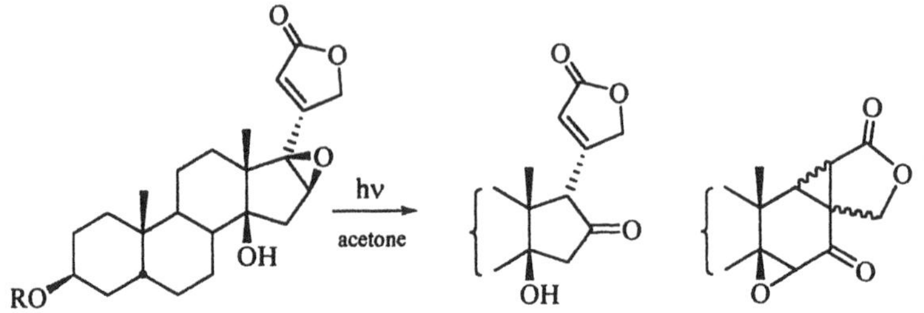

Scheme 22

In case of efficient (i.e., proceeding with high quantum yields) and rapid ODPM reactions, the products from photochemical transformations can sometimes not be isolated because of further photochemical transformations. When the epoxyenone **24** is photolysed in acetone in order to generate an oxirane to carbonyl transformation, an appreciable amount of side product **26** was isolated originating from a photochemical rearrangement of the desired reaction product **25** (Sch. 22) [50].

7.4. SYNTHETIC POTENTIAL: REACTIVITY AND SELECTIVITY PATTERN

A general approach to the synthesis of linear polyquinanes with the desired *cis-trans-cis*-ring junction of the central tricyclopentanoid structure from the corresponding annulated bicyclo[2.2.2]octenones was reported by Singh et al. [51]. An extensive set of compounds were reacted in a primary thermal Diels-Alder reaction between cyclohexadienones and cyclic as well as bicyclic dienophiles such as spiroannulated cyclopentadienes, pentafulvene, indene, cycloalkenes and -alkadienes and norbornadiene. The cycloadducts **27** were rearranged by acetone-sensitization following the di-π-methane route. Neither the intramolecular [2+2]-photocycloaddition (a prominent side-reaction in bridged and conformationally rigid dienes) nor the 1,3-acyl shift competed with the triquinane (**28**) formation. The dimethylated starting material **29** resulted in the triquinane **30** in 55% yield. A set of structurally diverse products which originate from the corresponding annulated bicyclo[2.2.2]ocetenones are also shown in Sch. 23. The protoncatalyzed ring-opening of the cyclopropane unit has been studied as a further route to accomplish the synthesis of linear polyquinanes.

The propellane-type terpene *modhephene* is a further example of a structurally highly complex target molecule which can be traced back

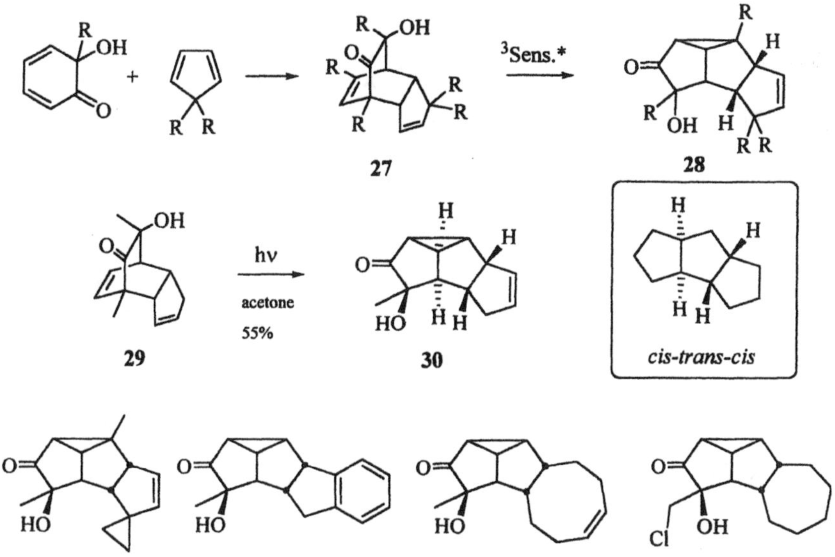

Scheme 23

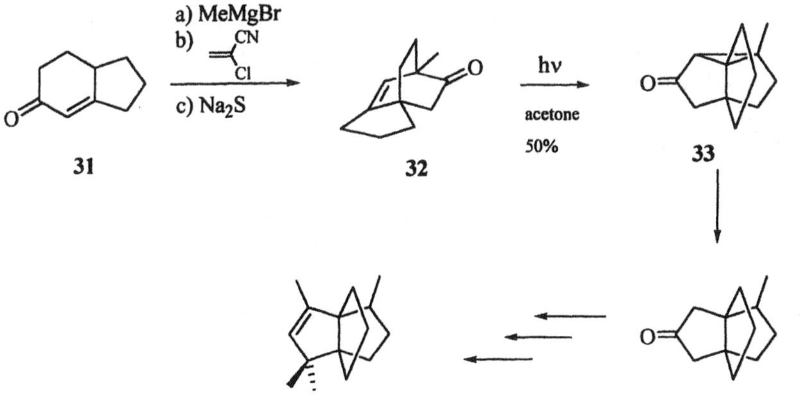

Scheme 24

retrosynthetically to a rather simple Diels-Alder cyclo-adduct. The central bridging five-membered ring must be bridgehead-connected in the substrate prior to the di-π-methane rearrangement [52]. The required bicyclo[2.2.2] octenone **32** is generated from the bicyclooctenone **31** by Grignard addition of methyl magnesium bromide and subsequent addition of a masked ketene (Sch. 24). Triplet-sensitized photolysis of **32** led to the formation of the

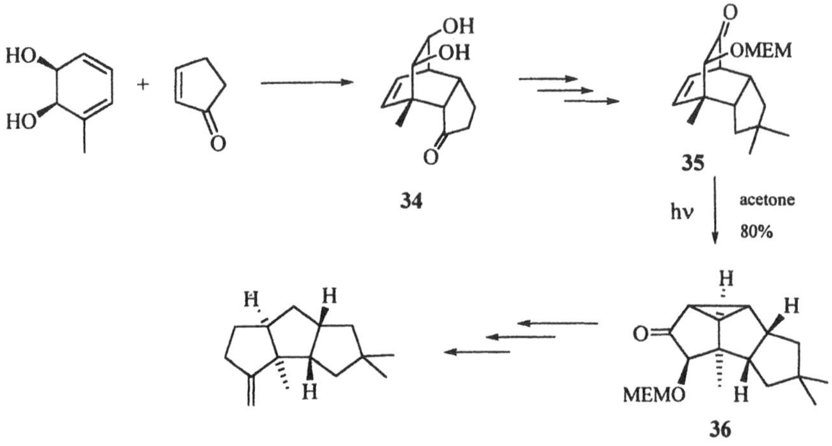

Scheme 25

propellane system **33** in moderate yield. The total synthesis of the natural product *modhephene* was accomplished in five further steps.

A chemoenzymatic route to the enantiomerically pure linear triquinane (−)-*hirsutene* can be accomplished starting from toluene via an ODPM rearrangement as the photochemical key step [53]. Microbial dihydroxylation of toluene provided the enantiomerically pure 1,2-dihydrocatechol which was reacted with cyclopentenone to the *endo*-product **34** in a high-pressure [4+2]-cycloaddition. Futher modifications gave the substrate **35** for the ODPM rearrangement which was performed by triplet sensitization and led to the triquinane **36** in 80% yield (at 71% conversion). The total synthesis of the (−)-enantiomer of *hirsutene* was completed by reductive cyclopropane ring-opening, carbonyl transposition, and methylenation (Sch. 25).

Also angularly annulated triquinanes can be obtained by the ODPM route. This was demonstrated for the terpene *isocomene* by Uyehara et al. [54]. In contrast to the approaches described above, annulation of the third five-membered ring is disconnected from the ODPM rearrangement and follows a classical enolate alkylation procedure (Sch. 26). The bridgehead-alkylated bicyclo[2.2.2]octenone **37** is covered into the tricyclic ketone **38** which, using a traditional α-alkylation procedure and subsequent funtionalization/defuntionalization (cyclopropane ring-cleavage) steps led to the angularly annulated triquinane **39**. Thus, the correct alignment of the activated C_3-chain and the degree of substitution at the α-carbonyl atom determines the type of annulation, i.e., from the gemdimethyl substrate **40**, the [3.3.3]-propellane ring system (as existing in *modhephene*, *vide supra*) becomes available.

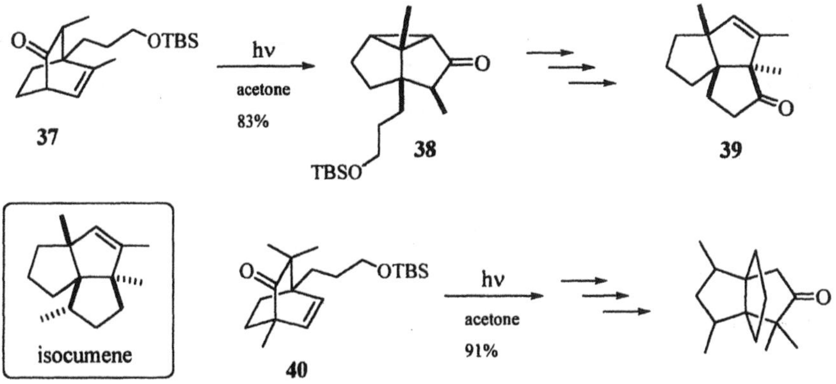

Scheme 26

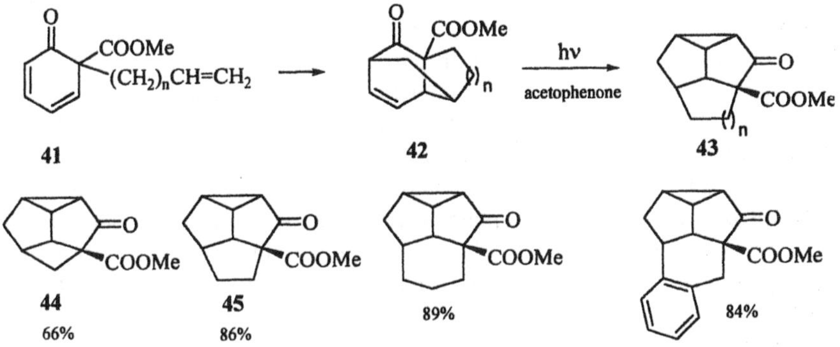

Scheme 27

Cyclohexadienones are not only ODPM-active substrates, but can also easily converted into more reactive tricyclic β,γ-unsaturated ketones by intramolecular thermal [4+2]-cycloaddition (Sch. 27). As versatile starting materials for this methodology, 6-alkenyl-2-4-cyclohexadien-1-ones 41 were converted into the annulated tricyclic β,γ-unsaturated ketones 42 and subsequently irradiated in the presence of acetophenone as triplet sensitizer to yield the tetracyclic products 43 [55]. Even the highly strained tetracyclic ketone 44, combining a diquinane with annulated cyclopropane and cyclobutane ring systems was isolated with 66% after long irradiation times (48 h). The *perhydrotriquinacene* skeleton (as in 45) is available in excellent yield as well as ring-enlarged or benzoannulated derivatives.

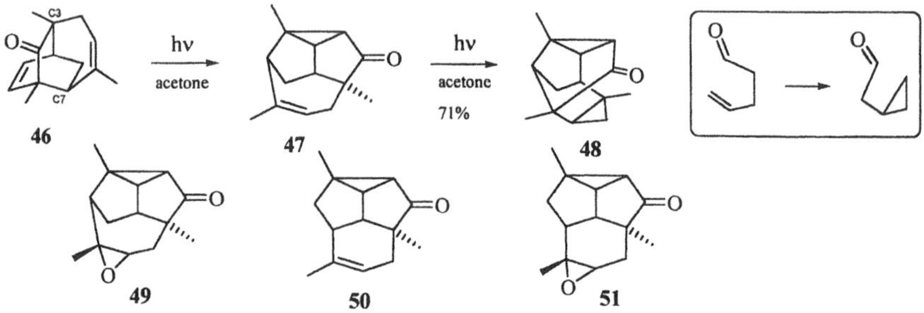

Scheme 28

Bridging bicyclo[2.2.2]octenones by unsaturated hydrocarbon chains at positions C3 and C7 or C8, respectively, leads to ODPM substrate which, upon triplet sensitization, result in bridged diquinanes [56]. Depending on the relative position of the hydrocarbon linker, more or less strained products are formed which can undergo further photochemical conversion (Sch. 28). An interesting light-induced reaction was documented with the intermediate **47** (by ODPM from the C_3-bridged bicyclo[2.2.2]octenone **46**) leading to an 1,4-acyl shift with formation of a half-cage structure **48**, i.e., a type of oxa-di-π-ethane rearrangement. No such reaction was observed with the corresponding epoxide **49**. Photochemical stability was also observed for the 1,3-linked diquinane **50**, resulting from the triplet-sensitized ODPM rearrangement of the C3–C8 bridged bicyclo[2.2.2]octenone. Also the corresponding epoxide **51** was generated by the same process.

Biased homoconjugated ketones (like **52**) offer two pathways for ODPM rearrangement: direct and vinylogue version [57]. The second variant take advantage of the allylic radical, that is formed in the first step of the ODPM, which has two reactive sites for C–C combination (Sch. 29). The preferred formation of the "normal" rearrangement product (3:1 ratio **53:54**) is expected on basis of the least motion principle. Higher amounts of the vinylogue products were formed from the nonmethyl-substituted starting material (9:4 ratio). A "concerted triplet mechanism" has also been discussed by the authors as a direct way to normal and vinylogue products. A thermal path was also detected which transforms normal ODPM-products in vinylogue products by a 1,3-sigmatropic shift with retention of configuration [58].

The ultimate challenge for an efficient synthetic pathway to complex chiral target molecules is the generation of enantiomerically pure products either by use of enantiomerically pure starting materials or by application of

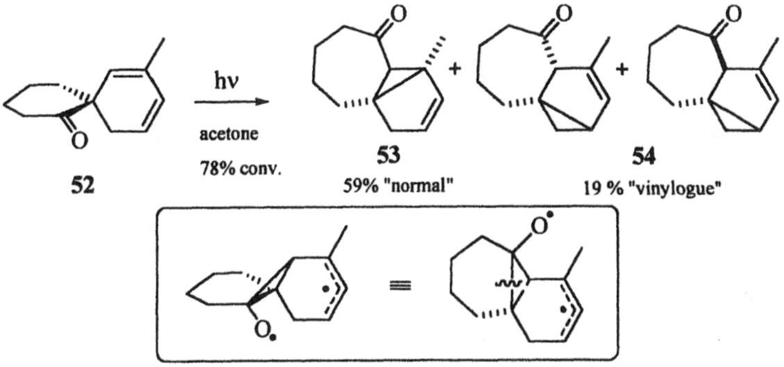

Scheme 29

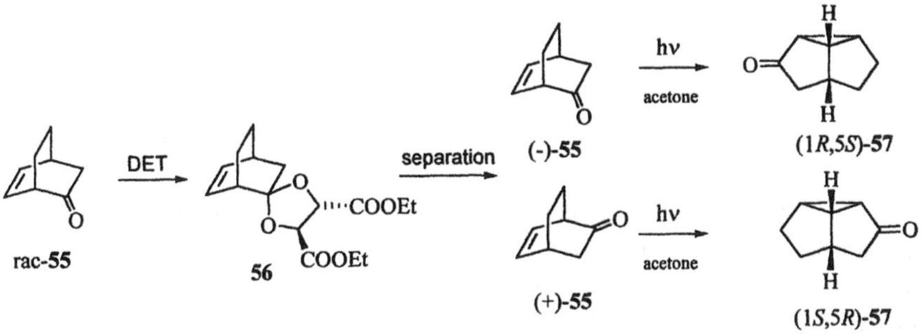

Scheme 30

asymmetric induction using prochiral substrates. The parent bicyclo[2.2.2]octenone is available in enantiomerically pure form by a simple resolution procedure from the racemic mixture involving the acetal diastereoisomers formed with tartaric acid (Sch. 30) [59]. Irradiation of the separated parent compound enantiomers under conditions of triplet sensitization leads to the expected diquinanes in high enantiomeric purity [59]. Another approach to enantiomerically pure ODPM substrates is described for bicyclo[2.2.1]heptenone [60].

This concept has been applied for the synthesis of the structurally complex and highly oxyfunctionalized triquinane (−)-*coriolin* (Sch. 31) [61]. Two carbonyl groups, both in the right position for 1,2-acyl shift were present in the trimethyl-functionalized bicyclo[2.2.2]octenone **58**. With a site-selectivity of 85% the expected regioisomeric tricyclic dione **59** was formed as a mixture of epimers (Sch. 31). Subsequent transformations involving the annulations of the third five-membered ring as well as epoxidation and hydroxylation steps led to the desired natural product

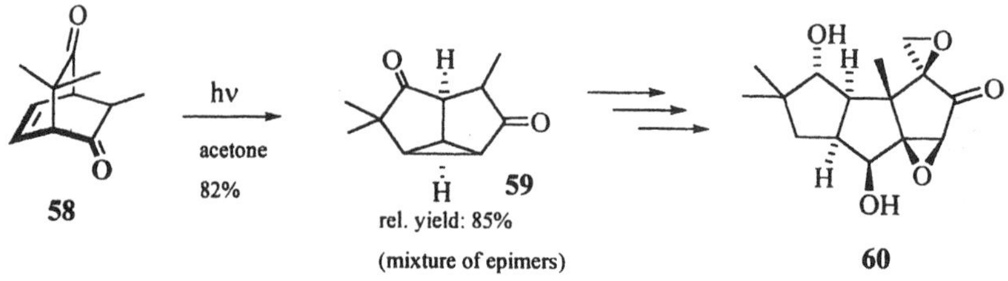

Scheme 31

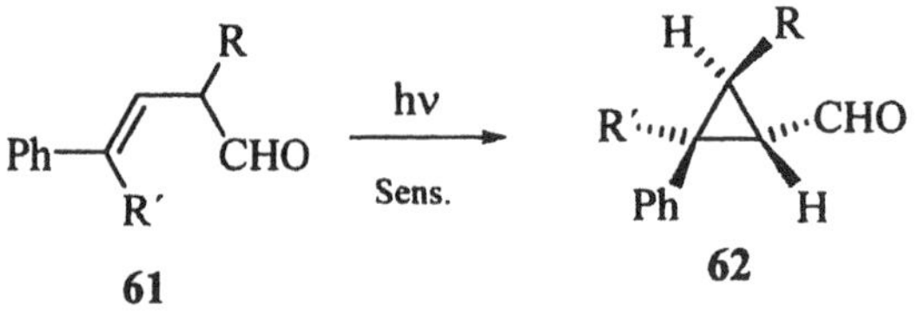

R=iPr, R'=H	21%
R=iPr, R'=Ph	23%
R=Et, R'=H	22%
R=Et, R'=Ph	24%

Scheme 32

coriolin. An impressive collection of additional target molecules has been prepared following the concept described above, e.g., iridoid terpene structures [62], the aglucons of forsythide [63], loganine [64], and silphiperfolenone (vide infra) [65].

Throughout the successful applications of ODPM rearrangement for the total synthesis of complex polyoxyfunctionalized target molecules, three major limitations were constantly reported and accepted: that no heterocyclic three-membered rings can be obtained, that β,γ-unsaturated aldehydes are ODPM-unreactive in the majority of cases and that the reaction has to be initiated by photon absorption, i.e., always occurs with the electronically excited carbonyl substrates. All three "rules" have been recently questioned by Armesto et al. A thorough investigation of the photochemistry of β,γ-unsaturated aldehydes revealed, that these species, contrary to literature opinion, are reactive substrates also for the ODPM rearrangement [66]. In the course of photoinduced electron transfer 2-nitrogen-DPM (2-ADPM) rearrangement (thus involving the radical cation of the substrate and not the electronically excited carbonyl analogue), also aziridines were isolated [67]. Crucial for the success of ODPM (as likewise ADPM) rearrangements appears to be at least one aryl substituent at the alkene part of the DPM substructure. In Sch. 32 the ODPM synthesis of cyclopropane carbaldehydes **62** from the β,γ-unsaturated aldehydes **61** is highlighted. As sensitizers acetophenone or *m*-methoxyacetophenone, respectively, were used.

7.5. REPRESENTATIVE EXPERIMENTAL PROCEDURES

7.5.1. Synthesis of Tricyclo[3.3.0.0.0]octane-3-one [59]

A solution of 10 g of the β,γ-unsaturated bicyclic ketone **55** (cf. Sch. 30) in 1000 mL of acetone was purged with argon and irradiated in a water-cooled quartz vessel placed in a Rayonet RPR-208 photochemical reactor equipped with RUL-3000 lamps ($\lambda \sim 300$ nm). Irradiation was continued for 72 h and the reaction was monitored by tlc. After 72 h of irradiation, the conversion was ~98% and the only detectable compound was the ODPM rearrangement product, tricyclo[3.3.0.0.0]octane-3-one **57**. The solvent was distilled off and the residue was chromatographed over silica gel using a benzene/ether solvent mixture. The product which was eluted (8.6 g) was further distilled under vacuum (50 °C/1 mm) to get the pure ODPM rearrangement product (tricyclic ketone **57**) in 81% yield with 99.5% purity. The quantum efficiency determined was found to be $\Phi = 1.0$.

7.5.2. Synthesis of 4,4,9,9-Tetramethyltetracyclo [6.4.0.0.0]dodecane-7,12-dione [44]

A solution of 1.25 g of 8,8,9,9-Tetramethyl-1,3,4,6,7,8-hexahydro-1,4-ethanonaphthalene-2,5-dione **11** (a δ-oxo-β,γ-unsaturated ketone; cf. Sch. 17) in 300 mL cyclohexane was degassed by bubbling nitrogen through the solution and was irradiated for 3 h in a Rayonet photochemical reactor ($\lambda \sim 300$ nm). The reaction was monitored by tlc and only one product was detectable. Evaporation of the solvent and recrystallization of the crude photolysate afforded the ODPM rearrangement product **12** in 78% yield. The quantum yield of the reaction was determined as $\Phi = 0.67$. Solvents like acetonitrile, benzene, or acetone and sensitizers like acetone or xanthone, respectively, can be employed for this reaction without altering the yield of the ODPM rearrangement product.

7.6. TARGET MOLECULES: NATURAL AND NONNATURAL PRODUCT STRUCTURES

There are numerous examples of compounds from the polyquinane family described in the literature as potential targets for ODPM-synthesis (Fig. 1). The annulation pattern can be adjusted in a highly variable way to give linear, angularly, or propellane-type annulated products and bridged systems like the triquinacence skeleton. From the basis ring structures, not only can further annulation been performed to give polycyclic products,

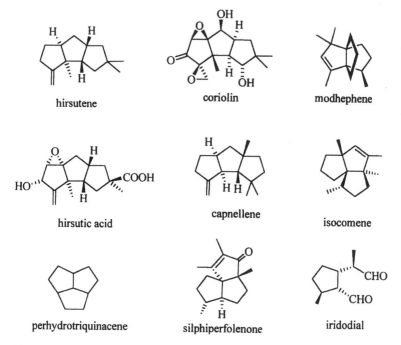

hirsutene   coriolin   modhephene

hirsutic acid   capnellene   isocomene

perhydrotriquinacene   silphiperfolenone   iridodial

Figure 1

but also ring-cleavage processes can be performed to reduce the ring number with concomitant increase in functional groups like in the synthesis of iridoid terpenes. Future, will definitely show more applications, presumably also in alkaloid synthesis, for which ODPM rearrangements currently are rarely applied [68].

REFERENCES

1. Tenny PL, Boykins DL Jr, Lutz RE. J Am Chem Soc 1966; 88:1835.
2. Swenton JS. J Chem Educ 1969; 46:217.
3. Givens RS, Oettle FW. J Chem Soc, Chem Commun 1969; 1164.
4. Dauben WG, Kellog MS, Seeman JI, Spitzer WA. J Am Chem Soc 1970; 92:1786.
5. Houk KN. Chem Rev 1976:76:1.
6. Schuster DI. Photochemical rearrangements of enones. In: deMayo P, ed. Rearrangements in Ground and Excited States. New York: Academic Press, 1980; 3:167.
7. Hixon SS, Mariano PS, Zimmerman HE. Chem Rev 1973; 73:531.

8. Zimmerman HE. Org Photochem 1991; 11:1.
9. Zimmerman HE, Armesto D. Chem Rev 1996; 96:3065.
10. Demuth M. Pure & Appl Chem 1986; 58:1233.
11. Demuth M. Org Photochem 1991; 11:37.
12. Demuth M. Oxa-di-π-methane photoisomerizations. In: Trost BM, ed. Comprehensive Organic Synthesis. Pergamon Press, 1991:5:215.
13. Büchi G, Burgess EM. J Am Chem Soc 1960; 82:4333.
14. Engel PS, Schexnayder MA, Ziffer H, Seeman JI. J Am Chem Soc 1974; 96:924.
15. Engel PS, Schexnayder MA. J Am Chem Soc 1975; 97:145.
16. Baggiolini E, Schaffner K, Jegger O. Chem Commun 1969; 1103.
17. Givens RS, Chae WK. J Am Chem Soc 1978; 100:6278.
18. Houk KN, Northington DJ, Duke RE. J Am Chem Soc 1972; 94:6233.
19. Eckersley T, Parker SD, Rogers NAJ. Tetrahedron Lett 1976; 4393.
20. Tegmo-Larsson IM, Gonzenbach HU, Schaffner K. Helv Chim Acta 1976; 59:1376.
21. Sadler DE, Wendler JE, Olbrich G, Schaffner K. J Am Chem Soc 1984; 106:2064.
22. Marsh G, Kearns DR, Schaffner K. J Am Chem Soc 1971; 93:3129.
23. Gonzenbach H-U, Tegmo-Larsson T-M, Grosclaude J-P, Schaffner K. Helv Chim Acta 1977; 60:1091.
24. Schexnayder MA, Engel PS. J Am Chem 1975; 97:4825.
25. (a) Hancock KG, Grider RO. J Am Chem Soc 1974; 96:1158; (b) Hancock KG, Grider RO. Tetrahedron Lett 1972; 1367; (c) Hancock KG, Grider RO. Tetrahedron Lett 1971; 4281.
26. Schexnayder MA, Engel PS. Tetrahedron Lett 1975; 1153.
27. Dauben WG, Lodder G, Ipaktschi. Top Curr Chem 1975; 54:73.
28. Henne A, Siew PY, Schaffner K. Helv Chim Acta 1979; 62:1952.
29. Henne A, Siew PY, Schaffner K. J Am Chem Soc 1979; 101:3671.
30. Dalton JC, Ming Shen, Snyder JJ. J Am Chem Soc 1976; 98:5023.
31. Schuster DI, Calcaterra LT. J Am Chem Soc 1982; 104:6397.
32. Domb S, Schaffner K. Helv Chim Acta 1970; 53:677.
33. Daubern WG, Lodder G, Robbins JD. J Am Chem Soc 1976; 98:3030.
34. Dauben WG, Lodder G, Robbins JD. Nouv J Chim 1977; 1:243.
35. Seeman JI, Ziffer H. Tetrahedron Lett 1973; 4413.
36. Schaffner K. Tetrahedron 1976; 32:641.
37. (a) Hart H, Collins PM, Warring AJ. J Am Chem Soc 1966; 88:1005; (b) Hart H, Murray RK. J Org Chem 1970; 35:1535.
38. Uppili S, Ramamurthy V. Org Lett 2002; 4:87.
39. Uppili S, Takagi S, Sunoj RB, Lakshminarasimhan P, Chandrasekhar J, Ramamurthy V. Tetrahedron Lett 2001; 42:2079.
40. Mirbach MJ, Henne A, Schaffner K. J Am Chem Soc 1978; 100:7127.
41. Bruno K, Sadler DE, Schaffner K. J Am CHem Soc 1986; 108:5527.
42. van der Weerdt AJA, Cerfontain H. Rec Trav Chim Pays-Bas 1977; 96:247.
43. Koppes MJCM, Cerfontain H. Rec Trav Chim Pays-Bas 1985; 104:272.
44. Kilger R, Körner W, Maragretha P. Helv Chim Acta 1984; 67:1493.

45. Sugihara Y, Sugimura T, Murata I. J Am Chem Soc 1981; 103:6738.
46. Sugihara Y, Suigmura T, Murata I. Chem Lett 1980; 1103.
47. Sadeghpoor R, Ghandi M, Najafi HM, Farzaneh F. Chem Commun 1998; 329.
48. Lee T-H, Rao PD, Liao CC. Chem Commun 1999; 801.
49. Mehta G, Srikrishna A. Tetrahedron Lett 1979; 3187.
50. Ishii K, Hashimoto T, Sakamoto M, Taira Z, Asakawa Y. Chem Lett 1988; 609.
51. Singh VK, Deota PT, Bedekar AV. J Chem Soc, Perkin Trans 1 1992; 903.
52. (a) Mehta G, Subrahmanyam D. J Chem Soc, Chem Commun 1985; 768; (b) Mehta G, Subrahmanyam D. J Chem Soc, Perkin Trans 1 1991; 395.
53. Banwell MG, Edwards AJ, Harfoot GJ, Jolliffe KA. J Chem Soc, Perkin Trans 1 2002; 2439.
54. Uyehara T, Murayama T, Sakai K, Onda K, Ueno M, Sato T. Bull Chem Soc, Jpn 1998; 71:231.
55. Schultz AG, Lavieri FP, Snead TE. J Org Chem 1985; 50:3086.
56. Hayakawa K, Schmid H, Fráter G. Helv Chim Acta 1977; 60:561.
57. Zizuashvili J, Abramson S, Shmueli U, Fuchs B. J Chem Soc, Chem Commun 1982; 1375.
58. Oren J, Schleifer L, Shmueli U, Fuchs B. Tetrahedron Lett 1984; 25:981.
59. Demuth M, Raghavan PR, Carter C, Nakano K, Schaffner K. Helv Chim Acta 1980; 63:2434.
60. Corey EJ, Guzman-Perez A, Loh TP. J Am Chem Soc 1994; 116:3611.
61. Demuth M, Ritterskamp P, Weight E, Schaffner K. J Am Chem Soc 1986; 108:4149.
62. Demuth M, Schaffner K. Angew Chem Int Ed Engl 1982; 21:820.
63. Liao CC, Wei CP. Tetrahedron Lett 1989; 30:2255.
64. Demuth M, Chandrasekhar S, Schaffner K. J Am Chem Soc 1984; 106:1092.
65. Demuth M, Hinsken W. Helv Chim Acta 1988; 71:569.
66. (a) Armesto D, Ortiz MJ, Agarrabeitia AR, Aparicio-Lara S. Synthesis 2001; 1149; (b) Armesto D, Ortiz MJ, Agarrabeitia AR. J Photoscience 2003; 10:9.
67. Armesto D, Caballero O, Amador U. J Am Chem Soc 1997; 119:12659.
68. McClure CK, Kiessling AJ, Link JS. Org Lett 2003; 5:3811.

8

Photocycloaddition of Cycloalk-2-enones to Alkenes

Paul Margaretha
University of Hamburg, Hamburg, Germany

8.1. HISTORICAL BACKGROUND

The photocycloaddition of (cyclic) α,β-unsaturated ketones to alkenes affording cyclobutanes as products comprises the four reaction types shown in Sch. 1, i.e., (a) intermolecular enone + alkene cycloaddition; (b) cyclo-isomerization of alkenylsubstituted enones; (c) photocyclodimerization of enones, one ground state enone molecule acting as "alkene"; and (d) cycloisomerization of *bis*-enones.

The first such reaction published in 1908 by Ciamician and Silber was the light induced carvone → carvonecamphor isomerization, corresponding to type b [1]. Between 1930 and 1960 some examples of photodimerizations (type c) of steroidal cyclohexenones and 3-alkylcyclohexenones were reported [2–5]. In 1964, Eaton and Cole accomplished the synthesis of cubane, wherein the key step is again a type b) photocycloisomerization [6]. The first examples of type a) reactions were the cyclopent-2-enone + cyclopentene photocycloaddition (Eaton, 1962) and then the photoaddition of cyclohex-2-enone to a variety of alkenes (Corey, 1964) [7,8]. Very soon thereafter the first reviews on "photocycloaddition of α,β-unsaturated ketones to alkenes" appeared [9,10]. Finally, one early example of a type d) isomerization was communicated in 1981 [11]. This chapter will focus mainly on intermolecular "enone + alkene" cycloadditions, i.e., type a), reactions and also comprise some recent developments in the intramolec-ular, i.e., type b) cycloisomerizations.

a)

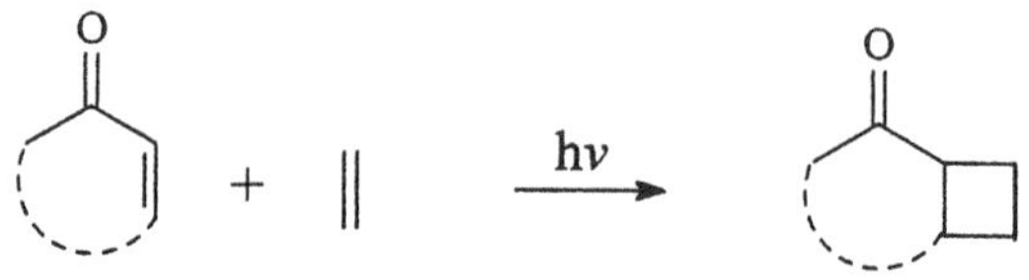

[2+2]-Photocycloaddition of a cycloalk-2-enone to an alkene

b)

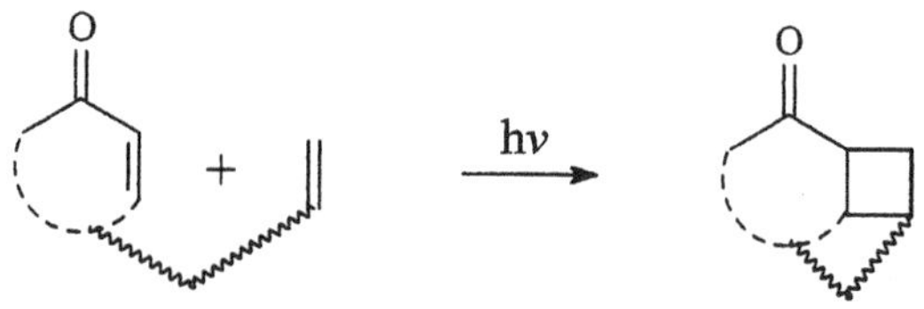

Photocycloisomerization of an alkenylcycloalk-2-enone

c)

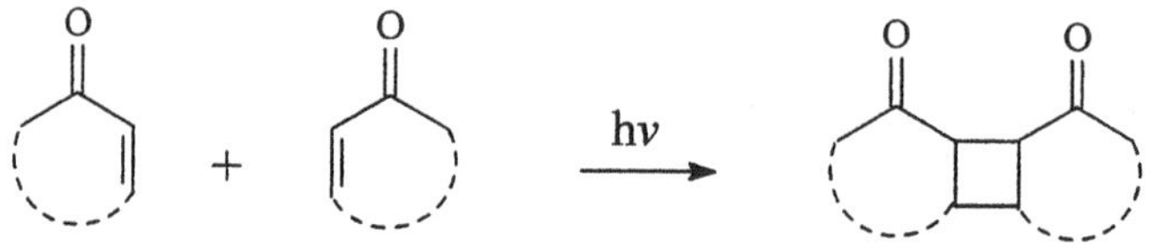

[2+2]-Photocyclodimerization of a cycloalk-2-enone

d)

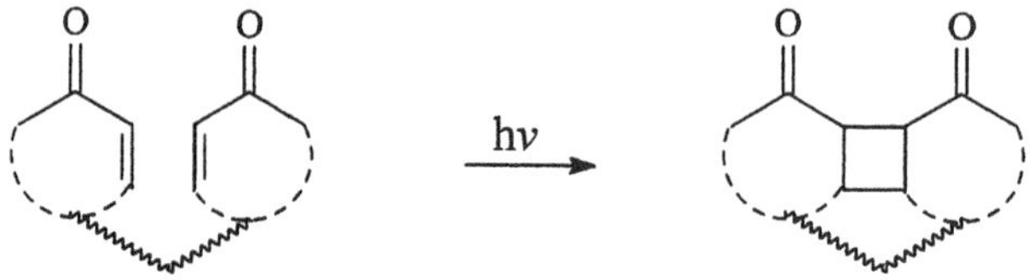

Photocycloisomerization of a "*bis*-cycloalk-2-enone"

Scheme 1

An early synthetic application of such intermolecular enone + alkene photocycloadditions was presented by de Mayo in 1964. 5,5-Dimethyl-3-hydroxycyclohex-2-enone (**1**), i.e., the enol of dimedone, adds to e.g., cyclohexene, and the primarily formed cyclobutane **2** undergoes a

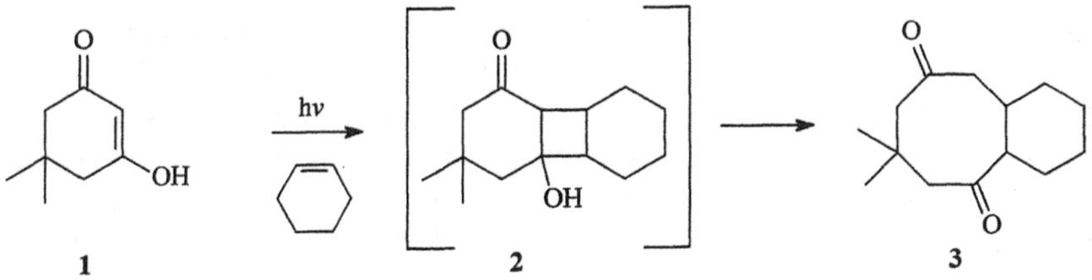

Scheme 2

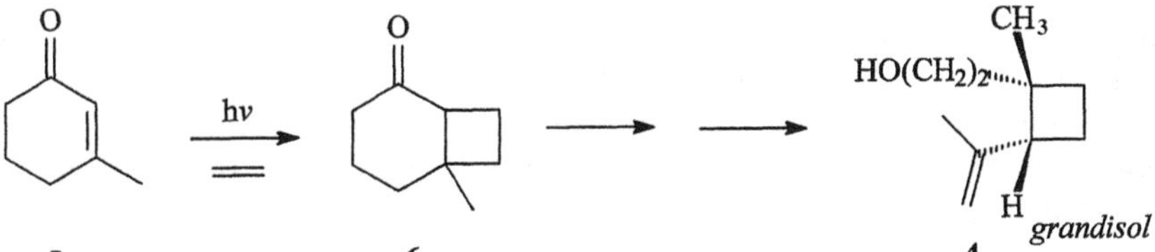

Scheme 3

consecutive retroaldol ring opening to yield bicyclo[6.4.0]octanedione **3** (Sch. 2) [12]. The first industrial application of such a reaction (Sch. 3) was the synthesis of *grandisol* (**4**), the male sex pheromon of the boll weevil, and included the photocycloaddition of 3-methylcyclohex-2-enone (**5**) to ethylene affording bicyclo[4.2.0]octan-2-one **6** [13].

8.2. STATE OF THE ART, MECHANISTIC MODELS

In the preparative application of [2+2]-photocycloadditions of cyclic enones to (substituted) alkenes, two factors concerning product formation are of decisive relevance, namely the regioselectivity and the (overall) rate of conversion. Regarding the regioselectivity in the addition to mono- and 1,1-disubstituted alkenes, Corey had shown that the preferred addition mode of cyclohex-2-enone to isobutene or 1,1-dimethoxyethylene was the one leading to—both *cis*- and *trans*-fused—bicyclo[4.2.0]octan-2-ones with the substituents on C(7) [8]. In contrast, in the reaction with acrylonitrile, the alternate orientation was observed to occur preferentially. Similar results were also reported by Cantrell for the photocycloaddition of 3-methyl-cyclohex-2-enone to differently substituted alkenes [14]. No significant differences in the overall rates of product formation for the different alkenes were observed in these studies. In order to explain these observed regioselectivities it was assumed that an *Umpolung* effect occurs in the charge distribution of the C–C-double bond in the (reactive) triplet excited cyclohexenone [8]. Shortly thereafter, a mechanistic sequence was proposed by de Mayo involving the formation of an exciplex between the triplet excited enone and alkene (which should be determining for the regiochemical outcome), followed by that of a triplet 1,4-biradical, which in turn after intersystem crossing affords the final product(s) via 1,4-cyclization [15]. It was also assumed that the triplet excited enone should exhibit an electrophilic behavior and therefore react faster with electron-rich than electron-deficient alkenes. This mechanistic description was corroborated in studies of photocycloadditions of 4-oxacyclohex-2-enones to 2-chloro-difluoromethyl-3-chloro-3,3-difluoropropene [16].

In a relevant paper describing triplet lifetime measurements by using nanosecond flash photolysis, Schuster and Turro tackled the problem of describing "relative rates" for such reactions because product yields in multistep processes depend on overall quantum efficiencies and not on the rate of a single specific step [17,18]. The results were, that (a) the rate constants for quenching of enone triplets are much larger for electron deficient alkenes than for those bearing electrondonor substituents, and (b) no correlation whatsoever was seen between these quenching rates and the (overall) quantum yield for product formation. These results were found to be inconsistent with expectations based on the exciplex model described above, and the direct formation of triplet biradicals from triplet enone and alkene was therefore suggested as a plausible reaction step. Slightly later, based on results of photocycloaddition between cyclopent-2-enone and acrylonitrile, Weedon concluded that it was unnecessary to introduce an exciplex into the mechanism of the photocycloaddition reaction of cyclic enones with alkenes in order to explain the reaction regiochemistry [19,20].

This mechanistic sequence (Sch. 4) wherein the triplet excited enone adds to the alkene, either via an exciplex intermediate or directly, to afford triplet 1,4-biradicals, which (after undergoing intersystem crossing) either cyclize to product(s) or revert to ground state reactants, is confirmed by both semi-empirical and ab-initio calculations [21–24]. The origin of regioselectivity is supposed to stem from the primary binding step, the enone triplet being considered as a (nucleophilic) alkyl radical at C(3) linked to an (electrophilic) α-acyl radical at C(2) [25]. Thus additions of C(2) to the less substituted terminus of electron rich alkenes and of C(3) to the least substituted terminus of electron deficient alkenes should occur preferentially [26].

In contrast to these intermolecular enone + alkene photocyclo-additions, the regioselectivity in the corresponding intramolecular cycloisomerizations of alkenylcycloalkenones is controlled primarily by

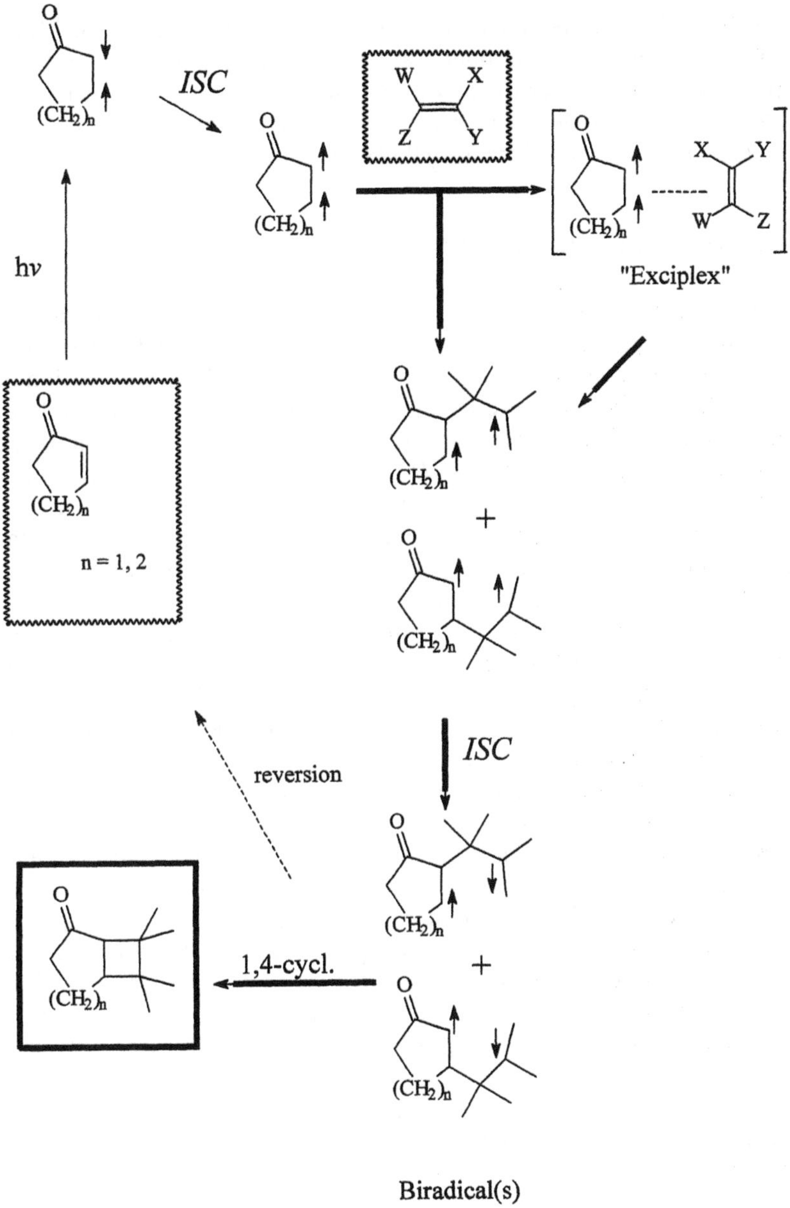

Scheme 4

the number of atoms between the two reactive C=C bonds [27]. Many examples wherein these two bonds are separated by one, two, or three additional atoms are governed by the so-called *"rule of five,"* which states that *"if triplet cyclizations can lead to rings of different size, the one formed by 1,5-cyclization is preferred kinetically"* [28,29].

8.3. SCOPE AND LIMITATIONS

Regarding the cycloalkenone itself, the reaction is limited to five- and six-membered unsaturated ketones, as smaller rings tend to undergo (mono-molecular) ring opening reactions and for larger rings, Z/E isomerization of the cycloalkenone becomes predominant. This latter type of (monomolecu-lar) deactivation of the triplet enone is also observed for six membered S-heterocycles, e.g., 2,3-dihydro-4H-thiin-4-ones as these rings rather resemble cyclohept-2-enones due to the longer bond length of the C–S bond [30–32]. In a recent review on enone–olefin cycloadditions, it has already been pointed out that the regiochemical control of such intermolecular photocycloadditions is the most critical aspect of the reaction from the standpoint of synthetic utility [33]. This is illustrated in Sch. 5 for the regiochemical outcome of the cycloaddition of various enones to 2-methylpropene: while cyclopent-2-enone (**7**) and cyclohex-2-enone (**8**) afford 3:1 mixtures of regioisomers **9 + 10** and **11 + 12** [8,34], the corresponding 4-heteracycloalk-2-enones **13** and **14** add regiospecifically to afford exclusively the socalled *head to tail* regioisomers **15** and **16**, respectively [30,35–39]. This reflects the higher degree of *Umpolung* in the C–C double bond of the triplet enone due to the adjacent heteroatom, which in turn facilitates exciplex formation with electron rich alkenes [40].

When the ionization potential of the alkene is even lower, e.g., for tetramethoxyethylene IP = 6.82 eV, charge transfer from the alkene to the triplet excited enone in the exciplex affording a contact ion pair, which in turn gives oxetanes as final products becomes the predominant path (Sch. 6) [41]. Thus, cyclohex-2-enone (**8**) affords oxaspirononene **17** selectively. The amount of the contact ion pair increases with growing difference in redox potentials for the enone reduction on the one side and the alkene oxidation on the other one. Easily reducible enones, e.g., 6-fluorocycloalk-2-enones **18**, already afford predominantly oxetanes **19** with 2,3-dimethylbut-2-ene (IP = 8.27 eV) [42,43].

Appropriately substituted triplet 1,4-biradicals can undergo *direct* 1,5- and 1,6-cyclization to triplet cyclopentenylcarbenes or triplet 1,2-cyclohexa-dienes, respectively (Sch. 7) [44–46]. The typical substitution pattern for such competitive reactions is the presence of a sp-hybridized C-atom linked to one

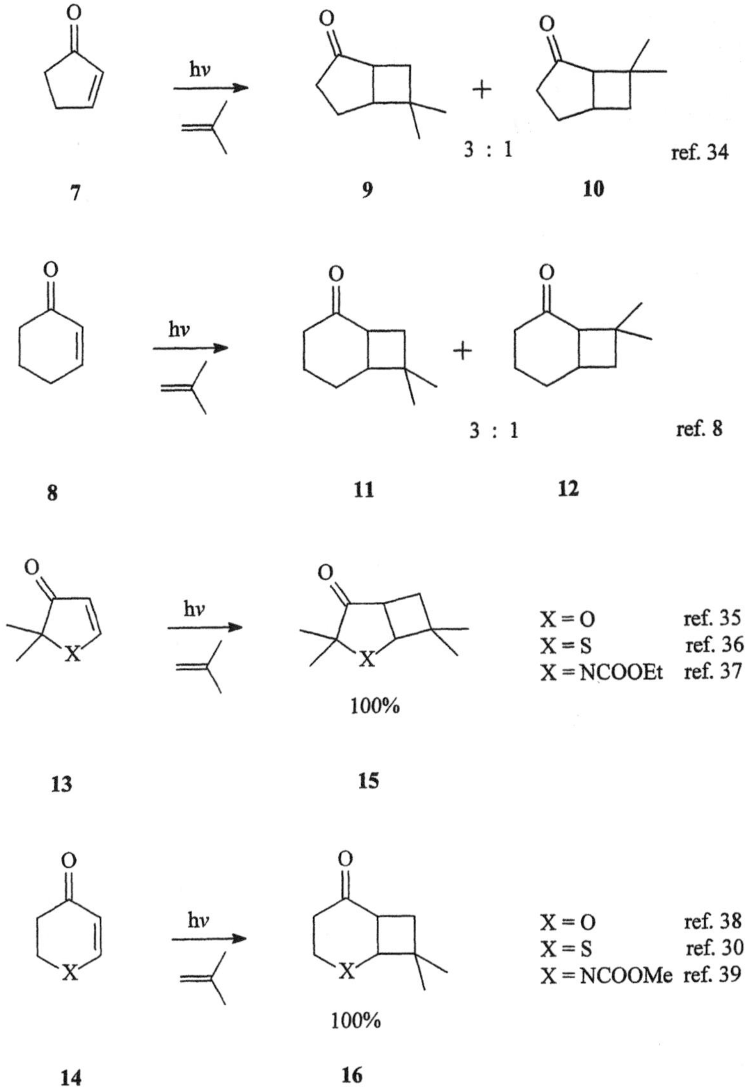

Scheme 5

of the tetramethylene radical centers. Selected examples are the formation of tricyclic furans **20** from 2-alkynylcyclohex-2-enones **21** or the "photo-dehydro-Diels-Alder-cycloaddition" of 2,5,5-trimethylcyclohex-2-enone (**22**) to 2-methylbut-1-en-3-yne to afford hexahydronaphthalenone **23**.

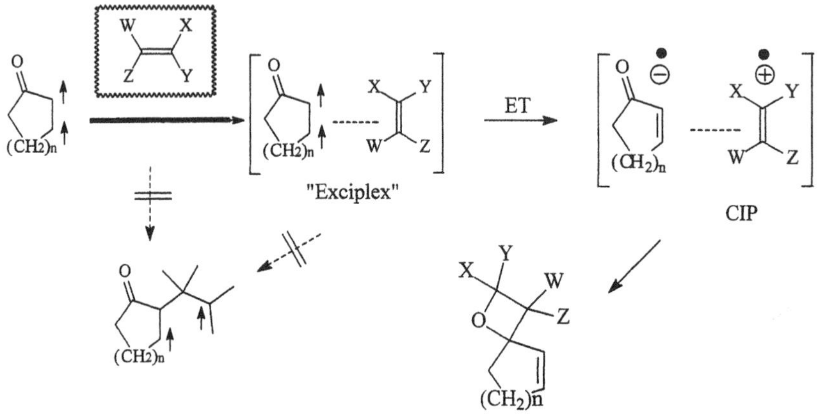

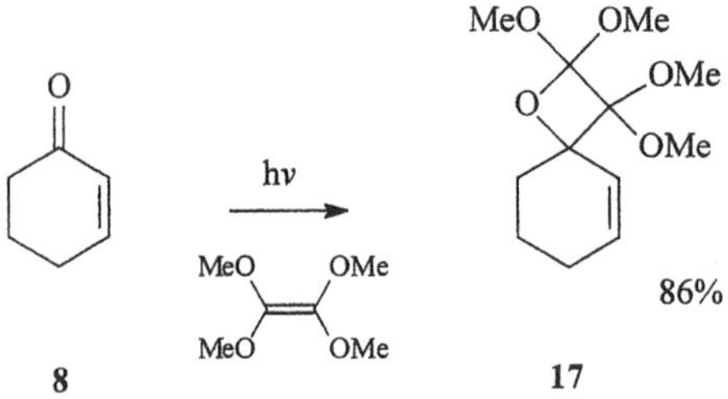

ref. 41

8 17

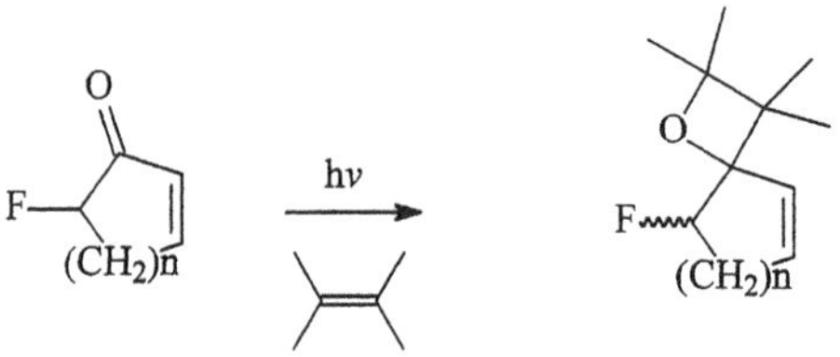

n = 1 100% ref. 43
n = 2 90% ref. 42

18 19

Scheme 6

The intermediate singlet 1,4-biradical can undergo hydrogen atom transfer to give products where only one C–C single bond has been formed, i.e., alkylated cycloalkanones **24** (Sch. 8). This is most often observed when the biradical has a *gem*-dimethyl substitution pattern, and is again favored, when the enone itself is easily reducible, e.g., for α′-fluoro- or α′-trifluoromethylcycloalkenones [47].

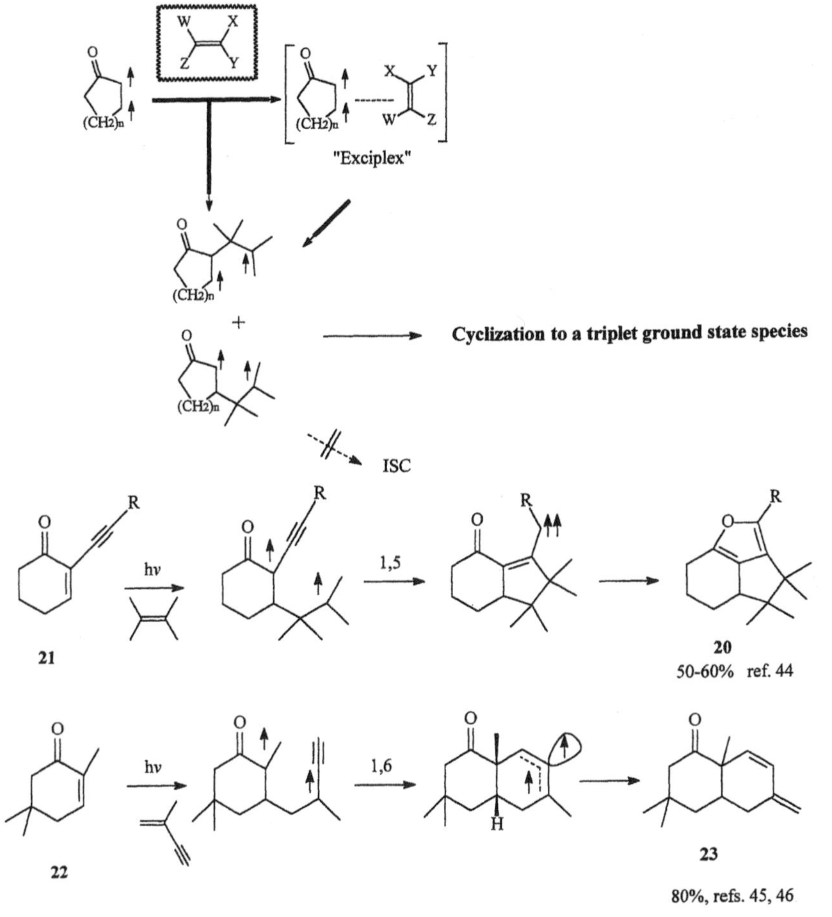

"Exciplex"

+   ⟶   Cyclization to a triplet ground state species

⫻⤑ ISC

21

hν

1,5

20
50-60% ref. 44

22

hν

1,6

23

80%, refs. 45, 46

Scheme 7

A different competitive reaction on the singlet 1,4-biradical level is observed for 2-acylcyclohex-2-enones **25**, which on irradiation in the presence of 2,3-dimethylbut-2-ene afford both 1-acylbicyclo[4.2.0]octan-2-ones **26** as well as 3,4,4a,5,6,7-hexahydro-8*H*-isochromen-8-ones **27** (Sch. 9). These concurrent photochemical [4 + 2]-cycloadditions represent the first examples of dihydropyran formation from enones and alkenes to be reported in the literature [48].

Finally it should be mentioned that many other compounds containing C–C- and C–O double bonds afford cyclobutanes on irradiation in the presence of alkenes. Examples of monocyclic compounds include endiones,

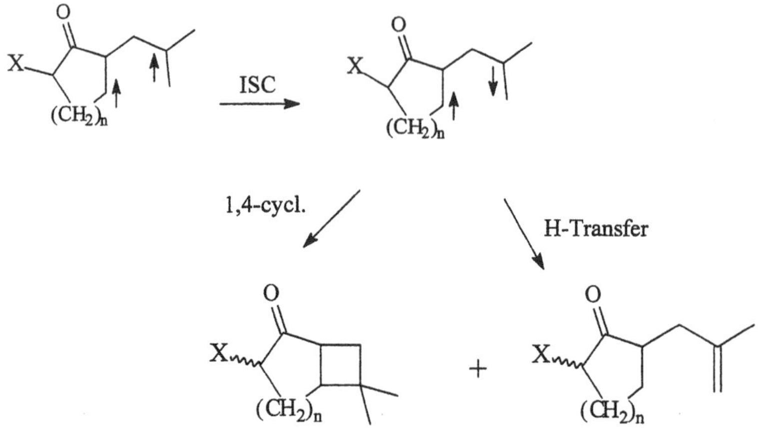

$$X = H, \ n = 2 \qquad\qquad 70 : 30$$
$$X = CF_3, \ n = 2 \qquad\quad 61 : 39 \qquad\qquad \text{ref. 47}$$
$$X = F, \ n = 2 \qquad\qquad 29 : 71$$

Scheme 8

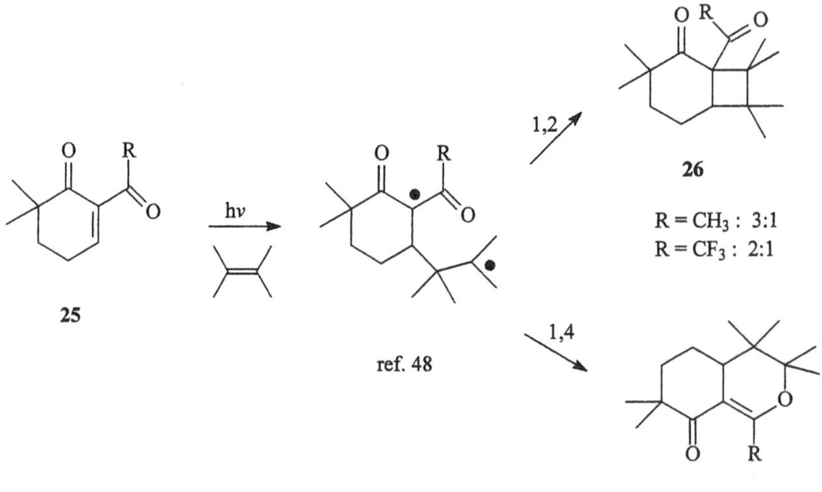

26

R = CH$_3$: 3:1
R = CF$_3$: 2:1

ref. 48

27

Scheme 9

e.g., homobenzoquinones [49], unsaturated lactones and lactams [50,51], cyclic cinnamic acid derivatives, i.e., coumarins, thiocoumarins, and quinolones [52–54], as well as the isomeric isocoumarins and isothiocoumarins [55]. This chapter will be limited to "true" enones, i.e., cyclic α,β-unsaturated ketones.

8.4. SYNTHETIC POTENTIAL, REACTIVITY, AND SELECTIVITY PATTERNS

As already mentioned (see Chaps. 8.2 and 8.3), the addition of an unsymmetrical enone to an unsymmetrical alkene usually affords a mixture of the possible regioisomers. This is illustrated in Sch. 10 for the reaction of cyclopentenone **28** with allylic alcohol as alkene where a mixture of eight cycloadducts is obtained. A very efficient way in controlling the regiochemical outcome for such a reaction has been proposed by Crimmins by using a siloxane as temporary tether, i.e., cyclopentenone **29** in an intramolecular reaction which now proceeds regioselectively to give first tricycle **30** and then the desired cycloadduct **31** [56].

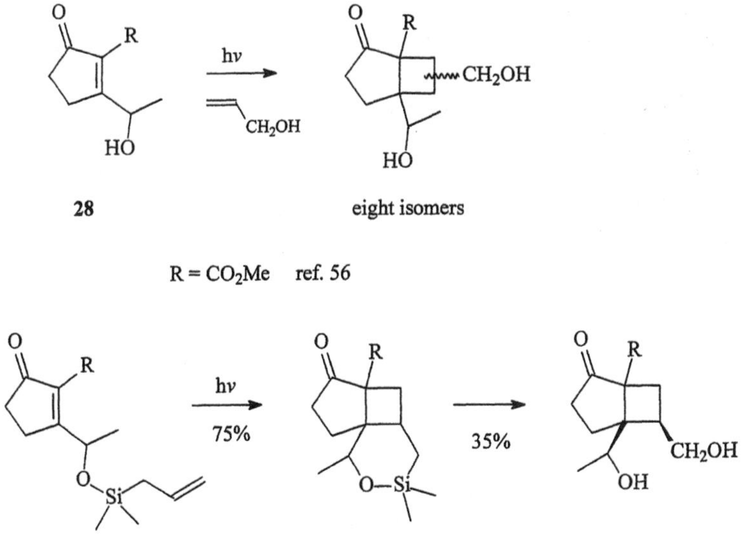

28   eight isomers

R = CO₂Me   ref. 56

29   **30**   **31**

Scheme 10

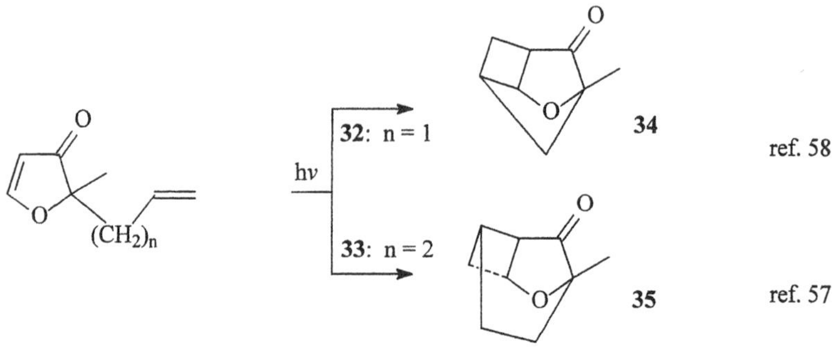

Scheme 11

Although the outcome of such intramolecular photocycloisomeriza-tions is usually predictable, depending on the number of atoms between the two reactive C=C bonds, slight modifications of the starting material sometimes lead to unforeseen inversions in the regiochemical outcome. This is illustrated in Sch. 11 for the allyl- and butenylfuranones **32** and **33**. Whereas the former affords the expected *straight* cycloadduct **34**, a complete reversal of the regiochemistry is observed for the latter which affords selectively the *crossed* adduct **35** [57,58].

The stereochemical outcome of a photochemical [2 + 2] photocycload-dition depends highly on the approach of the two reactants as well as the sterochemical features of the product(s). In this context stereoselective [2 + 2]-photocycloaddition reactions have emerged as powerful C–C bond forming transformations leading to useful and versatile building blocks [59,60]. Some examples of such reactions are presented below.

Simple diastereoselectivities are observed in the reaction of cyclo-alkenones with cyclic alkenes, e.g., cyclopentene, wherein the *transoid* tricycle is formed preferentially. Whereas cyclopent-2-enone (**7**) affords **36** selectively, the more flexible cyclohex-2-enone (**8**) gives a 3:1 mixture of diastereoisomers **37** and **38** [7,61]. A stereogenic center in the cycloalkenone also has a strong impact on the product ratio as shown for the reactions of 4-alkylcyclohex-2-enones **39** and **40** with acyclic alkenes wherein the major diastereoisomer formed is the one in which the enone-alkyl group is *trans* to the new ring forming C–C bonds, i.e., **41t** and **42t**, respectively (Sch. 12) [62,63]. Cycloadducts **42** have been further converted to the pheromone *periplanone-B*.

A combination of these two effects, i.e., the preferential formation of one of two possible *transoid* tricyclic adducts (**43** vs. **44**), is illustrated in Sch. 13 for the cycloaddition of furanones **45** to cyclopentene [64].

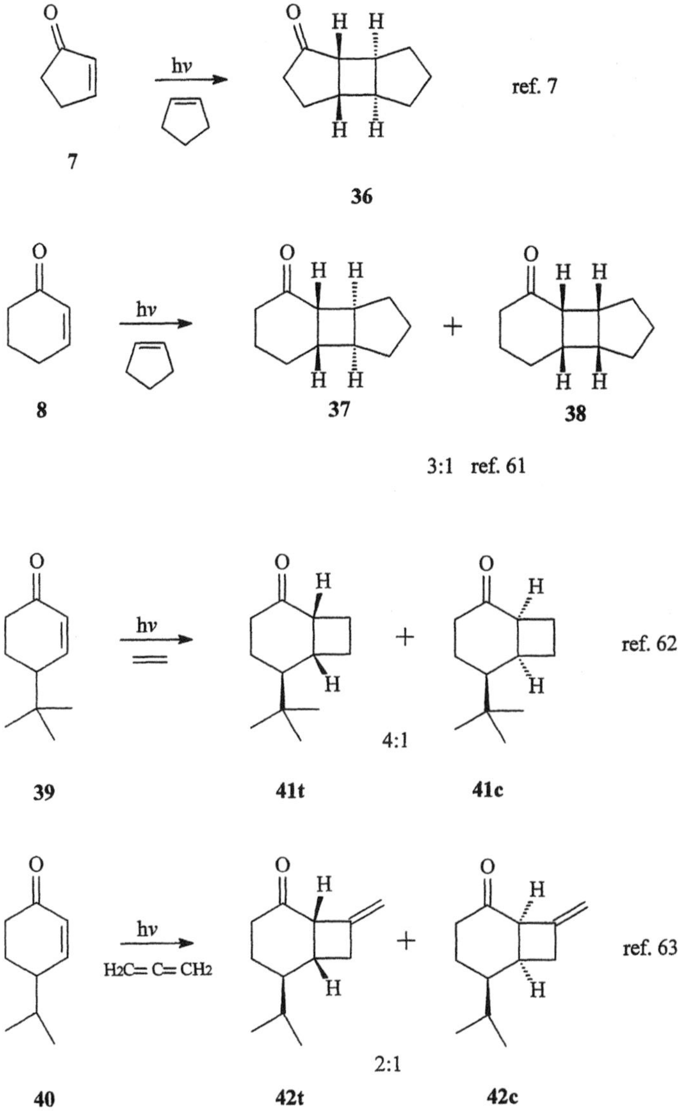

Scheme 12

As already mentioned, the stereochemical conditions of approach of the two reactants is an important parameter in such photocycloadditions. Very elegant supporting evidence therefore comes from an investigation of the photocycloaddition of diastereomeric hexahydro-indenones **46** and **47** to

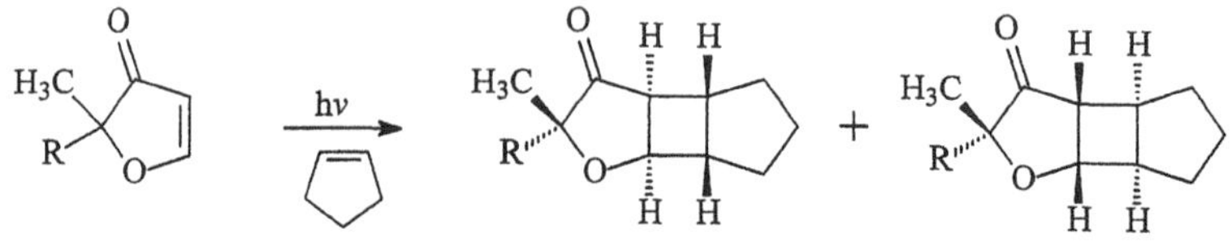

45   **43**   **44**

$$R = C_2H_5 \qquad 1 : 1.4$$
$$R = i.C_3H_7 \qquad 1 : 4 \qquad \text{ref. 64}$$
$$R = t.C_4H_9 \qquad 1 : 23$$

Scheme 13

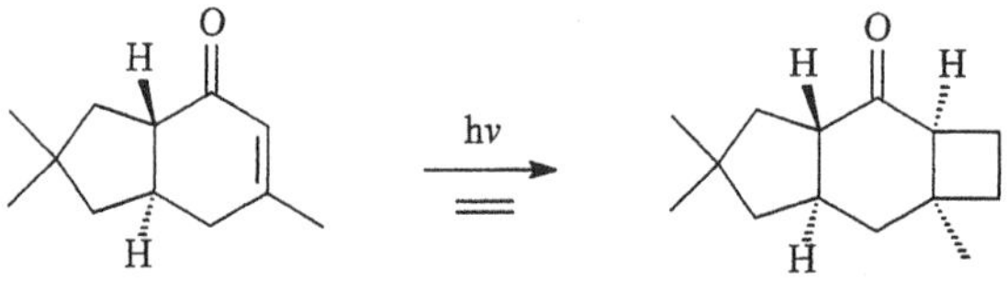

46   **48**

ref. 65

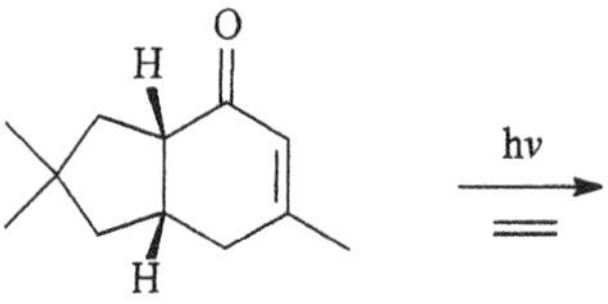

no reaction

47

Scheme 14

ethene (Sch. 14) in the course of which only the *trans*-fused bicyclic enone reacted to give tricycle **48** [65].

Enantioselectivity in intermolecular enone + alkene photocycloadditions is quite difficult to accomplish, due to the fact that stereochemical information inherent in the starting materials can easily be lost on the stage of the triplet 1,4-biradical intermediate. Nevertheless highly enantioselective photocycloadditions of quinolones to alkenes mediated by a chiral lactam

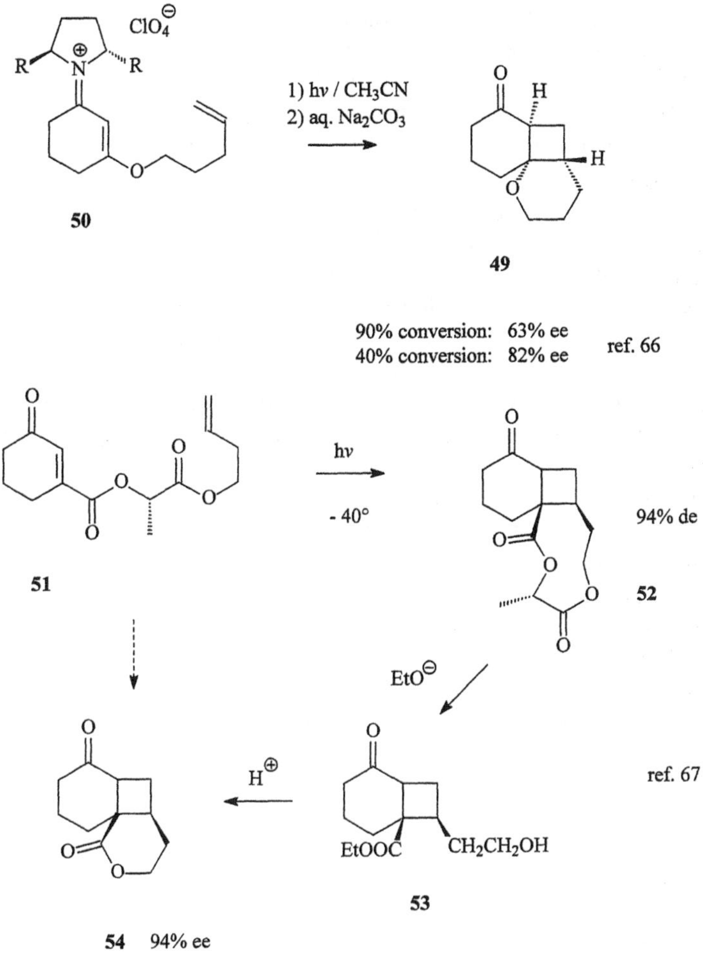

50

1) hv / CH₃CN
2) aq. Na₂CO₃

49

90% conversion: 63% ee
40% conversion: 82% ee   ref. 66

51

hv

- 40°

94% de

52

EtO$^{\ominus}$

54 94% ee

H$^{\oplus}$

EtOOC   CH₂CH₂OH

53

ref. 67

Scheme 15

host which interacts with the (quinolone) guest molecule via hydrogen bonding have been recently reported, but this method is limited to enones containing a N-atom [54]. This "problem" connected to triplet state intermediates has been elegantly circumvented by Mariano in a general startegy for absolute stereochemical control in such photocycloadditions by using eniminium salts of enones instead of the enones themselves, as these react from their singlet excited states [66]. Quite good enantiomeric excesses of tricycle **49** at moderate degrees of conversion of the chiral eniminium salt **50** (Sch. 15) have been obtained in such reactions.

Alternatively, enantioselective intramolecular photocycloisomerizations result from the use of hydroxy acids as chiral tethers, as illustrated in the conversion of 3-oxocyclohexenylcarboxylate **51** to the tricyclic lactone **52** which proceeds with 94% ee [67]. The chiral tether can be recycled by saponification to **53** and this hydroxyacid again converted to lactone **54** without loss of optical purity.

Some typical experimental procedures for the preparation of cyclobutanes will be presented (see Chap. 8.5) and the synthesis of cyclobutane containing natural products will be discussed (see Chap. 8.6). Very often photochemically generated cyclobutanes are used as synthetic intermediates wherein the cyclobutane ring is chemically altered. An excellent treatise on such transformations of cyclobutanes has appeared recently including both elimination- and ring-opening reactions, these latter proceeding by hydrogenolysis, addition reactions, cycloreversion, or rearrangement [68].

The central bond in (photochemically) easily accesible bicyclo[n.2.0]-alkan-2-ones ($n = 3, 4$) can be cleaved by retro-aldol reaction (cf. Sch. 2), but also by treatment with Broenstedt- and Lewis-acids. This is shown in Sch. 16 for the synthesis of (±)descarboxyquadrone (**55**) starting from indenone **56** via the cycloadduct **57** which reacts with HCl to **58** [69], or by the preparation of the AB-ring core of Taxol **59** from cyclopentenone **60** via cycloadduct **61** which rearranges to **62** in the presence of TiCl$_4$ [70].

Furthermore this same central bond in enone + alkene photocycload-ducts can be cleaved by thermally induced cycloreversion (Sch. 17). The re-arrangement of tricycle **63** – obtained from 3,5,5-trimethylcyclohex-2-enone (**64**) and methyl cyclobut-1-ene-1-carboxylate – to give the *trans*-decalin **65** proceeds via a sequence wherein the cycloreversion is followed by a transannular ene-reaction [71]. The photocycloisomer **66** of cyclopentenone **67** rearranges further to **68** containing a 5,8,5-sesquiterpene core [72].

The external bonds in bicyclo[4.2.0]octan-2-ones can be cleaved both by reductive and oxidative processes, both reactions affording disubstituted cyclohexanones. This is illustrated (Sch. 18) by the SmI$_2$-promoted formation of 3-alkyl-3-ethylcyclohexanones **69** from 6-alkylbicyclo[4.2.0]-octan-2-ones **70** [73], and by the formation of 2-alkyl-3-acylcyclohexanones **71** from 7-trimethylsilyloxybicyclo[4.2.0]octan-2-ones **72** via single electron oxidation [74].

Representative Experimental Procedures

Three typical procedures will be presented (Sch. 19). The first represents one of the (only) two "enone + alkene" photocycloaddition reactions which have published as a procedure in *Organic Synthesis* [75], the second one is of

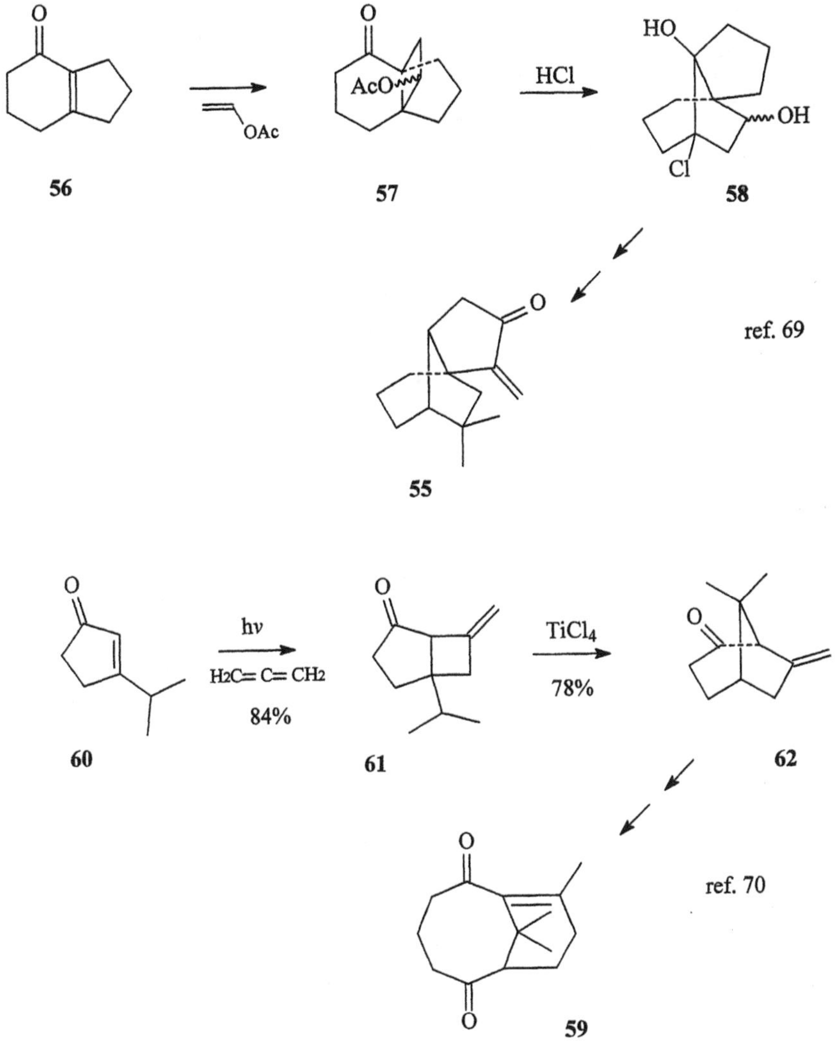

Scheme 16

singular importance, as it refers to a reaction wherein the photolysis of the "enone" affords the same photocycloadduct, both in the absence or in the presence of the alkene [76], and finally the third one exemplifies a photocycloaddition which affords a mixture of two diasteroisomers, but wherein one of the products can be quantitively converted to the second one by base-catalysis [55].

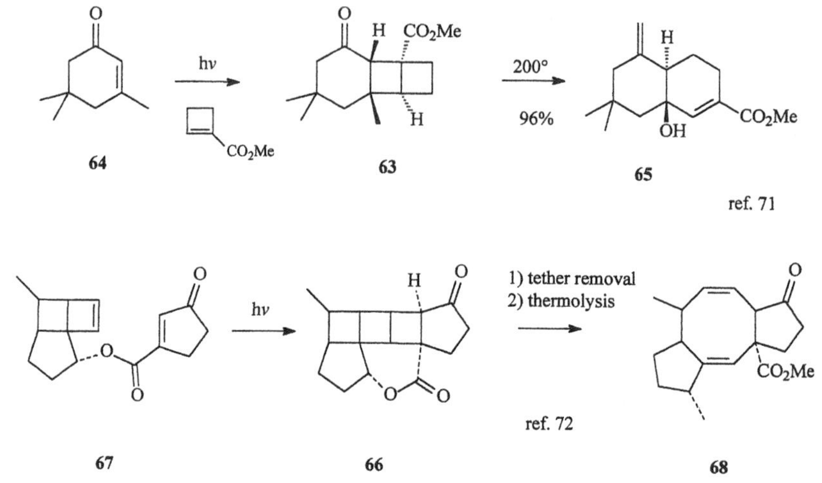

ref. 71

ref. 72

Scheme 17

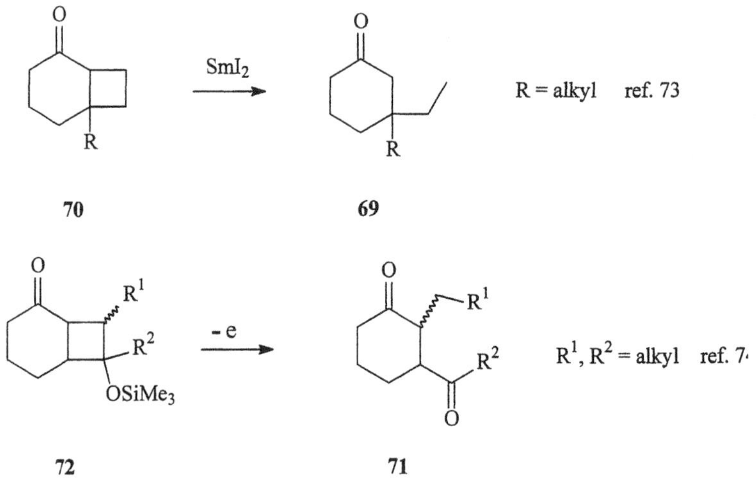

R = alkyl   ref. 73

R¹, R² = alkyl   ref. 7.

Scheme 18

1. Photocyclization of an Enone to an Alkene: 6-Methylbicyclo[4.2.0]octan-2-one [75]

The irradiation apparatus, consisting of a triple walled Dewar constructed of Pyrex, is charged with a solution of 25.0 g (0.277 mol) of 3-methyl-cyclohex-2-enone (**5**) in reagent-grade CH_2Cl_2. A gas outlet tube to an

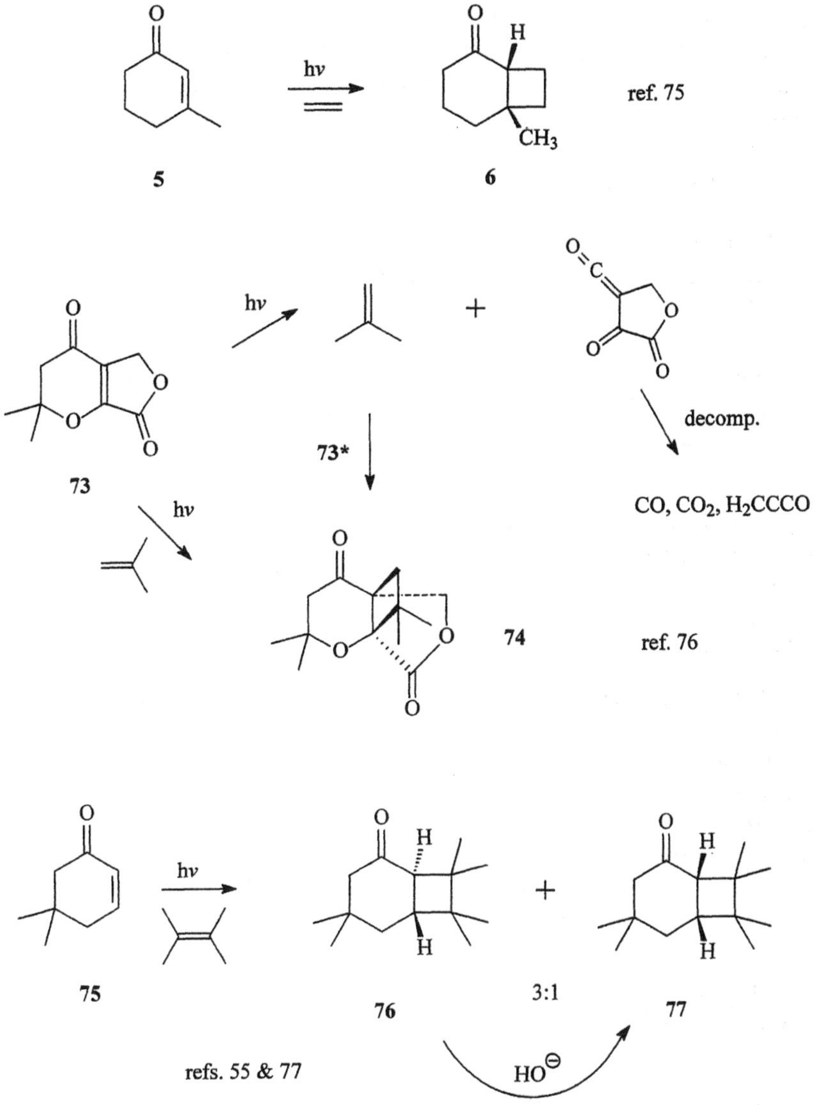

ref. 75

ref. 76

Scheme 19

efficient hood is placed in one 14/20 standard taper joint; in the other, there is a stopper that can be removed for periodic sampling. The cooling water is turned on and the apparatus is immersed in a dry-ice/propan-2-ol bath while the chilled solution is saturated with ethylene, a flow of ca. 100 mL/min for 2–3 h being adequate. The lamp, either a GE H1000-A36-15 or a Westinghouse H-36GV-1000, both 1000-W street lamps, is inserted into the irradiation well and turned on. Progress of the irradiation is conveniently followed by gas chromatography. After about 8 h, most of the starting material has reacted. At this time, the lamp is turned off and the apparatus removed from the cooling bath. The reaction mixture is degassed with a slow stream of N_2 while it warms to r.t., dried over $MgSO_4$, and concentrated with a rotary evaporator at a temperature below 30°C. The product is isolated by distillation to afford 27–28 g (86–90%) of 6-methylbicyclo[4.2.0]octan-2-one (6), b.p.: 62–65 °C (3.5 mm) of >90% purity.

2. Irradiation of 2,2-Dimethyl-2H-furo[3,4-b]pyran-4,7(3H,5H)-dione (73) in the Absence and Presence of 2-Methylpropene [76]

2a) Irradiation of 75. An argon-degassed solution of 1.82 g (0.01 mol) **75** in 20 mL t.BuOH is irradiated for 80 h in a *Rayonet RPR-100* photoreactor equipped with 350 nm lamps in combination with a liquid filter with cut-off at $\lambda = 340$ nm. Evaporation of the solvent and recrystallization of the residue from diethyl ether/pentane affords 300 mg (13%) of **74** (see below).

2b) Irradiation of 73 and 2-Methylpropene. An argon-degassed solution of 1.82 g (0.01 mol) **75** in 20 mL t.BuOH is saturated with 2-methylpropene at r.t. and irradiated for 16 h in a *Rayonet RPR-100* photoreactor equipped with 350 nm lamps in combination with a liquid filter with cut-off at $\lambda = 340$ nm. Evaporation of the solvent and recrystallization of the residue from diethyl ether/pentane affords 1.81 g (76%) of 3,3,10,10-tetramethyl-2,8-dioxatetracyclo[4.3.2.0^{1,1}]undecane-5,9-dione (74), m.p.: 128–130 °C.

3. Photocycloaddition of 5,5-Dimethylcyclohex-2-enone (75) to 2,3-Dimethylbut-2-ene [55,77]

An argon-degassed solution of 1.24 g (0.01 mol) **75** and 8.4 g (0.1 mol) 2,3-dimethylbut-2-ene in benzene (100 mL) is irradiated with a 250-W Hg-lamp in a double-walled immersion well using a filter solution (7 g $Pb(NO_3)_2 + 750$ g $NaBr/100$ mL water, cut-off at $\lambda = 340$ nm) as coolant for 12 h. After evaporation of the solvent and excess alkene the residue (2.0 g) consisting of a 3:1 mixture of *trans*- and *cis*-fused bicyclo-octanones **76** and **77** is refluxed for 2 h in a solution of 4 g KOH in methanol (100 mL). After cooling to r.t. the mixture is neutralized with 5% aq. HCl, the methanol evaporated and the residue extracted three times with ether (25 mL). The combined ethereal extracts are washed with water and brine and dried over $MgSO_4$. Evaporation of the solvent and Kugelrohr-distillation of the residue at $150\,°C/0.01\,mm$ affords $1.65\,g$ (77%) pure $1\alpha,6\alpha$-4,4,7,7,8,8-hexamethylbicyclo[4.2.0]octan-2-one [77]. ^{1}H-NMR: $J_{H(1),H(6)} = 8.9\,Hz$, $J_{H(3),H(3)} = 17.7\,Hz$. The *trans*-diastereoisomer 76 can be obtained in pure form by preparative gas chromatography. ^{1}H-NMR: $J_{H(1),H(6)} = 13.3\,Hz$, $J_{H(3),H(3)} = 13.6\,Hz$.

8.5. NATURAL- AND NONNATURAL TARGET MOLECULES

Several sesquiterpene skeletons include cyclobutane rings [78]. Some representative examples are summarized in Sch. 20. As can be easily seen, syntheses of caryophyllanes 78 and panasinsanes 79 are based on photocycloadditions to 2-methylpropene, those of tritomaranes (kelsoanes) 80, protoilludanes 81, and sterpuranes 82 are based on photocyclo-additions to ethylene, while the bourbonane skeleton 83 is accessible via

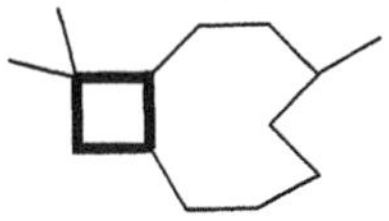

78 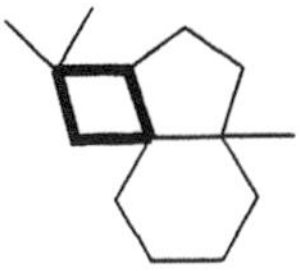   79

80 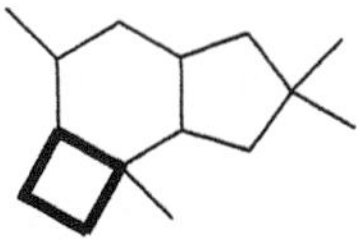   81 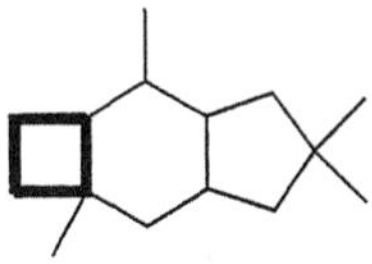   82

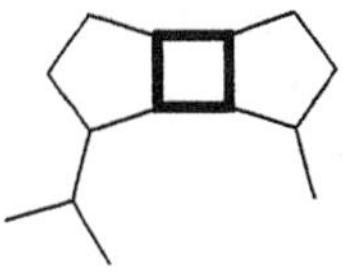

83

Scheme 20

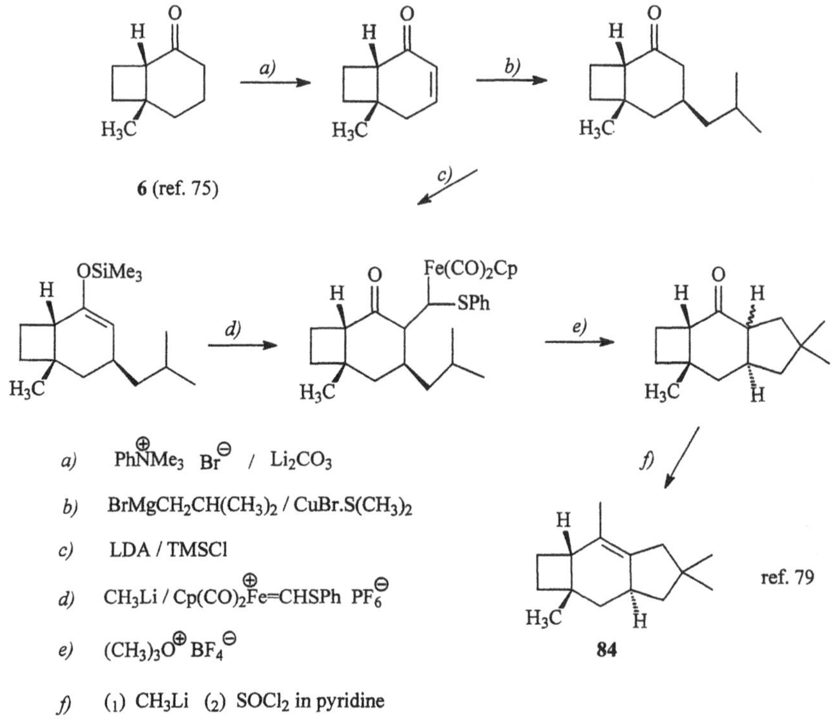

6 (ref. 75)

a) PhN⁺Me₃ Br⁻ / Li₂CO₃

b) BrMgCH₂CH(CH₃)₂ / CuBr.S(CH₃)₂

c) LDA / TMSCl

d) CH₃Li / Cp(CO)₂Fe⁺=CHSPh PF₆⁻

e) (CH₃)₃O⁺ BF₄⁻

f) (1) CH₃Li (2) SOCl₂ in pyridine

ref. 79

84

Scheme 21

photocycloaddition of an accordingly substituted cyclopentenone to 3-methylcyclopentene. A recent synthesis of the fungal metabolite *sterpurene* (**84**) in 15% overall yield starting from bicyclo-octanone **6** is illustrated in Sch. 21 [79].

Other examples of cyclobutane-ring containing natural products include the lycopodium alkaloid *annotinine* (**85**), *fomannosin* (**86**), the biologically active sesquiterpene metabolite of a wood-rotting fungus, and, as already illustrated in Sch. 3, the male sex pheromon of the oll weevil, *grandisol* (**4**) (Sch. 22). While for the two former compounds the syntheses of the four membered rings are achieved by nonphotochemical paths [80,81] several photochemical routes exist for the synthesis of **4**, a more recent one, using 3-methylcyclopent-2-enone (**87**) as starting material, being illustrated in Sch. 23 [82].

A complete overview of syntheses of "nonnatural" cyclobutanes via enone + alkene photocycloadditions would be far beyond the scope of this chapter, recent reviews on this topic having been published recently [33,83].

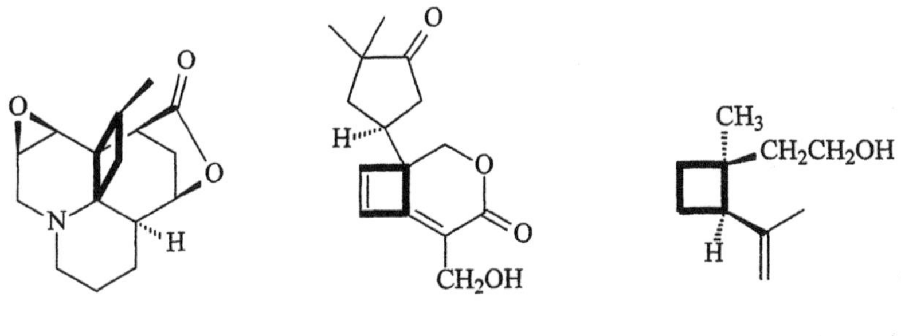

Annotinine (**85**) *Fomannosin* (**86**) *Grandisol* (**4**)

Scheme 22

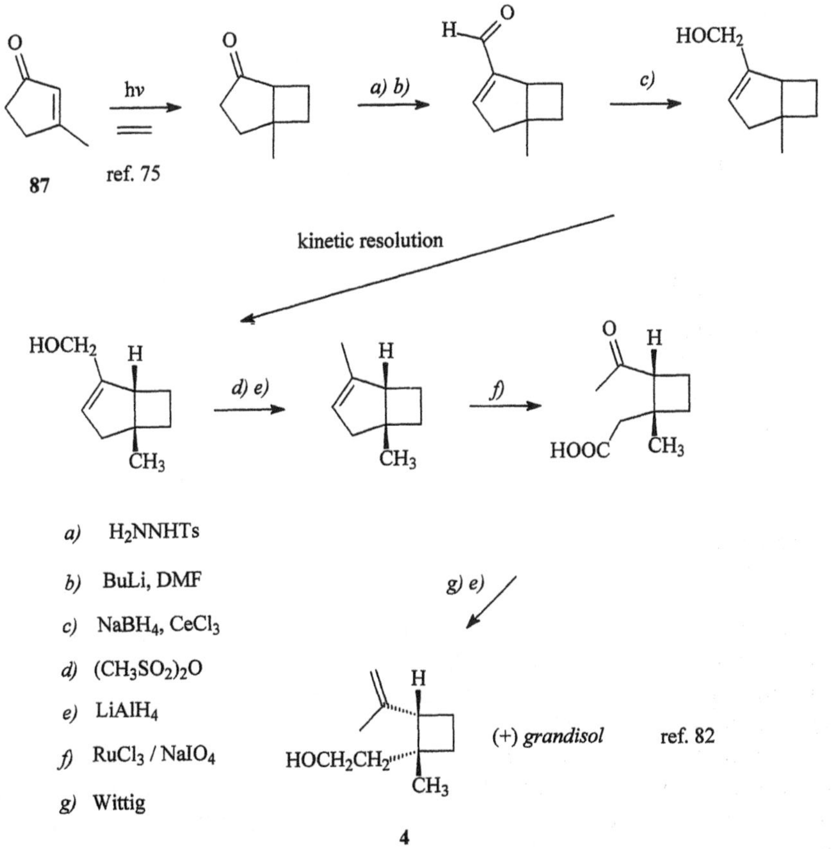

kinetic resolution

a) H$_2$NNHTs

b) BuLi, DMF

c) NaBH$_4$, CeCl$_3$

d) (CH$_3$SO$_2$)$_2$O

e) LiAlH$_4$

f) RuCl$_3$ / NaIO$_4$

g) Wittig

g) e)

(+) *grandisol*   ref. 82

4

Scheme 23

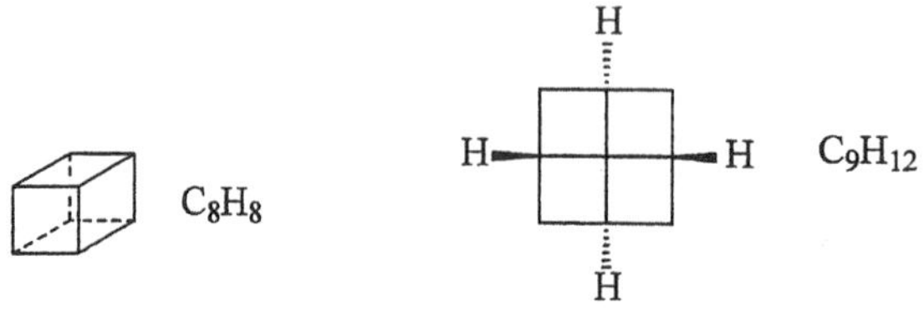

88   **89**

Scheme 24

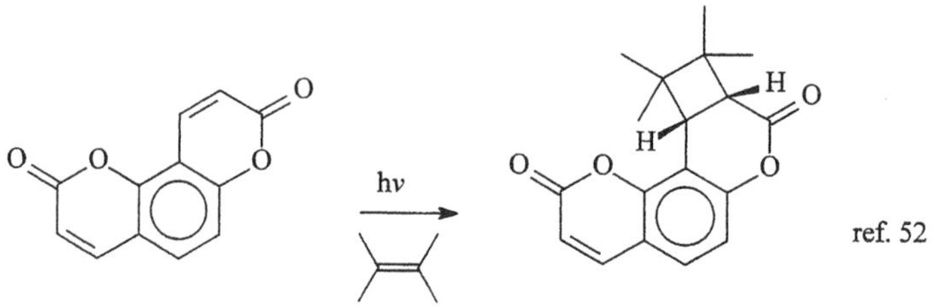

ref. 52

90   **91**

ref. 85

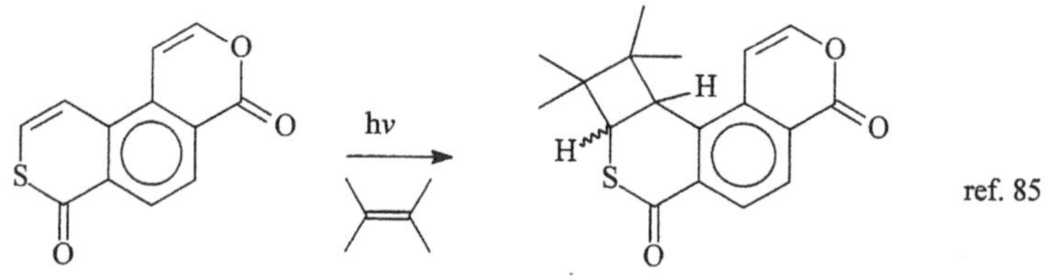

92   **93**

Scheme 25

Therefore only some (subjectively) representant examples will be mentioned. The syntheses of both *cubane* (**88**) and *[4,4,4,4]-fenestrane* (**89**) include an (intramolecular) photocycloisomerization step of an appropriate alkenyl-cycloalkenone (Sch. 24) [6,84].

Recent synthetic applications of intermolecular photocycloadditions have dealt with the question of chemoselective additions of alkenes to one (out of two) reactive excited double bonds (Sch. 25). In this context, it is of interest that 2,3-dimethylbut-2-ene adds selectively to benzodipyran **90** (affording **91**) and thiopyranobenzopyran **92** to give **93**, respectively [52,85].

REFERENCES

1. Ciamician G, Silber P. Ber 1908; 41:1928.
2. Treibs W. Ber 1930; 63:2378.
3. Treibs W. J Prakt Chem 1933; 138:299.
4. Butenandt A, Wolff A. Ber 1939; 72:1121.
5. Butenandt A, Poschmann LK, Failer G, Schiedt U, Biekert E. Liebigs Ann Chem 1952; 575:123.
6. Eaton PE, Cole TW. J Am Chem Soc 1964; 86:982.
7. Eaton PE. J Am Chem Soc 1962; 84:2454.
8. Corey EJ, Bass JD, LeMahieu R, Mitra RB. J Am Chem Soc 1964; 86:5570.
9. Chapman OL, Lenz G. In: Chapman OL, ed. Organic Photochemistry. Vol. 1. NY: Marcel Dekker, Inc., 1967:294.
10. Schönberg A, Schenck GO, Neumüller OA. In: Preparative Organic Photochemistry. Berlin: Springer Verlag, 1968:112.
11. Margaretha P, Grohmann K. Helv Chim Acta 1982; 65:556.
12. Hikino H, de Mayo P. J Am Chem Soc 1964; 86:3582.
13. Zurflueh RC, Siddel JB. Zoecon Corp, Chem Abstr 1971; 75:151423q.
14. Cantrell TS, Haller WS, Williams JC. J Org Chem 1969; 34:509.
15. de Mayo P. Acc Chem Res 1971; 4:41.
16. Margaretha P. Helv Chim Acta 1974; 57:1866.
17. Schuster DI, Heibel GE, Brown PB, Turro NJ, Kumar CV. J Am Chem Soc 1988; 110:8261.
18. Schuster DI. In: Horspool WM, Song PS, eds. CRC Handbook of Organic Photochemistry and Photobiology. Boca Raton: CRC Press, 1995:652.
19. Krug P, Rudolph A, Weedon AC. Tetrahedron Lett 1993; 34:7221.
20. Weedon AC. In: Horspool WM, Song PS, eds. CRC Handbook of Organic Photochemistry and Photobiology. Boca Raton: CRC Press, 1995:634.
21. Garcia-Esposito E, Bearpark MJ, Ortuno RM, Robb MA, Branchadell V. J Org Chem 2002; 67:6070.
22. Bertrand C, Bouquant J, Pete JP, Humbel S. Theochem 2001; 122:5866.
23. Wilsey S, Gonzalez L, Robb MA, Houk KN. J Am Chem Soc 2000; 122:5866.
24. Suishu T, Shimo T, Somekawa K. Tetrahedron 1997; 53:3545.
25. Broeker JL, Eksterowicz JE, Belk AJ, Houk KN. J Am Chem Soc 1995; 117:1847.
26. Meyer L, Alouane N, Schmidt K, Margaretha P. Can J Chem 2003; 81:417.
27. Mattay J. In: Horspool WM, Song PS, eds. CRC Handbook of Organic Photochemistry and Photobiology. Boca Raton: CRC Press, 1995:618.
28. Srinivasan R, Carlough KH. J Am Chem Soc 1967; 89:4932.
29. Agosta WC, Wolff S. J Org Chem 1980; 49:3139.
30. Anklam E, Lau S, Margaretha P. Helv Chim Acta 1985; 68:1129.
31. Kowalewski R, Margaretha P. Helv Chim Acta 1992; 75:1925.
32. Margaretha P. In: Ramamurthy V, Schanze KS, eds. Organic Photochemistry. Vol. 1. New York: Marcel Dekker, Inc., 1997:85.
33. Crimmins TM, Reinhold TL. Organic Reactions 1993; 44:297.

34. Crowley KJ, Erickson KL, Eckell A, Meinwald J. JCS Perkin Trans I 1973; 2671.
35. Margaretha P. Tetrahedron 1973; 29:1317.
36. Anklam E, Ghaffari-Tabrizi R, Hombrecher H, Lau S, Margaretha P. Helv Chim Acta 1984; 67:1403.
37. Patjens J, Margaretha P. Helv Chim Acta 1989; 72:1817.
38. Margaretha P. Liebigs Ann Chem 1973; 727.
39. Guerry P, Blanco P, Brodbeck H, Pasteris O, Neier R. Helv Chim Acta 1991; 74:163.
40. Margaretha P. Chimia 1975; 29:203.
41. Cruciani G, Rathjen HJ, Margaretha P. Helv Chim Acta 1990; 73:856.
42. Desobry V, Margaretha P. Helv Chim Acta 1975: 58:2161.
43. VoThi G, Margaretha P. Helv Chim Acta 1976; 59:2236.
44. Agosta WC, Margaretha P. Acc Chem Res 1996; 29:179.
45. Witte B, Margaretha P. Org Letters 1999; 1:173.
46. Meyer L, Elsholz B, Reulecke I, Schmidt K, Margaretha P. Helv Chim Acta 2002; 85:2065.
47. Cruciani G, Semisch C, Margaretha P. J Photochem Photobiol A 1988; 44:219.
48. Oliveira Ferrer L, Margaretha P. Chem Commun 2001; 481.
49. Kokubo K, Yamaguchi H, Kawamoto T, Oshima T. J Am Chem Soc 2002; 124:8912.
50. Alibes R, de March P, Figueredo M, Font J, Racamonde M. Tetrahedron Lett 2001; 42:6695.
51. Wrobel MN, Margaretha P. Helv Chim Acta 2003; 86:515.
52. Bethke J, Margaretha P. Helv Chim Acta 2002; 85:544.
53. Bethke J, Jakobs A, Margaretha P. J Photochem Photobiol A 1997; 104:83.
54. Bach T, Bergmann H, Grosch B, Harms K. J Am Chem Soc 2002; 124:7982.
55. Kinder MA, Meyer L, Margaretha P. Helv Chim Acta 2001; 84:2373.
56. Crimmins MT, Guise LE. Tetrahedron Lett 1994; 35:1657.
57. Bach T, Kemmler M, Herdtweck E. J Org Chem 2003; 68:1994.
58. Gebel RC, Margaretha P. Helv Chim Acta 1992; 75:1663.
59. Fleming SA, Bradford CL, Gao JJ. In: Ramamurthy V, Schanze KS, eds. Organic Photochemistry. Vol. 1. New York: Marcel Dekker, Inc., 1997:187.
60. Bach T. Synthesis 1998; 683.
61. Schuster DI, Kaprinidis N, Wink DJ, Dewan JC. J Org Chem 1991; 56:561.
62. Cargill RL, Morton GH, Bordner J. J Org Chem 1980; 45:3929.
63. Schreiber SL, Santini C. J Am Chem Soc 1984; 106:4038.
64. Baldwin SW, Mazzuckelli TJ. Tetrahedron Lett 1986; 27:5975.
65. Moens L, Baizer MM, Daniel Little R. J Org Chem 1986; 51:4497.
66. Chen C, Chang V, Cai X, Duesler E, Mariano PS. J Am Chem Soc 2001; 123:6433.
67. Faure S, Piva-Le-Blanc S, Bertrand C, Pete JP, Faure R, Piva O. J Org Chem 2002; 67:1061.
68. Wong HNC. In: de Meijere A, ed. Houben-Weyl, Methods of Organic Chemistry. Vol. E17e. Stuttgart: Thieme, 1997:435.

69. Kakiuchi K, Ue M, Tadaki T, Tobe Y, Odaira Y. Chem Letters 1986; 507.2.
70. Shimada Y, Nakamura M, Suzuka T, Matsui J, Tatsumi R, Tsutsumi K, Morimoto T, Kurosawa H, Kakiuchi K. Tetrahedron Lett 2003; 44:1401.
71. Wender PA, Hubbs JC. J Org Chem 1980; 45:365.
72. Lo PCK, Snapper ML. Org Letters 2001; 3:2819.
73. Kakiuchi K, Minato K, Tsutsumi K, Morimoto T, Kurosawa H. Tetrahedron Lett 2003; 44:1963.
74. Mitani M, Osakabe Y, Hamano J. Chemistry Letters 1994; 1255.
75. Cargill RL, Dalton JR, Morton GH, Caldwell WE. In: Freeman JP, ed. Org. Synth. Coll. Vol. VII. NY: Wiley, 1990:315.
76. Lau S, Margaretha P. Helv Chim Acta 1988; 71:498.
77. Takeda K, Shimono Y, Yoshii E. J Am Chem Soc 1983; 105:563.
78. Joulain D, König WA. In: The Atlas of Spectral Data of Sesquiterpene Hydrocarbons. Hamburg: E.B. Verlag, 1998.
79. Ishii S, Zhao S, Mehta G, Knors CJ, Helquist P. J Org Chem 2001; 66:3449.
80. Hughes DW, Gerard RV, MacLean DB. Can J Chem 1989; 67:1765.
81. Semmelhack MT, Tomoda S, Nagaoka H, Boettger SD, Hurst KM. J Am Chem Soc 1982; 104:747.
82. Hamon DPG, Tucker KL. Tetrahedron Lett 1999; 40:7569.
83. Margaretha P. In: de Meijere A, ed. Houben-Weyl, Methods of Organic Chemistry. Vol. E17e. Stuttgart: Thieme, 1997:149.
84. Agosta WC. In: Patai S, Rappoport Z, eds. The Chemistry of Alkanes and Cycloalkanes. New York: J. Wiley, 1992:927.
85. Kinder MA, Margaretha P. Photochem Photobiol Sci 2003, 2:1220.

9
Photocycloaddition of Alkenes (Dienes) to Dienes ([4+2]/[4+4])

Scott McN. Sieburth
Temple University, Philadelphia, Pennsylvania, U.S.A.

9.1. HISTORICAL BACKGROUND

Among the earliest photochemical reactions was photodimerization of anthracene **1** to give dimer **2** (Sch. 1), reported by Fritzsche in 1867 [1–3]. While not technically a reaction of 1,3-dienes, it was the first [4+4] cycloaddition reaction, and predates the first [2+2] photocycloaddition by a decade [3]. More than fifty years would pass before investigations of 1,3-diene photochemistry would be undertaken, and cycloaddition reactions were not reported until the early 1960s. The initial investigations, mercury-sensitized reactions of dienes, reported fragmentations [4]. Seminal investigations by Srinivasan at the University of Rochester in 1960 [5] and they by Hammond at the California Institute of Technology [6] led to the first examples of photo-[2+2], [4+2] and [4+4] cycloadditions by 1,3-dienes. Continuing studies of diene photochemistry have refined our under-standings of the interrelated aspects of the chemistry that include conformation, exiplex and intermediate lifetimes. Several reviews have concentrated on 1,3-diene photochemistry [7–9].

9.2. STATE OF THE ART MECHANISTIC MODELS

Woodward-Hoffman rules define photochemical [4+2] cycloadditions as disallowed based on orbital symmetry considerations, whereas [4+4] cycloadditions are allowed [10]. This is only true, however, when the

Scheme 1

photoreaction involves a singlet excited state. Consequently, photo-initiated [4+2] reactions likely proceed in an asynchronous, stepwise fashion; [4+4] cycloadditions can be concerted, but other reaction paths exist. The monograph by Michl and Bonačić-Koutecký details the many aspects of electronic transitions [11]. The use of frontier molecular orbitals and their applications to photocycloadditions has been described by Fleming [12].

Photoexcitation of 1,3-dienes can occur by direct irradiation of the diene chromophore, in which case a singlet excited state is initially produced. Intermolecular association of the diene with another molecule to form an exiplex can lead to concerted cycloadditions or, when the eximer has a strong charge-transfer character, solvent separated radical ion pairs can result [13]. The use of a sensitizer to channel the light energy to the reactants leads to a triplet excited state and the generation of biradical intermediates [14]. When electron deficient species are introduced, photo-chemically-initiated electron-transfer results in ionic intermediates, particularly in polar solvents, and ionic reaction pathways [15–18]. Access to this reaction manifold photochemically is often preferred over the use of ground state oxidants because the reaction conditions are generally less harsh.

Recent reports detail a computational analysis of the photodimeriza-tion of 1,3-butadiene that located conical intersections for concerted [4+4] and [2+2] cycloaddition paths [19,20] and have correlated these results with the products observed experimentally [21].

9.3. SCOPE AND LIMITATIONS

The photochemistry of 1,3-dienes can be highly dependent on the diene structure and reaction conditions. Important variables include the ground state conformation [22,23], the reaction concentration, the use (or not) and properties of a triplet sensitizer [14] or an electron acceptor [18], and solvent polarity. The simplest dienes also often yield the most complex chemistry. For example, 1,3-butadiene **3** undergoes unimolecular isomerization in "dilute solution" to give only cyclobutene **4** and bicyclobutane **5** (Sch. 2), and polymerization in "concentrated" solution [24]. At intermediate concentrations a set of dimers is formed. The overall conversion to dimers, less than 10% of the diene consumed, produce [2+2] **6**, [4+2] **7**, [4+4] **8** and bicyclic dimers **9** in a ratio of 30:0:8:50, plus one unidentified dimer [24]. When the dimerization of **3** was conducted using a sensitizer, this ratio changed to 98:2:0:0 with benzophenone, and 37:63:0:0 with camphor quinone [25]. Reactions that are highly selective in site, regio- and stereoselectivity are also well known (see below).

A significant factor leading to the somewhat complex photochemistry of 1,3-butadiene **3**, is its *s-trans*:*s-cis* conformer ratio, substantially disfavoring the *s-cis* conformation that is needed for [4+2] and [4+4] cycloaddition reactions. The position of this equilibrium is dependent, of course, on the diene substitution.

An instability of some cycloadducts can lead to identification problems because of nonphotochemical transformation. The *cis*-1,2-divinylcyclobutane **10**—a [2+2] product of 1,3-butadiene dimerization—will undergo Cope rearrangement at the relatively low temperature of 120 °C to give 1,5-cyclo-octadiene **11** (Sch. 3). In contrast, the tricyclic dienes related to **12**—potentially derived by [4+4] cycloaddition of 1,3-cyclohexadiene—are thermally unstable relative to the cyclobutane **13** and will often undergo Cope rearrangement under exceptionally mild conditions, sometimes at room temperature. The instability of **12** is generally ascribed to nonbonding interactions, and similar Cope rearrangements have been observed in a broad set of closely related structures [26–30].

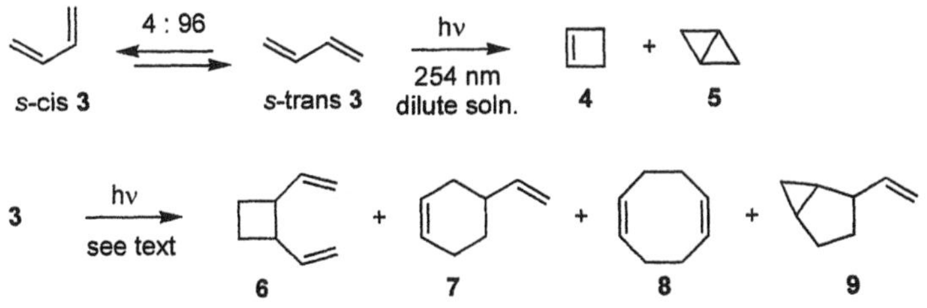

Scheme 2

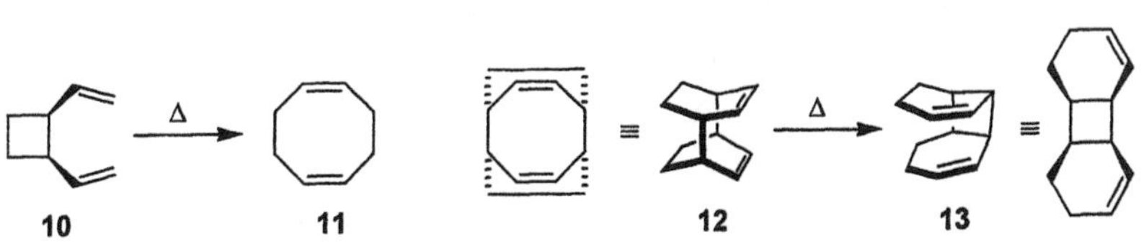

Scheme 3

9.4. SYNTHETIC POTENTIAL: REACTIVITY AND SELECTIVITY PATTERNS

Intermolecular [4+2] Cycloaddition

Irradiation of simple 1,3-dienes in the presence of a sensitizer leads to dimers, with the product ratio dependent on the triplet energy of the sensitizer. For example, acetophenone **14** converts 1,3-butadiene **3** almost exclusively to the [2+2] adduct **6** (80:20: *trans:cis*), (Sch. 4) [31], whereas the use of pyrene **15** gives mostly the [4+2] adduct **7** [25].

Similarly, isoprene **16** gave a sensitizer-dependent mixture of [2+2], [4+2] as well as [4+4] dimers (Sch. 5). This outcome is complicated by the isomers resulting from unsymmetric methyl substitution. Once again, [2+2]

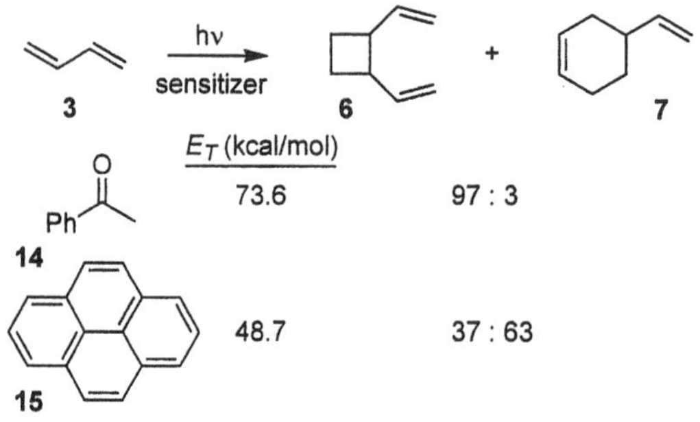

Scheme 4

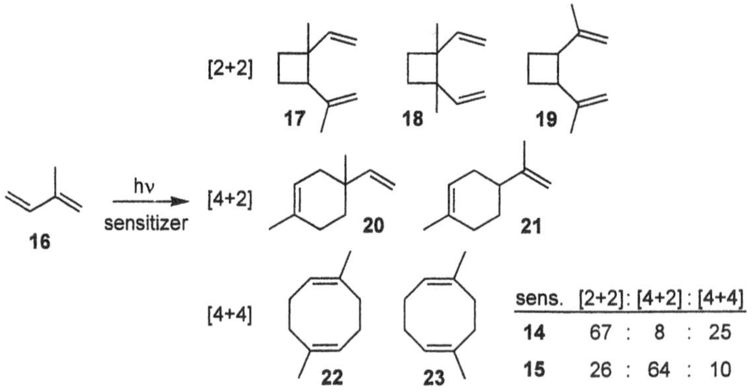

Scheme 5

adducts were dominant with acetophenone **14** as the sensitizer, and [4+2] adducts were in the majority when pyrene **15** was employed [25].

The sensitizer-dependence of these product ratios has been attributed to triplet energies that differ for the *s-cis* and *s-trans* conformers (Sch. 6). Conversion of the dienes to triplet biradicals **24** and **25** lock in the initial conformations. Subsequent bond formation with a second diene leads to two allylic radicals **26** and **27**, again preserving the geometries. Spin inversion and bond formation then forms the cyclobutane **6** from **26** and the cyclohexene **7** from **27**. The higher population of *s-cis* conformer for isoprene **16** is reflected in the higher proportion of [4+2] cycloadducts **20** and **21** as well as the appearance of [4+4] cycloadducts **22** and **23** in the product mixture (Sch. 5) [32].

Sensitized irradiation of dienes with 1-acetoxy acrylonitrile **29** also gave a sensitizer-dependent product mixture of [4+2] **30** and [2+2] **31** (Sch. 7), with a good correlation with the results of Hammond for 1,3-diene dimerization (Sch. 4), and consistent with triplet sensitization of the diene [33–35].

When the diene is locked into an *s-cis* conformation, such as cyclopentadiene **32**, cycloadditions lead to higher proportions of [4+2]

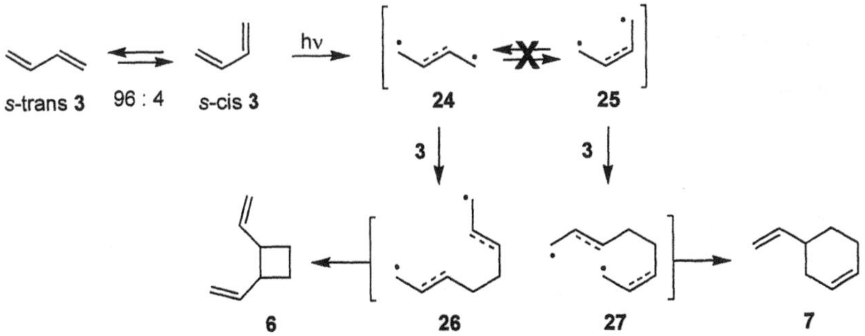

Scheme 6

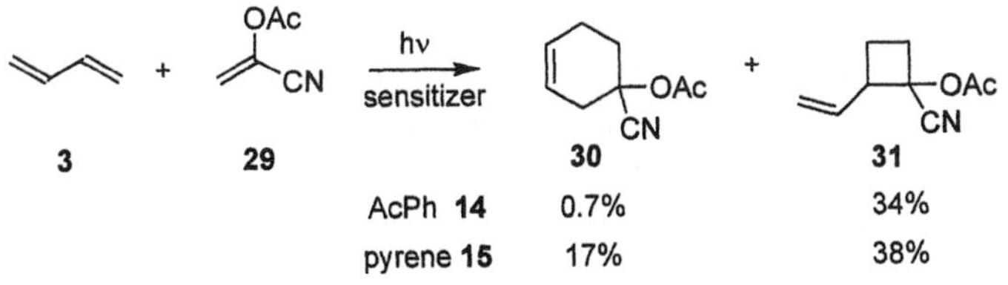

| | AcPh **14** | 0.7% | 34% |
| | pyrene **15** | 17% | 38% |

Scheme 7

products as well as [4+2] isomers not readily formed by the thermal cycloaddition (Sch. 8). Thermal Diels–Alder reaction yields only the *endo* adduct 33, whereas sensitized photoreaction yields a mixture of *endo* 33, *exo* 34, and [2+2] adduct 35 [36].

Numerous photodimerization studies of 1,3-cyclohexadiene 36 have been reported (Sch. 9). Thermal cycloaddition yields a 4:1 mixture of *endo*/ *exo* [4+2] adducts 37 and 38 in modest yield. Irradiation of the diene in cyclohexane near its λ_{max} of 254 nm yields very little dimer, but irradiation at 313 nm leads to a mixture of dimers, favoring the [2+2] adducts 39 [37]. The use of γ-radiation produces similar mixtures [38,39]. A triplet sensitizer leads to largely the [2+2] adducts plus *exo* 38 and little of the *endo* [4+2] isomer 37 [40]. When the photochemistry is conducted in the presence of the electron acceptors anthracene 41, LiClO$_4$-42 or pyrylium 43, only [4+2]

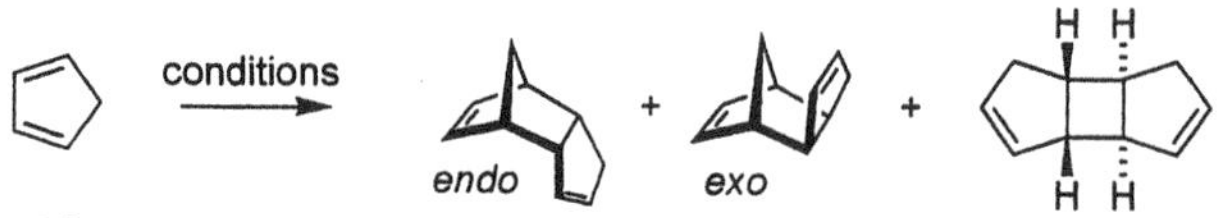

conditions	33 : 34 : 35
Δ	1 : 0 : 0
hv, acetone	1 : 1 : 1

Scheme 8

conditions	37 : 38 : 39 : 40	yield
Δ	4 : 1 : – : –	32%
hv, 313 nm, cyclohexane	1.5 : 1 : 5 : 2	
hv, triplet sensitizer	– : 1 : 4 : –	92%
hv, 41	3 : 1 : – : –	60%
hv, > 350 nm, LiClO$_4$, 42	1 : 1 : – : –	60%
hv, > 350 nm, 43	6 : 1 : – : –	70%
hv, 44	10 : 1 : 1 : –	82%

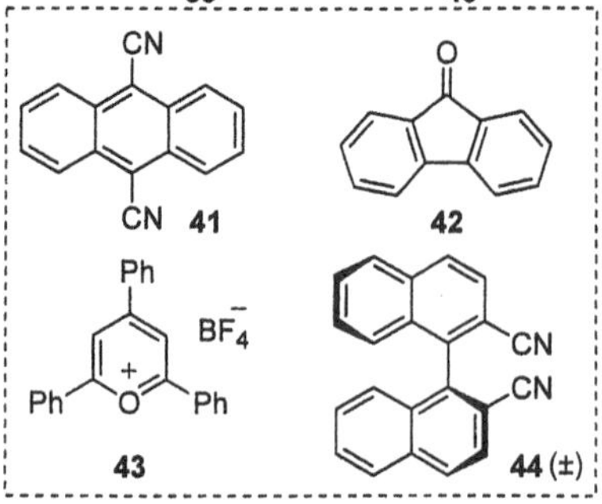

Scheme 9

adducts are produced. With anthracene **41** the reaction outcome is similar to the thermal cycloaddition, but the yield is nearly twice as high [41]. Dimerization of cyclohexadiene sensitized by **41** has been compared with dimerization using 1,8-dicyanonaphthalene **67b** [42]. Use of the $LiClO_4$-**42** mixture as an electron acceptor results in no *endo/exo* selectivity, whereas pyrylium ion **43** is higher yielding and more selective for *endo* **37** than the thermal reaction [43]. Binapthyl **44** is better still, with good yields and a high selectivity for the *endo* isomer [44]. The use of high pressure (up to 200 MPa) has only a small effect on the *endo/exo* ratio [45]. An extensive study of quantum yields as a function of concentration, sensitizer and solvent has appeared [46].

Sensitized addition of cyclic dienes with chlorinated alkenes, employing an excess of the latter, yields mixtures of [4+2] and [2+2] adducts (Sch. 10). A substantial proportion of [4+2] adduct **46a** is formed when cyclopentadiene **32** is the diene, but cyclohexadiene **36** yields almost entirely the [2+2] adduct **46b**. Use of acyclic 1,3-dienes leads only to [2+2] products. The regioselectivity of the cycloadditions is consistent with a biradical intermediate **48** [47]. Sensitized irradiation of cyclopentadiene with 1-acetoxy acrylonitrile **49** also gives a [4+2] and [2+2] mixture, but with a higher proportion of the [4+2] adduct than the reactions using chloroalkenes **45** [33–35].

Benzophenone-sensitized photoreaction of cyclopentadiene/1,3-butadiene mixtures gave three cross products (Sch. 11). Two [4+2] adducts **52** and **53** were formed, with the vinyl norbornene **52** as the dominant [4+2] isomer. Cyclobutane **54**, a mixture of two isomers, was the major product. One isomer of **54** rearranges to the norbornene **52** while the other isomer rearranges to the [4+4] adduct **55** [48].

Photo-induced radical-cation [4+2] cycloadditions between cyclohexadiene **36** and 1,3-dioxole **56a** yields **57a** and **58a** (Sch. 12). The ratio of

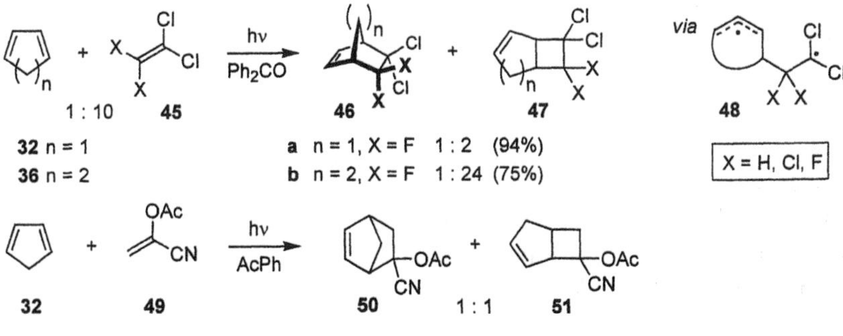

Scheme 10

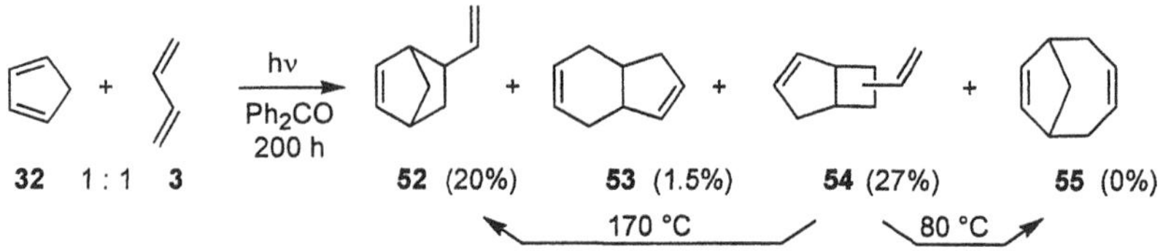

32 1 : 1 **3**   **52** (20%)   **53** (1.5%)   **54** (27%)   **55** (0%)

                                        170 °C   80 °C

Scheme 11

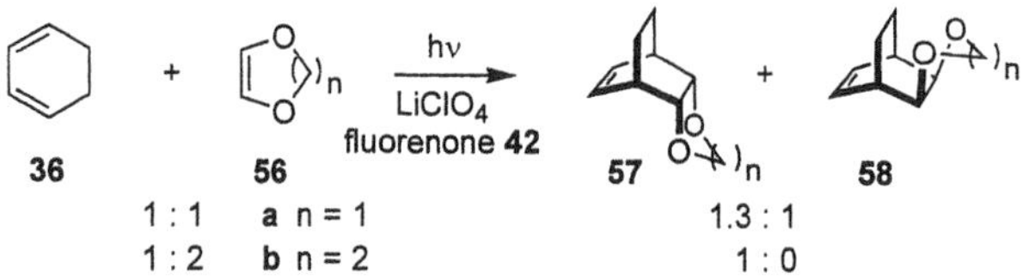

36   **56**
          1 : 1   **a** n = 1   1.3 : 1
          1 : 2   **b** n = 2   1 : 0

Scheme 12

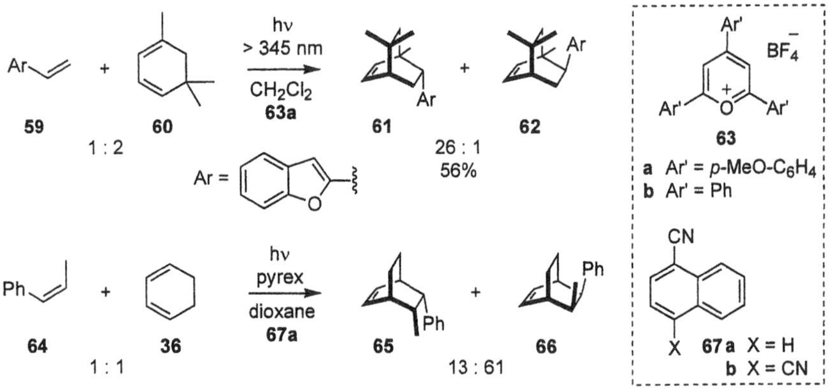

Scheme 13

products is mildly dependent on the sensitizer, and the product ratio can resemble that derived from nonphotochemical generation of radical cations. Dimers of both **36** and **56a** were also detected [43]. For dihydro-1,4-dioxin **56b**, the yields are low but only the *endo* adduct **57b** was formed [17].

Arene-substituted alkenes can undergo radical-cation [4+2] cycloadditions with 1,3-dienes when they are irradiated in the presence of an electron acceptor (Sch. 13) [49]. Good levels of regio and stereoselectivity are seen: both (*E*) and (*Z*) β-methyl styrenes (e.g., **64**) maintain stereo-chemical integrity [50]. For mixtures of alkenes and dienes with dicyanonaphthalene **67a** as the electron acceptor, a ternary reaction complex involving a singlet excited state has been implicated [50]. A closely related study of the cycloaddition of indene and cyclohexadiene promoted by **67a** also indicated that a ternary complex was present in the transition state [42].

Furan **69** can also act as the diene component (Sch. 14). A mixture of furan **69** and indene **68** (but not styrene), using dicyanonaphthalene **67b** as an electron acceptor, yields the [4+2] adduct **70** as the major product [51]. Indole **72** can also function as the dienophile using triphenylpyrylium ion **63b** as the electron acceptor. Good yields of **73** and **74** require trapping of the easily oxidized products as the acetamides [52].

Both thermal and photo-[4+2] reactions between C$_{60}$ **75** and (*E,Z*)-diene **76** (Sch. 15) gives largely adduct **77** in which the substituents of the newly formed cyclohexene substituents are *trans*, instead of the expected *cis* isomer, a result consistent with stepwise bond formation [53].

Styrenes can also participate as the diene component in photo-chemically-mediated radical cation [4+2] cycloadditions, as well as in

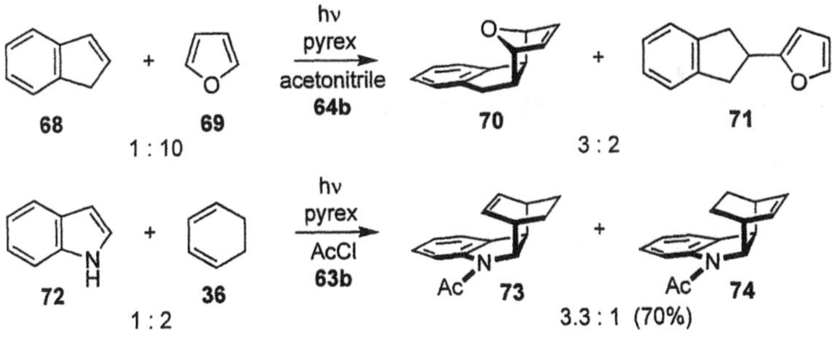

Scheme 14

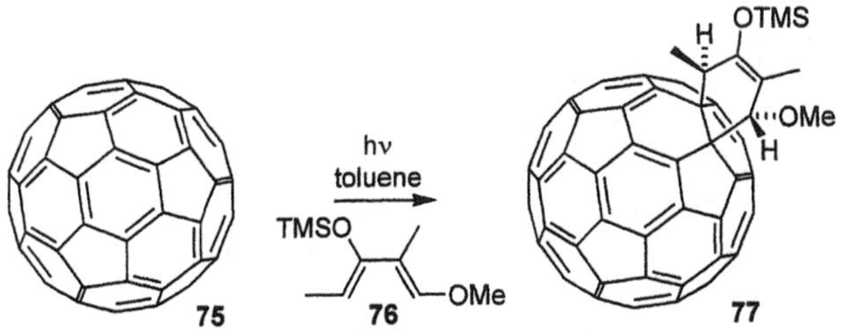

Scheme 15

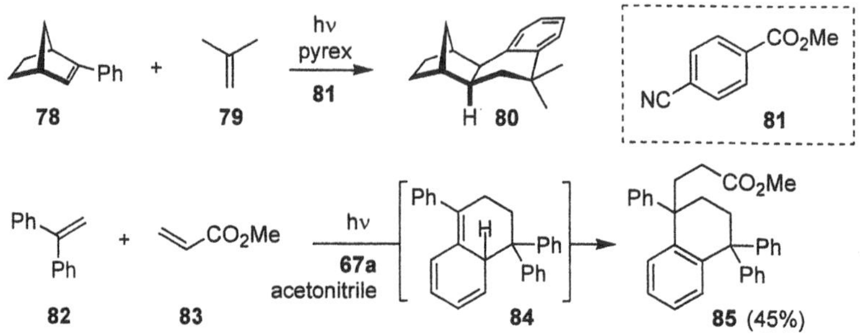

Scheme 16

dimerization reactions (Sch. 16) [54]. Irradiation of **78** in the presence of excess isobutylene **79**, using nitrile ester **81** as the electron acceptor, gives **80** in 45% yield. The reaction rate increases with solvent polarity [55,56]. The related electrophilic dimerization/cyclization of **82** leads to intermediate **84** which subsequently isomerizes and regenerates the aromatic ring. Dimerization of **82** using dicyanonaphthalene **67a** in the presence of methyl acrylate **83** generates **84**, which is intercepted by an ene reaction with the acrylate to give ester **85** in surprisingly good overall yield [57].

Intramolecular [4+2] Cycloaddition of Dienes

Sensitized intramolecular reaction of two 1,3-dienes (**86**) (Sch. 17) yields predominantly the [2+2] adduct **87**, with small amounts of [4+4] adduct **89** and little, if any, [4+2] product **88** [58,59], consistent with Hammond results for intermolecular reactions of acyclic dienes (Sch. 4). Benzophenone-sensitized reaction yields a mixture of two isomers of **87**. Heating this mixture to 200 °C converts both isomers of **87** to cyclo-octadiene **89** [58]. Unsensitized photoreaction of **86** in the presence of copper(I) triflate gives a significant amount of [4+2] adduct **88**. Extended irradiation time converts much of **87** and **89** into **88**, as well as producing secondary products [59]. Copper triflate-mediated photocycloaddition of a related tethered diene-monoalkene, gave only the [2+2] adduct [59].

Generation of the highly strained tricyclic **91** (Sch. 18) can be accomplished by transannular photo-[4+2] cycloaddition of 1,3,6-cyclo-octatrience **90** by direct (unsensitized) irradiation, albeit in modest yield [60]. Structures such as **91** can, however, be unstable toward a thermal retro-Diels–Alder reaction [61].

Radical cation-mediated intramolecular [4+2] cycloaddition proceeds efficiently only when the diene and dienophile are separated by three atoms

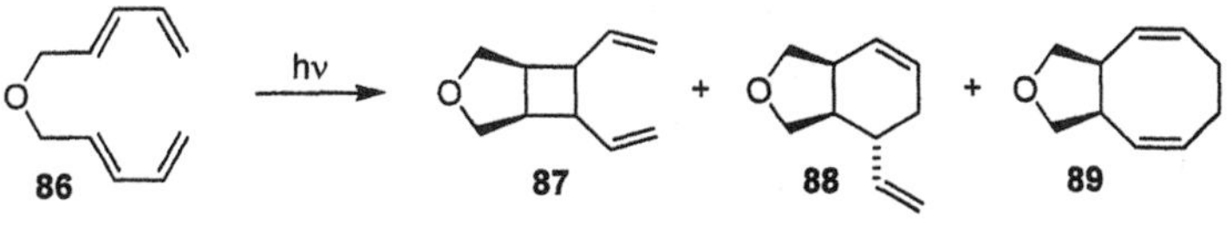

86 **87** **88** **89**

	87	88	89
acetone, pyrex	92% 3 isomers	< 1%	4%
Ph_2CO, pyrex.	50% 2 isomers	— 200 °C →	(75%)
CuOTf, quartz, 6 h	77%	14%	4%
CuOTf, quartz, 120 h	—	45%	—

Scheme 17

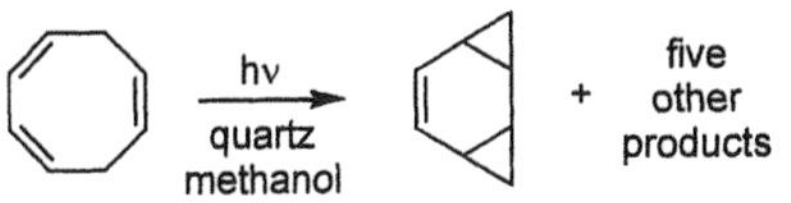

90 **91** (11%)

Scheme 18

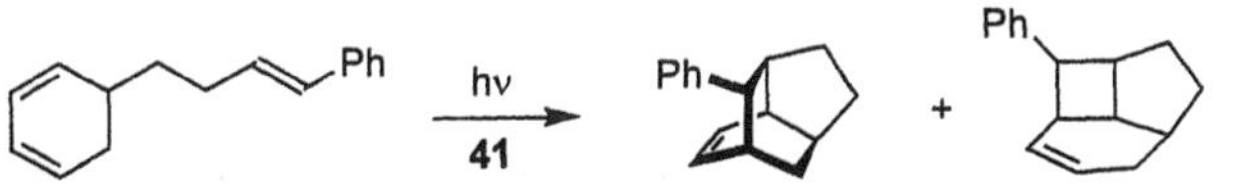

92 **93** (88%) **94** (12%)

Scheme 19

and when the diene is cyclic, i.e., **92** (Sch. 19). The yield for [4+2] cycloadduct **93** is high when the electron acceptor is dicyanoanthracene **41**. Use of benzophenone or an acyclic diene gives only [2+2] adducts [62].

When 2-pyridones are symmetrically tethered through nitrogen with a two-carbon chain (**95**) (Sch. 20), benzophenone-sensitized cycloaddition leads to the [4+2] adduct **96** in good yield. Other tether lengths largely lead to [2+2] products (see Sch. 43 for a [4+4] adduct using a four-carbon tether) [63].

Photo-[4+2] Cycloadditions of Enones

Intramolecular [4+2] addition of the enone to the proximal 1,3-diene in **97**—a product of thermal [6+4] dimerization—leads to the strained

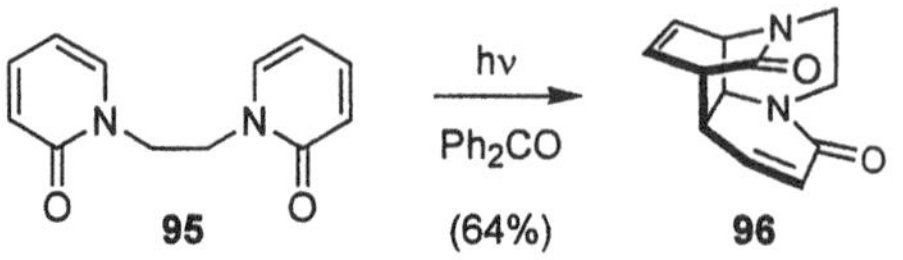

Scheme 20

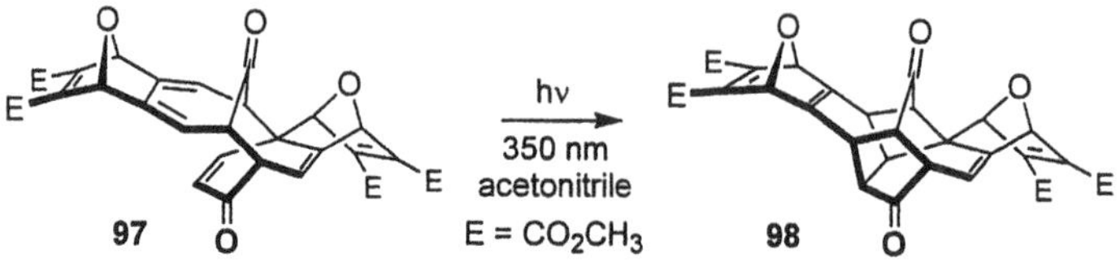

Scheme 21

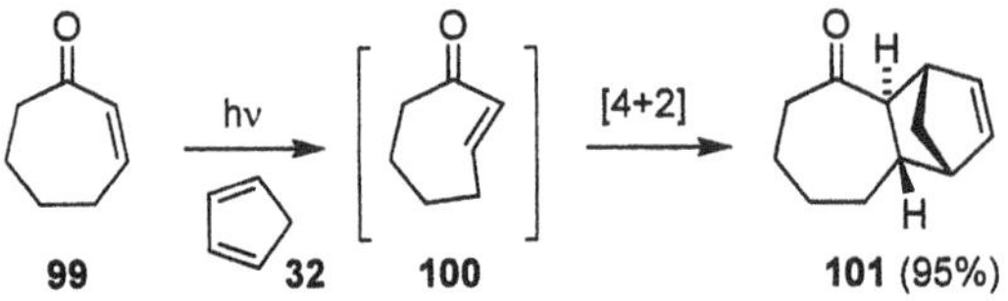

Scheme 22

structure **98** [64]. This is an unusual example of an enone photo-[4+2] reaction because the product retains the original enone stereochemistry.

Cyclic enones with ring sizes of six-to-eight carbons can be photochemically induced to undergo [4+2] cycloadditions via isomerization to a strained *trans* isomer (Schs. 22–24). Irradiation of 2-cycloheptenone **99** leads to [2+2] dimerization of an intermediate *trans*-2-cycloheptenone **100**, but if this irradiation is conducted with an excess of cyclopentadiene **32** at −50 °C, a single [4+2] adduct **101** is isolated in very high yield [65,66]. The somewhat less strained *trans*-2-cyclo-octenone can be generated and trapped by *subsequent* addition of a cyclopentadiene [67,68]. Extension of this reaction to intramolecular examples has recently been reported [69].

The photo-[4+2] cycloaddition of furan with Pummerer's ketone **102** [70,71] gives evidence for the intermediacy of the highly twisted enone intermediate **103**, and a biradical cycloaddition pathway (Sch. 23). The structures of the *endo* and *exo* products **104** were confirmed by X-ray crystallography [72,73]. In a related comparison of cyclohexenone and cyclopentenone photochemistry, conditions that gave [4+2] adducts for the cyclohexenone produced only [2+2] adducts from cyclopentenone [74].

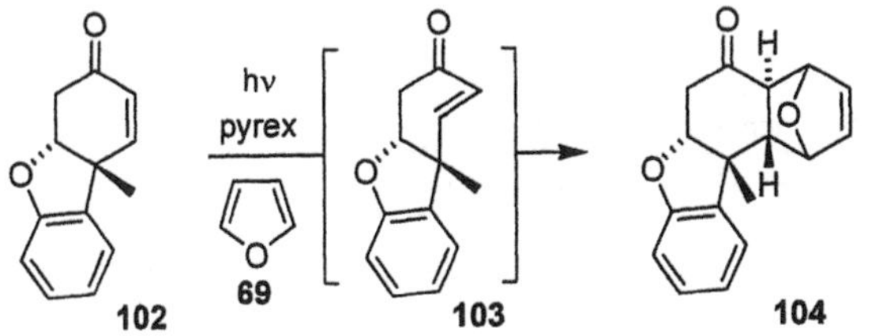

Scheme 23

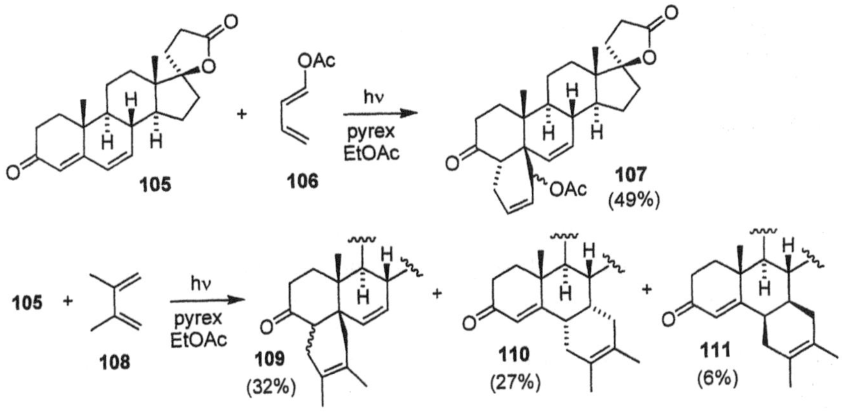

Scheme 24

Photo-[4+2] reactions of the dienone steroid **105** illustrates interesting regio-, stereo-, and site selectivities (Sch. 24) [75–78]. Reaction with 1-acetoxy-1,3-diene **106** gives *trans* adduct **107** in good yield, epimeric at the acetate. The *trans* cycloaddition was attributed to a triplet pathway rather than a twisted enone intermediate [75]. Reaction with 2,3-dimethyl-1,3-butadiene **108** leads to four [4+2] adducts, with reaction at both alkenes groups of the dienone. Note that the products of reaction at the γ,δ-alkene are both *cis*.

Additional [4+2] Cycloaddition Examples

The twisted, chiral diene **112** undergoes photosensitized [4+2] cycloadditions but only with heteroalkenes (Sch. 25). Napthoquinone **113**

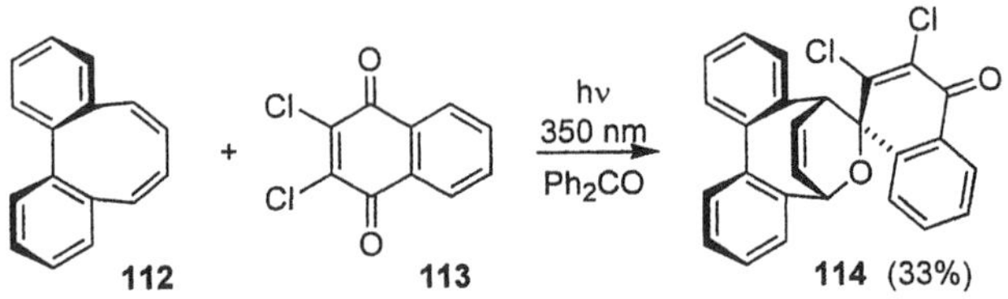

Scheme 25

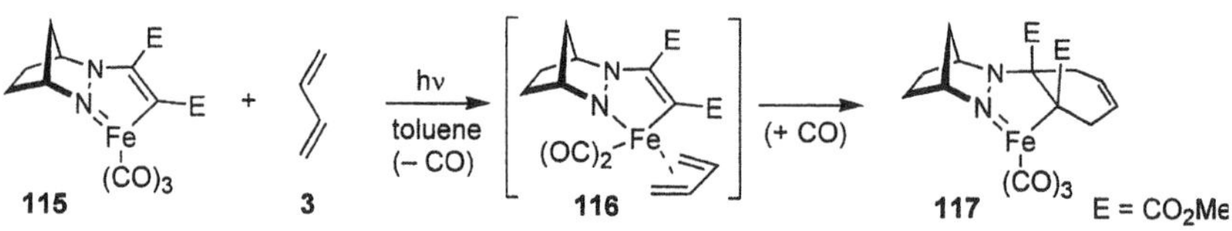

Scheme 26

yields **114** as a single stereoisomer. The absolute configuration of the biphenyl unit is not altered during this photocycloaddition [79].

An interesting example of photochemically-mediated [4+2] cycloaddition was found for **115** (Sch. 26), which does not undergo cycloaddition with butadiene **3** thermally. Irradiation is believed to transiently expel carbon monoxide, leading to co-ordination of butadiene to the iron (**116**). The diene is then delivered intramolecularly, to form **117** [80].

Intermolecular [4+4] Cycloaddition

Direct irradiation of 2,3-diphenyl-1,3-butadiene **118** (Sch. 27) leads to the [4+4] dimer **119**, in unspecified yield, in addition to tricyclic **120**, cyclobutene **121** (15%), and diphenylacetylene **122**, which may derive from **121**, The [4+4] adduct **119** is isolated by distillation, and may arise by rearrangement of an unidentified 1,2-divinylcyclobutane [2+2] adduct. Pyrolysis of **120** at 170 °C converts it to **119** [81].

The conformationally rigid *s-cis* diene **123**-(Sch. 28) yields, after triplet sensitized irradiation and distillation, the [4+4] adduct **127**. Inspection of the dimeric products before distillation, however, found that the cyclooctadiene **127** was not present, only two [2+2] adducts **124** and **125**, and one [4+2] adduct **126**. The *cis* divinylcyclobutane **124** rearranges during distillation [82].

Unsymmetrically substituted diene **128** (Sch. 29) undergoes dimerization under electron-transfer conditions with dicyanoanthracene as the catalyst. This dimerization is promoted by polar solvents and yields are also

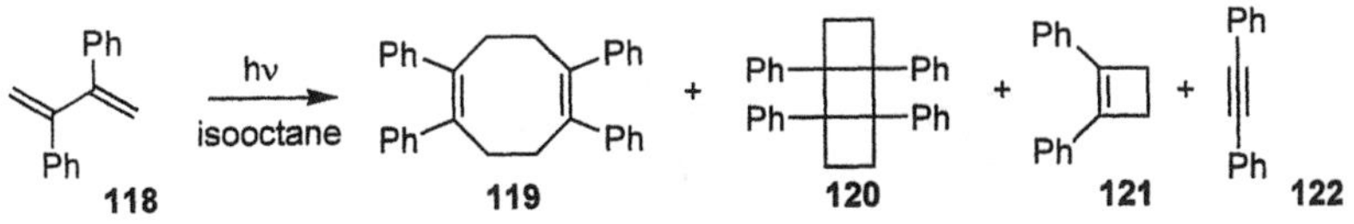

Scheme 27

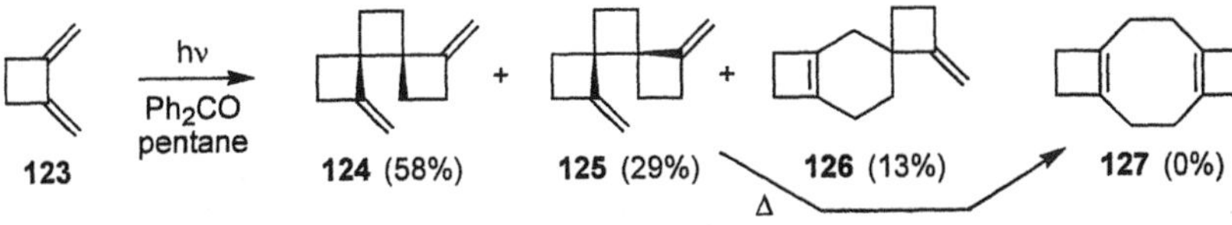

Scheme 28

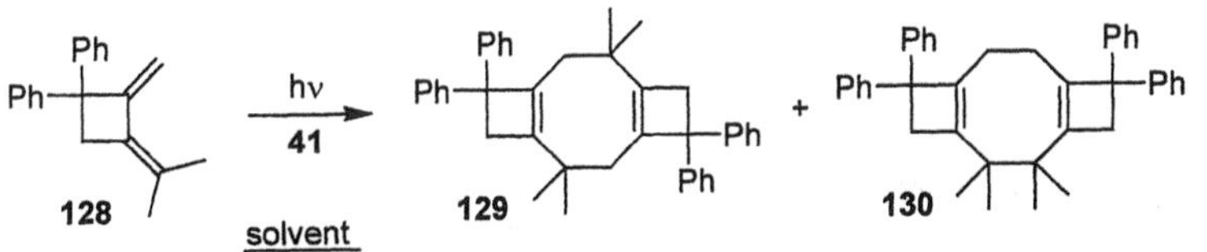

solvent		
benzene	0%	0%
CH$_2$Cl$_2$	3%	56%
acetonitrile	1%	32%

Scheme 29

Scheme 30

affected by the irradiation time. A [4+4] cycloaddition of **128** with the cyanoanthracene also competes with the dimerization [83].

Furans do not undergo [4+4] dimerization, however irradiation of a mixture of 2,5-dimethylfuran with an excess of cyclopentadiene leads to a mixture of two cross products **132** and **133** (Sch. 30) [84]. The alkene position in **133** was not determined, but **133** could have been formed by Cope rearrangement of the *cis* [4+4] adduct **134** that was not observed.

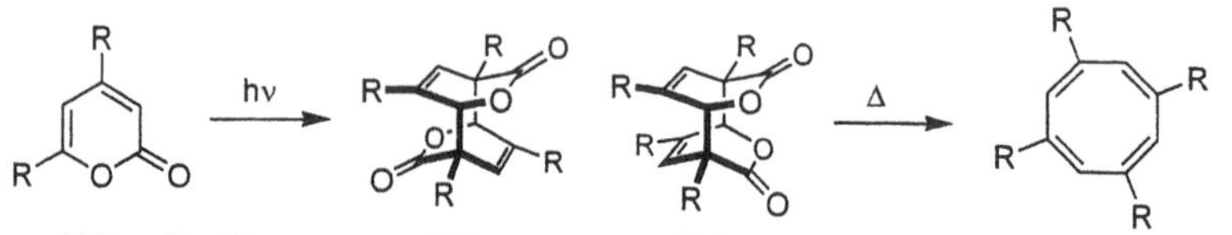

135 a R = Me
   b R = *t*-Bu
   c R = Ph

136   **137**   **138**

Scheme 31

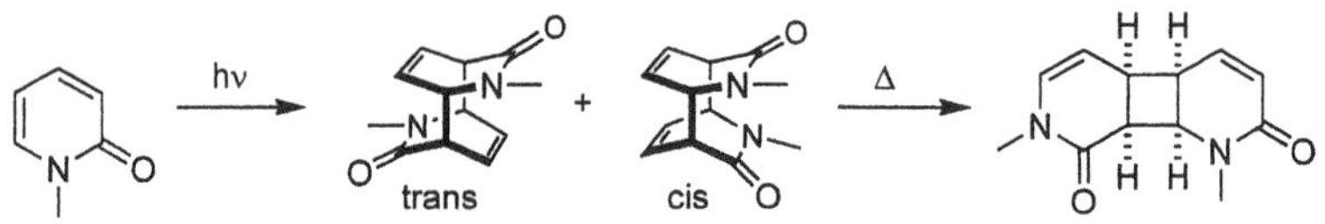

139   trans   cis   **142**
            140 56 : 44 **141**

Scheme 32

In contrast to furans, when the heterocyclic dienes take the form of 2-pyrones (**135**, Sch. 31) or 2-pyridones (**139**, Sch. 32), photo-[4+4] dimerization becomes a major reaction path [85–87]. In the case of 4,6-dimethyl-2-pyrone **135a**, irradiation forms a mixture of *trans* **136a** and *cis* **137a** plus a portion of [2+2] adduct (possibly resulting from a Cope rearrangement of **137a**) [88]. Heating this mixture leads to extrusion of carbon dioxide and formation of the 1,3,5,7-tetramethyl cyclo-octatetraene **138a** (18% overall) [89]. The utility of this application was demonstrated by the preparation of tetra-*tert*-butyl cyclo-octatetraene **138b** (72%!), although the mechanism in this case may involve an alternative cyclobutadiene pathway [90]. For the 4,6-diphenyl-2-pyrone **135c**, solid state photodimerization yields exclusively the *trans* [4+4] adduct **136c**, whereas in solution only [2+2] adducts were formed [91].

Pyridone photodimerization has been extensively studied. It proceeds via a short-lived singlet excited state, and tolerates a broad range of substitution [87]. Little solvent effect is observed for this dimerization, with comparable yields and *trans/cis* ratios using solvents ranging from water to benzene [92]. When the pyridone carries appropriate substitution, alignment in micelles can lead to very high proportions of the *cis* isomer [93]. The *cis* isomer **141** is thermally labile, undergoing Cope rearrangement under very mild conditions (ca. 60 °C) to give cyclobutane **142** [92].

Selective [4+4] cycloaddition with mixtures of 2-pyridones has been shown to be possible using 4-alkoxy-2-pyridone **144** as one component

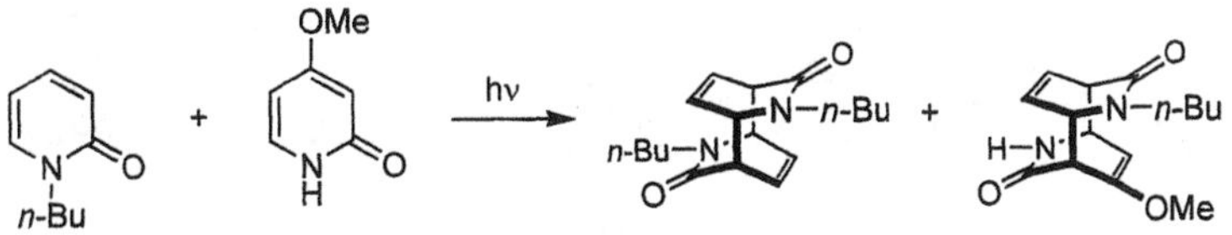

143 144 145 146

1 : 1 1.8 : 1 (28%)

1 : 7 1 : 13 (59%)

Scheme 33

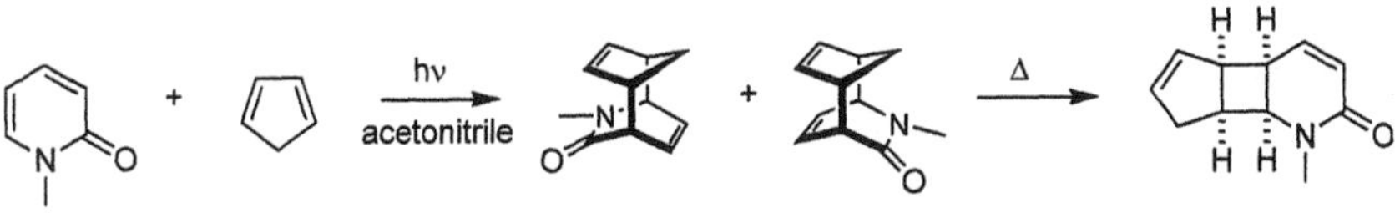

139 1 : 10 32 147 (10%) 148 (12%) 149

Scheme 34

(Sch. 33) [94] because 4-alkoxy substituent is incompatible with dimerization [95]. An excess of **144** is required to compete with dimerization of **143**, but the unreacted **144** is recovered in high yield.

Mixtures of pyridones and 1,3-dienes lead to cross-[4+4] products (Sch. 34), although dimerization of the pyridone can be a dominating reaction. With an excess of cyclopentadiene **32**, a 1:1 mixture of the *trans* **147** and *cis* **148** isomers are formed. The *cis* **148** subsequently rearranges to give **149** [96,97].

When the diene is acyclic, [4+4] cycloaddition remains the primary reaction pathway even when the result is a highly reactive and unstable *trans*-alkene product, e.g., **150** and **151** (Sch. 35). With 1,3-butadiene, these intermediates are intercepted by an additional equivalent of the diene, to give the 2:1 adducts **152** and **153**. When a diene is used that cannot achieve an *s-cis* conformation such as **154**, Diels–Alder reaction with [4+4] adducts **155** and **156** is impossible and these compounds relieve strain via Cope rearrangement to give cyclobutanes **157** and **158**, respectively [98]. An intramolecular version of this reaction has been reported [99].

Intramolecular [4+4] Cycloaddition

Proximity of two *s-cis*-1,3-dienes in the same molecule allows for [4+4] photocycloaddition. An extensively studied example is the tetraene **159** (Sch. 36) [100–106]. For the simplest example, R = H, this $C_{10}H_{10}$ molecule

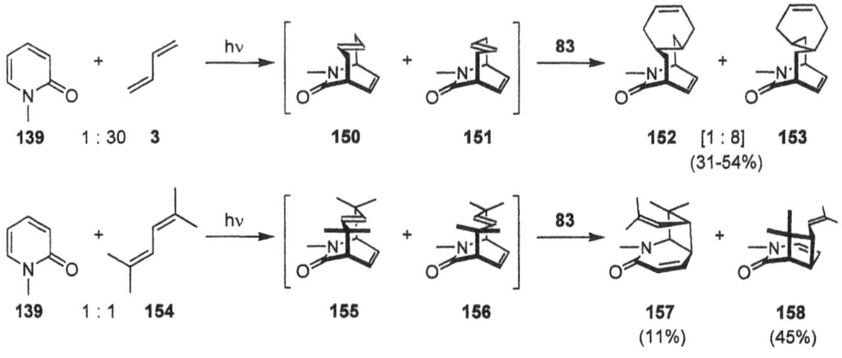

Scheme 35

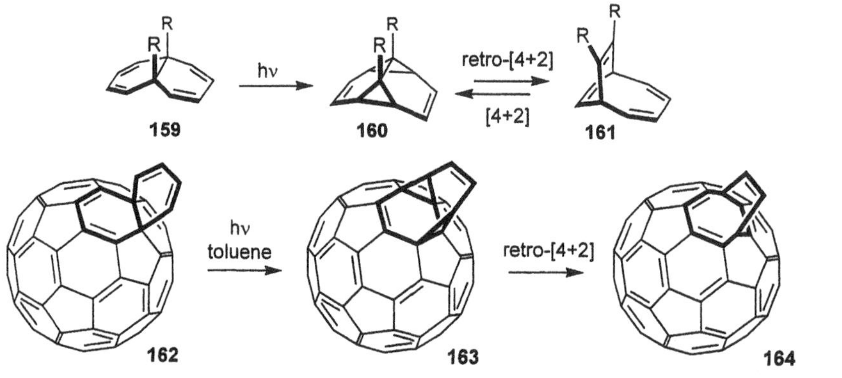

Scheme 36

is a constitutional isomer of semibullvalene. Irradiation leads to the highly strained **160**. This molecule has been found to be in thermal equilibrium with **161** (and related structures when R ≠ H). Recently, this mechanism has been invoked for the transformation of the C_{60} adduct **162** (see Sch. 15) into the expanded structure **164** [107–109]. The intermediacy of **163** has been questioned, however, with a di-π-methane mechanism proposed as an alternative pathway [110,111].

Tetraene **165**, with the dienes separated by two two-atom linkages, undergoes photocycloaddition to give the strained **166** (Sch. 37) [112]. Double extrusion of carbon monoxide from **167** yields tetraene **168** with the dienes separated by two three-carbon chains. This tetraene yields pentacyclic **169** in 50% overall yield [113].

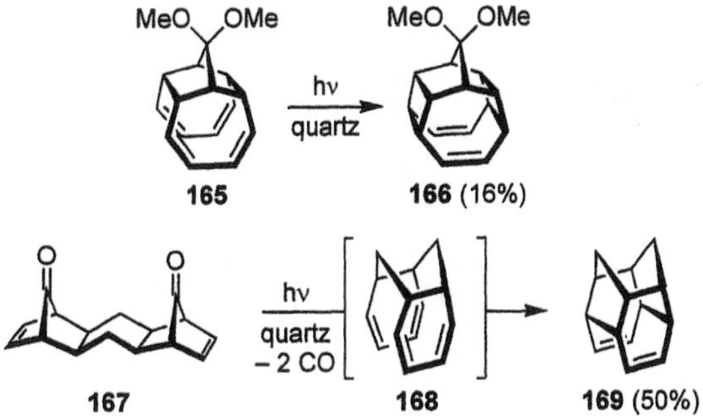

Scheme 37

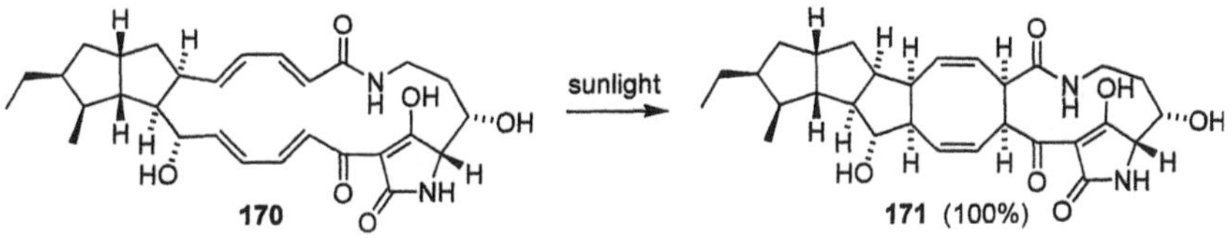

Scheme 38

The electronically and structurally more complex alteramide A **170** (Sch. 38) undergoes a quantitative [4+4] cycloaddition after 24 h of sunlight irradiation [114].

Intramolecular [4+4] cycloaddition of 2-pyrones with furans has been studied by West and coworkers (Sch. 39). Irradiation of **172**, with a furan tethered to C6 of the pyrone, leads to a mixture of *trans* **173** and *cis* **174** [115]. Interestingly, the Cope rearrangement of the *cis* isomer **174** yields an equilibrium mixture with cyclobutane **175**. The effect of a stereogenic center on the tether was also determined in this study. With the furan tethered to C3 of the triflate-substituted pyrone **176**, the *trans* isomer **177** was formed with good selectivity and isolated in very good yield [116].

Pyridone–pyridone [4+4] photocycloaddition can be intramolecular with both 3- and 4-atom tethers (Sch. 40). Photocycloaddition of **179** yields the *trans* **180** and *cis* **181** in a ratio that is similar to the intramolecular reactions (Sch. 32) [117]. The *cis* isomer undergoes Cope rearrangement producing cyclobutane **182** above 50 °C. Cope product **182**, unlike the [4+4] adducts **180** and **181**, has a conjugated π-system and on irradiation undergoes photocleavage back to **179** and photocycloaddition again. Two cycles of heat and irradiation converts the 2:1 mixture of [4+4] adducts to an 18:1 mixture, without a substantial change in yields [118].

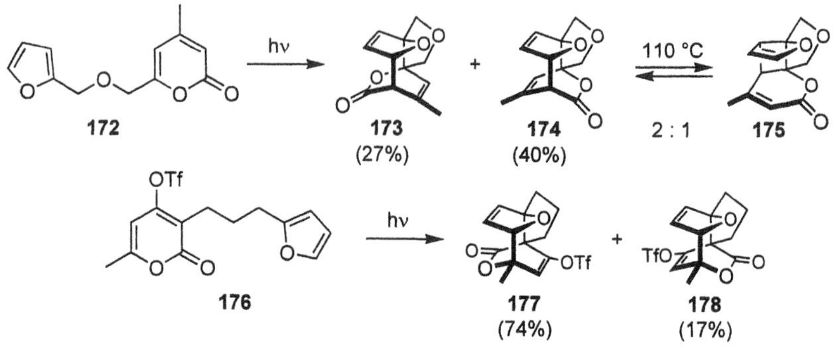

Scheme 39

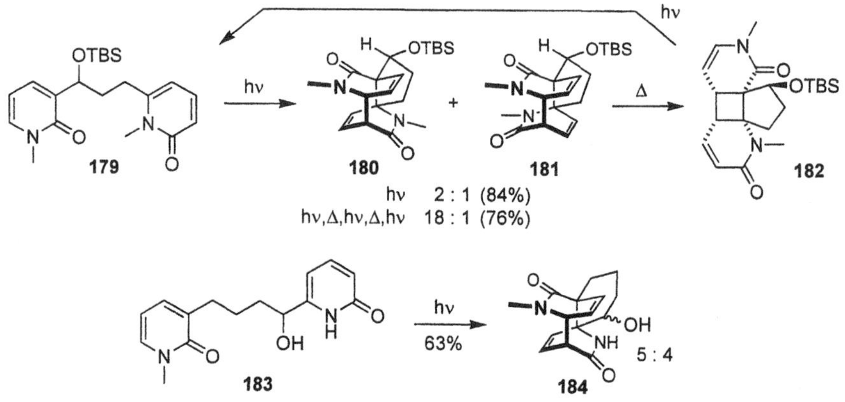

Scheme 40

When the tether is four carbons (**183**), only the *trans* isomer **184** is observed. The product **184** is substantially more strained than **180**, in large part because the cyclohexane derived from the tether is rigidly held in a boat conformation [119].

The *cis/trans* ratio for three carbon-tethered pyridones **185** (Sch. 41) was found to be solvent dependent when both pyridones lack nitrogen substitution [120]. For substrate **185a** with a single *N*-methyl group, only the *trans* isomer **186a** was formed regardless of the solvent. Without the *N*-methyl group, **185b**, the *trans/cis* ratio was solvent dependent, as well as concentration dependent for solvents of intermediate polarity such as ethyl acetate. The high *cis* selectivity of **185b** is due to a hydrogen bonded dimer **188** that aligns the substrates [121].

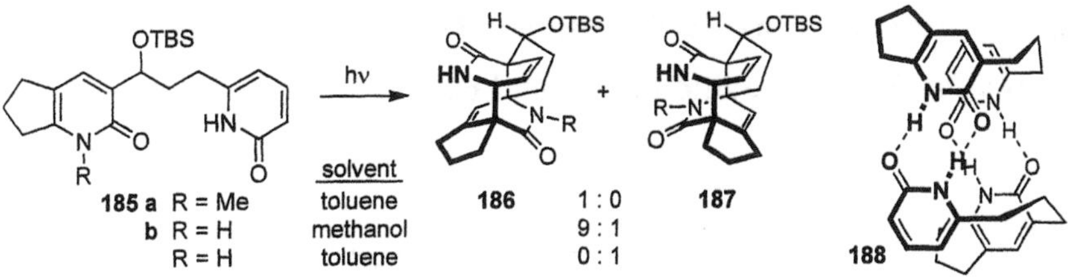

185 a	R = Me	toluene	186	1 : 0	187
b	R = H	methanol		9 : 1	
	R = H	toluene		0 : 1	

Scheme 41

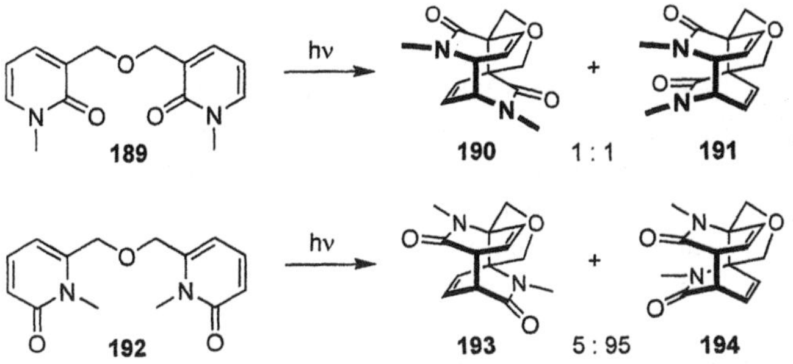

Scheme 42

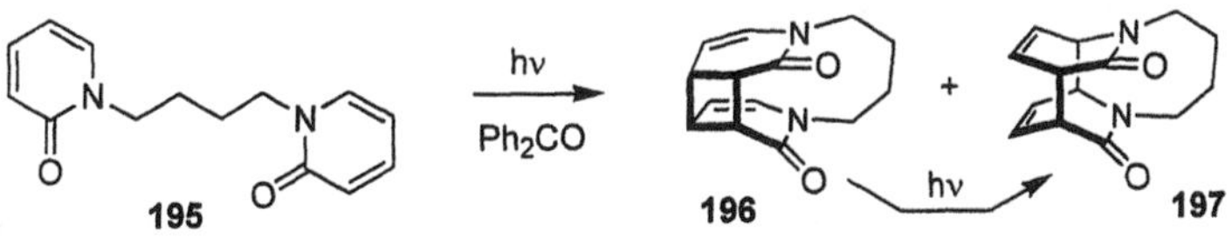

Scheme 43

The natural propensity of 2-pyridone photodimerization to favor the head-to-tail [4+4] adducts can be reversed by the use of a tether (Sch. 42). For head-to-head **189**, photocycloaddition leads to a 1:1 mixture of the *cis* and *trans* adducts [122]. For tail-to-tail isomer **192**, a high proportion of the *cis* adduct **194** is generated, although product mixtures for both reactions change under extended irradiation times, favoring the *cis* isomers [123].

When the symmetric tether is four carbons in length and attached at nitrogen (**195**), (Sch. 43), benzophenone-sensitized reaction yields a 2:1 mixture of [2+2] adduct **196** and [4+4] adduct **197**. Continued irradiation converts **196** to **197** [63]. Other tether lengths lead to only [2+2] and [4+2] adducts (see Sch. 20).

9.5. REPRESENTATIVE EXPERIMENTAL PROCEDURES

Photo-[4+2]-cycloaddition of 59 and 60 [49]

A solution of 2-vinylbenzofuran **59** (288 mg, 2 mmol), 1,5,5-trimethyl-1,3-cyclohexadiene **60** (488 mg, 4 mmol) and **63a** (20 mg, 2 mol%) in methylene chloride (120 mL) was cooled to 10–15 °C and irradiated with xenon arc lamp using a filter ($\lambda > 345$ nm) for 4 h. The product was filtered through a short column of silica gel to remove the pyrylium salt and then concentrated. Flash chromatography (cyclohexane) gave **61** (295 mg, 56%).

CuOTf-Catalyzed Irradiation of 86 [59]

A homogeneous solution of the tetraene ether **86** (1.28 g, 8.5 mmol) in 85 mL of THF (0.1 M) with 43 mg (0.17 mmol, 2 mol%) of CuOTf was irradiated for five days under argon by using a quartz immersion well. Evaporation of the solvent gave a brown, oily residue which was purified by filtration over basic alumina to give a 2.5:1 mixture of **88** and **198** (63%) which were separated by HPLC.

Photo-[4+2]-cycloaddition of Enone 102 and Furan [73]

A solution of **102** (526 mg, 2.5 mmol) in freshly distilled furan (50 mL) was degassed with argon and irradiated with a 450 W medium pressure mercury lamp using a pyrex filter for 18 h. After evaporation of the furan, the residue was recrystallized from ethanol. Preparative thin layer chromatography (3:1 methylene chloride/hexane) gave **104a** and **104b** which were recrystallized from ethyl acetate/ethanol (1:5).

Photodemerization of 1-Methyl-2-pyridone 139 [92]

A fine stream of nitrogen was bubbled through a solution of **139** (546 mg, 5 mmol) in water (10 mL) in a pyrex cell, irradiated by a 400 W high-pressure mercury lamp for 15 h. The resulting mixture was evaporated in vacuo below 40 °C. The residue was diluted with a small amount of ethanol and the precipitate was removed by filtration. The filtrate was concentrated and chromatographed on silica gel (99:1 methylene chloride/methanol) to give **140** (51%), **141** (11%), **199** (1%), and **200** (7%). Note that **199** and **200** are only observed when using water as the solvent.

Photocycloaddition of Tetraene 168 [113]

A benzene solution (500 mL) of the dione **167** (500 mg, 2.12 mmol) was carefully purged with a slow stream of nitrogen for 15 min. The solution was then irradiated with a Hanovia 450-W medium-pressure mercury vapor lamp in a quartz immersion vessel. After being irradiated for 30 min, solvent was evaporated and the residue chromatographed over silica gel (hexane) furnishing 180 mg (50%) of the crystalline diene **169**.

9.6. TARGET MOLECULES: NATURAL AND NON-NATURAL PRODUCT STRUCTURES

Natural products containing six-membered rings are ubiquitous, therefore the following set of target molecules are limited to those containing eight-membered rings that would be accessible via [4+4] cycloadditions. Notably, 1,5-cyclo-octadienes **88** can be converted to bicyclo[3.3.0]octanes **201** transannular electrophilic [124] and radical cyclization (Sch. 44) [125].

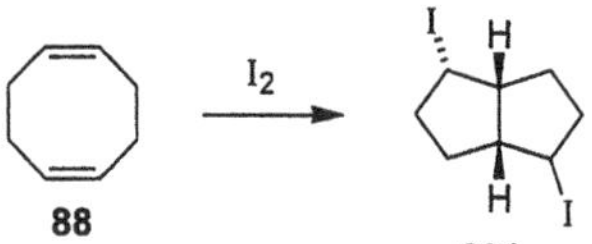

Scheme 44

(a) Polyquinane Natural Products

Pentalenolactone [126], silphinene [127], coriolin [128], and modhephene [129].

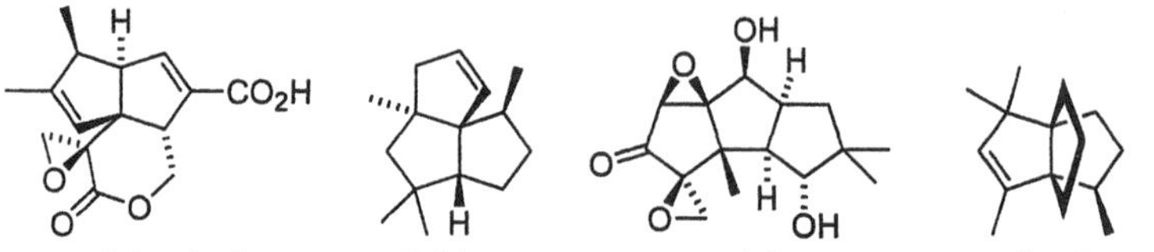

pentalenolactone   silphinene   coriolin   modhephene

(b) Monocarbocyclic Cyclo-octanoid Natural Products

Spartidiendione [130] and vulgarolide [131].

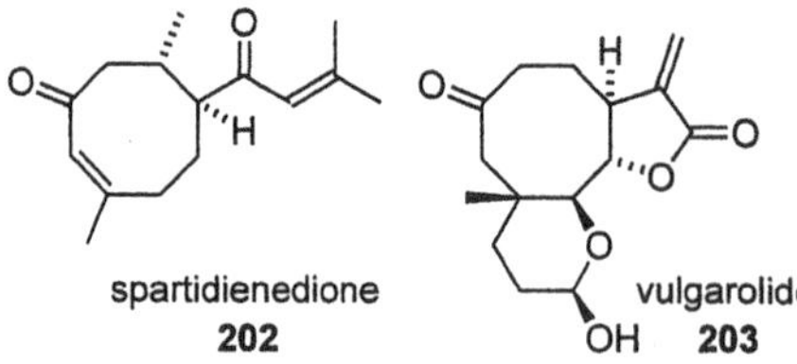

spartidienedione   vulgarolide
202   **203**

(c) Bicarbocyclic Cyclo-octanoid Natural Products

Dactylol [132], pachylactone [133], asteriscanolide [134], and salsolene oxide [135].

| dactylol | pachylactone | asteriscanolide | salsolene oxide |
| **204** | **205** | **206** | **207** |

(d) Tricarbocyclic Cyclo-octanoid Natural Products: dicyclopenta[a,d]cyclo-octanes

Ophiobolins [136], fusicoccin [137], epoxydictymet [138].

| ophiobolin A | fusicoccin A | epoxydictymene |
| **208** | **209** | **210** |

7,8-epoxy-4-basemen-6-one [139], and pleuromutilin [140].

| 7,8-epoxy-4-basmen-6-one | pleuromutilin |
| **211** | **212** |

(e) Tetracarbocyclic Cyclo-octanoid Natural Products

Longipenol [141] and kalmanol [142].

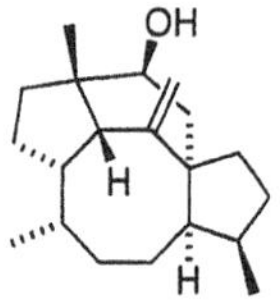

longipenol
213

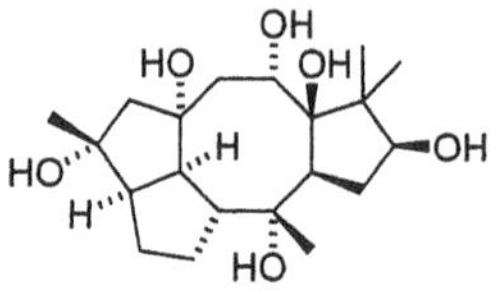

kalmanol
214

(f) Fused Cyclohexyl–Cyclo-octanoid Natural Products

Neolemnane [143], variecolin [144], taxol [145].

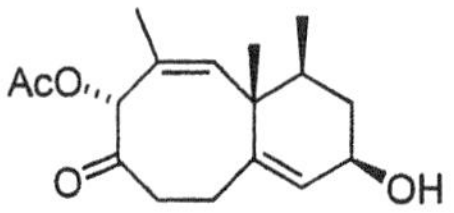

neolemnane
215

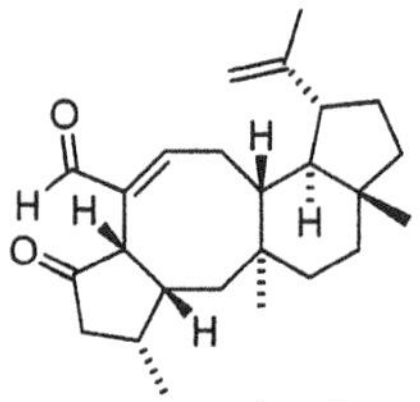

variecolin
216

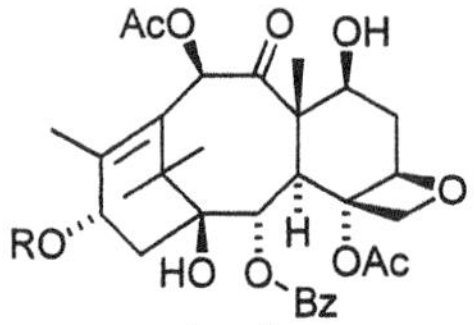

taxol
217

REFERENCES

1. Fritzsche J. J Prakt Chem 1867; 101:333.
2. Greene FD, Misrock SL, Wolfe JR Jr. J Am Chem Soc 1955; 77:3852.
3. Roth HD. Angew Chem Int Ed Engl 1989; 28:1193.
4. Gee G. Trans Faraday Soc 1938; 34:713.
5. Srinivasan R. J Am Chem Soc 1960; 82:5063.
6. Hammond GS, Turro NJ, Fischer A. J Am Chem Soc 1961; 83:4674.
7. Dilling WL. Chem Rev. 1969; 69:845.
8. Laarhoven WH. Photocyclizations and intramolecular cycloaddition of conjugated olefins. In: Org Photochem. Padwa A, ed. New York: Marcel Dekker, Inc., 1987:129–224.
9. Nuss JM, West FG. The photochemistry of dienes and polyenes: application to the synthesis of complex molecules. In: The Chemistry of Dienes and Polyenes. Rappoport Z, ed. New York: John Wiley, 1997:263–324.
10. Woodward RB, Hoffmann R. Angew Chem Int Ed Engl 1969; 8:781.
11. Michl J, Bonačić-Koutecký V. Electronic Aspects of Organic Photochemistry. New York, John Wiley and Sons, 1990.

12. Fleming I. Frontier Orbitals and Organic Chemical Reactions. New York: John Wiley and Sons, 1976.
13. Mattes SL, Farid S. Acc Chem Res 1982; 15:80.
14. Turro NJ, Dalton JC, Weiss DS. Photosensitization by energy transfer. In: Org Photochem. Chapman OL, ed. New York: Marcel Dekker, Inc., 1969:1–62.
15. Bauld NL, Bellville DJ, Pabon R, Chelsky R, Green G. J Am Chem Soc 1983; 105:2378.
16. Bauld NL, Bellville DJ, Harirchian B, Lorenz KT, Pabon RA Jr, Reynolds DW, Wirth DD, Chiou HS, Marsh BK. Acc Chem Res 1987; 20:371.
17. Mattay J, Trampe G, Runsink J. Chem Ber 1988; 121:1991.
18. Müller F, Mattay J. Chem Rev 1993; 93:99.
19. Olivucci M, Ragazos IN, Bernardi F, Robb MA. J Am Chem Soc 1993; 115:3710.
20. Bearpark MJ, Deumal M, Robb MA, Vreven T, Yamamoto N, Olivucci M, Bernardi F. J Am Chem Soc 1997; 199:709.
21. Deumal M, Bearpark MJ, Smith BR, Olivucci M, Bernardi F, Robb MA. J Org Chem 1998; 63:4594.
22. Jacobs HJC, Havinga E. Adv Photochem 1979; 11:305.
23. Dauben WG, Rabinowitz J, Vietmeyer ND, Wendschuh PH. J Am Chem Soc 1972; 94:4285.
24. Srinivasan R, Sonntag FI. J Am Chem Soc 1965; 87:3778.
25. Liu RSH, Turro NJ, Hammond GS. J Am Chem Soc 1965; 87:3406.
26. Yang NC, Libman J. J Am Chem Soc 1972; 94:9228.
27. Eaton PE, Chakraborty UR. J Am Chem Soc 1978; 100:3634.
28. Paquette LA, Doecke CW, Klein G. J Am Chem Soc 1979; 101:7599.
29. Tobe Y, Hirata F, Nishida K, Fujita H, Kimura K, Odaira Y. J Chem Soc Chem Commun 1981; 786.
30. McGee KF Jr, Al-Tel TH, Sieburth S. McN Synthesis 2001; 1185.
31. DeBoer CD, Turro NJ, Hammond GS. Org Synth 1973; C V 5:528.
32. Hammond GS, Turro NJ, Liu RSH. J Org Chem 1963; 28:3297.
33. Dilling WL, Kroening RD. Tetrahedron Lett 1968; 5601.
34. Dilling WL. J Am Chem Soc 1967; 89:2741.
35. Dilling WL. J Am Chem Soc 1967; 89:2742.
36. Turro NJ, Hammond GS. J Am Chem Soc 1962; 84:2841.
37. Bahurel YL, MacGregor DJ, Penner TL, Hammond GS. J Am Chem Soc 1972; 94:637.
38. Schutte R, Freeman GR. J Am Chem Soc 1969; 91:3715.
39. Penner TL, Whitten DG, Hammond GS. J Am Chem Soc 1970; 92:2861.
40. Valentine D, Turro NJ, Hammond GS. J Am Chem Soc 1964; 86:5202.
41. Jones CR, Allman BJ, Mooring A, Spahic B. J Am Chem Soc 1983; 105:652.
42. Calhoun GC, Schuster GB. J Am Chem Soc 1986; 108:8021.
43. Mattay J, Gersdorf J, Mertes J. J Chem Soc, Chem Commun 1985; 1088.
44. Vondenhof M, Mattay J. Chem Ber 1990; 123:2457.
45. Chung W-S, Turro NJ, Mertest J, Mattay J. J Org Chem 1989; 54:4881.
46. Mella M, Fasani E, Albini A. Tetrahedron 1991; 47:3137.

47. Turro NJ, Bartlett PD. J Org Chem 1965; 30:1849.
48. Sartori G, Turba V, Valvassori A, Riva M. Tetrahedron Lett 1966; 4777.
49. Botzem J, Haberl U, Steckhan E, Blechert S. Acta Chem Scand 1998; 52:175.
50. Akbulut N, Hartsough D, Kim J-I, Schuster GB. J Org Chem 1989; 54:2549.
51. Mizung K, Kaji R, Okada H, Otsuji Y. J Chem Soc, Chem Commun 1978; 594.
52. Gieseler A, Steckhan E, Wiest O, Knoch F. J Org Chem 1991; 56:1405.
53. Mikami K, Matsumoto S, Okubo Y, Fujitsuka M, Ito O, Suenobu T, Fukuzumi S. J Am Chem Soc 2000; 122:2236.
54. Mattes SL, Farid S. J Am Chem Soc 1986; 108:7356.
55. Neunteufel RA, Arnold DR. J Am Chem Soc 1973; 95:4080.
56. Maroulis AJ, Arnold DR. J Chem Soc Chem Commun 1979; 351.
57. Arnold DR, Borg RM, Albini A. J Chem Soc Chem Commun 1981; 138.
58. Wender PA, Correia CRD. J Am Chem Soc 1987; 109:2523.
59. Hertel R, Mattay J, Runsink J. J Am Chem Soc 1991; 113:657.
60. Roth WR, Peltzer B. Angew Chem Int Ed Engl 1964; 3:440.
61. Whitlock HW Jr, Schatz PF. J Am Chem Soc 1971; 93:3837.
62. Wölfle I, Chan S, Schuster GB. J Org Chem 1991; 56:7313.
63. Nakamura Y, Zsindely J, Schmid H. Helv Chim Acta 1976; 59:2841.
64. Wollenweber M, Fritz H, Rihs G, Prinzbach H. Chem Ber 1991; 124:2465.
65. Corey EJ, Tada M, LaMahieu R, Libit L. J Am Chem Soc 1965; 87:2051.
66. Eaton PE, Lin K. J Am Chem Soc 1965; 87:2052.
67. Eaton PE, Lin KJ. Am Chem Soc 1964; 86:2087.
68. Lange GL, Neidert E. Can J Chem 1973; 51:2207.
69. Dorr H, Rawal VH. J Am Chem Soc 1999; 121:10229.
70. Pummerer R, Puttfarcken H, Schopflocher P. Chem Ber 1925; 58B:1808.
71. Barton DHR, Deflorin AM, Edwards OE. J Chem Soc 1956; 530.
72. Mintas M, Schuster DI, Williard PG. J Am Chem Soc 1988; 110:2305.
73. Mintas M, Schuster DI, Williard PG. Tetrahedron 1988; 44:6001.
74. Cantrell TS. J Org Chem 1974; 39:3063.
75. Lenz GR. Tetrahedron Lett 1977; 2483.
76. Lenz GR. J Org Chem 1979; 44:1382.
77. Lenz GR. J Org Chem 1979; 44:1597.
78. Lenz GR. J Org Chem 1979; 44:4299.
79. Burger U, Lottaz P-A, Millasson P, Bernardinelli G. Helv Chim Acta 1994; 77:850.
80. Battagliz R, Frazier CC III, Kisch H, Krüger C. Z Naturforsch 1983; 38b:648.
81. White EH, Anhalt JP. Tetrahedron Lett 1965; 3937.
82. Borden WT, Reich IL, Sharpe LA, Weinberg RB, Reich HJ. J Org Chem 1975; 40:2438.
83. Ikeda H, Aburakawa N, Tanaka F, Fukushima T, Miyashi T. Eur J Org Chem 2001; 3445.
84. Cantrell TS. J Org Chem 1981; 46:2674.
85. West FG. Photocyclization and photocycloaddition reactions of 4- and 2-pyrones. In: Advances in Cycloaddition. Lautens M, ed. Greenwich, CT: JAI, 1997:1–40.

86. Sieburth S McN. The inter- and intramolecular [4+4] photocycloaddition of 2-pyridones and its application to natural product synthesis. In: Advances of Cycloaddition. Harmata M, ed. Greenwich, CT: JAI, 1999:85–118.
87. Sieburth S McN. Photochemical reactivity of pyridones. In: Handbook of Organic Photochemistry and Photobiology. Horspool WM, ed. CRC Boca Raton FL, 2003; Chapter 103, pp. 103-1–103-18.
88. de Mayo P, Yip RW. The photo-dimerization of 2,4-dimethylcoumalin: the synthesis of 1,3,5,7-tetramethylcyclooctatetraene. Proc Chem Soc London 1964; 84.
89. de Mayo P, McIntosh CL, Yip RW. 1,3,5,7-Tetramethylcyclooctatetraene. Org Photochem Synth 1971; 1:99.
90. Miller MJ, Lyttle MH, Streitwieser A Jr. J Org Chem 1981; 46:1977.
91. Rieke RD, Copenhafer RA. Tetrahedron Lett 1971; 879.
92. Nakamura Y, Kato T, Morita Y. J Chem Soc, Perkin Trans 1 1982; 1187.
93. Kato T, Nakamura Y, Morita Y. Chem Pharm Bull 1983; 31:2552.
94. Sieburth S McN, Lin C-H, Rucando D. J Org Chem 1999; 64:950.
95. De Selms RC, Schleigh WR. Tetrahedron Lett 1972; 3563.
96. Sato E, Ikeda Y, Kanaoka Y. Heterocycles 1989; 28:117.
97. Kanaoka Y, Ikeda Y, Sato E. Heterocycles 1984; 21:645.
98. Sato E, Ikeda Y, Kanaoka Y. Liebigs Ann Chem 1989; 781.
99. Sieburth S McN, Zhang F. Tetrahedron Lett 1999; 40:3527.
100. Vogel E, Grimme W, Meckel W, Riebel HJ, Oth JFM. Angew Chem Int Ed Engl 1966; 5:590.
101. von E Doering W, Rosenthal JW. Tetrahedron Lett 1967; 349.
102. Masamune S, Seidner RT, Zenda H, Wiesel M, Nakatsuka N, Bigam G. J Am Chem Soc 1968; 90:5286.
103. Babad E, Ginsburg D, Rubin MB. Tetrahedron Lett 1968; 2361.
104. Grimme W, Riebel HJ, Vogel E. Angew Chem Int Ed Engl 1968; 7:823.
105. Altman J, Babad E, Ginsburg D, Rubin MB. Israel J Chem 1969; 7:435.
106. Altman J, Babad E, Rubin MB, Ginsburg D. Tetrahedron Lett 1969; 1125.
107. Arce M-J, Viado AL, An Y-Z, Khan SI, Rubin Y. J Am Chem Soc 1996; 118:3775.
108. Hsiao T-Y, Santhosh KC, Liou K-F, Cheng C-H. J Am Chem Soc 1998; 120:12232.
109. Qian W, Rubin Y. J Org Chem 2002; 67:7683.
110. Iwamatsu S-i, Vijayalakshmi PS, Hamajima M, Suresh CH, Koga N, Suzuki T, Murata S. Org Lett 2002; 4:1217.
111. Suresh CH, Vijayalakshmi PS, Iwamatsu S-i, Murata S, Koga N. J Org Chem 2003; 68:3522.
112. Golobish TD, Burke JK, Kim AH, Chong SW, Probst EL, Carroll PJ, Dailey WP. Tetrahedron 1998; 54:7013.
113. Srikrishna A, Sunderbabu G. J Org Chem 1987; 52:5037.
114. Shigernori H, Bae M-A, Yazawa K, Sasaki T, Kobayashi J. J Org Chem 1992; 57:4317.
115. West FG, Chase CE, Arif AM. J Org Chem 1993; 58:3794.

116. Chase CE, Bender JA, West FG. Synlett 1996; 1173.
117. Sieburth S McN, Hiel G, Lin C-H, Kuan DP. J Org Chem 1994; 59:80.
118. Sieburth S McN, Lin C-H. J Org Chem 1994; 59:3597.
119. Sieburth S McN. J Chem Soc Chem Commun 1994; 1663.
120. Sieburth S McN, McGee KF Jr, Al-Tel TH. J Am Chem Soc 1998; 120:587.
121. Sieburth S McN, McGee KF Jr. Org Lett 1999; 1:1775.
122. Sieburth S McN, Siegel B. J Chem Soc, Chem Commun 1996; 2249.
123. Yin TY, Madsen-Duggan CB, Siegel B. SUNY Stony Brook research notes.
124. Uemura S, Fukuzawa S-i, Toshimitsu A, Okano M, Tezuka H, Sawada S. J Org Chem 1983; 48:270.
125. Curran DP, Shen W. Tetrahedron 1993; 49:755.
126. Martin DG, Slomp G, Mizsak S, Duchamp DJ, Chidester CG. Tetrahedron Lett 1970; 4901.
127. Bohlmann F, Jakupovic J. Phytochemistry 1980; 19:259.
128. Takahashi S, Naganawa H, Iinuma H, Takita T, Maeda K, Umezawa H. Tetrahedron Lett 1971; 1955.
129. Zalkow LH, Harris RN III, Van Derveer D. J Chem Soc, Chem Commun 1978; 420.
130. Norte M, Cataldo F, Sánchez A, González AG. Tetrahedron Lett 1993; 34:5143.
131. Appendino G, Gariboldi P, Valle MG. Gazz Chim Ital 1988; 118:55.
132. Schmitz FJ, Hollenbeak KH, Vanderah DJ. Tetrahedron 1978; 34:2719.
133. Ishitsuka M, Kusumi T, Kakisawa H, Kawakami Y, Nagai Y, Sato T. Tetrahedron Lett 1983; 24:5117.
134. San Feliciano A, Barrero AF, Medarde M, Miguel del Corral JM, Aramburu A. Tetrahedron Lett 1985; 26:2369.
135. Weyerstahl P, Marschall H, Wahlburg H-C, Kaul VK. Liebigs Ann Chem 1991; 1353.
136. Singh SB, Smith JL, Sabnis GS, Dombrowski AW, Schaeffer JM, Goetz MA, Bills GF. Tetrahedron 1991; 47:6931.
137. Ballio A, Brufani M, Casinovi CG, Cerrini S, Fedeli W, Pellicciari R, Santurbano B, Vaciago A. Experientia 1968; 24:631.
138. Enoki N, Furusaki A, Suehiro K, Ishida R, Matsumoto T. Tetrahedron Lett 1983; 24:4341.
139. Wahlberg I, Eklund AM, Nishida T, Enzell CR, Berg JE. Tetrahedron Lett 1983; 24:843.
140. Arigoni D. Pure Appl Chem 1968; 17:331.
141. Prestwich GD, Tempesta MS, Turner C. Tetrahedron Lett 1984; 25:1531.
142. Burke JW, Doskotch RW, Ni C-Z, Clardy J. J Am Chem Soc 1989; 111:5831.
143. Izac RR, Fenical W, Tagle B, Clardy J. Tetrahedron 1981; 37:2569.
144. Hensens OD, Zink D, Williamson JM, Lotti VJ, Chang RSL, Goetz MA. J Org Chem 1991; 56:3399.
145. Kingston DGI, Molinero AA, Rimoldi JM. Prog Chem Org Nat Prod 1993; 61:1.

10

Photoinduced Electron Transfer Cyclizations via Radical Ions

Michael Oelgemöller
Bayer CropScience K.K., Yuki-City, Ibaraki, Japan

Jens Otto Bunte and Jochen Mattay
Universität Bielefeld, Bielefeld, Germany

10.1. HISTORICAL BACKGROUND

Although *photoinduced electron transfer reactions occupy a central position in the chemistry of life* [1] their application to organic synthesis started only a few decades ago [2]. At first glance this is somewhat surprising since analogous reactions via neutral radicals are already known for a long time [3]. The reason for this delayed development of photoinduced electron transfer (PET) reactions may be simply related to the role of photochemistry in chemistry in general. Although the beginnings of modern organic photochemistry were laid more than 100 years ago [4], the fundamental understanding of electron transfer in general [5] and of PET processes in particular [1] developed relatively late in the middle of the twentieth century. Therefore, many photochemical transformations that are initiated by electron transfer were originally thought to proceed via homolytic steps.

Meanwhile, the basis of photoinduced electron transfer as well as the mechanisms of fundamental processes such as PET-sensitization, co-sensitization or medium and salt effects are widely understood [6]. As a consequence the number of applications of PET reactions in organic synthesis has increased dramatically as noticeable from a series of summarizing reviews and book chapters [2,7]. Among these applications are for example cycloadditions [8], fragmentations [9], macrocyclizations [10], and addition reactions [9], respectively. In order to keep this chapter within an acceptable length we will restrict ourselves to *intramolecular* cyclization reactions of radical ions. Readers interested in other aspects of photo-reactions involving radical ions may refer to the above mentioned reviews [2,7–10].

10.2. STATE OF THE ART MECHANISTIC MODELS

A number of excellent reviews on the theoretical aspects of photoinduced electron transfer have appeared in the literature and the interested reader may refer to these original articles. Due to the fundamental importance of the electron transfer step for the title process and in order to facilitate reading this chapter a brief introduction on PET will be given [2,6].

Upon electronic excitation the redox properties of either the electron donor (D) or the acceptor (A) are enhanced. The feasibility of an electron transfer can be estimated from a simple free reaction energy consideration as customary in the frame of the Rehm–Weller approach (Eq. (1)) [11], where $E_{1/2}^{Ox}(D)$ and $E_{1/2}^{Red}(A)$ represent the oxidation and reduction potential of the donor or the acceptor, respectively. ΔE_{excit} stands for the electronic excitation energy, whereas ΔE_{coul} indicates the coulombic interaction energy of the products formed (most commonly: radical ions). This simplified approach allows a first approximation on the feasibility of a PET process without considering the more complex kinetics as controlled by the Marcus theory [6c]. For exergonic processes ($\Delta G < 0$) a PET process becomes thermodynamically favorable.

$$\Delta G = F(E_{1/2}^{Ox}(D) - E_{1/2}^{Red\cdot}(A)) - \Delta E_{excit} + \Delta E_{coul}$$

In *intermolecular* PET processes, radical ions are formed either as close pairs or as free species from neutral molecules (Sch. 1) [2,6]. Most commonly, carbonyl compounds or related derivatives as for example enol ethers, cyclopropyl ketones, and siloxycyclopropanes are used for *intra-molecular* cyclization reactions. With the exception of cycloadditions the ring-building key step is always an *intramolecular* bond formation. In PET

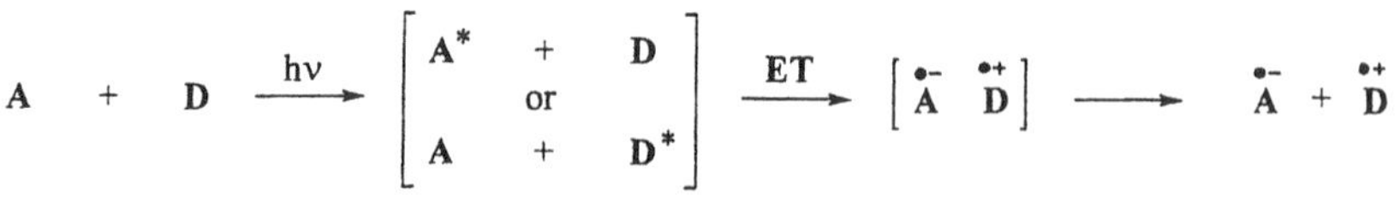

$$A + D \xrightarrow{h\nu} \begin{bmatrix} A^* & + & D \\ & or & \\ A & + & D^* \end{bmatrix} \xrightarrow{ET} \begin{bmatrix} \overset{\bullet-}{A} & \overset{\bullet+}{D} \end{bmatrix} \longrightarrow \overset{\bullet-}{A} + \overset{\bullet+}{D}$$

Scheme 1 Photoinduced electron transfer between neutral donor and acceptor molecules.

reactions this step can be an addition (or recombination) reaction of a radical or radical ion or a nucleophilic capture of a cation or radical cation, respectively.

In the *intramolecular* version, donor and acceptor are linked via a tether. P. J. Wagner splendidly described the photocyclization behavior of these so-called *donor-bridge-acceptor molecules as "a bug that opens its mouth when struck by light, can hold it open for only a short time, and during that time can swallow anything within striking range of its open mouth including its own tail"* [12]. The donor moiety is either incorporated into the newly formed ring-system (most common) or eliminated prior to cyclization. Typical electron donors for this approach are thioethers (Me_2S:$E_{Ox.} = 1.2\,V$ vs. SCE [13]) or tertiary amines (R_3N:$E_{Ox.} = 0.7$–$1.3\,V$ vs. SCE [13]). Alternatively, electron-rich alkenes, alkynes, or arenes have been applied. Frequently used acceptors are carbonyl compounds as, for example, ketones, esters, or imides.

According to Eq. (1), the thermodynamic driving force of a PET process increases with the solvent polarity and therefore, photoreactions can be simply switched from energy transfer to electron transfer by changing the solvent [8]. However, back electron transfer (BET) often diminishes the yields of radical ions formed and therefore various efforts have been undertaken to circumvent this energy loss process [14]. Among these approaches two processes have been widely used and will thus be described in more detail.

Scheme 2 outlines the general mechanism of PET-sensitization. The basic process is an electron transfer from the donor (D) to the acceptor (A, in Sch. 2: Sens), which proceeds most commonly via the electronically excited acceptor (A^* or Sens*). Since the acceptor is regenerated through a

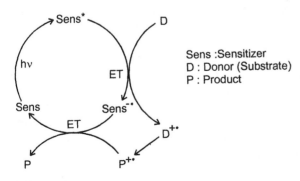

Sens :Sensitizer
D : Donor (Substrate)
P : Product

Scheme 2

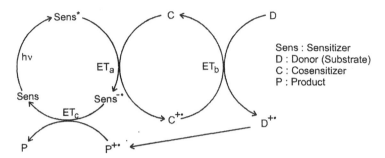

Scheme 3   Co-sensitized PET process.

secondary electron transfer process from the product radical cation, it acts as a sensitizer (Sens = A).

For oxidative purposes, electron-deficient arenes e.g., dicyanoanthracene (DCA), dicyanonaphthalene (DCN), or triphenylpyrylium salts are frequently used. For reductive PET reactions (not shown in Sch. 2) electron donor substituted arenas or amines are generally applied. The latter substrates are consumed during the reaction and are thus *sacrificial* co-substrates rather than sensitizers. Nevertheless, the strategy of a *sacrificial electron transfer* provides an effective way to avoid back electron transfer.

A more economic way to circumvent the BET pathway is co-sensitization as illustrated in Sch. 3 [2,15]. In this strategy the substrate is not directly oxidized (or reduced) by the electronically excited sensitizer but by the radical ion of the co-sensitizer (ET_b). This *thermal* electron transfer is not affected by back electron transfer since (i) sensitizer and substrate radical ions are separated from each other (back electron transfer from Sens$^{\cdot-}$ to D$^{\cdot+}$) and (ii) a back electron shift from C to D$^{\cdot+}$ is thermodynamically unfavourable. As a consequences co-sensitized reactions often proceed with higher efficiencies and different selectivities if compared with "simple" sensitized PET reactions.

Ion pair separation can also be facilitated by utilizing salt effects [2,6,16]. The basic principle is exemplary illustrated in Sch. 4 for a special salt effect induced by the addition of e.g., lithium perchlorate (generally in a polar solvent like acetonitrile). Applying this procedure, the primarily formed radical ion pairs (either as contact or solvent separate ion pairs) are subsequently replaced by the formation of a new and tight ion pair between the acceptor radical anion (A$^{\cdot-}$) and lithium cation (Li$^+$). PET reactions often proceed solely under these condition, e.g., when using ketones as PET-sensitizers [16].

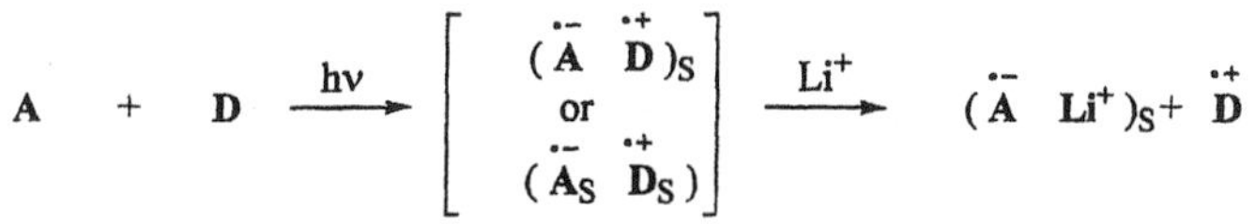

$$A \ + \ D \ \xrightarrow{h\nu} \ \left[\begin{array}{c} (\overset{\bullet-}{A} \ \overset{\bullet+}{D})_S \\ or \\ (\overset{\bullet-}{A_S} \ \overset{\bullet+}{D_S}) \end{array} \right] \ \xrightarrow{Li^+} \ (\overset{\bullet-}{A} \ Li^+)_S + \overset{\bullet+}{D}$$

Scheme 4 Ion separation of contact and solvent separated ion pairs due to the *special salt effect*.

10.3. SCOPE AND LIMITATIONS

A large number of PET photocyclizations have appeared in the literature. Since the synthesis of complex polycyclic or macrocyclic ring-system is still a challenge in synthetic organic chemistry [10,17], we have tried to select recent examples of preparative value for this paragraph. Specific target-structures are highlighted in chapter 10.6 (*vide infra*).

10.3.1. Photocyclizations of Donor-Bridge-Acceptor Molecules

10.3.1.1. PET Cyclization of Aromatic Oxoesters

PET cyclizations of donors-substituted phenylgloxylates have been intensively studied by the groups of Neckers and Hasegawa. For the thioether derivatives 1 the efficiency of the cyclization depended on the linker chain length (Sch. 5) [18]. With increasing carbon tether ($n = 2$–10) the yield dropped steadily and secondary reductive dimerizations or Norrish Type II processes became competitive. Consequently, the C_{11}-linked substrate solely underwent dimerization and cleavage [18b].

Hasegawa and Yamazaki investigated related amino-substituted phenylgloxylates 3 [19]. In line with the corresponding thioethers, cyclization exclusively occurred at the *remote* α position to nitrogen (Sch. 6). This remarkably high regioselectivity has been assigned to the increased acidity of the α-CH of the radical cation in combination with the close donor-acceptor geometry prior to cyclization. The photoreaction of the dibenzylated compound ($R^1 = Ph$, $R^2 = Bn$) proceeded most efficiently due to the benzylic stabilization, and the quantum yield for its disappearance (Φ_d) was determined to be unity. However, the higher efficiency for cyclization caused a lower diastereoselectivity and a ca. 1:1 isomeric mixture was obtained in total yield of 73%.

Photochemical reactions of alkenyl phenylglyoxylates have been additionally studied but PET induced photocyclization only occurred when alkenyl group was situated at a proper distance and in a suitable configuration [20]. Otherwise, dimerizations, Norrish Type II cleavage and

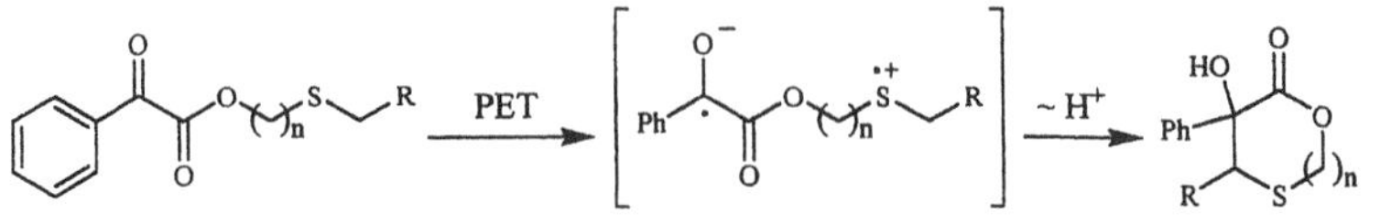

1 (n = 2-10)   PET: hv, benzene   2 (25-100%)

Scheme 5

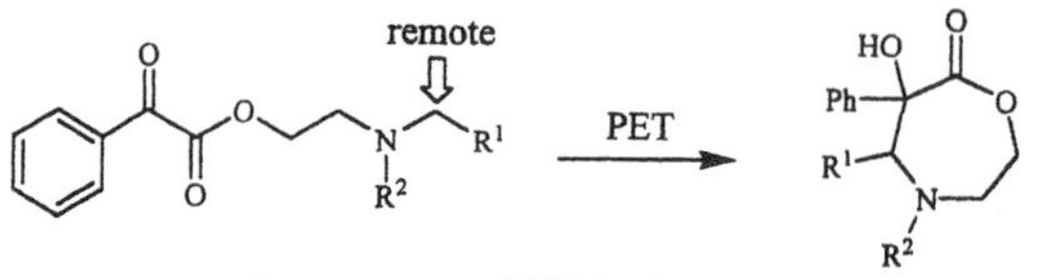

3   PET: hv, benzene   4 (38-76%)

Scheme 6

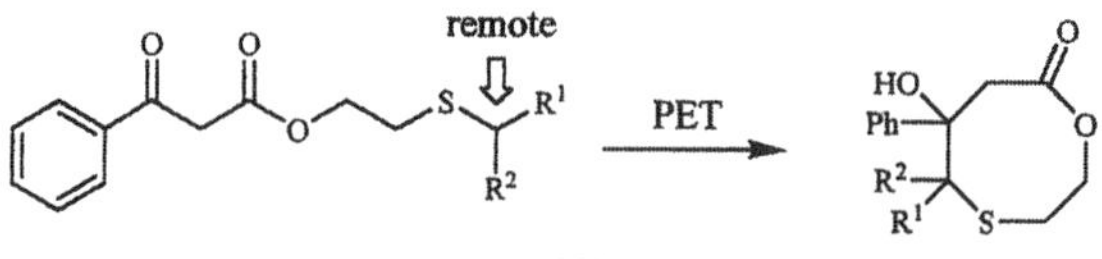

5   PET: hv, benzene   6 (35-65%)

Scheme 7

Paternò-Büchi reactions became dominate. Consequently, synthetic applications based on these derivatives are limited.

In recent publications, Hasegawa et al. were able to show that aromatic β-oxoesters are also capable for macrocyclization reactions initiated via photoinduced electron transfer. Upon irradiation in benzene, the thioether linked substrates **5** gave the eight-membered thia-lactones **6** in moderate yields (Sch. 7) [21]. Cyclization occurred exclusively at the *remote* position α to the heteroatom.

In contrast to that, regioisomeric azalactones (**8**) and lactones (**9**) were obtained when the corresponding amino analogues **7** were irradiated in benzene (Sch. 8) [22]. The photoproducts arising from hydrogen transfer from the *near side* α to nitrogen were formed in only yields, and their formation was totally suppressed by using twofold *N*-benzyl-substituted starting materials. Obviously, the rigidity of the chromophore-bridge-donor system is partly compensated for most of these derivatives which allows a competing proton transfer from both sides of the nitrogen radical cation.

PET reactions of aliphatic dialkylamino acetoacetates were much more sensitive towards the substitution pattern at the nitrogen atom and

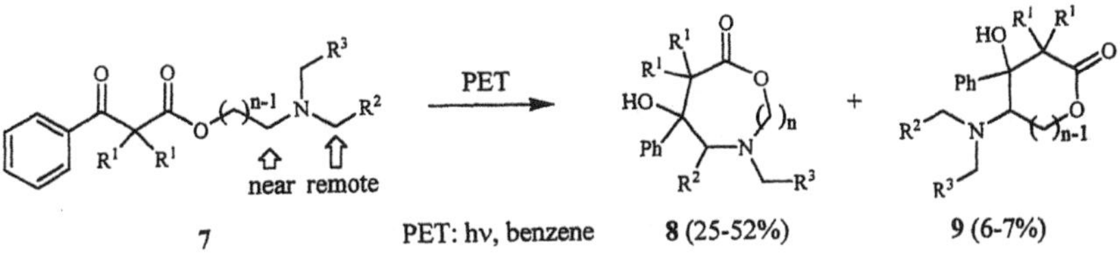

Scheme 8

solely derivatives carrying at least one benzyl group underwent effective cyclization [23]. A similar behavior was observed for aromatic γ-oxoesters and only substrates incorporating terminal benzylamino substituents gave the corresponding azolactones [24].

10.3.1.2. PET Cyclizations of Phthalimides

As noticeable from a number of reviews, the photochemistry of the phthalimide system is attracting ongoing attention [25]. From the simple alkylthio-substituted phthalimide derivatives (10) [26], the corresponding unbranched azathiacyclols (11) were formed as major products (Sch. 9) in moderate to good yields, thus reflecting the higher kinetic acidity of the primary vs. secondary α-CH group. The only exception was the butylene-linked substrate, which exclusively gave the branched product 12 by means of geometrically favored ϵ-CH$_2$-activation.

Using the phthalimide/methylthioether-pair, a variety of photochemical transformations resulting in medium- and macrocyclic sulfur containing amines, lactams, lactones, and crownether analogues, respectively, with a maximum ring-size of 38 atoms have been described [26,27]. Scheme 10 summarizes some selected examples.

Photoinduced electron transfer reactions of aminoalkyl-substituted phthalimides are highly exergonic, but since amines are potent hydrogen donors, photoreductions are commonly observed side-reactions. The product spectrum parallels that of the thioether case (*vide supra*) although yields were in general lower [28]. Higher conversions and yields up to 39% were obtained for dibenzylated amines [28c].

The progress of *intramolecular* PET-reactions involving alkenyl phthalimides in essentially influenced by the solvent [29]. Upon irradiation in MeCN, $[\pi^2 + \sigma^2]$-addition to the C(O)–N bond takes place and benzazepinediones are obtained. In alcohol, the intermediary formed radical cation is trapped in an *anti*-Markovnikov fashion depending on the polarity as well as the nucleophilicity of the solvent [30]. Recently, Xue et al. described an interesting modification of the latter process using tetrachloro-phthalimides with remote hydroxyalkyl substituents (13) [31]. During

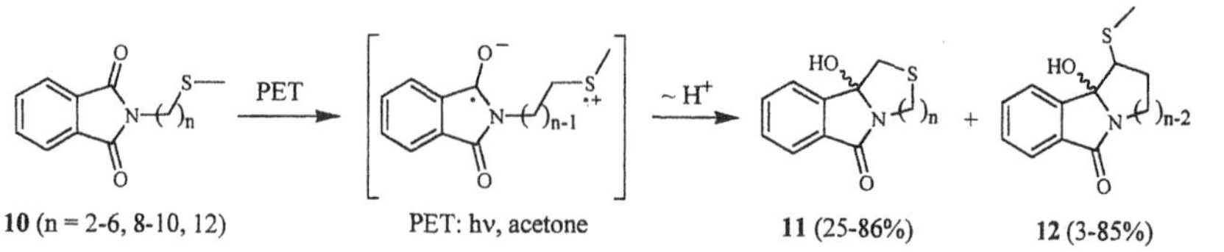
PET
~H+
10 (n = 2-6, 8-10, 12)
PET: hν, acetone
11 (25-86%)
12 (3-85%)

Scheme 9

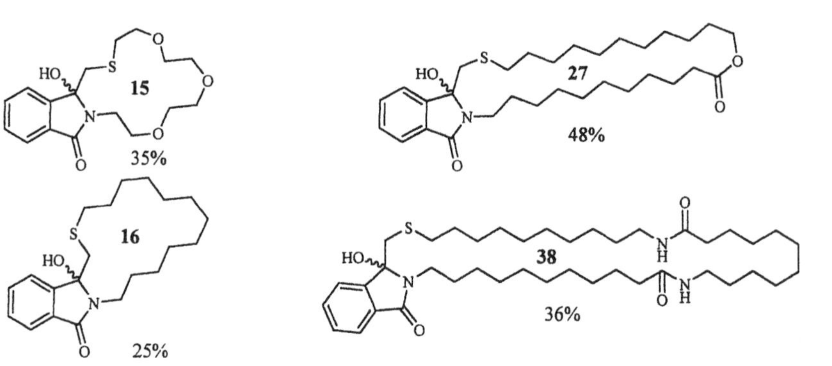
15
35%
16
25%
27
48%
38
36%

Scheme 10

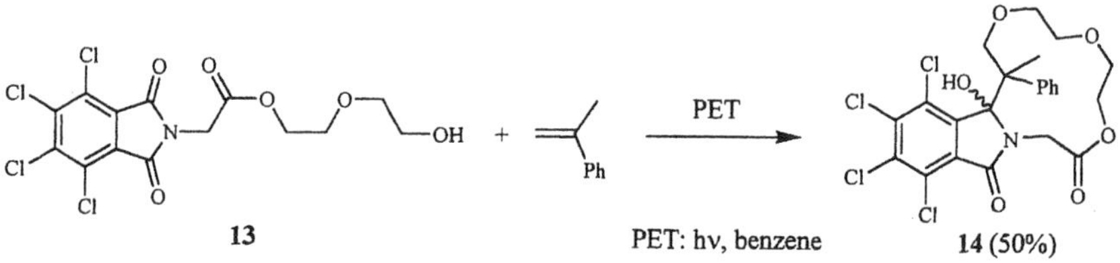
13
PET
PET: hν, benzene
14 (50%)

Scheme 11

photolysis in benzene and in the presence of alkenes, the alkene radical cation intermediate was trapped by the terminal hydroxy function, followed by an *intramolecular* radical-radical combination to give macrocyclic lactones, e.g., **14** in 50% yield (Sch. 11).

Griesbeck et al. have developed the decarboxylative photocyclization (PDC) of phthalimido ω-alkylcarboxylates (**15**) as a versatile route to macrocyclic ring systems (**16**). The carboxylate serves as electron donor and CO_2 is eliminated during the course of the reaction. Applying this concept, the syntheses of medium- and macrocyclic amines, polyethers, lactams, lactones, as well as cycloalkynes were accessible but the limitations

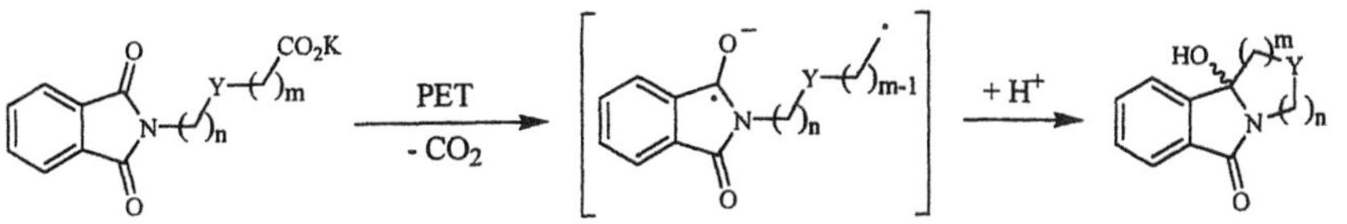

Scheme 12

concerning functional groups or maximum ring size have not been exactly explored yet (Sch. 12) [32,33].

The efficiency and selectivity of the cyclization can be improved by incorporating a suitable leaving group in α-position to the electron donor. The trimethylsilyl-, trialkylstannyl, or carbonxylate function has been utilized for this strategy [34].

10.3.2. Photocyclizations Initiated via Oxidative PET

An efficient photoinduced electron transfer initiated cyclization of α-trimethylsilylmethyl amines has been systematically evaluated by Pandey et al. [35]. Using this strategy, various mono- or bicyclic amines have been constructed via α-trimethylsilylmethyl amine radical cations (Sch. 13). The photosystem employed to generate this reactive intermediate utilized 1,4-dicyanonaphthalene (DCN) as the light harvesting electron acceptor. Extrusion of the α-trimethylsilyl group and subsequent cyclization to a tethered π-functionality gave the corresponding N-heterocycle.

In a series of publications, Pandey et al. described a variety of synthetic applications of this PET-mediated cyclization and, for example, the syntheses of (±)-isoretronecanol [35a], (±)-epilupinine [35a], and 1-N-iminosugars [35c–e] have been realized.

A similar radical cyclization reaction of unsaturated amino acid derivatives has been recently reported by Steckhan et al. [36]. The PET-catalyzed cyclization reaction proceeds under mild conditions using 9,10-dicyanoanthracene (DCA) as sensitizer and biphenyl (BP) as co-sensitizer (Sch. 14). The diastereoselectivity of the cyclization step was found to be moderate to high depending on the substitution pattern of the starting material. In almost all cases examined, the *trans*-diastereoisomer was predominately formed.

An elegant example of the latter reaction was its application to peptide chemistry for inducing structural changes within the peptide chain. Peptide **23** readily cyclized to the 4-methylproline-containing derivative **24** in 64% yield (Sch. 15). The diastereoisomeric ratio was found as 2:1 in favor of the *cis*-isomer for this example. The *cis*-selectivity reflects the approach of the

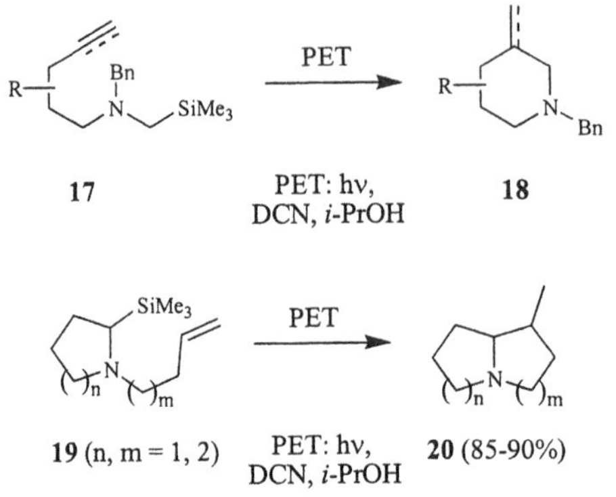

17
18

PET: hv,
DCN, *i*-PrOH

19 (n, m = 1, 2)
PET: hv,
DCN, *i*-PrOH
20 (85-90%)

Scheme 13

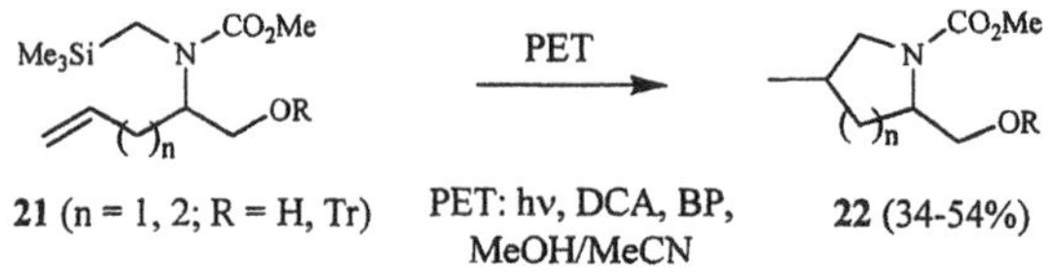

21 (n = 1, 2; R = H, Tr)
PET: hv, DCA, BP,
MeOH/MeCN
22 (34-54%)

Scheme 14

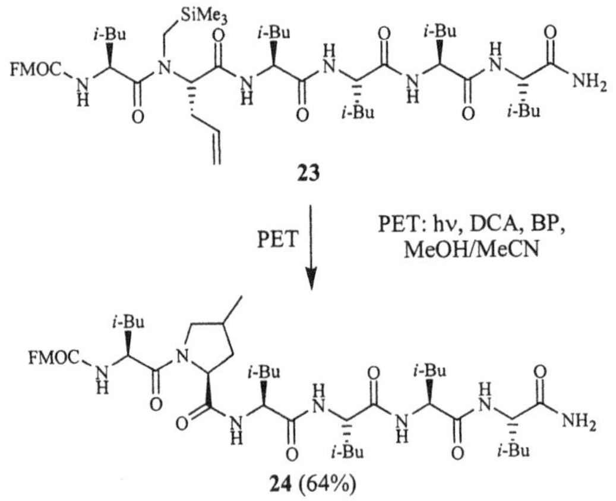

23

PET
PET: hv, DCA, BP,
MeOH/MeCN

24 (64%)

Scheme 15

intermediate radical towards the double bond which is determined by the predominant conformation of the peptide chain. Based on changes in the CD spectra it was concluded that a change of the secondary structure had occurred during the PET transformation of **23** to **24**.

PET activations of cyclic tertiary amines utilizing 1,4-dicyanonaphthalene (DCN) as electron acceptor have been studied by Pandey et al. [37]. The iminium cationic intermediates are generated via a electron–proton–electron (E–P–E) transfer sequence in a highly regiospecific fashion and as an example, tetrahydro-1,3-oxazines (26) were synthesized from substrates 25 with high control of regio- and stereoselectivity (Sch. 16).

The synthetic value of these transformations was further demonstrated through the synthesis of optically active α-alkyl piperidines or α-amino acid derivatives, respectively [37].

PET cyclizations of β,γ-unsaturated oximes (27) have been established by Armesto and Horspool [38]. This 9,10-dicyanoanthracene (DCA) sensitized reaction provides an efficient route for the synthesis of dihydroisoxazole derivatives (28) in reasonable to good yields (Sch. 17). Key step in the mechanistic scenario is an electron transfer from the oxime group to DCA in its excited singlet state. The radical cationic intermediate thus generated undergoes subsequent *exo*-cyclization with the olefinic moiety. Further, proton and electron transfer steps complete the reaction. A major limitation of this process is the restriction to molecules incorporating two aryl groups [38a].

Mattay et al. examined the regio- and stereoselective PET-cyclization of unsaturated silyl enol ethers using DCA or alternatively DCN as sensitizers [39]. The regiochemistry (6-*endo* vs. 5-*exo*) of the cyclization could be controlled via the solvent applied. In the absence of a nucleophilic solvent such as alcohol the cyclization of the siloxy radical cation to 30 dominates, whereas the presence of a nucleophile favors a reaction

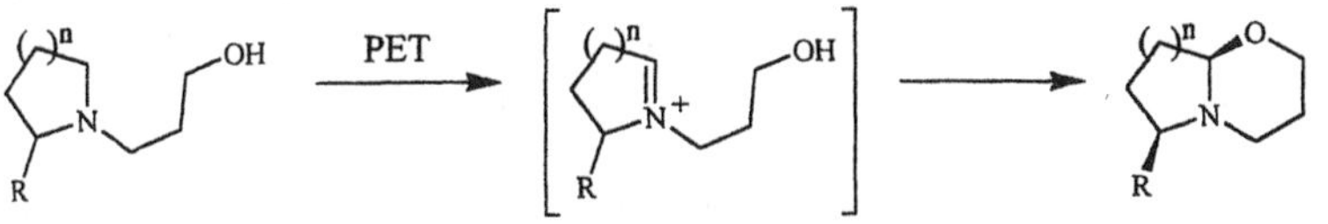

25 (n = 1, 2; R = Alkyl)   PET: hv, DCN, H₂O/MeCN   26 (88-92%)

Scheme 16

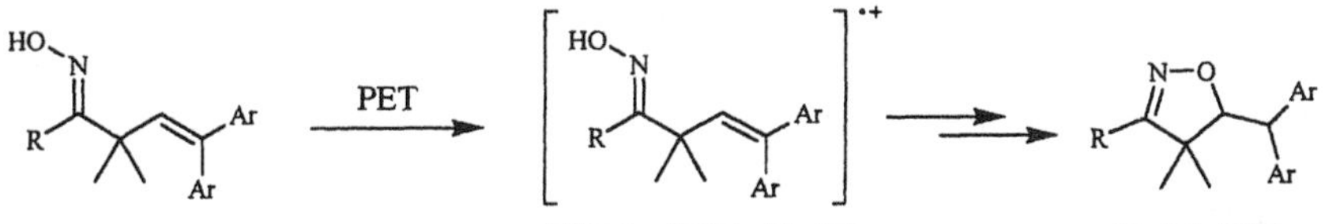

27   PET: hv, DCA, MeCN   28 (35-64%)

Scheme 17

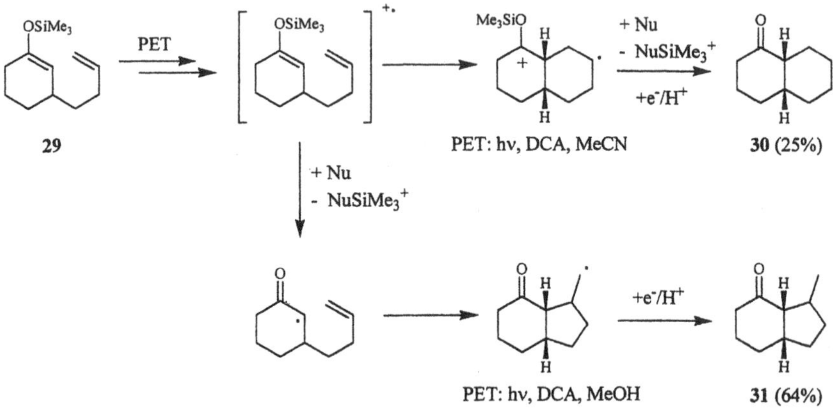

29

PET: hv, DCA, MeCN

30 (25%)

+ Nu
- NuSiMe₃⁺

PET: hv, DCA, MeOH

31 (64%)

Scheme 18

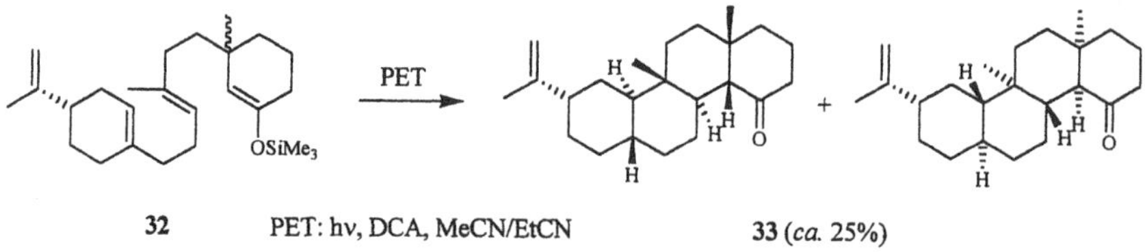

32   PET: hv, DCA, MeCN/EtCN   33 (ca. 25%)

Scheme 19

pathway to **31** via the corresponding α-keto radical (Sch. 18). The resulting stereoselective cis ring junction is due to a favored reactive chair-like conformer with a pseudeo-axial arrangement of the substituents. However, recent investigations on the mechanism of the latter pathway indicated that solely the differing rate for saturation of the intermediate causes this solvent sensitivity [40].

As an extension of this work, Mattay et al. have used this methodology for the construction of unnatural steroid analogues [41]. The polycyclic framework was build-up via a cascade cyclization of the silyl enol ether **32** using DCA as sensitizer (Sch. 19). The two major products **33** were formed with remarkable high *trans, anti*-stereoselectivity.

A versatile strategy for efficient *intramolecular* α-arylation of ketones was achieved by the reaction of silyle enol ethers with PET-generated arene radical cations. This strategy involved one-electron transfer from the excited methoxy-substituted arenes to ground-state DCN [42]. Pandey et al. reported the construction of five- to eight-membered benzannulated as well as benzospiroannulated compounds using this approach (Sch. 20) [42a]. The course of the reaction can be controlled via the silyl enol ether obtained

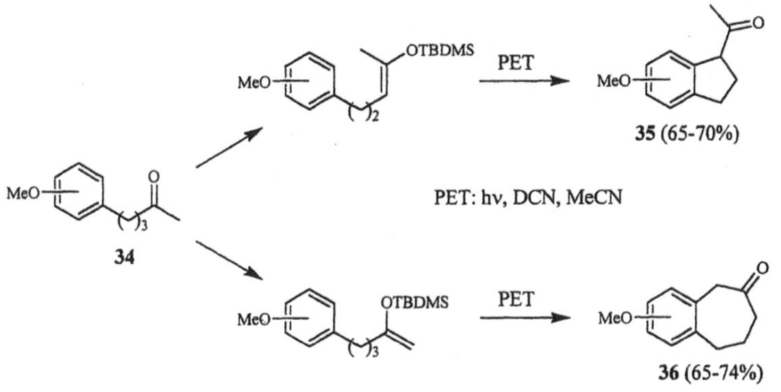

Scheme 20

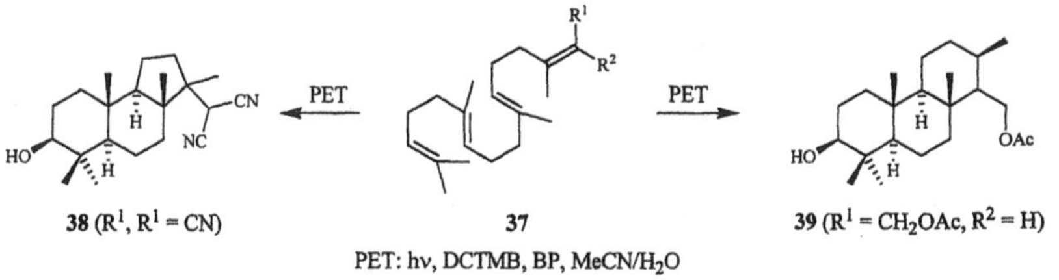

Scheme 21

from the starting ketone **34**. The *thermodynamically* controlled silyle enol ether gave the branched cyclization products **35**, whereas *kinetically* controlled silyl enol ethers lead to the unbranched substrates **36**.

Demuth et al. recently developed an efficient radical cationic cyclization of functionalized polyalkenes using 1,4-dicyano-tetramethylbenzene (DCTMB) as an acceptor and biphenyl (BP) as co-sensitizer [43]. The transformation is in general highly stereo- and chemoselective and the substitution pattern of the polyalkene allows the construction of either five- or six-membered rings (Sch. 21). So far, this methodology has been applied for the synthesis of several natural products such as stypoldione, hydroxyspongianone and abietanes, respectively [43].

A new synthetic pathway for the construction of cyclic acetals (**41**) from homobenzylic ethers (**40**) has been recently developed by Floreancig et al. [44]. The electron transfer initiated oxidative cyclization is efficiently catalyzed by *N*-methylquinolinium hexafluorophosphate (NMQPE$_6$) in the presence of oxygen (Sch. 22). For gram-scale preparations solid $Na_2S_2O_3$ has to be used as an additive to suppress decomposition caused by the reactive oxygen species involved.

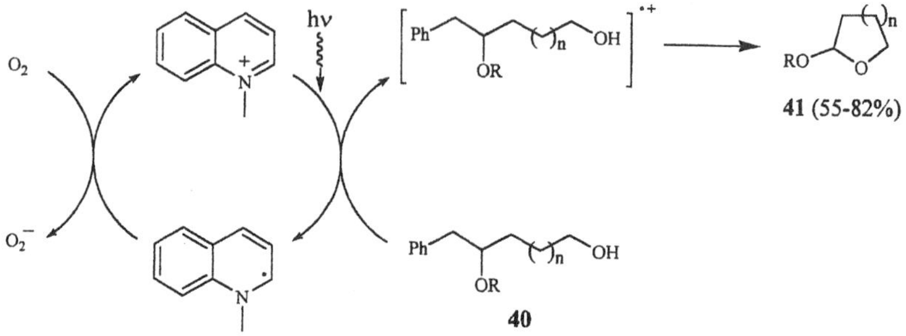

Scheme 22

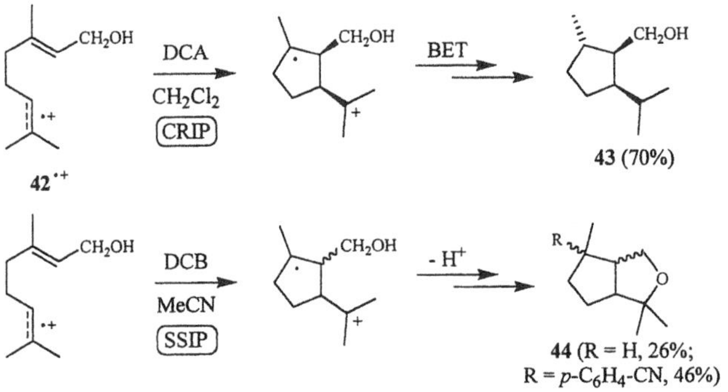

Scheme 23

The electron transfer photochemistry of geraniol (**42**) employing DCA and 1,4-dicyanobenzene (DCB) as sensitizers has been examined by Roth et al. [45]. The course of the reaction is controlled by the energy of the electron transfer step and thus by the sensitizer applied (Sch. 23).

The marginal driving force ($\Delta G = -0.2\,\text{eV}$) of the DCA sensitized reaction in CH_2Cl_2 causes the formation of contact radical ion pairs (CRIPs) The close proximity to the sensitizer forces the predominant formation of the *cis*-fused radical cation and ensures fast back electron transfer to the cyclized radical cation. Further, transformation give the cyclopentane derivative **43** in good yield. In contrast, the photoreactions involving DCB are exergonic with $\Delta G = -0.7\,\text{eV}$ in MeCN which favors the generation of solvent-separated radical ion pairs (SSRIPs) but lowers the stereoselectivity of the primary cyclization step. Due to the reduced back electron transfer efficiency, secondary cyclization takes place. Subsequently, reduction or coupling with the sensitizer radical anion gives mixtures of the diastereoisomeric 3-oxabicyclo[3.3.0]octanes **44**.

10.3.3. Photocyclizations Initiated via Reductive PET

The synthesis of bicyclic cyclopentanols (46) via photoreductive cyclization of δ,ε-unsaturated ketones (45) has been realized by Belotti et al. using hexamethylphosporic triamide (HMPA) or triethylamine (TEA) as electron donor [46]. The photocyclization proceeded remarkably efficient when HMPA was used as donor and solvent (Sch. 24). Furthermore, only one stereoisomer was obtained carrying methyl and hydroxy groups in *trans*-configuration. In contrast, the yields for cyclization dropped significantly when TEA in a polar solvent such as acetonitrile was used. As an example, the yield for 46 ($n = 1$) decreased from 81% in HMPA to 50% in TEA/MeCN.

The PET-cyclization of 2-bromomethyl-2-(3-butenyl)benzocyclic-1-alkanones developed by Hasegawa et al. can be regarded as a related version of the latter reaction [47]. In sharp contrast to thermal reactions involving SmI_2 [47a], irradiations of 47 in acetonitrile and in the presence of 1,3-dimethyl-2-anisylbenzimidazoline (DMABI) afforded the 5-*exo* cyclization products (48) in good yields (Sch. 25).

Mattay et al. used triethylamine as electron donor in tandem fragmentation/cyclization reactions of α-cyclopropylketones (49) [48]. The initial electron transfer to the ketone moiety is followed by subsequent cyclopropylcarbinyl-homoallyl rearrangement yielding a distonic radical anion (50). With an appropriate unsaturated side chain attached to the molecule both annelated and spirocyclic ring systems are accessible in moderate yields. Scheme 26 shows some representative examples.

Pandey et al. recently developed two useful photosystems for one-electron reductive chemistry and applied them to the activation of aldehyde

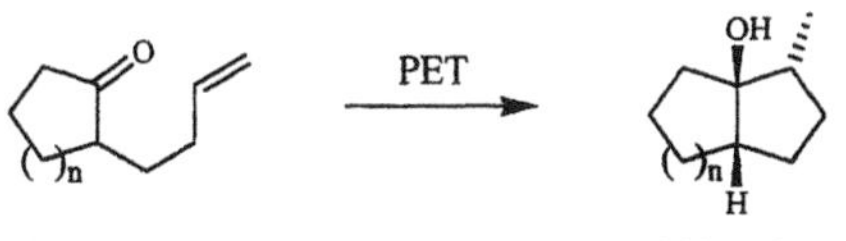

45 (n = 1, 2)   PET: hv, HMPA   46 (76-81%)

Scheme 24

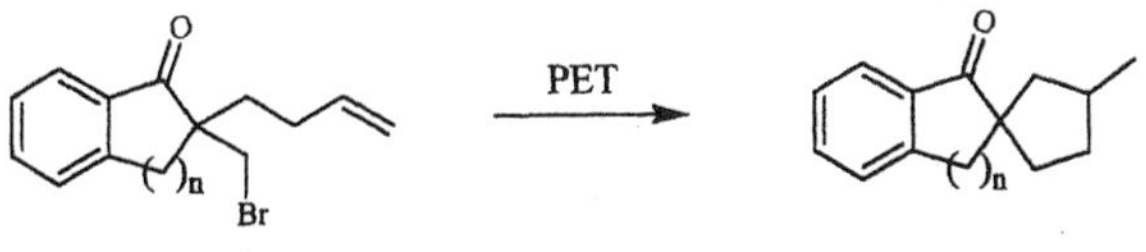

47 (n = 1-3)   PET: hv, DMABI, MeCN   48 (57-60%)

Scheme 25

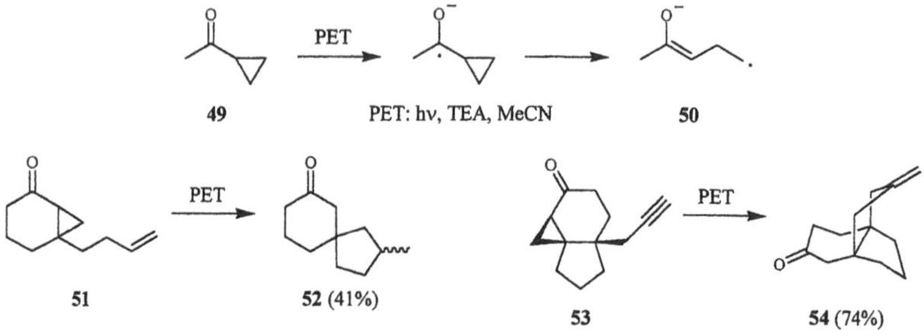

Scheme 26

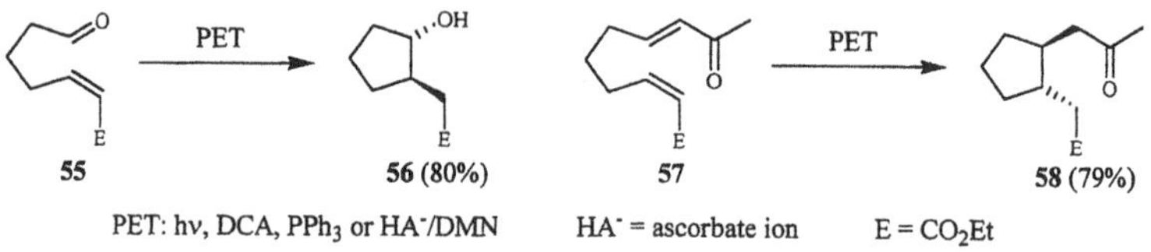

Scheme 27

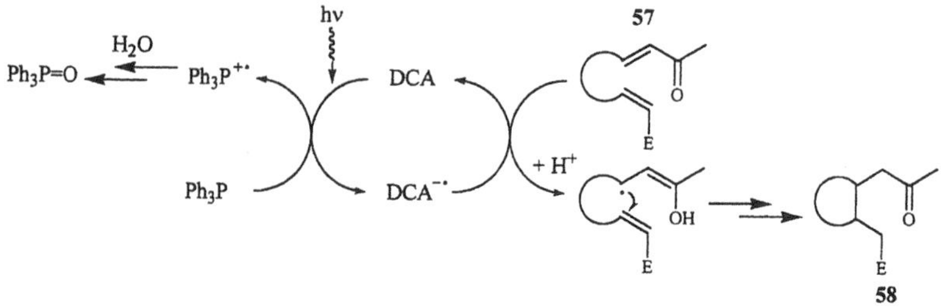

Scheme 28

or ketones tethered with activated olefins (**55**) [49a] and α,β-unsaturated ketones (**56**) [49b], respectively (Sch. 27).

Key-step in the mechanistic scenario is a primary electron transfer process involving a *sacrificial electron donor* as exemplary shown for the triphenylphosphine case in Sch. 28. The 9,10-dicyanoanthracene radical anion (DCA·⁻) thus generated undergoes a secondary *thermal* electron transfer to the unsaturated ketone. The resulting carbon-centered radical or radical anionic intermediate, subsequently cyclizes stereoselectively with a proximate olefin. The observed 1,2-*anti*-stereochemistry of the C–C bond formation step contrasts with the commonly observed *syn*-stereoselectivity of 5-hexenyl radical cyclizations. As sacrificial electron donors, the

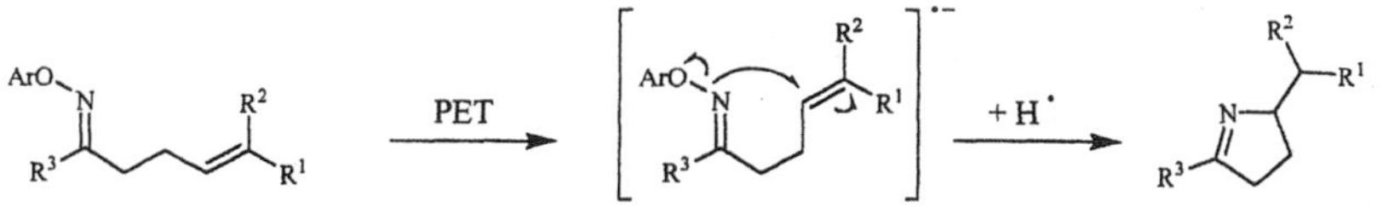

59 (Ar = p-C$_6$H$_4$CN)

PET: hv, DMN, 1,4-
cyclohexadiene, MeCN

60 (13-78%)

Scheme 29

ascorbate ion, combination with 1,5-dimethoxynaphthalene (DMN) as primary electron donor, or triphenylphoshine have been applied.

The versatile applicability of this photocyclization concept was demonstrated through the synthesis of important structural frameworks as for example triquinanes [49a] or furanosides [49b], respectively.

Photochemical transformations of γ,δ-unsaturated oximes ethers (59) to 3,4-dihydro-2H-pyrrole derivatives (60) have been developed by Mikami and Narasaka [50]. Irradiations were performed in the presence of 1,5-dimethoxynaphthalene (DMN) as sensitizer and 1,4-cyclohexadiene as hydrogen source. Mono- and bicyclic products were obtained selectively by 5-exo cyclization in moderate to good yields (Sch. 29).

10.4. SYNTHETIC POTENTIAL: REACTIVITY AND SELECTIVITY PATTERN

Depending on the structure of the starting material and the course of the reaction, at least one new stereogenic center is produced during the PET cyclization. Due to the involvement of excited-states and thus different spin multiplicities (single vs. triplet), the analysis and predictability of the regio- and stereoselectivity of the bond formation step has often been regarded as difficult and problematic. Recent investigations have led to a wider understanding in the stereochemistry of photochemical processes in general [51], and new concepts as spin-selectivity [52] or photochirogenesis [53] have been established.

10.4.1. Regioselectivity

For donor-bridge-acceptor systems, the regioselectivity of the C–C bond formation step critically depends on the position of the donor moiety along the tether chain and thus of the ring-size of the product. Conformational restrictions by certain functional groups within the tether can further influence the selectivity [12].

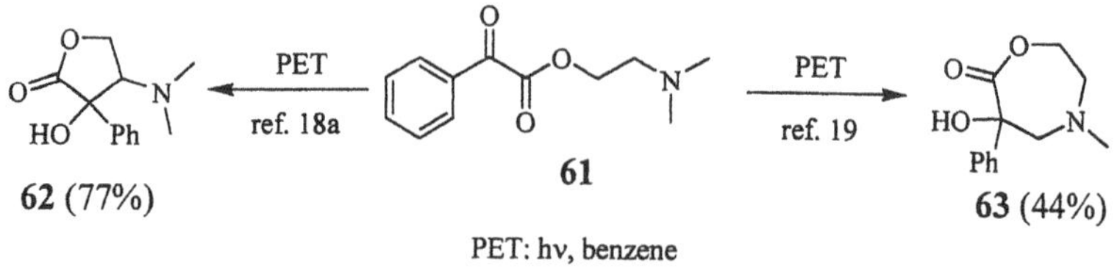
PET
ref. 18a
PET
ref. 19
61
62 (77%)
63 (44%)
PET: hν, benzene

Scheme 30

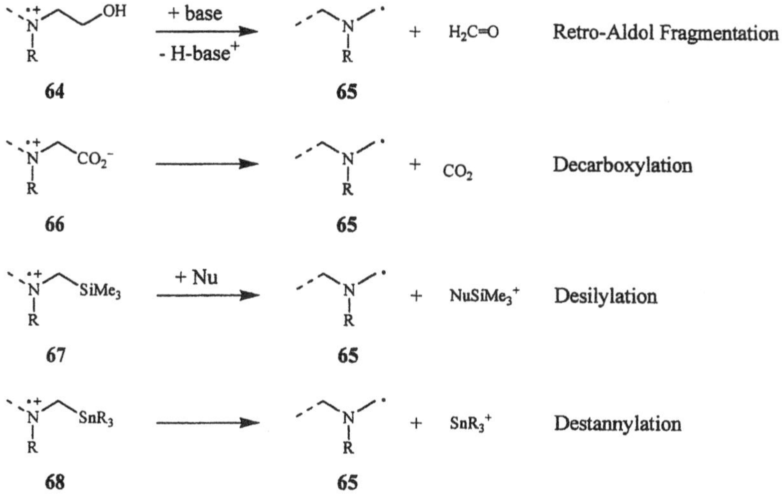
+ base
- H-base+
Retro-Aldol Fragmentation
64
65
Decarboxylation
66
65
+ Nu
Desilylation
67
65
Destannylation
68
65

Scheme 31

Contradictory results on the regioselectivity for cyclization have been reported for the amino substituted phenylglyoxylate **61** (Sch. 30). According to one report, the C–C bond formation of this particular compound occurred selectively and with 77% yield of **62** at the *near side* α to nitrogen, and this divergent behavior was attributed to the geometry imposed by the heteroatom to the conformations of the corresponding intermediates [18a]. In contrast to that report, it was described that **61** underwent solely *remote* photocyclization although the yield of **63** was significantly reduced with 44% [19]. The latter authors also extended their study to other derivatives and always observed similar regioselectivities, thus underlining the differences in kinetic acidity in combination with a close donor–acceptor geometry.

One strategy to overcome the regioselectivity problem and to enhance the efficiency for the photocyclization is the use of a suitable leaving group in α-position to the donor (Sch. 31). The intermediate donor radical-cation formed after electron transfer subsequently undergoes

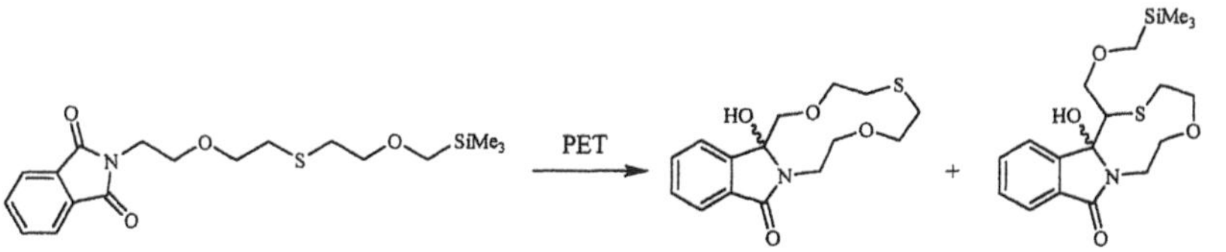

Scheme 32

secondary α-fragmentation to its corresponding radical. Among the α-fragmentation processes that have been widely used for amines are base-promoted retro-Mannich cleavage (**64 → 65**), unimolecular decarboxylation (**66 → 65**), silophile-induced desilylation (**67 → 65**), and destannylation (**68 → 65**), respectively [34b,54].

However, the efficiency of the latter strategy decreases when other potent electron donors are present in the donor-acceptor system. An illustrative example is the mixed ether/thiotether-linked phthalimide **69**, which yielded the branched macrocyclic compound **71** arising from competing sulfur oxidation as main product (Sch. 32) [55].

Another approach for regioselectivity control, which additionally circumvents the incorporation of the donor group into the product ring system, makes use of a terminal dithiane or dithiolane group as directing electron donor [56a]. However, the latter concept failed of the corresponding phenylglyoxylates possibly due to steric hindrance imposed by the dithianyl or dithiolanyl group, respectively [56b].

10.4.2. Diastereoselectivity

The stereochemistry generated via the cyclization step is in general controlled by the geometry of the corresponding intermediate and thus, especially by the ring-size of the product [51]. In addition to recent attempts to predict the stereochemical outcome via computer simulations [57], the analysis of the product-stereochemistry is still essential. This is especially important since the stereoselectivity seems often obscure and does not always match with intuitive expectations. Additionally, even small changes in the substitution pattern of the starting material can drastically influence or even reverse the seterochemistry of the C–C bond formation step.

A representative and interesting example for the dependency on the product's ring-size is the PET-cyclization of alkene-silylamine derivatives as described by Pandey et al. (Sch. 33) [35a]. 1,5-Cyclization (**19**, $n = 1$) predominantly gave the *cis*-isomer whereas the corresponding 1,6-cyclization (**19**, $n = 2$) yielded the opposite trans-isomer, respectively. Both findings were rationalized by chair-like transition states.

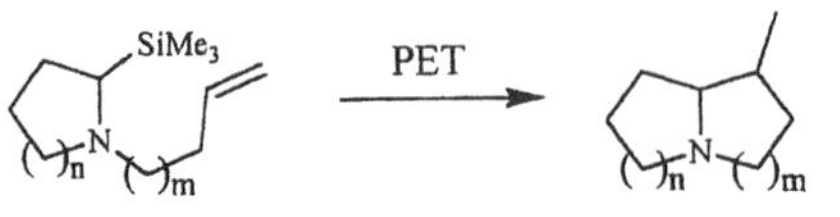

n	m	yield (%)	cis:trans
1	1	90	97:3
2	1	85	95:5
1	2	87	2:98
2	2	88	0:100

PET: hv, DCN, *i*-PrOH

Scheme 33

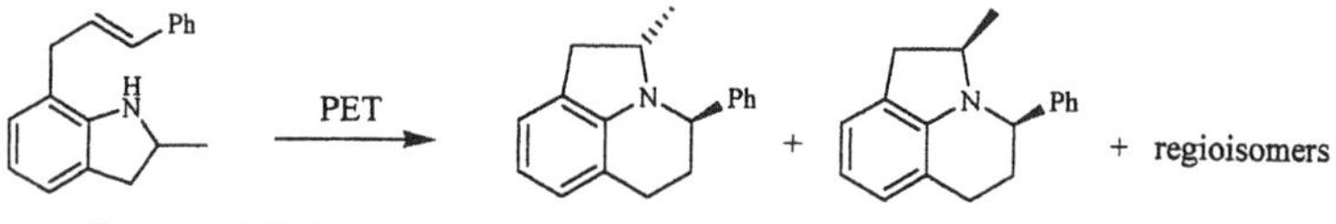

Scheme 34

Recently, Inoue et al. have established the concept of *multidimensional control of asymmetric photochemistry* [58]. Applying this strategy, the product stereoselectivity can be inverted by environmental factors such as the temperature, pressure or solvent, and this control has been interpreted in terms of the contributions of both enthalpy ($\Delta H^{\ddagger}$) and entropy ($\Delta S^{\ddagger}$). Although originally designed for *enantiodifferentiating photosensitization* of cycloalkenes [58], Miranda et al. were able to apply it to PET cyclizations of *o*-allylaniline derivatives [59]. In particular, irradiations at different temperatures revealed a significant entropy-controlled diastereoselectivity for compound **72** and the equipodal temperature was found at 292 K (Sch. 34).

10.4.3. Enantioselectivity

Highly selective cascade cyclizations of terpenoid polyalkenes via photo-induced electron transfer have been accomplished by Demuth et al. 1,4-Dicyanotetramethylbenzene (DCTMB) and biphenyl (BP) were applied as electron-acceptor couple and a spirocyclic dioxinone/(−)-menthone pair was used as chiral auxiliary [43c,h]. Although this directing group was placed in a remote location from the initiation site of the reaction, cyclization gave the enantiomerically pure cyclic terpenoids (**75**) through exclusive diastereofacial differentiation (Sch. 35). The efficient remote asymmetric induction is hereby most likely induced by enantioselective chiral folding of the polyalkene chain. Noteworthy, eight stereogenic centers were created in a single step and only 2 out of 256 possible isomers were formed.

The phenomenon of *memory of chirality* [60] was recently extended to PET-cyclization reactions. An interesting example was the highly

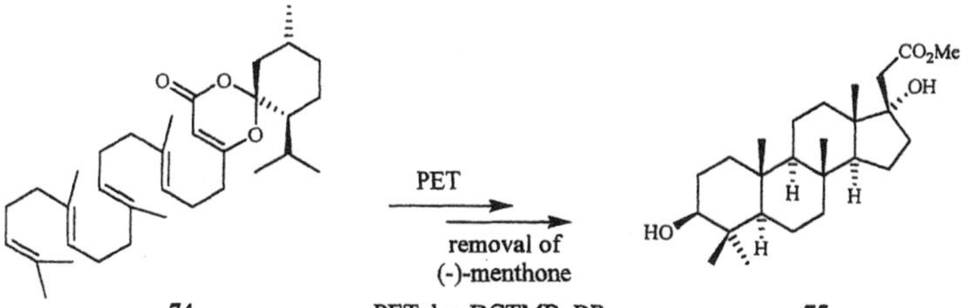

74 PET: hν, DCTMB, BP, $MeCN/H_2O$, -25°C 75

Scheme 35

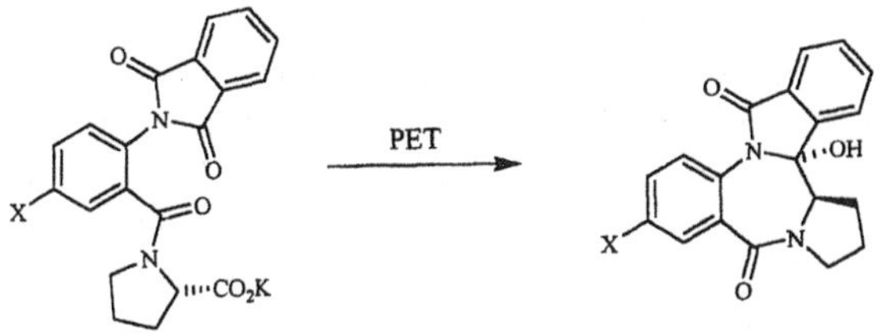

76 (X = H, Cl) PET: hν, acetone/H_2O 77 (X = H: 45%, 86% *ee*) (X = Cl: 50%, 79% *ee*)

Scheme 36

selective synthesis of pyrrolobenzodiazepines and photolysis of the proline derivatives **76** gave **77** with high enantiomeric excess (*ee*) values (Sch. 36) [61]. A reasonable explanation of this remarkably high memory effect was based on the restricted rotations about the amide C–N and the arene–N bonds in **76**. Consequently, when the *ortho*-phenylene linker was replaced by the conformationally more flexible ethylene linker, *chirality memory* vanished and a mixture of stereoisomeric products was obtained [61c].

10.5. REPRESENTATIVE EXPERIMENTAL PROCEDURES

10.5.1. General Remarks

The large variety of different PET-approaches for cyclizations results in a number of specific experimental procedures. Nevertheless, most of these procedures are relatively simple and vary mainly in the photochemical equipment applied and the composition of the starting material. As in any photochemical reaction, certain experimental features must be considered. To circumvent further photodecomposition of the desired product, a suitable wavelength should be selected which avoids absorption by the

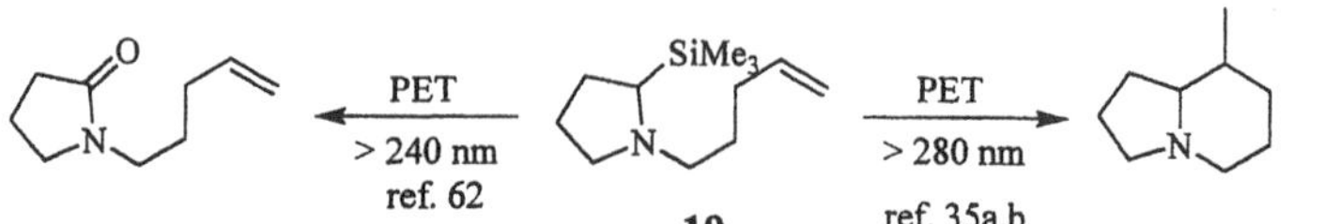

Scheme 37

product. In general, the choice of the right light source together with a proper filter can drastically influence the outcome of the photochemical transformation.

Illustrative examples for the latter precaution are the contradictory results on PET cyclizations of nonactivated alkene-silylamine systems as independently reported by the groups of Pandey [35a,b] and Mariano [62] (Sch. 37). In his study, Mariano obtained the pyrrolidinone **78** when he repeated the DCN-sensitized irradiation of **19** under condition which closely matched those of Pandey. Aware of this obvious discrepancy, Pandey launched a detailed study on the experimental conditions of this application. After he proofed the correctness of his earlier results and even extended the photoreaction to other nonactivated systems, he concluded that the only significant difference in the experimental conditions used by Mariano was a gap of ca. 40 nm in the applied wavelength.

In addition to common light-sources as e.g., medium or high pressure mercury lamps two alternative techniques deserve further mentioning. Griesbeck et al. have recently realized photoreactions in multigram quantities using a commercially available 60 cm 3 kW XeCl excimer lamp [63]. Furthermore, isolated examples of photocyclizations using solar radiation [27a,43] either direct or concentrated sunlight have appeared in the literature [64]. Due to the growing demand for alternative and environmentally friendly technologies, a growing number of future large-scale applications can be expected from both applications.

10.5.2. Radical Cationic Photocyclization of the Oligopeptide (23) [>345 nm Photolysis on a 0.1 mmol Scale] [36]

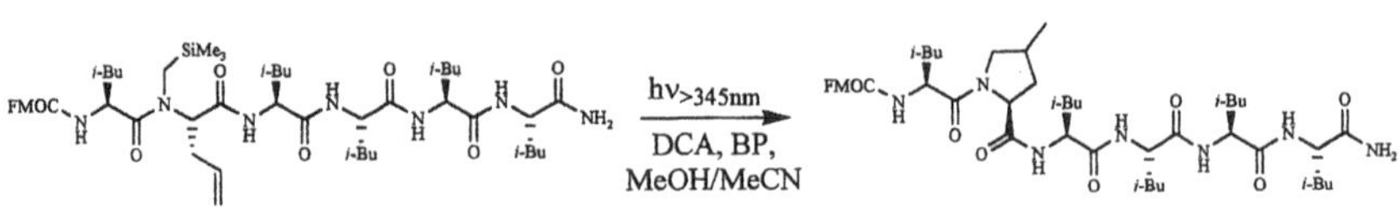

A solution of the oligopeptide **23** (95 mg, 0.1 mmol), biphenyl (4 mg), and 9,10-dicyanoanthracene (4 mg) in 50 mL of MeOH/MeCN (2:3) was irradiated for 1 h under argon with a 1000 W xenon arc lamp ($\lambda > 345$ nm). The crude product was purified by flash chromatography yielding 56 mg of a 2:1 mixture of *cis*- and *trans*-**24** (64%).

10.5.3. Radical Anionic Photocyclization of the α-Cyclopropyl Ketone (79) [300 ± 20 nm Photolysis on a 0.9 mmol Scale] [48b]

The PET-reaction of **79** (150 mg, 0.92 mmol) with TEA (1.0 g) in acetonitrile (12 mL) was conducted using a Rayonet photoreactor fitted with 300 ± 20 nm lamps. After roughly five days of irradiation, the solvent was evaporated and 40 mg of **80** were isolated (27%, purified by column chromatography).

10.5.4. Photodecarboxylative Cyclization of the N-Phthaloylanthranilic Acid Amide of L-Leucine (81) [308 nm XeCl Excimer Photolysis on a 32 mmol Scale] [63b]

Table 1.

Target structure of class	Photochemical key-step	yield [%]	Ref.
berberine (analog)	n = 1 or 2	30–31	[66]
(±)-epilupinine		n.g.	[35a]
(+)-isofagomine		60	[35c,d]
conine		80	[37a]
(+)-2,7-dideoxypancratistatin[67]	R = t-BuMe₂Si	68	[42b]
(±)-stypoldione	R = CH₂OAc, CO₂Et	20–30	[43b]
(±)-isoafricanol	+ *trans*-isomer	27 (*trans*: 55)	[68]
C-furanoside	R = t-BuMe₂Si	25	[49b]

A detailed description of the excimer set-up is given in reference [63a]. The potassium salt of 12 g (31.7 mmol) *N*-phthaloylanthranilic acid amide **81** was irradiated in ca. 2000 mL of water/acetone (40:1) for 3 h. The crude reaction mixture was extracted with ethyl acetate, the combined organic layers were dried over $MgSO_4$ and evaporated to give 5.5 g of the pyrrolobenzodiazepine **82** (52%, 4:6 *trans*:*cis*-mixture). Alternatively, a Rayonet photochemical reactor equipped with phosphor coated 300 ± 10 nm lamps can be used for reactions on a smaller scale.

10.6. TARGET MOLECULES: NATURAL AND NONNATURAL PRODUCT STRUCTURES

Numerous applications of PET cyclization reactions in organic synthesis appeared over the last years [65], and we have tried to summarize important target compounds in the following Table 1. Since most preparations follow multi-step procedures, we have additionally included the photochemical key-step. The interested reader may refer to the original article for further details.

REFERENCES

1. Roth HD. Top Curr Chem 1990; 156:1.
2. (a) Mattes SL, Farid S. In: Padwa A, ed. Organic Photochemistry. Vol. 6. New York: Marcel Dekker, Inc., 1982; 233; (b) Mattay J. Angew Chem 1987; 99:849; Angew Chem Int Ed Engl 1987; 26:825; (c) Mattay J. Synthesis 1989; 233; (d) Griesbeck AG, Schieffer S. In: Balzani A, Mattay J, eds. Handbook of Electron Transfer Chemistry. 2000:457.
3. (a) Giese B. Angew Chem 1983; 95:771; Angew Chem Int Ed Engl 1983; 22:753; (b) Giese B. Angew Chem 1985; 97:555; Angew Chem Int Ed Engl 1985; 24:553; (c) Linker T, Schmittel M. Radikale und Radikalionen in der Organischen Synthese. Weinheim: Wiley-VCH, 1998.
4. (a) Roth HD. Angew Chem 1989; 101:1220; Angew Chem Int Ed Engl 1989; 28:1193; (b) Roth HD. Pure Appl Chem 2001; 73:395; (c) Roth HD. EPA-Newslett 2001; 71:37.
5. Eberson L. Electron Transfer Reactions in Organic Chemistry. Berlin: Springer, 1987.
6. (a) Kovarnos GJ, Turro NJ. Chem Rev 1986; 86:401; (b) Mattay J, Vondenhof M. Top Curr Chem 1991; 159:219; (c) Kovarnos GJ. Fundamentals of Photoinduced Electron Transfer. New York: VCH Publishers, 1993; (d) Balzani V. ed. Electron Transfer in Chemistry. Vols. 1 and 2. Weinheim: Wiley-VCH, 2001.

7. (a) Hintz S, Heidbreder A, Mattay J. Top Curr Chem 1996; 177:77; (b) Schmoldt P, Rinderhagen H, Mattay J. In: Ramamurthy V, Schanze KS, eds. Molecular and Supramolecular Photochemistry. Vol. 9. New York: Marcel Dekker, Inc., 2003; 185; (c) Pandey G. Top Curr Chem 1993; 168:175; (d) Albini A, Fagnoni M, Mella M. In: Fabbrizzi L, Poggi A, eds. Chemistry at the Beginning of the Third Millenium. Heidelberg: Springer, 2000:83; (e) Albrecht A, Averdung J, Bischof EW, Heidbreder A, Kirschberg T, Müller F, Mattay J. J Photochem Photobiol A: Chem 1994; 82:219.

8. (a) Müller F, Mattay J. Chem Rev 1993; 93:99; (b) Mattay J, Albrecht E, Fagnoni M, Heidbreder A, Hintz S, Kirschberg T, Klessinger M, Mählmann J, Schlachter I, Steenken S. J Inf Rec 1996; 23:23.

9. Eaton DF. Pure Appl Chem 1984; 56:1191.

10. Griesbeck AG, Henz A, Hirt J. Synthesis 1996; 1261.

11. (a) Rehm D, Weller A. Isr J Chem 1970; 8:259; (b) Rehm D, Weller A. Ber Bunsenges Phys Chem 1969; 73:834.

12. Wagner PJ. Acc Chem Res 1983; 16:461.

13. Pienta NJ. In: Fox MA, Chanon M, eds. Photoinduced Electron Transfer. Amsterdam: Elsevier, 1988:421.

14. Fox MA. Adv Photochem 1986; 13:237.

15. Majima T, Pac C, Nakasone A, Sakurai H. J Am Chem Soc 1981; 103:4499. Pac et al. were probably the first who introduced "*redox photosensitization*" which is synonymous to "co-sensitization."

16. Mattay J, Trampe G, Runsink J. Chem Ber 1988; 121, 1991.

17. (a) Izatt RM, Christensen JJ, eds. Progress in Macrocyclic Chemistry. Vol. 1. New York: Wiley, 1973; Vol. 2. 1981; Vol. 3. 1987; (b) Thebtaranonth C, Thebtaranonth Y. Cyclization Reactions. Boca Raton: CRC Press, 1994; (c) Ross-Murphy SB, Stepto RFT. Large Ring Molecules. New York: John Wiley & Sons, 1996:599.

18. (a) Hu S, Neckers DC. Tetrahedron 1997; 53:2751; (b) Hu S, Neckers DC. Tetrahedron 1997; 53:7165.

19. Hasegawa T, Yamazaki Y. Tetrahedron 1998; 54:12223.

20. Hu S, Neckers DC. J Org Chem 1997; 62:6820.

21. Yamazaki Y, Miyagawa T, Hasegawa T. J Chem Soc, Perkin Trans 1 1997; 2979.

22. (a) Hasegawa T, Ogawa T, Miyata K, Karakizawa A, Komiyama M, Nishizawa K, Yoshioka M. J Chem Soc, Perkin Trans 1 1990; 901; (b) Hasegawa T, Miyata K, Ogawa T, Yoshihara N, Yoshioka M. J Chem Soc, Chem Commun 1985; 363.

23. Hasegawa T, Yasuda N, Yoshioka M. J Phys Org Chem 1996; 9:221.

24. Hasegawa T, Mukai K, Mizukoshi K, Yoshioka M. Bull Che Soc Jpn 1990; 63:3348.

25. (a) Oelgemöller M, Griesbeck AG. J Photochem Photobiol C: Photochem Rev 2002; 3:109; (b) Maurer H, Griesbeck AG. In: Horspool WM, Song PS, eds. CRC Handbook of Organic Photochemistry and Photobiology. Boca Raton: CRC Press, 1995:513; (c) Coyle JD. In: Horspool WM, ed. Synthetic Organic Photochemistry. New York: Plenum Press, 1984:259; (d) Mazzocchi PH. Org Photochem 1981; 5:421.

26. (a) Hatanaka Y, Sato Y, Nakai H, Wada M, Mizoguchi T, Kanaoka Y. Liebigs Ann Chem 1992; 1113; (b) Sato Y, Nakai H, Wada M, Mizoguchi T, Hatanaka Y, Migata Y, Kanaoka Y. Liebigs Ann Chem 1985; 1099.

27. (a) Sato Y, Nakai H, Wada M, Mizoguchi T, Hatanaka Y, Kanaoka Y. Chem Pharm Bull 1992; 40:3174; (b) Wada M, Nakai H, Aoe K, Kotera K, Sato Y, Hatanaka Y, Kanaoka Y. Tetrahedron 1983; 39:1273; (c) Wada M, Nakai H, Sato Y, Hatanaka Y, Kanaoka Y. Chem Pharm Bull 1983; 31:429.

28. (a) Machida M, Takechi H, Kanaoka Y. Chem Pharm Bull 1982; 30:1579; (b) Coyle JD, Smart LE, Challiner JF, Haws EJ. J Chem Soc, Perkin Trans 1 1985; 121; (c) Coyle JD, Newport GL. Synthesis 1979; 381.

29. (a) Maruyama K, Ogawa T, Kubo Y, Araki T. J Chem Soc, Perkin Trans 1 1985; 2025; (b) Machida M, Oda K, Kanaoka Y. Tetrahedron 1985; 41:4995; (c) Mazzocchi PH, Fritz G. J Org Chem 1986; 51:5362; (d) Maruyama K, Kubo Y. J Org Chem 1981; 46:3612.

30. (a) Mazzocchi PH, Shook D, Liu L. Heterocycles 1987; 26:1165; (b) Maruyama K, Kubo Y. Chem Lett 1978; 851.

31. Xue J, Zhu L, Fun HK, Xu JH. Tetrahedron Lett 2000; 41:8553.

32. (a) Griesbeck AG, Kramer W, Oelgemöller M. Synlett 1999; 1169; (b) Griesbeck AG, Henz A, Kramer W, Lex J, Nerowski F, Oelgemöller M, Peters K, Peters EM. Helv Chim Acta 1997; 80:912.

33. (a) Griesbeck AG, Heinrich T, Oelgemöller M, Molis A, Heidtmann A. Helv Chim Acta 2002; 85:4561; (b) Griesbeck AG, Nerowski F, Lex J. J Org Chem 1999; 64:5213; (c) Yoo DJ, Kim EY, Oelgemöller M, Shim SC. Heterocycles 2001; 54:1049.

34. (a) Yoon UC. Mariano PS. Acc Chem Res 2001; 34:523; (b) Yoon UC, Jin YX, Oh SW, Cho DW, Park KH, Mariano PS. J Photochem Photobiol A: Chem 2002; 150:77; (c) Griesbeck AG, Oelgemöller M, Lex J, Haeuseler A, Schmittel M. Eur J Org Chem 2001; 1831.

35. (a) Pandey G, Reddy GD, Chakrabarti D. J Chem Soc, Perkin Trans 1 1996; 219; (b) Pandey G, Chakrabarti D. Tetrahedron Lett 1996; 37:2285; Tetrahedron Lett 1998; 39:8371; (c) Pandey G, Kapur M. Tetrahedron Lett 2000; 41:8821; (d) Pandey G, Kapur M. Synthesis 2001; 1263; (e) Pandey G, Kapur M. Org Lett 2002; 4:3883.

36. Jonas M, Blechert S, Steckhan E. J Org Chem 2001; 66:6896.

37. (a) Pandey G, Gadre SR. ARKIVOC 2003; 45; (b) Pandey G, Das P, Reddy PY. Eur J Org Chem 2000; 657.

38. (a) Horspool WM, Hynd G, Ixkes U. Tetrahedron Lett 1999; 40:8295; (b) Armesto D, Ramos A, Oritz MJ, Horspool WM, Macheño MJ, Caballero O, Mayoral EP. J Chem Soc, Perkin Trans 1 1997; 1535.

39. (a) Ackermann L, Heidbreder A, Wurche F, Klärner FG, Mattay J. J Chem Soc, Perkin Trans 2 1999; 863; (b) Hintz S, Mattay J, van Eldik R, Fu WF. Eur J Org Chem 1998; 1583; (c) Hintz S, Fröhlich R, Mattay J. Tetrahedron Lett 1996; 37:7349.

40. Bunte JO, PhD Thesis, Universität Bielefeld, 2004.

41. (a) Bunte JO, Rinne S, Schäfer C, Neumann B, Stammler H-G, Mattay J. Tetrahedron Lett 2003; 44:45; (b) Bunte JO, Rinne S, Mattay J. Synthesis 2004; 619.

42. (a) Pandey G, Karthikeyan M, Murugan A. J Org Chem 1998; 63:2867; (b) Pandey G, Murugan A, Balakrishnan M. Chem Commun 2002; 624.

43. (a) Goeller F, Heinemann C, Demuth M. Synthesis 2001; 1114; (b) Xing X, Demuth M. Eur J Org Chem 2001; 537; (c) Heinemann C, Demuth M. J Am Chem Soc 1999; 121:4894; (d) Heinemann C, Xing X, Warzecha K-D, Ritterskamp P, Görner H, Demuth M. Pure Appl Chem 1998; 70:2167; (e) Warzecha K-D, Görner H, Demuth M. J Chem Soc, Faraday Trans 1998; 94:1701; (f) Warzecha K-D, Demuth M, Görner H. J Chem Soc, Faraday Trans 1997; 93:1523; (g) Görner H, Warzecha K-D, Demuth M. J Phys Chem A 1997; 101:9964; (h) Heinemann C, Demuth M. J Am Chem Soc 1997; 119:1129; (i) Warzecha K-D, Xing X, Demuth M. Pure Appl Chem 1997; 69:109; (j) Warzecha K-D, Xing X, Demuth M, Goodard R, Kesseler M, Krüger C. Helv Chim Acta 1995; 78:2065.

44. (a) Kumar VS, Aubele DL, Floreancig PE. Org Lett 2001; 3:4123; (b) Kumar VS, Floreancig PE. J Am Chem Soc 2001; 123:3842.

45. (a) Weng H, Scarlata C, Roth HD. J Am Chem Soc 1996; 118:10947; (b) Roth HD, Weng H, Zhou D, Herbertz T. Pure Appl Chem 1997; 69:809.

46. Belotti D, Cossy J, Pete JP, Portella C. J Org Chem 1986; 51:4196.

47. (a) Hasegawa E, Takizawa S, Iwaya K, Kurokawa M, Chiba N, Yamamichi K. Chem Commun 2002; 1966; (b) Hasegawa E. J Photosci 2003; 10:61.

48. (a) Fagnoni M, Schmoldt P, Kirschberg T, Mattay J. Tetrahedron 1998; 54:6427; (b) Kirschberg T, Mattay J. J Org Chem 1996; 61:8885; (c) Tzvetkov NT, Schmidunann M, Müller A, Mattay J. Tetrahedron Lett 2003; 44. 5979.

49. (a) Pandey G, Hajra S, Ghorai MK, Kumar KR. J Am Chem Soc 1997; 119:8777; (b) Pandey G, Hajra S, Ghorai MK, Kumar KR. J Org Chem 1997; 62:5966.

50. Mikami T, Narasaka K. Chem Lett 2000; 338.

51. (a) Griesbeck AG, Fiege M. In: Ramamurthy V, Schanze KS eds. Molecular and Supramolecular Photochemistry. Vol. 6. New York: Marcel Dekker, Inc., 2000; 33; (b) Everitt SRL, Inoue Y. In: Ramamurthy V, Schanze KS eds. Molecular and Supramolecular Photochemistry. Vol. 3. New York: Marcel Dekker, Inc., 1999; 71; (c) Hasegawa T, Yamazaki Y, Yoshioka M. Trends in Photochem & Photobiol 1997; 4:27.

52. (a) Griesbeck AG, Synlett 2003; 451; (b) Griesbeck AG. J Photosci 2003; 10:49.

53. Griesbeck AG, Meierhenrich UJ. Angew Chem 2002; 114:3279; Angew Chem Int Ed Engl 2002; 41:3147.

54. (a) Yoon UC, Mariano PS. J Photosci 2003; 10:89; (b) Su Z, Mariano PS, Falvey DE, Yoon UC, Oh SW. J Am Chem Soc 1998; 120:10676; (c) Ikeno T, Harada M, Arai N, Narasaka K. Chem Lett 1997; 169.

55. Yoon UC, Oh SW, Lee JH, Park JH, Kang KT, Mariano PS. J Org Chem 2001; 66:939.

56. (a) Wada M, Nakai H, Sato Y, Kanaoka Y. Chem Pharm Bull 1982; 30:3414; (b) Hu S, Neckers DC. J Photochem Photobiol A: Chem 1998; 114:103.

57. Hasegawa T, Yamazaki Y, Yoshioka M. J Photosci 1997; 4:61.

58. (a) Inoue Y, Wada T, Asaoka S, Sato H, Pete JP. Chem Commun 2000; 251; (b) Inoue Y, Sugahara N, Wada T. Pure Appl Chem 2001; 73:475; (c) Inoue Y, Ikeda H, Kaneda M, Sumimura T, Everitt SRL, Wada T. J Am Chem Soc 2000; 122:406.

59. Benali O, Miranda MA, Tormos R, Gil S. J Org Chem 2002; 67:7915.

60. Fuji K, Kawabata T. Chem Eur J 1998; 4:373.

61. (a) Griesbeck AG, Kramer W, Lex J. Synthesis 2001; 1159; (b) Griesbeck AG, Kramer W, Lex J. Angew Chem 2001; 113:586; Angew Chem Int Ed Engl 2001; 40:577; (c) Griesbeck AG, Kramer W, Bartoschek A, Schmickler H. Org Lett 2001; 3:537.

62. Hoegy SE, Mariano PS. Tetrahedron Lett 1994; 35:8319.

63. (a) Griesbeck AG, Maptue N, Bondock S, Oelgemöller M. Photochem Photobiol Sci 2003; 2:450; (b) Griesbeck AG, Kramer W, Oelgemöller M. Green Chem 1999; 1:205.

64. (a) Esser P, Pohlmann B, Scharf HD. Angew Chem 1994; 106:2093; Angew Chem Int Ed Engl 1994; 33:2009; (b) Pohlmann B, Scharf H-D, Jarolimek U, Mauermann P. Solar Energy 1997; 61:159; (c) Funken K-H, Ortner J. Z Phys Chem 1999; 213:99.

65. A number of additional examples can be find. In: Mattay J, Griesbeck AG, eds. Photochemical Key Steps in Organic Chemistry. Weinheim: VCH, 1994.

66. (a) Kessar SV, Singh T, Vohra R. Tetrahedron Lett 1987; 28:5323; (b) For a very similar approach, see: Mazzocchi PH, King CR, Ammon HL. Tetrahedron Lett 1987; 28:2473.

67. For an alternative approach via an aryl enamide photocyclization, see: Rigby JH, Maharoof USM, Mateo ME. J Am Chem Soc 2000; 122:6624.

68. Cossy J, Bouz-Bouz S, Mouza C. Synlett 1998; 621.

11
Photo-oxygenation of the [4+2] and [2+2] Type

Maria Rosaria Iesce
Università di Napoli Federico II, Napoli, Italy

11.1. HISTORICAL BACKGROUND

The first example of an 1,4-endoperoxide by photo-oxygenation was reported in 1928 by Windaus and Brunken [1] who obtained the peroxide of ergosterol by irradiation in the presence of oxygen and a dye (eosin) (Sch. 1).

More than 40 years later did two different groups [2,3] isolate [2+2]-cycloadducts (1,2-dioxetanes) by photo-oxygenation of alkenes shortly after Kopecky had synthetized the first stable dioxetanes by base-catalyzed cyclization of α-bromohydroperoxides (Schs. 2,3) [4].

Since these pioneering works a considerable number of conjugated dienes or alkenes have been photo-oxygenated and have led to the corresponding endoperoxides [5] or 1,2-dioxetanes [6]. A large part of these compounds prove stable enough to be isolated and characterized. In many cases they have been trapped by suitable reagents or postulated on the basis of the resulting oxygenated products. However, the synthetic potential of [4+2] and [2+2] photo-oxygenation reactions was recognized early and an enormous number of applications are reported in literature.

Photo-oxygenation is a process which usually combines a substrate, light, oxygen in the presence of a sensitizer. The electronically excited sensitizer can either activate the substrate by energy or hydrogen transfer which in turn reacts with molecular oxygen (type I) or activates the oxygen to singlet excited-state oxygen which then reacts with the substrate (type II) [7a]. A recently defined Type III-process is also possible involving substrate

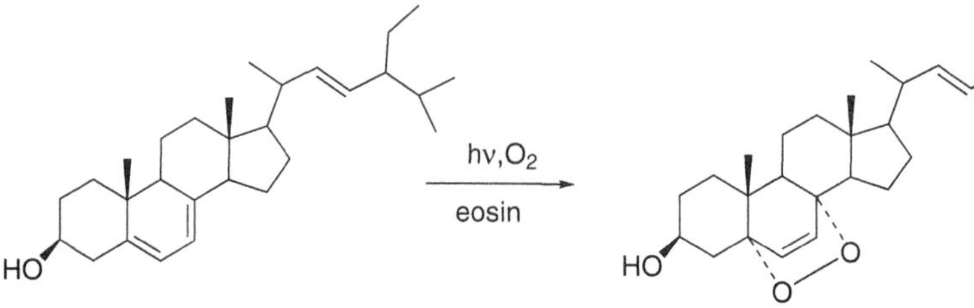

Scheme 1

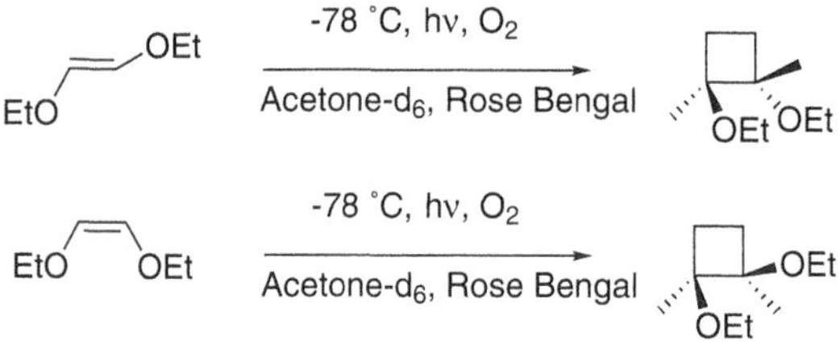

Scheme 2

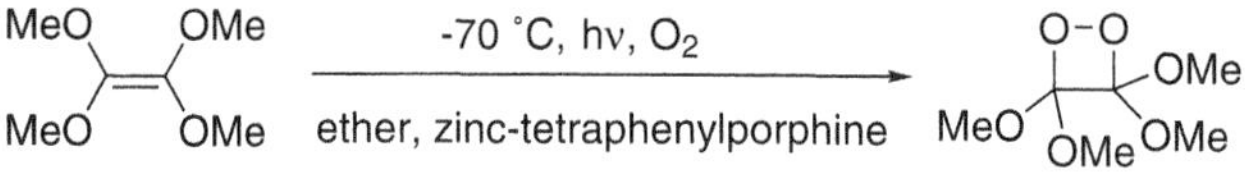

Scheme 3

radical cations or superoxide radical anion [7b–d]. Proper choise of the reaction conditions can discriminate among the various possibilities. Thus, dyes, sun lamps, halogenated or deuterated solvents, and low temperatures favor II type reactions. The latter reactions, due to the mild experimental conditions as well as the generally high selectivity of the oxygenating species (singlet oxygen), represent a potent means to introduce oxyfunctionalizations in organic material. Singlet oxygen can react with unsaturated substrates by [4+2] or [2+2] or ene-mode (Sch. 4) or with heteroatoms as sulfur or undergoes physical quenching [8].

Its involvement can be easily confirmed through experimental techniques which include quenching with 1,4-diazabicyclo[2.2.2]octane (DABCO) or β-carotene, increasing of reaction rate in deuterated solvents,

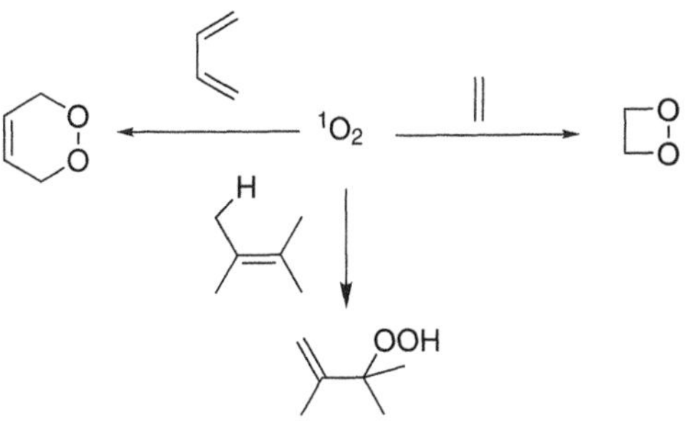

Scheme 4

identical product distribution with different sensitizers, competitive inhibition with singlet oxygen acceptors as tetramethylethylene [5a,8a]. The frequently used sensitizer-solvent systems are tetraphenylporphine (TPP) or 5,10,15,20-*tetrakis*(pentafluorophenyl)porphine (TPFPP) in dichloromethane, Rose Bengal (RB) or methylene blue (MB) in methanol, polymer-bound Rose Bengal (Sensitox) in acetone [5b,c,8a]. The use of halogenated or deuterated solvents and low temperature are recommended to accelerate the peroxide formation and minimize that of probable radical-derived by-products. The use of filters to remove ultraviolet light or a lamp with low ultraviolet content is useful in preventing UV-initiated rearrangement of the peroxides [5b,c,8a].

Most reactions reported in this chapter are of Type II involving singlet oxygen in the first electronically excited state ($^1\Delta_g$) and, in particular, photo-oxygenation of [4+2] and [2+2] type will be illustrated. Excellent books [5,6,8] and reviews [9] cover the extraordinarialy wide literature on the mechanistic and synthetic aspects of these reactions, and in the recent years attention has been given to the stereochemical aspects, especially in view of the application of these reactions in asymmetric synthesis [10].

11.2. [4+2] PHOTO-OXYGENATION

11.2.1. State of Art Mechanistic Models

Due to the topographical similarity to the Diels–Alder reaction, the addition of singlet oxygen to conjugated systems is generally described as a concerted reaction ([4π+2π]-cycloaddition). Actually, the frontier MO theory applied to the reaction with naphthalenes and other polycyclic aromatic compounds shows good correlations between the highest occupied molecular orbital (HOMO) on the diene system and the lowest unoccupied molecular orbital (LUMO) on singlet oxygen [11]. However, over the years the assumption

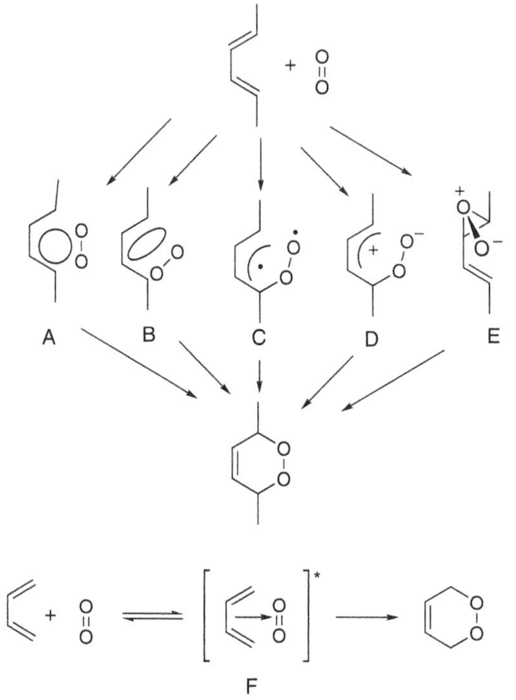

Scheme 5

of a simple concerted reaction has been shown limited in perspective and a variety of mechanisms has been proposed which vary from concerted (A) to nonconcerted (B) and include diradical (C) and/or open-chain zwitterionic intermediates (D) or perepoxide (E) (Sch. 5) [5,8b].

The intermediacy of a perepoxide which had been proposed on the basis of MINDO/3 [12] has been excluded by successive theoretical calculations on the 1,4-cycloaddition of singlet oxygen to *s-cis*-1,3-butadiene [13]. On the basis of this investigation, a diradical intermediate has been found on the lowest energy pathway to the 1,4-adduct [13]. However, a very recent work [14] suggests a concerted—even if highly nonsynchronous—mechanism in the formation of the cyclohexadiene endoperoxide. The debate still continues and the multiplicity of versions reflects a general problem in scrutinizing reactions of electronically excited molecules such as singlet oxygen by theoretical methods.

From the experimental perspective, most of the data (the lack of appreciable solvent effects [15], the stereospecificity observed in different reactions [16], the effects of the substituents [17]) are consistent with the concerted mechanism. However, the formation of methanol-trapping products [18] or the nonstereospecificity observed in some cases [18b] indicate that stepwise mechanisms proceeding through open zwitterion or biradical intermediates are feasible as well. The reversible formation of exciplex intermediates has been also proposed (F, Sch. 5) [19]. In particular, kinetic data of singlet oxygen [4+2] cycloaddition with furans and cyclopentadienes suggest that this step is followed by a symmetrical and unsymmetrical concerted collapse to the [4+2] cycloadduct depending on the substrate (only symmetrical furans react symmetrically) and/or solvatation [19b]. The reversible formation of an exciplex as the primary reaction intermediate also accounts for the near-zero activation energy found in the singlet oxygen [4+2] cycloaddition [20] and for the solvent insensitivity [15], as well as for the stereochemical results [10]. So, this species appears to describe best and in a unifying manner the reaction mechanism.

11.2.2. Chemoselectivity

1,4-Addition of singlet oxygen can compete with the ene-reaction in the presence of allylic hydrogens or with the 1,2-addition particularly in the presence of heteroatom substituents which have a pronounced activation effect on dioxetane formation. However, the reactivity pattern can often be predicted with a detailed knowledge of electronic and structural factors which promote one type of reactivity at the expense of the other. [4+2]-Cycloaddition is favored by even little amounts of *cis*-conformer but it can be surpassed by ene reaction in the presence of an easy to ionize olefinic linkage (tri- or tetrasubstituted double bond). Due to its electrophilic character, singlet oxygen prefers *to react with electron rich bonds* [21]. Thus, only α-mircene 1 gives endoperoxide 2 (79% yield) [17], while for 3 the attack at the more substituted bond occurs leading to ene-adducts 4 and 5 (25–35% and 50%, respectively, by I.R.) [22] (Sch. 6).

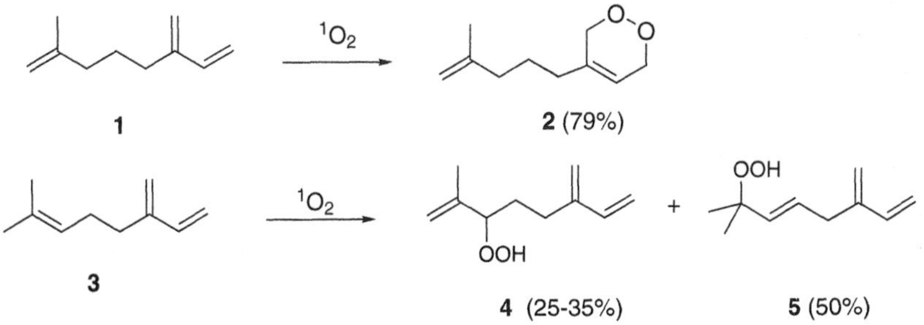

Scheme 6

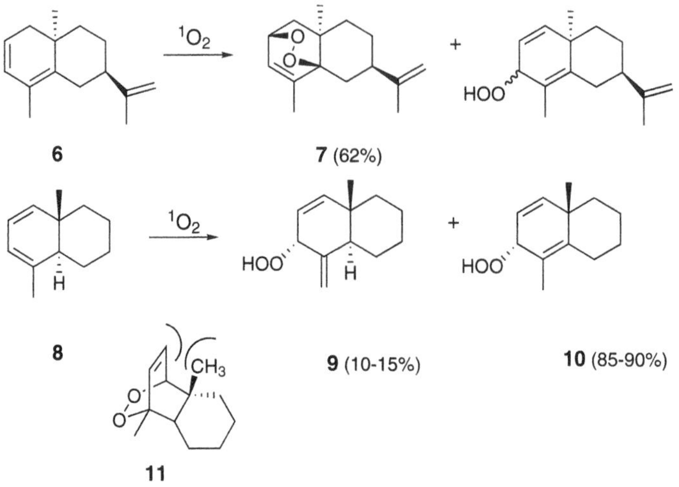

Scheme 7

In spite of the small dimensions of singlet oxygen, *steric interaction* also plays an important role. Singlet oxygen attacks both compounds **6** and **8** at the less hindered face. However, only the oxygenation of compound **6** affords the endoperoxide **7** [23b] while in the other case it leads to ene-adducts **9** and **10** (Sch. 7) [23a].

The absence of an endoperoxide formed by approach to the sterically less hindered α-face of **8** has been assigned to a severe 1,3-interaction as shown in **11** (Sch. 7). It appears more likely that the high chemoselectivity is due to a *correct orthogonal alignment of the allylic hydrogen to the olefinic plane*, and this is a condition which strongly favors the ene reaction [9a]. The *distance between the terminal carbons C1–C4* appears equally important in determining the rate of 1,4-addition [16a,24]. Therefore, too large a distance would be responsible of the poor 1,4-reactivity of the dienes **12** and **15** as these preferentially undergo [2+2] addition which is favored by the presence of the heteroatom substituents (Sch. 8) [24]. Even, the 1,4-unsubstituted diene **19** is recovered unchanged.

11.2.3. Stereochemistry

Owing to the inherent lack of stereochemical features in the singlet oxygen molecule, stereoselectivity is directed by the substrate. In view of the formal analogy to the classical Diels–Alder reaction, a stereochemical requirement imposed by the concerted mechanism is that the configuration of the diene system is retained in the cycloadduct, as occurs for diene

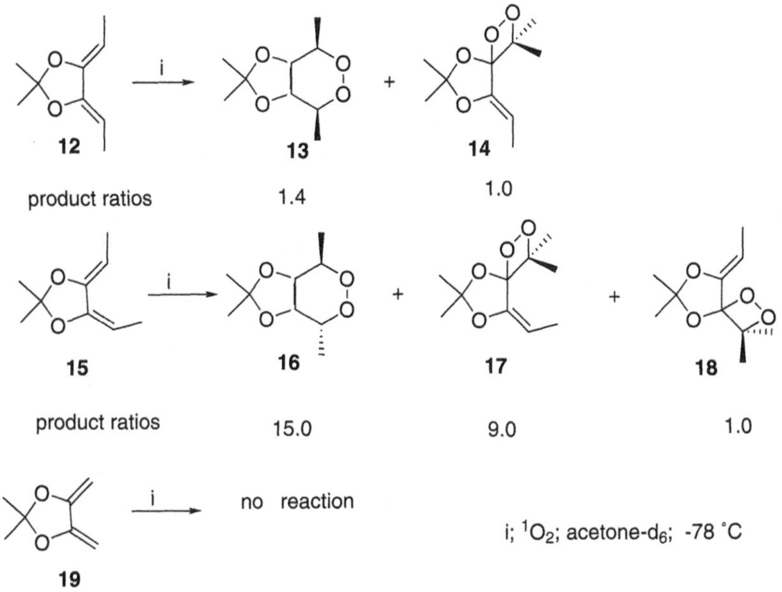

12 → **13** + **14**

product ratios   1.4   1.0

15 → **16** + **17** + **18**

product ratios   15.0   9.0   1.0

19   →   no reaction

i; $^1O_2$; acetone-d_6; -78 °C

Scheme 8

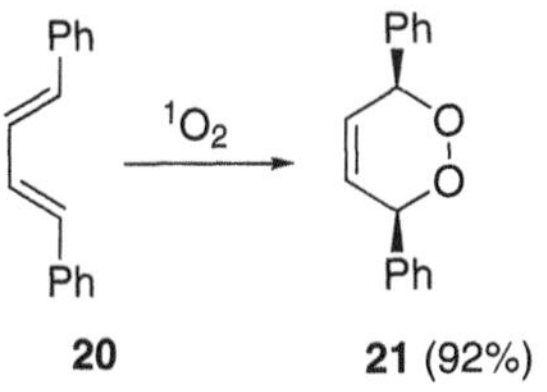

20   **21 (92%)**

Scheme 9

20 which gives the peroxide **21** stereospecifically, with 92% of yield [17] (Sch. 9).

This control, however, can be tested only in acyclic dienes, which lack of the conformational rigidity of cyclic substrates. Loss of the stereo-chemical information could be due to isomerization of the starting diene [18b,19c], to two-step processes [18], or to the reversible formation of an exciplex which, depending on the substrate structure, may have a zwitterionic and/or diradical character [19].

For plane-nonsymmetric dienes, either acyclic or cyclic molecules, there is an additional stereochemical feature which is the π-facial selectivity.

The attack of singlet oxygen occurs generally at the less hindered of the two different faces, as highlighted in Sch. 7, and this sensitivity to steric control accounts for the high π-facial selectivity generally observed in the photo-oxygenation of polycyclic substrates [8d]. However, efforts have been made to evaluate the π-facial directing effects of functional groups on the attack. These effects may be steric but also electronic (attraction, repulsion, H-bonding interaction), and their knowledge may be used for prediction and control of the stereoselectivity and, hence, is essential for preparative purposes, particularly in the synthesis of natural products.

In the reaction of compounds **22**, which have diene moieties embedded in rigid polycyclic frames, the carbonyl-face selectivity is completely reversed through protection of the carbonyl groups as mono- or *bis*-acetals. This suggests the use of protecting groups as stereodirectors (Sch. 10) [25].

The *tert*-butyldimethyl silyl group also behaves as stereodirecting group since it enhances the diastereofacial selectivity significantly in the endoperoxidation of dienes **25**. (Sch. 11) [26].

Indeed, the photo-oxygenation of **25a** leads to endoperoxides **26a** and **27a** (total yield 80%) with the β-isomer **26a** resulting from the attack at the face opposite to the silyl group as the dominant product (54:1). In comparison, the reaction carried out on the desilylated **25b** affords the corresponding β- and α-endoperoxides **26b** and **27b** in a ratio of 3:2 [26].

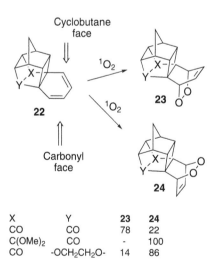

X	Y	23	24
CO	CO	78	22
C(OMe)$_2$	CO	-	100
CO	-OCH$_2$CH$_2$O-	14	86

Scheme 10

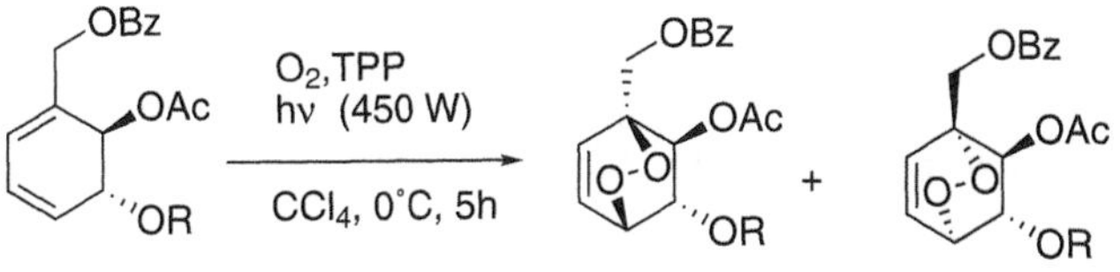

25

R = TBDMS (a)
R= H (b)

Scheme 11

26

78.5%
48%

27

1.5%
32%

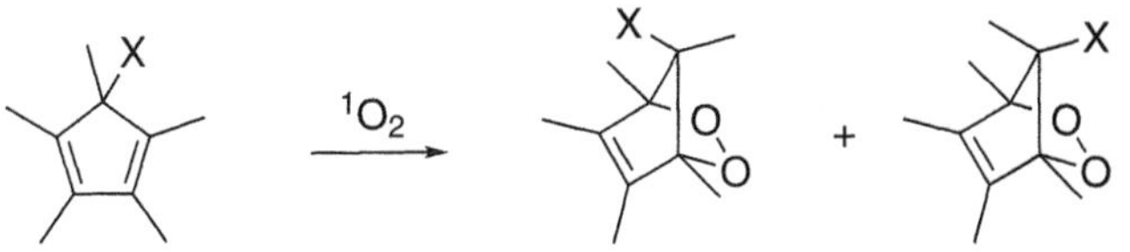

28   *anti* adduct   *syn* adduct

X=	*anti:syn*
H	20:80
SMe,Et	>90:10
SOMe, SO$_2$Me, *i*-Pr,CHO, CH$_2$OH	>95: 5

Scheme 12

Adam has found that in compounds **28** the reaction occurs with a diastereoselectivity *anti:syn* ≥ 90:10, whatever the substituent, and, hence, the stereocontrol is exclusively by steric bias (Sch. 12) [10d].

The loss of stereochemical control in the oxygenation of the isodicyclopentadiene derivative **29**, even compared with typical carbon dienophiles, cannot be due to steric factors by the norbornane ring, since remote, and has been rationalized in terms of electronically facilitated reversal of the exciplex on that π-face which is most conducive to bonding in the first instance (Sch. 13) [27].

In particular, it has been suggested that the higher selectivity of **28** than **29** is due to its higher activation and, hence, higher reactivity [10d].

In chiral molecules as **30**, in addition to steric and electronic factors, energies of different rotamers can come into play and have to be considered. So, by oxygenation of a series of chiral naphtalene derivatives **30**, a significant stereoelectronic control has been highlighted, particularly by a

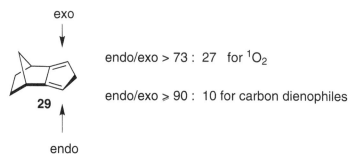

exo
endo/exo > 73 : 27 for $^1O_2$
endo/exo ⩾ 90 : 10 for carbon dienophiles
29
endo

Scheme 13

X
Me
$^1O_2$
$CDCl_3$, -30 °C
+
30
like-**31**
unlike-**31**

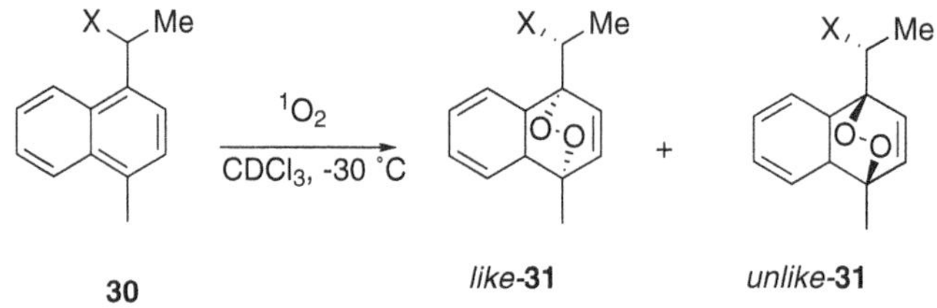

X=	*l*-**31** / *u*-**31**
OH	85:15
OMe	66:34
CH_2OH	90:10
CO_2Me	22:78
t-Bu	34:66

2
Me
8a 6- H 7
H
X
unlike
like
A
B
δ+
δ−
O=O

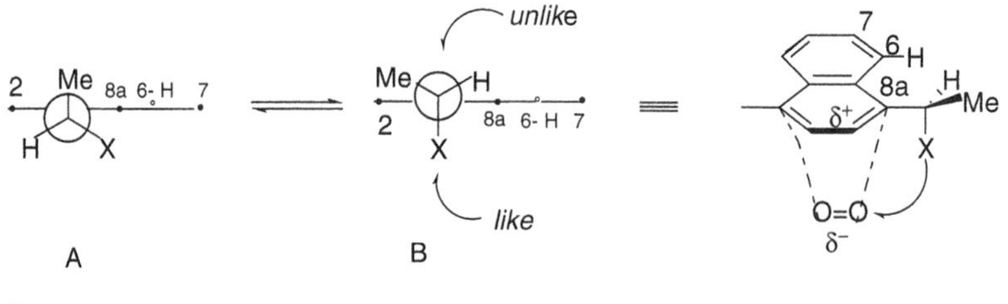

Scheme 14

hydroxy group, which has been explained on the basis of the mechanistic picture in Sch. 14 [10d].

Since in naphtalene substrates a steric interaction exists between the substituent at the stereogenic center and the *peri* H, the preferred conformations A and B bear the small hydrogen substituent proximate to this atom so that the attack of oxygen can be controlled by steric and electronic factors. The small dimensions of the OH group and its attractive interaction with the incoming oxygen can account for the favored *like*-attack [10b,d]. The determinant role of H-bonding is confirmed by the low selectivity observed when protecting the free hydroxy group (Sch. 14, X = OMe) or when the reaction is performed in methanol (*l*-31: *u*-31 = 55:45) [10b]. The *unlike*-selectivity for X = CO_2Me (but also for Br, Cl, $SiMe_3$, COOH) might be due to electrostatic repulsion in the *like* approach between the negatively charged $^1O_2$ and the electronegative substituent. The steric bias alone is not particularly effective as shown by the poor (66:34) selectivity in the photo-oxygenation of *tert*-butyl-substituted derivative.

Different from the naphthyl alcohol, the presence of a hydroxy-substituted stereogenic center in acyclic diene 32 appears to scarcely affect the diastereoselectivity (Sch. 15) [10d].

Due to the absence of appreciable allylic strain, the diastereoselectivity might be primarily determined by stereoelectronic effects with a preference for one of the conformations. The moderate control may be accounted for by a reduced sensitivity of singlet oxygen towards stereoelectronic factors.

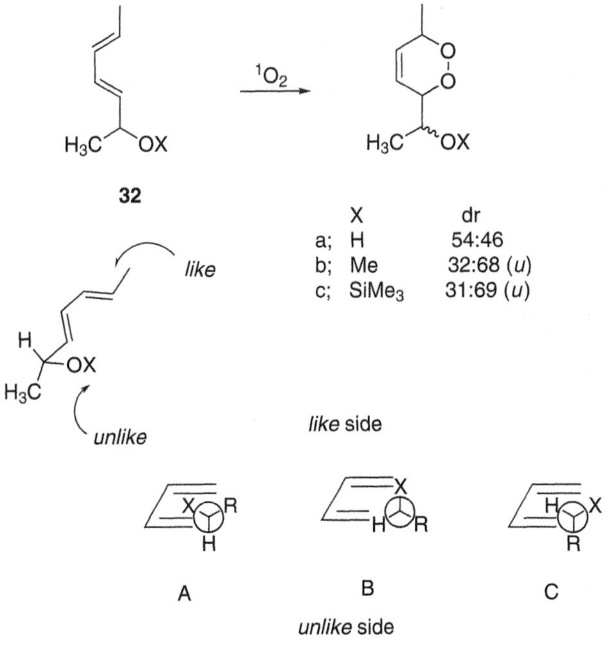

X	dr
a; H	54:46
b; Me	32:68 (*u*)
c; SiMe$_3$	31:69 (*u*)

Scheme 15

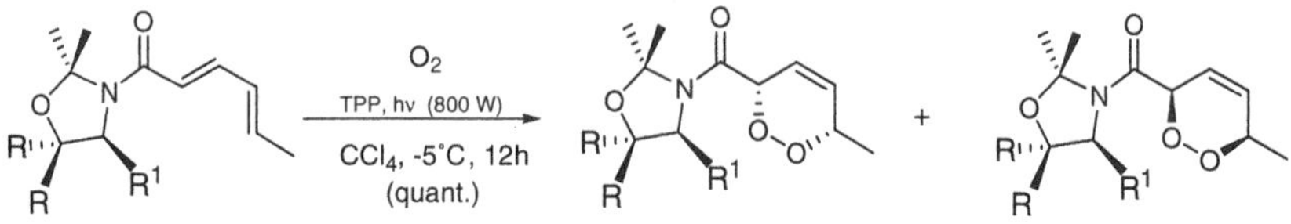

33 *l*-34 *u*-34

amide	R	R^1	dr **l-34 : u-34**
a	H	CH_2Ph	68 : 32
b	H	CH_2Nph	67 : 33
c	H	*i*-Pr	76 : 24
d	H	Ph	91 : 9
e	Me	CH_2Ph	⩾95 : 5

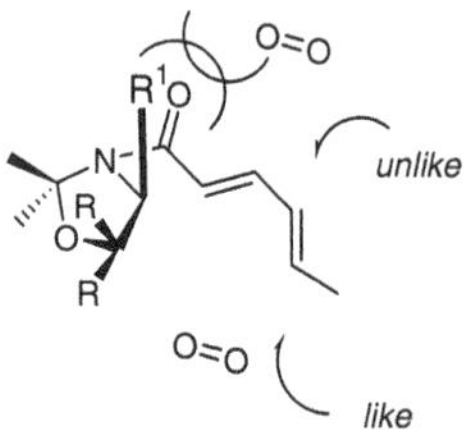

Scheme 16

So, due to the high reactivity, it captures the various rotamers of the substrate, all of which are populated in the ground state (Sch. 15) [10d]. More satisfactory results have been achieved by steric control using optically active 2,2-dimethyloxazolidines as chiral auxiliaries (Sch. 16) [10c].

As shown in the corresponding Table, the oxygenation of the amides **33** occurs with a π-facial diastereoselectivity which increases with increasing branching at the α-carbon atom of the R^1 group ($CH_2Ph \approx CH_2NPh <$ *i*-Pr < Ph). Complete stereochemical control (diastereomeric ratio ⩾ 95:5) is achieved for **33e** where the two methyl adjacent to the C-4 position force the phenyl group of the benzyl substituent to be placed directly over the reacting diene moiety [10c].

11.2.4. Scope and Limitations

11.2.4.1. Cyclic Dienes

Endoperoxidation is the preferred process for cyclic dienes, and simple five, six, seven, and eight membered ring aliphatic dienes lead, by photo-oxygenation, only to the corresponding endoperoxides which, except for that of cyclopentadiene, are quite stable (Sch. 17) [28].

There is, however, about a tenfold decrease in the rate constant when the ring size is increased by one methylene unit [29], and an ideal C1–C4 distance is responsible for the impressive reactivity of cyclopentadiene in comparison to larger cyclic homologues [16a]. Electron-donating substituents enhance the reactivity, and for 1,3-cyclopentadiene derivatives

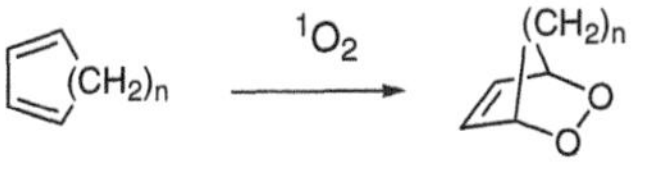

n	Yield (%)
1	86[28a]
2	21[28b]
3	29[28c]
4	53[28d]

Scheme 17

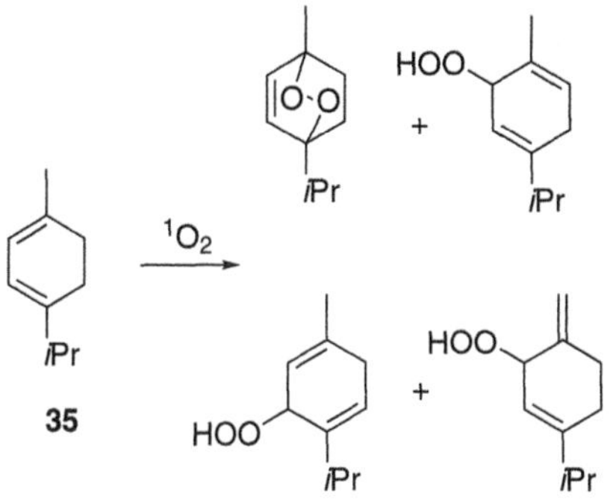

[4+2]/ene-adducts = 90% : 4.9%

Scheme 18

the [4+2]-cycloadduct is still the only product even in the presence of alkyl [10d,30] or aryl [30] substituents. For higher homologues the extent of the reaction depends on the factors reported above as the substitution site and/ or steric shielding. Thus, the two alkyl groups at positions 1 and 4 favor the formation of the endoperoxide from α-terpinene **35** [31a] while it is less dominant from compound **36** with one alkyl group at position 2 [31a,b] or in the 1,4-unsubstituted diene **37** [31c] (Schs. 18–20).

Endoperoxides derived from cyclohexa-1,3-dienes are the most abundant in this category due to the greater stability compared to their analogues and to the substrate availability [5c,8d]. This subunit is present in a great number of terpene and steroid products. Most of steroid endoperoxides are 5,8-derivatives in which the oxygen bridge has the α-configuration [1,5c,8d]. The steric demand of the methyl at C-10 appears

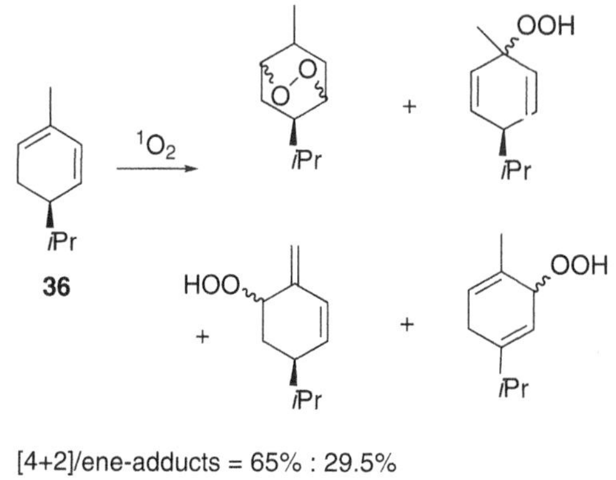

[4+2]/ene-adducts = 65% : 29.5%

Scheme 19

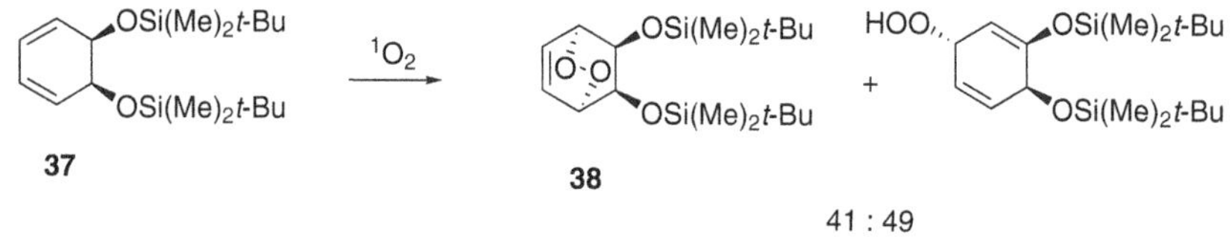

41 : 49

Scheme 20

to determine upon which face of the steroid singlet oxygen attacks, so, the formation of β-endoperoxides is found in steroids with α-methyl group at C-10 [5c,8d,32].

Photo-oxygenation of 1,3,5-cycloheptatrienes **39** which can undergo valence tautomerization to norcaradiene isomers, leads to the bicyclic endoperoxides **40** and/or tricyclic derivatives **41** depending on the steric and electronic nature of the substituents (Sch. 21) [33].

In particular, in the reaction of 7-substituted trienes **39**, if R^1 is more bulky than R^2, the norcaradiene endoperoxide **41** is favored, especially with π-electron acceptor substituents (CO_2R, COR, $CONH_2$), while the tropilidene endoperoxide **40** is the only one formed with substituents with π-electron donor ability, such as OMe [33c–e]. Starting from asymmetric compounds, complex mixtures of endoperoxides are formed depending on the location and type of substituent [5c,33a,b]. A good example of high regioselectivity (but inverse) π-facial selectivity has been found in the [4+2] photo-oxygenation of colchicine **42** [34a] and its isomer isocolchicine **44** [34b] which gave stereoselectively the tropilidene endoperoxides *syn*-**43** and *anti*-**45**, respectively (Sch. 22).

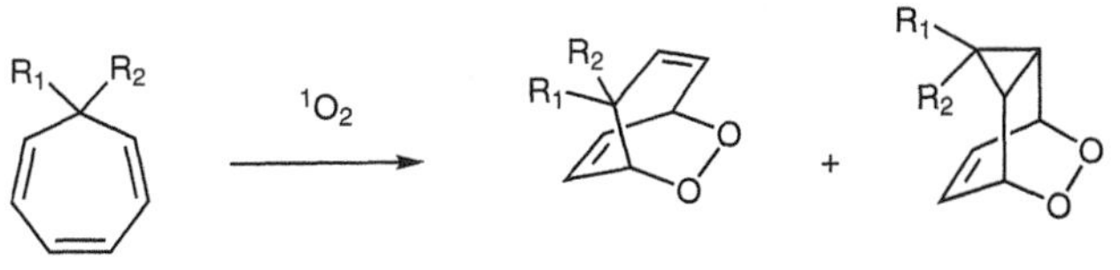
R1
R2
$^1O_2$
R2
R1
O
O
+
R1
R2
O
O

39 40 41

	R_1	R_2	**40**	**41**
a,[33e]	H	H	92	8
b,[33d]	CO_2Me	H	-	100
c,[33c]	Ph	H	14	86
d,[33d]	OMe	H	100	-

Scheme 21

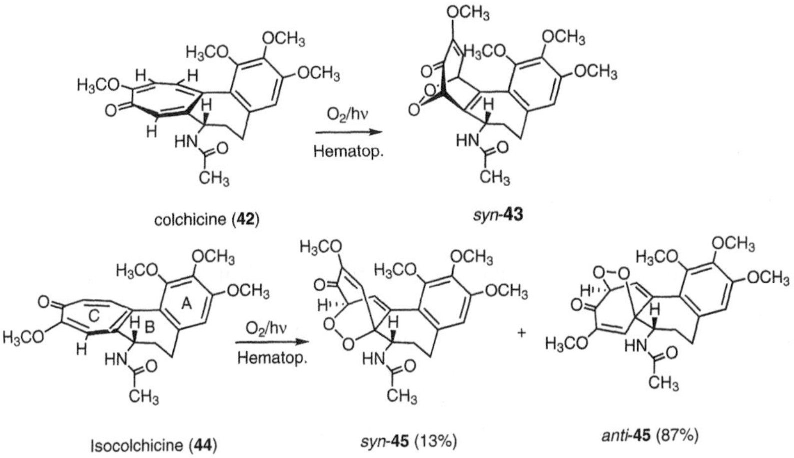
colchicine (42)
O2/hν
Hematop.
syn-43
Isocolchicine (44)
O2/hν
Hematop.
syn-45 (13%)
+
anti-45 (87%)

Scheme 22

The diastereoselectivity in the reaction of **42** was proven to be independent of the polarity of the solvent used and is therefore considered a consequence of steric factors [34a]. Photo-oxygenation of 1,3,5-cyclooctatriene [35] affords both the bicyclic and tricyclic endoperoxides (isolated yield 20 and 5%, respectively) as found for the heptatriene **39a**, while for cyclooctatetraene [36] only the bicyclic peroxide is obtained in 26% yield.

11.2.4.2. Acyclic Dienes

Photo-oxygenation of acyclic dienes is complicated by the possibility that all three types of singlet oxygen reactions occur. It is, however, possible

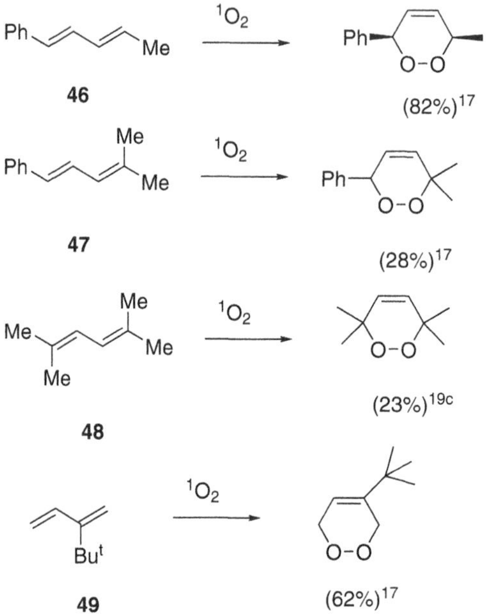

Scheme 23

according to the above considerations to predict and/or to favor formation of the 1,4-adduct. The decreasing yield of 1,4-adducts from **46** [17] to **47** [17] or **48** [19c] can be easily explained on the basis of steric factors which play a role in the s-*cis*/s-*trans* conformational isomerism (Sch. 23).

A *cis*-1-substituent on a diene increases the strain energy of the *cis* conformation required for the 1,4-cycloaddition so that this reaction mode becomes difficult and other oxygenations such as [2+2]- or ene-addition, take place (Sch. 24).

On the contrary, a bulky substituent at 2-position as in **49** makes the *trans* conformation less stable and, hence, favors the 1,4-addition [17]. From a stereochemical perspective, some results agree with a concerted mechanism as found, for example, in the stereospecific formation of the above **21** or of the *cis*-cycloadduct **51** (isolated yield 85%) from *E,E*-1,4-dimethylbuta-1,3-diene **50** [37] (Sch. 25).

However, *E,Z*-**50** leads to a mixture of both the expected *trans*- and, also, *cis*-adduct **51** [37], and the oxygenation of either (*E,E*)- or (*E,Z*)-1-aryl-1,3-pentadienes **52** gives mainly *cis*-endoperoxides **53a–c** (72–84%) (Sch. 26) [38].

These results have been explained through a photoinduced isomerization of the (*Z,Z*)-isomers in the presence of both tetraphenylporphine and

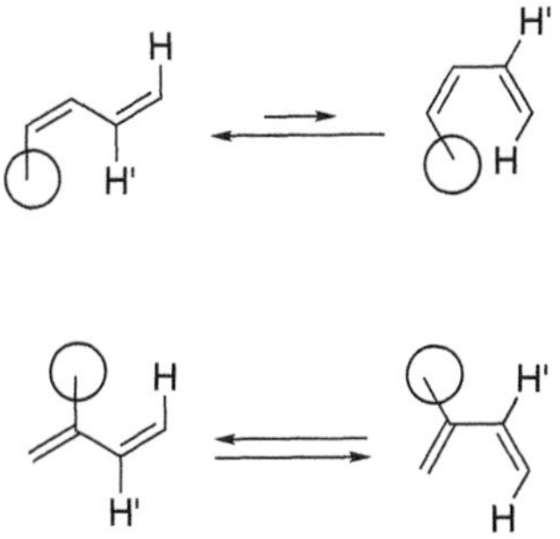

Scheme 24

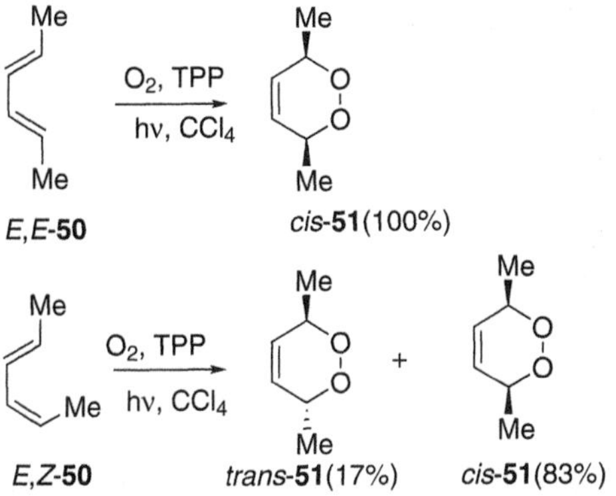

E,E-**50**   *cis*-**51**(100%)

E,Z-**50**   *trans*-**51**(17%)   *cis*-**51**(83%)

Scheme 25

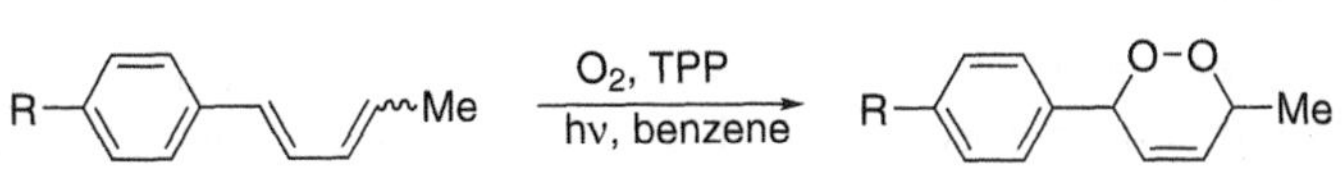

52a; R=H   *cis*-**53a-c**
52b; R=Me   72-84%
52c; R=OMe

                    isomerization $^1O_2$-addition
          E,Z-diene ⇌ *E,E*-diene ⇌ endoperoxide

Scheme 26

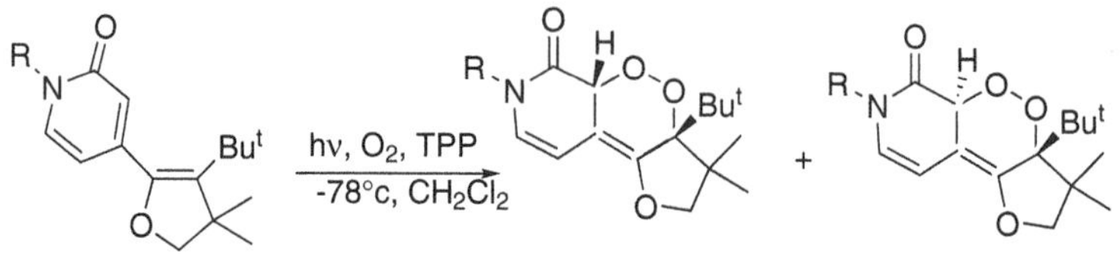

54

cis-55
(65%)

trans-55
(35%)

R=PhCH$_2$

Scheme 27

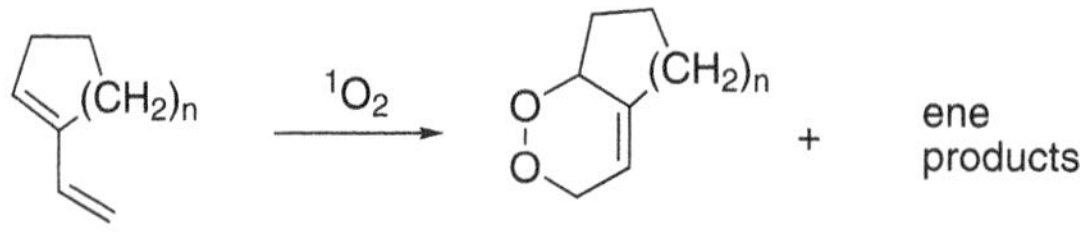

56

ene
products

n	[4+2]: ene
1	16 : 84
2	77 : 23
3	22 : 78
4	50 : 50
5	47 : 53
6	67 : 33

Scheme 28

singlet oxygen [37,38], or of sensitizers with low oxidation potential [38]. The isomerization is followed by the concerted Diels–Alder addition. It has recently been found that singlet oxygen adds to 1,3-diene **54**, whose E–Z isomerization cannot take place, to give a mixture of stereoisomeric 1,4-endoperoxides **55** (Sch. 27) [39].

To explain these results, the authors assume that in this case a primary zwitterion (or perepoxide) is formed, which attacks the double bond at both faces [39].

In the photo-oxygenation of vinylcycloalkenes **56**, the ratio of [4+2] cycloaddition to ene reaction is significantly influenced by the ring size (Sch. 28) [40].

This appears to be related to the availability of allylic hydrogens capable of approaching a position orthogonal to the olefinic plane, thus favoring the ene reaction [40]. Attempts to control the diastereoselectivity with a hydroxy group next to C1 lead to moderate facial selectivity (Sch. 29) [10d], and similar results have been found in semicyclic dienes with the hydroxy group next to C2 [40] of the diene moiety.

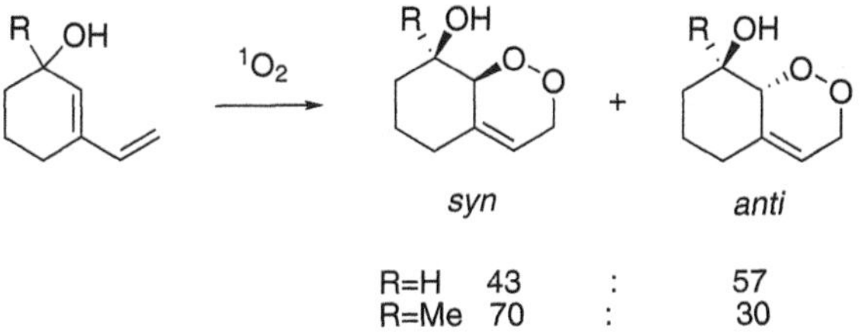

syn   anti

R=H   43   :   57
R=Me 70   :   30

Scheme 29

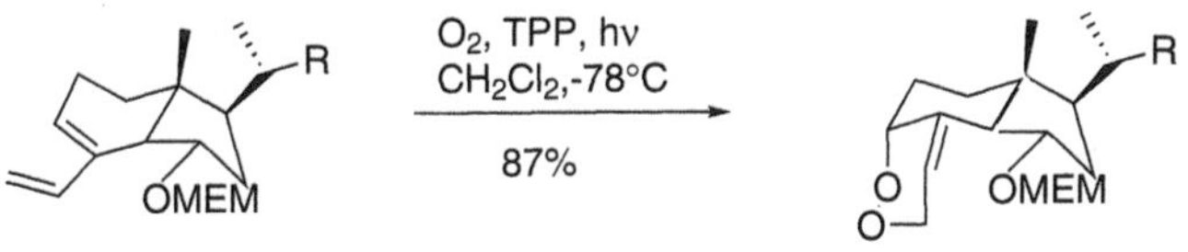

O_2, TPP, hv
CH_2Cl_2, -78°C

87%

57   **58**

R = $(CH_2)_3CH(CH_3)_2$

Scheme 30

High steric control by remote groups has been observed in the oxygenation of diene **57**, which gives the peroxide **58** with 87% yield and excellent stereoselectivity (>95%) (Sch . 30) [41]. The α-face diastereo-preference has been attributed to the presence of the OMEM group and the C-12 angular methyl group [41].

11.2.4.3.   Aromatic Compounds

The capability of aromatic compounds to react with singlet oxygen increases with increasing number of condensed rings [42]. Thus, benzene derivatives react only if electron-donating substituents are present (methoxy, *N,N*-dimethylamino, methyl groups) or when the benzene moiety is incorporated into a cyclophane system [42]. In the presence of alkyl groups, cycloaddition is followed by ene reaction on the allylic moiety thus formed as in **59** (90% yield) (Sch. 31) [43]. The peroxides from benzene derivatives are unstable and only recently has the preparation and X-ray crystallographic analysis of a stable endoperoxide of a pentamethyl[2.2]metacyclophane been reported [44].

In contrast to anthracene or higher members of the acene series [45], naphtalene does not react with singlet oxygen [19a]. The reaction occurs readily when electron-donating substituents are introduced on the 1,4

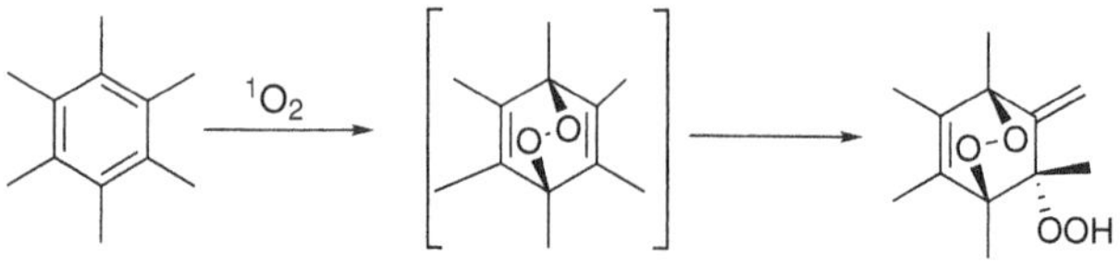

59 (90%)

Scheme 31

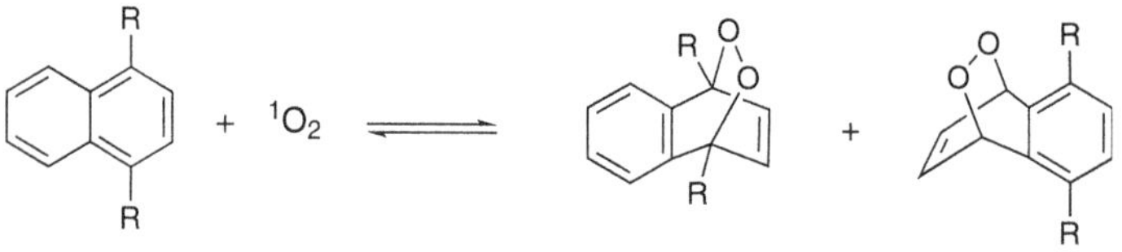

a;[47] R = Me   100%   -

b;[48b] R=CH₂CH[CONHCH₂CH(OH)CH₂(OH)]₂   80%   20%

Scheme 32

positions of the naphtalene core and leads regiospecifically to 1,4-endo-peroxides (Sch. 32) [46].

However, the regioselectivity can be modified by both electronic and steric effects. 5,8-Endoperoxides have been observed when at least one additional methyl group is placed on the 6,7 [47] or 8 positions [10d] or when bulky groups are present in 1,4-positions [48] (Sch. 32). In the latter cases it has been suggested that steric hindrance induced by side chains contributes to steer the incoming singlet oxygen to the non-substituted cycle [48].

Oxidation of anthracene or higher acenes also occurs in the absence of a dye, since the same substrates act as sensitizers in singlet oxygenation [9c,42a]. These compounds react mainly at the *meso*-positions of the aromatic nucleus [49a], but electron-donating groups have a strong directing effect when placed at the 1- and 4-positions (Sch. 33) [49b].

The thermal stability of the endoperoxides of aromatic compounds enhances going from the, only spectroscopically detectable, endoperoxides of benzene derivatives to those from acenes, which are stable enough to be stored for several days. The main rearrangement is reversion to singlet oxygen and the starting materials and depends on the aromatic system and on the type of substituents in the *meso* positions when they carry the

a;[49a] $R^1 = R^2 = H$ $R^3 = R^4 = Ph$ 95%   -

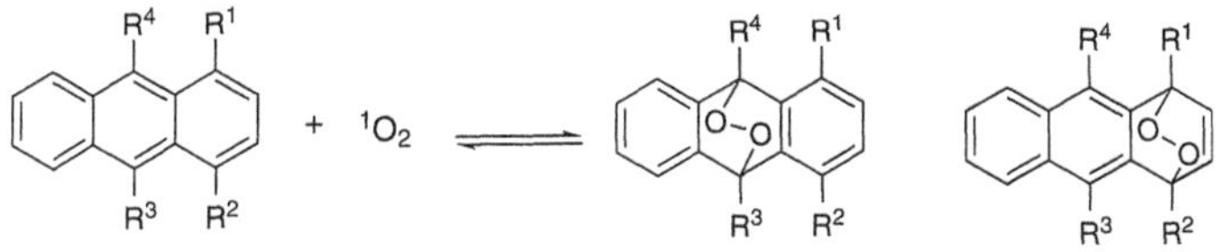

b;[49b] $R^1 = R^2 = OMe$ $R^3 = R^4 = Ph$ -   80-90%

Scheme 33

peroxide bridge. Thus, naphtalene endoperoxides invariably revert to the parent compounds and oxygen [47,50a] whereas for higher acenes the extent of reversion decreases from anthracene to higher acenes, and aryl substituents at the *meso* position lead to increased oxygen release compared to alkyl or hydrogen substituents [50b,c]. Thermal retrocycloaddition yields singlet oxygen rather than the ground-state molecule so that naphtalene or anthracene peroxides are currently used to generate singlet oxygen by means of an alternative route to photosensitization [6b,8a,48]. This aspect is of particular interest, especially for mechanistic studies in biological systems, and it accounts for the continuous search for new derivatives [48,51]. An interesting water-soluble naphtalene endoperoxide has recently been prepared by MB-photosensitization in the presence of [^{18}O]-labelled molecular oxygen, which appears particularly suitable as a source of [^{18}O]-labelled singlet oxygen in aqueous media as required in order to study the reactivity of singlet oxygen towards biomolecules [51].

11.2.4.4. Vinylarenes

The presence of an exocyclic double bond on aromatic [52] or heteroaromatic [53] compounds dictates the locoselectivity of the cycloaddition occurring preferentially at the alkene terminus, and sometimes with retention of stereochemistry [52b]. Thus, the [4+2] photo-oxygenation of a vinylbenzene derivative, such as **60**, leads to dihydrobenzo-1,2-dioxin in the primary step, which is followed by rapid oxygenation of the cyclohexadiene chromophore to give diendoperoxides as **61** and **62** (Sch. 34) [52c].

The [4+2] cycloaddition occurs preferentially, also in the presence of allylic hydrogens [52c] or alkoxy groups [52a,b,d]. In the oxygenation of vinylnaphtalenes *trans*- and *cis*-**63**, the lower yield of the endoperoxide *cis*-**64** than that of *trans*-**64** might be due to the steric interaction between the

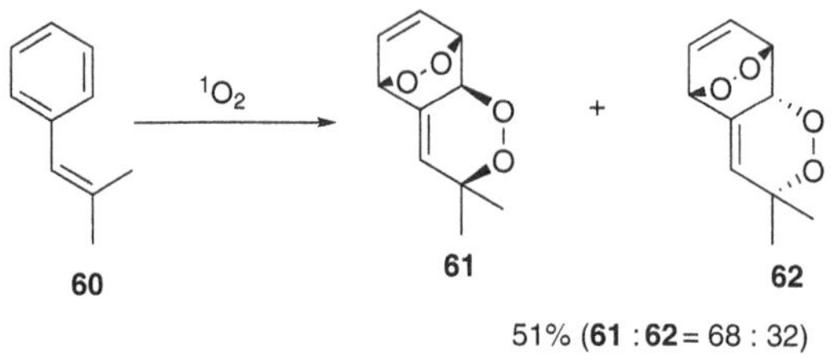

$^1O_2$
60
61
62
51% (61 : 62 = 68 : 32)

Scheme 34

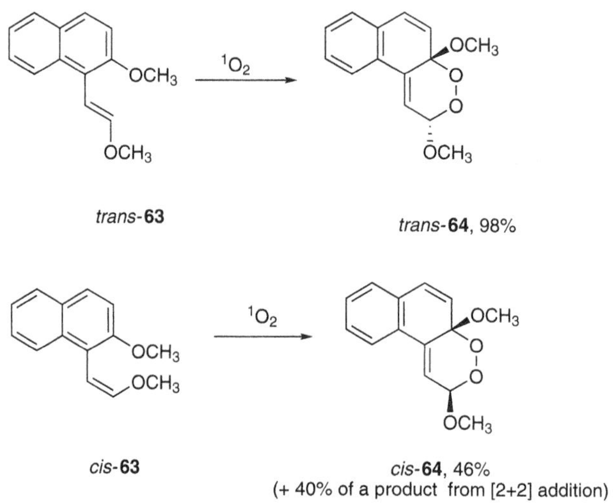

OCH3
$^1O_2$
trans-63
trans-64, 98%
cis-63
cis-64, 46%
(+ 40% of a product from [2+2] addition)

Scheme 35

two methoxy groups which makes the [4+2]-cycloaddition difficult and, therefore, favors a competing [2+2] reaction on the electron-rich double bond [52a]. (Sch. 35).

11.2.4.5. Heterocyclic Systems

Photo-oxidation of heterocycles leads to unstable endoperoxides whose existence has been generally proven by low temperature NMR. In addition to the classical transformations of peroxides (reduction, hydrolysis, deoxygenation, generally performed at low temperature), the bicyclic peroxides from heterocycles can afford characteristic rearranged products

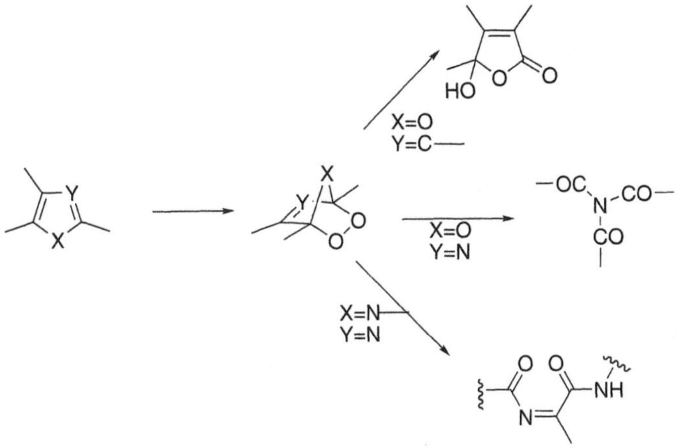

Scheme 36

depending on the heteroatom, substitution pattern, experimental conditions [5c,54]. The reactions can sometimes be complicated by the formation of products deriving from both 1,4- and 1,2-oxygen addition, especially in the photo-oxygenation of pyrroles, imidazoles, indoles [54]. However, since the works of Schenck who isolated the highly explosive endoperoxide of the parent furan and evidenced the first properties of this class of peroxides [55], the synthetic scope and high versatility of the photo-oxygenation of heterocycles is widely accepted. There are a number of instances in which butenolides or ene-diones from furans have been incorporated into the framework of more complex molecules (Sch. 36) [8c,56].

Likewise, oxazoles have been used as protecting-activating groups for carboxylic acids [57], and imidazoles have been transformed to amino acid diamides and dipeptides (Sch. 36) [54a].

11.2.4.5.1. Furans. Scarpati showed that the presence of an electron-withdrawing group at the β position enhances the thermal stability of furan endoperoxides, which prove sufficiently stable to be isolated and characterized spectroscopically and chemically (Sch. 37) [58].

Stability appears to follow the order Me >Ph >H >OMe when these substituents are in α-positions [58–60]. A variety of multifunctional compounds [*bis*-epoxides [59,60d,g], epoxylactones [59,60d], butenolides [59,60d,e] (e.g., **67**), enediones [60a], enol esters [60f,h] (e.g., **68**), *cis*-diacyloxiranes [58,60f,i] (e.g., **69**)] have been obtained from furan endoperoxides depending on the substitution pattern [59,60a,d,g] and/or the reaction conditions [60e,f]. Rearrangements to peroxidic species as dioxetanes [60a,61] or 3*H*-1,2-dioxoles [60a] or oxetanylhydroperoxides

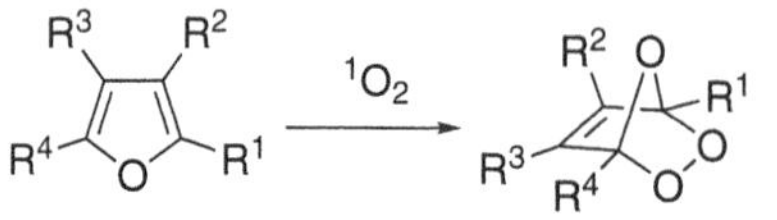

Dec. Temp. of **66** (in °C)

$R^1 = R^4 = Me$ $R^2 = CO_2Me$ $R^3 = H$ $+78$[58]
$R^1 = R^4 = Ph$ $R^2 = R^3 = CO_2Me$ $+4$[60g]
$R^1 = R^2 = R^3 = R^4 = H$ -15[59]
$R^1 = OMe$ $R^2 = R^3 = R^4 = H$ -80[60a]

Scheme 37

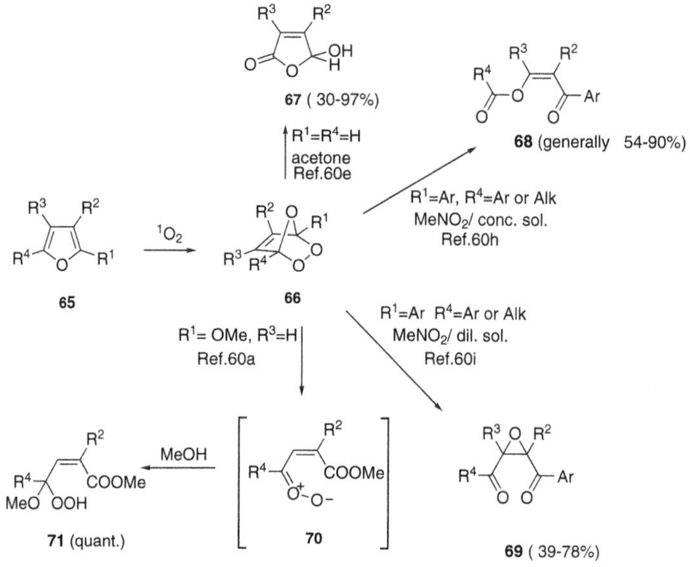

Scheme 38

[60b] are documented. Evidence for the intermediacy of carbonyl oxides **70** in thermal conversion of 1-alkoxyendoperoxides (only for 5-unsubstituted derivatives) has been obtained by trapping reactions with methanol, which lead to hemiperacetals **71**, but also by trapping with electron-rich and electron-poor alkenes, heterocumulenes, carbonyl compounds, oximes [60a,62]. Scheme 38 shows some of the applications from furan

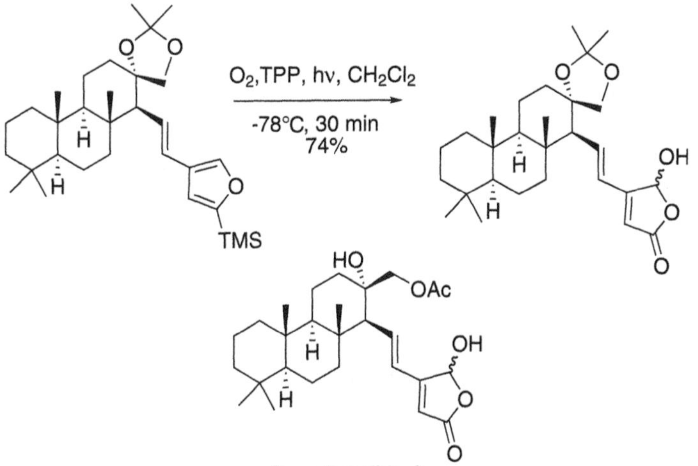

O₂,TPP, hv, CH₂Cl₂

-78°C, 30 min
74%

TMS

Spongianolide A

Scheme 39

endoperoxides. The reactions are all one-pot and do not require isolation of the starting unstable endoperoxides, and are highly stereoselective.

The oxidation of furans has often been used to express the latent functionality present within this heterocyclic framework, and it results particularly attractive for the synthesis of the widespread butenolides [8c,56]. Dye-sensitized photo-oxygenation of furfural, furoic acid and α- or α,α'-unsubstituted furans leads directly to butenolides [8c,56,60d,63]. The presence of a silyl group in α-position enhances the rate of endoperoxide formation and controls the regiochemical outcome of the second-step leading to the related butenolide via a rapid intramolecular silatropic shift [64]. An application of this reaction is shown in Sch. 39 as a step in the synthesis of spongianolide A, an antitumoral natural sesterterpenoid [65].

11.2.4.5.2. Pyrroles. 1,4-Addition of singlet oxygen to pyrroles leads to the formation of highly unstable endoperoxides whose existence has been proven by low-temperature NMR [66]. The oxidation process often takes place in low yield to give multiple products, along with considerable decomposition, depending on substitution pattern [66,67]. The reaction can be complicated by the formation of products deriving from 1,2-oxygen addition [67–69] or of hydroperoxides [70], and this complexity is exalted in condensed derivatives as indoles for which the main oxygenation routes are [2+2] cycloaddition and ene-reaction [71]. Endo-peroxides of α- or α,α'-unsubstituted pyrroles afford characteristic

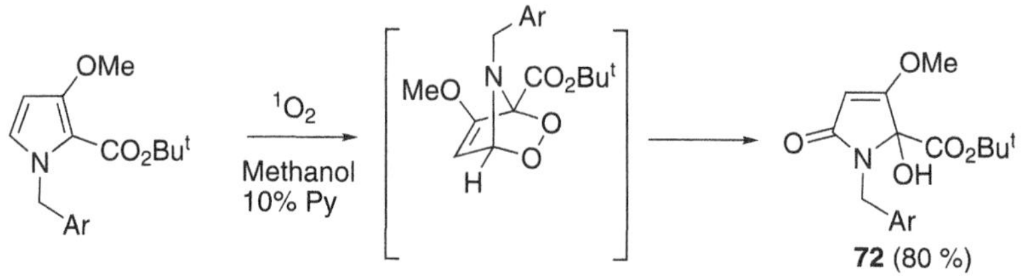

Scheme 40

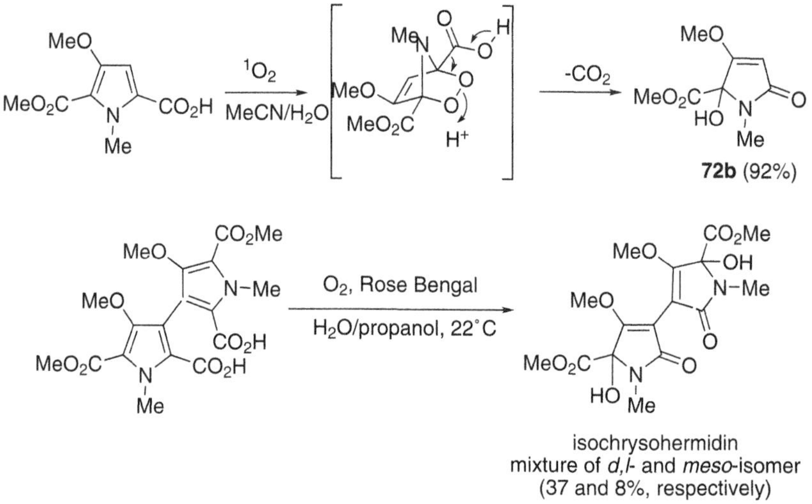

Scheme 41

5-hydroxy-Δ^3-pyrrolinones, e.g., **72**, which are formed by a β-elimination process (Sch. 40) [66–68].

Photo-oxygenation takes place under improved control when both electron-releasing and electron-attracting groups are substituted on the hetero ring [70,72]. A singlet oxygen mediated oxidative decarboxylation of pyrrole-2-carboxylic acids has been observed and used to obtain *d,l*- and *meso*-isochrysohermidin, effective interstrand DNA cross-linking agents (Sch. 41) [72a].

A noteworthy application of pyrrole photo-oxygenation has been recently reported by Wasserman (Sch. 42) starting from the *tert*-butyl ester of 3-methoxy-2-pyrrolecarboxylic acid **73** [70]. Oxygenation of this compound leads to the intermediate hydroperoxide **74** which undergoes

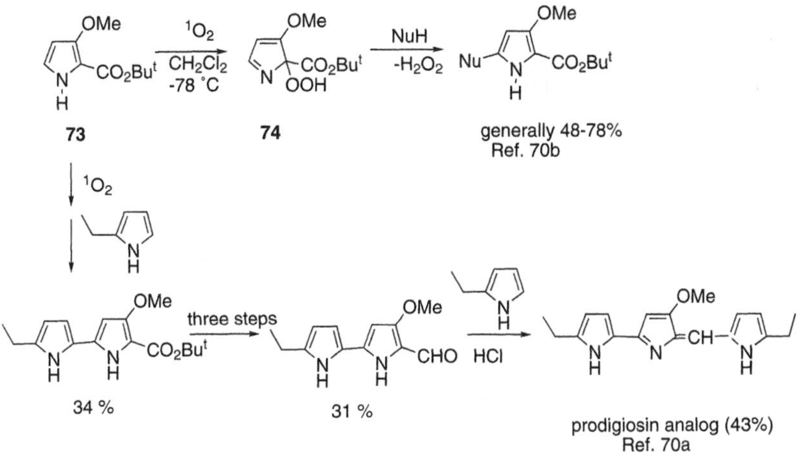

Scheme 42

addition-elimination reactions with a variety of nucleophiles yielding 5-substituted pyrroles [70b].

Formation of the hydroperoxide **74** is suggested by the authors to involve a proton shift to the zwitterion deriving from the initial addition of singlet oxygen [70c]. This sequence has been successfully used to prepare α,α'-bipyrroles, precursors of bioactive natural products in the prodigiosin family and ring A analogs (Sch. 42) [70a].

11.2.4.5.3. Thiophenes. Unlike pyrroles or furans, thiophenes undergo hardly any [4+2] addition of singlet oxygen (the parent compound is unreactive) and alkyl groups favor the reaction. The corresponding highly unstable endoperoxides decompose easily leading to *cis*-sulfines **75** and mainly to *cis*-dicarbonyl compounds **76** with the extrusion of sulfur (Sch. 43) [73].

11.2.4.5.4. Heterocycles with More than One Heteroatom. Good substrates for photo-oxygenation are oxazoles which lead exclusively to 1,4-adducts. These particularly unstable peroxides can be detected at very low temperatures by spectroscopic means [74,75], and lead to various products depending on solvent and substituents [54,57,74,75]. The main rearrangement leads to triamides, e.g., **78**, and this reaction has many potential applications in organic synthesis since it provides a means of generating a latent carboxyl group (Sch. 44) [76].

The formation of triamides transforms all three C atoms of the oxazole to activated carboxylate derivatives. By carefully choosing the substituents, it is possible to limit the reaction with nucleophiles to one of

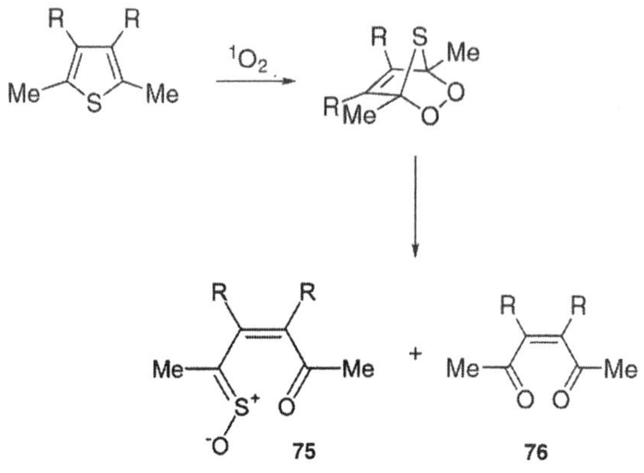

Scheme 43

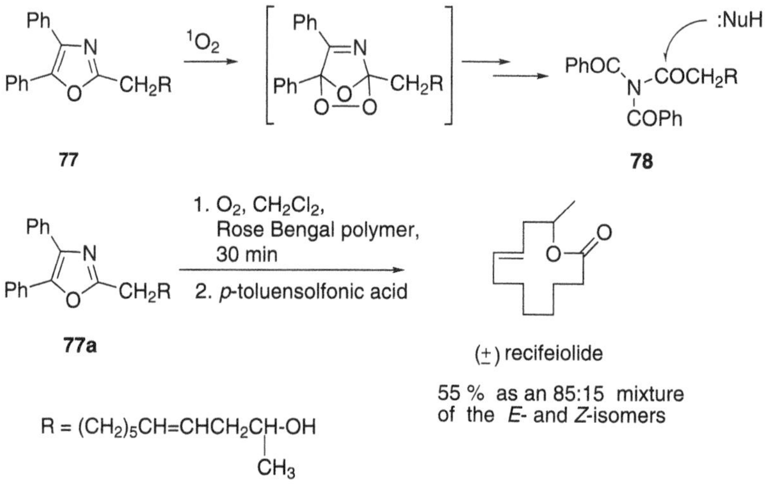

R = (CH₂)₅CH=CHCH₂CH-OH
                          |
                         CH₃

Scheme 44

the three CO groups. Thus, with 2-alkyl-4,5-diphenyloxazoles **77**, the triamides undergo selective nucleophilic attack at the less hindered alkyl CO (Sch. 44). The oxidation-acylation sequence may be successfully employed for the preparation of macrolides in a mild and selective way. In Sch. 44 the synthesis of the naturally occurring (±)recifeiolide is reported starting from the oxazole **77a** [76a]. An application of the oxygenation of oxazoles as a

i; Sharpless, (+)-DET. ii; 1) Ac$_2$O; 2) ^{1}O$_2$. iii; MeOH, TsOH,benzene

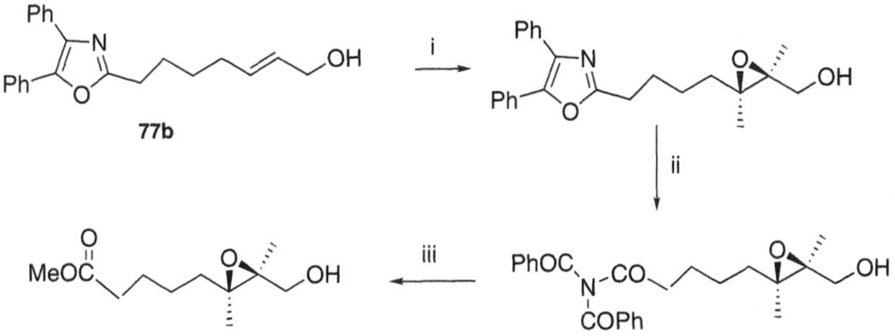

Scheme 45

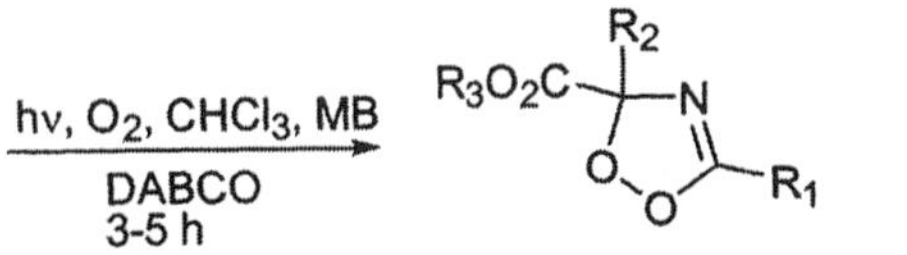

Scheme 46

means of protecting/activating carboxylates is for the ester **79** in the synthesis of leukotrienes (Sch. 45) [76b].

While the conventional carboxylate protecting groups interfere with the course of the reaction, the enantioselective epoxidation on the oxazole **77b** followed by photo-oxygenation and hydrolysis of triamide **78b** affords the ester **79** in good yield with the desired stereochemistry [76b]. Starting from alkoxyoxazoles, rearrangement to dioxetanes [75b] or dioxazoles [75] has been observed, and a simple and high-yield route for stable 3H-1,2,4-dioxazoles **80** has been described (Sch. 46) [77].

Fused ring oxazoles lead, in apolar solvents, to products which derive from cleavage at the 2–3 and 4–5 positions of the endoperoxides to cyano-anhydrides **81** followed by hydrolysis or CO loss. This sequence has been efficiently used to prepare ω-cyano carboxylic acids **82** (Sch. 47) [78].

An interesting recent application of oxazole dye-sensitized photo-oxidation is to photoimaging processes [79]. This is based on the efficient transformation (using diverse dyes and, hence, light of a wide range of wavelengths) of oxazole rings attached to a polymer backbone to the corresponding N,N-dibenzoylcarboxamide derivatives [79].

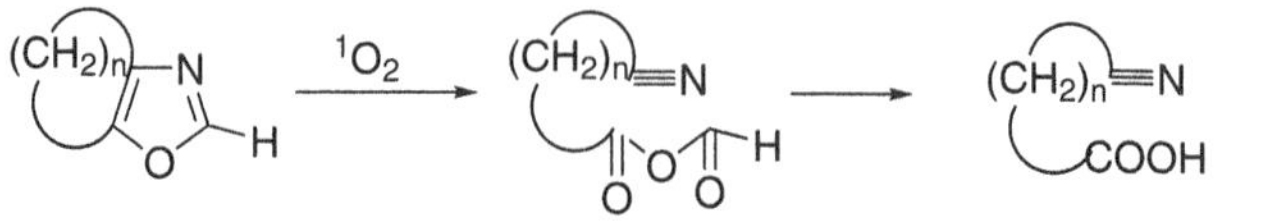

n= 4,5,6   **81**   82 (80-90 %)

Scheme 47

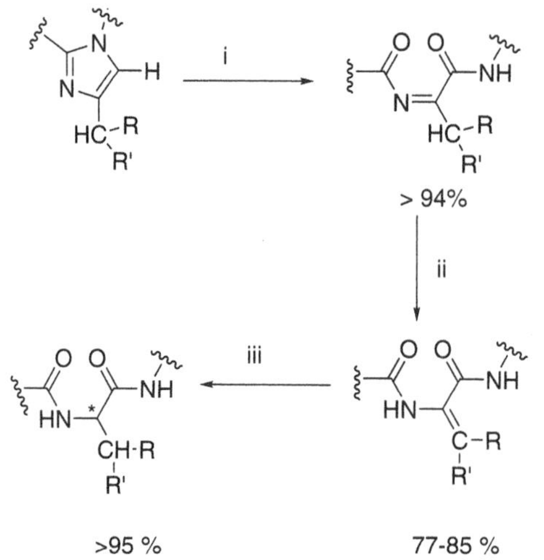

i, ${}^1O_2$, DBU. ii; KO-t-Bu, THF/t-BuOH. iii; H_2, Rh(NBD)Cl_2-(R,R)-DIPAMP

Scheme 48

Isoxazoles, 1,2,5-oxadiazoles and 1,3,4-oxadiazoles are apparently inert under dye-sensitized photo-oxygenation [74a], while a thiotriamide has been found in the photo-oxygenation of arylthiazole presumably via the undetected endoperoxide in a fashion similar to oxazoles [67,80].

The highly unstable endoperoxides of imidazoles, sometime spectroscopically detected [81a], rearrange to dioxetanes or hydroperoxides or rapidly fragment or polymerize so that from the synthetic perspective, the reaction has a poor scope [54,67,81]. One interesting application has been pointed out by Lipshutz as a general asymmetric synthesis of N-protected amino acids/dipeptides via appropriately substituted imidazoles (Sch. 48) [82].

The three-step protocol (photo-oxygenation in the presence of DBU/ imine to enamine isomerization/hydrogenation with optically active Rh(I)/ DIPAMP combination as catalyst) leads in high yields and excellent ee to the final products. Despite the scarce synthetic interest, attention is paid to the mechanistic aspects of the photo-oxygenation of imidazoles and, this is related to the presence of the heterocyclic system in molecules of biological interest, such as histidine or purines and their role as predominant target in photosensitized oxidation of DNA [83].

11.2.5. Synthetic Potential

[4+2] Photo-oxygenation is a powerful synthetic tool to introduce, simultaneously, oxygenated *syn*-1,4-functionalizations in 1,3-diene systems, from alicyclic to cyclic, from carbocyclic to heterocyclic or aromatic. The primary peroxidic products (endoperoxides) can undergo a variety of transformations (Sch. 49), some of them stereospecific, which can either or not involve the cleavage of the O–O bond and, often, can be performed in situ and at low temperature overcoming the thermal instability problem of the peroxide intermediate [5b,9b].

Many reactions are common to the diverse classes of endoperoxides. In other cases, as above reported, peculiar transformations can occur especially for peroxides deriving from heterocycles. Here follows a brief description of the most general applications with specific references and some recent examples.

11.2.5.1. Reduction (a,b,c)

Reducing agents for unsaturated endoperoxides fall into one of three categories: those capable of selective reduction of the O–O bond (thiourea

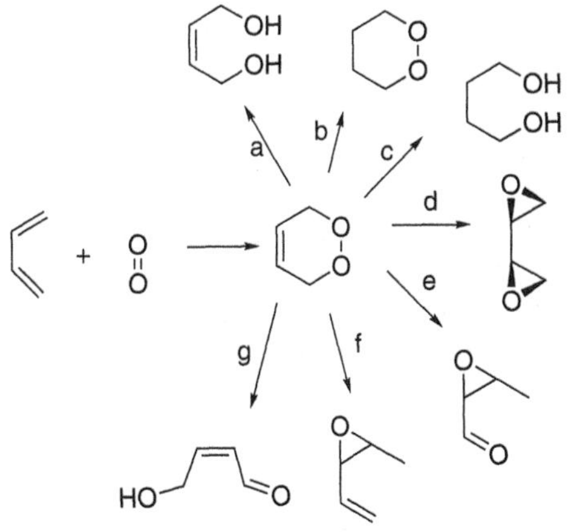

Scheme 49

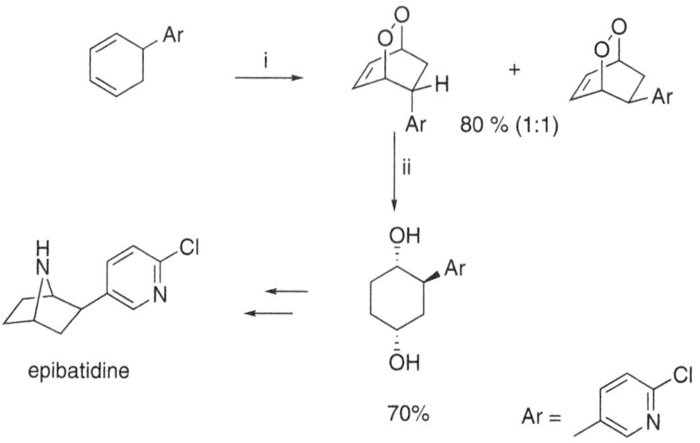

i; hv, O$_2$, 5,10,15,20-tetraphenyl-21H,23H-porphine. ii; H$_2$, Rh/Al$_2$O$_3$

Scheme 50

[30b,35,84] or sodium tetrahydroborate [85]); those capable of selective reduction of the C–C bond (diimide) [86]; those unselective as PtO$_2$, Ni, Pd/C [33e,35] or Rh/Al$_2$O$_3$ [87]. The latter treatment, which has been employed in the total synthesis of epibatidine, a potent natural analgesic with a non-opioid mode of action, proves milder than the use of palladium which also leads to hydrogenolysis of the pyridine-chlorine bond (Sch. 50) [87].

Compared with NaBH$_4$ whose action is restricted to strained ring systems [85], LiAlH$_4$ is a powerful reductant for O–O bond but its ability to act as a base or to reduce other functional groups decreases its general utility [88]. Reduction with thiourea of the *trans*-endoperoxide **38**, obtained with high stereoselectivity by photo-oxygenation of diene **37**, followed by epoxidation are the first steps in the synthesis of pinitol, a synthetic analogue of mio-inositol (Sch. 51) [31c].

The use of diimide to selectively reduce the C=C bond in bicyclic systems has received great attention as an efficient route to prostaglandin endoperoxides [86b]. It can be employed also at low temperatures and photooxygenation at −78 °C followed by diimide reduction at the same temperature has been used to trap unstable singlet oxygen adducts [73b,86d].

11.2.5.2.   Thermolysis (d,e)

Thermal rearrangement of endoperoxides of bicyclic dienes gives *syn*-diepoxides [9b,33a], which are sometimes accompanied by epoxyketones

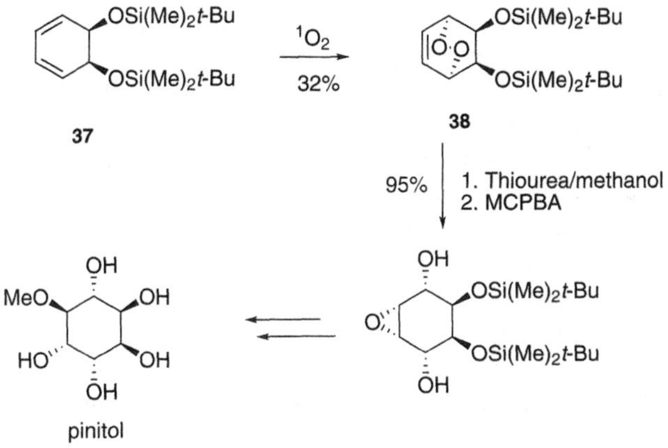

Scheme 51

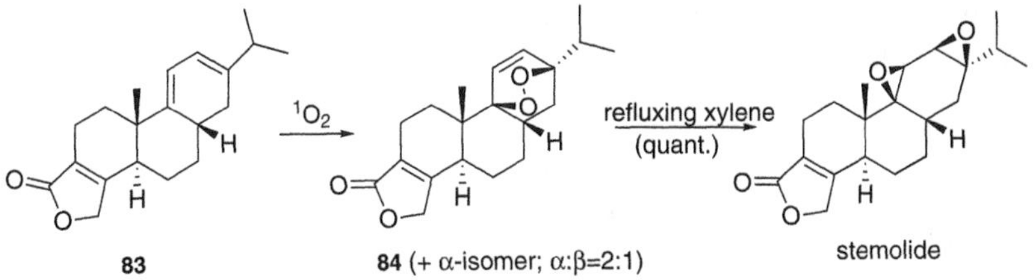

Scheme 52

[9b,89] or, when endoperoxides gain significant resonance [43,45], by loss of oxygen. The same products can also be formed by photolysis [medium- or high-pressure mercury lamps, filter pyrex ($\lambda > 366$ nm)] but a greater proportion of epoxyketones is observed [90].

Quantitative conversion in refluxing xylene of the β-peroxide **84** is the final step in the total synthesis of the diterpenoid *bis*-epoxide stemolide (Sch. 52) [91].

The β-peroxide **84** was obtained by methylene blue-sensitized photo-oxygenation of diene **83** together with its α-isomer (2:1 9,13-α:β) and was separated by silica gel chromatography. Diepoxides are the thermally rearranged products of α,α′-unsubstituted or alkylsubstituted furan endoperoxides [59,60d,g].

11.2.5.3. Metal-Promoted Decompositions (d)

The use of cobalt(II) tetraphenylporphyrin (CoTPP) as catalyst in the decomposition of unsaturated bicyclic endoperoxides represents the preferred method for synthesizing diepoxides [9b]. The reaction is rapid at low temperature and leads stereospecifically to *syn*-diepoxides. The yields are much better than pyrolysis and epoxy ketones are generally absent [9b,92]. Monocyclic endoperoxides by the same treatment lead to furans in quantitative yields [93]. A unusual CoTPP-catalyzed decomposition mode has been recently observed from endoperoxide **85** which has a strained double bond moiety [94]. In addition to the diepoxide **86** (70%), the reaction leads to the unprecedented bicyclic ketoaldehydes **87** (30% of a 9:1 diastereomeric mixture), important precursors for the synthesis of the furofuran system **88** (Sch. 53) [94].

The CoTPP-catalyzed decomposition of endoperoxides as **89** leads to divinylepoxides as **90** which can be converted through a 3,3-sigmatropic rearrangement to dihydrooxepine derivatives as **91** (Sch. 54) [33a]. The result appears interesting due to the discovery of the oxepine nucleus in a number of biologically active natural products.

Diepoxides have also been obtained, in addition to two spiro compounds, by treating peroxides from abietic acid with $FeCl_3$ in aqueous acetonitrile [95].

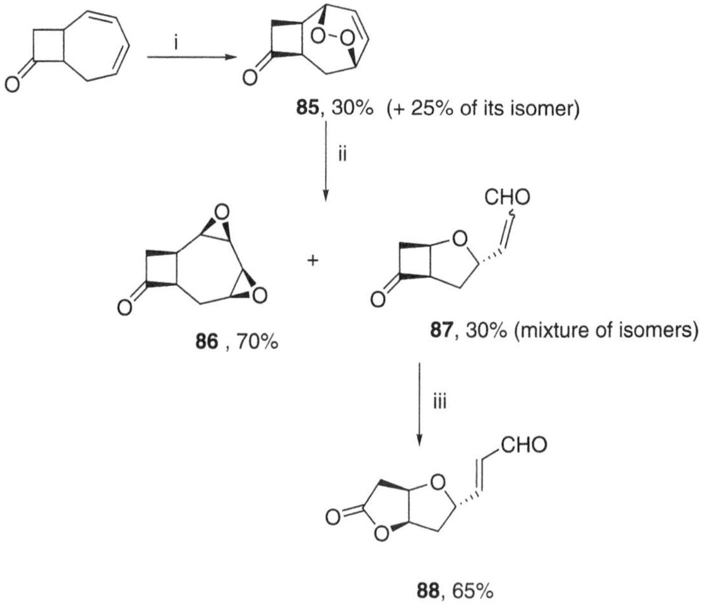

85, 30% (+ 25% of its isomer)

86 , 70%   +   **87**, 30% (mixture of isomers)

88, 65%

i; TPP, O$_2$, hv, CHCl$_3$/CCl$_4$. ii; CoTPP, CHCl$_3$, r.t.. iii; *m*-CPBA, NaHCO$_3$, CH$_2$Cl$_2$

Scheme 53

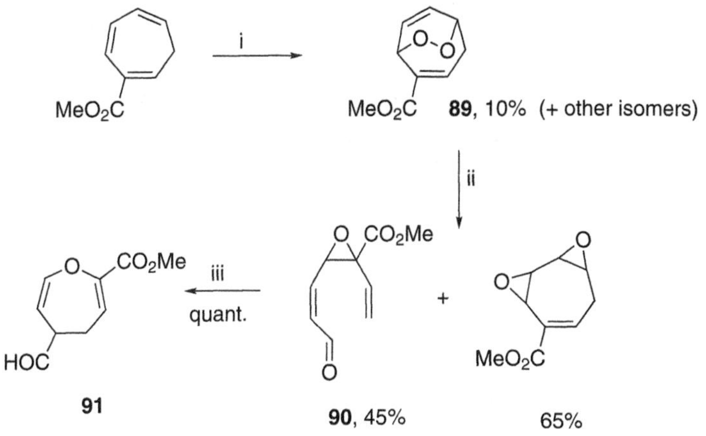

i; TPP, O_2, hν,CCl_4. ii; CoTPP, CCl_4, 0°C. iii; 45 °C

Scheme 54

11.2.5.4. Deoxygenation by Trivalent Phosphorus Compounds or Dialkyl Sulfides (f)

The reaction of triphenylphosphine [or $P(OMe)_3$] with unsaturated endoperoxides proceeds by initial cleavage of the peroxide linkage followed by S_N2' displacement to give triphenylphosphine oxide [or $O=P(OMe)_3$] and the unsaturated epoxide [9b]. Mild alkyl sulfides work well with furan endoperoxides, and low temperature photo-oxygenation followed by in situ treatment with the sulfides is a general and very convenient alternative to other methods for the furan ring opening to ene diones [60a,96]. Treatment with Me_2S is used as a key step in the furan-to-pirone transformation in the synthesis of cardioactive steroids as bufalin (Sch. 55) [96c].

11.2.5.5. Acid- and Base-Catalyzed (g) Reactions

The most prevalent base-catalyzed reaction of an endoperoxide is the Kornblum-DeLaMare decomposition [97a] which leads to a hydroxy ketone by removal of a proton from the carbon adjacent to the peroxy linkage [9b,90b,97]. The formation, in a basic solvent as acetone, of hydroxyfuranones **69** (Sch. 38) in the photo-oxygenation of α,α'-unsubstituted furans might occur via a similar rearrangement [60e]. Attempts to induce asymmetry in endoperoxides with enantiotopic hydrogens or with chiral bases have led to moderate success [98]. The Et_3N-catalyzed rearrangement of substituted cycloheptatriene endoperoxides **92** leads to

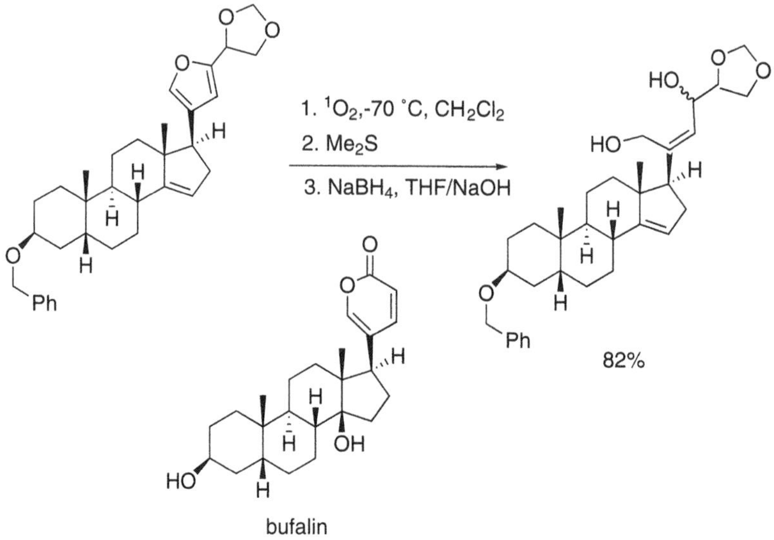

bufalin

Scheme 55

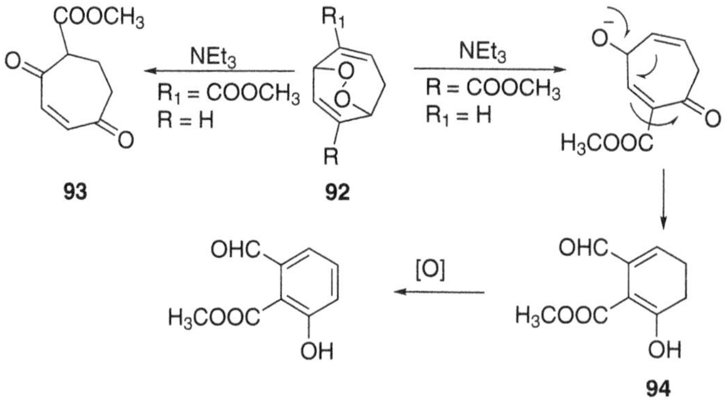

93   92   94

Scheme 56

diketone **93** or ring contraction product **94** depending on the location of the substituent, namely on the location in the seven or six membered ring (Sch. 56) [99].

Broensted or Lewis acids promote heterolysis of the C–O bond and formation of a peroxy allylic or benzylic cation which can collapse to dioxetane [52d] or be trapped by aldehydes or ketones [100] to form

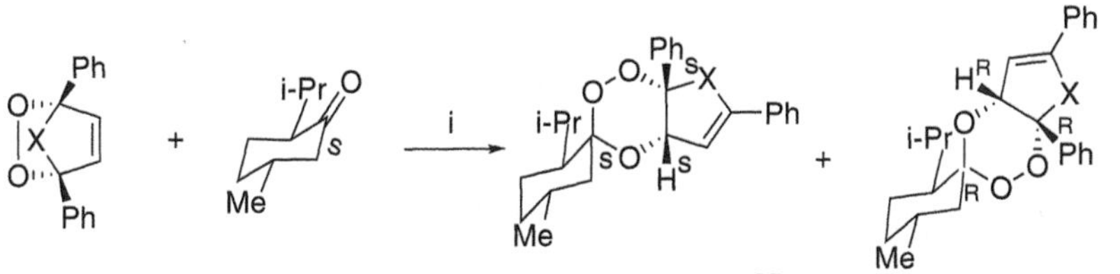

X = CH$_2$ Yield 21% (1 : 1)
X = CH$_2$CH$_2$ Yield 75% (2 : 1)

i; ketone : peroxide = 5 : 1, CH$_2$Cl$_2$, -78 °C, Me$_3$SiOTf (cat. for X= CH$_2$; 1.1 equiv. for X=CH$_2$CH$_2$)

Scheme 57

trioxanes or by alcohols [9c] to give hydroperoxy ethers. Interesting applications of the reaction of endoperoxides with aldehydes and ketones in the presence of catalytic amount of Me$_3$SiOTf have been described by Jefford and provide a convenient means of preparing a wide variety of tricyclic, bicyclic, and monocyclic trioxanes [100]. The reaction with chiral cyclohexanones results diastereoselectively controlled. So, by treating the cyclopentadiene endoperoxide with (−) menthone trioxanes **95** are formed, only two of the eight diastereoisomeric trioxanes which in principle can be formed (Sch. 57). The stereochemical outcome has been rationalized in terms of the respective enantiomeric silylperoxy cations which are completely differentiated by the *si* and *re* faces of the ketone function [100a].

Methoxyhydroperoxides can be directly obtained by running the oxygenation in methanol as solvent [55,59,60a]. A peculiar trapping reaction of 1-methoxyfuran endoperoxides has been observed with 4-nitrobenzaldehyde oxime which leads regio- and stereoselectively to the adducts **96** (Sch. 58) [60c].

The same endoperoxides with different oximes (alkyl or aryl substituted) afford *N*-hydroperoxyalkylnitrones **97** stereoselectively [62]. It has been shown that the first step is an oxygen nucleophilic trapping reaction of the intermediate carbonyl oxides by oxime which lead to α-oxime ether hydroperoxides and these, depending on the nature of the substituents, thermally rearrange to the hydroperoxynitrones or sometime can be isolated [62a]. Nitrones **97** have been used as starting materials for oxazinones **98** [101a], hydroperoxyisoxazolines **99** [101b], *cis*-hydroperoxyoxaziridines **100** [101c].

Endoperoxides of *N*-carbomethoxypyrroles react with nucleophiles in the presence of stannous chloride to give 2-substituted pyrroles [102a]. The

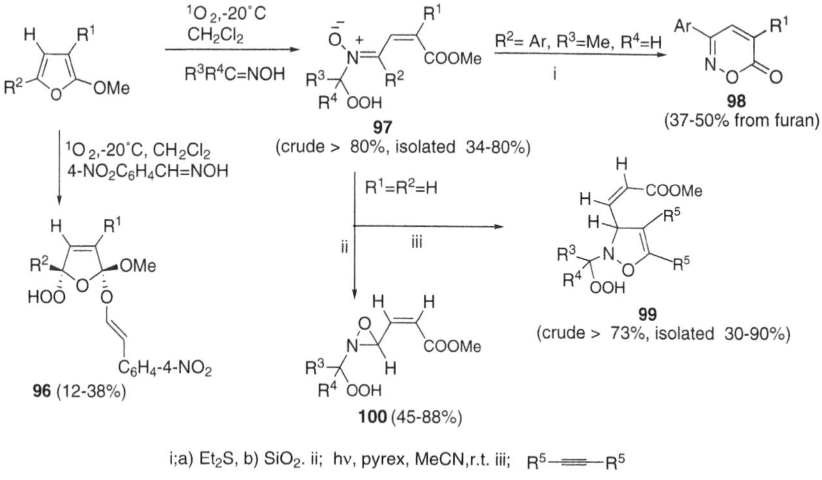

Scheme 58

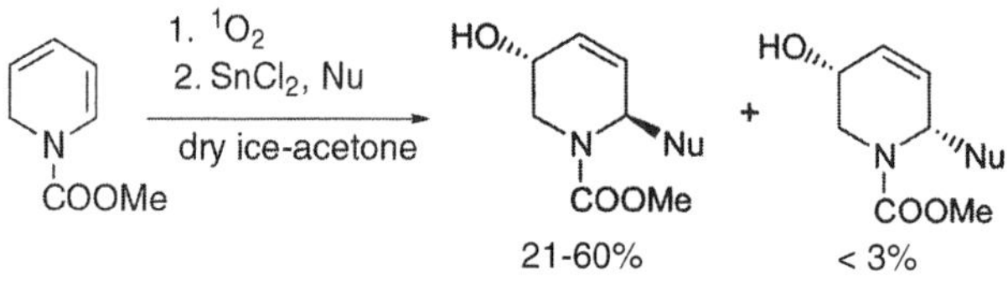

Scheme 59

same reaction can be carried out on the endoperoxides of dihydropyridines [102b,c]. Starting from 2-substituted-1,2-dihydropyridines stereochemically defined products can be prepared in good yields since the endoperoxides are formed with high stereoselectivity by the approach of singlet oxygen from the opposite side of the substituent and the nucleophile attacks the less hindered carbon (Sch. 59) [102b]. The method has been employed for the synthesis of complex molecules which are of interest in the framework of indole alkaloids [102c].

11.2.6. Target Molecules

The high diastereoselectivity found in the reaction of $^1O_2$ with chiral cyclic molecules together with the *syn*-functionalization can be usefully directed to

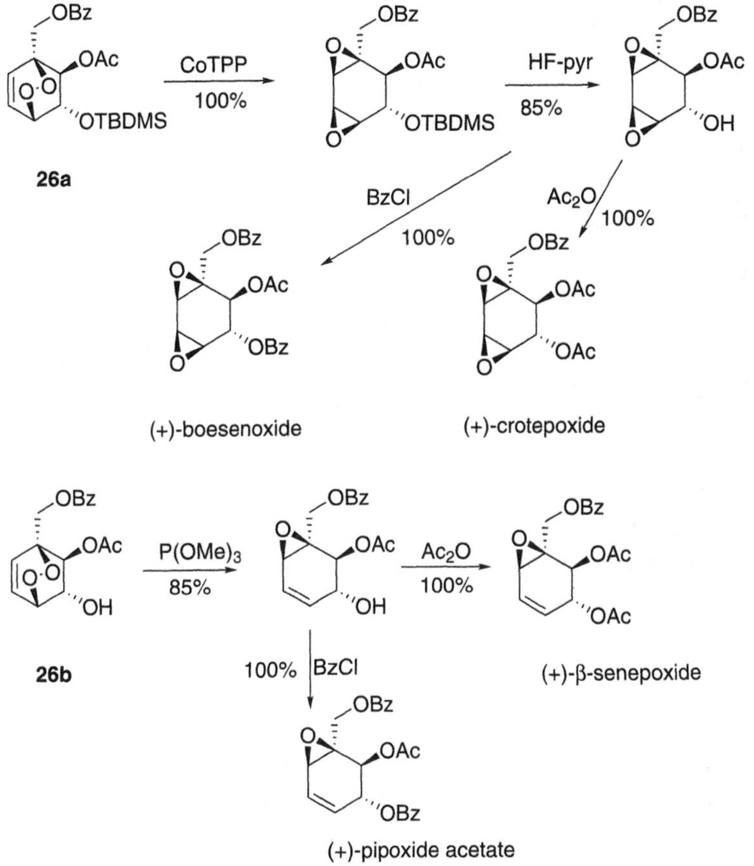

Scheme 60

enantiospecific syntheses of natural products. Two representative examples are reported.

Endoperoxides **26a** and **26b**, which can be prepared and easily separated from their isomers (Sch. 11) [27], are the starting material, by means of simple and high-yield reactions as described above, for the enantiospecific syntheses of naturally occurring cyclohexane diepoxides and cyclohexene epoxides, which display significant tumor-inhibitory activity. Thus, (+)crotepoxide, (+)boesenoxide and (−)*iso*-crotepoxide or (−)senepoxide, (+)β-sepoxide, (+)pipoxide, and (−)tingtanoxide can be obtained by singlet oxygenation of the suitable dienes **25a** and **25b** followed by treatment of the resultant endoperoxides with either cobalt-*meso*-tetraphenylporhyrin or trimethyl phosphite (Sch. 60) [27].

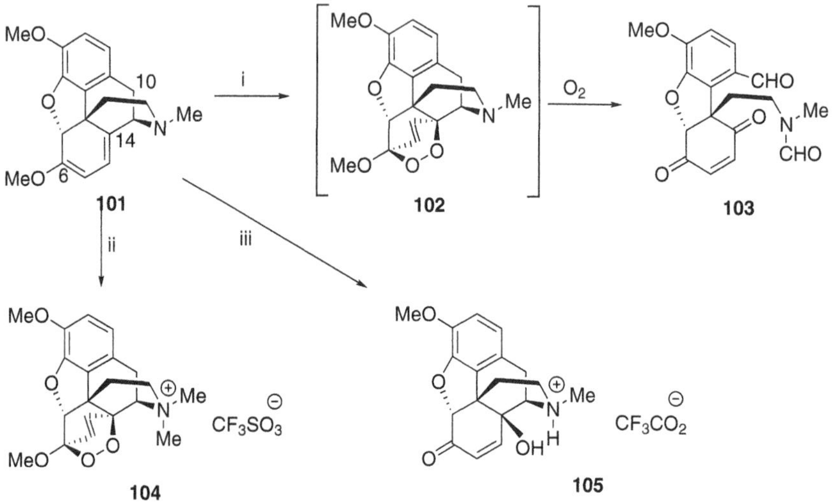

i; hv, TPP, O₂, CH₂Cl₂, yield 62%
ii; a) CF₃SO₃Me, MeNO₂, 0 °C, yield 92%; b) hv,TPP, O₂, CH₂Cl₂,yield 85%
iii; hv, TPP, O₂, CH₂Cl₂ (acidified solution, pH 4), yield 61%

Scheme 61

The TPP-sensitized photo-oxidation of thebaine **101** as ammonium salts (entries ii and iii) affords the endoperoxide **104** and the 14-hydroxy-codeinone **105**, stereoselectively and in high yields. Instead, product **103** is the main product under neutral conditions and its formation has been rationalized via an endoperoxide intermediate **102** that undergoes oxidation at the nitrogen atom (Sch. 61) [103].

All the compounds can be used for further transformations to give pharmacologically important morphine derivatives. The oxygenation represents a valid alternative to the multi-stage nonstereoselective methods described for the preparation of this class of compounds [103].

11.2.7. Synthetic Procedures: An Example

An example of the simplicity and potential of the photo-oxygenation reaction is described for the preparation of (Z)-1-hydroperoxy-N-[(Z)-3-methoxycarbonyl)-2-propenylidene]cyclohexylamine N-oxide (**P**) starting from 2-methoxyfuran and cyclohexanone oxime, which both can be purchased and used without further purification.

i; O$_2$, tetraphenylporphine,
hv, CH$_2$Cl$_2$, -20°C

65% isolated product

A 0.02 M solution of 2-methoxyfuran (1 mmol) in dry CH$_2$Cl$_2$ in the presence of cyclohexanone oxime (5 mmol) and tetraphenylporphine (3.6×10^{-4} mmol) was irradiated with an external halogen lamp (650 W) at -20 °C. The temperature was maintained by immersion of the reactor (a Pyrex round-bottom flask) into a Pyrex vessel containing ice and salt or acetone and in the latter case the bath was thermostated by a Cryocool. During irradiation dry oxygen was bubbled through the solution. When the reaction was complete (90 min, TLC) the solvent was removed at rt and the residue was chromatographed on a short column of silica gel, eluting with light petroleum/diethyl ether (8:2, 1:1) to give successively the unreacted oxime and the nitrone (65%, m.p. 76–78 °C [62b] from diethyl ether/hexane).

11.3. [2+2] PHOTO-OXYGENATION

11.3.1. Scope and Limitations

The photo-oxygenation of electron rich alkenes such as vinyl ethers, ketene acetals, enamines, vinyl sulfides as well as of olefins without allylic hydrogens or with inaccessible allylic hydrogens is a general route for 1,2-dioxetanes [6]. A considerable number of stable dioxetanes have been prepared from alkyl- and/or aryl-substituted alkenes or from enol ethers, in many cases dioxetane formation has been postulated on the basis of the characteristic cleavage products, frequently accompanied with emission of light [6]. Dicarbonyl compounds deriving from the cleavage of 1,2-dioxetanes are often found in the photo-oxygenation of some heterocycles as pyrroles [54a] or indoles [71].

Bisadamantylidenedioxetane 106 is the most stable thermally among hitherto known dioxetanes (Sch. 62) [104], and a single adamantylidene group has been proven to exert already a strong stabilizing effect [105].

Actually, steric hindrance between geminal substituents of a dioxetane (3,3-steric interaction) appears to especially stabilize the peroxidic ring [6d]. Conversely, the presence of an easily oxidized substituent, that has low

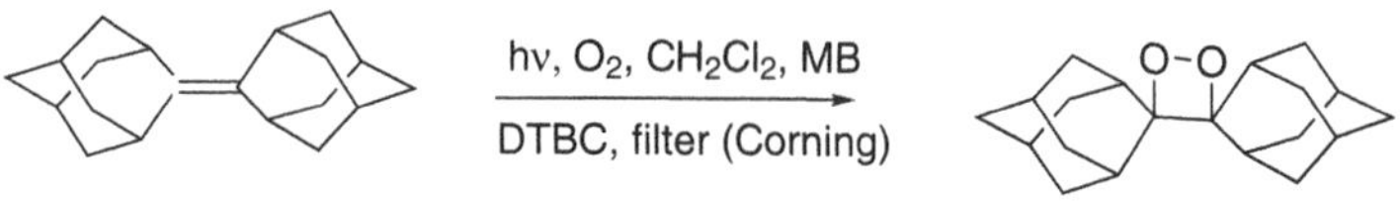

hv, O$_2$, CH$_2$Cl$_2$, MB
DTBC, filter (Corning)

106 (66%)

DTBC= 2,6-di-*t*-butyl-p-cresol

Scheme 62

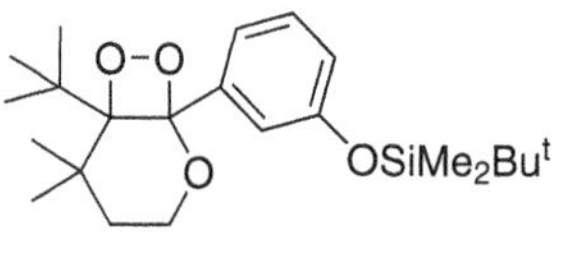

107 (74%)[108a]

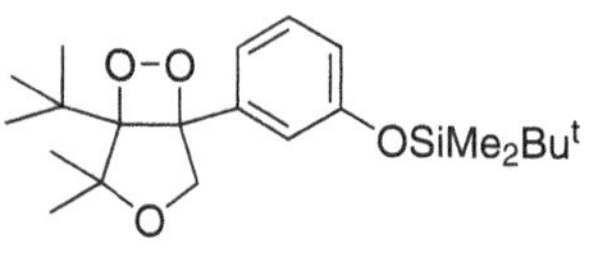

108 (96.2%)[108b]

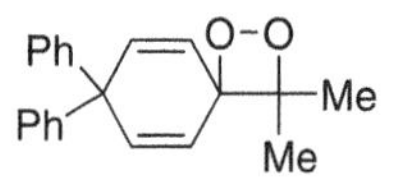

82% (m.p. 72-73 °C)[6b]

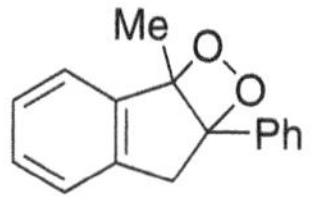

54% (m.p. 69-70 °C)[109]

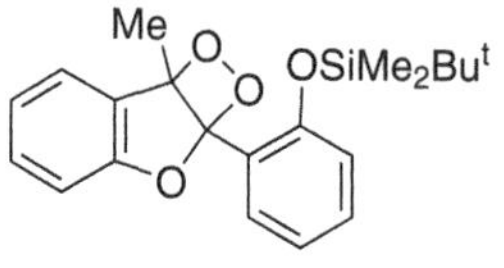

37% (m.p. 97-99 °C)[110]

Scheme 63

oxidation potential [106], or of a heteroatom [107] at the double bond considerably decreases the thermal stability of the peroxidic intermediate. Sulfur-substituted dioxetanes, for example, are much less thermally stable than the *O*-analogues and have been detected by spectroscopic means at very low temperature, if at all [107]. Some examples of stable dioxetanes prepared by photo-oxygenation are shown in Sch. 63.

There is a high interest to know the structural features which endow the thermal stability of dioxetanes since this property makes the dioxetane easy to handle and, if accompanied with high chemiluminescence, usable as enzyme label in bioanalytical, clinical application [111]. In this field,

Matsumoto's group has recently synthetized compounds **107** and **108** which present both properties at a good level [108].

Since dioxetanes are usually thermally labile compounds and prone to respond to redox chemistry, the dye-sensitized photo-oxygenation is the most general procedure for their preparation due to the mild reaction conditions. With stable derivatives photo-electron transfer reactions can also be carried out using 9,10-dicyanoanthracene as sensitizer [112]. Particular care is paid to solvents which must be metal-free in order to avoid catalyzed "dark" decomposition [113] or to temperature which should be maintained subambient. Most dioxetanes are pale yellow ($\lambda_{max} \approx 280$ nm) and have a weak tail end absorption up to 450 nm [6], so that irradiation is conveniently performed through filters with $K_2Cr_2O_7$ or $CuSO_4$. Purification by low-temperature column or flash chromatography on silica gel can be performed with stable dioxetanes [6d]. Pure samples of dioxetanes can be explosive under vacuum.

11.3.2. State of Art Mechanistic Models

Several mechanisms have been postulated, including a Woodward-Hoffmann-allowed [2s+2a] concerted cycloaddition [114a,b], or two-step reactions proceeding through 1,4-biradical [114c], perepoxide [114d], charge transfer, or exciplex intermediate [114e], or an open zwitterion [114f,g] (Sch. 64).

The intermediacy of polar species appears the most accepted [115], and evidence has been claimed on the basis of solvent effects, [116,117] skeletal

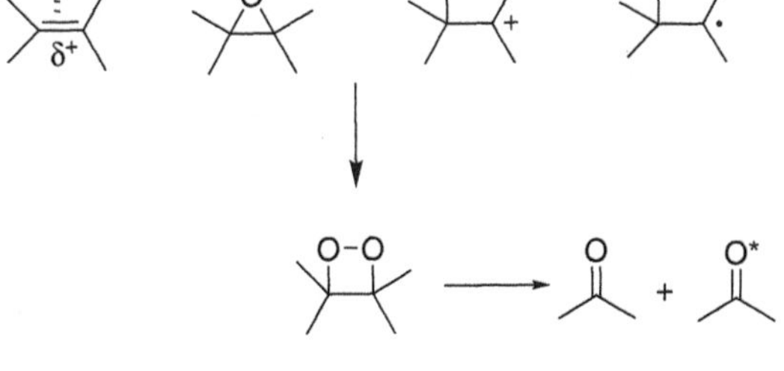

Scheme 64

rearrangements [118], or by trapping reactions with methanol [116,119a] or aldehydes [119b] or with electrophilic agents such as sulfoxides [120a] or sulfenate and sulfinate esters [120b]. There is some controversy over the relationship between perepoxide and open zwitterion, particularly when both the ene and [2+2] reaction pathways are viable. Primary and secondary isotope effects and kinetic calculations are consistent with the primary formation of a perepoxide intermediate which rearranges in a faster step to an open zwitterion [121]. Contemporary theoretical data indicate that there is a high energy barrier to dioxetane from peroxirane, so that the latter corresponds to a dead-end pathway and cannot lead to dioxetane [122]. That distinct mechanisms operate, depending on the symmetry of the substrate, the electron-donating capabilities of the substituents, and the solvent is the conclusion, based on the comparison with experimental results, of *ab initio* molecular orbital (MO) studies of the oxygenation of alkenes, enol ethers, and enamines [123].

11.3.3. Chemoselectivity and Diastereoselectivity

In the photo-oxygenation reactions, the formation of ene-products or 1,4-endoperoxides is highly competitive with dioxetane mode in the presence of allylic hydrogens or a vinyl substituent, respectively. In both cases the proportion of dioxetane is enhanced appreciably with increasing solvent polarity, especially in methanol [109,117,119b], or at low temperature [107b,119b,124] or in the presence of heteroatom substituents [6,107]. In the oxygenation of cyclic enolethers (dihydrofurans or pyrans) it has been found that five ring size favors the dioxetane to ene mode [107b].

The singlet oxygenation of isomeric Z,Z- E,E- and Z,E-alkoxydienes **109** leads to a mixture of products with a predominance of dioxetanes in all the three reactions (Sch. 65) [125].

For all of them [2+2] addition is the predominant route and endoperoxide **113** is formed only starting from the E,Z-isomer. The results have been rationalized in terms of an allylic-stabilized open zwitterionic intermediates whose rotation can compete with closure to dioxetanes [125a].

Solvent acidity plays a critical role in the reaction of singlet oxygen with *trans*-4-propenylanisol **115**, and the authors invoke a proton transfer from methanol or acid to the ionic intermediates to explain the enhanced dioxetane concentrations in methanol and in nonprotic solvents with benzoic acid (Sch. 66) [117].

In the oxygenation of 5,6-dihydro-1,4-oxathiin **116**, despite the presence of allylic hydrogens and the sulfide moiety, the sole reaction product is the dicarbonyl compound **118** (isolated yield 90%) [126a] deriving from the corresponding dioxetane **117**, which has been spectroscopically

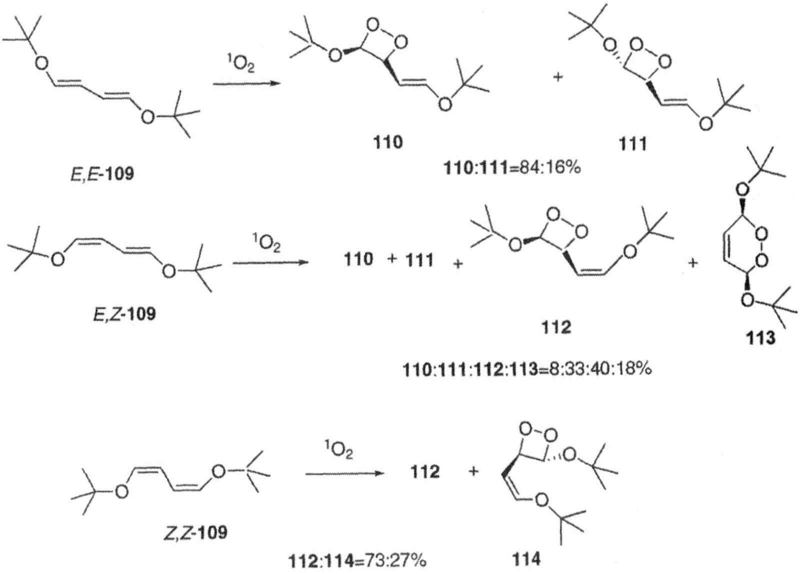

Scheme 65

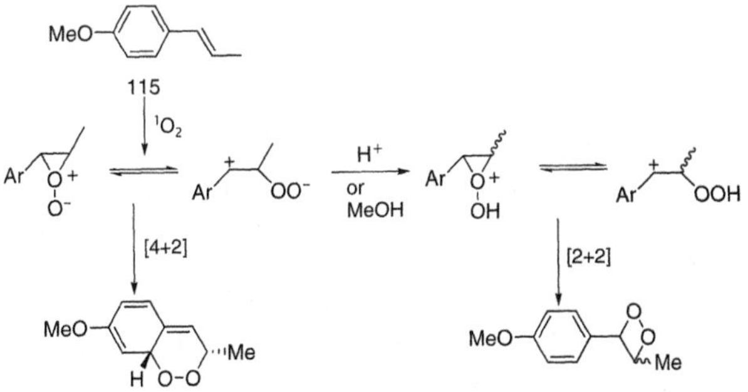

solvent	products %	
	[4 + 2]	[2 + 2]
C_6D_6	73	24
$CDCl_3$	80	18
CD_3CN	70	29
C_6D_6 + acid (cat.)	10	86
$CDCl_3$ + acid (cat.)	4	93
CD_3OD	0	97

Scheme 66

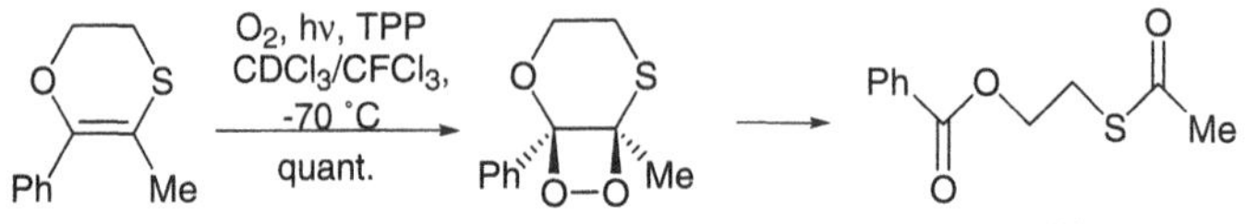

116   **117**   **118**

Scheme 67

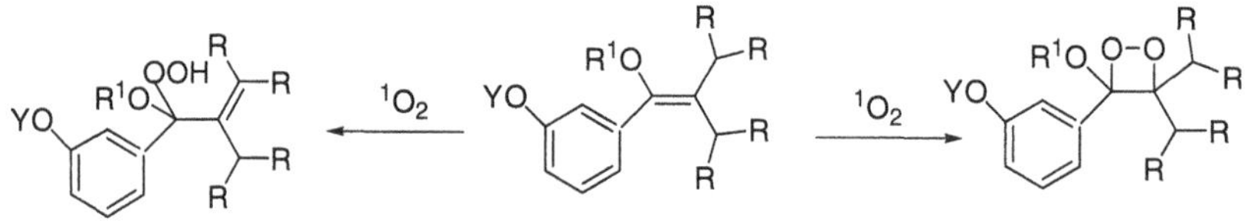

120   **119**   **121**

alkene	R	R	R^1	25°C	reaction temperature 0°C	-78°C
				1,2-Ad. : Ene- r.	1,2-Ad. : Ene-r.	1,2-Ad. : Ene-r.
119a	-CH$_2$-CH$_2$-		Me	22 : 78	32 : 68	67 : 33
119b	Me	Me	Me	29 : 71	69 : 31	95 : 5
119c	Me	Me	Et	80 : 20	91 : 9	99 : 1
119d	Me	Me	i-Pr	98 : 2	99 : 1	100 : 0
119e	Me	Me	t-Bu	100 : 0	100 : 0	100 : 0

Scheme 68

evidenced [126b]. The strong preference to 1,2-addition has been assigned to the mesomeric effect of both sulfur and oxygen (Sch. 67) [126b].

Dioxetane mode becomes highly selective when allylic hydrogens cannot align appropriately to give the ene-reactions due to steric and/or stereoelectronic interactions. For olefins **119** the ratio of 1,2-addition vs. ene reaction is increased by lowering the temperature (from 25 to −78 °C) and by increasing the bulkiness of the alkoxy group (Sch. 68) [124].

This has been explained on the basis of the greater stability of the rotamer (A) than that of the rotamer (B) due to the "gear effect" between the two isopropyls (or cyclopropyls) at geminal positions, which generates distinct energy minima with respect to rotation of the groups (Sch. 69).

The rotation of the two gears is also affected by the bulkiness of the vicinal alkoxy group as highlighted by the order of mode selectivity OMe < OEt < i-PrO < t-BuO. It is worth noting that the order of stability is instead OMe < OEt< i-PrO ≫ t-BuO suggesting that a steric interaction between vicinal substituents, a 3-alkoxy group and 4-isopropyl group (3,4-interaction), stabilizes to some extent the dioxetane ring

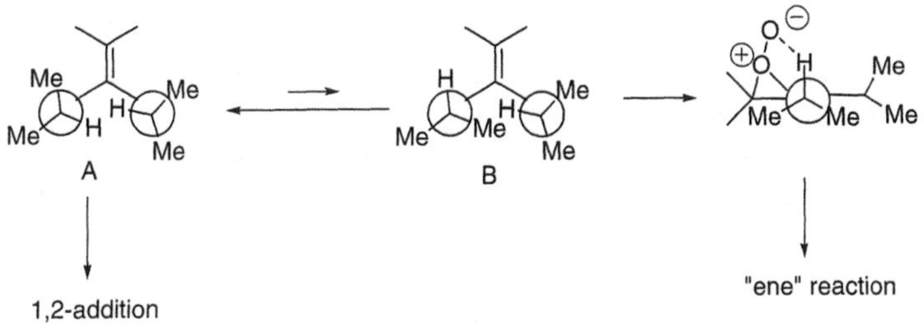

1,2-addition

"ene" reaction

Scheme 69

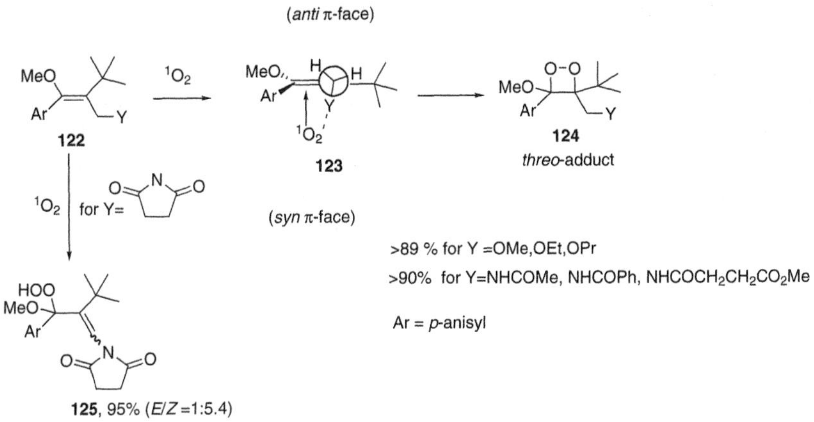

(*anti* π-face)

122

123

124
threo-adduct

(*syn* π-face)

>89 % for Y =OMe,OEt,OPr

>90% for Y=NHCOMe, NHCOPh, NHCOCH$_2$CH$_2$CO$_2$Me

Ar = *p*-anisyl

125, 95% (*E/Z* =1:5.4)

Scheme 70

except for overly large *t*-BuO group, for which the steric interaction acts negatively [124].

Electronic effets have been suggested to control the chemo- and diastereoselectivity observed in the oxygenation of olefins **122** with allylic oxygen [127a] and nitrogen [127b] (Sch. 70).

In particular, the selectivity has been attributed to the coordination of the allylic nucleophilic heteroatom with the incoming electrophilic singlet oxygen (steering effect). As depicted in **123**, this interaction guides the oxygen to the *syn*-face where no allylic hydrogens are oriented to suffer the prototropic ene reaction, and leads to *threo*-adduct **124** [127]. In nitrogen derivatives, the character of the heteroatom also plays a major role [127b]. Indeed, the *syn* (*threo*) π-facial selectivity has been observed only for primary amines and amides. Secondary amines give α-oxidation to imines likely due to a charge transfer which deprives the dioxygen of electrophilicity needed to react with the double bond. Imides afford preferentially ene-adducts as **125** since singlet oxygen attacks from the *anti*-face where allylic hydrogen(s) is suitably oriented (Sch. 70) [127b].

A highly diastereoselective dioxetane formation has been found in the oxygenation of the adamantylidene-substituted chiral allylic alcohol **126a** (Sch. 71) [10b,128].

The reaction leads to dioxetane *threo*-**127a** with a ⩾ 95:5 diastereo-selectivity. The *threo*-selective hydroxy-group directivity is explained in terms of hydrogen bonding in the exciplex between singlet oxygen and the hydroxy group, as in **128**, which is also appropriately conformationally aligned in the substrate by 1,3-allylic strain [10b,128]. In CD$_3$OD/CCl$_4$ the diasteromeric ratio drops to 89:11 due to the reduced strength of hydrogen bonding caused by methanol (intermolecular hydrogen bonding) [128]. For acetate **126b** hydrogen bonding is absent and singlet oxygen coordinates with the allylic hydrogen atom at the acetoxy-bearing chirality center leading, via abstraction of this atom, to the ene-adduct (Sch. 71). Optically active oxazolidinones, which work well with dienes **33** in [4+2] oxygenation (Sch. 16) [10c], have proven to be highly effective chiral auxiliaries also in the oxygenation of the enecarbamates **129** (Sch. 72) [10a].

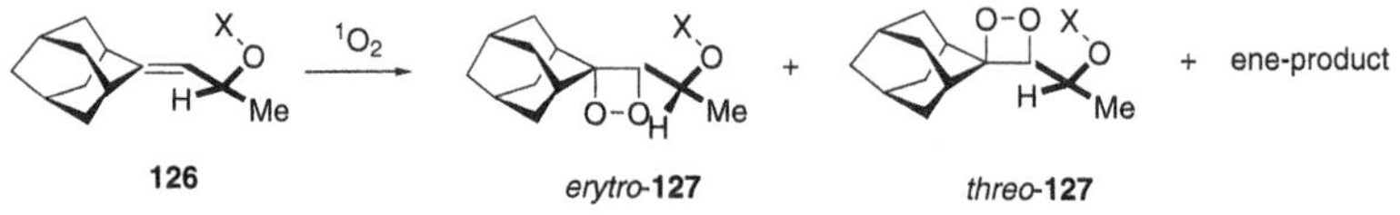

X	solvent	*threo/eritro*	[2+2]/ene
a H	CDCl$_3$	⩾95:5	47:53
a H	CD$_3$OD/CCl$_4$	89:11	⩾95:5
b Ac	CDCl$_3$ (or CD$_3$OD)		⩽5:95

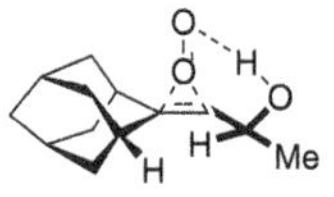

Scheme 71

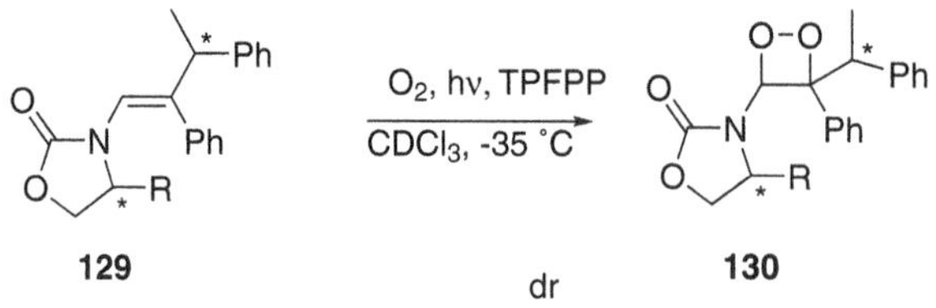

	dr
R = H	50 : 50
R = Me or ^{i}Pr	>95 : 5

Scheme 72

The reaction proceeds with high stereoselectivity (dr >95:5) due to an effective steric repulsion by the substituent (even methyl) at the oxazolidinone moiety [10a].

11.3.4. Synthetic Potential

11.3.4.1. Fragmentation

The [2+2] cycloaddition of singlet oxygen to alkenes followed by thermal or catalyzed decomposition of the dioxetane intermediate to carbonyl fragments represents a valid and simple alternative to ozonolysis. However, the synthetic utility is limited to electron-rich olefins due to the competition with the other reaction modes (ene and [4 + 2] cycloaddition). So, enamines can be converted to ketones and amides [129a] or diketones and hydroxyketones [129b], ketenimines to isocyanates [129c], thioketene acetals to thioketones [107e], nitronates to ketones [129d].

Photo-oxidative double bond cleavage can be usefully adopted for the construction of large heterocyclic rings containing carbonyl groups. Thus, photo-oxygenation of an indole **131** leads to a nine-membered ring system via a dioxetane intermediate [6c,130a], and oxygenation of the enamine **132** is a convenient route to compound **133** in the synthesis of protopine-type alkaloids (Schs. 73, 74) [130b].

Compound **135** which is formed as the main product by the oxygenation of pyrrole **134** leads, by acid-catalyzed dehydrocyclization,

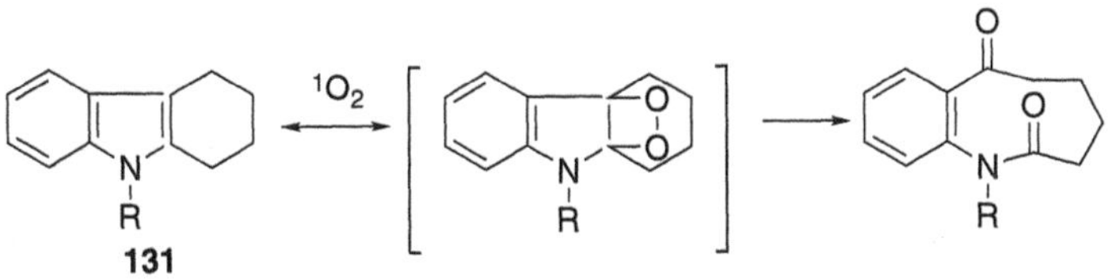

131

Scheme 73

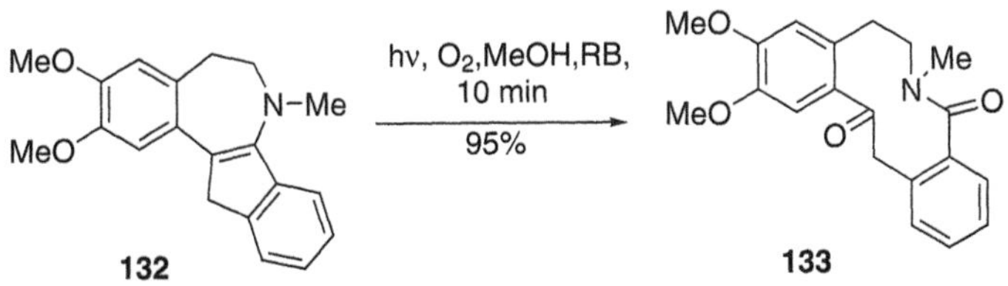

132   **133**

Scheme 74

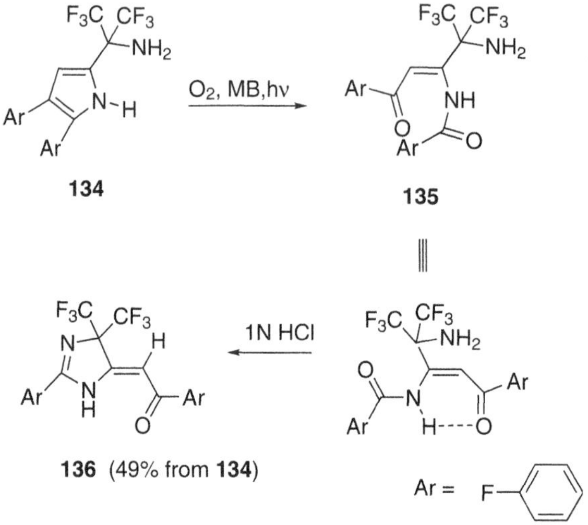

134

135

136 (49% from 134)

Scheme 75

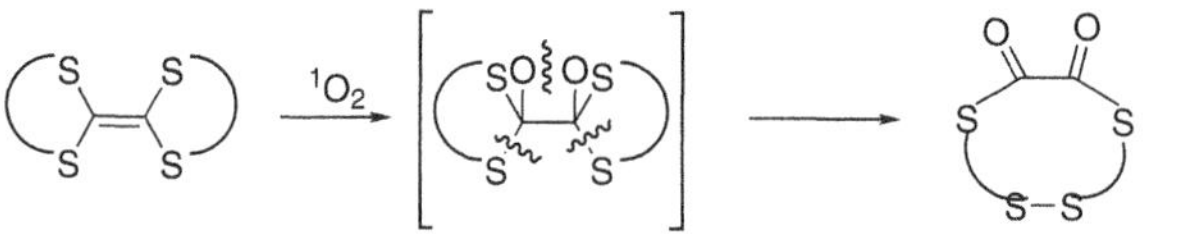

137 (14%)

Scheme 76

to 4,4-*bis*(trifluoromethyl)imidazoline **136**, a key intermediate for the preparation of acyl CoA:cholesterol acyltransferase (ACAT) inhibitors (Sch. 75) [69].

In addition to the cleavage to two carbonyl compounds, sulfur- or nitrogen-substituted dioxetanes can undergo a different mode of fragmentation, namely fission of the C–S [131a,c] or C–N [131b] bond subsequent to the O–O bond cleavage. This sequence might be involved in the formation of the cyclic disulfide **137** (Sch. 76) [131c].

11.3.4.2. Nucleophilic Transformations

Dioxetanes are converted to *vic*-diols with reductants as lithium aluminum hydride [132a] or thiols [132b] (Sch. 77).

Epoxides can be formed by treatment with phosphines [133a] via the corresponding phosphorane **138** (Sch. 78) [133b], and analogous reactions

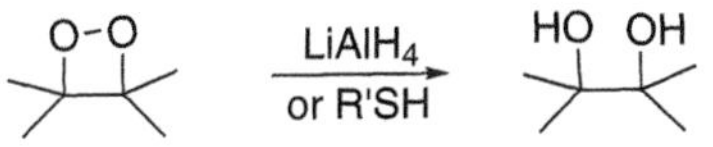

Scheme 77

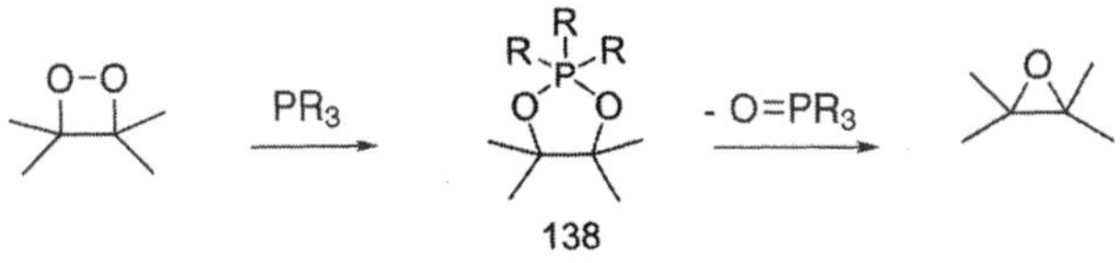

Scheme 78

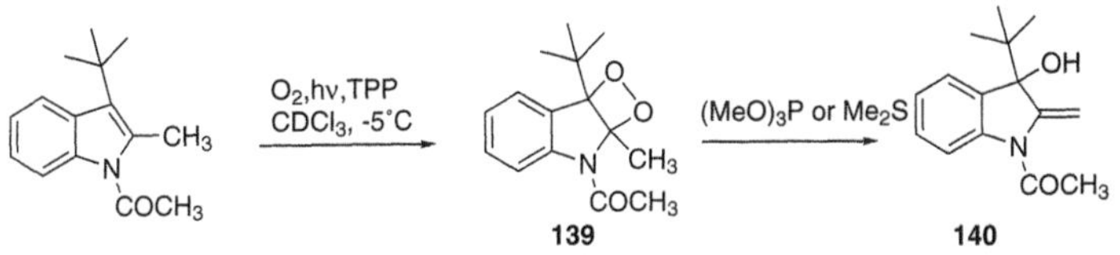

Scheme 79

can also occur with sulfides [133c]. A peculiar ring-opening product, the allylic alcohol **140**, has been obtained by treatment of dioxetane **139** with P(OMe)₃ or Me₂S (Sch. 79) [71a].

Investigation of the reaction of 3,3-disubstituted 1,2-dioxetanes with various heteroatom nucleophiles establishes the S$_N$2 reactivity of these strained peroxides [134]. As reported in Sch. 80 for dioxetane **141**, the sterically exposed oxygen of the peroxide bond becomes the site of nucleophilic attack to produce an anionic or zwitterionic adduct **142**. Different reaction channels become available for the intermediate which depend on the chemical nature of the nucleophile. So epoxy alcohol **143**, β-hydroxy hydroxylamine **144**, diol **145**, cyclic carbonate **146**, and cyclic sulfite **147** can be obtained (Sch. 80) [134].

Intramolecular nucleophilic attacks of nitrogen [106a] and sulfur [126] at O–O bond have been recently proposed to explain the formation of unexpected products in the oxygenation of 2(o-aminophenyl)-4,5-dihydro-furans **148** (Sch. 81) and 5,6-dihydro-1,4-oxathiins **149** (Sch. 82).

In the latter cases, the stereoselective formation of keto sulfoxides **153** and **154** has been observed instead of the expected dicarbonyl compounds **151** (as found for 3-methyl-2-phenyl-5,6-dihydro-1,4-oxathiin (**116**), Sch. 67), only in the presence of an electron withdrawing group at the double

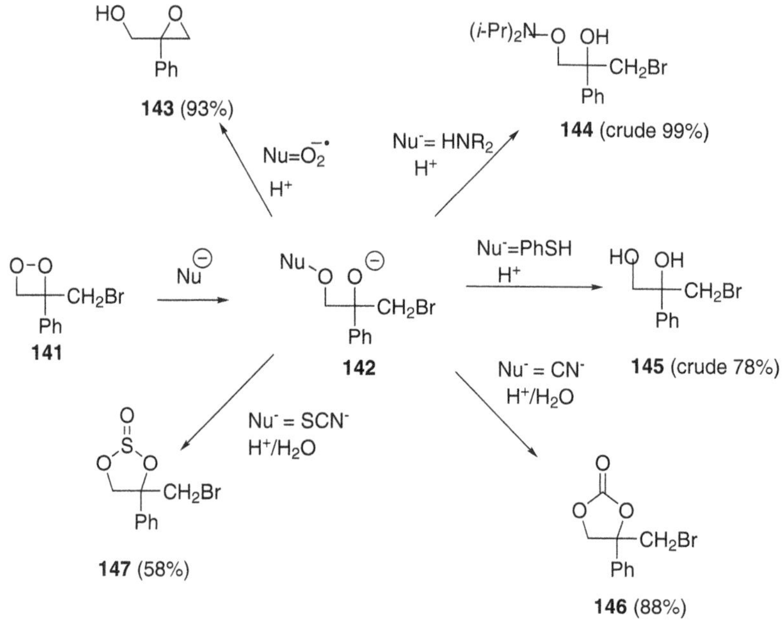

Scheme 80

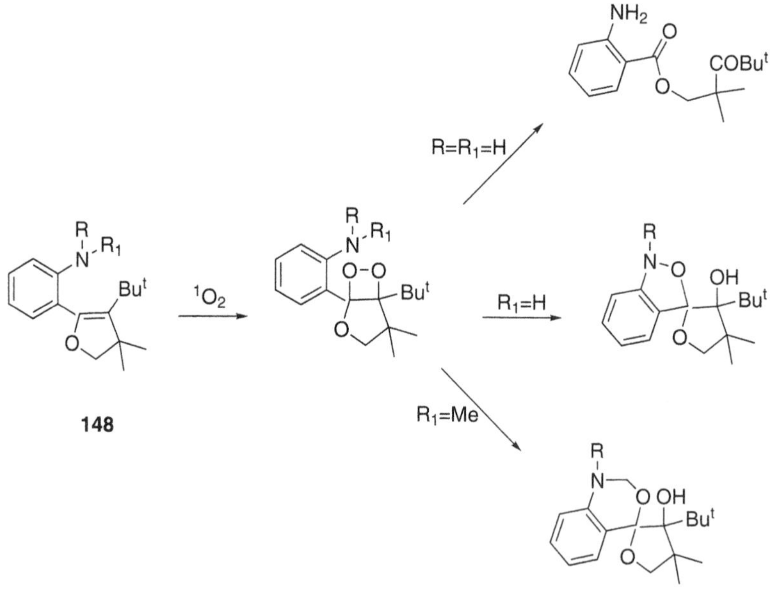

Scheme 81

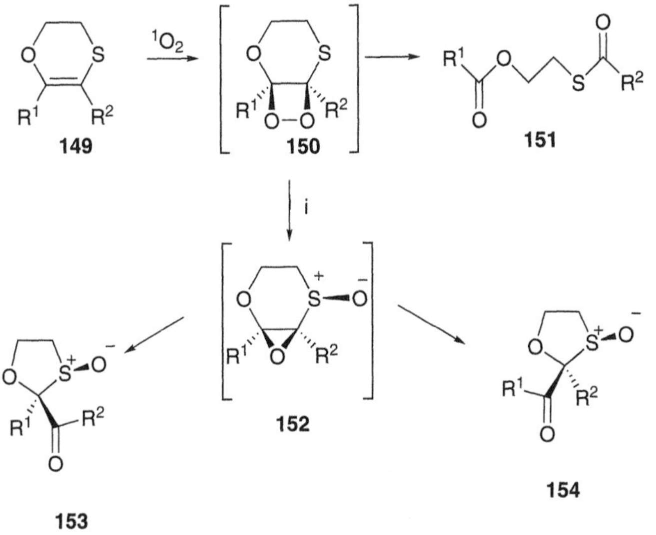

i; only for R² = electron-withdrawing [COMe, CO₂Me, CONR₂]

Scheme 82

bond. It has been suggested that in the dioxetane **150** the increased electron demand by the O–O bond favors the intramolecular nucleophilic attack of the ring sulfur leading to final products **153** and **154** via the labile undetected sulfoxide epoxides **152** (Sch. 82) [126].

11.3.4.3. Rearrangements

Some alkoxy-substituted dioxetanes yield α-peroxy ketones and esters via zwitterionic intermediates which are formed by heterolytic C–O bond cleavage as proven by trapping with acetaldehyde [135]. This behavior which has been fully explored by Jefford [135b], has been exploited for the total synthesis of the antimalarial (+)artemisinin [136a–c] and related analogs [136d–f]. One recent example is reported in Sch. 83 [136f].

The oxygenation of the appropriate alkenes, such as **155** generates the intermediate dioxetanes **156** in a stereoselective mode (attack on the *si* face) and these undergo acid-catalyzed opening to give directly the final products **157** presumably via the ion **158** [136f]. 3-Silyloxydioxetanes **159** rearrange at room temperature to α-(trialkylsilyl)peroxy ketones **160** via silyl migration [137a,b]. Similar products (**163**) have also been found in the oxygenation of silyl enol ethers **161** derived from indan-1-ones together with dioxetanes **162**, so that their direct formation has been

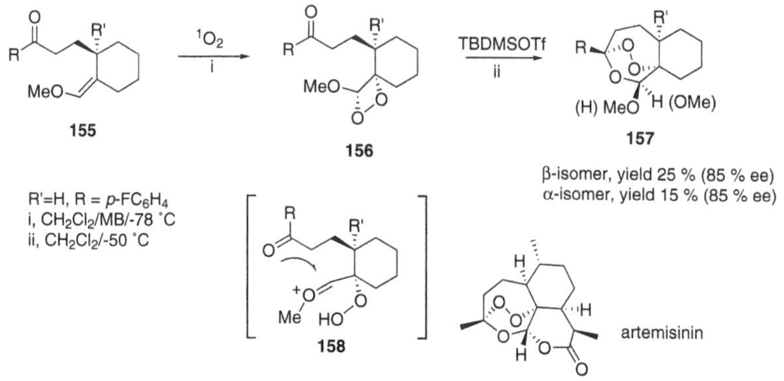

Scheme 83

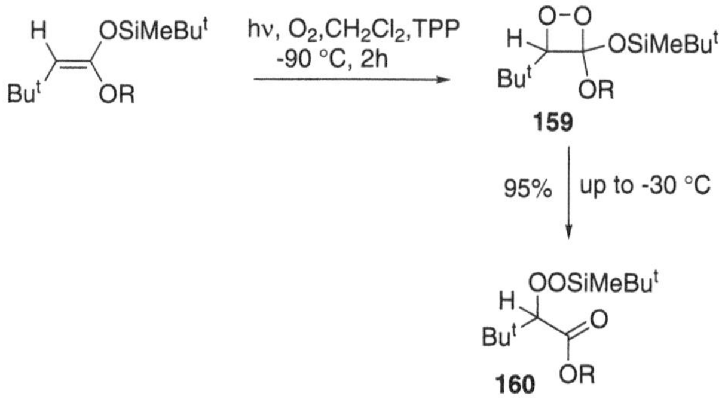

155

R'=H, R = p-FC$_6$H$_4$
i, CH$_2$Cl$_2$/MB/-78 °C
ii, CH$_2$Cl$_2$/-50 °C

156

i ^{1}O$_2$

TBDMSOTf ii

157

β-isomer, yield 25 % (85 % ee)
α-isomer, yield 15 % (85 % ee)

158

artemisinin

Scheme 84

H   OSiMeBu^t

Bu^t OR

hv, O$_2$,CH$_2$Cl$_2$,TPP
-90 °C, 2h

159

95%   up to -30 °C

160   OR

rationalized by silyl migration in the 1,4-zwitterion intermediates (Schs. 84, 85) [137b,c].

The photo-oxygenation of 2-phenylnorbornene **164** in the presence of hydrogen peroxide leads to the 1,2-*bis*-hydroperoxide **165** in essentially quantitative yield via trapping of the zwitterionic peroxidic intermediate by the reagent (Sch. 86) [138]. Suitable treatment of this compound leads to tricyclic peroxides **166** and **167**, some of which [**166a** and **167** (for R,R^1-CH(CH$_3$)(CH$_2$)$_4$-)] have shown moderate antimalarial activity.

11.3.5. Chemiluminescence

A unique property of dioxetanes is to decompose thermally to electronically excited carbonyl products with subsequent emission of blue light

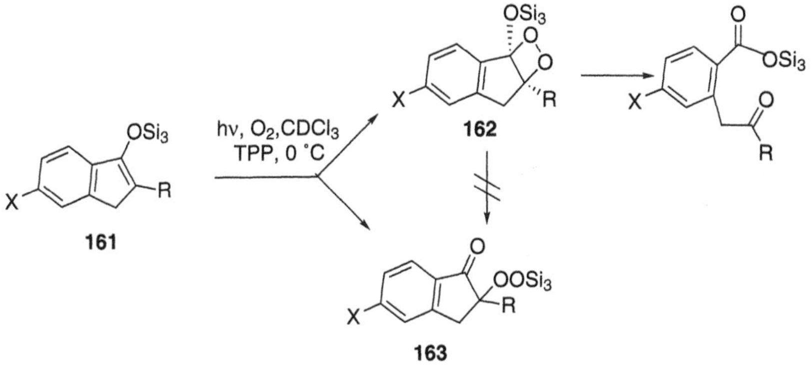

Scheme 85

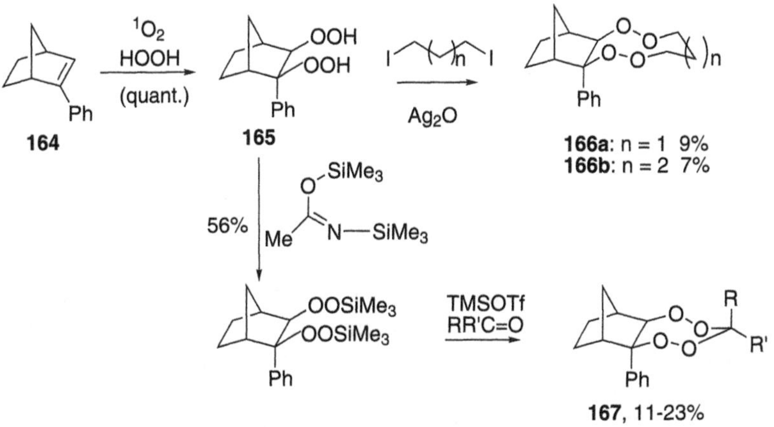

Scheme 86

(λ_{max} 400–430 nm) [6,139]. The two carbonyl compounds are formed partly in the S_1 state and partly in the T_1 state, and the light emission results from deactivation of the S_1 excited species [6] (Sch. 87).

A different decomposition mechanism has been found in dioxetanes containing a substituent of low oxidation potential such as aminoaryl or aryloxy groups [106,140], in oxyphenylethenyl or oxyphenylethynyl-substituted dioxetanes [141] or in nitrogen- and sulfur-substituted dioxetanes [107c]. In these derivatives an intramolecular electron transfer from the easily oxidized group to O–O bond occurs, generating an intermediate which cleaves to an excited singlet product (CIEEL mechanism = chemically

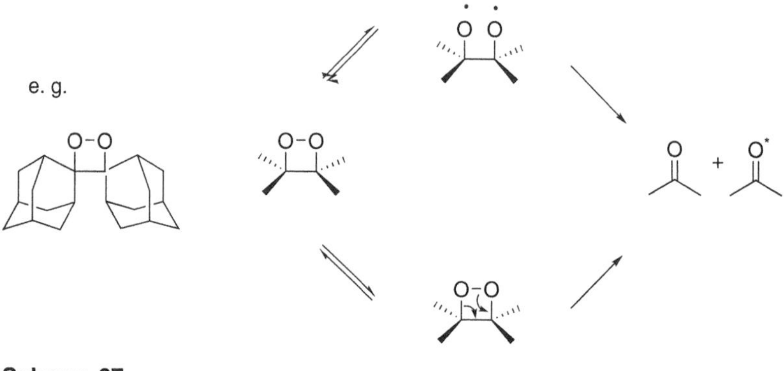

e. g.

Scheme 87

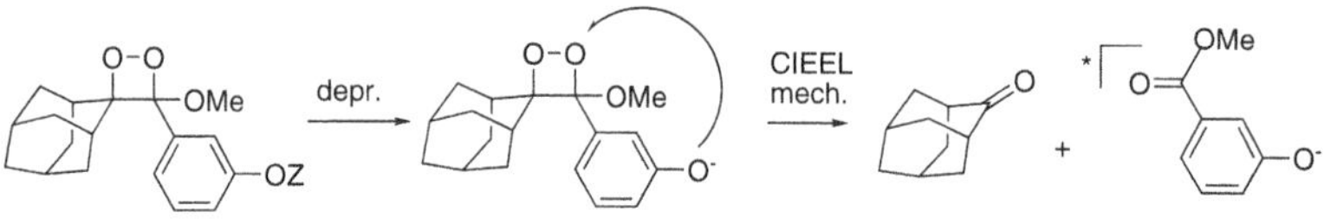

Z = Na$_2$O$_2$P(O)-

Scheme 88

initiated electron exchange luminescence) (Sch. 88) [140]. The presence of electron-donor groups [142], especially the silyloxy-substituted aromatic group [142d], as well as the conformation of the aromatic electron donor [143] influences significantly the rate of decomposition.

The chemiluminescent properties of dioxetanes have led to the development of thermally stable derivatives to be used as luminogenic substrates in enzyme labels which find applications in the area of immunoassays [111,144]. Although functionalized adamantylideneadamantanes have been described as labels for thermochemiluminescent immunoassays [145], CIEEL-active dioxetanes are more effective. They generally contain an adamantyl group to provide stability and a fluorgenic protected phenolic substituent which can be triggered chemically or enzymatically [142,146]. In the protected form (ester) these dioxetanes are stable with half-lives up to 19 years at r.t.; deprotection (e.g., for phosphate esters by hydrolysis with alkaline phosphatase) generates an unstable phenoxide-substituted dioxetane which decomposes rapidly to emit light by CIEEL mechanism. Scheme 88 reports the commercially available AMPPD® (or PPD®) which is used for biochemical and clinical applications [144].

The novel triggerable bicyclic dioxetanes also possess remarkable thermal stability and high CIEEL efficiency. The most common strategy is to immobilize the target DNA on a solid support, attach a DNA probe containing an enzyme (alkaline phosphatase) to the target DNA, followed by exposure to an enzyme-(phosphatase) triggerable stable dioxetane. The light emitted from the decomposition of the newly generated unstable dioxetane is then measured on a luminometer. The advantages of the use of luminescent labels if compared with the use of radioactive isotopes are that they are safer, faster, and more sensitive. Thus, in less than 20 years since their discovery, dioxetanes have developed from being merely laboratory curiosities to becoming stable derivatives employed worldwide in immunological and biochemical analysis [111], and the considerable amount of literature in this field bears testimony to the general interest of the scientific community.

11.4. CONCLUSION

Although the applications here illustrated are only a part of the impressive number of examples described in literature, it becomes clear that the utility of singlet oxygen [4+2] and [2+2] photo-oxygenations and how the various transformations of the primary oxygenated products can be used to oxyfunctionalize organic molecules in a regio- and, often, stereoselective manner, is enormous. Versatility, selectivity, and practical simplicity (workup and product isolation often entail merely evaporation of the solvent) make these reactions particularly attractive for the synthetic chemist, even in the frame of green chemistry due to the use of light and environmentally friendly oxygen.

REFERENCES

1. Windaus A, Brunken J. Justus Liebigs Ann Chem 1928; 460:225.
2. Bartlett PD, Schaap AP. J Am Chem Soc 1970; 92:3223.
3. Mazur S, Foote CS. J Am Chem Soc 1970; 92:3225.
4. Kopecky KR, Van de Sande JH, Mumford C. Can J Chem 1968; 46:25; Kopecky KR, Mumford C. Can J Chem 1969; 47:709.
5. (a) Adam W, Griesbeck AG. In: Horspool WM, Song P-S, eds. CRC Handbook of Organic Photochemistry and Photobiology. Boca Raton: CRC, 1992:311; (b) Clennan LA, Foote CS. In: Ando W, ed. Organic Peroxides. London: Wiley, 1992:255; (c) Bloodworth AJ, Eggelte HJ. In: Frimer AA, ed. Singlet Oxygen. Vol. II. Boca Raton: CRC Press, 1985:94.

6. (a) Saha-Möller CR, Adam W. In: Padwa A, ed. Comprehensive Heterocyclic Chemistry II. Vol. 1B. New York: Pergamon, 1996:1041; (b) Adam W, Heil M, Mosandl T, Saha-Moller CR. In: Ando W, ed. Organic Peroxides. New York: Wiley, 1992:221; (c) Baumstark AL, Rodriguez A. In: Horspool WM, Song P-S, eds. CRC Handbook of Organic Photochemistry and Photobiology. Boca Raton: CRC, 1992:335; (d) Baumstark AL. In: Frimer AA, ed. Singlet Oxygen. Vol. II. Boca Raton: CRC Press, 1985:1; (e) Schaap AP, Zaklika KA. In: Wasserman HH, Murray RW, eds. Singlet Oxygen. London: Academic Press, 1979:174.

7. (a) Gollnick K. Adv Photochem 1968; 6:1. (b) Foote CS. Photochem Photobiol 1991; 54:659; (c) Lopez L. Top Curr Chem 1990; 156:119; (d) Fox MA. In: Fox MA, Chanon M, eds. Photoinduced Electron Transfer. Vol. D. Amsterdam: Elsevier, 1988:1.

8. (a) Foote CS, Clennan EL. In: Foote CS, Valentine JS, Greenberg A, Liebman JF, eds. Active Oxygen in Chemistry. London: Chapman & Hall, 1995:105; (b) Clennan EL. Advances in Oxygenated Processes. Vol. I. Greenwich: JAI Press, 1988:85; (c) Matsumoto M. In: Frimer AA, ed. Singlet Oxygen. Vol. II. Boca Raton: CRC Press, 1985:205; (d) Denny RW, Nickon A. Org React 1973; 20:133.

9. (a) Clennan EL. Tetrahedron 1991; 47:1343; (b) Balci M. Chem Rev 1981; 91; (c) Wasserman HH, Ives JI. Tetrahedron 1980; 37:1825.

10. (a) Adam W, Bosio SG, Turro NJ. J Am Chem Soc 2002; 124:8814; (b) Adam W, Saha-Möller CR, Schambony SB, Schmid KS, Wirth T. Photochem Photobiol 1999; 70:476; (c) Adam W, Güthlein M, Peters E-M, Peters K, Wirth T. J Am Chem Soc 1998; 120:4091; (d) Adam W, Prein M. Acc Chem Res 1996; 29:275.

11. van den Heuvel CJM, Verhoeven JW, de Boer Th. J Recl Trav Chim Pays-Bas 1980; 99:280.

12. Dewar MJS, Thiel W. J Am Chem Soc 1977; 99:2338.

13. Bobrowski M, Liwo A, Oldziej S, Jeziorek D, Ossowski T. J Am Chem Soc 2000; 122:8112.

14. Sevin F, McKee ML. J Am Chem Soc 2001; 123:4591.

15. (a) Gollnick K, Griesbeck A. Tetrahedron Lett 1984; 25:725; (b) Stevens B, Perez SR. Mol Photochem 1974; 6:1.

16. (a) Rigaudy J, Capdevielle P, Combrisson S, Maumy M. Tetrahedron Lett 1974, 2757; (b) Rio G, Berthelot J. Bull Soc Chim Fr 1969, 1664; (c) Gollnick K, Schenck GO. In: Hamer J, ed. 1,4-Cycloaddition Reactions. New York: Academic Press, 1967; 255.

17. Matsumoto M, Dobashi S, Kuroda K, Kondo K. Tetrahedron 1985; 41:2147.

18. (a) Kwon B-M, Foote CS, Khan SI. J Org Chem 1989; 54:3378; (b) O'Shea KE, Foote CS. J Am Chem Soc 1988; 110:7167.

19. (a) Aubry J-M, Mandard-Cazin B, Rougee M, Bensasson RV. J Am Chem Soc 1995; 117:9159; (b) Clennan EL, Mehrsheikh-Mohammadi ME. J Am Chem Soc 1984; 106:7112; (c) Gollnick K, Griesbeck A. Tetrahedron 1984; 40:3235.

20. Gorman AA, Lovering G, Rodgers MAJ. J Am Chem Soc 1979; 101:3050.

21. Paquette LA, Liotta DC, Baker AD. Tetrahedron Lett 1976; 2681.
22. Kenney RL, Fisher GS. J Am Chem Soc 1959; 81:4288.
23. (a) Sasson I, Labovitz J. J Org Chem 1975; 40:3670; (b) Barrett HC, Buchi G. J Am Chem Soc 1967; 89:5665.
24. Clennan EL, Nagraba K. J Org Chem 1987; 52:294.
25. Mehta G, Uma R. Tetrahedron Lett 1995; 36:4873.
26. Shing TKM, Tam EKW. J Org Chem 1998; 63:1547.
27. Hathaway SJ, Paquette LA. Tetrahedron 1985; 41:2037.
28. (a) Schenck GO, Dunlap DE. Angew Chem 1956; 68:248; (b) Schenck GO. Angew Chem 1952; 64:12; (c) Cope AC, Liss TA, Wood GA. J Am Chem Soc 1957; 79:6287; (d) Horinaka A, Nakashima R, Yoshikawa M, Matsuura T. Bull Chem Soc Jpn 1975; 48:2095.
29. (a) Wilkinson F, Brummer JG. J Phys Chem Ref Data 1981; 10:809; (b) Monroe BM. J Am Chem Soc 1981; 103:7253.
30. Rio G, Chiarify M. Bull Soc Chim Fr 1970; 3585.
31. (a) Matusch R, Schmidt G. Angew ChemInt EdEngl 1988; 27:717; (b) Matusch R, Schmidt G. Helv Chim Acta 1989; 72:51; c) Carless HAJ, Billinge JR, Oak OZ. Tetrahedron Lett 1989; 30:3113.
32. Bladon P. J Chem Soc 1955; 2176.
33. (a) Sengul ME, Balci M. J Chem Soc Perkin Trans 1 1997; 2071; (b) Atasoy B, Balci M. Tetrahedron 1986; 42:1461; (c) Adam W, Rebollo H. Tetrahedron Lett 1982; 23:4907; (d) Adam W, Balci M, Pietrzak B. J Am Chem Soc 1979; 101:6285; (e) Adam W, Balci M. J Am Chem Soc 1979; 101:7537.
34. (a) Brecht R, Haenel F, Seitz G, Frenzen G, Pilz A, Massa W, Wocadlo S. Liebigs Ann/Recueil 1997; 851; (b) Brecht R, Buttner F, Bohm M, Seitz G, Frenzen G, Pilz A, Massa W. J Org Chem 2001; 66:2911.
35. (a) Adam W, Erden I. Tetrahedron Lett 1979; 2781; b) Adam W, Gretzke N, Hasemann L, Klug G, Peters E-M, Peters K, von Schnering HG, Will B. Chem Ber 1985; 118:3357.
36. Adam W, Klug G, Peters E-M., Peters K, von Schnering HG, Tetrahedron 1985; 41:2045.
37. Gollnick K, Griesbeck A. Tetrahedron Lett 1983; 24:3303.
38. Motoyoshiya J, Okuda Y, Matsuoka I, Hayashi S, Tagaguchi Y, Aoyama H. J Org Chem 1999; 64:493.
39. Matsumoto M, Nasu S, Takeda M, Murakami H, Watanabe N. Chem Commun 2000; 821.
40. Herz W, Juo R-R. J Org Chem 1985; 50:618.
41. Izzo I, Meduri G, Avallone E, De Riccardis F, Sodano G. Eur J Org Chem 2000; 439.
42. (a) Albini A, Freccero M. In: Horspool WM, Song P-S, eds. CRC Handbook of Organic Photochemistry and Photobiology. Boca Raton: CRC, 1995:346; (b) Saito I, Matsuura T. In: Wasserman HH, Murray RW, eds. Singlet Oxygen. London: Academic Press, 1979:511.
43. van den Heuvel CJM, Hofland A, Steinberg H, de Boer TJ. Recl Trav Chim Pays-Bas 1980; 99:275.

44. Sawada T, Mimura K, Thiemann T, Yamato T, Tashiro H, Mataka S. J Chem Soc, Perkin Trans I 1998; 1369.
45. Rigaudy J. Pure Appl Chem 1968; 16:169.
46. (a) Pierlot C, Hajjam S, Barthélémy C, Aubry JM. J Photochem Photobiol B, 1996; 36:31; (b) Wasserman HH, Larsen DL. J Chem Soc, Chem Commun 1972; 253; (c) Hart H, Oku A. J Chem Soc, Chem Commun 1972; 254.
47. van den Heuvel CJM, Steinberg H, de Boer TJ. Recl Trav Chim Pays-Bas 1980; 99:109.
48. (a) Pierlot C, Poprawski J, Marko J, Aubry J-M. Tetrahedron Lett 2000, 41:5063; (b) Pierlot C, Aubry J-M. J Chem Soc, Chem Commun 1997; 2289.
49. (a) Rigaudy J, Scribe P, Breliere C. Tetrahedron 1981; 37:2585; (b) Viallet A, Rouger J, Cheradame H, Gandini A. J Photochem 1979; 11:129.
50. (a) Wasserman HH, Larsen DL. J Chem Soc, Chem Commun 1972; 253; (b) Turro NJ, Chow MF, Rigaudy J. J Am Chem Soc 1981; 103:7218; (c) Sparfel D, Gobert F, Rigaudy J. Tetrahedron 1980; 36:2225.
51. Martinez GR, Ravanat J-L, Medeiros MHG, Cadet J, Di Mascio P. J Am Chem Soc 2000; 122:10212.
52. (a) Matsumoto M, Kuroda K, Suzuki Y. Tetrahedron Lett 1981; 3253; (b) Matsumoto M, Kuroda K. Tetrahedron Lett 1979; 1607; (c) Matsumoto M, Dobashi S, Kuroda K. Tetrahedron Lett 1977; 3361; (d) Schaap AP, Burns PA, Zaklika KA. J Am Chem Soc 1977; 99:1270.
53. (a) Zhang X, Khan SI, Foote CS. J Org Chem 1993; 58:7839; (b) Matsumoto M, Dobashi S, Kondo K. Tetrahedron Lett 1975; 4471.
54. (a) Lipshutz BH. Chem Rev 1986; 86:795; (b) Wasserman HH, Lipshutz BH, In: Wasserman HH, Murray RW, eds. Singlet Oxygen. London: Academic Press, 1979:429.
55. Koch E, Schenck GO. Chem Ber 1966; 99:1984.
56. Esser P, Pohlmann B, Scharf H-D. Angew Chem Int Ed Engl 1994; 33:2009.
57. Wasserman HH, McCarthy KE, Prowse KS. Chem Rev 1986; 86:845.
58. Graziano ML, Iesce MR, Scarpati R. J Chem Soc, Perkin Trans I 1980; 1955.
59. Gollnick K, Griesbeck A. Tetrahedron 1985; 41:2057.
60. (a) For a review on alkoxyfurans: Scarpati R, Iesce MR, Cermola F, Guitto A. Synlett, 1998; 17; For other furans, (b) Iesce MR, Cermola F, De Lorenzo F, Orabona I, Graziano ML. J Org Chem 2001; 66:4732; (c) Iesce MR, Cermola F, Guitto A, Giordano F. J Chem Soc, Perkin Trans I 1999; 475; (d) Graziano ML, Iesce MR, Cinotti A, Scarpati R. J Chem Soc, Perkin Trans I 1987; 1833; (e) Graziano ML, Iesce MR. Synthesis 1985; 1151; (f) Graziano ML, Iesce MR, Chiosi S, Scarpati R. J Chem Soc, Perkin Trans I 1983; 2071; (g) Graziano ML, Iesce MR, Scarpati, R. J Chem Soc, Perkin Trans I 1982; 2007; (h) Graziano ML, Iesce MR, Carli B, Scarpati R. Synthesis 1982; 736; (i) Graziano ML, Iesce MR, Scarpati R. Synthesis 1984; 66.
61. Adam W, Ahrweiler M, Sauter M. Angew Chem 1993; 105:104.
62. (a) Iesce MR, Cermola F, Guitto A. J Org Chem 1998; 63:9528; (b) Iesce MR, Cermola F, Guitto A, Giordano F, Scarpati R. J Org Chem 1996; 61:8677.

63. Ogliaruso MA, Wolfe JF. In: Patai S, Rappoport Z, eds. Synthesis of Lactones and Lactams. New York: Wiley, 1993:109–110, 337, 341.
64. Adam W, Rodriguez A. Tetrahedron Lett 1981; 22:3505.
65. Hata T, Tanaka K, Katsumura S. Tetrahedron Lett 1999; 40:1731.
66. Lightner DA, Bisacchi GS, Norris RD. J Am Chem Soc 1976; 98:802 and references therein.
67. George MV, Bhat V. Chem Rev 1979; 79:447.
68. Wasserman HH, Frechette R, Rotello VM, Schulte G. Tetrahedron Lett 1991; 32:7571.
69. Li H-Y, Drummond S, DeLucca I, Boswell GA. Tetrahedron 1996; 52:11153.
70. (a) Wasserman HH, Petersen AK, Xia M, Wang J. Tetrahedron Lett 1999; 40:7587; (b) Wasserman HH, Xia M, Wang J, Petersen AK, Jorgensen M. Tetrahedron Lett 1999; 40:6145; (c) Wasserman HH, Power P, Petersen AK. Tetrahedron Lett 1996; 37:6657.
71. (a) Zhang X, Foote CS, Khan SI. J Org Chem 1993; 58:47; (b) Nakagawa M, Yokoyama Y, Kato S, Hino T. Tetrahedron 1985; 41:2125.
72. (a) Boger DL, Baldino CM. J Am Chem Soc 1993; 115:11418; (b) Boger DL, Baldino CM. J Org Chem 1991; 56:6942.
73. (a) Gollnick K, Griesbeck A. Tetrahedron Lett 1984; 25:4921; (b) Adam W, Eggelte HJ. Angew Chem Int Ed Engl 1978; 17:765.
74. (a) Gollnick K, Koegler S. Tetrahedron Lett 1988; 29:1003; (b) Gollnick K, Koegler S. Tetrahedron Lett 1988; 29:1007.
75. (a) Graziano ML, Iesce MR, Cimminiello G, Cermola F, Scarpati R, Parrilli M. J. Chem Soc Perkin Trans II 1991; 1085; (b) Graziano ML, Iesce MR, Cimminiello G, Scarpati R, Parrilli M. J Chem Soc, Perkin Trans I 1990; 1011.
76. (a) Wasserman HH, Gambale RJ, Pulwer MJ. Tetrahedron 1981; 37:4059; (b) Pridgen LN, Shilcrat SC, Lantos I, Tetrahedron Lett 1984; 25:2835.
77. Graziano ML, Iesce MR, Carotenuto A, Scarpati R. Synthesis 1977; 572.
78. (a) Wasserman HH, Lenz G. Heterocycles 1976; 5:409; (b) Wasserman HH, Druckrey E. J Am Chem Soc 1968; 90:2440.
79. (a) Ito H, Ichimura K. Macromolecular Chemistry and Physics 1998; 199:2547; (b) Ito H, Ikeda T, Ichimura K. Macromolecules 1993; 26:4533.
80. Matsuura T, Saito I. Bull Chem Soc Jpn 1969; 42:2973.
81. (a) Ryang H-S, Foote CS. J Am Chem Soc 1979; 101:6683; (b) Graziano ML, Iesce MR, Scarpati R. J Chem Soc, Chem Commun 1979; 7.
82. Lipshutz BH, Morey MC. J Am Chem Soc 1984; 106:457.
83. See for example: (a) Adam W, Saha-Moller CR, Schonberger A. J Am Chem Soc 1997; 119:719; (b) Sheu C, Foote CS. J Am Chem Soc 1995; 117:474; (c) Steenken S. Chem Rev 1989; 89:503.
84. Secen H, Sutbeyaz Y, Balci M. Tetrahedron Lett 1990; 31:1323.
85. Harada N, Uda H, Ueno H, Utsumi S. Chem Lett 1973; 1173.
86. (a) Posner GH, Tao X, Cumming JN, Klinedinst D, Shapiro TA. Tetrahedron Lett 1996; 37:7225; (b) Salomon RG. Acc Chem Res 1985; 18:294;

(c) Bloodworth AJ, Eggelte HJ. J Chem Soc, Perkin Trans 2 1984; 2069; (d) Adam W, Eggelte HJ, Rodriguez A. Synthesis 1979; 383; (e) Adam W, Erden I. J Org Chem 1978; 43:2737.

87.   Ko SY, Lerpiniere J, Linney ID, Wrigglesworth R. J Chem Soc, Chem Commun 1994; 1775.

88.   Rigaudy J, Maumy M, Capdevielle P, Breton L. Tetrahedron 1977; 33:53.

89.   Adam W, Balci M. Tetrahedron 1980; 36:833.

90.   (a) Rigaudy J. In: Horspool WM, Song P-S, eds. CRC Handbook of Organic Photochemistry and Photobiology. Boca Raton: CRC, 1995:325; (b) Carless HAJ, Atkins R, Fekarurhobo GK. Tetrahedron Lett 1985; 26:803.

91.   van Tamelen EE, Taylor EG. J Am Chem Soc 1980; 102:1202.

92.   Akbulut N, Menzek A, Balci M. Tetrahedron Lett 1987; 28:1689.

93.   O'Shea KE, Foote CS. J Org Chem 1989; 54:3475.

94.   (a) Sengul ME, Simsek N, Balci M. Eur J Org Chem 2000; 1359; (b)Takahashi K, Kishi M. Tetrahedron 1988; 44:4737.

95.   Muellner U, Huefner A, Haslinger E. Tetrahedron 2000; 56:3893.

96.   (a) D'Onofrio F, Piancatelli G, Nicolai M. Tetrahedron 1995; 51:4083; (b) Graziano ML, Iesce MR, Carli B, Scarpati R. Synthesis 1983, 125; (c) Tsai TYR, Wiesner K, Can J Chem 1982; 60:2161.

97.   (a) Kornblum N, DeLaMare HE. J Am Chem Soc 1951; 73:880; (b) Balogh V, Beloeil J-C, Fetizon M. Tetrahedron 1977; 33:1321.

98.   Hagenbuch J-P, Vogel P. J Chem Soc, Chem Commun 1980:1062.

99.   Sengul ME, Ceylan Z, Balci M. Tetrahedron 1997; 53:10401.

100. (a) Jefford CW, Jin S, Bernardinelli G. Helv Chim Acta 1997; 80:2440; (b) Jefford CW, Kohmoto S, Jaggi D, Timari G, Rossier J-C, Rudaz M, Barbuzzi O, Gerard D, Burger U, Kamalaprija P, Mareda J, Bernardinelli G, Manzanares I, Canfield CJ, Fleck SL, Robinson BL, Peters W. Helv Chim Acta 1995; 78:647; (c) Jefford CW, Jin S, Bernardinelli G. Tetrahedron Lett 1991; 32:7243; (d) Jefford CW, Jaggi D, Boukouvalas J, Kohmoto S. J Am Chem Soc 1983; 105:6497.

101. (a) Iesce MR, Cermola F, Guitto A. Synlett 1999; 417; (b) Iesce MR, Cermola F, Guitto A. Synthesis 1998; 333; (c) Iesce MR, Cermola F, Guitto A. Synthesis 1997; 657.

102. (a) Natsume M, Muratake H. Tetrahedron Lett 1979; 3477; (b) Natsume M, Sekine Y, Ogawa M, Soyagimi H, Kitagawa Y, Tetrahedron Lett 1979; 3473; (c) Natsume M, Utsunomiya I, Yamaguchi K, Sakai S-I. Tetrahedron Lett 1985; 41:2115.

103. Lopez D, Quinoa E, Riguera R. J Org Chem 2000; 65:4671.

104. Schuster GB, Turro NJ, Steinmetzer H-C, Schaap AP, Faler G, Adam W, Liu JC. J Am Chem Soc 1975; 97:7110.

105. (a) Adam W, Arias Encarnacion LA, Zinner K. Chem Ber 1983; 116:839; (b) Adam W, Arias Encarnacion LA. Chem Ber 1982; 115:2592.

106. See for example: (a) Matsumoto M, Murakami H, Watanabe N. J Chem Soc, Chem Commun 1998; 2319; (b) Schaap AP, Gagnon SD, Zaklika KA.

Tetrahedron Lett 1982; 23:2943; (c) McCapra F, Beheshti I, Burford A, Hann RA, Zaklika KA. J Chem Soc, Chem Commun 1977; 944.

107. See for example: (a) Gollnick K, Knutzen-Mies K. J Org Chem 1991; 56:4027; (b) Gollnick K, Knutzen-Mies K. J Org Chem 1991; 56:4017; (c) Handley RS, Stern AJ, Schaap AP. Tetrahedron Lett 1985; 26:3183; (d) Ando W, Takata T. In: Frimer AA, ed. Singlet Oxygen. Vol. III. Boca Raton: CRC Press, 1985; (e) Geller GG, Foote CS, Pechman DB. Tetrahedron Lett 1983; 24:673; (f) Adam W, Arias LA, Scheutzow D. Tetrahedron Lett 1982; 23:2835.

108. (a) Matsumoto M, Murayama J, Nishiyama M, Mizoguchi Y, Sakuma T, Watanabe N. Tetrahedron Lett 2002; 43:1523; (b) Matsumoto M, Watanabe N, Kasuga NC, Hamada F, Tadokoro K. Tetrahedron Lett 1997; 38:2863.

109. Burns PA, Foote CS. J Am Chem Soc 1974; 96:4339.

110. Adam W, Fell R, Schulz MH. Tetrahedron 1993; 49:2227.

111. For a review: Mayer A, Neuenhofer S. Angew Chem Int Ed Engl 1994; 33:1044.

112. Lopez L, Troisi L, Rashid SMK, Schaap AP. Tetrahedron Lett 1989; 30:485.

113. Wilson T, Landis ME, Baumstark AL, Bartlett PD. J Am Chem Soc 1973; 95:4765.

114. (a) Bartlett PD. Pure Appl Chem 1971; 27:597; (b) Zaklika KA, Kashar B, Schaap AP. J Am Chem Soc 1980; 102:386; (c) Harding LB, Goddard WA III. J Am Chem Soc 1980; 102:439; (d) Frimer AA, Bartlett PD, Boschung AF, Jewett JG. J Am Chem Soc 1977; 99:7977; (e) Gorman AA, Gould IR, Hamblett I. J Am Chem Soc 1982; 104:7098; (f) Manring LE, Foote CS. J Am Chem Soc 1983; 105:4710.

115. Jefford CW. Chem Soc Rev 1993; 59.

116. Asveld EWH, Kellogg RM. J Am Chem Soc 1980; 102:3644.

117. Greer A, Vassilikogiannakis G, Lee KC, Koffas TS, Nahm K, Foote CS. J Org Chem 2000; 65:6876.

118. (a) Wilson SL, Schuster GB. J Am Chem Soc 1983; 105:679; (b) McCapra F, Beheshti I. J Chem Soc, Chem Commun 1977; 517; (c) van der Heuvel CJM, Steinberg H, de Boer TJ. Recl Trav Chim Pays-Bas 1985; 104:145.

119. (a) Jefford CW, Rimbault CG. J Am Chem Soc 1978; 100:295, 6437, 6515; (b) Jefford CW, Kohmoto S, Boukouvalas J, Burger U. J Am Chem Soc 1983; 105:6498.

120. (a) Schaap AP, Recher SG, Faler GR, Villasenor SR. J Am Chem Soc 1983; 105:1691; (b) Clennan EL, Chen M-F, Xu G. Tetrahedron Lett 1996; 37:2911.

121. Vassilikogiannakis G, Stratakis M, Orfanopoulos M. J Org Chem 1998; 63:6390.

122. Maranzana A, Ghigo G, Tonachini G. J Am Chem Soc 2000; 122:1414.

123. Yoshioka Y, Yamada S, Kawakami T, Nishino M, Yamaguchi K, Saito I, Bull Chem Soc Jpn 1996; 69:2683.

124. Watanabe N, Suganuma H, Kobayashi H, Mutoh H, Katao Y, Matsumoto M. Tetrahedron 1999; 55:4287.

125. (a) Clennan EL, Nagraba K. J Am Chem Soc 1988; 110:4312; (b) Clennan EL, L'Esperance RP. J Am Chem Soc 1985; 107:5178.

126. (a) Cermola F, De Lorenzo F, Giordano F, Graziano ML, Iesce MR, Palumbo G. Org Lett 2000; 2:1205; (b) Cermola F, Iesce MR. J Org Chem 2002; 67:4937.

127. (a) Matsumoto M, Kobayashi H, Matsubara J, Watanabe N, Yamashita S, Oguma D, Kitano Y, Ikawa H. Tetrahedron Lett 1996; 37:397; (b) Matsumoto M, Kitano Y, Kobayashi H, Ikawa H. Tetrahedron Lett 1996; 37:8191.

128. Adam W, Saha-Moller CR, Schambony SB. J Am Chem Soc 1999; 121:1834.

129. (a) Wasserman HH, Ives, JL. J Am Chem Soc 1976; 98:7868; (b) Ando W, Saiki T, Migita T. J Am Chem Soc 1975; 97:5028; (c) Inoue Y, Turro NJ. Tetrahedron Lett 1980; 21:4327; (d) Williams JR, Unger LR, Moore RH. J Org Chem 1978; 43:1271.

130. (a) Otsuji Y, Ohmura N, Nakanishi S, Mizuno K. Chem Lett 1972; 1197; (b) Orito K, Kurokawa Y, Itoh M. Tetrahedron 1980; 36:617.

131. (a) Ando W, Watanabe K, Suzuki J, Migita T. J Am Chem Soc 1974; 96:6766; (b) Wasserman HH, Terao S. Tetrahedron Lett 1975; 1735; (c) Adam W, Liu J-C. J Am Chem Soc 1972; 94:1206.

132. (a) Kopecky KR, Filby JE, Mumford C, Lockwood PA, Ding J-Y. Can J Chem 1975; 53:1103; (b) Adam W, Epe W, Schiffmann D, Vargas F, Wild D. Angew Chem Int Ed Engl 1988; 27:429.

133. (a) Bartlett PD, Baumstark AL, Landis ME. J Am Chem Soc 1973; 95:6486; (b) Baumstark AL, McCloskey CJ, Williams TE, Chrisope DR. J Org Chem 1980; 45:3593; (c) Wasserman HH, Saito I. J Am Chem Soc 1975; 97:905.

134. Adam W, Heil M. J Am Chem Soc 1992; 114:5591.

135. (a) Jefford CW, Boukouvalas J, Kohmoto S, Bernardinelli G. Tetrahedron 1985; 41:2081; (b) Jefford CW, Favarger F, Ferro S, Chambaz D, Bringhen A, Bernardinelli G, Boukouvalas J. Helv Chim Acta 1986; 69:1778.

136. For a review: (a) Butler AR, Wu Y-L. Chem Soc Rev 1992; 85. For artemisin: (b) Schmid G, Hofheinz W. J Am Chem Soc 1983; 105:624; (c) Xu X-X, Zhu J, Huang D-Z, Zhou W-S. Tetrahedron 1986; 42:819. For analogues; (d) Avery MA, Jennings-White C, Chong WKM. Tetrahedron Lett 1987; 28:4629; (e) Rong Y-J, Wu Y-L. J Chem Soc, Perkin Trans I 1993; 2149; (f) O'Neill PM, Miller A, Bickley JF, Scheinmann F, Oh CH. Posner GH. Tetrahedron Lett 1999; 40:9133.

137. (a) Adam W, Kades E, Wang X-H. Tetrahedron Lett 1990; 31:2259; (b) Adam W, Wang X. J Org Chem 1991; 56:4737; (c) Einaga H, Nojima M, Abe M. J Chem Soc, Perkin Trans I 1999, 2507.

138. McCullough KJ, Nonami Y, Masuyama A, Nojima M, Kim H-S, Wataya Y. Tetrahedron Lett 1999; 40:9151.

139. Wilson T. In: Frimer AA, ed. Singlet Oxygen. Vol. II. Boca Raton: CRC Press, 1985:37.

140. (a) Cilento G, Adam W. Photochem Photobiol 1988; 48:361; (b) Richardson WH, Stiggall-Estberg DL. J Am Chem Soc 1982; 104:4173; (c) Schuster GB, Schmidt SP. Adv Phys Org Chem 1982; 18:187; (d) Lee C, Singer LA. J Am Chem Soc 1980; 102:3823; (e) Schuster GB. Acc Chem Res 1979; 12:366; (f) Harding LB, Goddard WA III. J Am Chem Soc 1977; 99:4520.

141. Matsumoto M, Ishihara T, Watanabe N, Hiroshima T. Tetrahedron Lett 1999; 40:4571.

142. (a) Matsumoto M, Watanabe N, Kobayashi H, Azami M, Ikawa H. Tetrahedron Lett 1997; 38:411; (b) McCapra F. Tetrahedron Lett 1993; 34:6941; (c) Edwards B, Sparks A, Voyta JC, Bronstein I. J Biolumin and Chemilumin 1990; 5:1; (d) Schaap AP, Chen T-S, Handley RS, DeSilva R, Giri BP. Tetrahedron Lett 1987; 28:1155.

143. (a) Matsumoto M, Watanabe N, Shiono T, Suganuma H, Matsubara J. Tetrahedron Lett 1997; 38:5825; (b) Closs GL, Calcaterra LH, Green NJ, Penfield KW, Miller JR. J Phys Chem 1986; 90:3673.

144. Bronstein I, Olesen CEM. In: Wiedbrauk DL, Farkas DH, eds. Molecular Method for Virus Detection. New York: Academic, 1995:149.

145. Beck S, Koster H. Anal Chem 1990; 62:2258.

146. Hummelen JC, Luider TM, Wynberg H. Methods Enzymol 1986; 133:531.

12

Photo-oxygenation of the Ene-Type

Edward L. Clennan
University of Wyoming, Laramie, Wyoming, U.S.A.

12.1. HISTORICAL BACKGROUND

The earliest report of a singlet oxygen ene reaction (Sch. 1) was made by G. O. Schenck in a German patent in 1943 [1]. The synthetic potential of the reaction was recognized very early [2]; it provides an exquisite regioselective, and in many cases stereoselective, entry into synthetically useful feedstock using an environmentally benign oxidant [3,4]. In recognition of its pivotal synthetic utility it is has been referred to as the "Schenck Ene Reaction" in honor of its discoverer [2,3]. In this manuscript we will try to provide an overview of the synthetic scope and limitations of this reaction. We will also provide a more limited mechanistic discussion to the extent that it will enhance the synthetic discussion. Several reviews have expertly dealt with the equally fascinating mechanistic aspects of this intriguing reaction and should be consulted for more specialized details [5–9].

Over the years several different mechanisms for the singlet oxygen ene reaction have been considered as depicted in Sch. 2. Biradical (mechanism B in Sch. 2) [10,11] and zwitterionic (mechanism C in Sch. 2) mechanisms have been considered as very unlikely for a variety of reasons including the facts that Markovnikov directing effects are not observed [4], radical scavengers have no influence on the reaction, only minor solvent effects are observed [12–14], and *cis–trans* isomerizations of the substrates do not occur [6]. A dioxetane intermediate (mechanism D in Sch. 2) was considered very early [6] but this suggestion was dismissed when isolated dioxetanes were shown to cleave to the carbonyl compounds rather than rearrange to the allylic

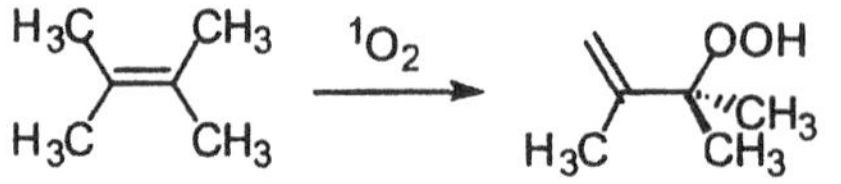

Scheme 1

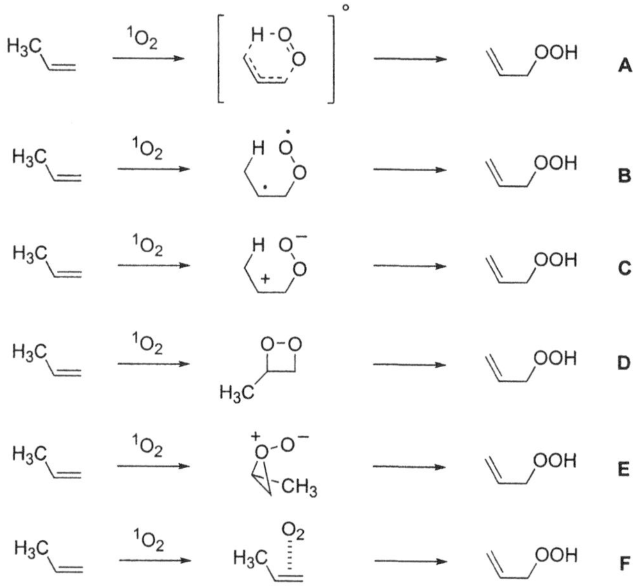

Scheme 2

hydroperoxides [15] Historically, the three mechanisms that have received the most serious consideration are the concerted ene [11,16] (mechanism A in Sch. 2), the perepoxide [17,18] (mechanism E in Sch. 2), and the exciplex [19,20] (mechanism F in Sch. 2) pathways.

12.2. STATE OF THE ART MECHANISTIC MODELS

The mechanistic description of the singlet oxygen ene reaction is still a source of debate. The lack of a mechanistic consensus is related to the glaring absence of agreement between experimental and computational results.

Experimental Evidence. A three-step pathway via an exciplex, **1**, and perepoxide, **2**, is supported by a large body of experimental work (Sch. 3). Gorman and coworkers [20] pointed out that the small negative activation

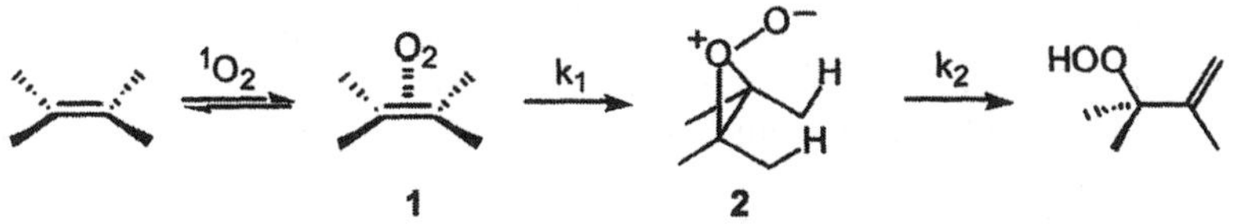

Scheme 3

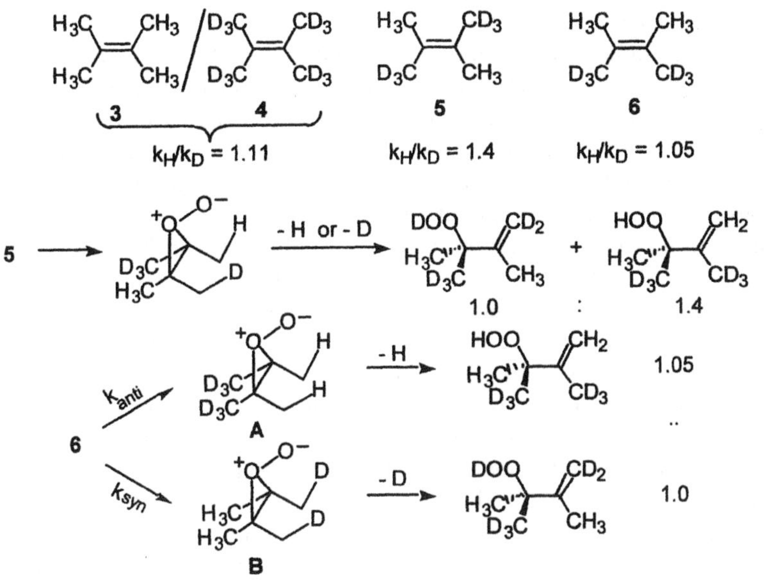

Scheme 4

enthalpies and highly negative activation entropies exhibited by singlet oxygen ene reactions [19] are characteristic of reactions involving exciplex, **1**, intermediates. They also provided compelling kinetic evidence that this exciplex was formed reversibly in the singlet oxygen ene reaction of tetramethylethylene.

The most often cited experimental work in support of perepoxide, **2**, is the intra-/intermolecular isotope effect [21,22] study of Stephenson [23] with isotopically labeled tetramethylethylenes **3**, **4**, **5**, and **6** (Sch. 4). The different isotope effects observed with **3/4** (intermolecular $k_H/k_D = 1.11$) and with **5** (intramolecular $k_H/k_D = 1.4$) was used to argue for distinct rate and product determining steps. It was anticipated that a concerted reaction should show identical kinetic (intermolecular) and product (intramolecular) isotope effects since the rate and product determining steps are one and the same. In addition, the unusual dependence of the symmetry of isotopic labeling on the magnitude of the isotope effect (see reaction of **5** and **6** in Sch. 4) is exactly what one would anticipate for an intermediate with the symmetry of perepoxide, **2**.

Computational Evidence. Recent high level computational results suggest that the perepoxide is a transition state rather than an intermediate in direct disagreement with the conclusions drawn from the experimental results [24–26]. Houk, Singleton, and coworkers [27] in an attempt to reconcile the disagreement between these dichotomous viewpoints have suggested that singlet oxygen ene reactions proceed via two transition states without an intervening intermediate; a topographical arrangement made possible because these two maxima on the reaction path to products are saddle points on a three dimensional energy surface. The first transition state does not involve hydrogen abstraction but leads to the second perepoxide-like transition state. The perepoxide-like transition state lies near a valley-ridge inflection where a bifurcation to the isomeric allylic hydroperoxide products occur. The two transition states are sufficiently isolated/separated from one another on the potential energy surface that minor structural perturbations such as isotopic substitution can selectively influence one but not the other transition state. The net result is a concerted reaction that behaves as if it proceeds via two kinetically distinguishable steps.

Despite the mechanistic controversy a recent review [4], pointed out that the reaction surface represented by the transformations shown in Sch. 3 is suitable as a template for the design of new regio- and stereospecific singlet oxygen ene reactions. In particular the structural constraints inherent in the perepoxide, **2**, provides a satisfying rationale for the suprafacial character of the reaction (i.e., addition of oxygen and abstraction of hydrogen occur from the same face of the alkene). This synthetically very useful stereochemical feature was elegantly demonstrated with the chiral alkene in Sch. 5 which only reacted to give the suprafacial products **A** and **B** and not the antarafacial products **C** and **D** [23].

12.3. SCOPE AND LIMITATIONS

The singlet oxygen ene reaction is a bimolecular process with a rate that depends on concentrations of both the substrate and singlet oxygen [28]. Convenient rates (i.e., sufficient concentrations of singlet oxygen) are achievable in many common organic solvents using photosensitization (Sch. 6). This can be accomplished, even at the very low concentrations of the sensitizers (10^{-4} M or less) which are used to prevent direct interaction with the substrates, since the quantum yields (ϕ_Δ) of singlet oxygen formation are

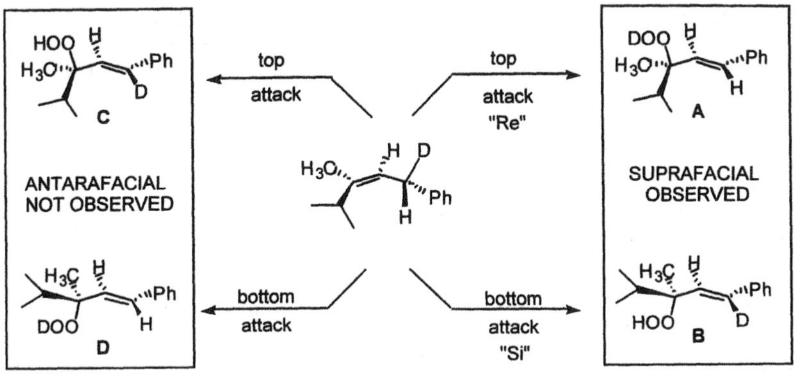

Scheme 5

$$\text{Sens} \xrightarrow{h\nu} {}^1\text{Sens}^*$$
$${}^1\text{Sens}^* \longrightarrow {}^3\text{Sens}^*$$
$${}^3\text{Sens}^* \xrightarrow{{}^3O_2} \text{Sens} + {}^1O_2$$
$${}^1\text{Sens}^* \xrightarrow{{}^3O_2} \text{Sens}/{}^3\text{Sens}^* + {}^1O_2$$

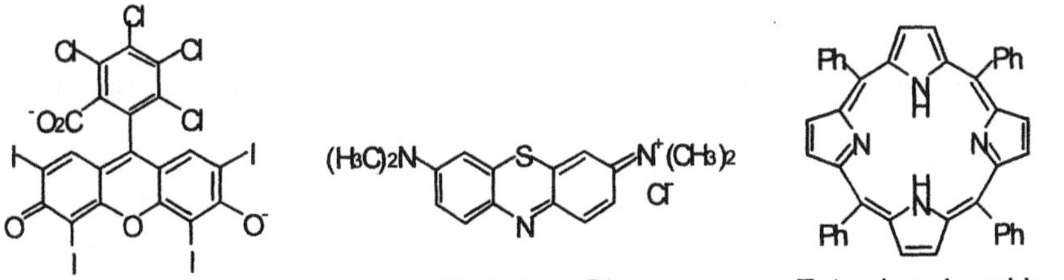

Rose Bengal
$\phi_\Delta(CH_3OH) = 0.8$

Methylene Blue
$\phi_\Delta(CH_3OH) = 0.51$

Tetraphenylporphine
$\phi_\Delta(C_6H_6) = 0.66$

Scheme 6

generally very high, as shown for three of the more commonly utilized sensitizers in Sch. 6 [29]. Product formation is even faster in deuterated solvents that do not have C–H or O–H bonds which deactivate singlet oxygen by an electronic to vibronic coupling mechanism [30–32]. However, even at high concentrations of singlet oxygen, synthetically useful ene reactions are successful only if the rates of competing reactions can be suppressed.

Rate constants for a large number of potentially competing singlet oxygen reactions have been reported by Wilkinson et al. [33]. A selection of representative examples is given in Table 1. Examination of this table reveals

Table 1   Rate Constants for the Deactivation of Singlet ($^1\Delta_g$) Oxygen.

Compound	Process	k $(M^{-1}s^{-1})$
H₃C, CH₃ / H₃C, CH₃ (tetramethylethylene structure)	Ene reaction[a]	3×10^7
MeO—C₆H₄—CH=CH—CH=CH—C₆H₄—OMe (structure)	4 + 2 Cycloaddition[b]	1.1×10^6
H₃CO, OCH₃ (structure)	2 + 2 Cycloaddition[c]	4.4×10^7
H₃CO, OCH₃ (structure)	2 + 2 Cycloaddition[c]	2.6×10^7
N,N-bicyclic diamine (DABCO structure)	Physical quenching[d]	2.9×10^8
Et₂S	Physical quenching[e]	2.1×10^7

[a]In acetone [30].
[b]In CCl₄ [34].
[c]Faler GR. Ph.D. thesis, Wayne State Univ., Detroit, MI, 1977, 157 p.
[d]In benzene [35].
[e]In CH₃CN [36].

that the rate constants for all these reactions *can be* similar in magnitude. Consequently, the successful use of the singlet oxygen ene reaction requires a very judicious choice of substrate and a thorough understanding of the factors that influence the rate constants of the reaction. These rate determining structural factors can be grouped into three categories; (1) conformational effects, (2) steric effects, and (3) electronic effects.

1.   *Conformational Effects.* The development of the incipient π-bond on the reaction surface for the formation of the allylic hydroperoxide provides a stringent conformational requirement on the alignment of the allylic hydrogen. The ene reaction does not occur in alkenes such as adamantane adamantylidene, **7**, in which the allylic hydrogens are structurally prevented from adopting a conformation which places them perpendicular to the alkene linkage. In the case of **7** exclusive 2+2 cycloaddition to give the dioxetane becomes the preferred mode of interaction with singlet oxygen [37]. The importance of conformational control on product composition is also evident during photo-oxidation of **8**. Singlet oxygen approaches **8** from the least hindered bottom face to give exclusively allylic hydroperoxide **9** [5] The formation

of **10** is not observed because of the lack of any low energy conformation that provides at the starred (*) carbon a perpendicular aligned allylic hydrogen on the bottom face of the molecule.

2. *Steric Effects.* Despite the fact that singlet oxygen is small, steric effects can reduce the rate of the singlet oxygen ene reaction and allow competitive reactions to occur. For example, the preferential approach to the bottom face of **8** is a result of the steric interaction with the angular methyl group on the top face of the molecule. A steric effect is also responsible for the reversal of the *exo/endo* hydroperoxide ratio in the photo-oxidations of the norbornene derivative **11** and **12** [38,39].

3. *Electronic Effects.* Singlet oxygen is an electrophilic oxidant that exhibits a clear preference for reactions with nucleophilic substrates. This preference is strikingly evident in a comparison of the rates constants for ene reactions of simple methyl substituted alkenes; 2,3-dimethyl-2-butene $(k = 2.2 \times 10^7 \, M^{-1} s^{-1})$ [19] reacts more than 30 times faster than the tri-substituted alkene 2-methyl-2-butene $(k = 7.2 \times 10^5 \, M^{-1} s^{-1})$ [19] and more than 500 times faster than the di-substituted alkene Z-2-butene $(k = 4.8 \times 10^4 M^{-1} s^{-1})$ [19]. The practical implications of these electronic effects are

dramatically illustrated by a comparison of the photo-oxidations of β-myrcene, **13**, and α-myrcene, **14** [40]. Ene reaction at the trisubstituted double bond in **13** competitively inhibits 4+2 cycloaddition with endoperoxide formation occurring only sub-sequent to allylic hydroperoxide formation. In contrast, the ene reaction at the disubstituted double bond is appreciably slower and can no longer compete with endoperoxide formation.

Through space electronic effects have also been observed during singlet oxygen ene reactions. This was elegantly demonstrated with a series of 7-isopropylidenebenzonorbornenes, **15a–d** [41]. In these cases the preference for anti addition in **15a** was reversed by decreasing the electron density in the aryl ring. Initially it was suggested that the predilection for anti addition in **15a** was a result of anchimeric assistance from the aromatic ring to the developing anti-perepoxide (**2** in Sch. 3). However, more recently the results have been attributed to a through space destabilizing electronic interaction between the π-cloud of the electron rich aromatic rings and the approaching singlet oxygen [42].

	15a	15b	15c	15d
anti/syn	79/21	80/20	48/52	46/54

Although the stringent conformational, electronic, and steric require-ments of the singlet oxygen ene reaction can be viewed as severe limitations they can also be taken advantage of to design regiospecific and stereospecific reactions as discussed in Sec. 12.4. In addition the singlet oxygen ene reaction is compatible with a remarkably wide range of functional groups including carbonyl, halogen, cyano [43], hydroxyl, and nitrogen containing groups (e.g., $-NH_2$, $-NH_3^+$, -NHAc, NHBoc, NHPhth, $NBoc_2$) [44], which dramatically extends its scope and synthetic utility. However, some problems are encountered as a result of the electronic character of these groups when they are directly attached to the alkene linkage. For example, α,β-unsaturated aldehydes [45], esters [46], ketones [47–49], amides [43], and sulfoxides [50] will all react with singlet oxygen albeit with reduce reactivity. On the other hand, electron rich, vinyl ethers [51,52], sulfides [53–57], and enamines [58] all have a tendency to react via 2+2 cycloaddition rather than via the ene mode with singlet oxygen. Adam et al. [53], have circumvented this problem to certain extent by demonstrating (Sch. 7) that decreasing electron density at sulfur in vinyl sulfides enhances the ene reactivity at the expense of 2+2 cycloaddition.

Cumulated double bonds such as those found in allenes, in contrast to isolated or conjugated double bonds, are very unreactive in the singlet oxygen

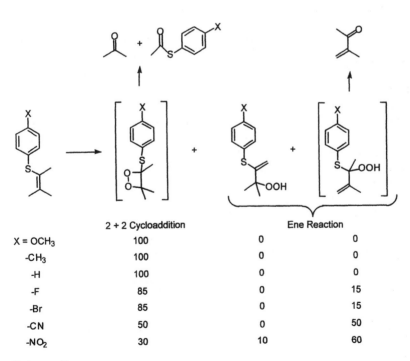

	2 + 2 Cycloaddition	Ene Reaction	
X = OCH_3	100	0	0
$-CH_3$	100	0	0
-H	100	0	0
-F	85	0	15
-Br	85	0	15
-CN	50	0	50
$-NO_2$	30	10	60

Scheme 7

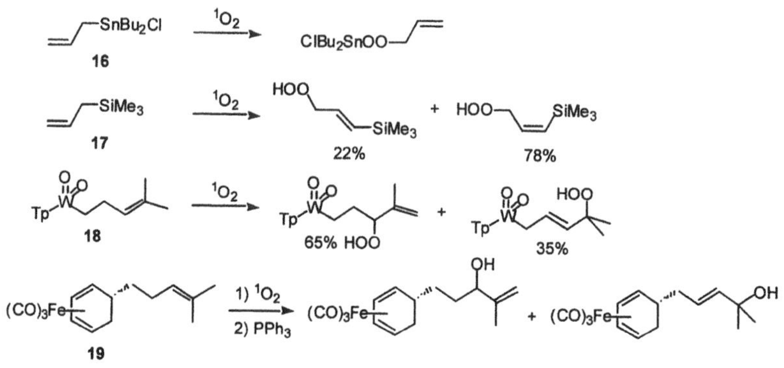

Scheme 8

ene reaction [59]. Products are formed during extended photo-oxidations of tetramethylallene [59] and some cyclic allenes [60], however, the complicated reaction mixtures render these reactions synthetically unattractive.

The difficulty in using the singlet oxygen ene reaction to functionalize organometallic ligands is a result of the propensity of very polar metal-carbon bonds to react rapidly with ground state triplet oxygen. Nevertheless, many organometallic complexes containing metals such as tin, iron, tungsten, or metalloids such as silicon can serve as singlet oxygen substrates when present in the right co-ordination environment (Sch. 8).

The singlet oxygen ene reaction of allylic stannane, **16**, is a synthetically useful procedure that leads to quantitative formation of the metal ene (M-ene) product [61–64]. However, other allylic stannanes, (e.g., **20** and **21**) with less electropositive tin centers generate both hydrogen ene (H-ene) and novel cycloaddition products along with the M-ene product. The complete absence of a M-ene product in the photo-oxygenation of **22** has also led to the speculation that the M-ene reaction is less important for allylic stannanes in which the tin is bonded to a 2° rather than to a 1° carbon [65–67]. On the other hand, the use of more polar solvents can be used to enhance M-ene product formation [68].

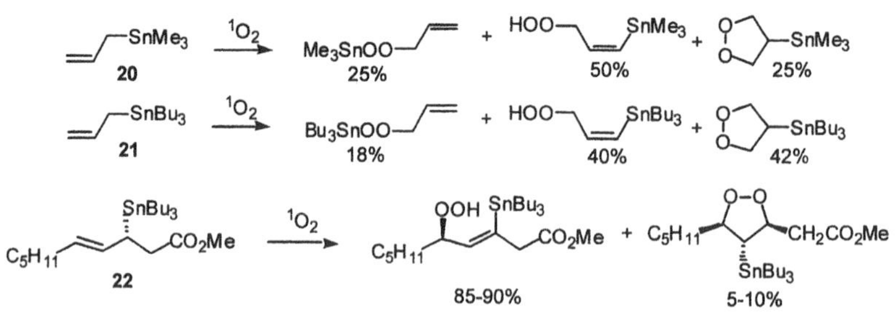

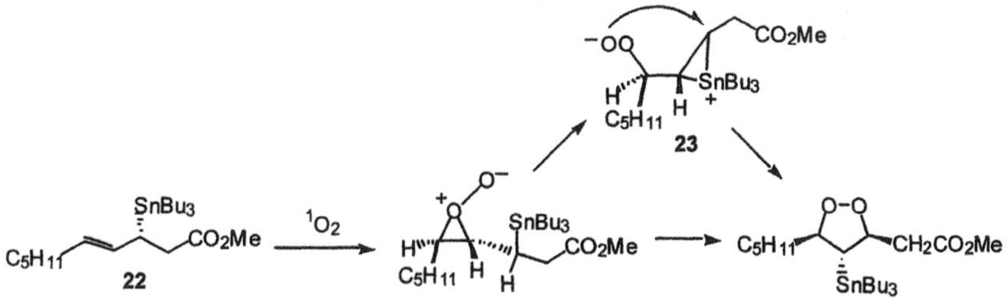

Scheme 9

Scheme 10

The formation of the dioxolanes in the photo-oxygenations of allylic stannanes with electron rich tin centers (i.e., compare **16** and **20**) can be attributed to the ability of tin to stabilize and migrate to an electron deficient β carbon (Sch. 9). The reduced yield of dioxolane in the reaction of **22** in comparison to **20** or **21** can be attributed to a steric effect operating in conjunction with an electronic effect of the carbomethoxy group in the bridged (or perhaps open) intermediate **23** which promotes hydrogen abstraction in lieu of sterically more demanding nucleophilic attack (Sch. 9).

1,2-Dioxolane formation is not observed in allyl silanes as a result of silicons' diminished ability in comparison to tin to stabilize a β positive charge. Nevertheless, the propensity of the carbon–silicon bond to donate electron density to electron deficient sites leads to population of a perpendicular conformation (Sch. 10) and ultimately to an unusually high yield of the sterically less stable Z-allylic hydroperoxide (e.g., **17**).

Allyl and homoallyl *tris*(pyrazolyl)borato-substituted (Tp) alkenyltungsten (Sch. 11 and **18**) react more rapidly with singlet oxygen than simple alkenes with the same substitution pattern [69]. In addition, allylic tungsten complexes react in a similar fashion to allyl silanes and stannanes showing an unusual proclivity for Z- rather than E-allylic hydroperoxide formation (Sch. 11). Z-Selectivity is presumably derived from hyperconjugative stabilization of the developing positive charge on the β-carbon atom.

Adam and Schuhmann [70] have reported the clever use of the iron tricarbonyl fragment as a protecting group (Sch. 12). Photo-oxygenation of **24** at low temperatures ($-80\,^\circ$C) resulted, after reduction of the allylic

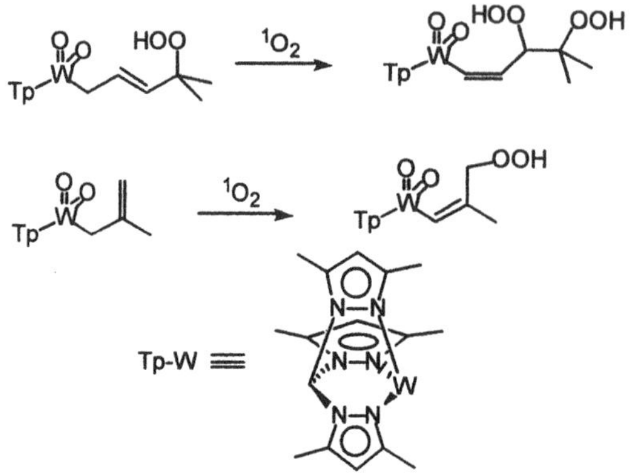

Scheme 11

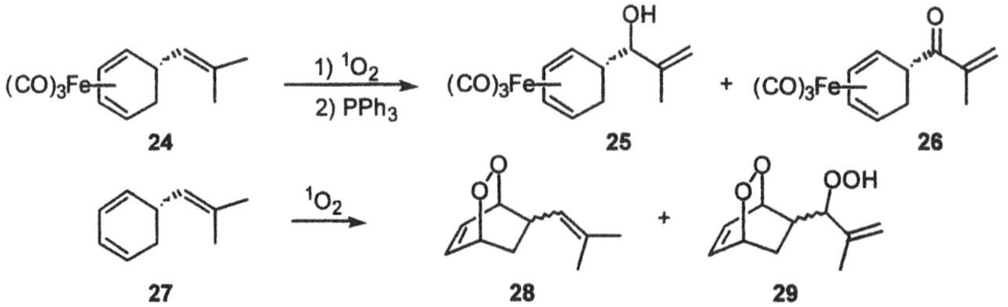

Scheme 12

hydroperoxide with PPh$_3$, in isolation of **25** in 83% yield as a 57:43 mixture of diastereomers. A trace of the enone, **26**, was obtained presumably as a result of the dehydration of the hydroperoxide. In contrast, direct photo-oxidation of **27** resulted in a complex mixture of endoperoxides **28** and **29**. The iron tricarbonyl fragment could be removed from **25** by treatment with cerium(IV) ammonium nitrate (CAN) in high yield. A major drawback to the procedure is the requirement to conduct the photo-oxygenation at low temperatures to prevent an intramolecular redox reaction between the iron and the hydroperoxy group that leads to extensive decomposition. This detrimental side reaction becomes increasingly severe as the distance between the iron center and the hydroperoxy group decreases.

Two additional examples of singlet oxygen ene reactions which further extends the scope of this valuable procedure are illustrated with **30** and **31**. The reaction of **30** demonstrates that the singlet oxygen ene reaction can be used to generate allenyl hydroperoxides in "significantly twisted" 1,3-dienes that have the vinyl hydrogen optimally aligned with the double bond for abstraction [71–74]. The reaction of **31** demonstrates that fullerene adducts can undergo self-sensitized singlet oxygen ene reactions [75,76]. Both steric and electronic effects of the fulleroid core on the ene regiochemistry have been noted.

12.4. SYNTHETIC POTENTIAL: REACTIVITY AND SELECTIVITY PATTERN

Complete stereochemical control to achieve perfect asymmetric allylic hydroperoxide formation using the singlet oxygen ene reaction is often the desired and/or the ultimate goal but to achieve it requires mastery of both the diastereoselectivity and enantioselectivity of the reaction. Unfortunately, despite significant progress to control the diastereoselectivity of the singlet oxygen ene reaction, little progress to control the enantioselectivity has been made. Efforts to achieve enantioselective formation of allylic hydroperoxides are hampered by the inability to generate "chiral singlet oxygen." As a consequence one is restricted to using a chiral environment in efforts to generate allylic hydroperoxides enantioselectivity. Attempts to employ chiral auxiliaries have been reported [77], however, these are intrinsically diastereoselective processes despite the fact that the overall result is formation of an excess of one enantiomer after removal of the chiral directing group.

Diastereoselective Singlet Oxygen Ene Reactions. The stereochemical features of the singlet oxygen ene reaction which need to be controlled in order to achieve diastereoselectivity are depicted in Sch. 13. They include: [1] the end of the double bond to which oxygen becomes bound (end selectivity), [2] the side of double bond from which hydrogen is abstracted (side selectivity), [3] the face of the double bond approached by singlet oxygen and from which hydrogen is removed (facial selectivity), and [4] the identity of the diastereotopic allylic hydrogen removed (abstraction selectivity).

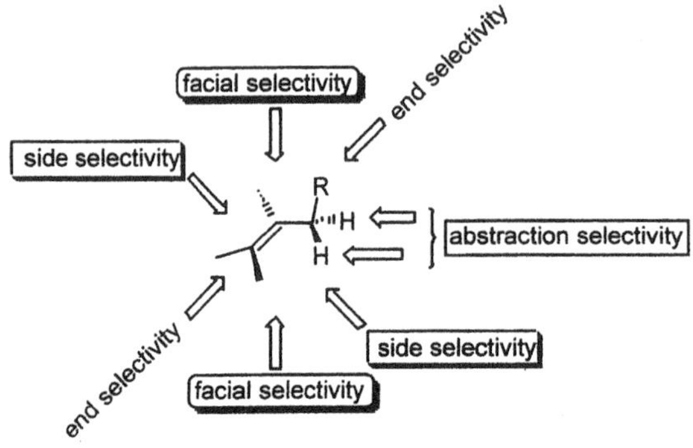

Scheme 13

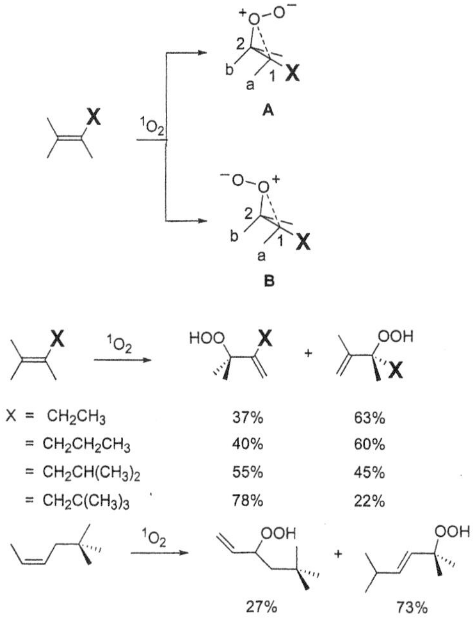

X =	CH$_2$CH$_3$	37%	63%
=	CH$_2$CH$_2$CH$_3$	40%	60%
=	CH$_2$CH(CH$_3$)$_2$	55%	45%
=	CH$_2$C(CH$_3$)$_3$	78%	22%

27%   73%

Scheme 14

(1) *End Selectivity.* Despite the fact that singlet oxygen is electrophilic, Markovnikov directing effects do not play a role in these reactions (see **32** in Sch. 16) and end selectivity is determined to a large extent by steric effects. The steric effect is illustrated by addition of singlet oxygen to an alkene with a substituent X in Sch. 14. As the size of X increases the ratio of perepoxides **A**/**B** decreases and the length of the C$_1$-oxygen bond increases promoting abstraction of hydrogen from the end of the alkene bearing the

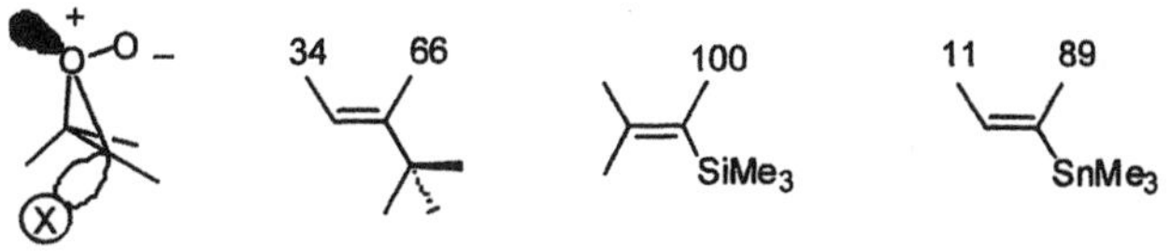

Scheme 15

substituent X. This phenomenon has been referred to in the literature as either the "geminal effect" [78] or the "large group nonbonding effect" [79].

A large number of different types of groups X including alkyl [78–82], carbonyl [43,45–49,83], aryl [84], sulfinyl [50], silyl [85–88], stannyl [89,90], cyano [43], and iminyl groups [91,92] are capable of promoting end selectivity. However, it is clear that superimposed upon the steric influence of these groups is an albeit minor stereoelectronic effect depicted in Sch. 15 [86]. The HOMO of the perepoxide is a linear combination of the σ-bond to the substituent and the lone-pair orbital on the perepoxide oxygen (Sch. 15). Any substituent that decreases the energy difference between the σ-bond and the lone-pair orbital will promote cleavage of the adjacent C–O bond. Electron withdrawing groups in particular are anticipated to lower the σ-bond energy bringing it into consonance with the lone pair energy. As a result of opposing steric (effective size; $Me_3Sn- < Me_3Si- < Me_3C-$) and electronic effects (highest energy σ-orbital; C–Sn > C–Si > C–C) in the alkenes shown in Sch. 15 a maximum in regioselectivity is observed for the vinyl silane [89].

(2) *Side Selectivity.* Several side selective singlet oxygen ene reactions are depicted in Sch. 16. In each case a surprising selectivity for abstraction of hydrogen from the most substituted side of the alkene is observed (99% in **32**; 81% in **33**; 93% in **34**; 92% in **35**; 100% in **36**; and 76% in **37**). This phenomenon, which was first recognized in enol ethers by Conia [93] and Foote [94], has become known as the "*cis* effect" [95,96]. Two different models have been suggested to explain the "*cis* effect." In the Stephenson/ Fukui model [97,98] a favorable HOMO-LUMO interaction between the trailing oxygen in the incipient perepoxide and the allylic hydrogens is maximized on the more congested side of the olefin. In an interesting suggestion, Houk et al. [99] suggested that the barrier to rotate the allylic hydrogen to the perpendicular geometry favorable for abstraction plays a dominate role in dictating the regiochemistry of the reactions. Their calculations demonstrated that the eclipsed geometry of a methyl group is destabilized and its rotation barrier lowered when buttressed by a *cis* substituent. Regardless of which model for the *cis* effect is correct, it is useful to remember that steric effects (vide supra) also operate and, in some highly conjested alkenes, can vitiate the stereoelectronic effect responsible for side selectivity [100].

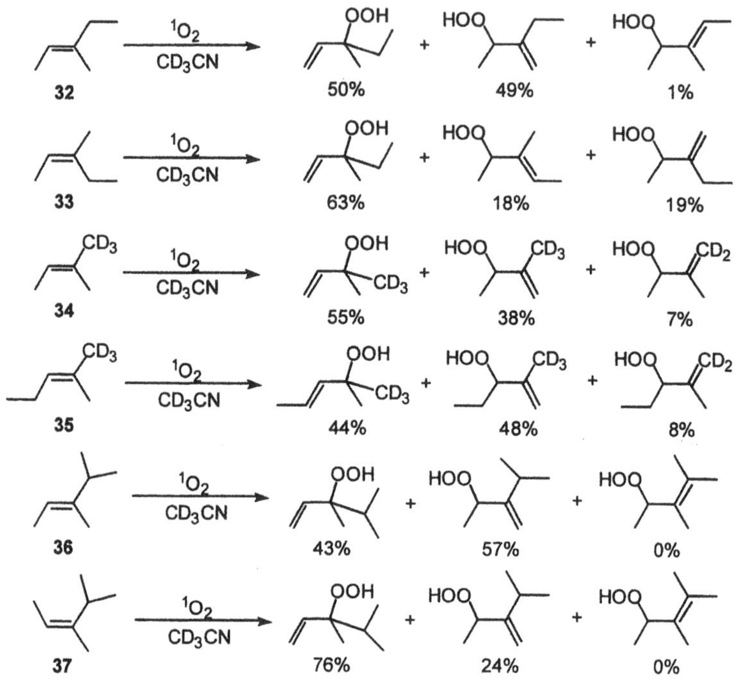

Scheme 16

(3) *Facial Selectivity.* Face selectivity, as anticipated by the preceding discussion, is a function of both the availability of a suitably aligned hydrogen and steric effects. Recent efforts to improve facial selectivity have also discovered that groups capable of hydrogen bonding can control the diastereoselectivity of the reaction [3,101–103]. These hydrogen bonding groups are most effective when they are placed on the allylic carbon close to the ene reaction center [104]. In the case of acyclic alkenes the steering effect of the hydrogen bonding group must also be coupled with a conformational bias for a single conformer of the alkene to generate synthetically useful diastereoselectivities. Adam and coworkers have used allylic strain ($A_{1,3}$ strain) to bias conformational populations in a series of allylic alcohols and amines [44,105–108]. A comparison of the allylic alcohol **38** and its acylated derivative **39** provides a direct measure of the remarkable steering efficiency of the hydrogen bond. The predominant formation of the SS* diastereomer in the reaction of **38** is consistent with the threo transition state with minimal allylic strain and energetically favorable hydrogen bonding. On the other hand the predominant formation of the S*R* diastereomer in the singlet oxygen ene reaction of **39** is consistent with

an erythro transition state which maximizes the *cis* effect (vide supra) while providing the minimal allylic strain. Alternative hydrogen-bonding sites in hydrogen-bond accepting solvents compete with singlet oxygen and diminish this useful steering effect [109,110].

(4) *Abstraction Selectivity.* Abstraction selectivity (Sch. 13) usually occurs to give the thermodynamically more stable *trans*-allylic hydroperoxide. This *trans*-selective regiochemistry is illustrated for an allylic amine in Sch. 17 which gives exclusively the *E*-rather than the *Z*-hydroperoxy styrenes [111]. Exceptions to this *trans*-selective regiochemistry has been observed in both allylic silanes and stannanes (see **17** in Sch. 8 and Sch. 17). In these cases the stereoelectronic preference for the allylic substituent to reside in a conformation placing it perpendicular to the alkene linkage provides ready access to the inside (i) rather than outside (o) hydrogen (Sch. 17) and as a consequence least motion collapse with abstraction of H_i leads specifically to the *E* rather than *Z* product.

Enantioselective Singlet Oxygen Ene Reactions. Progress in designing enantioselective singlet oxygen ene reactions lags far behind efforts to influence the diastereoselectivity of the reaction. Nevertheless, several enantioselective singlet oxygen ene reactions have been reported using supramolecular systems [112–114] or solid-state photo-oxidations [115]. Kuroda et al. [112,113] reported formation of allylic hydroperoxides with moderate enantiomeric excesses (e.e.) of 10–20% during photo-oxidations

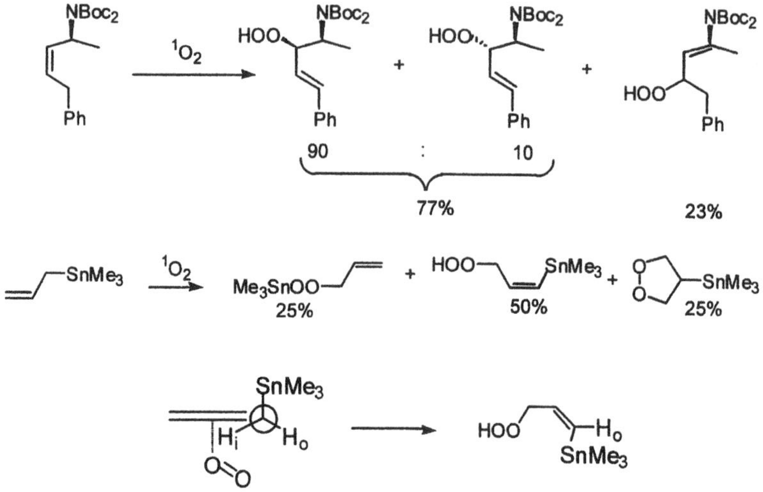

Scheme 17

of linoleic acid using porphyrin sensitizers covalently linked to β-cyclodextrin. Joy et al. [114] reported that photo-oxidation of **40** in (+)-ephedrine doped zeolite Y also generated an allylic hydroperoxide with a modest e.e. of approximately 15%. The lack of significant progress in designing effective enantioselective singlet oxygen ene reactions is in part a result of the small size of singlet oxygen and represents a significant future challenge for workers in the area.

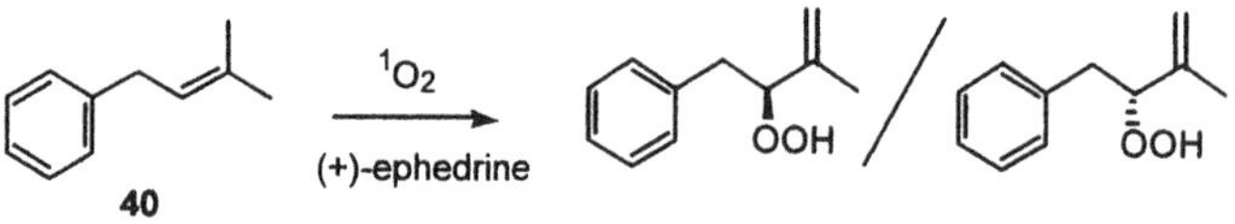

12.5. REPRESENTATIVE EXPERIMENTAL PROCEDURES

Solution Photo-oxidations. Singlet oxygen reactions can be conducted in water or in a wide variety of volatile organic solvents that are easily removed from the sensitive peroxidic products. Photo-oxidations in water suffer from both the low solubility of oxygen (1.0 mM) [116] and short lifetime of the reactive $^1\Delta_g$ state of singlet oxygen. Photo-oxidation rates are enhanced in organic solvents with higher attainable oxygen concentrations (CH$_3$CN 8.1 mM; hydrocarbons ≈ 10 mM; fluorocarbons ≈ 25 mM) [116]

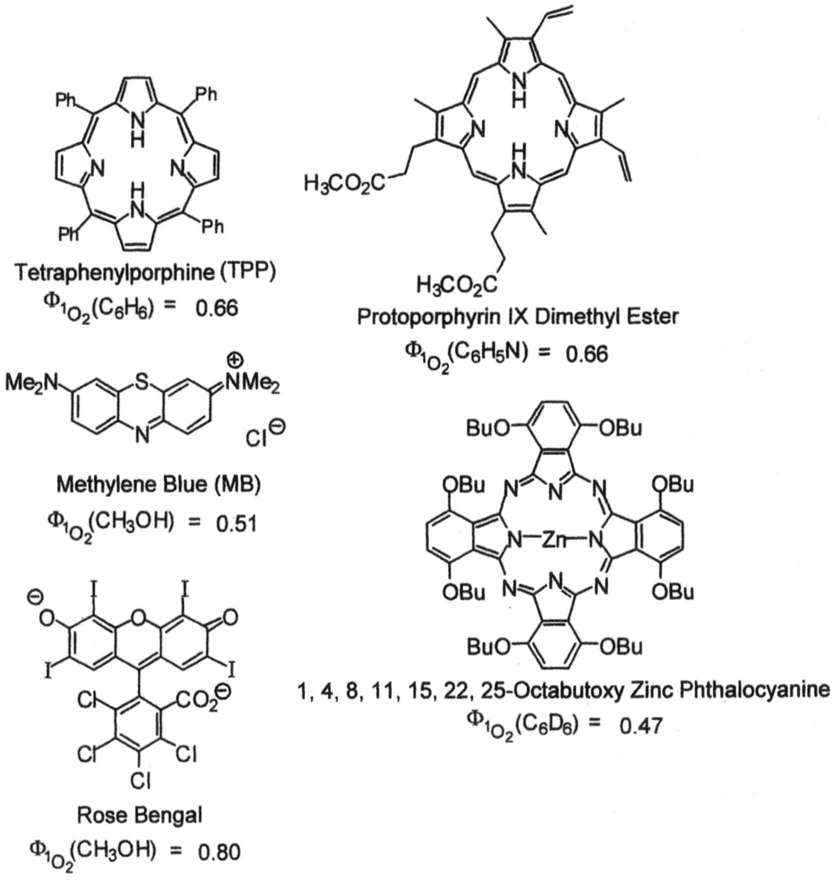

Tetraphenylporphine (TPP)
$\Phi_{^1O_2}(C_6H_6) = 0.66$

Protoporphyrin IX Dimethyl Ester
$\Phi_{^1O_2}(C_6H_5N) = 0.66$

Methylene Blue (MB)
$\Phi_{^1O_2}(CH_3OH) = 0.51$

1, 4, 8, 11, 15, 22, 25-Octabutoxy Zinc Phthalocyanine
$\Phi_{^1O_2}(C_6D_6) = 0.47$

Rose Bengal
$\Phi_{^1O_2}(CH_3OH) = 0.80$

Scheme 18

and in solvents devoid of C–H or O–H bonds which serve to deactivate singlet oxygen and reduce its lifetime and its maximum achievable steady state concentration [31].

Photosensitization is the method of choice for the generation of singlet oxygen on the preparative scale because of its simplicity and efficiency [28]. A large number of sensitizers are available, however, the most frequently encountered and generally useful ones are depicted in Sch. 18 [29]. These sensitizers will absorb visible light from a medium-pressure sodium street lamp and transfer their excitation energy to triplet oxygen. Alternatively, tungsten, xenon, or mercury lamps can also be used, but proper wavelength filtering is necessary in order to circumvent detrimental peroxide decomposition. Fortunately, a large number of glass and solution filters are available to remove UV light from a variety of light sources [117–119]. An immersion system provides maximum efficiency in the utilization of the lamp output and is especially attractive for preparative reactions [28].

Heterogeneous Photo-oxidations. The use of supramolecular systems to enforce "enzyme-like" organization on the activated complexes is a very attractive strategy (vide supra) for stereoselective synthesis given the small size, and high reactivity, of singlet oxygen. In addition, these supramolecular systems often provide the added benefit of a green process for synthesis of oxidized organic materials [120]. Several manuscripts have reported the use of organized media such as zeolites [114,121–133], micelles and vesicles [134–139], and polymer films [140–142] to influence the singlet oxygen ene reaction with varying degrees of success and should be consulted for experimental details.

Photo-oxidation of 2,3,5,5-tetramethyl-2-hexene, **41**. A 0.5 mL acetone-d_6 solution 5×10^{-2} M in **41** and 2×10^{-5} M in Rose Bengal was irradiated with a 750 W, 120 V tungsten-halogen lamp under continuous oxygen agitation at $-80\,^{\circ}$C through a 0.5% $K_2Cr_2O_7$ filter solution. The progress of the reaction was monitored by proton NMR. 2-(1-Hydroperoxy-1-methylethyl)-4,4-dimethyl-1-pentene, **42**, and 3-hydroperoxy-2,3,5,5-tetramethyl-1-hexene, **43**, were formed in 78 and 22% yield, respectively. The hydroperoxides were quantitatively reduced to the corresponding alcohol after treatment with one-equivalent of Ph$_3$P and were subsequently isolated by flash column chromatography.

Photo-oxidation of citronellol in polystyrene beads [120]. A sample of 3.0 g of polystyrene beads (commercial, cross-polymerized with 1% of divinylbenzene) was treated with a solution of 2 mg of tetraphenylporphyrin and 780 mg (5 mmol) of citronellol in 20 mL of ethyl acetate in a petri-dish (30 cm diameter). After 2 h in a ventilated hood, the solvent has evaporated and the petri-dish was covered with a glass plate and irradiated for 5 h with a 150 W halogen lamp. The solid support was then washed with 3×20 mL of ethanol, the combined ethanol fractions were rota-evaporated and 900 mg of the hydroperoxide mixture (96%) was isolated as a slightly yellow oil. The hydroperoxides were quantitatively reduced to the corresponding allylic alcohols by treatment with sodium sulfite. One of these products is used in the industrial synthesis of rose oxide.

12.6. TARGET MOLECULES: NATURAL AND NONNATURAL PRODUCT STRUCTURES

Synthetic applications of the singlet oxygen ene reaction have been addressed in several reviews [3,103,143]. The ability to reduce the allylic hydroperoxides to synthetically valuable allylic alcohols with a variety of reducing agents (LiAlH$_4$, NaBH$_4$, PPh$_3$, etc.,) provided the incentive behind the rapid development of this reaction. Its value as a key step in the synthesis of the diterpene Garryine, **44**, was demonstrated in the early 1960s by both Nickon [144] and Masamune [145]. More recent examples include its use in the synthesis of the lactone dihydromahubanolide B, **45**, isolated from the trunks of *Clinostemon mahuba* [146], in the syntheses of (±)*proto*-Quercitol, **46** [147], and (±)*gala*-Quercitol, **47** [147], via novel tandem singlet oxygen ene and [4+2] cycloadditions, and in the synthesis of *talo*-Quercitol, **48** [148].

ACKNOWLEDGMENT

We thank the National Science Foundation for their generous support of this research. We thank Axel G. Griesbeck for the procedure for preparative photo-oxygenation in polystyrene.

REFERENCES

1. Schenck GO. German Patent 933,925, 1943; In a paper entitled, Zur Theorie der photosensibilisierten Reaktion mit molekularem Sauerstoff. which appeared in Naturwissenschaften 1948; 35:28–29.
2. Schönberg A. Preparative Organic Photochemistry. Berlin: Springer, 1958.
3. Prein M, Adam W. Angew Chem Int Ed Engl 1996; 35:477–494.
4. Clennan EL. Tetrahedron 2000; 56:9151–9179.
5. See footnote on page 332 in Gollnick K. Adv Photochem 1968; 6:1–122.
6. Gollnick K, Kuhn HJ. In: Wasserman HH, Murray RW, eds. Singlet Oxygen. Vol. 40. New York: Academic Press, 1979:287–427.
7. Stephenson LM, Grdina MJ, Orfanopoulos M. Acc Chem Res 1980; 13:419–425.
8. Frimer AA, Stephenson LM. In: Frimer AA, ed. Singlet O_2. Reaction Modes and Products. Vol. II. Boca Raton, FL: CRC Press, 1985:67–91.
9. Stratakis M, Orfanopoulos M. Tetrahedron 2000; 56:1595–1615.
10. Harding LB, Goddard WA III. J Am Chem Soc 1980; 102:439–449.
11. Yamaguchi K, Yabushita S, Fueno T, Houk KN. J Am Chem Soc 1981; 103:5043–5046.
12. Manring LE, Foote CS. J Am Chem Soc 1983; 105:4710–4717.
13. Gollnick K, Griesbeck A. Tetrahedron Lett 1984; 25:725–728.
14. Orfanopoulos M, Stratakis M. Tetrahedron Lett 1991; 32:7321–7324.
15. Kopecky KR, Mumford C. Can J Chem 1969; 47:709–711.
16. Davies AG, Schiesser CH. Tetrahedron 1991; 47:1707–1726.
17. Sharp DB. Abstracts 138th Meeting of the American Chemical Society, New York, NY, 1960; 79.
18. Kopecky KR, Reich HJ. Can J Chem 1965; 43:2265–2270.
19. Hurst JR, Wilson SL, Schuster GB. Tetrahedron 1985; 41:2191–2197.
20. Gorman AA, Hamblett I, Lambert C, Spencer B, Standen MC. J Am Chem Soc 1988; 110:8053–8059.
21. Song Z, Chrisope DR, Beak P. J Org Chem 1987; 52:3938–3940.
22. Song Z, Beak P. J Am Chem Soc 1990; 112:8126–8134.
23. Orfanopoulos M, Stephenson LM. J Am Chem Soc 1980; 102:1417–1418.
24. Yoshioka Y, Yamada S, Kawakami T, Nishino M, Yamaguchi K, Saito I. Bull Chem Soc Jpn 1996; 69:2683–2699.
25. Okajima T. Nippon Kagaku Kaishi 1998; 2:107–112.
26. Sevin F, McKee ML. J Am Chem Soc 2001; 123:4591–4600.

27. Singleton DA, Hang C, Szymanski MJ, Meyer MP, Leach AG, Kuwata KT, Chen JS, Greer A, Foote CS, Houk KN. J Am Chem Soc 2003; 125:1319–1328.
28. Foote CS, Clennan EL. In: Foote CS, Valentine JS, Greenberg A, Liebman JF, eds. Active Oxygen in Chemistry. Vol. 2. New York: Blackie Academic & Professional, 1995:105–140.
29. Wilkinson F, Helman WP, Ross AB. J Phys Chem Ref Data 1993; 22:113–262.
30. Ogilby PR, Foote CS. J Am Chem Soc 1983; 105:3423–3430.
31. Rodgers MAJ. J Am Chem Soc 1983; 105:6201–6205.
32. Hurst JR, Schuster GB. J Am Chem Soc 1983; 105:5756–5760.
33. Wilkinson F, Helman WP, Ross AB. J Phys Chem Ref Data 1995; 24:663–1021.
34. Takahasi Y, Wakamatsu K, Kikuchi K, Miyashi T. J Phys Org Chem 1990; 3:509–518.
35. Iu K-K, Thomas JK. J Photochem Photobiol A Chem 1993; 71:55–60.
36. Nahm K, Foote CS. J Am Chem Soc 1989; 111:1909–1910.
37. Wieringa JH, Strating J, Wynberg H, Adam W. Tetrahedron Lett 1972; 169–172.
38. Jefford CW, Laffer MH, Boschung AF. J Am Chem Soc 1972; 94:8904–8905.
39. Jefford CW, Boschung AF. Helv Chim Acta 1974; 57:2242–2257.
40. Matsumoto M, Kondo K. J Org Chem 1975; 40:2259–2260.
41. Paquette LA, Bellamy F, Wells GJ, Böhm MC, Gleiter R. J Am Chem Soc 1981; 103:7122–7133.
42. Wu Y-D, Li Y, Na J, Houk KN. J Org Chem 1993; 58:4625–4628.
43. Adam W, Griesbeck A. Synthesis 1986; 1050–1052.
44. Brünker H-G, Adam W. J Am Chem Soc 1995; 117:3976–3982.
45. Adam W, Catalani LH, Griesbeck A. J Org Chem 1986; 51:5494–5496.
46. Orfanopoulos M, Foote CS. Tetrahedron Lett 1985; 26:5991–5994.
47. Ensley HE, Carr RVC, Martin RS, Pierce TE. J Am Chem Soc 1980; 102:2836–2838.
48. Kwon BM, Kanner RC, Foote CS. Tetrahedron Lett 1989; 30:903–906.
49. Ensley HE, Balakrishnan P, Ugarkar B. Tetrahedron Lett 1983; 24:5189–5192.
50. Akasaka T, Misawa Y, Goto M, Ando W. Tetrahedron 1989; 45:6657–6666.
51. Bartlett PD, Landis ME. In: Wasserman HH, Murray RW, eds. Singlet Oxygen. New York, NY: Academic Press, 1979:243–286.
52. Bartlett PD, Schaap AP. J Am Chem Soc 1970; 92:3223–3224.
53. Adam W, Kumar AS, Saha-Möller CR. Tetrahedron Lett 1995; 36:7853–7854.
54. Ando W, Suzuki J, Arai T, Migita T. Tetrahedron 1973; 29:1507–1513.
55. Ando W, Suzuki J, Arai T, Migita T. J Chem Soc Chem Commun 1972; 477–478.
56. Adam W, Liu J-C. J Chem Soc Chem Commun 1972; 73–74.
57. Adam W, Liu J-C. J Am Chem Soc 1972; 94:1206–1209.
58. Foote CS, Dzakpasu AA, Lin W-P. Tetrahedron Lett 1975; 1247–1250.
59. Gollnick K, Schnatterer A. Tetrahedron Lett 1985; 26:5029–5032.
60. Erden I, Song J, Cao W. Org Lett 2000; 2:1383–1385.
61. Dang H-S, Davies AG. Tetrahedron Lett 1991; 32:1745–1748.

62. Dang H-S, Davies AG. J Chem Soc Perkin Trans 2 1991; 2011–2020.
63. Dang H-S, Davies AG. J Chem Soc Perkin Trans 2 1992; 1095–1101.
64. Dang H-S, Davies AG. J Organomet Chem 1992; 430:287–298.
65. Dussault PH, Woller KR, HIllier MC. Tetrahedron 1994; 50:8929–8940.
66. Dussault PH, Zope UR. Tetrahedron Lett 1995; 36:2187–2190.
67. Dussault PH, Eary CT, Lee RJ, Zope UR. J Chem Soc Perkin Trans 1 1999; 2189–2204.
68. Kinart WJ, Kinart CM, Tylak I. J Organomet Chem 1999; 590:258–260.
69. Adam W, Putterlik J, Schuhmann RM, Sundermeyer J. Organometallics 1996; 15:4586–4596.
70. Adam W, Schuhmann RM. J Org Chem 1996; 61:874–878.
71. Mori H, Ikoma K, Masui Y, Isoe S, Kitaura K, Katsumura S. Tetrahedron Lett 1996; 37:7771–7774.
72. Mori H, Ikoma K, Katsumura S. J Chem Soc Chem Commun 1997; 2243–2244.
73. Mori H, Ikoma K, Isoe S, Kitaura K, Katsumura S. J Org Chem 1998; 63:8704–8718.
74. Mori H, Matsuo T, Yamashita K, Katsumura S. Tetrahedron Lett 1999; 40:6461–6464.
75. An Y-Z, Viado AL, Arce M-J, Rubin Y. J Org Chem 1995; 60:8330–8331.
76. Chronakis N, Vougloukalakis GC, Orfanopoulos M. Org Lett 2002; 4:945–948.
77. Adam W, Brünker H-G, Nestler B. Tetrahedron Lett 1991; 32:1957–1960.
78. Clennan EL, Chen X, Koola JJ. J Am Chem Soc 1990; 112:5193–5199.
79. Orfanopoulos M, Stratakis M, Elemes Y. J Am Chem Soc 1990; 112:6417–6419.
80. Orfanopoulos M, Stratakis M, Elemes Y. Tetrahedron Lett 1989; 30:4875–4878.
81. Clennan EL, Koola JJ, Oolman KA. Tetrahedron Lett 1990; 31:6759–6762.
82. Stratakis M, Orfanopoulos M. Synth Commun 1993; 23:425–430.
83. Adam W, Griesbeck A. Angew Chem Int Ed Engl 1985; 24:1070–1071.
84. Adam W, Richter MJ. Tetrahedron Lett 1993; 34:8423–8426.
85. Fristad WE, Bailey TR, Paquette LA. J Org Chem 1978; 43:1620–1623.
86. Fristad WE, Bailey TR, Paquette LA, Gleiter R, Böhm MC. J Am Chem Soc 1979; 101:4420–4423.
87. Fristad WE, Bailey TR, Paquette LA. J Org Chem 1980; 45:3028–3037.
88. Adam W, Richter MJ. J Org Chem 1994; 59:3335–3340.
89. Adam W, Klug P. J Org Chem 1993; 58:3416–3420.
90. Adam W, Klug P. J Org Chem 1994; 59:2695–2699.
91. Akasaka T, Takeuchi K, W A. Tetrahedron Lett 1987; 28:6633–6636.
92. Akasaka T, Takeuchi K, Misawa Y, Ando W. Heterocycles 1989; 28:445–451.
93. Rousseau G, LePerchec P, Conia JM. Tetrahedron Lett 1977; 2517–2520.
94. Lerdal D, Foote CS. Tetrahedron Lett 1978; 3227–3230.
95. Orfanopoulos M, Grdina SMB, Stephenson LM. J Am Chem Soc 1979; 101:275–276.
96. Schulte-Elte KH, Rautenstrauch V. J Am Chem Soc 1980; 102:1738–1740.
97. Stephenson LM. Tetrahedron Lett 1980; 21:1005–1008.
98. Inagaki S, Fujimoto H, Fukui K. Chem Lett 1976; 749–752.

99. Houk KN, Williams JC Jr, Mitchell PA, Yamaguchi K. J Am Chem Soc 1981; 103:949–951.
100. Orfanopoulos M, Stratakis M, Elemes Y, Jensen F. J Am Chem Soc 1991; 113:3180–3181.
101. Adam W, Wirth T. Acc Chem Res 1999; 32:703–710.
102. Adam W, Brünker H-G, Kumar AS, Peters E-M, Peters K, Schneider U, von Schnering HG. J Am Chem Soc 1996; 118:1899–1905.
103. Adam W, Saha-Möller CR, Schambony SB, Schmid KS, Wirth T. Photochem Photobiol 1999; 70:476–483.
104. Remote hydrogen bonding groups are capable of directing the facial selectivity of singlet oxygen in some cases. For example exceptional diastereoselectivity has been reported in the photo-oxidation of a cyclohexadiene bearing a homoallylic alcohol substituent. Linker T, Fröhlich L. J Am Chem Soc 1995; 117:2694–2697.
105. Adam W, Nestler B. J Am Chem Soc 1992; 114:6549–6550.
106. Adam W, Nestler B. J Am Chem Soc 1993; 115:5041–5049.
107. Adam W, Gevert O, Klug P. Tetrahedron Lett 1994; 35:1681–1684.
108. Adam W, Brünker H-G. J Am Chem Soc 1993; 115:3008–3009.
109. Vassilikogiannakis G, Stratakis M, Orfanopoulos M, Foote CS. J Org Chem 1999; 64:4130–4139.
110. Stratakis M, Orfanopoulos M, Foote CS. Tetrahedron Lett 1996; 37: 7159–7162.
111. Adam W, Brünker H-G. Synthesis 1995; 1066–1068.
112. Kuroda Y, Hiroshige T, Sera T, Shiroiwa Y, Tanaka H, Ogoshi H. J Am Chem Soc 1989; 111:1912–1913.
113. Kuroda Y, Sera T, Ogoshi H. J Am Chem Soc 1991; 113:2793–2794.
114. Joy A, Robbins RJ, Pitchumani K, Ramamurthy V. Tetrahedron Lett 1997; 38:8825–8828.
115. Gerdil R, Barchetto G. J Am Chem Soc 1984; 106:8004–8005.
116. Sawyer DT, Chiericato G Jr, Angelis CT, Nanni EJ Jr, Tsushiya T. Anal Chem 1982; 54:1720–1724.
117. Calvert JG, Pitts JN. Photochemistry. New York, NY: John Wiley & Sons, 1966:443–450.
118. Parker CA. Photoluminescence of Solutions. With Applications to Photochemistry and Analytical Chemistry. New York: Elsevier, Inc., 1968.
119. Gould IR. In: Scaiano JC, ed. CRC Handbook of Organic Photochemistry. Vol. I. Boca Raton, FL: CRC Press, Inc., 1989:37–117.
120. Griesbeck AG, Bartoschek A. J Chem Soc Chem Commun 2002; 1594–1595.
121. Li X, Ramamurthy V. Tetrahedron Lett 1996; 37:5235–5238.
122. Li X, Ramamurthy V. J Am Chem Soc 1996; 118:10666–10667.
123. Shailaja J, Sivaguru J, Robbins RJ, Ramamurthy V, Sunoj RB, Chandrasekhar J. Tetrahedron 2000; 56:6927–6943.
124. Robbins RJ, Ramamurthy V. J Chem Soc Chem Commun 1997; 1071–1072.
125. Ramamurthy V, Lakshminarasimhan P, Grey CP, Johnston LJ. J Chem Soc Chem Commun 1998; 2411–2424.

126. Clennan EL, Sram JP. Tetrahedron Lett 1999; 40:5275–5278.
127. Stratakis M, Froudakis G. Org Lett 2000; 2:1369–1372.
128. Clennan EL, Sram JP. Tetrahedron 2000; 56:6945–6950.
129. Pettit TL, Fox MA. J Phys Chem 1986; 90:1353–1354.
130. Joy A, Scheffer JR, Ramamurthy V. Org Lett 2000; 2:119–121.
131. Tung C-H, Wang H, Ying Y-M. J Am Chem Soc 1998; 120:5179–5186.
132. Pace A, Clennan EL. J Am Chem Soc 2002; 124:11236–11237.
133. Clennan EL, Sram JP, Pace A, Vincer K, White S. J Org Chem 2002; 67: 3975–3978.
134. Ehrenberg B, Anderson JL, Foote CS. Photochem Photobiol 1998; 68:135–140.
135. Lissi E, Rubio MA. Pure & Appl Chem 1990; 62:1503–1510.
136. Lissi EA, Encinas MV, Lemp E, Rubio MA. Chem Rev 1993; 93:699–723.
137. Hovey MC. J Am Chem Soc 1982; 104:4196–4202.
138. Horsey BE, Whitten DG. J Am Chem Soc 1978; 100:1293–1295.
139. Ohtani B, Nishida M, Nishimoto S, Kagiya T. Photchem Photobiol 1986; 44:725–732.
140. Lee PC, Rodgers MAJ. J Phys Chem 1984; 88:4385–4389.
141. Niu EP, Mau AWH, Ghiggino KP. Aust J Chem 1991; 44:695–704.
142. Tung C-H, Guan J-Q. J Am Chem Soc 1998; 120:11874–11879.
143. Wasserman HH, Ives JL. Tetrahedron 1981; 37:1825–1852.
144. Nickon A, Mendelson WL. J Am Chem Soc 1963; 85:1894–1895.
145. Masamune S. J Am Chem Soc 1964; 86:290–291.
146. Adam W, Klug P. Synthesis 1994; 567–572.
147. Salamci E, Secen H, Sütbeyaz Y, Balci M. J Org Chem 1997; 62:2453–2457.
148. Maras A, Secen H, Sütbeyaz Y, Balci M. J Org Chem 1998; 63:2039–2041.

13

Photogenerated Nitrene Addition to π-Bonds

H.-W. Abraham
Institut für Chemie der Humboldt-Universität zu Berlin,
Berlin, Germany

13.1. HISTORICAL REMARKS

Nitrenes of the general formula R–N are reactive intermediates containing a monovalent nitrogen atom with a sextet of electrons in its outer shell. A variety of names such as azenes [1], imine radicals [2], imene [3], and imido intermediates (imidogens) [4] have all been used by various authors, however the term "nitrene" is generally accepted in the literature. The R group may be an alkyl, aryl, acyl, a hetaryl, sulfonyl, phosphazyl, or an amino group. Nitrenes were first mentioned as reactive intermediates more than 100 years ago by Tiemann [5]. Nitrenes are isoelectronic to carbenes and this relationship heated up the investigation of nitrenes themselves and their addition to olefins in parallel with carbenes in the 1960s. As a result there are a number of reviews dealing with the generation of and chemistry of nitrenes [3a,4,6]. However, given all this study, only carbenes have been produced as stable entities, this development has yet to be reported for nitrenes.

13.2. MECHANISTIC MODELS

13.2.1. Generation of Nitrenes

In general, the reactive nitrenes can be generated by thermolytic or photolytic elimination of stable compounds from suitable precursors such as heterocycles, ylides, and azides. The most commonly used method for the generation of nitrenes is the thermolysis or photolysis of the corresponding azides, however, not all nitrenes are available by the thermolysis route. The decomposition temperature of azides is often high and a competing reaction of the azide or the intermediate nitrene may take place. Therefore, azide photolysis is perhaps the most popular method for making nitrenes [6d]. Nitrenes generated via the excited states of heterocyclic or azide precursors are formed in the ground state, therefore nitrene generation is the only photochemical key step in nitrene chemistry. The accessibility of the intermediate nitrenes depends on three parameters:

1. An electronic transition in the far UV with high extinction coefficients is necessary in order to conduct an effective photoreaction. The excitation energy of the direct photolysis depends on the group connected to the azide substituent; for example, in the gas phase, methyl azide exhibits an ultraviolet absorption maximum at 288 nm [7]. Chromophores such as aryl groups that are present in aryl, hetaryl, and aroyl azides offer the more convenient generation of the corresponding nitrenes. Such chromophoric groups often transfer excitation energy to the azide group, thus initiating the nitrogen elimination.

   As an alternative to the direct photolysis of azides, sensitized nitrogen elimination can be performed using triplet generators such as acetophenone and benzophenone yielding nitrenes in the triplet state [6b]. Singlet energy transfer yields singlet nitrenes [8a], whereas electron transfer yields singlet or triplet nitrene products depending on the sensitizer used [8b].

2. The quantum yield of the nitrogen elimination from azides is generally high [6e,9,10a], and in some cases may be far greater than unity at high azide concentrations [9,10b], the very high quantum yields reported for phenyl azide have yet to be confirmed by further experiments [10a].

3. The lifetime of the nitrene is governed by its stability; this in turn governs its reactivity with respect to added substrates. The most reactive nitrenes, i.e., those having electron acceptor substituents such as sulfonyl and acyl nitrenes are extremely short-lived [6b], In contrast, amino nitrenes that are stabilized by resonance can be stored for several hours at $-78\,°C$ [11].

13.2.2. Properties of Nitrenes

In general, nitrenes can be exploited in the amination of target molecules by insertion and addition reactions. The properties of nitrenes strongly depend on the substituent that is connected to the N-atom; the most reactive

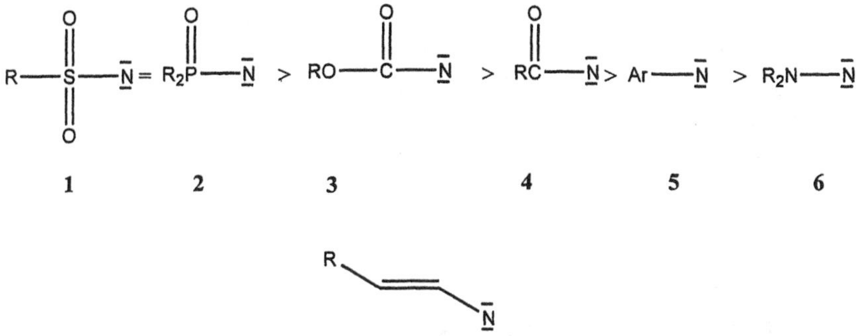

Scheme 1

nitrenes are the sulfonyl and acyl nitrenes, amino nitrenes are much less reactive (Sch. 1). The reactivity is not only determined by the energy content of the nitrene but is also dependent on the spin multiplicity of the unshared electrons. It is commonly assumed, that the triplet state is the ground state for nitrenes [10a], indeed, low temperature ESR spectra reveal the triplet state of aryl nitrenes [12], carbethoxycarbonyl nitrene, and sulfonylnitrenes [6b], whereas singlet ground states can only be indirectly inferred from the reactivity of the nitrenes. Thus, certain aroyl nitrenes and triazinyl nitrenes are assumed to react from the singlet state [10a].

The irradiation of azides mainly results in the formation of the singlet nitrene and the reactions of the singlet nitrene have to compete with the intersystem crossing process leading to the production of the triplet nitrene. Therefore intermolecular reactions with singlet state nitrenes may be possible at high nitrene quencher concentrations.

Knowledge regarding the multiplicity of the nitrenes is important because the reactivity is quite different in the singlet vs. the triplet state of the reactive intermediates:

- Singlet nitrenes such as intermediates 3–7 (Sch. 1) are known to undergo intra- and intermolecular insertion into C–H- and O–H-bonds as well as cycloaddition towards π-bonds.
- Alkylnitrenes often rearrange to form imines.
- Nitrenes can be trapped by compounds having lone electron pairs such as dimethylsulfoxide, sulfides, or pyridine.
- Triplet nitrenes mostly react by hydrogen abstraction (inter- and intramolecularly). Cycloaddition reactions with π-bonds occur nonstereoselectively.
- Triplet nitrenes are able to dimerize yielding azo compounds.

In order to use photogenerated nitrenes for selective cycloaddition towards π-bonds, a number of possible side reactions must to be suppressed. In particular, the choice of the solvent is very important as it can serve as reaction partner. Singlet nitrenes are expected to be more suitable than triplet nitrenes for the above cycloaddition, to date, the addition of nitrenes **1**, **2**, and **6** to double bonds has not been demonstrated.

13.3. SCOPE AND LIMITATIONS

13.3.1. Alkyl Nitrenes

The only known example of a useful preparative intramolecular cycloaddition reaction is the formation of *cis,cis*-trialkyltriaziridines by the reaction of the nitrene with the N=N-double bond (Sch. 2) [13]. In this example, the fixed position of the reactants enables the fast intramolecular reaction.

13.3.2. 1-Alkenyl Nitrenes

Alkenyl as well as alkyl nitrenes can be produced both by thermolysis and photolysis. However, photolysis is preferred due to the more favorable reaction conditions such as low temperature and the ability to use of particular solvents, which avoid side reactions such as the polymerization observed under thermolytic conditions. However, in some cases, alkenyl nitrenes can only be generated by photolysis [14]. Photochemically generated alkenyl nitrenes normally react by intramolecular addition to the π-bond forming 2H-azirines at rather high yields (Sch. 3) [6d,15], in addition, the

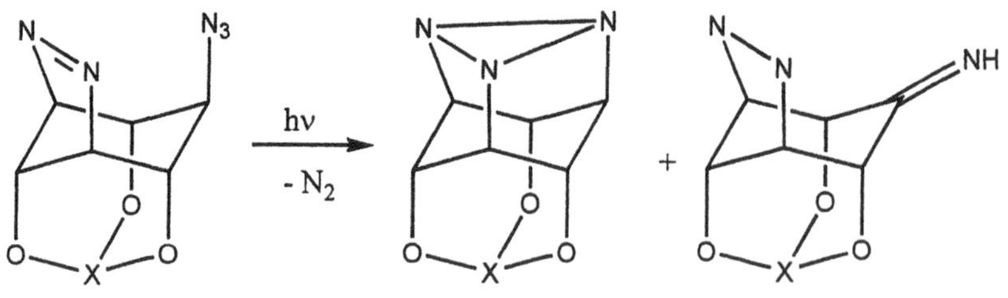

X = CH, CCH₃, PO, P

8	**9** (20 - 65%)	**10** (10 - 20%)

Scheme 2

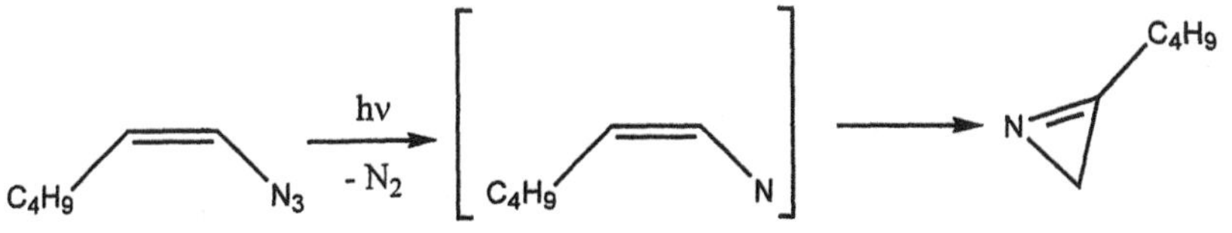

11   **12**   (81%)

Scheme 3

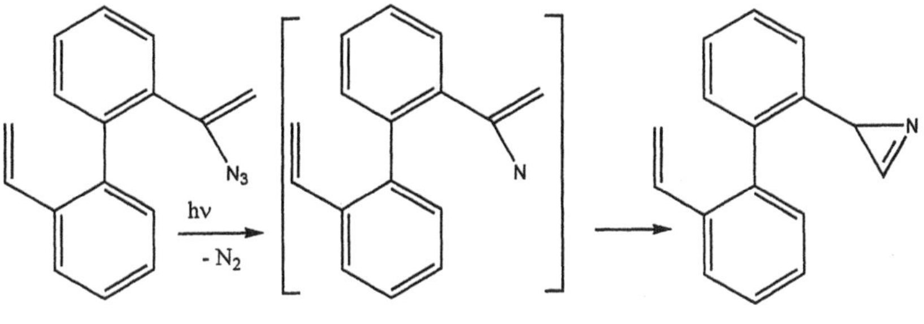

13   **14**   (80%)

Scheme 4

irradiation of the azide **13** results only in intramolecular nitrene addition (Sch. 4).

13.3.3. Aryl Nitrenes

The reactions of singlet and triplet aryl nitrenes differ significantly [16]. Triplet nitrenes generated by photolysis of azides, sensitized by aldehydes or ketones react intermolecularly resulting in the formation of anilines and diazo compounds. In general, singlet aryl nitrenes cannot be quenched by the addition of alkenes, as nitrenes react intramolecularly through addition to the aromatic π-system yielding azepines (compound **16**) and anilines (compound **17**) upon addition of nucleophiles (Sch. 5). When diethylamine is used as nucleophile, 2-diethylamino-3H-azepines are produced with good yields (Sch. 6).

13.3.4. Heteroaryl Nitrenes

Heteroaryl nitrenes can be produced by thermolysis and photolysis, however, the reaction pathways are fundamentally different [17]. Under thermolytic conditions, ring cleavage is obtained, however, photolytically

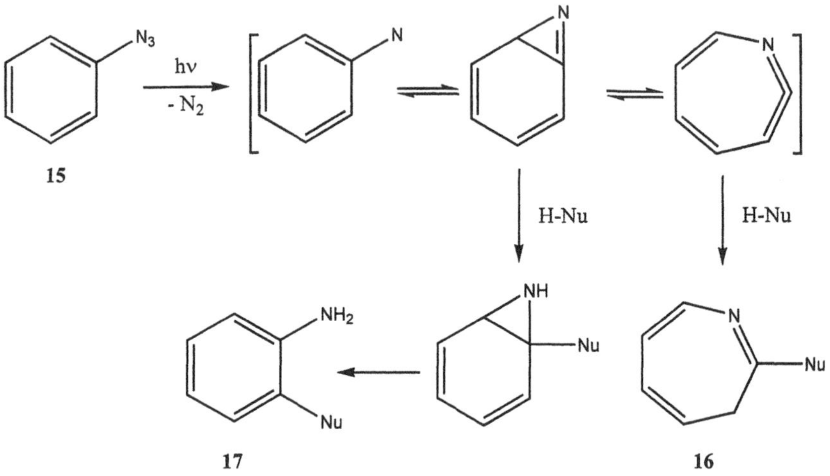

Scheme 5

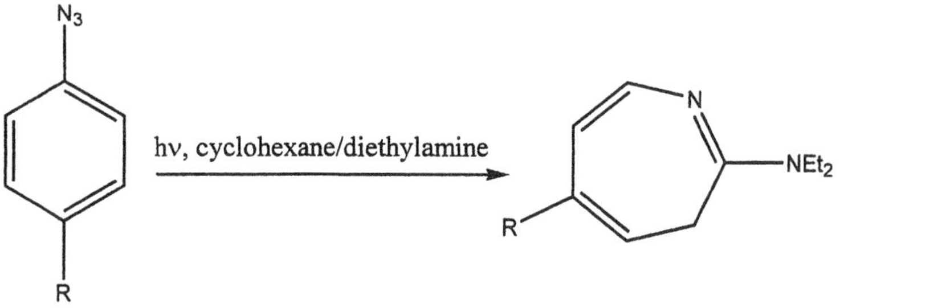

15 R = H
18 R = CO-NMe₂

16 R = H, Nu = NEt₂ (80%)
19 R = CO-NEt₁ (91%)

Scheme 6

generated heteroaryl nitrenes behave like aryl nitrenes, in that singlet nitrenes exhibit addition reactions and triplet nitrenes undergo H-abstraction reactions. It is not clear whether heteroaryl nitrenes exist in a singlet or a triplet ground state. There are some results indicating, that the ground state of the triazinyl nitrene is a singlet, and accordingly, addition reactions have mainly been observed [10a] where the heterocycle is involved, such as for compound **21** (Sch. 7) [18]. Azidoisoquinolines such as compound **23** (Sch. 8) react intramolecularly analogous to arylnitrenes, resulting in the formation of benzodiazepines [19].

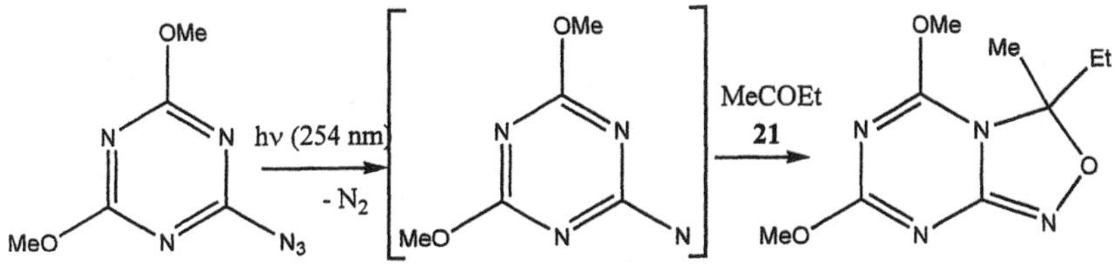

Scheme 7

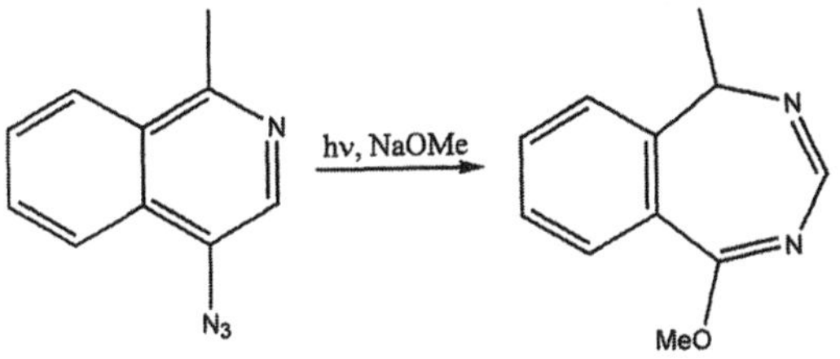

Scheme 8

13.3.5. Acyl Nitrenes

Acyl nitrenes can be photolytically generated from heterocycles by incorporating suitable leaving groups or from the corresponding azides (Sch. 9) [6d]. However, only the azide method has been employed where acyl nitrene is used for preparative purposes. Among the acyl nitrenes depicted in Sch. 10, only intermediates **3** and **26** can be generated both by the thermolysis and photolysis of the corresponding azide. All the other nitrenes are only produced via photolysis due to the isocyanate being thermolytically formed (Curtius rearrangement). In the excited state, almost equal proportions of the nitrene and isocyanate are formed; however the nitrene is not the intermediate of the rearrangement product.

The nitrene **28** is not produced from the azide precursor, but from heterocycles via photolysis and thermolysis as shown in Sch. 11 [20]. Iminoacyl nitrenes react intramolecularly giving benzimidazoles with good yields (Sch. 11), and, dependending on the precursor used and the reaction conditions, varying amounts of carbodiimides are obtained. The reactivity of the acyl nitrenes is influenced by the substituent connected to the acyl group (see Sch. 10), however all acyl nitrenes are quite reactive and therefore rather unselective. Apart from cycloaddition reactions with π-bonds, insertion reactions into σ-bonds, additions to lone pair electrons of

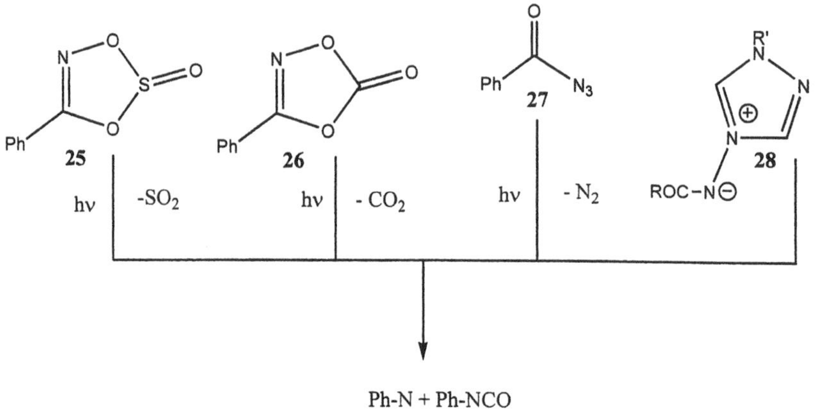

Scheme 9

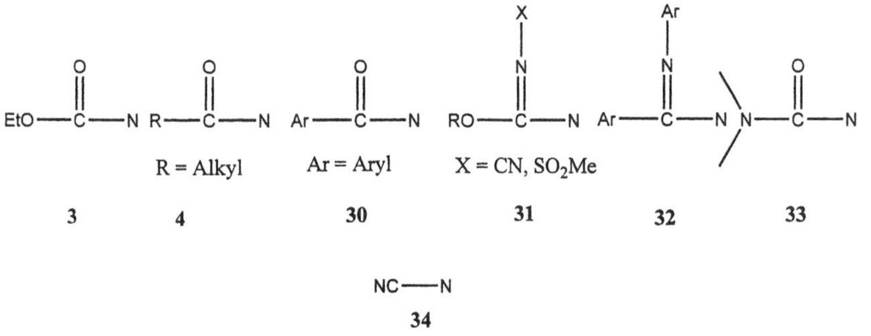

Scheme 10

heteroatoms, and H-abstraction reactions (with triplet nitrenes) can also be expected. In so far there are a number of reaction possibilities with the solvent, the choice of the solvent for the cycloaddition reaction plays an important role in selective addition reactions [21]. Apart from the alkoxy-carbonyl azides, an additional drawback of using acylazides is the high yield of isocyanate formed in parallel with the nitrene. These competing reactions via the singlet excited state of the corresponding azides cannot be avoided. In addition, substituents present on the aryl group of aroyl azides only partially control the wasteful formation of the isocyanate. It also seems that electron donor substituents favor isocyanate formation. For example, *p*-dimethylaminophenyl azide exclusively yields the corresponding isocyanate upon irradiation in acetonitrile solution (Sch. 12) [22].

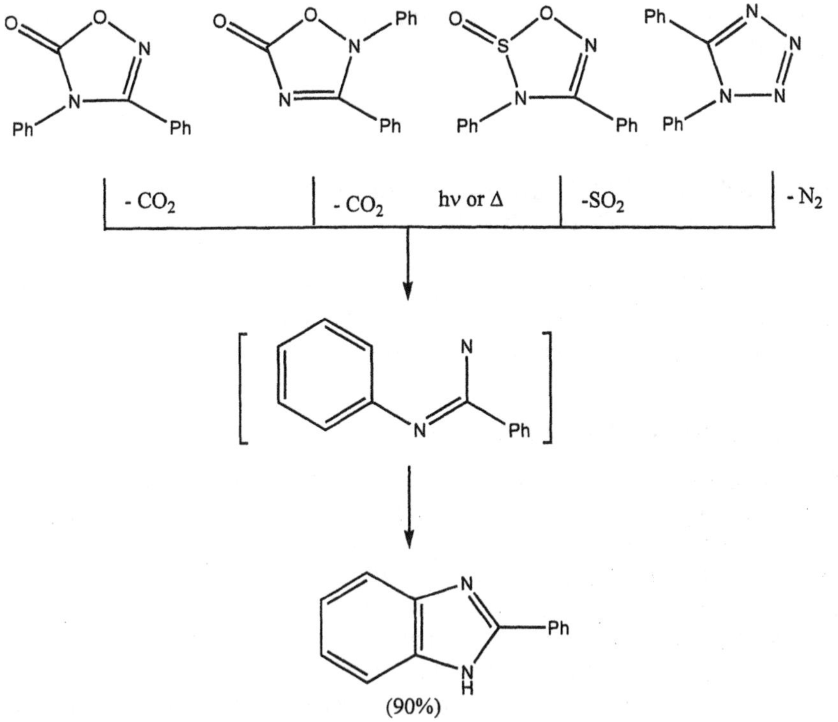

Scheme 11

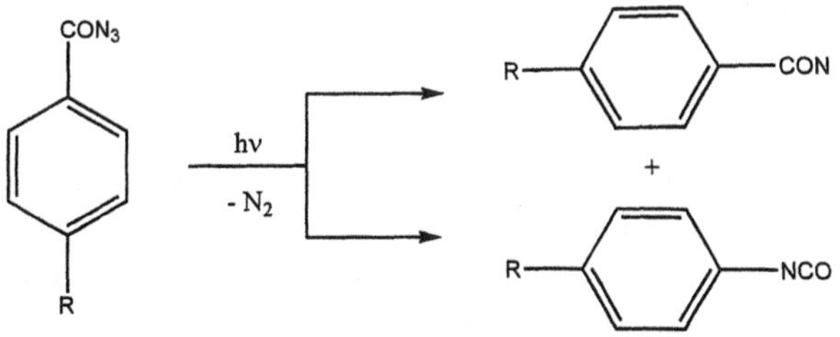

	R	% isocyanate	% nitrene
41	CN	53	47
42	Br	58	36
27	H	49	46
43	OMe	56	30
44	NMe$_2$	100	0

Scheme 12

Accordingly, the choice of the acyl azide and of the solvent must be carefully planned.

13.4. SYNTHETIC POTENTIAL: REACTIVITY AND SELECTIVITY PATTERNS

Acyl azides which produce the nitrenes **3** and **4** upon photolysis are the most suitable for synthetic applications of cycloaddition reactions. Therefore, this subchapter will mainly deal with alkoxycarbonyl and aroyl nitrenes.

13.4.1. Alkoxycarbonyl Nitrenes

Olefinic double bonds react with most nitrenes to form aziridines. However, alkoxycarbonyl nitrenes are not very suitable for making aziridines as they are too reactive, forming various C–H insertion products often resulting in the aziridine yield being low (an exception is given in Sch. 13) [23,24]. Furthermore under photolytical conditions, one-third of the nitrene is generated in the triplet state that reacts nonstereospecifically [25].

   The outcome of the nitrene addition reaction depends on the type of π-bond involved. In contrast to electron deficient olefins [26] and nonpolar olefins forming aziridines, electron rich olefins react with alkoxycarbonyl nitrenes to give oxazolines (Sch. 14) [22]. The same type of cycloaddition reaction leading to the production of five-membered rings has also been observed with nitriles [27] (such as compound **35** in Sch. 14) and isocyanates [28] as illustrated in Sch. 15.

13.4.2. Alkoxyimidoyl Nitrenes

For the more selective production of aziridines, alkoxyimidoyl nitrenes of the type represented by species **26** may be used [29]. The photolysis of the corresponding azides results in greater yields than the thermolysis and the less reactive alkoxyimidoyl nitrenes add *cis*-stereoselectively to olefins (see Sch. 16).

13.4.3. Alkylcarbonyl Nitrenes

Alkylcarbonyl nitrenes have only rarely been used for intermolecular cycloaddition reactions. 2,2-Dimethylpropanoyl nitrene reacts in the usual manner with cyclohexene giving the aziridine (45%) as well as the

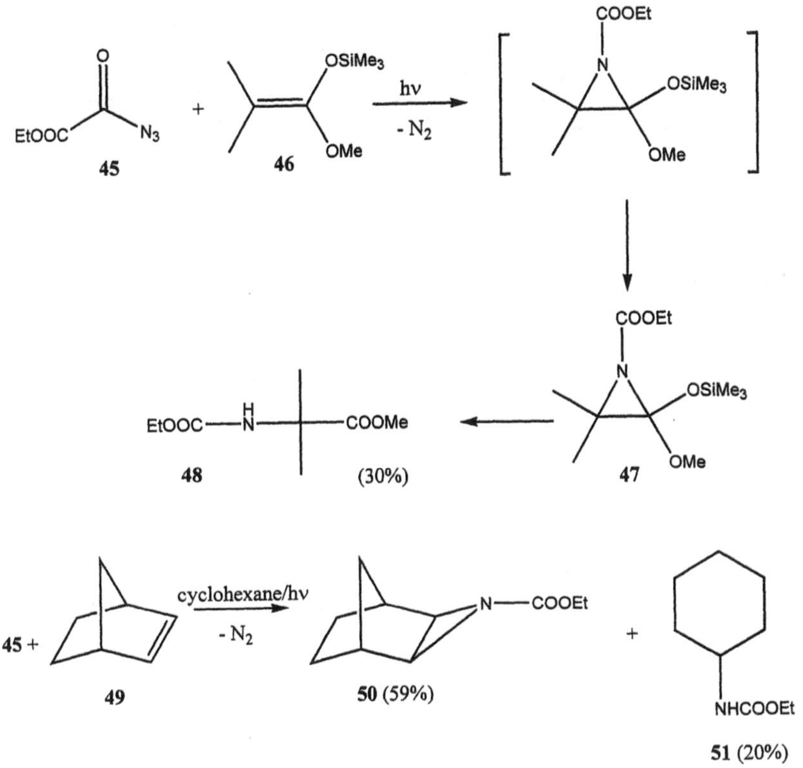

Scheme 13

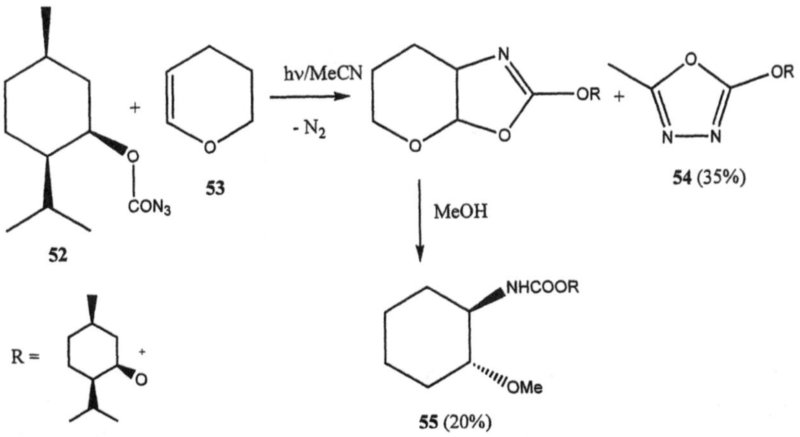

Scheme 14

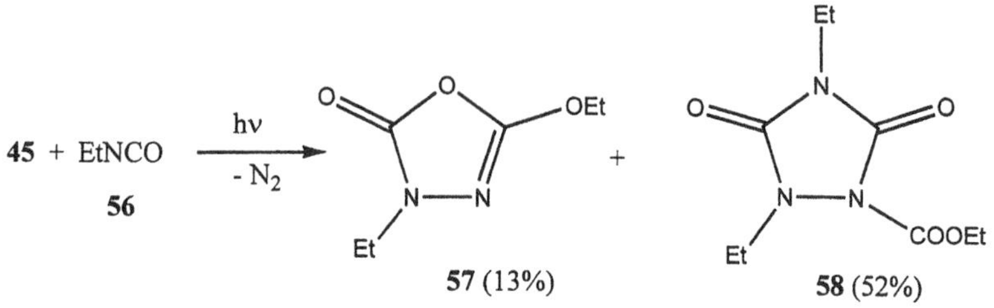

Scheme 15

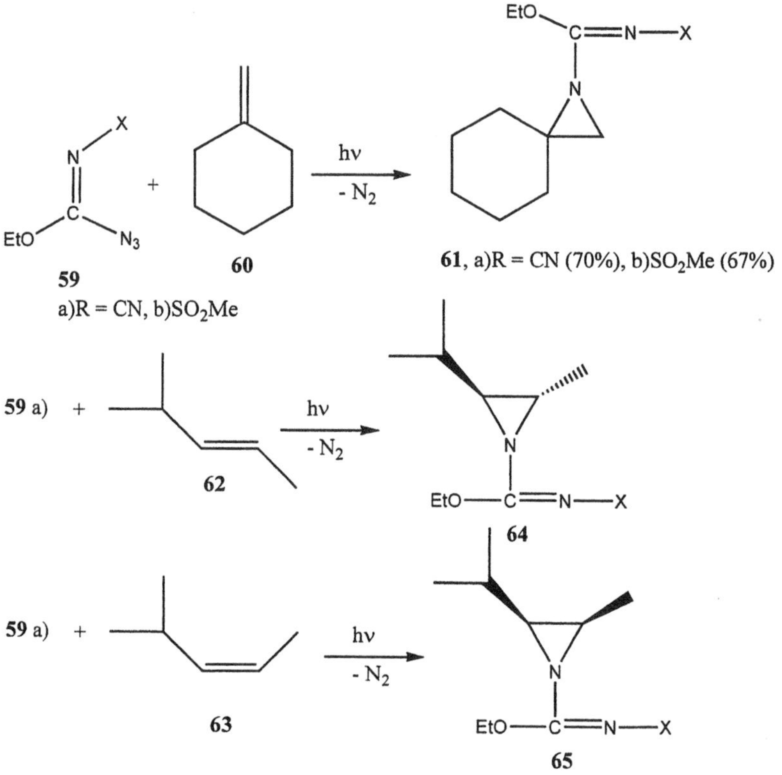

Scheme 16

isocyanate (41%) [30]. *tert*-Butylcarbonyl azide (compound **66**, Sch. 17) reacts with *cis*-2-butene in dichloromethane (in a 1:1 ratio) giving stereoselectively 38% of the *cis*-aziridine and 50% of the corresponding isocyanate. In contrast, with the *trans*-olefin (compound **70**), the triplet

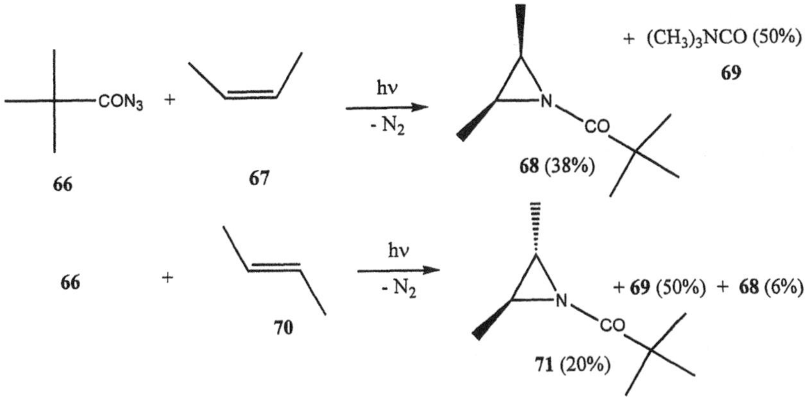

66 CON₃ + 67 $\xrightarrow[-N_2]{h\nu}$ 68 (38%) + (CH₃)₃NCO (50%) 69

66 + 70 $\xrightarrow[-N_2]{h\nu}$ 71 (20%) + 69 (50%) + 68 (6%)

Scheme 17

products, compound **68** and 2,2-dimethylpropionyl amide were obtained (probably by an H-abstraction reaction) (Sch. 17) [31].

13.4.4. Addition of Aroyl Nitrenes to Nonpolar and Electron Deficient π-Bonds

Aroyl nitrenes can be generated from their corresponding azides by three methods:

- Direct photolysis.
- Sensitized decomposition by singlet or triplet energy transfer, yielding singlet or triplet nitrenes that may undergo intersystem crossing.
- Electron transfer from excited electron donors in the singlet or the triplet state to aroylazides (Sch. 18) [6e].

Both the singlet energy and electron transfer methods suffer from competing isocyanate formation. Energy and electron transfer in the triplet state avoids this rearrangement reaction which can result in higher yields of cycloaddition products. However, in both sensitization modes, side reactions of the triplet nitrene such as H-abstraction reactions have to be taken into account.

Both singlet and triplet nitrenes add to C=C double bonds, but by different mechanisms. The stereospecific addition of singlet aroyl nitrenes occurs in a single step. Triplet nitrenes add in two steps and therefore isomerization by rotation about the former C=C-bond may lead to a total loss of the geometrical information in the starting material. The standard

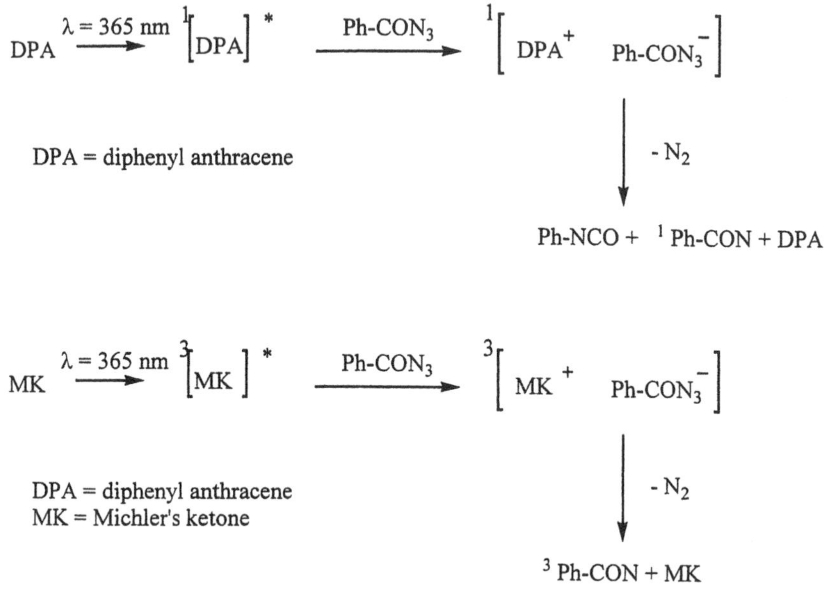

$$DPA \xrightarrow{\lambda = 365 \text{ nm}} {}^{1}\left[DPA\right]^{*} \xrightarrow{Ph\text{-}CON_3} {}^{1}\left[DPA^{+} \quad Ph\text{-}CON_3^{-}\right]$$

DPA = diphenyl anthracene

$\downarrow \; -N_2$

$$Ph\text{-}NCO + {}^{1}Ph\text{-}CON + DPA$$

$$MK \xrightarrow{\lambda = 365 \text{ nm}} {}^{3}\left[MK\right]^{*} \xrightarrow{Ph\text{-}CON_3} {}^{3}\left[MK^{+} \quad Ph\text{-}CON_3^{-}\right]$$

DPA = diphenyl anthracene
MK = Michler's ketone

$\downarrow \; -N_2$

$${}^{3}Ph\text{-}CON + MK$$

Scheme 18

example for this cycloaddition reaction is the formation of aziridines by the irradiation of compound **27** in olefins such as cyclohexene or cis- or trans-4-methyl-2-pentene (compounds **62** and **63**) (Sch. 19). Two results in this reaction are important, these are that aziridines are formed exclusively and the cycloadditions occur stereoselectively [32]. The last finding can be explained by the formation of a singlet nitrene that attacks the olefin before the intersystem crossing process takes place. However, it is not clear whether the benzoyl nitrene exists as a singlet or a triplet in the ground state. From its reactivity both a singlet [33] and a triplet state [9,34] can be concluded. In addition, substituted benzoyl nitrenes attack cyclic olefins such as compound **76** [22] and compound **77** [21] (Sch. 20) and form three-membered rings. Aziridinofullerenes such as compound **79** have been synthesized by the photolysis of the corresponding azide in the presence of C_{60} in 1,1,2,2-tetrachloroethane [35], thus the electron deficient nature of the π-bonds of C_{60} has been confirmed. The 6-6-ring fused fulleroaziridines can be thermally transformed into the fullero-oxazoles represented by compound **81** (Sch. 20).

The formation of an aziridine has also been assumed to occur during the photolysis of the dichloromethane solution of benzoyl azide in the presence of diketene giving the spiro intermediate (see Sch. 21) [36]. The

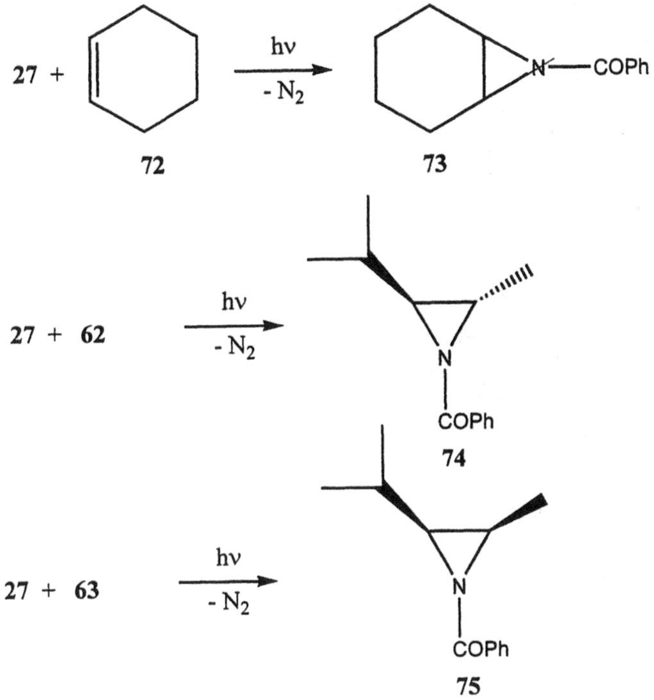

Scheme 19

rapid rearrangement of the primary product results in the pyrrolin-2-one (compound **82**).

13.4.5. Formation of Five-Membered Rings

It is often the case that the oxygen and nitrogen atom of the acyl nitrene are involved in the cycloaddition reaction. In this respect, acyl nitrenes are able to react like a 1,3-dipole [37]; however this is different to other species commonly classified as 1,3-dipoles as they are not octet stabilized.

Intermediate three-membered rings have never been detected in the cycloaddition reaction and a direct route to the five-membered rings is probable in most cases. For example, benzoyl nitrene adds to triple bonds and in contrast to the addition to nonpolar double bonds, a five-membered oxazole is obtained (Sch. 22) [21,38]. An azirine intermediate is not detected, however it cannot be excluded that the three-membered ring containing a π-bond rearranges rapidly giving rise to the oxazole.

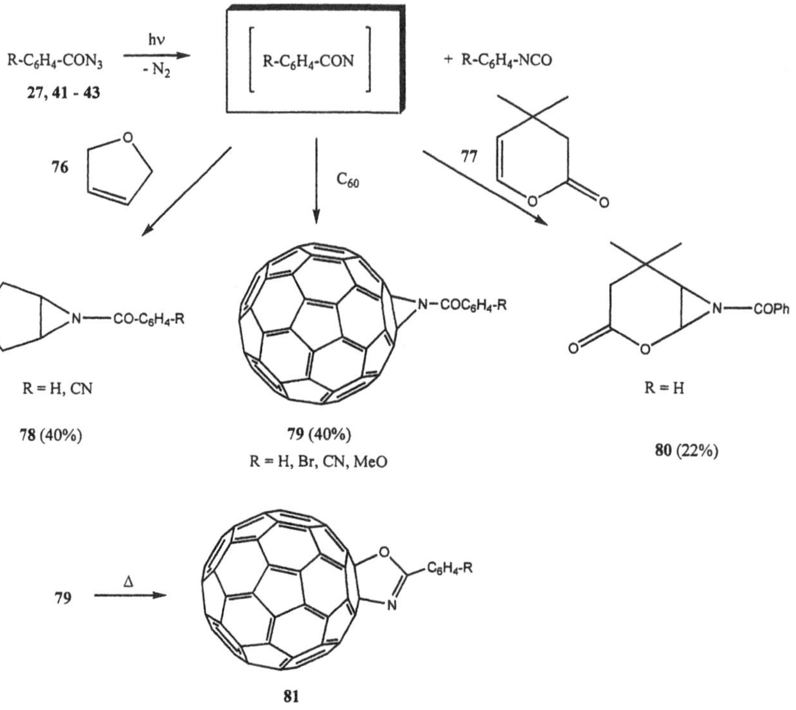

Scheme 20

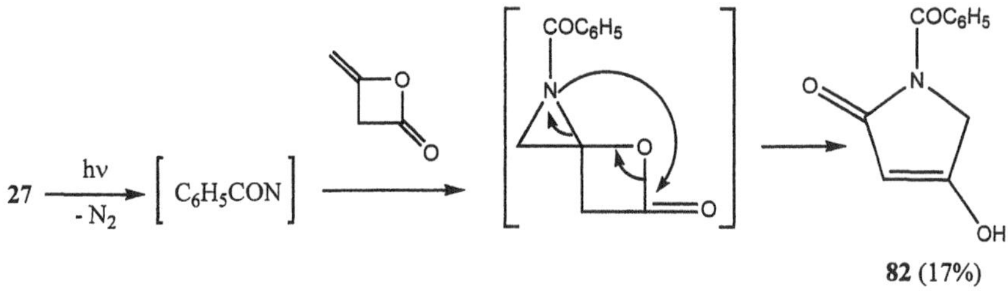

Scheme 21

The cycloaddition of aroyl nitrenes to electron rich double bonds such as in dihydropyran (compound **85**) [Sch. 23] also results in oxazoline formation without any reference to an intermediate aziridine [22,39]. The same holds true for polar π-bonds of ketones and aldehydes that are added to aroyl nitrenes forming dioxazolines such as compound **89** (Sch. 24) [40].

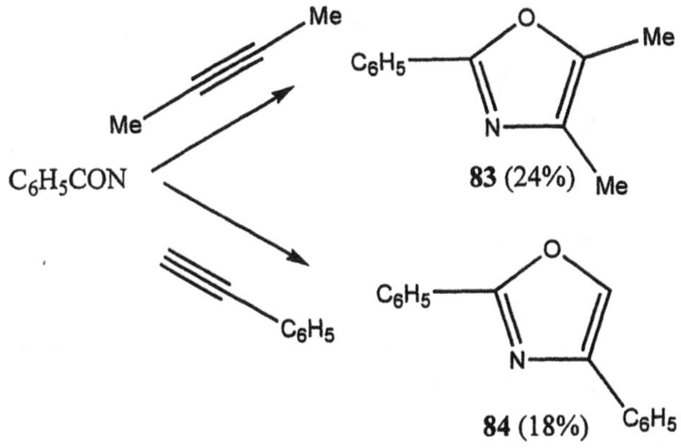

Scheme 22

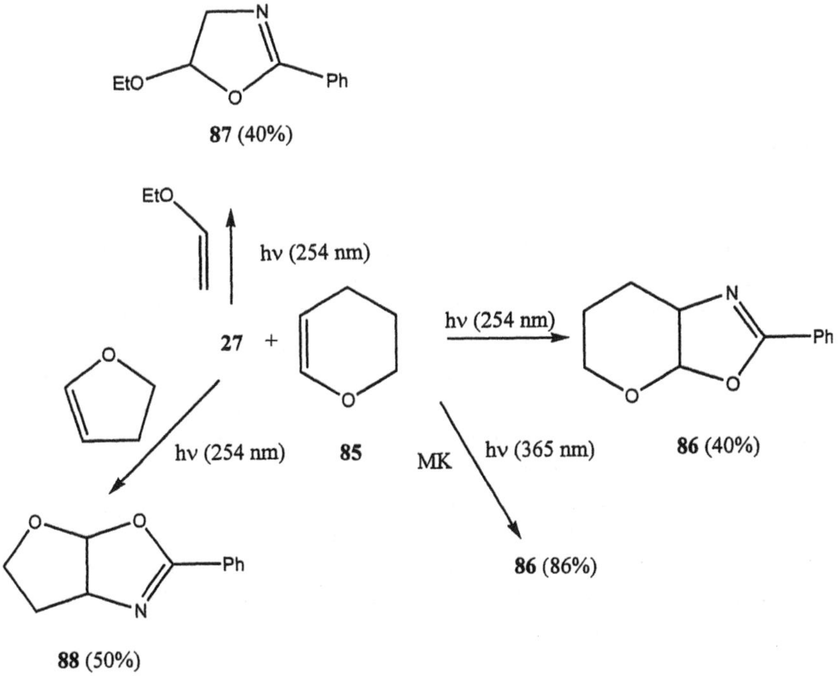

Scheme 23

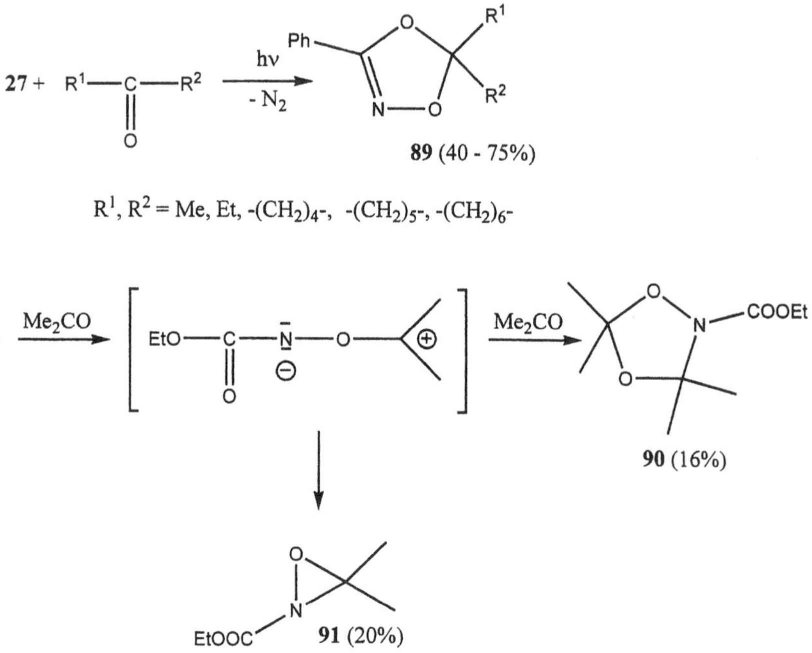

89 (40 - 75%)

$R^1, R^2 = $ Me, Et, -(CH$_2$)$_4$-, -(CH$_2$)$_5$-, -(CH$_2$)$_6$-

90 (16%)

91 (20%)

Scheme 24

In contrast to the above, the more reactive ethoxycarbonyl nitrene **4** is able to attack the carbonyl group of ketones forming three-membered rings such as compound **90** shown in Sch. 24. The ground state for this nitrene is the triplet state, meaning that the cycloaddition reaction occurs in at least two steps. Indeed is has been found that acetone can be attacked by the reactive intermediate **4** primarily at the carbonyl O-atom (see Sch. 24); the dipolar intermediate is able to add a second acetone molecule to yield the dioxazoline (compound **91**) [21].

In contrast to that, benzoyl nitrene reacts with acetone yielding the five-membered ring even upon triplet sensitization by excitation of the carbonyl compound with light at wavelengths of >300 nm [20]. Thus, the reactivity of the aroyl nitrene seems to be principally different from that of the ethoxycarbonyl nitrene.

The triple bond of nitriles is attacked by aroyl nitrenes to give rise to oxadiazoles as illustrated in Sch. 25 [22,41]. However, additions of acyl nitrenes to olefinic double bonds can be carried out in aceto-nitrile solution because the cycloaddition reaction to the solvent is much slower.

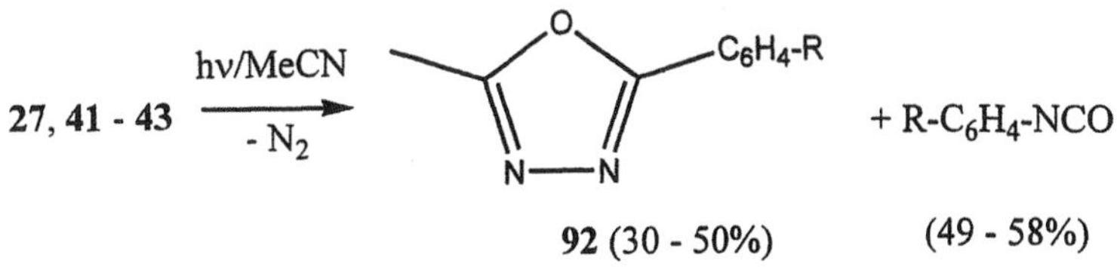

$$27, 41 - 43 \quad \xrightarrow[- N_2]{hv/MeCN} \quad \text{(structure 92)} \quad + \text{R-C}_6\text{H}_4\text{-NCO}$$

92 (30 - 50%)   (49 - 58%)

Scheme 25

By considering the product yields it has been found that the reactivity sequence of various bond types with respect to benzoyl nitrene is as follows:

$$O-H > C=C-OR > C=O = C\equiv C > C-H > C\equiv N \text{ [21]}.$$

Accordingly, cycloaddition reactions can be carried out in solvents such as alkanes, cycloalkanes and acetonitrile. The insertion reaction into the O–H-bond of alcohols is, in every case, faster than the cycloaddition reaction. Furthermore, 100% production of isocyanate was observed upon irradiation of benzoyl azide in dichloromethane solution, however, nitrenes that can be trapped by compounds with double bonds are also formed in this solvent [21].

The question arises, which bond of the reaction partners, containing C=C, C=O-, and C–H-bonds, is attacked by aroyl nitrenes. It has been shown that unsaturated carbonyl compounds such as compounds **93** and **95** react exclusively at the carbonyl group to give rise to dioxazolines (Sch. 26) [21]. Thus, the reactivity of the benzoyl nitrene is similar to sulfonium ylides that also react in a kinetically controlled manner with the carbonyl group instead with the olefinic double bond [42]. The yield of compound **93** can be increased to 80% by taking advantage of the sensitized decomposition of compound **27** using Michler's ketone [21]. The competing reaction of the nitrene with the solvent is not totally suppressed.

13.4.6. Stereochemistry

The cycloaddition of acyl nitrenes towards electron-rich double bonds creates two stereogenic centers. It has already been noted that singlet and triplet nitrenes exhibit a divergent stereoselectivity; whereas singlet nitrenes react with noncyclic alkenes retaining the configuration of the initial olefin, triplet nitrenes have to add in two steps yielding two diastereomers (see the kinetic Sch. 27) [6b]. Accordingly the concentration of the olefinic quencher of the singlet nitrene may affect the stereochemical outcome of the

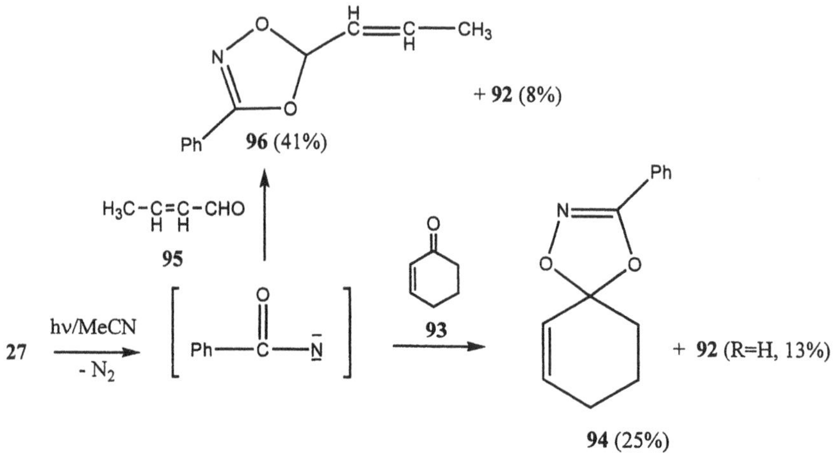

Scheme 26

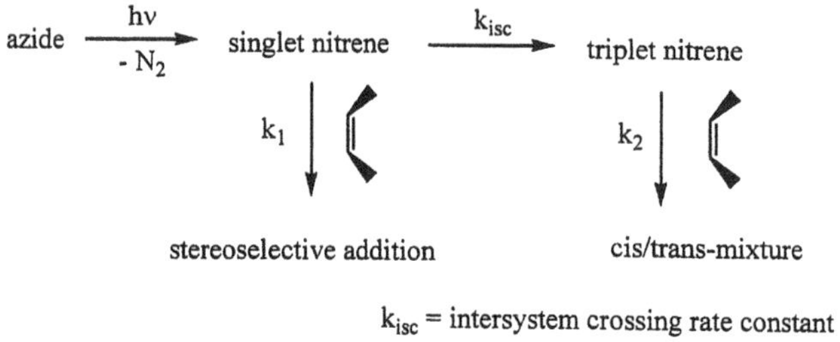

k_{isc} = intersystem crossing rate constant

Scheme 27

cycloaddition reaction because the intermolecular reaction (k_1x[alkene]) has to compete with k_{isc} and k_2x [alkene].

13.4.6.1. Chiral Acylazides

Some attempts have been made to modify diastereoselectivity by introducing chiral substituents into the azide precursor of the nitrene (see Sch. 28) [22,43,44]. The photocycloaddition of acyl nitrenes bearing chiral substituents to cycloalkenes having enantiotopic faces such as compound **85**, or prochiral ketones, can lead to the formation of two diastereomers. However, this chiral induction has not been observed in the reaction of the nitrenes

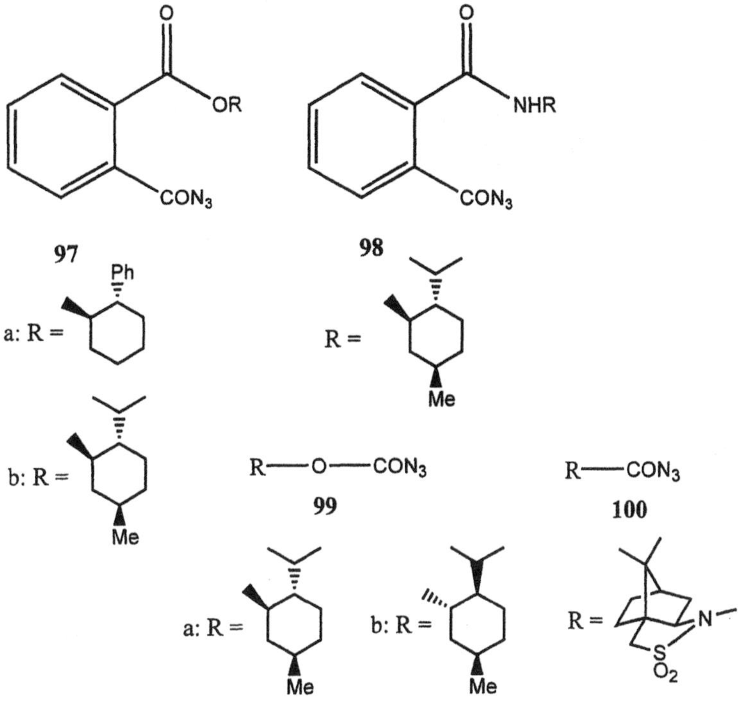

Scheme 28

derived from the azide compounds **97–99** with either compound **85** or unsymmetrical ketones [43,45] In contrast, the acyl nitrene derived from the azide compound **100** exhibits high diastereoselectivity (d.e. >90%) of the cycloaddition towards methyl-cyclohexene (Sch. 29) [44]. Unfortunately the chemical selectivity of the carbamoyl nitrene is rather low.

13.4.6.2. Chiral Alkenes

The photolysis of nonchiral acyl azides such as compounds **45** [22] (Sch. 13), **27**, **41**, and **43** [22,45] (Sch. 12) in the presence of the substituted dihydropyranes (compound **104**, used as a racemate) yields two diastereomers that can be characterized as "*exo-*" and "*endo-*" compounds (**105**) according to the position of the five-membered ring in relation to the alkoxy substituent (see Sch. 30). The d.e. values are about 30% and are not affected by the size of the alkoxy group of the dihydropyran. In contrast, substituents present in benzoyl azides do influence the diastereoselectivity (compounds **41**: 15%; **43**: 10%) [46].

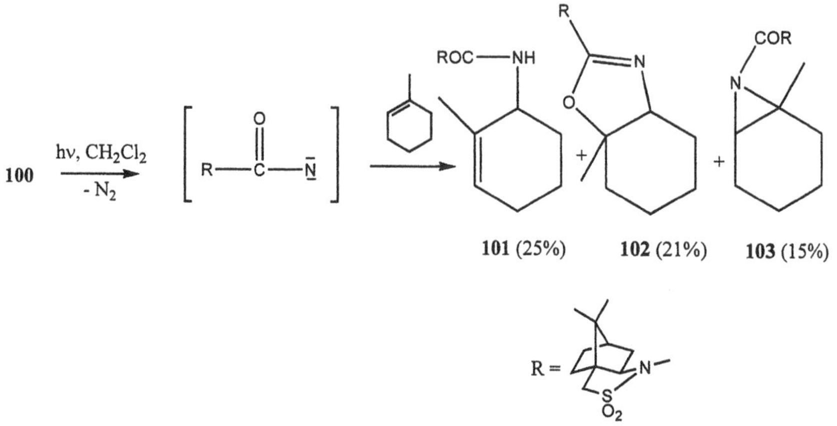

Scheme 29

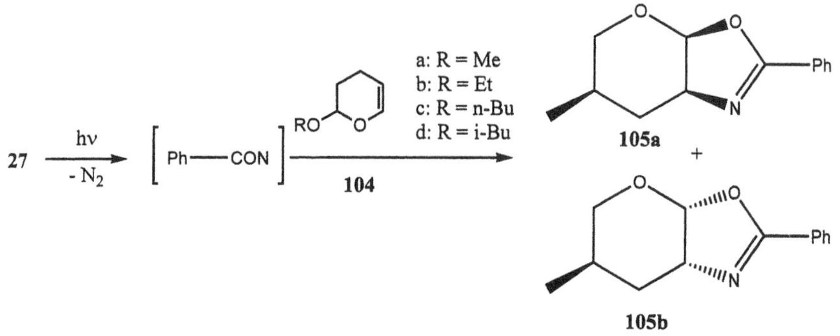

Scheme 30

An interesting result has been obtained with respect to the stereo-selectivity of the addition of benzoyl nitrene to compounds **104a–104d**: The ratio of the two diastereomers depends on the temperature, such that an inversion of the diastereoselectivity can be observed at a distinct temperature (*isoinversion principle* [47]). Such an effect may indicate a reversible first bond formation between the benzoyl nitrene and the alkene [45].

Ethoxycarbonyl nitrene is able to add to compound **104a** with a lower degree of stereoselectivity (d.e. 10%) compared with benzoylnitrene [22]. Nevertheless the addition of alkoxycarbonyl nitrene to unsaturated sugars such as compound **106** has been used in order to synthesize amino

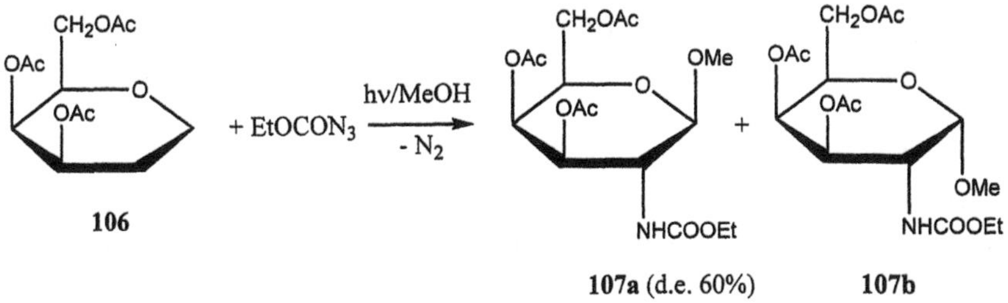

Scheme 31

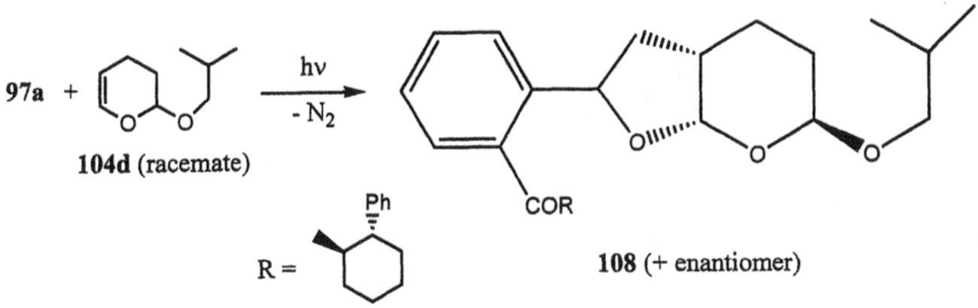

Scheme 32

substituted sugars via opening the ring of the intermediate adduct. In this case, the degree of diastereoselectivity is relatively high (Sch. 31) [48].

The combination of nitrene and substrate chirality improves the diastereoselectivity of the cycloaddition reaction. For example, due to the use of both chiral aroyl azides and chiral alkenes, the pure *exo*-product (compound **108**) was obtained (Sch. 32) [43].

13.5. REPRESENTATIVE EXPERIMENTAL PROCEDURES

13.5.1. 5-Methoxy-2-phenyl-3a,6,7,7a-tetrahydro-5H-pyrano[3,2-d]oxazole (105a) [22]

Procedure A

Benzoyl azide (1 g, 6.8 mmol) and 6-methoxy-5,6-dihydropyran (11.4 g, 100 mmol) dissolved in acetonitrile (100 mL) are placed in 10 quartz test tubes and irradiated using light of wavelength 254 nm (mercury low pressure lamp) for 9 h using a merry-go-round photoreactor. After removing the solvent and the olefin excess by vacuum evaporation, the reaction mixture was separated by column chromatography (silica gel, benzene/acetonitrile 4:1) to give 0.41 g *exo*-**105a** (26%) and 0.13 g *endo*-**105a** (9%) as oils.

Procedure B

Sensitized photolyses are carried out under nitrogen. Michler's ketone (270 mg, 1 mmol) was added to the acetonitrile acetonitrile solution of benzoyl azide (1 g, 6.8 mmol) and 6-methoxy-5,6-dihydropyran (11.4 g, 100 mmol). This solution was irradiated using light of wavelength 350 nm (Rayonet photoreactor, RPR-3500 lamps). Working-up the reaction mixture according to procedure A yields 0.127 g (80%) of the two diastereomers of compound **105a**.

13.5.2. Phenyl-1,4-dioxa-2-aza-spiro[4.5]dec-2-ene (89, $R^1, R^2 = (CH_2)_5$) [21]

Cyclohexanone (9.8 g, 100 mmol) and benzoyl azide (1.47 g, 10 mmol) dissolved in acetonitrile (100 mL) are irradiated according to procedure A for 6 h. After evaporating the solvent and the excess of cyclohexanone, methanol was added and the products were separated by column chromatography (silica gel; hexane/ethyl acetate 8:2). 0.98 g (45%) of the cycloadduct is obtained as oil as well as 0.19 g (12%) of the cycloadduct of benzoyl nitrene towards the solvent (2-methyl-5-phenyl-oxadiazol) and 0.6 g (40%) phenyl-*N*-methylcarbamate.

REFERENCES

1. (a) Bunyan PJ, Cadogan JIG. J Chem Soc 1963; 42; (b) Smith PAS, Hall JH, Kan RO. J Am Chem Soc 1962; 84:485.
2. (a) Heacock JF, Edmison MT. J Am Chem Soc 1960; 82:3460; (b) Rice FO, Luckenbach TA. J Am Chem Soc 1960; 82:2681.
3. (a) Horner L, Christmann A. Angew Chem 1963; 75:707; (b) Lüttringhaus A, Jander J, Schneider R. Chem Ber 1959; 92:1756.
4. Abramovitch RA, Davis BA. Chem Rev 1964; 64:149.
5. Tiemann F. Chem Ber 1891; 24:4162.
6. (a) L'Abbe G. Chem Rev 1968; 68:345; (b) Lwowski W. Acyl azides and nitrenes. In: Scriven EEFV, ed. Azides and Nitrenes. Chapter 4. New York: Academic Press, 1984:205; (c) Lwowski W. Nitrenes. In: Jones M, Moss RA, eds. Reactive Intermediates. Vol. 3. New York: J. Wiley & Sons, 1985:305; (d) Backes J. Nitrene. In: Houben-Weyl, Methoden der Organischen Chemie. Vol. E16c. Stuttgart: Georg Thieme Verlag, 1992:67–275; (e) Abraham W. Trends in Photochem Photobiol. Vol. 4. Res.Trends, 1997:13.

7. Currie CL, Darwent BB. Can J Chem 1963; 41:1552.
8. (a) Murate S, Nakatsuji R, Tomioka H. J Chem Soc Perkin Trans 2 1995; 793; (b) Abraham W, Buchallik M, Zhu QQ, Schnabel W. J Photochem Photobiol A: Chem 1993; 71:119.
9. Abraham W, Siegert S, Kreysig D. J Prakt Chem 1989; 331:177.
10. (a) Schuster GB, Platz S. Adv Photochem 1992; 16:69; (b) Bierwisch M, Abraham W. J Inf Recording 1998; 24:329.
11. Hinsberg WD III, Dervan PB. J Am Chem Soc 1979; 101:6142.
12. Wassermann E. Prog Phys Org Chem 1971; 8:319.
13. Klingler O, Prinzbach H. Angew Chem 1987; 99:579; Angew Chem Int Ed 1987; 26:566.
14. (a) Sakurai M, Komura T, Saruwatari M, Taniguchi H. Chem Lett 1987; 883; (b) Padwa A, Ku H, Mazzu A. J Org Chem 1978; 43:381; (c) Tamura Y, Kato S, Yoshimura Y, Nishimura T, Kita Y. Chem Pharm Bull 1974; 22:1291.
15. Hassner A, Fowler FW. J Am Chem Soc 1968; 90:2869.
16. Backes J. In: Houben-Weyl, ed. Methoden der Organischen Chemie. Vol. E16c. Stuttgart: Georg Thieme Verlag, 1992:149–228.
17. Sutherland DR, Pickard J. J Heterocycl Chem 1974; 11:457.
18. Kayama R, Shizuka H, Sekiguchi S, Matsui K. Bull Chem Soc Jpn 1975; 48:3309.
19. Sawanishi H, Sashida H, Tsuchiya T. Chem Pharm Bull 1985; 33:4564.
20. (a) Boyer JH, Ellis PS. J Chem Soc, Perkin Trans 1 1979; 483; (b) Boyer JH, Frints PJA. J Heterocycl Chem 1970; 7:59; (c) Rees CW. Pure Appl Chem 1979; 51:1243.
21. Clauss KU, Buck K, Abraham W. Tetrahedron 1995; 51:7181.
22. Buck K, Jacobi D, Plögert Y, Abraham W. J Prakt Chem 1994; 336:768.
23. Cipollone A, Loreto MA, Pellacani L, Sardelle PA. J Org Chem 1987; 52:2584.
24. Edwards OE, Elder JW, Lesage M, Retallak RW. Can J Chem 1975; 53:1019.
25. McConaghy JC Jr, Lwowski W. J Am Chem Soc 1967; 89:4450.
26. Hiyama T, Taguchi H, Nozaki H. Bull Chem Soc Jpn 1974; 47:2909.
27. Lwowski W, Hartenstein A, deVita C, Smick RL. Tetrahedron Lett 1964; 36:2497.
28. Hai SMA, Lwoeski W. J Org Chem 1973; 38:2442.
29. Subbaraj A, Subba Rao O, Lwowski W. J Org Chem 1989; 54:3945.
30. Tisue GT, Linke S, Lwowski W. J Am Chem Soc 1967; 89:6303.
31. Felt GR, Lwowski W. J Org Chem 1976; 96:96.
32. Hayashi Y, Swern D. J Am Chem Soc 1973; 95:5205.
33. (a) Autrey T, Schuster GB. J Am Chem Soc 1987; 109:5814; (b) Sigman ME, Autrey T, Schuster GB. J Am Chem Soc 1988; 110:4297.
34. Abraham W, Glänzel A, Buchallik M, Grummt U-W. Z Chem 1990; 30:405.
35. Averdung J, Mattay J, Jacobi D, Abraham W. Tetrahedron 1995; 51:2543.
36. Kato T, Suzuki Y, Sato M. Chem Pharm Bull 1979; 27:1181.
37. Huisgen R. In: Padwa A, ed. 1,3-Dipolar Chemistry. Vol. 1. New York: Wiley Interscience, 1984:7.
38. Huisgen R, Blaschke H. Tetrahedron Lett 1964; 5:1409.

39. Abraham W, Buck K, Clauß K-U. J Inf Recording 1996; 22:389.
40. Eibler E, Skura J, Sauer J. Tetrahedron Lett 1976; 17:4325.
41. Eibler E, Sauer J. Tetrahedron Lett 1974; 15:2569.
42. Brückner R. Reaktionsmechanismen. Heidelberg, Berlin, Oxford: Spektrum Akademischer Verlag, 1996.
43. Roeske Y, Abraham W. Synthesis 2001; 1125.
44. Del Signore G, Fioravanti S, Pellacani L, Tardella PA. Tetrahedron 2001; 57:4623.
45. Roeske Y. Ph.D. thesis, Humboldt-University, Berlin, 2001.
46. Jacobi D. Diploma thesis, Humboldt-University, Berlin, 1994.
47. Buschmann H, Scharf H-D, Hoffmann N, Esser P. Angew Chem 1991; 103:480; Angew Chem Int Ed 1991; 30:477.
48. Kozlowska-Gramisz E, Descotes G. Can J Chem 1982; 60:598.

14

C=C Photoinduced Isomerization Reactions

Tadashi Mori
Osaka University, Suita, Japan

Yoshihisa Inoue
Osaka University and ICORP/JST, Suita, Japan

14.1. INTRODUCTION

The photoisomerization of alkenes is one of the most well investigated fundamental photoreactions. In particular, the geometrical photoisomerization around the C=C double bond has attracted much attention from theoretical, mechanistic, and synthetic points of view. In the late 1960s and early 1970s Nobel laureate G. Wald demonstrated that the photochemical *Z–E* isomerization of rhodopsin is an essential process in vision [1–5]. Rhodopsin contains an 11-(*Z*)-retinal chromophore **1Z** tethered to an opsin protein through a Schiff base linkage. This chromophore readily isomerizes to the (*E*)-isomer with light stimulus at wavelengths around 500 nm, which is followed by sequential structural changes, eventually leading to its release form the protein. The photoisomerization takes place in about 200 fs and in a quantum yield of up to 0.67 (Sch. 1) [6–10].

In the rather short history of organic photochemistry, the geometrical *E–Z* photoisomerization has been exceptionally intensively studied for half a century and a number of reviews have been published [11–18]. Although the geometrical isomerization of alkenes can be effected thermally, catalytically, and photochemically, one of the unique features of photoisomerization is that the photostationary *E/Z* ratio is independent from the ground-state thermodynamics but is instead governed by the excited-state potential surfaces, which enables the thermodynamically less-stable isomers

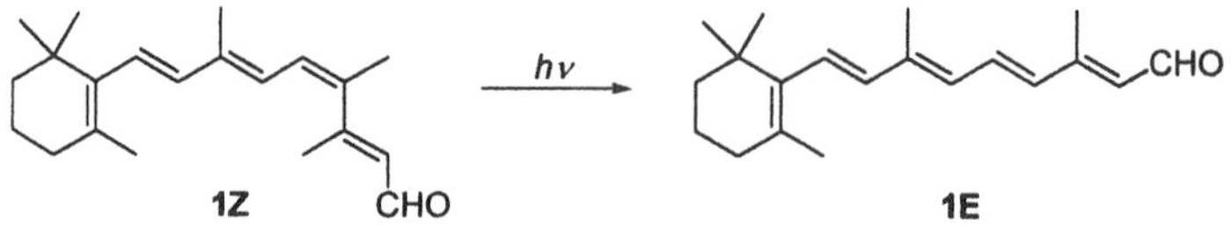

Scheme 1

to be obtained in excess. Therefore, in this chapter the focus will mainly be on the preparative aspects of the photoinduced geometrical isomerization of alkenes. Valence isomerization, rearrangement, and fragmentation, which are also observed upon photoirradiation of acyclic and small-membered cyclic alkenes are only briefly mentioned, as they have been reviewed elsewhere [12,13,19,20]. The synthetically useful Z-to-E photoisomerization of cyclic alkenes will also be dealt with, since the photobehavior of cycloalkenes is usually classified as a subcategory of alkene photochemistry and has not been comprehensively reviewed from a synthetic viewpoint [19–24]. Photochemically produced medium-sized (E)-cycloalkenes are thermally unstable and short-lived in general, but can be trapped, for example, by alcohols in the presence and even in the absence of an acid [25,26]. In particular, (E)-cyclo-octenes can readily be prepared by direct and sensitized photoisomerizations and isolated at ambient temperatures [27,28]. Recent studies on the asymmetric photoisomerization of cycloalkenes are subsequently summarized in a separate section in connection with the explosion of interest in of asymmetric photochemistry [29]. Photochemical asymmetric synthesis is a rapidly growing, but still less-explored, area of current photochemistry. Subsequently, the synthetic aspects of the geometrical photoisomerization of conjugated aromatic alkenes, such as styrenes and stilbenes, will be described; finally representative preparative photoreactions of three typical photoisomerizations will be detailed, i.e., the syntheses of racemic and optically active (E)-cyclo-octene and (Z)-stilbene.

14.2. PHOTOCHEMISTRY OF ACYCLIC ALKENES

14.2.1. Direct Irradiation

For most acyclic alkenes the (E)-isomer is thermodynamically more stable than the (Z)-isomer. Hence, geometrical photoisomerization provides us with a conventional tool for directly obtaining (Z)-alkenes in good yields. Carotenoids, vitamin A, vitamin D, and polyenes are known often to undergo the preferential photoisomerization of one of the conjugated double bonds to give a single photoisomer [30–36]. For example, photoirradiation of β-ionol (**2E**) and its derivatives are know to give the

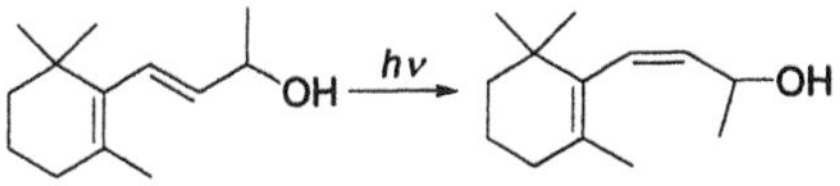

Scheme 2

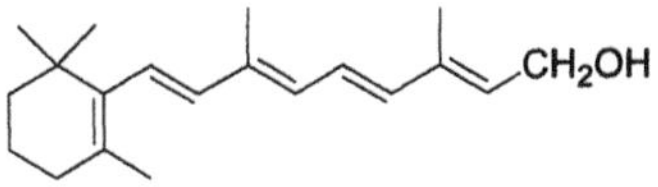

Scheme 3

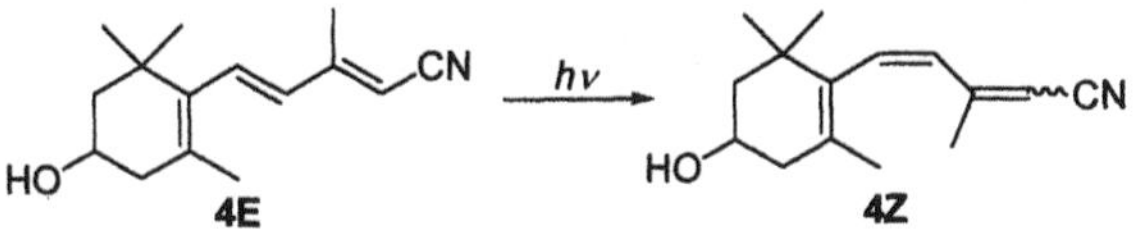

Scheme 4

corresponding (Z)-isomers in high Z/E ratios [37–43], and indeed 7-(Z)-β-ionol (**2Z**) is prepared in 86% yield by the photochemical one-way isomerization of **2E** in benzene (Sch. 2) [37]. All the isomers of vitamins A and A_2 series such as **3** have been prepared by triplet-sensitized photoisomerization of 3-hydroxy-β-ionylidene. For example, **4Z** has been prepared by the photolysis of **4E** (Schs. 3 and 4) [44–47]. However, such a position-selective photoisomerization is not always applicable to all conjugated alkenes. Recent studies revealed that the photobehavior of conjugated alkenes associated with supramolecular systems is very different from that of unbound alkenes. It has been found that the photoisomerization proceeds through the "Hula-twist" mechanism, keeping the volume change to a minimum during the photoisomerization [48–51]. Although simple acyclic alkenes can be similarly isomerized geometrically upon direct and sensitized irradiations, its practical application is less important for synthetic purposes; this is presumably because of the concurrent formation of byproducts and the less-preferable photostationary-state E/Z ratios of around unity.

Although the E–Z isomerization through the π,π^* state is undoubtedly the major decay process of excited alkenes, several byproducts are also derived from the $\pi,R(3s)$ Rydberg and π,σ^* states. These states lie closely in

5 → hv → **6** + **7** + **8** + **9** + **10** -OR + **11** -OR

R = H, Me

Scheme 5

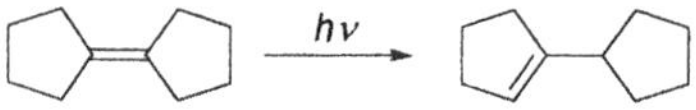

12   **13**

Scheme 6

energy in the case of simple acyclic alkenes and therefore the excitation is unfortunately nonselective. This is in sharp contrast to the photoisomerization of simple alkenes in the triplet manifold, in which one excited $^3\pi,\pi^*$ state is clearly low lying [52–55].

The major photoprocesses observed for simple acyclic alkenes upon direct irradiation are the E–Z isomerization and, at high concentrations, [2+2] cycloaddition. These processes are generally attributed to the $^1\pi,\pi^*$ state. However, the direct irradiation of simple alkenes, for which the E–Z isomerization is degenerate due to symmetrical substitution, is frequently accompanied by the formation of several rearrangement products which are derived from the $\pi,R(3s)$ Rydberg state [56–58]. Upon direct irradiation in hydrocarbon solvents, tetramethylethylene (**5**) affords the positional isomer **6** (18%), carbene-derived products **7** (17%) and **8** (24%), and a small amount of reduced product **9** (Sch. 5). In methanol or aqueous acetonitrile, the formation of ethers or alcohols (**10** and **11**) is preferred at the expense of the carbene-derived products, which is attributed to the radical cationic nature of the Rydberg state (Sch. 5) [59]. The double bond-migration process is enhanced for cycloalkenes; e.g., direct irradiation of cyclopentylidenecyclopentane **12** gives **13** in 85% yield (Sch. 6). The reaction is markedly sensitive to the degree of alkyl substitution and the solvent polarity [60]. Photochemical behavior of sterically congested tri- and di-*tert*-butylethylenes has also been reported [61]. For example, direct photolysis of tri-*tert*-butylethylene **14** in pentane afforded cyclobutane **15** (42%), cyclopropanes **16** (5%), and **17** (1%) (Sch. 7). In methanol, carbene-derived methanol adducts were formed together with rearrangement products. Direct photolyses of leaf alcohol, (Z)-3-hexen-1-ol, and related alkenes at 185 nm give the geometrical isomers in moderate yields. Esterification of the substrate, however, improves the yield from 51% for **18aE** to 84% for the corresponding acetate **18bE** (Sch. 8) [62].

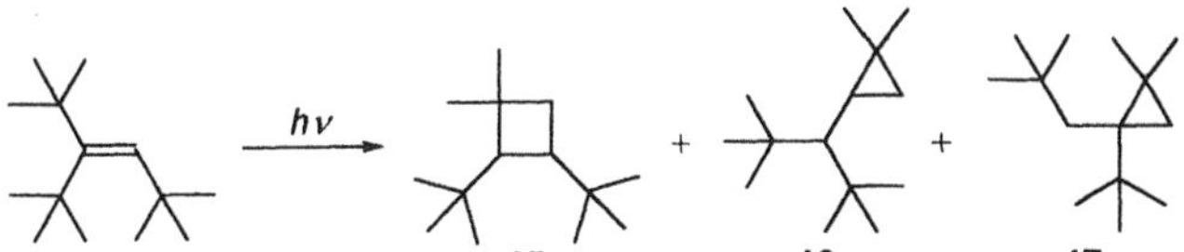

Scheme 7

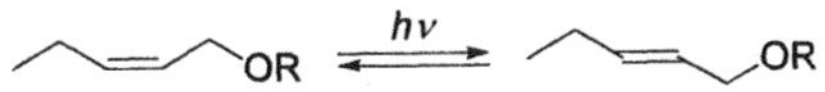

18aZ: R = H   **18E**
18bZ: R = COMe

Scheme 8

14.2.2. Sensitized Isomerization

Benzene-photosensitized $E–Z$ isomerization of 2-butene **19a** in the gas phase affords a photostationary E/Z ratio of around unity (Sch. 9) [63]. Isomerization occurs only when the triplet energy of the sensitizer is higher than 65 kcal/mol [64], the triplet energy-transfer mechanism is proposed on the basis of spectroscopic studies [65,66]. The $E–Z$ photoisomerization of 2-pentene **19b** at 313 nm was performed in the presence of acetone and acetophenone as sensitizers to give a decay ratio of 1.17 with acetone, indicating that the triplet energy transfer, rather than the Schenck mechanism via 1,4-biradical intermediate, is responsible for the photo-isomerization. On the other hand, the large decay ratio of up to 1.90 obtained with acetophenone indicates the operation of the Schenck mechanism (Sch. 9) [67]. In the photoisomerization of 2-octene **19c**, carbonyl and aromatic sensitizers with triplet energy of 50–74 kcal/mol are effective and the obtained E/Z ratios are dependent upon the sensitizer energy (Sch. 9) [68].

　　Re-examination of the benzene-sensitized isomerization of alkenes revealed that the photostationary E/Z ratio critically depends on the structure of alkenes and the triplet energy of sensitizers [69]. For instance, thermal $E–Z$ equilibrations of 2-octene **19c** and 3,4-dimethyl-2-pentene **20** give the thermodynamic E/Z ratios of 3.3 and 3.5, respectively, while the photostationary E/Z ratios upon benzene sensitization are 1.1 and 1.6, respectively (Schs. 9 and 10). For practical runs, the use of p-xylene or phenol, rather than benzene, in ether is recommended, as the yellowing of the irradiated solution, which retards the photoreaction, is substantially decreased.

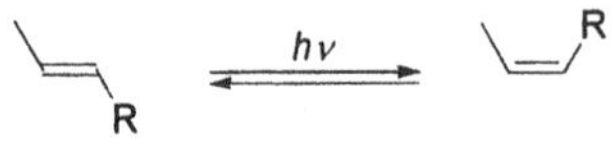

19aE: R = Me   **19Z**
19bE: R = Et
19cE: R = C$_5$H$_{11}$

Scheme 9

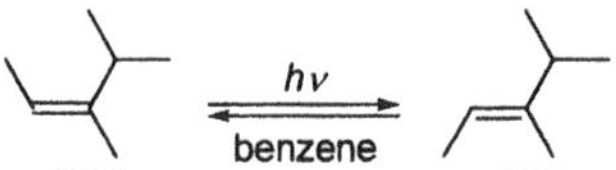

E/Z = 3.5 (thermal)
E/Z = 1.6 (photochemical)

20Z   **20E**

Scheme 10

Selective isomerization of the (E)-alkene to (Z)-alkene has also been carried out via infrared multiphoton excitation [70]. Upon IR laser irradiation at ~1000/cm of 2-butene **19a** and 2-pentene **19b**, essentially quantitative contrathermodynamic conversion (E-to-Z) can be achieved (Sch. 9). It is concluded that small differences in cross-section may be amplified in the multiphoton up-pumping process. However, the quantum yield was too small to measure ($<10^{-6}$).

14.3. DIRECT IRRADIATION OF CYCLOALKENES

Direct photolyses of small-sized cycloalkenes lead to rearrangement as well as decomposition. For example, the substituted cyclopropene **21** undergoes a rearrangement reaction upon direct irradiation, affording the allene **22**, cyclopentylacetylene **23**, and vinylcyclopentene **24** (Sch. 11) [71,72]. Upon direct irradiation, cyclobutene **25** gives the ring-opened 1,3-butadiene **26** as the major product, along with methylenecyclopropane **27**, ethylene, and acetylene (Sch. 12) [73–76], while cyclopentene **28** affords the rearranged methylenecyclobutene **29** and bicyclopentane **30** (Sch. 13) [77,78]. Most of the photoproducts derived from small-sized cycloalkenes involve the corresponding ring-contracted carbenes as common intermediates [12,13,19,20].

As is the case with acyclic alkenes, geometrical isomerization becomes the major photoreaction observed for medium-sized cycloalkenes upon direct excitation, although the carbene-derived products are still obtained as minor products. Thus, direct irradiation of cyclohexene **31Z** in aprotic

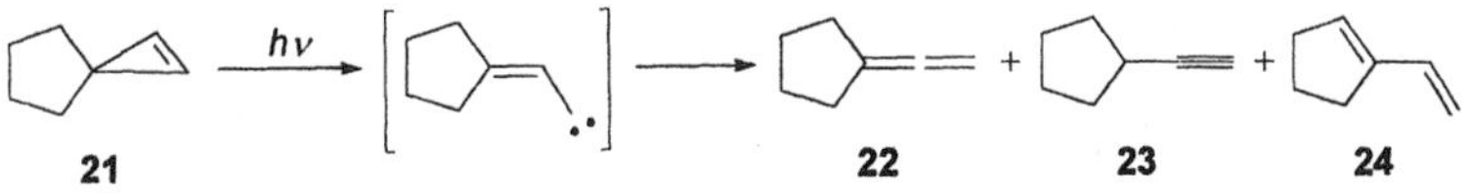

Scheme 11

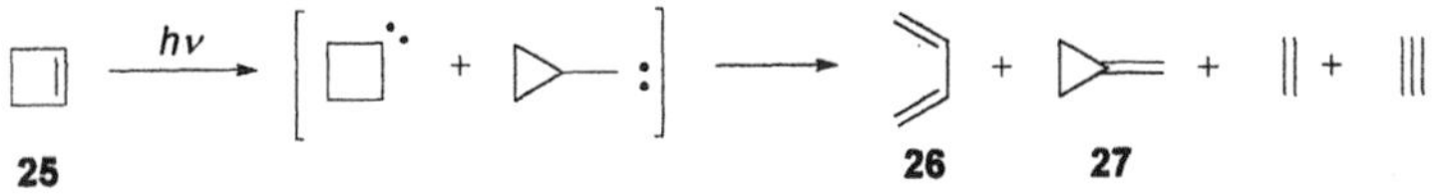

Scheme 12

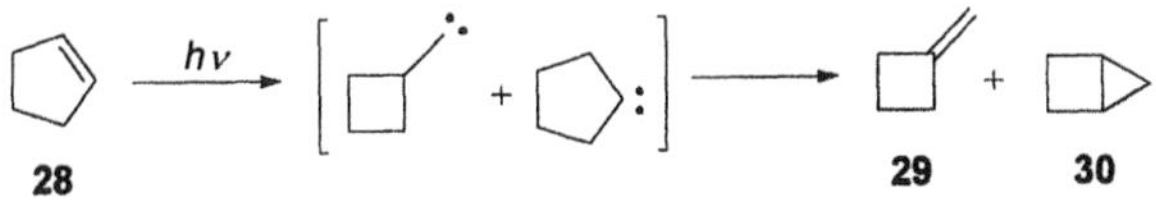

Scheme 13

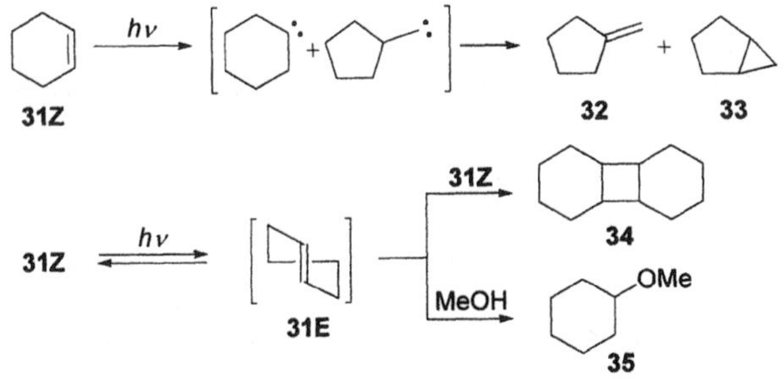

Scheme 14

solvents affords a stereoisomeric mixture of [2+2] cyclodimers **34** in a 1.6:2.3:1 ratio, along with methylenecyclopentane **32** and bicyclo[3.1.0]-alkane **33** (Sch. 14). The formation of cyclodimers involves the initial Z-to-E photoisomerization to give highly constrained (E)-isomer **31E**, which subsequently cycloadds to **31Z** in the ground state [79]. In contrast, direct irradiation of cyclohexene **31Z** and cycloheptene **36Z** in acidic methanol result in the formation of methanol adducts **35** and **37**, respectively, which is attributed to the proton-mediated addition of methanol to the respective (E)-isomers **31E** and **36E** produced photochemically (Sch. 15) [80].

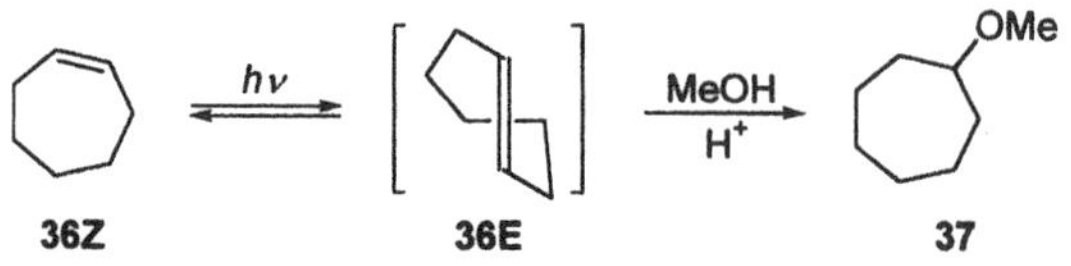

Scheme 15

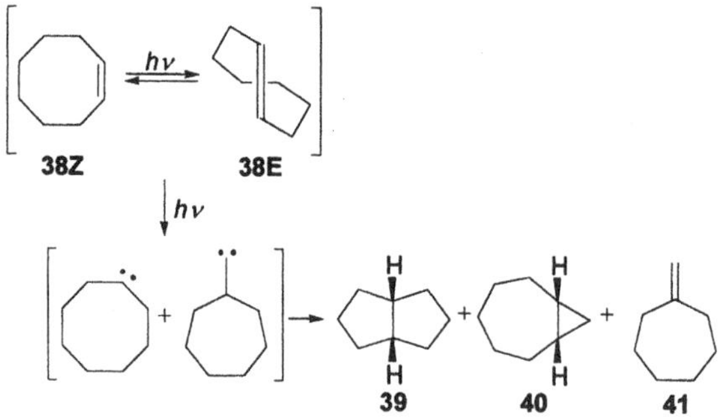

Scheme 16

(*E*)-Cyclo-octene **38E** is the smallest member of (*E*)-cycloalkenes, which is stable and isolable at ambient temperature. Although a couple of conventional thermal multistep synthetic routes to **38E** are known [81,82], direct irradiation of (*Z*)-cyclo-octene **38Z** at 185 nm provides more convenient access, affording the (*E*)-isomer **38E** as the major photoproduct in addition to small amounts of carbene-derived (*Z*)-bicyclo[3.3.0]octane **39**, (*Z*)-bicyclo[5.1.0]octane **40**, and methylenecycloheptane **41** (Sch. 16) [83]. A high photostationary-state E/Z ratio of 0.96 is attained upon direct irradiation at 185 nm [27,84], while sensitized photoisomerizations, in general, give E/Z ratios smaller than 0.3 [85]. Thus, a preparative-scale photoisomerization of (*Z*)-cyclo-octene **38Z** is possible; a pentane solution of **38Z** (0.12 M) was irradiated at 185 nm for 6 h under a nitrogen atmosphere, and the irradiated solution extracted with aqueous silver nitrate. **38E** of 99.6% purity was isolated in 26% yield after valve-to-valve distillation [27]. The geometrical photoisomerization of **38Z** proceeds through the $^1\pi,\pi^*$ state, while the carbenes are formed via the $\pi,R(3s)$ state [86].

Larger-sized (*Z*)-cycloalkenes, such as cyclononene **42aZ**, cyclodecene **42bZ**, cycloundecene **42cZ**, and cyclododecene **42dZ** also undergo geometrical photoisomerization to their (*E*)-isomers, but have yet to be isolated (Sch. 17) [87]. Upon prolonged irradiation, (*Z*)-cyclononene **42aZ** gives the carbene-derived (*Z*)-octahydro-1*H*-indene **43a** [88], and

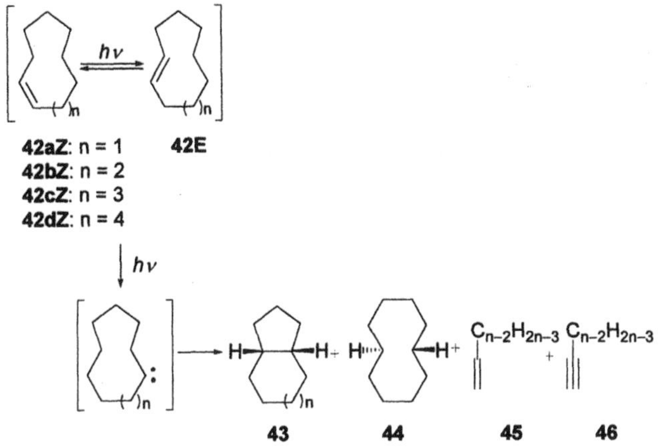

42aZ: n = 1   42E
42bZ: n = 2
42cZ: n = 3
42dZ: n = 4

43   44   45   46

Scheme 17

(Z)-cyclodecene **42bZ** the carbene-derived **43b** and saturated **44**. Direct irradiation of **42cZ** in pentane affords (Z)-bicyclo[6.3.0]undecane **43c** as the major product together with ring-opening ethylene and acetylene (**45c** and **46c**). (Z)-Cyclododecene **42dZ** similarly gives **43d**, **45d**, and **46d**. No detailed mechanistic or synthetic studies focused on the E–Z photoisomerization of these cycloalkenes have as yet been published in the literature [83].

14.4. PHOTOSENSITIZED ISOMERIZATION OF CYCLOALKENES

14.4.1. Cyclohexenes and Cycloheptenes

A convenient photochemical procedure for obtaining (E)-cycloalkenes and the products derived therefrom is the photosensitized isomerization of (Z)-isomers, where more conventional instrumentations, such as low/high pressure mercury lamps and quartz/Vycor/Pyrex vessels, can be used.

Photosensitization of cyclohexene derivatives leads to two types of reactions, i.e., E–Z isomerization and, if an allylic hydrogen is available, double-bond migration. Thus, photosensitizations of 1-methylcyclohexene **47aZ** and p-menth-1-ene **47bZ** with alkylbenzenes gives the exocyclic isomers **48** via a 1,3-hydrogen shift [89], as well as the [2+2] cycloadducts **49** and, in presence of methanol, adducts **50** through the initial E–Z photoisomerization and subsequent thermal reaction of the (Z)-isomer and alcohol, respectively (Sch. 18). Unsubstituted (Z)-cyclohexene **31Z** gives a stereo-isomeric mixture of [2+2] cyclodimers **34** upon triplet sensitization with

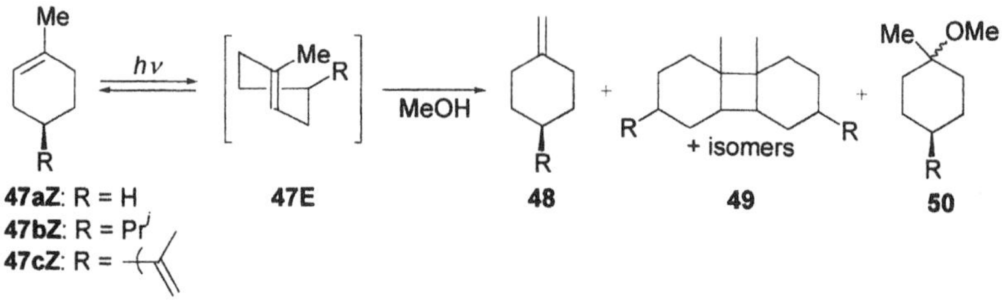
Me
hν
Me
R
MeOH
R
R
+ isomers
R
Me
OMe
R
47aZ: R = H
47bZ: R = Pr^i
47cZ: R =
47E
48
49
50

Scheme 18

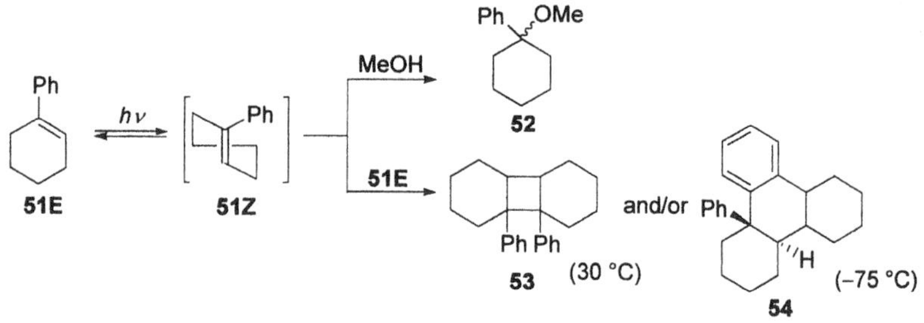
Ph
hν
Ph
MeOH
Ph
OMe
52
51E
51E
51Z
Ph Ph
53
(30 °C)
and/or Ph
H
54
(-75 °C)

Scheme 19

alkylbenzenes in aprotic solvent (Sch. 14) [79,90]. Xylene-sensitization of limonene (1-methyl-4-isopropenylcyclohexene, **47cZ**) in the presence of acidic methanol leads to the photoprotonation, affording 1.6:1 mixture of (*Z*)- and (*E*)-*p*-menth-8-en-1-yl methyl ether **50c** in 56% yield [91–94]. Cyclohexene **31Z** and 1-phenylcyclohexene **51E** also undergo the photosensitized addition of methanol in the presence of acid [95]. In the photoreaction of **51E** in methanol at 30 °C, 1-methoxy-1-phenylcyclohexane **52** and three [2+2] cyclodimers **53** in a ratio of 55:40:5 are formed (Sch. 19). At −75 °C, the [4+2] cyclodimer **54** was obtained upon direct and/or acetophenone-sensitized irradiations, the stereochemistry of which was determined by X-ray crystallography and taken as evidence for the formation of the highly strained (*Z*)-isomer **51Z** (note that **51Z** possesses the *trans* configuration and is strained) [96,97]. Upon direct excitation at 300 nm, the disappearance quantum yield of **51E** is moderate ($\Phi = 0.12$) [98], but the efficiency of the intersystem crossing is very low in general for cycloalkenes ($\Phi \sim 0.001$) [99]. Hence, the singlet mechanism is thought to operate.

Although (*E*)-cyclohexene has not been isolated, various chemical and spectroscopic observations strongly support its intervention as a reactive

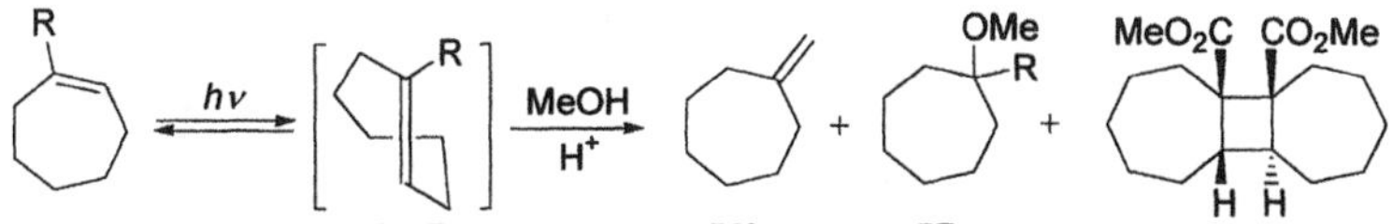

55aE: R = Ph   **55aZ**
55bZ: R = Me   **55bE**
55cE: R = CO$_2$Me   **55cZ**

56b   **57a**
         57b

+ isomers
58c

Scheme 20

transient intermediate upon direct and sensitized photolyses. A flash photolysis of 1-phenylcyclohexene **51E** in methanol clearly reveals the existence of the strained (*Z*)-isomer as a transient intermediate [100]. A large secondary kinetic deuterium isotope effect (k_D/k_H) of ~2 was observed for the thermal isomerization of **51Z** to **51E**, for which the significant changes in C=C torsion angle and vinyl bending angles at the transition state should be responsible [101].

(*Z*)-Cycloheptene **36Z** and (*E*)-1-phenylcycloheptene **55aE** also undergo photoinduced addition of methanol in the presence of acid through the initial geometrical photoisomerization (Schs. 15 and 20) [95]. (*Z*)-1-Methylcycloheptene **55bZ** is converted to the exocyclic isomer **56b** and methanol adduct **57b** upon alkylbenzene-sensitized irradiation in methanol, most probably via the (*E*)-isomer **55bE** [89]. Irradiation of (*E*)-1-methoxy-carbonylcylcoheptene **55cE** in pentane gives *head-to-head* cyclodimers **58c** in 86:14 ratio with high *trans-anti-cis* selectivity, suggesting the intermediacy of the (*Z*)-isomer [102]. Although (*E*)-cycloheptene **36E** has not been isolated, the *trans*-structure was characterized by various spectroscopic means, such as in-situ NMR measurement of photochemically prepared **36E** at low temperatures. Thus, singlet-sensitized irradiation of **36Z** at −90 °C affords the (*E*)-isomer, which quickly relapses to the (*Z*)-isomer upon warming to −30 °C. The energy barrier for the thermal *E*-to-*Z* isomerization of this seven-membered cycloalkene is as low as 10 kcal/mol [103]. Triplet sensitization with toluene or xylenes afforded the (*E*)-isomer in low yields with an *E/Z* ratio of 0.065 or 0.094, while singlet sensitization of **36Z** with methyl benzoate at −78 °C gives a high *E/Z* ratio of 0.24 [104,105]. Laser flash photolysis of (*E*)-1-phenylcycloheptene **55aE** at low temperatures demonstrated the transient absorption spectrum of the (*Z*)-isomer [106].

14.4.2. Cyclo-octenes

When (*Z*)-cyclo-octene **38Z** is irradiated in the presence of a large amount of benzene or naphthalene, the corresponding 1:1 cross-adduct is

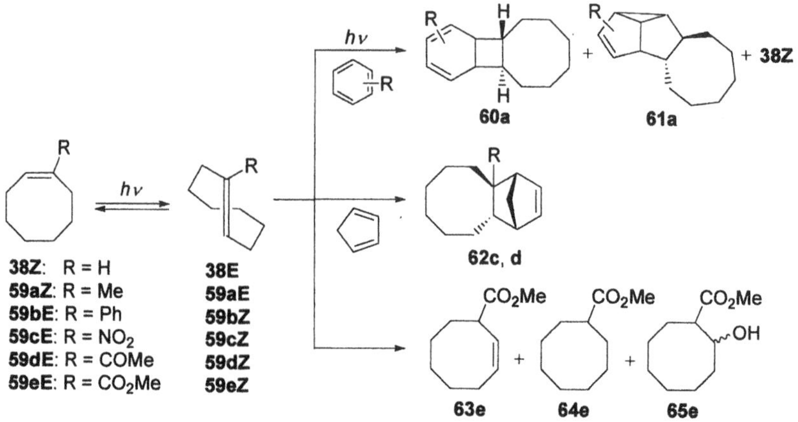

Scheme 21

produced [107–109]. Upon xylene sensitization, however, the adduct formation is retarded and the geometrical isomerization predominates to give the (E)-isomer **38E** in an acceptable isolated yield (0.5%, >97% purity) [110]. In the triplet-sensitized photoisomerization of **38Z**, the E/Z ratio is known to depend on the alkene concentration. The observed concentration dependence is attributed to the involvement of the short-lived excited singlet state at higher concentrations; the singlet-excited aromatic sensitizer not only mediates the geometrical isomerization but also adds to **38Z** to give the cross-adduct, such as **60a** and **61a** (Sch. 21) [111]. The triplet-sensitized photoisomerization of cyclo-octene suffers from critical temperature and solvent–viscosity effects [112]. In the triplet-sensitized isomerization of **38Z** in pentane, the E/Z ratio at the photostationary state is substantially affected by temperature, being doubled when the temperature is lowered from 20 to −78 °C, which is in marked contrast to the sensitized isomerization of acyclic alkene. The photostationary-state E/Z ratios obtained upon singlet-sensitized photoisomerization of **38Z** are considerably higher than those obtained upon triplet sensitization. Typically, benzene(poly)carboxylate sensitizers with singlet excitation energy of about 100 kcal/mol are employed for the geometrical isomerization of **38Z** [113] to give the photostationary E/Z ratios of 0.25–0.34 [85], but those with electron withdrawing group(s), such as trifluoromethylbenzoates, afford much higher photostationary E/Z ratios of up to 0.6 [28,114].

Geometrical photoisomerization of (Z)-1-methylcyclo-octene **59aZ** can be performed by direct, singlet-sensitized, and triplet-sensitized irradiations. The photostationary E/Z ratio is 0.30 upon direct irradiation at 214 nm in pentane, but becomes smaller upon triplet sensitization with alkylbenzenes (0.03–0.26) and also upon singlet sensitization with alkyl benzoates (0.01–0.24). It is interesting to note that the E/Z ratios obtained upon triplet sensitization of **59a** are comparable to or slightly higher than the corresponding values for unsubstituted cyclo-octene **38**, but the ratios obtained upon singlet sensitization of **59a** are much lower than those obtained for unsubstituted **38**, reflecting the difficulty of singlet exciplex formation with **59a** [115]. In the sensitized photoisomerization of (E)-1-phenylcyclo-octene **59bE**, fluorenone, and chrysene are not effective for isomerization because of low triplet energy, 9-cyanophenanthrene and thioxanthone work well as sensitizers, affording photostationary Z/E ratios of 32:68 and 42:58, respectively. Isolation of (Z)-1-phenylcyclo-octene was not successful [116]. (Z)-1-Nitrocyclo-octene **59cZ** produced upon irradiation of the (E)-isomer is trapped by thermal Diels-Alder reaction with cyclopentadiene to give tricyclic adduct **62c** in 40% yield [117]; at the photostationary state, the (Z)-isomer content reaches >28%. Irradiation of (E)-1-acetylcyclo-octene **59dE** at 337 nm in acetonitrile affords a photostationary Z/E mixture in a ratio of 88:12 with the (Z)-isomer also trapped by cyclopentadiene [118]. Irradiation of (E)-1-methoxycarbonyl-cyclo-octene **59eE** in pentane leads to deconjugation, giving (Z)-3-methoxycarbonylcylco-octene **63e** in 70% yield, with concurrent reduction and hydroxylation reactions giving **64e** and **65e** in 4% and 1% yields, respectively (Sch. 21) [102]. The intervention of the (Z)-intermediate was also suggested.

14.4.3. Miscellaneous Cycloalkenes

Photoisomerization of (Z)-cyclododecene **42dZ** sensitized by benzene affords an E/Z ratio of 39:61 at the photostationary state. Other triplet sensitizers such as benzophenone, acetophenone, and acetone can also be used for the geometrical photoisomerization (Sch. 17) [119]. Direct irradiation of (E)-3-methylenecyclodecene **66E** afforded the (Z)-isomer **66Z**; prolonged irradiation lead to the formation of (Z,Z)-1-methyl-1,3-cyclodecadiene **67**, (Z,Z)-2-methyl-1,3-cyclodecadiene **68**, and bicyclo[7.1.1] undec-1(10)-ene **69** (Sch. 22). Triphenylene-sensitized isomerization of **66E** affords the photostationary mixture of **66E** and **66Z** in a 94:6 ratio [120]. Doubly-bridged (Z)-bicyclo[8.8.0]octadec-1(10)-ene **70aZ** and (Z)-bicyclo[10.8.0]eicos-1(12)-ene **70bZ** photoisomerize to the corresponding (E)-isomers upon xylene-sensitized or direct irradiation. An E/Z ratio of 0.42 is attained upon photosensitization of **70bZ** [121], whilst the E/Z ratio is enhanced up to 9 and 2 upon direct irradiation at 254 nm of **70aZ** and **70bZ**, respectively (Sch. 23) [122].

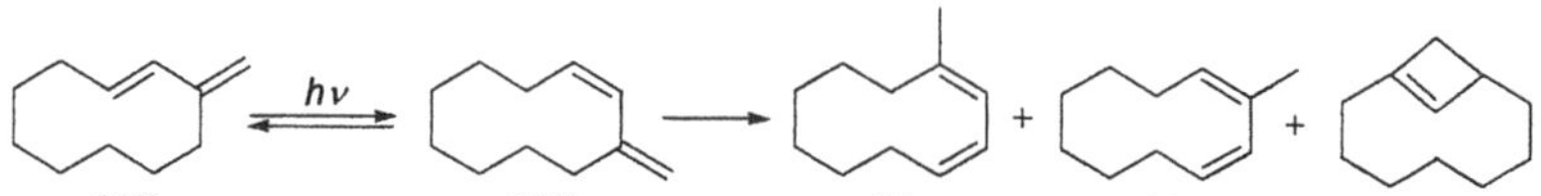

Scheme 22

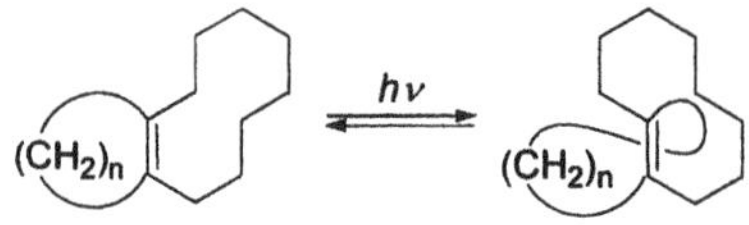

70aZ: (n = 8)
70bZ: (n = 10)   **70E**

Scheme 23

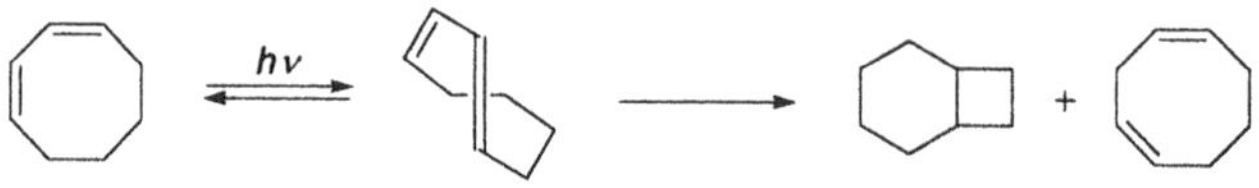

71ZZ   **71EZ**   **72   73**

Scheme 24

Irradiation of (Z,Z)-1,3-cyclo-octadiene **71ZZ** afforded the (E,Z)-isomer **71EZ** as the major product, along with small amounts of bicyclo[4.2.0]oct-7-ene **72** and (Z,Z)-1,4-cyclo-octadiene **73** [123]. The (E,Z)-1,3-cyclo-octadiene **71EZ** is produced from **71ZZ** in 62–89% yields (Sch. 24) [124]. Low-temperature irradiation of (Z,Z)-1,3-cycloheptadiene **74ZZ** in a glass matrix at $-196\,°C$ affords the (E,Z)-isomer **74EZ**, which is stable and spectroscopically characterized at that temperature [125]. Direct and triphenylene-sensitized irradiations of **74ZZ** in pentane or in neutral methanol at room temperature affords bicyclo[3.2.0]hept-6-ene **75** in excellent yields. Similar irradiations in acidic methanol gives 3-methoxy-cycloheptene **76** and 7-methoxybicyclo[4.1.0]heptane **77** in good yields through the trapping of photochemically produced (E,Z)-isomer **74EZ** by acidic methanol (Sch. 25). Singlet photosensitization of (Z,Z)-1,5-cyclo-octadiene **78ZZ** in pentane gives the (E,Z)-isomer **78EZ** together with tricyclo[3.2.0.0^{2,6}]octane **79** as a secondary product from the thermal cyclization of **78EZ** (Sch. 26). A nonvertical singlet-excitation mechanism

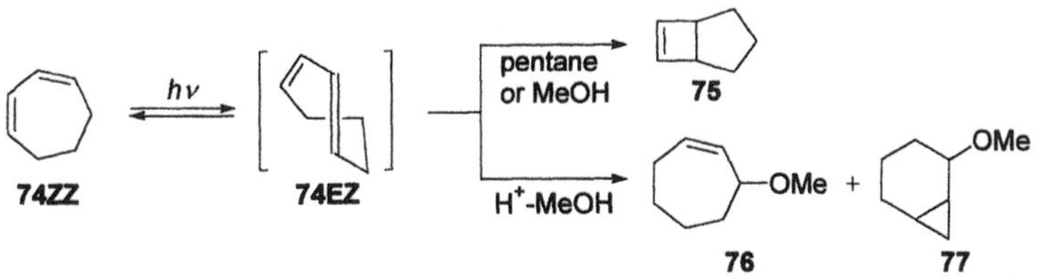

Scheme 25

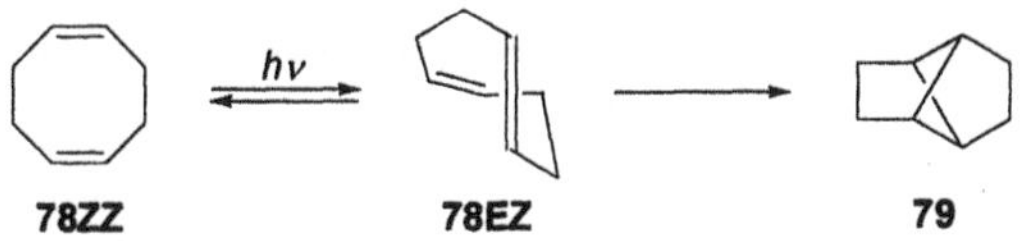

Scheme 26

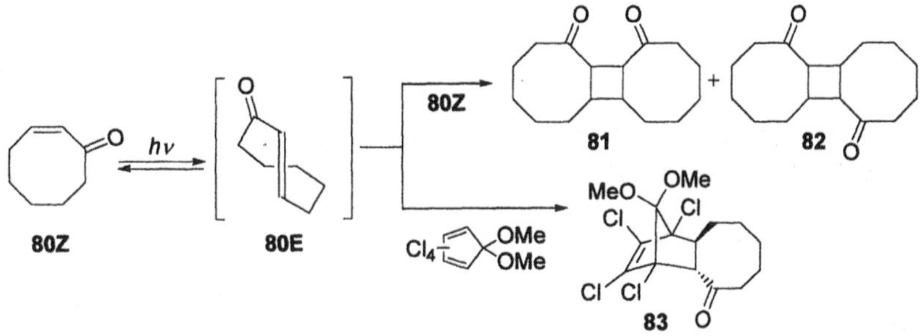

Scheme 27

involving an exciplex intermediate is proposed as is the case with cyclo-octene **38** [126].

Upon direct irradiation at 300 nm (Z)-2-cyclo-octenone **80Z** isomerizes to the (E)-isomer **80E**, affording an 80:20 E/Z mixture. The (E)-isomer **80E** is trapped by **80Z** or 5,5-dimethoxy-1,2,3,4-tetrachlorocyclopentadiene at room temperature to give the cyclodimers **81** and **82** or the Diels-Alder adduct **83**, respectively (Sch. 27) [127]. Similarly, (E)-2-cycloheptenone **84E** is produced upon irradiation of **84Z** at −50 °C, and is also trapped by cyclopentadiene to afford the adduct **85** (Sch. 28) [128]. The formation of the (E)-isomer as a transient intermediate has been proved by IR spectroscopy at −160 °C [129]. Laser flash photolysis of (Z)-2-cyclohep-tenone **84Z** revealed that the (E)-isomer has a lifetime of 45 s in cyclohexane

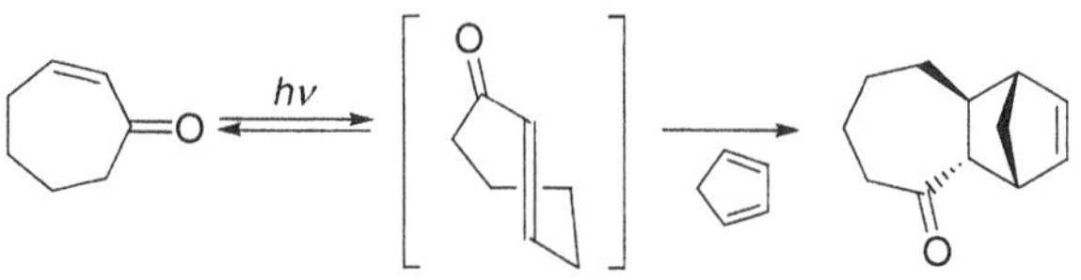

84Z   84E   85

Scheme 28

at room temperature [130]. A variety of alcohols and amines are known to add to the photochemically produced (*E*)-cycloheptenone **84E**, (*E*)-cyclo-octenone **80Z**, and (*E*)-cyclononenone [131]. The smaller-sized (*Z*)-cyclohexenone also affords a similar addition product in a low yield (0.7%) upon irradiation in methanol, suggesting the involvement of the (*E*)-isomer as an intermediate. Photoinduced methanol additions to 2-cycloheptenone, 2-cyclo-octenone, and 2,3-benzo-2,6-cycloheptadienone are believed to involve the (*E*)-isomers as precursors to the final products as based on deuterium-labeling experiments [132].

14.5.   ASYMMETRIC PHOTOISOMERIZATION OF CYCLOALKENES

14.5.1.   Cyclo-octene

There are a couple of comprehensive reviews on general asymmetric photochemistry in solution [133,134] and also on asymmetric photosensitization [135,136]. An account on multidimensional control of asymmetric photoreaction by environmental factors has also appeared recently [137]. This reflects a keen interest in chiral photochemistry and photochemical asymmetric synthesis [29]. In this section, we will concentrate mostly on the asymmetric photosensitization of cycloalkenes.

As a consequence of the restricted jump-rope rotation around the *trans* double bond, (*E*)-cyclo-octene **38E** is chiral. Optically active (*E*)-cyclo-octene has long been known, but the conventional multistep synthesis is rather tedious [138–140]. In contrast, direct-preparation of optically active (*E*)-cyclo-octene through asymmetric photosensitization is an attractive alternative. The first enantiodifferentiating *Z–E* photoisomerization of cyclo-octene **38Z** sensitized by simple chiral alkyl benzenecarboxylates was reported in 1978 to give low enantiomeric excesses (ee's) of <6% [141]; a variety of systems and conditions have been examined since then to raise the product ee. For an efficient transfer of chiral information

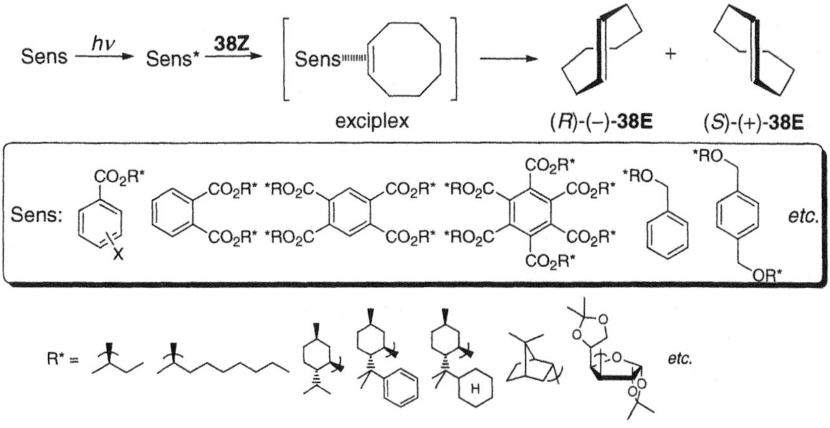

Scheme 29

from sensitizer to substrate, the existence of intimate interactions in the ground and/or excited state, forming ground-state complex or exciplex, is essential [85].

The enantiodifferentiating photoisomerization of cyclo-octene sensitized by chiral polyalkyl benzenepolycarboxylates has been studied extensively (Sch. 29) [85,113,114,141–148]. The singlet exciplex mechanism proposed is supported by the quenching of sensitizer fluorescence, direct observation of exciplex fluorescence and kinetic analyses. The enantiodifferentiating step is the rotational relaxation of (Z)-cyclo-octene to enantiomeric perpendicular singlets within the exciplex. The sensitizations with bulky chiral 8-phenylmenthyl and 8-cyclohexylmenthyl 1,2,4,5-benzenetetracarboxylates afford the (E)-cyclo-octene in up to a 50% ee even at room temperature and in 64% at −89 °C [113]. Interestingly, chiral triplet sensitizers are ineffective for enantiodifferentiating photoisomerization of **38Z**, giving poor ee's even at low temperatures [146]. Enantiodifferentiating photoisomerization can also be effected by other types of chiral photosensitizers, including aromatic amides [144], phosphoryl esters [145], and phosphoramides [142].

Crucially, not only temperature [148,149] but also other environmental factors, such as pressure [143], magnetic/electronic fields [150,151], and solvent [152] play vital roles in a variety of asymmetric photochemical reactions. In the enantiodifferentiating photoisomerization of (Z)-cyclo-octene **38Z**, (R)-(−)-(E)-**38E** is produced predominantly at temperatures higher than the isoenantiodifferentiating temperature (T_0), while the antipodal (S)-(+)-isomer prevails at temperatures lower than T_0 even if the same sensitizer (with the same chiral sense) is employed. In a representative case, the product chirality is inverted at $-19\,°C$ in the photosensitized isomerization of **38Z** sensitized by tetra-$(-)$-menthyl 1,2,4,5-benzenetetracarboxylate. When the sensitization is performed with saccharide esters, the product chirality is switched simply by changing the solvent from nonpolar to polar. In fact, the major product is (S)-$(+)$-**38E** in pentane but is switched to (R)-$(-)$-**38E** in diethyl ether in the photosensitization of **38Z** with *tetrakis*(diacetoneglucosyl)1,2,4,5-benzenetetracarboxylate at low temperatures [152].

Asymmetric photoisomerization of cyclo-octene was also investigated in supramolecular systems such as native [153] and modified cyclodextrins [154,155], chirally-modified zeolite [156], and DNA grooves [157].

14.5.2. Cyclo-octene Derivatives

Photoisomerization of (Z)-1-methylcyclo-octene **59aZ** sensitized by chiral benzene(poly)carboxylates suffers significantly from steric effects. The photosensitization of **59aZ** with bulky sensitizers leads to photostationary E/Z ratios and product ee's much lower than those obtained for less-hindered (Z)-cyclo-octene **38Z** [158]. Enantiodifferentiating photoisomerizations of (Z,Z)-1,3-cyclo-octadiene and (Z,Z)-1,5-cyclo-octadiene (**71Z** and **78Z**) have also been studied by using chiral benzene(poly)carboxylates at various temperatures [159]. Photosensitization of **71Z** with hexa-$(-)$-menthyl benzenehexacarboxylate affords the (E,Z)-product **71E** in 18% ee in pentane at $-40\,°C$. In the photoisomerization of **78Z** in pentane sensitized by $(-)$-menthyl benzoate, the $(-)$-(E,Z)-isomer **78E** is only slightly favored giving a 2% ee [126].

In the diastereodifferentiating E–Z photoisomerization of (Z)-3-benzoyloxycyclo-octene **86aZ**, the diastereomeric excess (de) of the photoproduct **86E** displays a unique concentration dependence (Sch. 30). Thus, when **86aZ** was irradiated at a low substrate concentration (1 mM) in

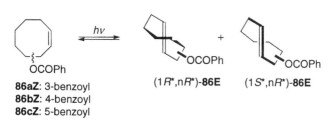

86aZ: 3-benzoyl
86bZ: 4-benzoyl
86cZ: 5-benzoyl

$(1R^*,nR^*)$-**86E** $(1S^*,nR^*)$-**86E**

Scheme 30

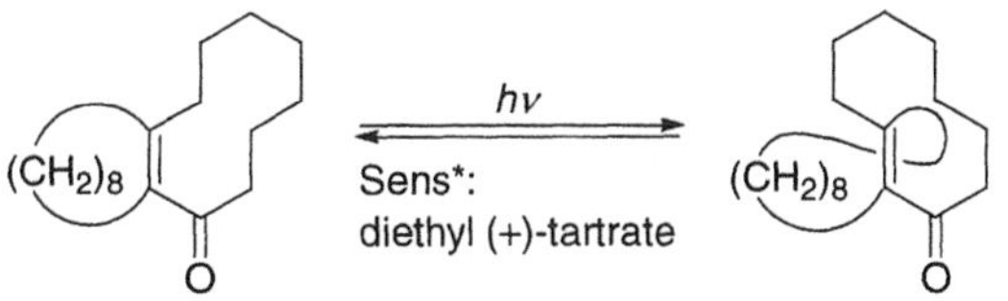

Scheme 31

pentane at 25 °C, the $(1R^*,3R^*)$-(E)-isomer **86aE** is predominantly obtained as a predominant product in 17% de, whereas the de decreases to almost zero at 10 mM, and finally the epimeric $(1S^*,3R^*)$-isomer becomes dominant at higher concentrations of up to 100 mM [160]. This phenomenon is attributed to the competing intra- and intermolecular photosensitization processes, each of which exhibits the opposite stereoselectivity. Interestingly, the diastereodifferentiating photoisomerizations of 4- and 5-benzoyloxy-cyclo-octenes (**86bZ** and **86cZ**) do not show any switching of product stereochemistry, at least in the same range of substrate concentration (Sch. 30) [161]. For the 4-benzoyl derivative **86bZ**, both of the intra- and intermolecular sensitizations give the same diastereomer in slightly different de's, whilst essentially no intramolecular sensitization takes place in the case of the 5-benzoyloxycyclo-octene **86cZ** due to steric inaccessibility.

In the presence of (+)-ephedrine, (E)-1-acetylcyclo-octene **59dE** photoisomerizes to enantiomeric (Z)-isomer, which is trapped by cyclo-pentadiene to give the Diels-Alder adduct **62d** in 22% ee (Sch. 21) [118]. Direct irradiation at >280 nm of doubly bridged (Z)-1(10)-bicyclo[8.8.0] octadecen-2-one **87Z** in diethyl (+)-tartrate as a solvent affords the (E)-isomer **87E** in favor of the levorotatory enantiomer where the E/Z ratio is 7 (Sch. 31) [162].

14.5.3. Cyclohexene and Cycloheptene

The smaller-sized cyclohexene and cycloheptene have also been subjected to enantiodifferentiating photoisomerization, although the corresponding (E)-isomers are short-lived transient species. Photosensitization of (Z)-cyclohexene **31Z** with chiral benzene(poly)carboxylates affords *trans-anti-trans*-, *cis–trans*-, and *cis-anti-cis*-cyclodimers **34** (Sch. 32). Interestingly, of the former two chiral products, only the *trans-anti-trans* isomer is optically active and its ee reaches up to 68%, whilst the *cis–trans* isomer is totally racemic under a variety of irradiation conditions, for which two competing, concerted, and stepwise cyclodimerization mechanisms are responsible. Thus, the enantiodifferentiating photoisomerization of **31Z** to the optically

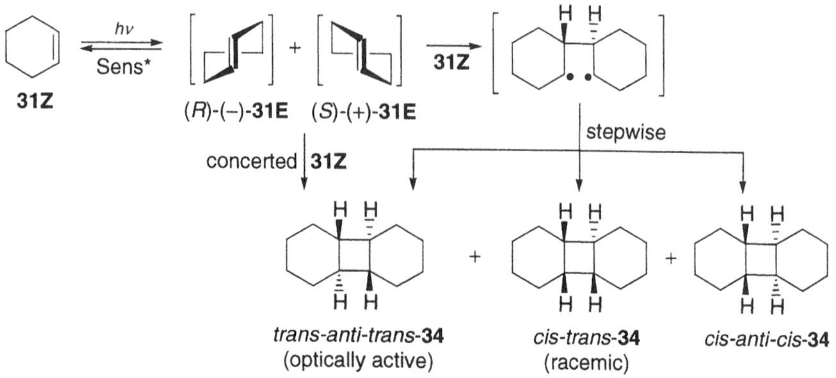

Scheme 32

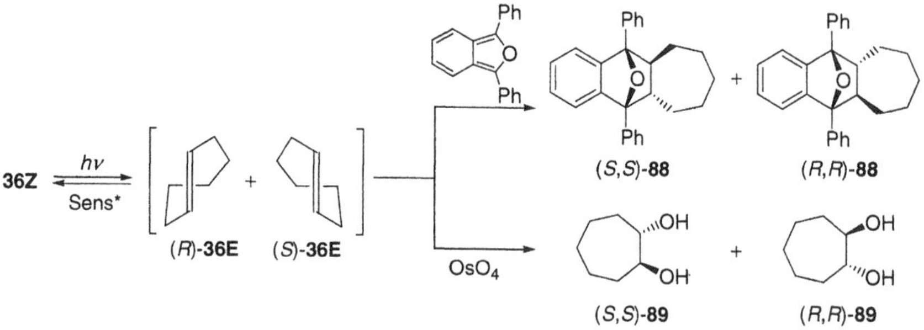

Scheme 33

active (E)-isomer **31E** and the subsequent stereospecific concerted cycloaddition to **31Z** in the ground state affords the optically active *trans-anti-trans* isomer, whereas the *cis–trans*, as well as *trans-anti-trans* and *cis-anti-cis*, isomers are formed in the stepwise cyclodimerization via a biradical intermediate, loosing the original chiral information in **31E** [90].

The enantiodifferentiating photoisomerization of (Z)-cycloheptene **36Z** was also studied in some detail [163]. Low temperature irradiations of **36Z** in the presence of chiral benzenepolycarboxylates as sensitizers give the labile (E)-isomer **36E**, which is subsequently trapped in the dark by either 1,3-diphenylisobenzofuran or osmium tetraoxide to afford the adduct **88** or the diol **89**, respectively (Sch. 33). The observed photostationary-state E/Z ratios of ~0.1 are comparable or slightly lower than that obtained for the singlet-sensitized photoisomerization of (Z)-cyclo-octene **38Z**. The ee of

36E, evaluated from that of the adduct **88** is higher than that of (*E*)-cyclo-octene **38E** obtained under the comparable conditions, and is sensitive to the solvent polarity and reaction temperature. Upon sensitization with tetra-(−)-bornyl 1,2,4,5-benzenetetracarboxylate in hexane at −80 °C, the ee of **36E** goes up to 77% ee, which is the highest value reported for the enantiodifferentiating photosensitized isomerization to-date.

14.6. PHOTOISOMERIZATION OF STILBENES AND STYRENES

14.6.1. Stilbenes

The photoisomerization of stilbenes is undoubtedly the most extensively studied and documented photoreaction, to which a number of excellent reviews from various viewpoints have been devoted [164–180]. The photoisomerization dynamics of stilbene have also been specifically reviewed [172]. The structure and dynamics of the lowest excited singlet state of (*E*)-stilbene have recently been discussed on the basis of the time-resolved Raman spectral data and isotope effects, by using the solvent-induced dynamic polarization model [164]. Laser flash photolysis studies have revealed the spectral, energetic, and lifetime profiles of the excited stilbenes in the ps and fs time scales [172,173,179]. A recent application of stilbene photochemistry to biophysical studies lead to the development of the Fluorescence-Photochrome Labeling technique for investigating the local medium effects in biological membranes and surface systems [165]. Reviews on photoisomerization of stilbene derivatives are also available [168,169,174,175,180]. Therefore, we will not repeat the photophysical and mechanistic aspects, but rather concentrate on the synthetic aspects of stilbene photochemistry. (*Z*)-Stilbenes undergo electrocyclic ring closure to dihydrophenanthrenes, which in turn are easily oxidized to the corresponding phenanthrenes. Synthetic applications and mechanistic details of this photocyclization are descried elsewhere [181–186]. This photocyclization/oxidation process has been employed in the syntheses of phenanthrene analogs such as benzoquinolines [187], benzoquinolizines [188], and helicenes [189–193]. Photoisomerization of diarylethenes has incited much interest because of the photochromic behavior, which has been reviewed from the mechanistic and synthetic points of view [194,195]. Recently, light/heat-driven chiroptical molecular motors/rotors, which possess a partial stilbene or analogous framework, that can turn around the olefinic double bond in only one direction upon successive photo/thermal activation have been developed [196–200].

The photostationary Z/E ratio of stilbene, $(Z/E)_{pss}$, is known to be significantly dependent on the excitation wavelength. Thus, the Z/E ratio is almost unity (48:52) upon irradiation at 254 nm, but is remarkably enhanced up to 93:7 upon irradiation at 313 nm. This apparently surprising change is readily interpreted in terms of the following equation, which relates the $(Z/E)_{pss}$ ratio with the relative extinction coefficient of the two isomers at the excitation wavelength and the relative efficiency (quantum yield) of the forward and reverse reactions: $(Z/E)_{pss} = \varepsilon_E/\varepsilon_Z \times \Phi_{Z \to E}/\Phi_{E \to Z}$ [201]. Preparative-scale direct irradiation should be done at low stilbene concentrations, since photodimerization of (E)-stilbene may compete with the photoisomerization as the concentration increases [202–206].

Triplet-sensitized photoisomerization of stilbenes has also been extensively studied, employing a variety of sensitizers. The $(Z/E)_{pss}$ ratios plotted against the triplet energy (E_T) of the sensitizer employed give a bell-shaped profile. In triplet sensitization of stilbene **90a**, sensitizers with E_T >62 kcal/mol afford practically constant $(Z/E)_{pss}$ ratios of around unity. However, as the E_T decreases, the energy transfer to (Z)-stilbene [possessing a higher E_T than the (E)-isomer] is decelerated and hence the $(Z/E)_{pss}$ ratio greatly increases at $E_T < 62$ kcal/mol to reach a peak value of ca. 12 at $E_T = 49$ kcal/mol. For sensitizers of yet lower E_T, the nonvertical energy transfer mechanism operates and the efficiency becomes comparable for both isomers, giving $(Z/E)_{pss}$ ratios again very close to unity [23].

Selective isomerization of (Z)-stilbene **90aZ** by nonresonant two-photon excitation has also been reported [207]. Upon laser flash photolysis in hexane at 266 nm, **90aZ** affords (E)-stilbene **90aE** and phenanthrene in 27 and 23% yields, respectively (Sch. 34). In contrast, laser excitation at 532 nm gives **90aE** in 16% yield without accompanying the formation of phenanthrene. The authors claimed that the reaction proceeds through the two-photon allowed excited state with A symmetry in nonresonant two-photon excitation. Alteration in product selectivity upon two-photon excitation has also been reported for the [2+2] cycloaddition of **90aZ** with tetramethylethylene [208] and for the competitive debromination/isomerization of (E)-4-bromostilbene [209]. Electron-transfer initiated $E–Z$ isomerization of stilbenes, though less important for the synthetic purposes, has also been documented [210–213]. For example, the stilbene radical cation, generated upon quenching of excited 2,4,6-triphenylpyrylium, undergoes Z-to-E isomerization [211].

14.6.2. Stilbene and Styrene Derivatives

A variety of *para*-substituted stilbenes have been subjected to direct photolysis (Sch. 34). Upon irradiation at 313 nm, 4-halostilbenes give

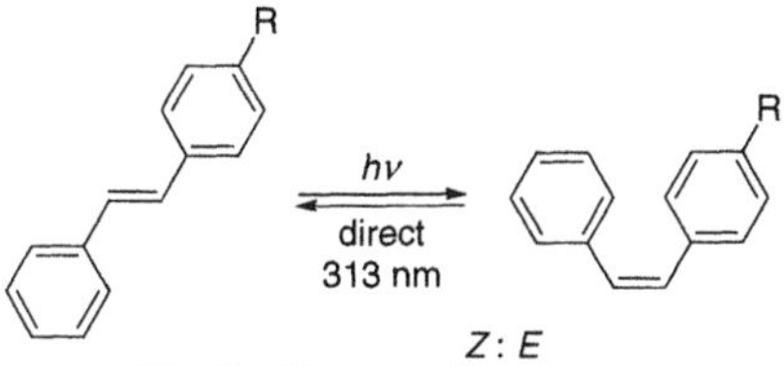

	Z : E
90a: R = H	93 : 7
90b: R = NH$_2$	81 : 19
90c: R = OMe	85 : 15
90d: R = F	89 : 11
90e: R = Cl	91 : 9
90f: R = Br	88 : 12
90g: R = CN	81 : 19
90h: R = NO$_2$	68 : 32

Scheme 34

high $(Z/E)_{pss}$ ratios of ~9:1, which are comparable to that observed for unsubstituted stilbene [214], whilst 4-amino- and 4-methoxystilbenes give appreciably lower Z/E ratios of 81:19 and 85:15, respectively. Three-substituted stilbenes have also been similarly studied. Photoisomerization of halostilbenes have been studied extensively in relation to the heavy atom effects on the singlet–triplet intersystem crossing process [201,215–217].

Photoisomerization of 4-methoxy-4'-nitrostilbene **91** suffers from a strong solvent effect (Sch. 35). The $(Z/E)_{pss}$ ratio upon excitation at 366 nm is 91:9 in nonpolar petroleum ether, but the Z/E preference is dramatically switched to give the Z/E ratio of 17:83 in polar dimethylformamide [218]. Laser flash photolysis studies revealed the multiplicity of the excited state involved. The singlet mechanism is operative in the photoisomerization of (E)-4-cyano-4'-methoxystilbene and (E)-4-methoxy-4'-nitrostilbene [219,220], while the photoisomerization of nitrostilbenes involves the triplet state [221].

Direct and sensitized photoisomerizations of 1,2-diphenylpropene **92** have also been examined (Sch. 36). The $(Z/E)_{pss}$ ratio upon sensitization in benzene varies critically with the sensitizer employed, thus giving 54:46 for acetophenone, 17:83 for eosin, 89:11 for fluorenone and 90:10 for duroquinone [222].

Direct irradiation at 254 nm of 1-phenylpropene **93a** in cyclohexane affords a Z/E mixture in a ratio of 65:35 [223], whilst photosensitized isomerization of **93a** gives $(Z/E)_{pss}$ ratios of 55:45 and 88:12 upon benzophenone and chrysene sensitization, respectively (Sch. 37). Similarly, a variety of β-alkylstyrenes has been subjected to triplet sensitization [224,225]. Upon direct irradiation of (E)-β-bromo-β-nitro-styrenes, the (Z)-isomers are obtained in 40–50% yield, but the major

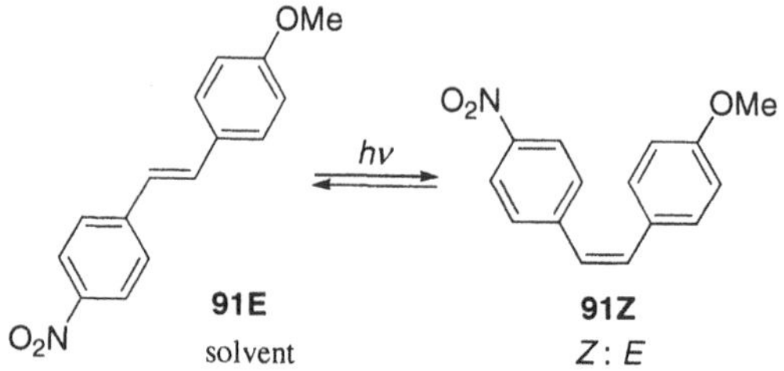

91E **91Z**

solvent	Z : E
petroleum ether	91 : 7
benzene	73 : 27
chloroform	56 : 44
ethanol	60 : 40
methanol	71 : 29
dimethylformamide	83 : 17

Scheme 35

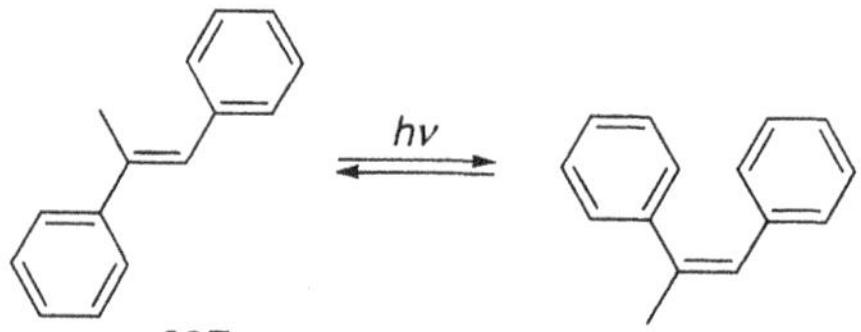

92E **92Z**

Scheme 36

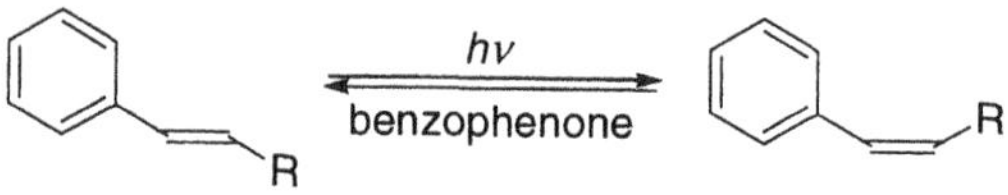

	Z : E
93aE: R = Me	63 : 37
93bE: R = Et	63 : 37
93cE: R = *i*-Pr	70 : 30
93dE: R = *t*-Bu	93 : 7
93eE: R = *t*-C_5H_{11}	94 : 6

Scheme 37

products are the corresponding oximes [226]. The temperature effects on the Z/E photoisomerization of substituted arylpropenes has recently been investigated in detail [227].

Photosensitized isomerization of fluoranthenylalkenes **94** has been studied in search of a "one-way" isomerization system [228–231].

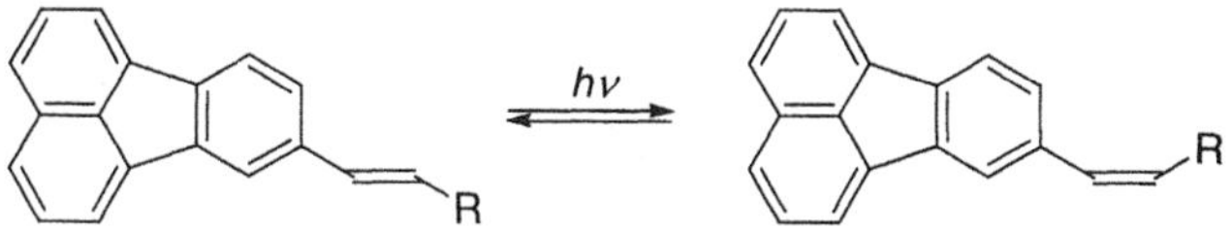

94aE: R = Bu^t
94bE: R = Ph

94Z

Scheme 38

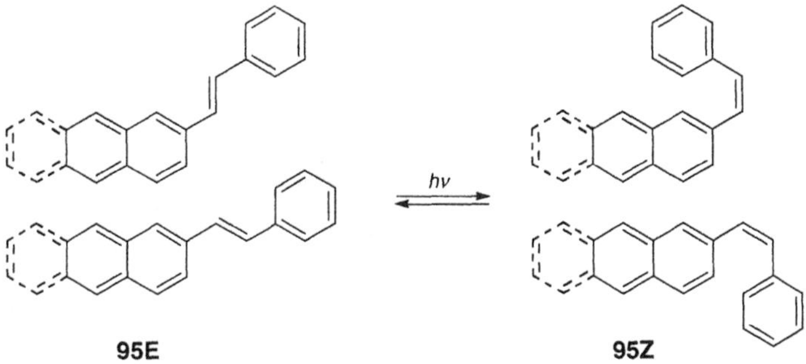

95E

95Z

Scheme 39

8-(3,3-Dimethyl-1-butenyl)fluoranethene **94a** undergoes a Z-to-E one-way isomerization [231], but 8-styrylfuruoranthene 94b affords a Z/E mixture upon camphorquinone sensitization in benzene (Sch. 38) [229], Interestingly, the $(Z/E)_{pss}$ ratio in the latter case is concentration dependent, varying from 57:43 at 0.5 mM to 18:82 at 8.5 mM, for which the equilibrium between (E)-triplet and twisted triplet species is thought to be responsible.

In the photoisomerization of 1-(2-anthryl)-2-phenylethene [232–234] and 1-(2-naphthyl)-2-phenylethene [235–237] **95**, the *s-cis/s-trans* conformational isomerism plays an important role. (Z)-1-(2-Naphthyl)-2-phenylethene is isolated in ~70% yield in fluorenone-sensitized isomerization of the (E)-isomer (Sch. 39) [238,239]. The photochemical behavior of related (E)-stilbenoids (**96–98**) has been reported (Sch. 40) [240]. The direct and sensitized photoisomerizations of (Z)-styrylferrocene in benzene have also been reported [241].

Photoisomerization of charge-transfer (CT) complexes of (E)- and (Z)-1,2-*bis*-(1-methyl-4-pyridinio)ethenes **99** with iodine can be effected by irradiating at the CT band in acetonitrile (Sch. 41). The CT excitation of the

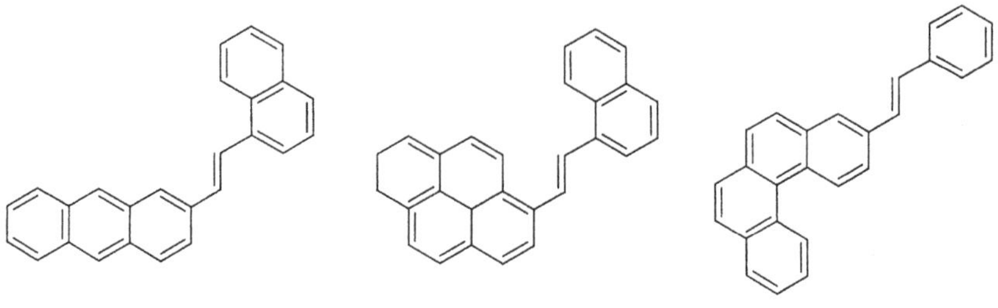

96   **97**   **98**

Scheme 40

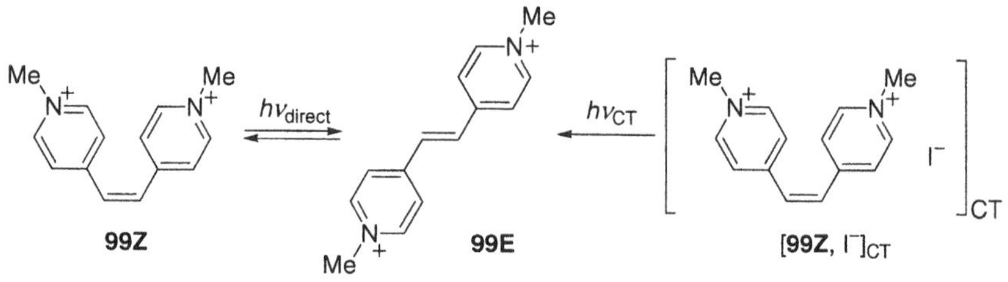

99Z   **99E**   [**99Z, I⁻**]$_{CT}$

Scheme 41

99Z·I₂ complex leads to the formation of (Z)- as well as (E)-radicals of **99**. In the absence of iodine, the free **99** salt isomerizes reversibly upon direct irradiation [242].

   In Wittig alkylidenations producing stilbene/styrene derivatives, the E/Z ratio of the product has been reported to increase by irradiation with a daylight lamp, reaching up to 99:1 for 1-aryl-2-alkylethenes and ca. 3:1 for 1,2-dialkylethenes [243]. Similar effects are known for γ-keto-α,β-unsaturated esters [244].

14.7.   REPRESENTATIVE SYNTHETIC PROCEDURES

14.7.1.   (E)-Cyclo-octene [28]

Direct irradiation of (Z)-cyclo-octene **38Z** at 185 nm affords a high $(E/Z)_{pss}$ ratio of unity in the absence of possible contamination by sensitizers and/or its decomposition products [27]. Nevertheless, the use of a photosensitization route is recommended, as only conventional equipment such as a low-pressure mercury lamp and quartz vessels are needed. Preparative-scale irradiation of (Z)-cyclo-octene **38Z** is conducted as follows. A pentane solution pentane (250 mL), containing 11 g (0.1 mol) of **38Z** and 1.0 g of dimethyl isophthalate (5 mmol) is placed in a donut-shaped annular quartz vessel (5 cm i.d. and 20 cm in height), purged with argon at −10 °C and irradiated at 254 nm with a 50-W low-pressure mercury lamp (Eikosha) in a water bath at 25 °C. Aliquots from the irradiated solution are occasionally monitored by GC for determination of the E/Z ratio. After 24 h irradiation, when the E/Z ratio approaches the photostationary state of ∼0.3, the irradiated solution is concentrated (to ca. 50 mL) and the concentrate extracted with 20% aqueous silver nitrate (3 × 25 mL) at ∼0 °C. The combined extracts are washed with pentane (2 × 25 mL) and then added dropwise to a stirred solution of ice-cooled 15% aqueous ammonia (100 mL). The resulting mixture is extracted with pentane (3 × 25 mL), washed with water (2 × 25 mL), and dried over anhydrous magnesium sulfate. After filtration, the solvent is removed by rotary evaporation at ca. 80–100 Torr. A further purification is performed by trap-to-trap distillation to afford 2.2 g (20% yield) of (E)-isomer **38E** of >99.5% GC purity.

14.7.2. Optically Active (E)-Cyclo-octene [147]

A pentane solution (200 mL) of (Z)-cyclo-octene **38Z** (3.3 g, 30 mmol) and (−)-tetrabornyl 1,2,4,5-benzenetetracarboxylate (0.5 g, 0.7 mmol) is irradiated at 254 nm for 72 h under argon in a methanol bath kept at −88 °C by using a Cryocool CC-100II (NESLAB). The irradiated solution is worked up as above and 0.2 g (6% yield) of chemically pure (E)-isomer **38E** is obtained. The sample is analyzed by chiral GC with a Supelco beta-DEX 225) column (0.25 mm × 30 m, 0.25 μm) to show 41% ee in favor of (R)-(−)-isomer. Using (−)-tetrakis(8-cyclohexylmenthyl)1,2,4,5-benzenetetracarboxylate as a sensitizer gives (S)-(+)-**38E** with a 64% ee in 4% yield [113]. Note that the product ee is highly temperature dependent and the above-mentioned temperature should be seriously taken [148,149]. It is also noted that as entropic factors govern the enantioselectivity the product ee is not always enhanced by lowering the irradiation temperature.

14.7.3. (Z)-Stilbene [245]

For general synthetic purposes, any of the direct and sensitized photo-isomerization routes can be used in the preparation of (E)-stilbenes from the corresponding (Z)-isomers. However, if really pure (Z)-stilbene **90aZ** is required, fluorenone sensitization is highly recommended. Otherwise, trace amounts of (E)-stilbene and bibenzyl remaining in **90aZ** sample prepared photochemically may lead to erroneous results/analyses, for example, in fluorescence measurements. Commercial (E)-stilbene **90aE**, which contains 3.5% (Z)-stilbene and 0.5% bibenzyl, is first purified by sublimation, chromatography on alumina and recrystallization from ethanol to give a sample containing <0.01% bibenzyl. 9-Fluorenone is also purified by chromatography and recrystallization from methanol. Purified (E)-stilbene (14.0 g, 78 mmol) and 9-fluorenone (6.0 g, 33 mmol) are dissolved in 200 mL of benzene (Aldrich, spectrometric grade), and the solution is placed in a donut-shaped annular Pyrex vessel (5 cm i.d. and 20 cm in height) and is irradiated at >280 nm with a medium-pressure mercury lamp under an argon atmosphere in a water bath at 25 °C. After 4 h of irradiation, when the E/Z ratio monitored by GC approaches to the photostationary state (Z/E ~ 5.5), the solution is concentrated to ca. 50 mL. The crude product is chromatographed on neutral alumina using pentane as an eluent. The first elute should be (Z)-stilbene, but the later fractions are contaminated by bibenzyl. Repeated column chromatography purification affords bibenzyl-free (Z)-stilbene **90aZ** (6.5 g, 46% yield).

14.8. CONCLUSION

As can be clearly recognized from the above discussion, the photochemistry of alkenes represents a distinctive area of photochemistry that provides valuable insights into the excited state, mechanism, reactivity, synthesis, and chiral induction. Photoisomerization of alkenes has long been a target of mostly mechanistic interest, but currently attracts more attention from the viewpoints of (asymmetric) synthesis and materials science. Direct irradiation of (Z)-cycloalkenes leads to a variety of photoreactions, including the Z–E isomerization, rearrangement, decomposition, cyclodimerization, and addition reactions through the π,π^*, π,σ^*, and $\pi,R(3s)$ excited states. Although both singlet and triplet sensitization can be employed for the photoisomerization of cycloalkenes, the singlet manifold is advantageous for obtaining higher photostationary-state E/Z ratios, particularly when the (E)-isomer is highly constrained. Environmental factors, such as temperature, concentration, pressure, and solvent can affect the photo-stationary E/Z ratio in sensitized photoisomerizations of cyclo-octene and derivatives.

Recent studies on the enantiodifferentiating photosensitization reveal that the singlet sensitization with appropriate enantiopure compounds leads to good-to-excellent enantiodifferentiating photoisomerization of cyclohexene, cycloheptene, and cyclo-octene through the corresponding diastereomeric exciplex intermediates. This strategy is applicable in general to the preparation of optically active (E)-cycloalkenes. The enantiodifferentiating rotational relaxation of alkene's double bond in the exciplex, and therefore the product ee, are known to be very sensitive not only to the sensitizer energy and structure, but also to the entropy-related environmental factors such as temperature, pressure, and solvent.

Mechanistic studies of the photoisomerization of stilbenes have fascinated photochemists for more than six decades; with these studies significantly contributing to synthetic photochemistry. Photochemical synthesis of the thermodynamically unfavorable (Z)-isomers can be achieved either by direct or sensitized irradiations. The photostationary Z/E ratio of stilbene obtained upon direct irradiation is simply governed by the irradiation wavelength, or the excitation ratio at the irradiation wavelength. In contrast, the Z/E ratio obtained upon triplet sensitization does not show a simple dependence on the triplet energy, affording the optimum ratio when the triplet-energy transfer to the (Z)-isomer is highly retarded in comparison to the (E)-isomer. The geometrical photoisomerization around the olefinic double bond is the essential part of the photochromic behavior of diarylethenes and light-driven chiroptical molecular motors.

Based on a rich heritage of photophysical and photochemical, as well as mechanistic and synthetic studies which have been pursued in the second half of the last century, the photochemistry of alkenes is expected to find further applications in chemical, physical, biological, and materials sciences and technologies.

REFERENCES

1. Wald G. Exp Eye Res 1974; 18:333.
2. Wald G. Origins Life 1974; 5:7.
3. Wald G. Biochem Physiol Visual Pigments, Symp 1973; 1.
4. Wald G. Prix Nobel 1968; 260.
5. Wald G. Science 1968; 162:230.
6. Herzfeld J, Lansing JC. Ann Rev Biophys Biomol Struct 2002; 31:73.
7. Kandori H, Shichida Y, Yoshizawa T. Biochemistry 2001; 66:1197.
8. Okada T, Palczewski K. Curr Op Struct Biol 2001; 11:420.
9. Spudich JL. Science (Washington, DC) 2000; 288:1358.
10. Yoshizawa T. Adv Biophys 1984; 17:5.
11. Saltiel J, Sears, DF Jr, Ko D-H, Park K-M. In: Horspool WM, ed. CRC Handbook of Organic Photochemistry and Photobiology. Florida: CRC Press, Inc., 1995:3.

12. Kropp PJ. In: Horspool WM, ed. CRC Handbook of Organic Photochemistry and Photobiology. Florida: CRC Press, Inc., 1995:16.
13. Collin GJ, De Mare GR. J Photochem 1987; 38:205.
14. Saltiel J, Charlton JL. In: de Mayo P, ed. Rearrangements in Ground and Excited States. New York: Academic Press, 1980:25.
15. Kropp PJ. Mol Photochem 1978; 9:39.
16. Saltiel J, D'Agostino J, Megarity DE, Metts L, Neuberger KR, Wrighton M, Zafiriou OC. Org Photochem 1973; 3:1.
17. Kropp PJ. Pure Appl Chem 1970; 24:585.
18. Turro NJ. Photochem Photobiol 1969; 9:555.
19. Steinmetz MG. Org Photochem 1987; 8:67.
20. Kropp PJ. Org Photochem 1979; 4:1.
21. Langer K, Mattay J. In: Horspool WM, ed. CRC Handbook of Organic Photochemistry and Photobiology. Florida: CRC Press, Inc., 1995:105.
22. Kropp PJ. Mol Photochem 1978–1979; 9:39.
23. Saltiel J, D'Agostino J, Megarity ED, Metts L, Neuberger KR, Wrighton M, Zafiriou OC. Org Photochem 1973; 3:1.
24. Horspool WM. Photochemistry 1971; 2:427.
25. Marshall JA. Science 1970; 170:137.
26. Kropp PJ. In: Horspool WM, ed. CRC Handbook of Organic Photochemistry and Photobiology. Florida: CRC Press, Inc., 1995:105.
27. Inoue Y, Takamuku S, Sakurai H. Synthesis 1977; 111.
28. Yamasaki N, Inoue Y, Yokoyama T, Tai A. J Photochem Photobiol A 1989; 48:465.
29. Photochirogenesis, 1st International Symposium on asymmetric photochemistry. Osaka, Japan, Sept 4–6, 2001.
30. Jacobs HJC, Havinga E. Adv Photochem 1979; 11:305.
31. Jacobs HJC, Gielen JWJ, Havinga E. Proc Workshop Vitam D 1979; 4th:61.
32. Havinga E. Chem Weekbl Mag 1979; 427.
33. Havinga E. Chimia 1976; 30:27.
34. Havinga E. Experientia 1973; 29:1181.
35. Sanders GM, Pot J, Havinga E. Fortschr Chem Org Naturst 1969; 27:131.
36. Zechmeister L. *Cis–trans* Isomeric Carotenoids, Vitamin A, and Arylpolyenes. New York: Academic Press, 1962.
37. Ramamurthy V, Liu RSH. Org Photochem Synth 1976; 2:70.
38. Ramamurthy V, Liu RSH. J Am Chem Soc 1976; 98:2935.
39. Ramamurthy V, Tustin G, Yau CC, Liu RSH. Tetrahedron 1975; 31:193.
40. Ramamurthy V, Liu RSH. J Am Chem Soc 1974; 96:5625.
41. Ramamurthy V, Liu RSH. Tetrahedron Lett 1973; 1393.
42. Ramamurthy V, Liu RSH. Tetrahedron Lett 1973; 441.
43. Ramamurthy V, Butt Y, Yang C, Yang P, Liu RSH. J Org Chem 1973; 38:1247.
44. Simmons CJ, Colmenares LU, Liu RSH. Tetrahedron Lett 1996; 37:4103.
45. Chen R-L, Liu RSH. Tetrahedron 1996; 52:7809.
46. Zhu Y, Ganapathy S, Trehan A, Asato AE, Liu RSH. Tetrahedron 1992; 48:10061.

47. Trehan A, Mirzadegan T, Liu RSH. Tetrahedron 1990; 46:3769.
48. Liu RSH. Ganguang Kexue Yu Guang Huaxue 2002; 20:81.
49. Liu RSH, Hammond GS. Chem Eur J 2001; 7:4536.
50. Liu RSH. Acc Chem Res 2001; 34:555.
51. Liu RSH, Hammond GS. Proc Natl Acad Sci USA 2000; 97:11153.
52. Mulliken RS. Acc Chem Res 1976; 9:7.
53. Watson FH Jr, McGlynn SP. Theor Chim Acta 1971; 21:309.
54. Watson FH Jr, Armstrong AT, McGlynn SP. Theor Chim Acta 1970; 16:75.
55. Merer AJ, Mulliken RS. Chem Rev 1969; 69:639.
56. Kropp PJ, Fravel HG, Jr, Fields TR. J Am Chem Soc 1976; 98:840.
57. Fravel HG Jr, Kropp PJ. J Org Chem 1975; 40:2434.
58. Fields TR, Kropp PJ. J Am Chem Soc 1974; 96:7559.
59. Reardon EJ Jr, Kropp PJ. J Am Chem Soc 1971; 93:5593.
60. Kropp PJ, Reardon EJ Jr, Gaibel ZLF, Williard KF, Hattaway JH Jr. J Am Chem Soc 1973; 95:7058.
61. Kropp PJ, Tise FP. J Am Chem Soc 1981; 103:7293.
62. Inoue Y, Goan K, Hakushi T. Bull Chem Soc Jpn 1985; 58:2217.
63. Tonaka M, Terao T, Sato S. Bull Chem Soc Jpn 1965; 38:1645.
64. Cundall RB, Fletcher FJ, Milne DG. J Chem Phys 1963; 39:3536.
65. Sigal P. J Chem Phys 1965; 42:1953.
66. Shekk YB. Khim Vys Energ 1973; 7:221.
67. Saltiel J, Neuberger KR, Wrighton M. J Am Chem Soc 1969; 91:3658.
68. Golub MA, Stephens CL. J Phys Chem 1966; 70:3576.
69. Snyder JJ, Tise FP, Davis RD, Kropp PJ. J Org Chem 1981; 46:3609.
70. Teng PP, Weitz E, Lewis FD. J Am Chem Soc 1982; 104:5518.
71. Padwa A, Chou CS, Rosenthal RJ, Terry LW. J Org Chem 1988; 53:4193.
72. Zimmerman HE, Bunce RA. J Org Chem 1982; 47:3377.
73. Lawless MK, Wickham SD, Mathies RA. J Am Chem Soc 1994; 116:1593.
74. Leigh WJ. Can J Chem 1993; 71:147.
75. Leigh WJ, Postigo JA. J Am Chem Soc 1995; 117:1688.
76. Leigh WJ, Postigo JA, Venneri PC. J Am Chem Soc 1995; 117:7826.
77. Inoue Y, Mukai T, Hakushi T. Chem Lett 1982; 1045.
78. Adam W, Oppenlaender T. J Am Chem Soc 1985; 107:3924.
79. Kropp PJ, Snyder JJ, Rawlings PC, Fravel HG Jr. J Org Chem 1980; 45:4471.
80. Inoue Y, Takamuku S, Sakurai H. J Chem Soc, Perkin Trans 2 1977; 1635.
81. Ziegler VK, Wilms H. Justus Libigs Ann Chem 1950; 567:1.
82. Cope AC, Pike RA, Spencer CF. J Am Chem Soc 1953; 75:3212.
83. Kropp PJ, Mason JD, Smith GFH. Can J Chem 1985; 63:1845.
84. Inoue Y, Takamuku S, Sakurai H. J Chem Soc, Chem Commun 1976, 423.
85. Inoue Y, Takamuku S, Kunitomi Y, Sakurai H. J Chem Soc, Perkin Trans 2 1980; 1672.
86. Inoue Y, Takamuku S, Sakurai H. J Phys Chem 1977; 81:7.
87. Inoue Y, Takamuku S, Sakurai H. Bull Chem Soc Jpn 1976; 49:1147.
88. Haufe G, Tubergen MW, Kropp PJ. J Org Chem 1991; 56:4292.
89. Kropp PJ, Krauss, HJ. J Am Chem Soc 1967; 89:5199.

90. Asaoka S, Horiguchi H, Wada T, Inoue Y. J Chem Soc, Perkin Trans 2 2000; 737.
91. Shim SC, Kim DS, Yoo DJ, Wada T, Inoue Y. J Org Chem 2002; ACS ASAP.
92. Kim DS, Shim SC, Wada T, Inoue Y. Tetrahedron Lett 2001; 42:4341.
93. Tise FP, Kropp PJ. Org Synth 1983; 61:112.
94. Kropp PJ. J Org Chem 1970; 35:2435.
95. Kropp PJ. J Am Chem Soc 1969; 91:5783.
96. Dauben WG, Van Riel HCHA, Robbins JD, Wagner GJ. J Am Chem Soc 1979; 101:6383.
97. Dauben WG, Van Riel HCHA, Hauw C, Leroy F, Juoussot-Dubien J, Bonneau R. J Am Chem Soc 1979; 101:1901.
98. Rosenberg HM. J Org Chem 1972; 37:141.
99. Zimmerman HE, Kamm KS, Werthemann DP. J Am Chem Soc 1975; 97:3718.
100. Bonneau R, Joussot-Dubien J, Salem L, Yarwood AJ. J Am Chem Soc 1976; 98:4329.
101. Caldwell RA, Misawa H, Healy EF, Dewar MJS. J Am Chem Soc 1987; 109:6869.
102. Gibson TW, Majeti S, Barnett BL. Tetrahedron Lett 1976; 4801.
103. Squillacote M, Bergman A, De Felippis J. Tetrahedron Lett 1989; 30:6805.
104. Inoue Y, Ueoka T, Kuroda T, Hakushi T. J Chem Soc, Chem Commun 1981; 1031.
105. Inoue Y, Ueoka T, Kuroda T, Hakushi T. J Chem Soc, Perkin Trans 2 1983; 983.
106. Bonneau R, Joussot-Dubien J, Yarwood J, Pereyre J. Tetrahedron Lett 1977; 235.
107. Bryce-Smith D, Gilbert A, Orger BH. J Chem Soc, Chem Commun 1966; 1966:512.
108. Bryce-Smith D. Pure Appl Chem 1968; 16:47.
109. Bryce-Smith D. Pure Appl Chem 1973; 34:193.
110. Swenton JS. J Org Chem 1969; 34:3217.
111. Inoue Y, Kobata T, Hakushi T. J Phys Chem 1985; 89:1973.
112. Inoue Y, Yamasaki N, Tai A, Daino Y, Yamada T, Hakushi T. J Chem Soc, Perkin Trans 2 1990; 1389.
113. Inoue Y, Yamasaki N, Yokoyama T, Tai A. J Org Chem 1993; 58:1011.
114. Inoue Y, Yamada T, Daino Y, Kobata T, Yamasaki N, Tai A, Hakushi T. Chem Lett 1989; 1933.
115. Tsuneishi H, Inoue Y, Hakushi T, Tai A. J Chem Soc, Perkin Trans 2 1993; 457.
116. Strickland AD, Caldwell RA. J Phys Chem 1993; 97:13394.
117. Yokoyama K, Kato M, Noyori R. Bull Chem Soc Jpn 1977; 50:2201.
118. Henin F, Muzart J, Pete JP, Rau H. Tetrahedron Lett 1990; 31:1015.
119. Nozaki H, Nishikawa Y, Kawanisi M, Noyori R. Tetrahedron 1967; 23:2173.
120. Dauben WG, Poulter CD, Suter C. J Am Chem Soc 1970; 92:7408.

121. Nakazaki M, Yamamoto K, Yanagi J. J Chem Soc, Chem Commun 1977; 346.
122. Nakazaki M, Yamamoto K, Yanagi J. J Am Chem Soc 1977; 101:147.
123. Nebe WJ, Fonken GJ. J Am Chem Soc 1969; 91:1249.
124. Inoue Y, Daino Y, Hagiwara S, Nakamura H, Hakushi T. J Chem Soc, Chem Commun 1985; 804.
125. Inoue Y, Hagiwara S, Daino Y, Hakushi T. J Chem Soc, Chem Commun 1985; 1307.
126. Goto S, Takamuku S, Sakurai H, Inoue Y, Hakushi T. J Chem Soc, Perkin Trans 2 1980; 1678.
127. Eaton PE, Lin K. J Am Chem Soc 1964; 86:2087.
128. Corey EJ, Tada M, LaMahieu R, Libit L. J Am Chem Soc 1965; 87:2051.
129. Eaton PE, Lin K. J Am Chem Soc 1965; 87:2052.
130. Bonneau R, de Violet PF, Joussot-Dubien J. Nouv J Chim 1976; 1:31.
131. Noyori R, Kato M. Bull Chem Soc Jpn 1974; 47:1460.
132. Hart H, Dunkelblum E. J Am Chem Soc 1978; 100:5141.
133. Rau H. Chem Rev 1983; 83:535.
134. Inoue Y. Chem Rev 1992; 92:741.
135. Inoue Y. J Synth Org Chem, Jpn 1995; 53:348.
136. Everitt SRL, Inoue Y. In: Schanze KS, ed. Organic Molecular Photochemistry. New York: Marcel Dekker, Inc., 1999:71.
137. Inoue Y, Wada T, Asaoka S, Sato H, Pete J-P. Chem Commun 2000; 251.
138. Cope AC, Ganellin CR, Johnson HW Jr, Van Auken TV, Winkler HJS. J Am Chem Soc 1963; 85:3276.
139. Corey EJ, Shulman JI. Tetrahedron Lett 1968; 3655.
140. Corey EJ, Carey FA, Winter RAE. J Am Chem Soc 1965; 87:934.
141. Inoue Y, Kunitomi Y, Takamuku S, Sakurai H. J Chem Soc, Chem Commun 1978; 1024.
142. Shi M, Inoue Y. Aust J Chem 2001; 54:113.
143. Inoue Y, Matsuyama E, Wada T. J Am Chem Soc 1998; 120:10687.
144. Shi M, Inoue Y. J Chem Soc, Perkin Trans 2 1998; 1725.
145. Shi M, Inoue Y. J Chem Soc, Perkin Trans 2 1998; 2421.
146. Tsuneishi H, Hakushi T, Inoue Y. J Chem Soc, Perkin Trans 2 1996; 1601.
147. Inoue Y, Yamasaki N, Yokoyama T, Tai A. J Org Chem 1992; 57:1332.
148. Inoue Y, Yokoyama T, Yamasaki N, Tai A. J Am Chem Soc 1989; 111: 6480.
149. Inoue Y, Yokoyama T, Yamasaki N, Tai A. Nature 1989; 341:225.
150. Raupach E, Rikken GLJA, Train C, Malezieux B. Chem Phys 2000; 261:373.
151. Rikken GLJA, Raupach E. Nature (London) 2000; 405:932.
152. Inoue Y, Ikeda H, Kaneda M, Sumimura T, Everitt SRL, Wada T. J Am Chem Soc 2000; 122:406.
153. Inoue Y, Kosaka S, Matsumoto K, Tsuneishi H, Hakushi T, Tai A, Nakagawa K, Tong L-H. J Photochem Photobiol A 1993; 71:61.
154. Inoue Y, Dong F, Yamamoto K, Tong L-H, Tsuneishi H, Hakushi T, Tai A. J Am Chem Soc 1995; 117:11033.

155. Inoue Y, Wada T, Sugahara N, Yamamoto K, Kimura K, Tong L-H, Gao X-M, Hou Z-J, Liu Y. J Org Chem 2000; 65:8041.
156. Wada T, Shikimi M, Inoue Y, Lem G, Turro N. J Chem Commun 2001; 1864.
157. Wada T, Sugahara N, Kawano M, Inoue Y. Chem Lett 2000; 1174.
158. Tsuneishi H, Hakushi T, Tai A, Inoue Y. J Chem Soc, Perkin Trans 2 1995; 2057.
159. Inoue Y, Tsuneishi H, Hakushi T, Tai A. J Am Chem Soc 1997; 119:472.
160. Inoue T, Matsuyama K, Inoue Y. J Am Chem Soc 1999; 121:9877.
161. Matsuyama K, Inoue T, Inoue Y. Synthesis 2001; 1167.
162. Nakazaki M, Yamamoto K, Maeda M. J Chem Soc, Chem Commun 1980; 294.
163. Hoffmann R, Inoue Y. J Am Chem Soc 1999; 121:10702.
164. Hamaguchi H-o, Iwata K. Bull Chem Soc Jpn 2002; 75:883.
165. Papper V, Likhtenshtein GI. J Photochem Photobiol A 2001; 140:39.
166. Vachev VD, Frederick JH. Structure and Dynamics of Electronic Excited States 1999; 137.
167. Laane J. International Reviews in Physical Chemistry 1999; 18:301.
168. Hazai L, Hornyak G. ACH Mod Chem 1998; 135:493.
169. Goerner H, Kuhn HJ. Adv Photochem 1995; 19:1.
170. Waldeck DH. J Mol Liq 1993; 57:127.
171. Morita A. Bussei Kenkyu 1993; 60:647.
172. Waldeck DH. Chem Rev 1991; 91:415.
173. Saltiel J, Sun Y-P. In: Duërr H, Bouas-Laurent H, eds. Photochromism, Molecules and Systems. Amsterdam: Elsevier, 1990:64.
174. Park NS, Waldeck DH. J Chem Phys 1989; 91:943.
175. Park NS, Sivakumar N, Hoburg EA, Waldeck DH. Springer Ser Chem Phys 1988; 48:551.
176. Troe J. J Phys Chem 1986; 90:357.
177. Hamaguchi H. J Mol Struct 1985; 126:125.
178. Doany FE, Greene BI, Liang Y, Negus DK, Hochstrasser RM. Springer Ser Chem Phys 1980; 14:259.
179. Hochstrasser RM. Pure Appl Chem 1980; 52:2683.
180. Schulte-Frohlinde D, Goerner H. Pure Appl Chem 1979; 51:279.
181. Steinmetz MG. Org Photochem 1967; 1:247.
182. Laarhoven WH. Organic Photochemistry 1989; 10:163.
183. Mallory FB, Mallory CW. Org React (NY) 1984; 30:1.
184. Laarhoven WH. Recl: JR Neth Chem Soc 1983; 102:241.
185. Laarhoven WH. Recl: JR Neth Chem Soc 1983; 102:185.
186. Mallory CW, Mallory FB. Org Photochem Synth 1971; 1:55.
187. Dybans RA. Org Photochem Synth 1971; 1:25.
188. Mariano PS, Krochmal E Jr, Leone A. J Org Chem 1977; 42:1122.
189. Kitahara Y, Tanaka K. Chem Commun 2002; 932.
190. Tanaka K, Osuga H, Kitahara Y. J Org Chem 2002; 67:1795.
191. Stammel C, Froehlich R, Wolff C, Wenck H, De Meijere A, Mattay J. European Journal of Organic Chemistry 1999; 1709.

192. Moradpour A, Kagan H, Baes M, Morren G, Martin RH. Tetrahedron 1975; 31:2139.
193. Martin RH, Baes M. Tetrahedron 1975; 31:2135.
194. Irie M. Chem Rev 2000; 100:1685.
195. Irie M, Uchida K. Bull Chem Soc Jpn 1998; 71:985.
196. Koumura N, Geertsema EM, van Gelder MB, Meetsma A, Feringa BL. J Am Chem Soc 2002; 124:5037.
197. Feringa BL, van Delden RA, ter Wiel MKJ. Molecular Switches 2001; 123.
198. Feringa BL. Acc Chem Res 2001; 34:504.
199. Koumura N, Geertsema EM, Meetsma A, Feringa BL. J Am Chem Soc 2000; 122:12005.
200. Koumura N, Zijistra RWJ, Van Delden RA, Harada N, Feringa BL. Nature (London) 1999; 401:152.
201. Saltiel J, Marinari A, Chang DWL, Mitchener JC, Megarity ED. J Am Chem Soc 1979; 101:2982.
202. Peters KS, Freilich SC, Lee J. J Phys Chem 1993; 97:5482.
203. Ito Y, Kajita T, Kunimoto K, Matsuura T. J Org Chem 1989; 54:587.
204. Lewis FD, Johnson DE. J Photochem 1977; 7:421.
205. Ulrich H, Rao DV, Stuber FA, Sayigh AAR. J Org Chem 1970; 35:1121.
206. Stegemeyer H. Chimia (Aarau) 1965; 19:536.
207. Miyazawa T, Koshihara S-y, Segawa Y, Kira M. Chem Lett 1995; 217.
208. Miyazawa T, Liu C, Koshihara S-y, Kira M. Photochemistry and Photobiology 1997; 66:566.
209. Miyazawa T, Liu C, Kira M. Chem Lett 1997; 459.
210. Kuriyama Y, Arai T, Sakuragi H, Tokumaru K. Chem Phys Lett 1990; 173:253.
211. Kuriyama Y, Arai T, Sakuragi H, Tokumaru K. Chem Lett 1988; 1193.
212. Tolbert LM, Ali MZ. J Org Chem 1985; 50:3288.
213. Mattes SL, Farid S. Org Photochem 1983; 6:233.
214. Guesten H, Klasinc L. Tetrahedron Lett 1968; 3097.
215. Goerner H. J Photochem Photobiol A 1995; 90:57.
216. Krueger K, Lippert E. Zeit Phys Chem Neu Folg 1969; 66:293.
217. Shmuel M, Ernst F. J Phys Chem 1964; 68:1153.
218. Schulte-Frohlinde D. Justus Libigs Ann Chem 1958; 615:114.
219. Goerner H, Schulte-Frohlinde D. Ber Bunsenges Physik Chem 1977; 81:713.
220. Pisanias MN, Schulte-Frohlinde D. Ber Bunsenges Physik Chem 1975; 79:662.
221. Bent DV, Schulte-Frohlinde D. J Phys Chem 1974; 78:446.
222. Hammond GS, Saltiel J, Lamola AA, Turro NJ, Bradshaw JS, Cowan DO, Counsell RC, Vogt V, Dalton C. J Am Chem Soc 1964; 86:3197.
223. Nakagawa CS, Sigal P. J Chem Phys 1973; 58:3529.
224. Arai T, Sakuragi H, Tokumaru K. Bull Chem Soc Jpn 1982; 55:2204.
225. Caldwell RA, Sovocol GW, Peresue RJ. J Am Chem Soc 1971; 93:779.
226. Schaffer GW. Can J Chem 1970; 48:1948.
227. Lewis FD, Bassani DM, Caldwell RA, Unett DJ. J Am Chem Soc 1994; 116:10477.

228. Furuuchi H, Kuriyama Y, Arai T, Sakuragi H, Tokumaru K. Bull Chem Soc Jpn 1991; 64:1601.
229. Furuuchi H, Arai T, Kuriyama Y, Sakuragi H, Tokumaru K. Chem Phys Lett 1989; 162:211.
230. Arai T, Tokumaru K. Photomed Photobiol 1988; 10:51.
231. Arai T, Kuriyama Y, Karatsu T, Sakuragi H, Tokumaru K, Oishi S. J Photochem 1987; 36:125.
232. Karatsu T, Itoh H, Nishigaki A, Fukui K, Kitamura A, Matsuo S, Misawa H. J Phys Chem A 2000; 104:6993.
233. Saltiel J, Zhang Y, Sears DF Jr. J Am Chem Soc 1997; 119:11202.
234. Saltiel J, Zhang Y, Sears DF Jr. J Am Chem Soc 1996; 118:2811.
235. Hamond GS, Shim SC, Van SP. Mol Photochem 1969; 1:89.
236. Saltiel J, Choi JO, Sears DF Jr, Eaker DW, O'Shea KE, Garcia I. J Am Chem Soc 1996; 118:7478.
237. Saltiel J, Tarkalanov N, Sears DF Jr. J Am Chem Soc 1995; 117:5586.
238. Goerner H, Eaker DW, Saltiel J. J Am Chem Soc 1981; 103:7164.
239. Saltiel J, Eaker DW. Chem Phys Lett 1980; 75:209.
240. Bartocci G, Elisei F, Spalletti A. Gazz Chim Ital 1994; 124:89.
241. Richards JH, Pisker-Trifunac N. J Paint Technol 1969; 41:363.
242. Ebbesen TW, Tokumaru K, Sumitani M, Yoshihara K. J Phys Chem 1989; 93:5453.
243. Matikainen JK, Kaltia S, Hase T. Synlett 1994; 817.
244. Sestrick MR, Miller M, Hegedus LS. J Am Chem Soc 1992; 114:4079.
245. Saltiel J, Waller AS, Sears DF Jr. J Am Chem Soc 1993; 115:2453.

15
Photoinduced CX Cleavage of Benzylic Substrates

Angelo Albini and Maurizio Fagnoni
Dipartimento Chimica Organica, Università di Pavia, Pavia, Italy

15.1. HISTORICAL REMARKS

Like most photochemical reactions, benzylic fragmentation has been known from the beginning of the twentieth century. The photoinduced cleavage of triphenylmethylcarbinol has been noted by Gomberg in Ann Arbor in 1913 [1] and the generation of triarylmethyl cationic dyes from their leuco form by Lifschitz in Zürich in 1919 [2]. A solution of the colorless derivative obtained from *p*-rosaniline and cyanide became colored in a few seconds when exposed to an iron arc lamp (Sch. 1, old and new notation).

The author showed, that only UV light was active and that a polar solvent was required and described the reaction as a benzylic heterolysis. This and related photochemical reactions with the leuco derivatives of crystal violet and malachite green were later studied in detail in their mechanism [3–9] and were among the first compounds proposed as chemical actinometers [10–12], although in the event these were found to be less suited than other derivatives, since the high sensitivity was counterbalanced by the high absorption of the photoproducts in the UV.

Simpler benzylic systems have been studied only later. The resurgence of organic photochemistry in the sixties touched this field, in particular with a seminal paper by Zimmerman and Sandel in 1963 [13], where it was shown that photoinduced solvolysis in substituted benzylic derivatives underwent a different substituent effect with respect to the ground state reaction. Activation by a *m*-methoxy group was found and rationalized in terms of LCAO MO electron distribution in the first electronic excited state.

453

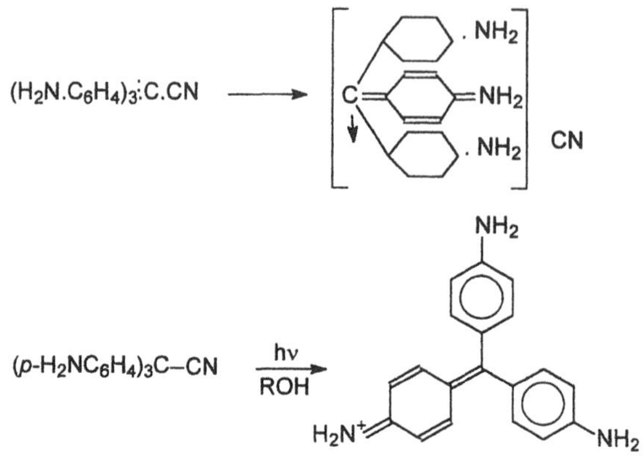

Scheme 1

The activity grew steadily in the field: the review by Cristol and Bindel in 1983 [14] featured over 120 references and the one by Fleming and Pincock in 1999 over 160 [15]. Another mode of benzylic cleavage, intermolecular benzylic hydrogen abstraction by excited ketones leading to bibenzyls has been discovered in 1909 by Paternò [16,17], and extended by other laboratories in the following years [18], in particular by Schönberg in the forties, when he was working in Egypt [19–21]. The intramolecular version of the reaction had been early discovered for the nitro derivatives, and indeed recognized by Sachs as a general property of o-nitrobenzylic derivatives (Sachs' rule) already in 1904 [22].

In turn, photosensitized electron transfer fragmentation of a benzylic bond has been among the first reaction reported when this branch of photochemistry began to take shape in the seventies, with the decarboxylation of phenylacetic acids reported by Libman [23] and the C–C cleavage in phenylethyl methyl ethers reported by Arnold [24].

15.2. STATE OF THE ART MECHANISTIC MODELS

The relative stability of benzylic radicals and cations makes photoinduced C–X cleavage of benzylic derivatives a widespread occurrence, and in fact according to different modes. These are i) fragmentation of an excited state, formed by direct irradiation or via energy transfer sensitization; ii) atom abstraction by a "chemical" sensitizer; iii) fragmentation of a benzylic radical cation or anion generated via electron transfer sensitization or quenching (Sch. 2).

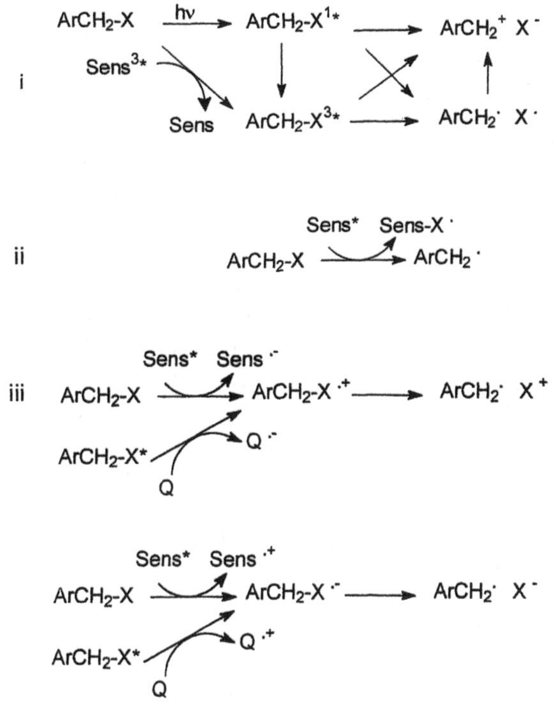

i
$$ArCH_2\text{-}X \xrightarrow{h\nu} ArCH_2\text{-}X^{1*} \longrightarrow ArCH_2^+ \ X^-$$

$$Sens^{3*}$$

$$Sens \quad ArCH_2\text{-}X^{3*} \longrightarrow ArCH_2^{\cdot} \ X^{\cdot}$$

ii
$$\overset{Sens^* \quad Sens\text{-}X^{\cdot}}{ArCH_2\text{-}X \longrightarrow ArCH_2^{\cdot}}$$

iii
$$\overset{Sens^* \quad Sens^{\cdot -}}{ArCH_2\text{-}X \longrightarrow ArCH_2\text{-}X^{\cdot +} \longrightarrow ArCH_2^{\cdot} \ X^+}$$

$$\overset{ArCH_2\text{-}X^*}{\quad\quad\quad Q}\nearrow^{Q^{\cdot -}}$$

$$\overset{Sens^* \quad Sens^{\cdot +}}{ArCH_2\text{-}X \longrightarrow ArCH_2\text{-}X^{\cdot -} \longrightarrow ArCH_2^{\cdot} \ X^-}$$

$$\overset{ArCH_2\text{-}X^*}{\quad\quad\quad Q}\nearrow^{Q^{\cdot +}}$$

Scheme 2

15.2.1. Fragmentation of an Excited State

The photoinduced C–X cleavage can involve either the singlet or the triplet state, the latter being reached through intersystem crossing or by energy transfer sensitization, and occurs either in a homolytic or in a heterolytic mode. Products arising from radicals (such as bibenzyls) or from ions, e.g., from the combination of benzyl cation with the solvent or another nucleophile present are obtained in various amounts depending on structure and conditions. This is not necessarily an indication of the pathway followed in the primary photochemical event, however.

Pincock and Fleming have recently summarized the mechanistic alternatives [15,25]. The probability of the fragmentation reaction has been evaluated first of all on the basis of thermochemical analyses. The energy of the singlet excited state of simple benzylic derivatives is >100 kcal/mol (106 for toluene) and drops to 90 for 1-methylnaphthalene and 73 for 9-methylanthracene. The S_1-T_1 gap is quite large for the $\pi\pi^*$ states of aromatics, and the corresponding values for the triplet are 83 (for toluene), 61, and 41 kcal/mol. As pointed out by these authors, comparison with the corresponding bond dissociation energies shows that homolytic cleavage is markedly exothermic for the singlet excited state of several benzene derivatives, e.g., BDE (PhCH$_2$-OH) 81 kcal/mol, (PhCH$_2$-Cl) 72, (PhCH$_2$-OMe) 71, and (PhCH$_2$-OAc) 62 kcal/mol. However, this is per se no indication of the efficiency of the reaction, since as an example benzyl alcohol and ethers do not cleave, while the chloride and the acetate do. Various theoretical approaches have been used for rationalizing, in qualitative or quantitative terms, how the excited state localized on the aromatic moiety interacts with the σ orbital of the C–X bond, leading to fragmentation, and thus account for the structure-dependent efficiency [26–28]. These are not discussed here. Certainly, as the excited state energy drops on going to the naphthalene and anthracene series the reaction probability diminishes and at any rate triplets are far less favored than the singlets from the mere thermochemical factor.

Fleming and Pincock have also evaluated the energy of the ion pair PhCH$_2^+$ X$^-$ on the basis of the oxidation potential of the benzyl radical (actually known in acetonitrile, but probably not changing much in other solvents) and the reduction potential of the leaving group. It turns out that for many common leaving groups heterolytic fragmentation is exothermal in polar media, e.g., ΔG (PhCH$_2$Cl^{1*} → PhCH$_2^+$ Cl$^-$) – 24 kcal/mol in MeCN, −36 in H$_2$O, ΔG (PhCH$_2$OH^{1*} → PhCH$_2^+$ HO$^-$) – 6 kcal/mol in MeCN, −22 in H$_2$O, ΔG (PhCH$_2$OAc^{1*} → PhCH$_2^+$ AcO$^-$) – 18 kcal/mol in MeCN, −34 in H$_2$O.

As a matter of fact, the energy of the ion pair in polar media is for many leaving groups lower than that of the corresponding radical pair, so that the occurring of a ionic reaction may involve homolytic cleavage as the photochemical step, followed by electron transfer between the paired radicals, rather than direct heterolysis, a distinction that is not always easy to determine experimentally. This is particularly relevant for rationalizing the substituent effect. Much of the original input to the studies of photoinduced benzylic fragmentation came from work by Zimmerman [13] evidencing an accelerating effect by a methoxy group in the *meta* position, and was rationalized as an increase in the heterolysis rate in the excited state (the *meta* accelerating effect differentiating excited state from ground state chemistry was a strong point of organic photochemistry when this took a unitary mechanistic shape in the sixties). Such a suggestion has been later supported by high-level calculations [29], but is not unanimously accepted. Pincock in particular found evidence for the reaction of the excited state being in most cases homolytic fragmentation, followed by electron transfer in the radicals pair leading to the ions pair [25]. For 3,5-dimethoxybenzyl derivatives yet another path has been detected, involving rearrangement to a methylenecyclohexadiene, which then gives thermally the product from the formal solvolysis of the reagent; this path plays a minor role, however [30,31].

15.2.2. Fragmentation via Photoinduced Atom Abstraction

Hydrogen abstraction from alkyl benzenes occurs efficiently by using aromatic ketones The mechanism of the reaction has been extensively studied, with ketones having both a $n\pi^*$ and a $\pi\pi^*$ state as the lowest triplet, and found to involve some degree of electron transfer, which grows with more easily reduced ketones [32,33]. The same reaction occurs intramolecularly, e.g., in the photoinduced hydrogen transfer in 2-methylbenzophenone to give the (trappable) enol [34–36].

The intramolecular hydrogen abstraction has been largely developed for the case of the nitro group [22,37], and is well known for some derivatives of o-nitrotoluene, for o-nitrobenzyl alcohols, ethers and esters [38–40] as well as on the acetals of o-nitrobenzaldehyde [41]. With these compounds, the $n\pi^*$ (singlet or triplet) state abstracts a hydrogen and the biradical relaxes to the aci-nitro form (the first intermediate is actually observed in bicyclic systems where formation of the nitronic acid is sterically hindered) [41], which under basic conditions dissociates to the corresponding anion [42].

15.2.3. Fragmentation via a Radical Ion

Benzylic derivatives are reasonably easily oxidized [43–49] and the corresponding radical cation can be conveniently generated by using an appropriate electron acceptor, whether as the sensitizer (an aromatic nitrile or ester, an aromatic ketone, a quinone, titanium dioxide powder) or as the quencher (an electron acceptor-substituted aromatic, an aliphatic polyhalide or polynitro derivative) in a polar medium. The stability of the benzyl radical makes fragmentation of a benzylic σ bond (C–H, C–C, or carbon-heteroatom) in the cation radical often an exothermic process, as it may be calculated through the appropriate thermochemical cycle. As an example, the toluene radical cation is a strong acid (calculated pKa in MeCN -12) [50]. The competition by back electron transfer may severely cut down the efficiency of the process, which actually does not necessarily involves a mere unimolecular fragmentation, since in most cases some form of nucleophile assistance can be evidenced. Kinetic data are now available for a number of benzylic fragmentation and show that the cleavage is fastest with benzylsilanes and stannanes as the electrofugal group, and quite fast also with a stabilized carbocation, such as a dialkoxy carbocation, while decarboxylation from phenylacetic acids and still more deprotonation from alkylbenzenes are slower [51–54]. This is a mild way for the generation of benzyl radicals.

Furthermore, there are several examples of benzylic radical fragmentation via the radical anion, including the potentially useful $S_{RN}1$ photoinduced substitution [55] as well as via the free radical arising from the photosensitized oxidation of an anionic reagent, such as an arylacetate or benzyltriphenylborate [56,57].

15.3. SCOPE AND LIMITATION

Most of the studies on benzylic systems are targeted at mechanistic issues. In fact, this is an interesting field in this respect and the detailed rationalization of many reactions has required much work (and presumably still will in the future) and in some cases generated a hot debate. Typical important issues are distinguishing singlet and triplet paths in benzylic fragmentation (as well as homolytic vs. heterolytic paths, actually a more difficult task), or distinguishing $S_{RN}1$ and S_N2 paths in benzylic substitution, or evidencing the (assisted) path for benzylic radical cation fragmentation. A large amount of fast kinetics studies have been devoted to these subjects.

However, there surely are synthetic aspects that deserve attention and are worth further development. This is one of the typical instance where the excited state reaction is only the beginning of the process, photochemistry is used for the generation of a highly reacting intermediate under mild conditions, in this case a benzyl cation, radical, or anion or a product arising from further elimination from the former ones, e.g., an o-quinone methide. This offers two main advantages with respect to thermal methods for obtaining the same species. First, the range of precursors from which the intermediate is generated is much wider, and does not require that these contain a weak σ bond. As an example, a benzyl radical can be generated via C–H cleavage from an alkylbenzene or C–Si cleavage from a benzyl-silane through electron transfer photosensitization, rather than be limited to the use a benzyl iodide as in thermal reactions. Second, the relative independence of the initiating photochemical step on conditions allows a much wider choice of the experimental parameters, e.g., of temperature, solvent characteristics, compatible traps. As a result, the photochemical method is much more versatile than any thermal counterpart.

A series of exemplary reactions that are based on a C–X benzylic bond fragmentation and have some synthetic potential are presented in the following section. Benzylic substitution has probably a limited preparative application, but benzylation reactions involving carbon–carbon bond formation appear to be more promising, if as yet applied to a limited extent with a primarily synthetic target. The topic is presented according to the following classification: substitution at the benzylic position; C–C bond formation; oxidation; reaction of arylated three- and four-membered rings; further applications of C–X benzylic fragmentation. It is possible that the last class, involving photoremovable protecting groups and photolabile linkers and resins based on the cleavage of a benzylic C–X bond, will develop into a method more important for organic synthesis than the use of such photochemical steps in synthetic plans.

15.4. SYNTHETIC POTENTIAL: REACTIVITY AND SELECTIVITY PATTERNS

15.4.1. Substitution at the Benzylic Position

$$Ar-\overset{|}{\underset{|}{C}}-X \quad \xrightarrow[\text{(NuH)}]{\text{SolvH}} \quad Ar-\overset{|}{\underset{|}{C}}-Solv\ (Nu)$$

$X = Cl, Br, OR, NR_2, NR_3{}^+, PR_3{}^+, SO_2R$
     SOR, CN, COOH,

$$-\overset{|}{\underset{|}{C}}-OH\ ,\quad \overset{}{\underset{\overset{||}{O}}{O-C}}-Y,\quad \overset{}{\underset{\overset{||}{O}}{O-S}}-Y$$

$$\overset{}{\underset{\overset{||}{O}}{O-P}}-Y_2$$

Equation 1

15.4.1.1. H/D Exchange

This reaction, contrary to most of the cases discussed below, involves liberation of an electrofugal rather than of a nucleofugal group. In fact, some benzylic derivatives behave as photoacids in the excited state. Wan has found that the reaction is effective when the cleavage yields a $4n$ π-electron carbanion. Thus, dibenzo[b,f]cycloheptatriene undergoes deuterium exchange in D_2O-MeCN (1:1), and both dibenzopyrane and dibenzo-thiopyrane do so under base catalysis, while diphenylmethane or dibenzocycloheptadiene, where no such carbanion can form, undergo no exchange under these conditions [58–62]. As noted above, the toluene radical cation is a strong acid. However, deprotonation is a relatively slow process and competition with back electron transfer may cut down the yield when the radical cations of alkylbenzenes are generated by electron transfer photosensitization. The reactions via benzyl radicals generated in this way are discussed in Sec. 16.4.2. The toluene radical cation has been generated and characterized by two photon ionization in water (τ between 100 ps and 20 ns), but then the fastest reaction is addition of water to the ring rather than deprotonation [63].

15.4.1.2. Substitution of Halides

As mentioned in the mechanistic section, both homolytic and heterolytic cleavage can occur [14]. Earlier work established the ground by demonstrating the formation of a carbocation, by determining the extent of epimerization [64], by evidencing the expected cationic rearrangements [65] as well as determining the extent of internal return of the anion and showing that the capture ratio by different nucleophiles was the same as when the cations were generated thermally [66,67]. In a LFP work on diphenylmethyl chloride and bromide (no major difference between the two derivatives), it was shown that both cleavages can occur from either the singlet or the triplet state. It was suggested that the determining factor is solvation [68]. When solvation occurs on the same time scale as the reaction from the excited state, charged fragments are stabilized and the heterolytic path is favored. In 4,4′-disubstituted diphenylmethyl chlorides the formation of ion pairs from radical pairs has been directly measured [69–71]. Work on naphthylmethyl- and fluorenyl halides give both radical recombination and cationic photoproducts, in a ratio dependent on the halide and on the medium [72–74]. Trifluoromethyl substituted phenols and naphthols underwent photohydrolysis to give the corresponding carboxylates under basic conditions [75,76].

4-Nitrocumyl chloride undergoes smooth light-induced substitution by nucleophiles such as phenolate and thiophenolate, benzenesulfinate, azide, cyanide or amines, and the process has been proven to involve a $S_{RN}1$ mechanism [55,77,78].

15.4.1.3. Substitution of Alcohols and Ethers

Photoheterolytic cleavage of benzyl alcohol in neutral media occurs only when it gives rise to highly stabilized cations, such as triphenylmethyl cations (compare Sec. 15.1; the reaction may be at least in part adiabatic in this case and yield the excited cation) [79–81], xanthyl [82] or fluorenyl cations [83]. However, acid catalysis is effective and methyl ethers are by far the main products from benzyl alcohols in acidified aqueous methanol [84–88]. Electron donating substituents in the *ortho* and, to a lesser extent, in the *meta* position enhance the quantum yield (acid catalysis may be

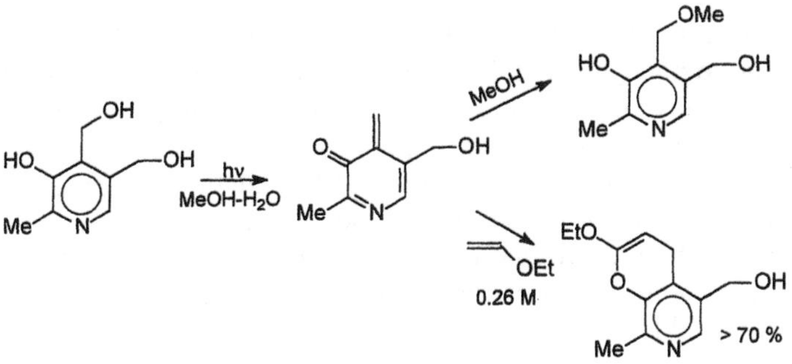

Scheme 3

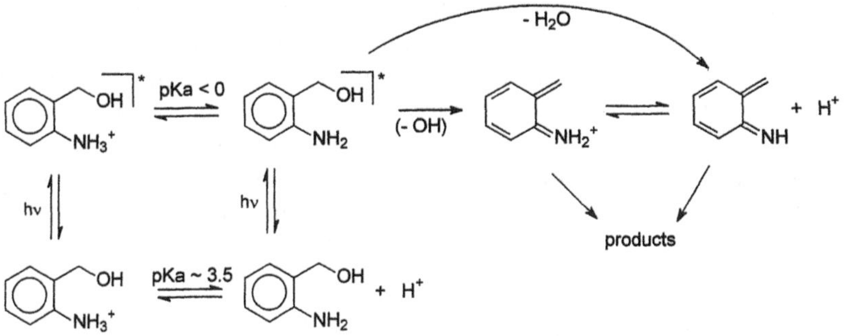

Scheme 4

important). With 2,6-dimethoxybenzyl alcohol the ionization quantum yield is close to unity in neutral aqueous solution.

o-Hydroxybenzyl alcohol also fragments efficiently, reasonably due to intramolecular assistance by the phenolic group; the accompanying deprotonation yield o-quinone methides that originate photochromism and can be trapped by Diels–Alder cycloaddition [89–92]. The reaction has been extended to a variety of substrates, including Vitamin B6 [93], 2-(2'-hydroxyphenyl)benzyl alcohols [94–97] as well as naphthyl and fluorenyl analogues [98,99], which are models for molecular conductors or switches or (using resolved binaphthyls) as chiral photochromic optical triggers for liquid crystals [100] (Sch. 3).

Indeed, also p- and m-hydroxybenzhydryl alcohols yield the corresponding quinone methides or zwitterions. o-Hydroxymethylanilines are excellent photo-precursors for the generation of o-quinone methide imides [101] (Sch. 4).

Benzylic ethers undergo C–O cleavage only when the fragments are stabilized. Thus, benzyl phenyl (as well as naphthyl) ethers undergo homolytic cleavage, in a typical photo-Fries reaction [102–104]. Triphenyl-methyl [105] and diphenylmethyl [106] phenyl ethers, on the contrary, fragment heterolytically and in this case a *meta* activation effect by electron withdrawing substituents such as nitro and cyano has been observed. Interestingly benzylated sugars have been found to undergo debenzylation by irradiation in methanol in the presence of iodine. Electron transfer from the latter generates the radical anion of the sulfonate that cleaves. When both benzyl groups and mesilates or triflates are present, the former ones are preferentially removed [107,108].

15.4.1.4. Substitution of Other Oxygen-Linked Groups

Benzylic esters cleave with a quantum yield varying from less than 0.01 to ca. 0.3. Several reports assessed a significant contribution of internal return after fragmentation [109–111]. Esters occupy a special place in the frame of photolysis of benzylic C–X bonds since the seminal paper by Zimmerman proposing the *meta* activation effect for electron-donating substituted benzyl acetates has been followed by a number of studies. Products resulting from both homolysis and heterolysis have been obtained, and it has been discussed whether the relevant competition takes place at the level of the excited state reaction—and is controlled by the electron distribution in that state [29] or the cleavage in any case for the most part homolytic, and then the ion pair results from electron transfer within the radical pair. In the original study, a comparison of the photohydrolysis of 4-methoxy, 3-methoxy and 3,5-dimethoxybenzyl acetates in water-dioxane showed an increase of benzyl alcohol from trace amounts to 29–35% and to 79%, respectively, with a corresponding decrease of the radical coupling products. More extensive examination of the substituent effect in the methanolysis of benzyl acetates [112,113], and even more clearly with naphthylmethyl acetates [114], and phenylacetates [115,116] for which the product distribution is simpler, showed a dependence of the ionic products on the oxidation potential of the radicals. The use of radical clocks allowed to obtain the rate of electron transfer from radical pair to ion pair and to note that their dependence on the radical E_{ox} fitted the Marcus equation. This supported the mechanism involving electron transfer from the radical pair to give the ion pair as the key step [25,117,118].

Benzylcarbonates [119] and carbamates [120] also suffer photosol-volysis. The latter compounds have been extensively investigated in view of their utility for the photoinduced generation of amines (as bases) [121] (see also Sec. 15.5). Intramolecular abstraction by nitro group [122–124] as well as α-cleavage of α-acylbenzylcarbamates [125,126] have also been used for amine liberation (Sch. 5).

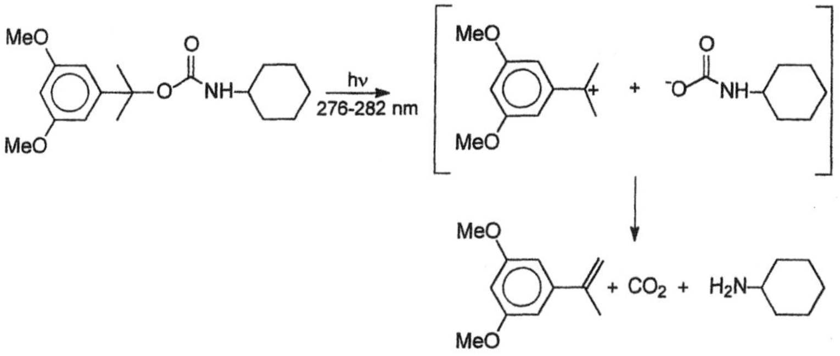

Scheme 5

Esters of inorganic acids have shown photolability. Thus, benzyl dialkyl phosphite is smoothly converted to the benzylphosphonate upon direct irradiation (85–95% yield) with only a minor amount of bibenzyl. This is a photochemical analogue of the Arbuzov reaction and proceeds with retention both at phosphorous and at the benzylic carbon. It appears to involve cleavage from the singlet excited state followed by rapid recombination of the proximate radicals [127–131]. The reaction has been extended to naphthyl-methyl and 1-arylethyl phosphates [132]. In contrast to the thermal Arbuzov reaction, secondary benzylphosphonate esters are smoothly prepared through the photochemical method (for synthetic applications see Sec. 16.6) [133,134]. However, the direct irradiation of 4-acetylbenzyl phosphites or energy transfer sensitization of naphthylmethyl phosphites lead to the decomposition from the triplet state, yielding a triplet radical pair. In this case, phosphonates are formed in a small yield (<10%) and the main products arise from reduction or coupling of the radicals (or their trapping, e.g., in benzene) [135].

Benzyl sulfonates generally suffer easy thermal solvolysis, unless electron-withdrawing groups stabilize them, and then photoinduced solvolysis may become interesting. Several 4-nitrobenzyl sulfonates have been found to cleave photochemically giving the corresponding radical [136,137] and some of such derivatives, in particular 4-nitrobenzyl 9,10-dimethoxyanthracene-2-sulfonate, have been applied as photoacid generators in polymers and in particular in photolithography [138,139].

15.4.1.5. Substitution of Nitrogen-Linked Groups

The arylmethyl-nitrogen bond appears to be photostable, as shown for 1-naphthylmethylaniline [25,119]. However, irradiation of N-alkyl-9-phenanthrenemethylamines in the presence of N-methylaniline causes cleavage of the CH$_2$-N bond and gives both bibenzyls and products from the (cationic) benzylation of the aniline [140]. Irradiation of substituted 1-benzyladenines in water gives the free base and the benzyl alcohols [141]. The protecting o-nitrobenzylgroup can be easily removed from amides (see Sec. 15.5), including NAD(P)$^+$ [142] and other biologically active amides [143,144]. Furthermore, o-hydroxybenzylamines photofragment to the o-quinone methides, just as the corresponding alcohols [145].

Benzylated quaternary ammonium salts are photoreactive [146–148]. Sensitization and quenching studies of (substituted) benzyl chlorides and bromides showed the existence of both singlet and triplet pathways; the solvolyzed products were not quenched, while of the radicalic products, bibenzyl gave a linear Stern–Volmer plot; toluene gave a nonlinear plot, suggesting that it arose in part from the singlet [149]. Naphthylmethyl-trimethylammonium tetrafluoborate was shown to react only from the singlet, reasonably because of the low energy of the triplet. Both solvolysis and reduction (from in cage hydrogen atom transfer between the amine radical cation and the naphthylmethyl radical resulting from homolytic cleavage) are observed. The proportion of solvolysis increases with electron-donating substituted derivatives, but more with the 4- than with the 3-methoxy derivative (in contrast to the concept of *meta* activation) [150]. o-Hydroxy-benzylammonium salts are a convenient source for o-quinone methides (Sch. 6), which may be useful for the trapping of biological substrates such as amino acids under mild conditions [151].

α,4-Dinitrocumene undergoes benzylic substitution through the S$_{RN}$1 process in the same way as the α-chloride (see above). In the case of 1-nitro-1-(4-nitrophenyl)-2-methyl-4-*tert*-butylcyclohexane the S$_{RN}$1 reaction with thiophenolate proceeds with retention of configuration [152] (Sch. 7).

15.4.1.6. Substitution of Other Groups

Benzyltriphenylphosphonium chloride photodecomposes to yield radical products [153], while the corresponding tetrafluoroborate gives both radical products and an ionic product (the *N*-benzylacetamide in acetonitrile); the last type of compounds predominates with most of the substituted benzyl derivatives studied, except with 3- and 4-methoxy, while only radical products were formed upon triplet sensitization [154]. The 4-cyanobenzylphosphonium salt gave only radicalic products [155], but

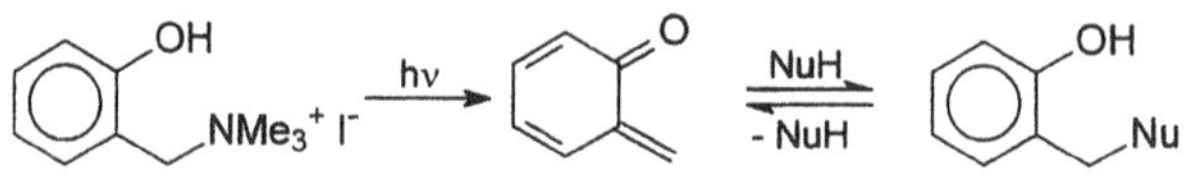

Scheme 6

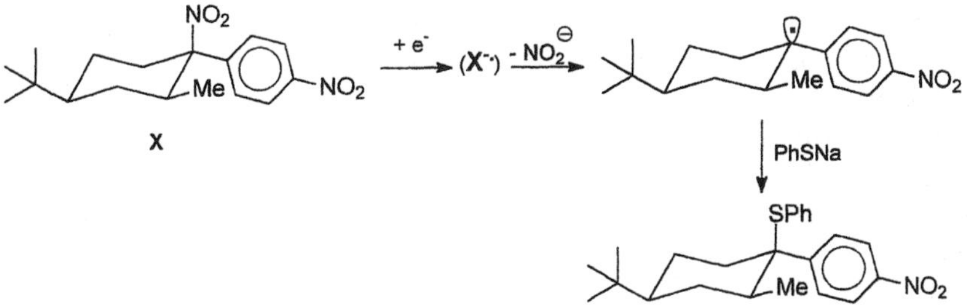

Scheme 7

various diarylmethylphosphonium derivatives were found to follow both paths in a proportion depending on the medium [156]. Similarly, both types of fragmentation products have been obtained from various benzylsulfonium tetrafluoroborates; these occurred to the same extent in methanol from the parent compound, but a large predominance of the methyl ether was obtained from the 3-methoxy substituted derivative [157]. Solvolysis predominated also from 1-naphthylmethylsulfonium tetrafluoroborate [73], while photolysis of [4-(9-anthracenyl)phenylmethyl]-(4-cyanophenylmethyl)phenylphosphonium trifluorosulfonate in acetonitrile led to intramolecular electron transfer and fragmentation to yield the 4-cyanobenzyl radical, which was trapped by the solvent [158]. Triphenyl-4-cyanobenzylarsonium tetrafluoborate irradiated in acetonitrile gave both the cation trapping product, N-(4-cyanobenzyl)acetamide, and radical coupling products [158]. Nitro substituted benzylphosphonic acids undergo photoinduced C–P bond cleavage ($p>m$, o) giving the carbanion that is protonated [159,160]. Among neutral substrates, benzyl aryl sulfides have been shown to undergo homolytic photocleavage with substituent-dependent efficiency, from $\Phi = 0.07$ for 4-cyano to $\Phi = 0.31$ for 4-methoxy [161–163]. Benzyl sulfones likewise give only products from radical recombination. The reaction proceeds via the singlet state for benzyl and 2-naphthylmethyl derivatives, via the triplet for 1-naphthyl derivatives [164–167]. Benzyl sulfoxide (and indeed also alkyl sulfoxides) likewise cleaves homolytically [168].

Fragmentation of a carbon–carbon bond has been observed in several classes of compounds. Detachment of a cyano group has been observed with the leuco derivatives of triphenylmethane dyes, both in pioneering studies (see Sec. 15.1) and more recently [169–171]. Less heavily phenylated acetonitriles do not cleave upon direct irradiation, but do so through the electron transfer path. Phenylacetic acid and related compounds photodecarboxylate to give benzyl radicals. The corresponding anions fragment to give benzyl anions and carbon dioxide [172]. As an example, phenylglyoxylate is decomposed to benzyl alcohol and dibenzocyclo-hexatriencarboxylate to the hydrocarbon in acetonitrile-water (deuterated products with D_2O) [62,173]. Retroaldol fragmentation occurs with 4-nitrophenethyl alcohol under base catalysis, yielding 4-nitrotoluene via the benzyl anion [174]. 2-Phenyl-2-(*m*- or *p*-nitrobenzyl)-1,3-dioxolanes undergo heterolytic cleavage to give 3- or 4-nitrotoluene, respectively, via the anion and hydroxyethylbenzoate from the dialkoxy-substituted cation [175]. C_α-C_β fragmentation is largely documented via SET oxidation, and yields benzyl radicals whenever a stabilized electrofugal group is present. A typical example is the fragmentation of bibenzyl derivatives into a benzyl radical, which is reduced, and a benzyl cation, which is trapped by the solvent (Sch. 8).

The mode of the reaction can be predicted on the basis of simple thermochemical considerations on the stability of the fragments [176–179]. As one may expect, β-alkoxy-, β,β-dialkoxy- and β-aminophenethyl derivatives are likewise cleaved via the radical cation [180–184]. Another extensively investigated class is that of 1,2-diarylpinacols, aminoalcohols and diamines [185–188], particularly efficient when electron donating substituents are present on the aromatic ring [189,190] (Sch. 9).

The cofragmentation of a benzylic radical cation and a fragmentable radical anion, such as that of alkyl halides, is another way of obtaining an efficient process [191–193] (Sch. 10).

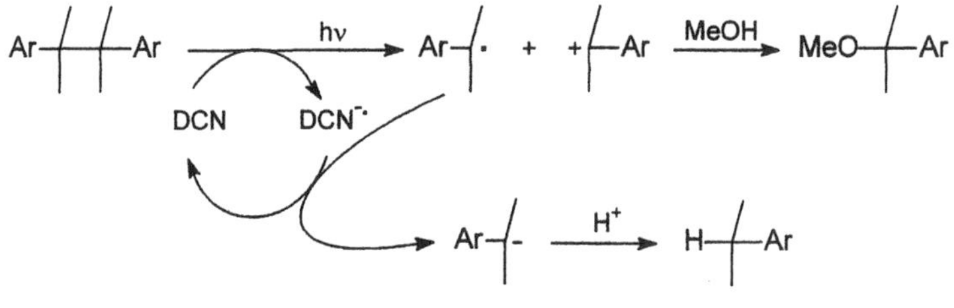

Scheme 8

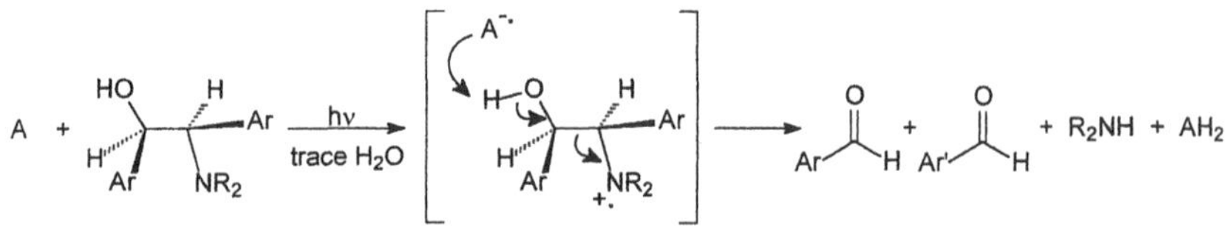

Scheme 9

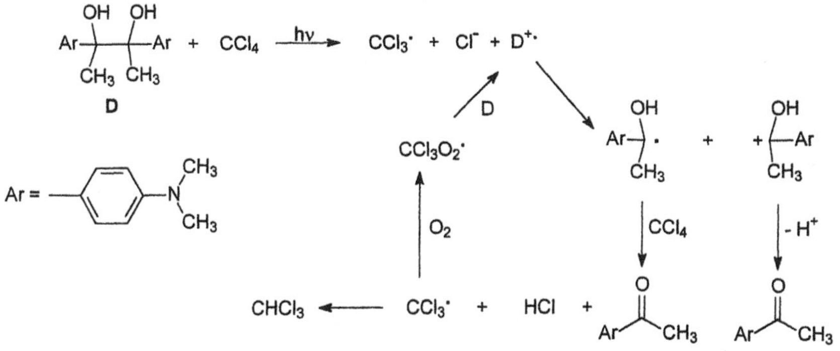

$$\text{Ar}-\overset{\overset{\displaystyle OH}{|}}{\underset{\underset{\displaystyle CH_3}{|}}{C}}-\overset{\overset{\displaystyle OH}{|}}{\underset{\underset{\displaystyle CH_3}{|}}{C}}-\text{Ar} \; + \; CCl_4 \; \xrightarrow{h\nu} \; CCl_3^{\cdot} \; + \; Cl^- \; + \; D^{+\cdot}$$

D

$$\text{Ar} = \;-\!\!\left\langle\!\!\bigcirc\!\!\right\rangle\!\!-\overset{\overset{\displaystyle CH_3}{|}}{\underset{\underset{\displaystyle CH_3}{|}}{N}}$$

$CCl_3O_2^{\cdot}$

$\uparrow O_2$

$$CHCl_3 \longleftarrow CCl_3^{\cdot} \; + \; HCl \; + \; \text{Ar}\overset{\displaystyle O}{\overset{\|}{C}}CH_3$$

$$\text{Ar}-\overset{\overset{\displaystyle OH}{|}}{\underset{\underset{\displaystyle CH_3}{|}}{C}}\!\cdot \; + \; \cdot\overset{\overset{\displaystyle OH}{|}}{\underset{\underset{\displaystyle CH_3}{|}}{C}}-\text{Ar}$$

$\downarrow CCl_4 \qquad \downarrow -H^+$

$$\text{Ar}\overset{\displaystyle O}{\overset{\|}{C}}CH_3 \qquad \text{Ar}\overset{\displaystyle O}{\overset{\|}{C}}CH_3$$

Scheme 10

15.4.2. Formation of a Carbon–Carbon Bond at a Benzylic Position

15.4.2.1. Benzylation of Arenes

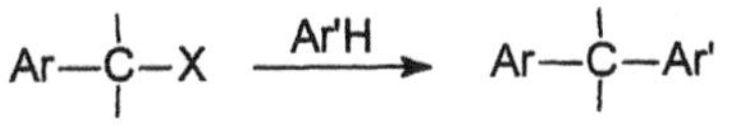

$$\text{Ar}-\overset{|}{\underset{|}{C}}-X \; \xrightarrow{Ar'H} \; \text{Ar}-\overset{|}{\underset{|}{C}}-Ar'$$

Equation 2

Again due to their nucleophilicity, benzyl radicals react with electron-withdrawing substituted aromatics. Typical is the case of arenenitriles, which are effective electron transfer sensitizers in the singlet state and promote fragmentation of electrofugal groups (H^+, Me_3Si^+ etc.,) from benzylic donors. The thus formed radicals either add to the radical anion or add to the neutral nitrile and the adducts are reduced by the persistent radical anion of the acceptor. In both cases an anion results, and protonation or cyanide loss then ensue, leading respectively to benzylated dihydroaromatic or aromatics. The aromatic product prevails in the benzene series, where the reaction is too slow with benzonitrile, but quite effective with polycyano derivatives [194,195]. As an example, p-xylene gives 85% of the corresponding bibenzyl with 1,2-dicyanobenzene and 100% with 1,4-dicyanobenzene; toluene gives 68% substitution with 1,2,4,5-tetracyanobenzene. Reaction at an unsubstituted position may also take place, as in the case of 1,3,5-tricyanobenzene, where 30% of 2,4,6-tricyanobibenzyl and 15% of 3,5-dicyanobibenzyl are formed. This has been rationalized on the basis of the larger spin density at position 2 in the acceptor radical

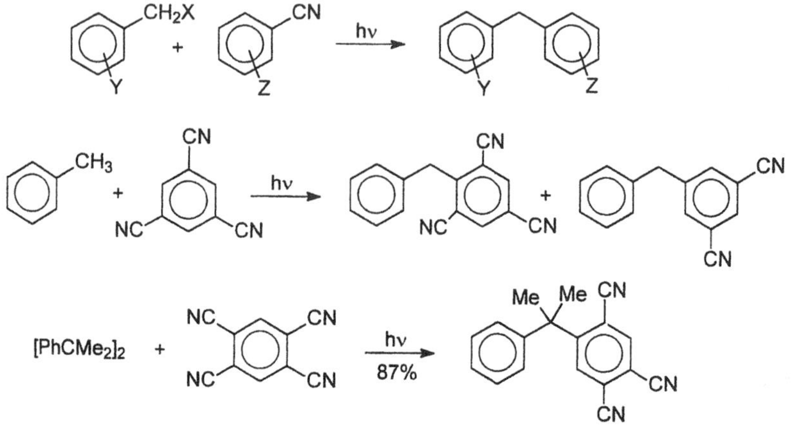

Scheme 11

anion [196]. In the latter product, the adduct rearomatizes through some unspecified mechanism (Sch. 11).

Formal substitution of a benzyl for a cyano group by a radical resulting from the deprotonation of the radical cation is observed also with other derivatives, e.g., acenaphthene in the reaction with dicyano and tetracyanobenzene [197] or 4,4′-dimethoxybibenzyl in the reaction with 1,4-dicyanonaphthalene. Furthermore, the benzyl radical may result from C–C bond (e.g., bibenzyls) or carbon-heteroatom (in particular Si, Ge, Sn, B) fragmentation. As an example, an α,α-dimethylbenzyl group is introduced when using tetramethylbibenzyl [198].

The outcome is more complex in the naphthalene series, where both the position of attack and the competition by rearomatization depend on the reagent chosen and conditions. Monocyanonaphthalenes are benzylated only by *p*-methoxytoluene, or by *m*- or *p*-methoxyphenylacetic acids, not by (poly)methylbenzenes [23,199]. 1,4-Dicyanonaphthalene reacts easily with toluene and a variety of precursors of the benzyl radical (alkylbenzenes [200–205], benzylsilanes, stannanes [206], and borates [56], as well as phenylacetic acids) [207] giving three types of products, viz. 1-benzyl-4-cyanonaphthalenes, 1- or 2-benzyl-1,2-dihydro-1,4-dicyanonaphthalenes and dibenzobicyclo[3.3.1]nonanes resulting from the formation of a second carbon–carbon bond (Schs. 12 and 13).

The formation of mixtures obviously detracts from the preparative interest of this reaction, but in some cases the reaction is relatively clean. As an example, the adduct in position 1 is the main product with benzyltrimethylsilane or with benzyl methyl ether [187] (the rearomatized

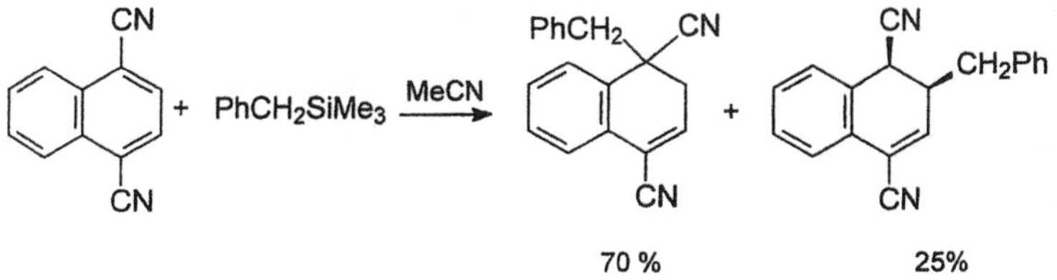

Scheme 12

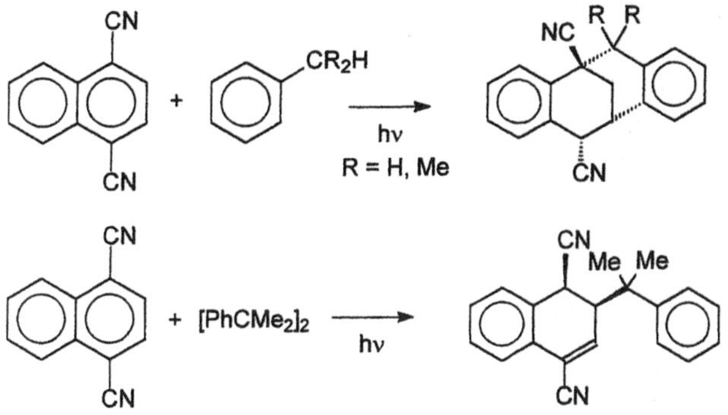

Scheme 13

products are obtained under basic conditions) [208,209]. Perhaps the most interesting point is the formation of bicyclononane derivatives, which, except for bulky group, occurs diastereoselectively. This is probably due to the reaction initiated directly from the donor acceptor complex. The reaction is thus at the borderline between radical ion photochemistry and exciplex chemistry. With some simple alkylbenzenes such as toluene and cumene this product is formed in up to 50% yield and, similarly to the cycloaddition of aromatics with alkenes, may be valued for the formation of a fairly elaborated structure in a single step. With cyanoanthracenes, *meso*-benzylated products are formed, whether rearomatized or as the dihydro derivatives [48,210].

As mentioned in Sec. 15.2.3, benzylic radicals are obtained also from the cleavage of a nucleofugal group from the radical anion. This may lead again to benzylation, and it has been shown that irradiation of 1,4-dimethoxynaphthalene in the presence of substituted benzyl halides leads to benzylated naphthalenes (mainly in position 2) via benzyl radical/arene radical cation combination, which is analogous to the benzyl radical/radical

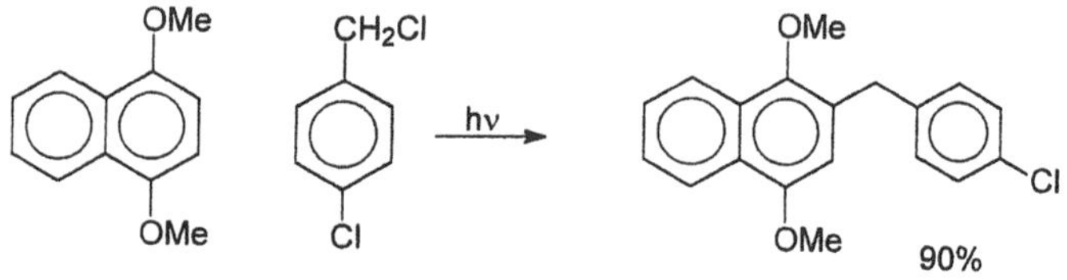

Scheme 14

anion combination discussed above, with reversed charge; yields are good in some cases [211] (Sch. 14).

Another instance is the benzylation of N,N-dimethylanilines and derivatives upon irradiation of benzyl cyanides, which cleave via the radical anion formed by electron transfer [212]. Yet another path for the benzylation of (hetero)aromatics is the $S_{RN}1$ reaction, as shown by the smooth photoassisted reaction of the enolate from ethyl phenylacetate with 2-bromopyridine (77% yield) [213].

15.4.2.2. Benzylation of Alkenes

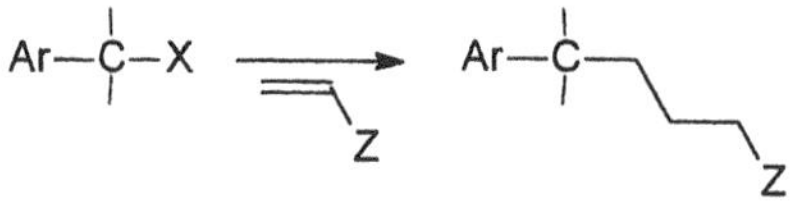

Equation 3

The persistent benzyl radicals often couple to bibenzyl rather than adding to electrophilic alkenes. When using 1,4-dicyanonaphthalene as an electron accepting sensitizer and toluene as the donor, most of the chemistry occurs via a radical ion pair, leading to various products resulting from the coupling of the two reagents (and a little bibenzyl), as indicated in the previous section. With a protic solvent such as *tert*-butanol, free solvated radical ions are formed, deprotonation occurs at the radical cation stage and the benzyl radical can be trapped. Under these conditions, as an example, dimethyl benzylsuccinate is formed from the dimethylmaleate (Sch. 15). When the alcohol is methanol, however, hydrogen transfer to the benzyl radical occurs and hydroxymethylation of maleate occurs in preference to benzylation [214].

Still in the electron transfer field, a useful benzylation procedure is based on the heterogeneous sensitization by titanium dioxide. In this case, methylbenzenes, benzylsilanes and phenylacetic acids are used as donors and electron-withdrawing substituted alkenes have the double role of

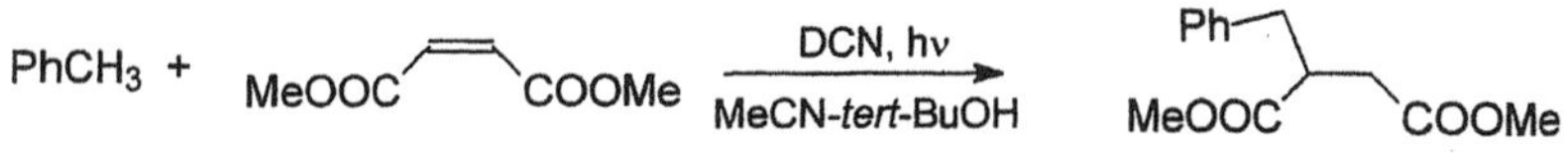

Scheme 15

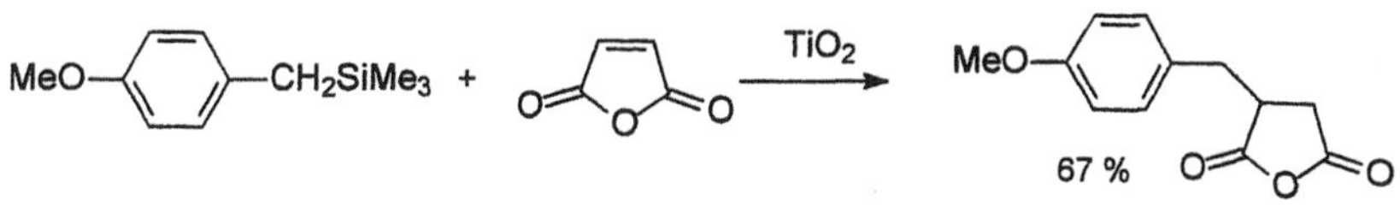

Scheme 16

electron acceptors, required for exploiting the otherwise quickly reversible photoinduced hole-electron separation on the semiconductor surface, and of electrophilic traps for the radicals. Particularly with *p*-methoxy substituted derivatives, yields of benzylated succinic acids, anhydrides or nitriles are good [215–217] (Sch. 16).

A sub case in this category is the benzylation of enolates through the $S_{RN}1$ process. In this way, both α-nitro- and α-chloro-4-nitrocumene are alkylated by the enolates of 2-nitropropane, diethylmalonate, or diethyl 2-butylmalonate. A particular case of benzylic C–C fragmentation is the electron transfer photosensitized Cope rearrangement of 2,5-diphenyl-1,5-hexadienes [218,219].

15.4.2.3. Benzylation of Carbonyl, Carboxamide, and Iminium Groups

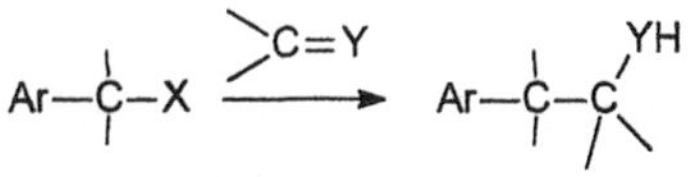

Equation 4

As mentioned in Sec. 15.2.2, hydrogen abstraction of benzylic hydrogens by the $n\pi^*$ triplet state of ketones is generally efficient and offers an access to benzylic radicals. However, the method is rarely of preparative interest since benzylic and ketyl radicals couple statistically giving a mixture of bibenzyls, alcohols, and pinacols, although the initial hydrogen abstraction may show some interesting intramolecular selectivity with polymethylbenzenes and substituted alkylbenzenes [220,221]. However, in

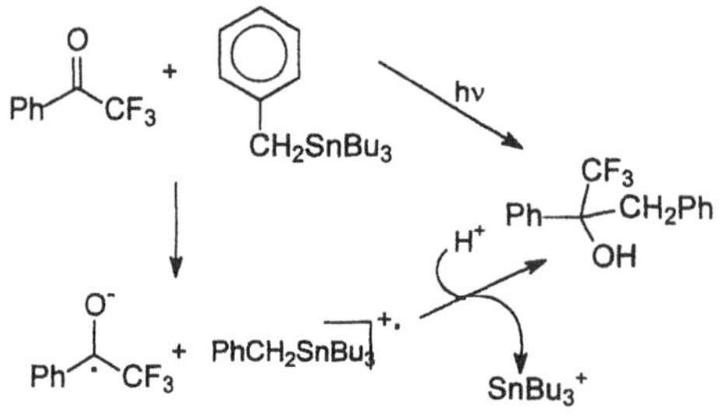

Scheme 17

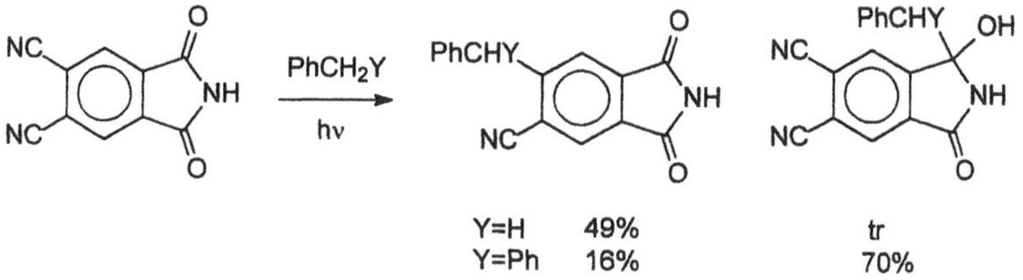

Y=H   49%   tr
Y=Ph   16%   70%

Scheme 18

favorable cases useful yields can be obtained. As an example, highly stabilized radicals may give good yields of the bibenzyls from homocoupling (see below). On the other hand, when the ketone is a good electron acceptor, e.g., trifluoroacetophenone, and donors such as benzylsilanes and stannanes are used, the reaction follows an electron transfer path and a higher percentage of a single product, the alcohol from heterocoupling, is obtained [222] (Sch. 17).

Phthalimides have often been used in the double role of electron acceptor sensitizers and radical traps, and are in fact benzylated by irradiation in the presence of toluene and several other precursors. 4,5-Dicyanophthalimide undergoes substitution of a benzyl for a cyano group when irradiated with toluene, but mainly attack at the imide carbon occurs with diphenylmethane. With various donors the competition between the two modes of reaction has been found to depend on the cage vs. out of cage radical cation cleavage, in the first case with the assistance of the radical anion [223] (Sch. 18).

Irradiation of chloranil with p-xylene [224] or with hexamethylbenzene [225] gives the corresponding 4-hydroxyphenyl benzyl ethers. Another good trap for benzyl radicals is the iminium group. Thus, irradiation of 1-methyl-2-phenylpyrrolinium perchlorate with toluene, benzyltrimethylsilane, or

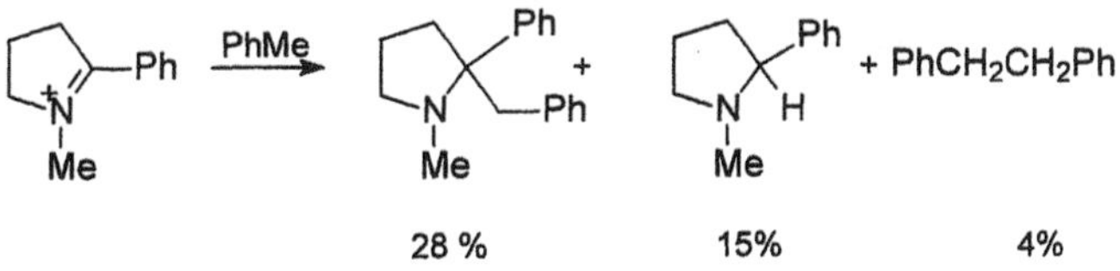

28 %   15%   4%

Scheme 19

benzyltributylstannane yields the 2-benzylated pyrrolidine in moderate yields [226] (Sch. 19). The intramolecular variation of this reaction with *o*-methylbenzylpyrrolinium salts or analogues has shown to be a useful method for the synthesis of various heterocycles (see Sec. 15.6).

15.4.3. Benzylic Oxidation or Functionalization

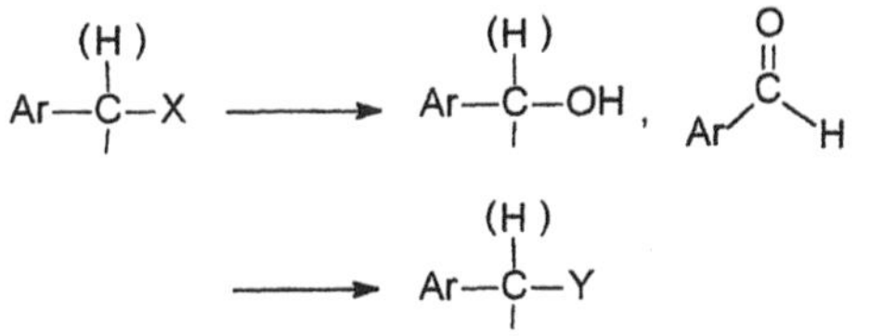

Equation 5

Toluene, durene, hexamethylbenzene, 1- and 2-methylnaphthalenes are oxidized to the corresponding benzaldehydes by irradiation in oxygen-equilibrated acetonitrile sensitized by 1,4-dicyanonaphthalene, 9-cyano-, 9,10-dicyano-, and 3,7,9,10-tetracyanoanthracene. The reaction involves proton transfer from the radical cation of the donor to the sensitizer radical anion or the superoxide anion, to yield the benzyl radical which is trapped by oxygen. In the case of durene, some tetramethylphthalide is also formed; with this hydrocarbon it is noteworthy that the same photosensitization, when carried out in an nonpolar medium, yields the well-known singlet oxygen adduct, not the aldehyde [227,228] (Sch. 20).

Nitroaromatics also sensitize the oxidation of methylarenes and it has been found that silica-grafted 2,4,6-trinitrobenzene is a convenient heterogeneous sensitizer, giving the aldehydes in 79–90% yields with 100% selectivity [229]. Bibenzyls, pinacols and pinacol ethers are likewise oxidized to ketones or respectively esters through carbon–carbon bond fragmentation upon dicyanonaphthalene sensitization [230]. A good method for benzylic oxidation is based on titanium dioxide photocatalysis [231–233].

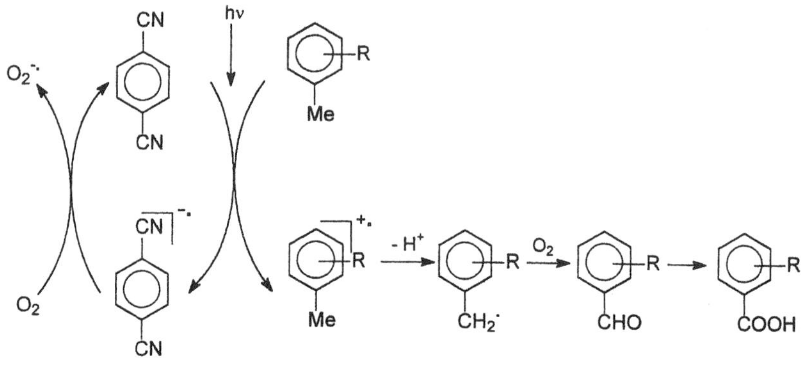

Scheme 20

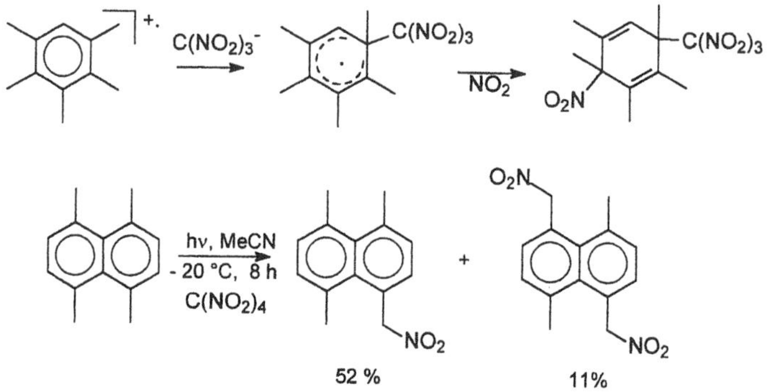

52 %   11%

Scheme 21

Further functionalizations are obtained via the electron transfer—radical cation fragmentation pathway; a typical example is side-chain nitration by irradiation of methyaromatics with tetranitromethane. Aromatics form charge-transfer complexes with $C(NO_2)_4$; irradiation leads to electron transfer and fragmentation of the $C(NO_2)_4$ radical anion to yield the triad $[Ar^{\cdot+}\ C(NO)_3^-\ NO_2]$, followed by combination between the arene radical cation and the trinitromethanide anion. Thus, cyclohexadienes are formed that generally eliminate and rearomatize at room temperature yielding ring-functionalized products [234] (Sch. 21).

However, with penta- and hexamethylbenzene [235,236] and 1,4,5,8-tetramethylnaphthalene [237] the main process is side-chain nitration, although ring functionalization predominates with less heavily substituted derivatives [238] (Sch. 22).

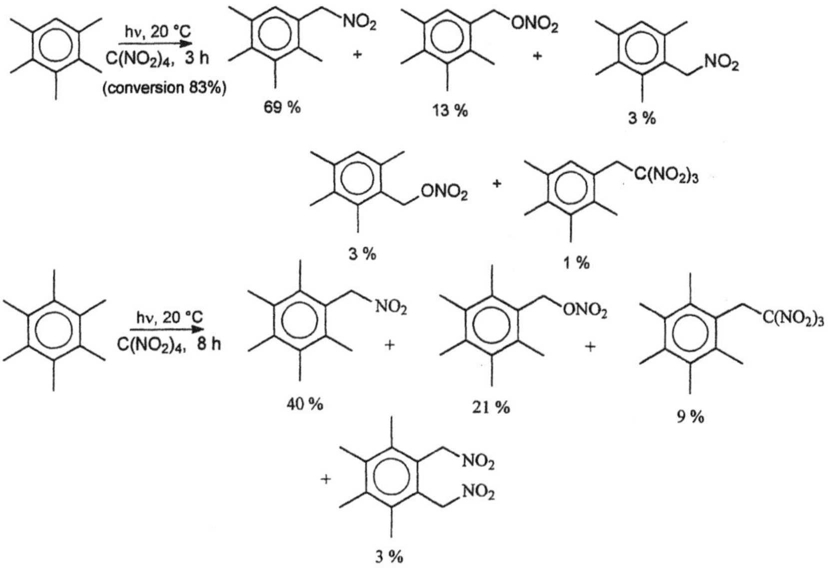

hv, 20 °C
C(NO₂)₄, 3 h
(conversion 83%)

NO₂
+
69 %

ONO₂
+
13 %

NO₂
3 %

ONO₂
+
3 %

C(NO₂)₃
1 %

hv, 20 °C
C(NO₂)₄, 8 h

NO₂
+
40 %

ONO₂
+
21 %

C(NO₂)₃
9 %

+
NO₂
NO₂
3 %

Scheme 22

A particular case of oxidative functionalization is the formation of bibenzyls. This occurs with medium to good yield when the benzyl radical is highly stabilized (e.g., benzhydryl, xanthyl) using aromatic ketones as the sensitizers [19–21]. Better results are obtained with titanium dioxide, particularly when starting form benzylsilanes, in view of the easy fragmentation of the radical cation [239].

15.4.4. Reactions of Arylated Three- and Four-Membered Cyclic Compounds

Arylcyclopropanes and their heterocyclic analogues are liable to electron transfer induced fragmentation of a carbon–carbon bond that in some cases leads to synthetically useful products. Thus, 1,2-diarylcyclopropanes [240–243] as well as 2,3-diaryloxirans [244–246] and -aziridines (in the last case, also 2-monophenyl derivatives) [247,248] are cleaved upon photoinduced electron transfer sensitization. The final result, after back electron transfer, is *trans–cis* isomerization of the ring. In the presence of a suitable trap, however, a cycloaddition reaction takes place, involving either the radical cation or the ylide. Thus, dioxoles, ozonides or azodioxoles, respectively, are formed in the presence of oxygen and oxazolidines have been obtained from cyclopropanes in the presence of nitrogen oxide (Sch. 23).

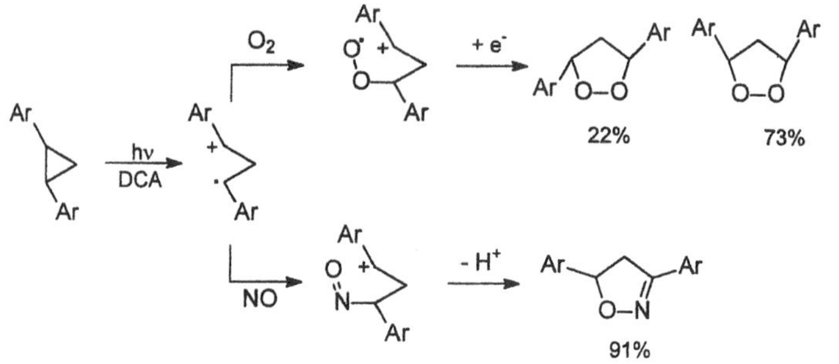

Scheme 23

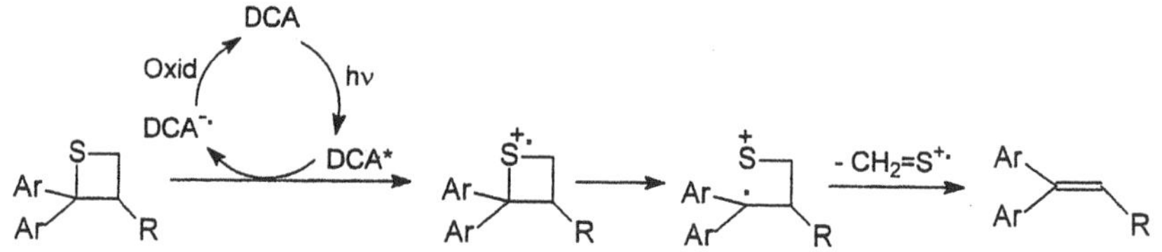

Scheme 24

1,5-Diphenylaziridines also cleave under similar conditions and the open-chain intermediate can be trapped to yield imidazoles and hetero-phanes [249]. Furthermore, 2,2-diarylthietanes have been found to cleave to the diphenylalkenes upon electron transfer sensitization (Sch. 24) [250].

15.4.5. Further Applications of Benzylic Photocleavage

Besides the direct application in a synthetically useful reaction, benzylic derivatives can also be useful in an indirect way. As an example, at several instances in this section it has been mentioned that photocleavage of benzylic derivatives generates acids or bases. Such a fragmentation has been applied e.g., in promoting polymerization.

More strictly connected with organic synthesis is photoinduced deprotection [251]. The most largely used photoremovable groups are based on the o-nitrobenzyl chromophore [22]. Intramolecular hydrogen abstraction leads to an acetal derivative, which under the reaction conditions collapses to o-nitrosobenzaldehyde and the liberated substrate (Sch. 25).

Substituting methoxy groups on the ring (e.g., using the 6-nitrover-atroyloxycarbonyl group as protecting reagent) pushes the absorption to longer wavelength and often ensures rapid deprotection with minimal

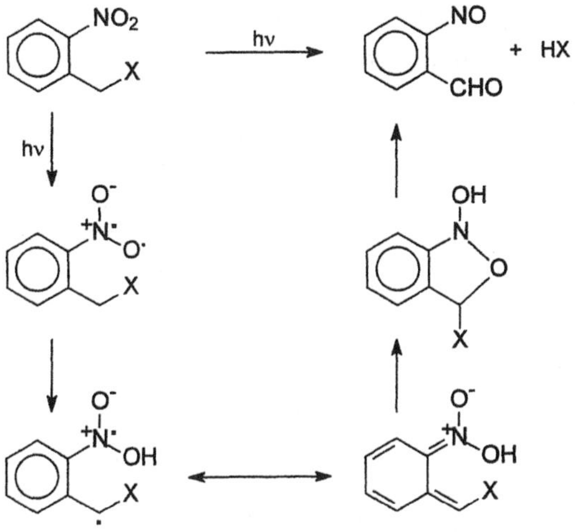

Scheme 25

competing reactions. The method has been used for the protection of the amino group in amino acids [252–254] (also for the direct protection of the imidazole side-chain of histidine) [255]; for the hydroxy group in carbohydrates [256,257] and ribonucleosides [258]; for the phosphate group in nucleotide synthesis [259]; for ketones (using *o*-nitrophenylglycol) [260,261]. Furthermore, the fragmentation of benzyl alcohols discussed in Sec. 15.4.1.3 can be used for deprotection [120], and it has been found that two methoxy groups in 3,5 greatly favor the cleavage [262], and the reactivity is further increased by methyl groups in the α-position [263]. Further (hetero)arylmethyl groups that have been used as photoremovable groups (for alcohols) are 9-phenylxanthen-9-yl [264], 1-pyrenmethyl, 9-phenanthreneylmethyl, 2-anthraquinonemethyl, 4-(7-methoxycoumarin)-methyl [265]. A third class of photolabile protecting group based on benzylic cleavage involves benzoin derivatives for carboxylic [266,267] and phosphoric acids [268–270]. The availability of different families of photoremovable protecting group had made possible to reach wavelength-based orthogonality, as shown by Bochet for the veratroyl group and the 3′,5′-dimethoxybenzoin group [271,272].

This mild method is particularly attractive for the cleavage from the solid support in combinatorial chemistry. A variety of orthogonal linkers have been devised, which are stable to both acidic and basic conditions and allow releasing the target molecules after removal of any protecting group, minimizing contamination. Most such linkers are again based on the *o*-nitrobenzyl structure and have been used for peptides

[273–276], β-lactams [277], polysaccharides [278,279], small molecules [280] and tagging moieties [281]. The initially tested linkers required a long irradiation time, which caused a considerable loss due to side-reactions. More reactive linkers have been evolved based on the insertion of substituents in the α position [282–284] or on the ring; the veratryl groups has allowed for the use of >350 nm radiation, short irradiation time (1–2 h) and minimization of reactive by-products [285–288].

Another family of photoremovable protecting agents is that of benzoin esters and related derivatives (such as benzoin phosphates, carbamates, carbonates), with the 3′,5′-dimethoxybenzoincarbonate (DMBC) as a particularly convenient group [289–293]. Also with these derivatives suitable linkers for combinatorial chemistry have been evolved [294,295]. A further degree of versatility is offered by the use of dithiane protected benzoin linkers, which may be advantageous in the binding steps and are then deprotected before proceeding to the photochemical liberation [296].

15.5. REPRESENTATIVE EXPERIMENTAL PROCEDURES

Historically Important Procedures

1,1,2,2-Tetraphenylethane [21]. Diphenylmethane (4 g) and *p*-benzoquinone (1.2 g) in thiophene-free, sodium-dried benzene (20 g) were exposed to sunlight for one month in Cairo. The reaction was carried out in a sealed glass tube filled with carbon dioxide. Quinhydrone began to separate out after only an hour; after a month the complete precipitate was filtered off. The benzene was evaporated under vacuum and the oily residue steam-distilled, in order to remove diphenylmethane and *p*-benzoquinone, and then ether extracted. The extract was dried (sodium sulfate) and the ether removed under vacuum. The residue was extracted with petroleum ether. A semisolid mass was obtained which was recrystallized from petroleum (100–110°) to give colorless crystals of the title compound (m.p. 210°, yield 0.7 g).

9,9′-Bithioxanthene [19]. Thioxanthene (1 g) and xanthone in equimolecular proportions in benzene solution were exposed to sunlight (6 h, July, Cairo) under the same experimental conditions. Afterwards, the bithioxanthene was filtered off, washed with benzene and recrystallized from xylene, m.p. 325°, yield 80%.

Modern Procedures

2(4-Methoxybenzyl)succinic acid. (compare Sch. 16) A 140 mg sample of Degussa P25 titanium dioxide was weighted in a round-bottomed Pyrex tube (2.4 cm internal diameter) and 40 mL of an acetonitrile solution of 0.02 M 4-methoxybenzyltrimethylsilane and 0.04 M maleic acid was added (no previous treatment of the solvent or of the commercial titania powder was required). The tube was sealed with a serum cap and the mixture was sonicated for 1 min. Two needles were inserted and purified nitrogen was passed for 15 min with magnetic stirring. The tubes (usually four of them) were put in the center of four 15 W phosphor coated lamps with 350 nm center of emission) and irradiated while maintaining nitrogen flushing and magnetic stirring. After completion of the reaction (30 h, as monitored by glc), the mixture was filtered through "slow" analytical filter paper, the solvent evaporated and the residue recrystallized to give a 72% yield of the title compound [217].

Dimethyl[1-phenyl-4-[(tert-butyldimethylsilyl)oxy]butyl]phosphonate. (see Sch. 26 below) A solution of 1-phenyl-4[(*tert*-butyldimethylsilyl)-oxy]butyl dimethyl phosphite (5.94 g, 16.0 mmol) in benzene (160 mL) was divided between six quartz test tubes. The solution was degassed by bubbling argon through it for 15 min. The solution was then irradiated by a 450 W Hanovia medium pressure UV lamp. The conversion of the starting material was complete in 8 h. Benzene was removed, and the yellowish liquid residue was dissolved in chloroform and applied to a silica gel column. This was eluted first with chloroform and then with 1% MeOH in CHCl₃. The title phosphonate was obtained as a colorless liquid (4.2 g, 72%) [133].

2,3-Dimethyl-2-nitro-3-(4-nitrophenyl)-butane. 4-Nitrocumyl chloride (2.0 g, 10 mmol), the lithium salt of 2-nitropropane (1.9 g, 20 mmol),

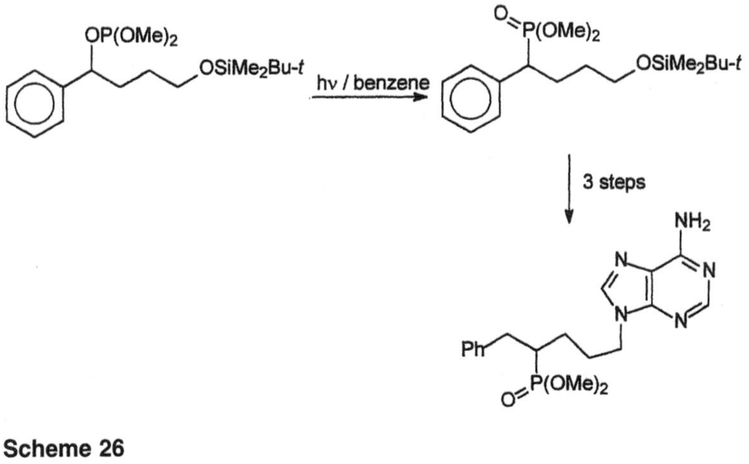

Scheme 26

and 100 mL of HMPA were exposed for 2 h to the light from two 20 W white fluorescent lamps mounted horizontally about 12 cm apart. The crude product obtained on workup was digested with diethyl ether for 90 min after which the mixture was allowed to cool at room temperature. Filtration gave 1.93 g of white crystals of the title compound. Evaporation of the filtrate left a yellow solid that was chromatographed on acid-washed alumina with benzene. This gave another 0.14 g of the product, total yield 81% [55].

Equation 6

6,11-Dicyano-12,12-dimethyl-5,11-methano-5,6,11,12-tetrahydrodi-benzo[a,e]cyclooctene. (see Sch. 13) An acetonitrile solution (70 mL) containing 250 mg (1.4 mM) of 1,4-dicyanonaphthalene and 2.4 g (20 mM) of isopropylbenzene was refluxed, cooled while flushing with argon, and irradiated with a Pyrex-filtered 150 W medium pressure mercury arc at 17°C until the starting nitrile was almost completely converted (4 h, tlc). After evaporation of the solvent, the photolysate was chromatographed on silica gel, eluting with cyclohexane in order to eliminate excess arene and then with cyclohexane-ethyl acetate 9 to 1 mixture to give 185 mg (45%) of the title compound [203].

Photoinduced liberation of a resin-bound substrate (with the dialkoxy-nitrobenzylamine moiety as photolabile group). The resin (50 mg) was suspended in 1–2 mL of pH 7.4 PBS buffer containing 5% of DMSO. Photolyses were carried out by irradiating the samples with a 500 W mercury lamp fitted with a 350–450 nm dichroic mirror on a vertical axis ate a 10 mW/cm power level measured at 365 nm. The samples were irradiated from above with gentle mixing from an orbital shaker table. The success of photolysis was demonstrated by analysis of the supernatant by reverse-phase HPLC and ^{13}C NMR analysis of the support (3 h, 95% cleavage). Both linker and bound molecule were stable to incubation with 95% TFA-5% water for 1 h at room temperature and, despite the photosensitivity, the use of foil-wrapped vessels allowed using the resin in the laboratory [288].

15.6. TARGET MOLECULES: NATURAL AND NONNATURAL PRODUCT STRUCTURES

Photoinduced substitution at the benzylic position rarely has a synthetic advantage with respect to the thermal counterpart. However, there are interesting exceptions. As mentioned in Sec. 15.4.1.4, benzylphosphites are smoothly converted to phosphonates through a photochemical reaction. This reaction is convenient in the case of secondary derivatives, since the thermal alternative (the Michaelis–Arbuzov reaction) is inefficient because secondary halides react only sluggishly with trialkyl phosphites. The availability of secondary benzylphosphonate diesters from the irradiation of the corresponding benzylphosphites has been exploited for the synthesis of 1-phenyl-4-silyloxybutyl dimethyl phosphonate that has been converted to the 4-bromide and further elaborated to various nucleoside phosphonic acids (with adenine, cytosine, guanine, 2,6-diaminopurine). These acyclic nucleoside-base phosphonates and other that have been similarly prepared are analogues of important antiviral agents and have tested for their pharmacological activity [132,133] (Sch. 26).

In other cases, the photoinduced fragmentation serves as a convenient source of highly reactive intermediates under mild conditions. Typical examples involve benzyl radicals or intermediates arising from further fragmentation, as illustrated in the following. The electron transfer photosensitized fragmentation of benzylsilanes and the easy nucleophilic addition of benzyl radicals to electrophilic double bonds, in particular to iminium salts suggested a synthetic path based on intramolecular reactions for the synthesis of heterocycles, e.g., of benzoindolizines from N-[o-(trimethylsilylmethylbenzyl)]pyrrolinium salts [297].

Likewise, benzyldihydroisoquinolinium derivatives can be used in a photochemical synthesis of tetrahydroisoquinolines. Thus, 2-(2-trimethyl-silylmethylphenylmethyl)-3,4-dihydroisoquinoliniun perchlorates have been successfully cyclized, as in the synthesis of the protoberbine alkaloids (+)xylopinine and (+)stylopine. The reaction proceeds via SET from the xylyl donor to the iminium moiety, fragmentation of the benzylsilane radical cation and carbon–carbon bond formation in the intermediate diradical. The synthesis is rather general and the yields compare favorably with those obtained from related substrates via a dipolar cycloaddition methodology [298] (Sch. 27).

Examination of various derivatives bearing substituents at the benzylic sites has shown, however, that stereoselectivity is modest and difficult to predict [299]. In this vein, 1-(2-trimethylsilylmethylphenylmethyl)-3,4-dihydroisoquinoliniun perchlorates have been used for the synthesis of spirocycloalkyltetrahydroisoquinolines, though the yields are lower in this

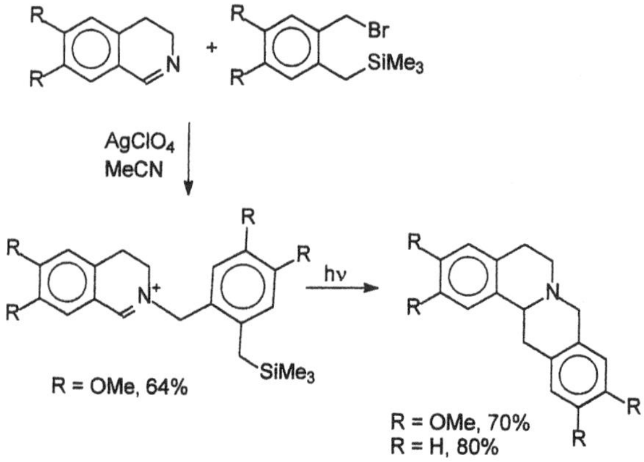

Scheme 27

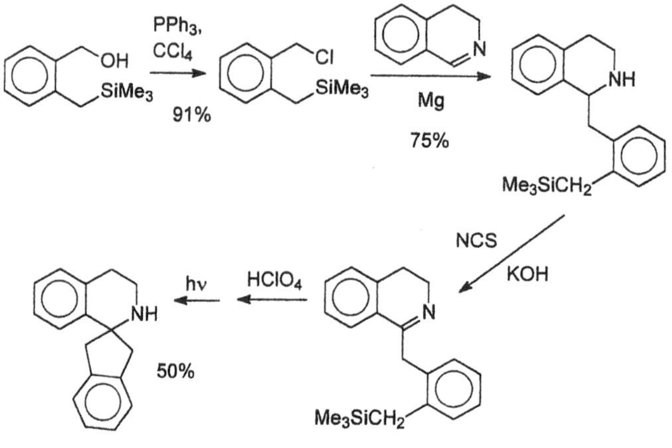

Scheme 28

case, due to competitive paths from the diradical not leading to cyclization (Sch. 28).

The photoinduced generation of *o*-quinonemethides from *o*-hydroxybenzylalcohols (see Sec. 15.4.1.3) has been exploited in the intramolecular cycloaddition to a double bond for the synthesis of the hexahydrocannabinol ring system. The reaction gives the desired product in >80% yield in acetonitrile-water, provided that the tethered alkenes are sufficiently electron-rich (trisubstituted). With less substituted derivatives nucleophilic addition by the solvent on the intermediate quinonemethide prevails over cycloaddition [300].

Once again, it may be that the utility of photoinduced C–X bond fragmentation in organic synthesis will be more significant as a mild method for deprotection or for combinatorial synthesis. Convenient photoremovable protecting groups are presently available, which require a limited irradiation time and operate selectively with relatively long wavelength irradiation, thus avoiding competing photoinitiated reactions of the substrate. A number of well-established photolabile protecting groups, which are orthogonal to thermally labile groups, has thus been added to the palette of synthetists. Furthermore, photoremovable protecting groups that are orthogonal one to another have been elaborated and an elegant example of selectively has been reported. Thus, it has been demonstrated that the nitroveratroyl group and the α-benzoyl-3,5-dimethoxybenzyl group can be selectively detached by irradiating at 254 and at 420 nm respectively [271,272]. It has been found that energy transfer from the initially excited chromophore can be minimized and the yield of selectively deprotected substrate can be high, opening the path to wavelength-driven orthogonality (Sch. 29).

Also in the field of combinatorial synthesis, photolabile linkers have now been evolved that are photolyzed at a high rate, although this is often quite solvent-dependent [285,288]. The α-methyl-2-nitro-5-methoxy-4-(3′-carboxypropyloxy)benzylamino structure has proved to be particularly

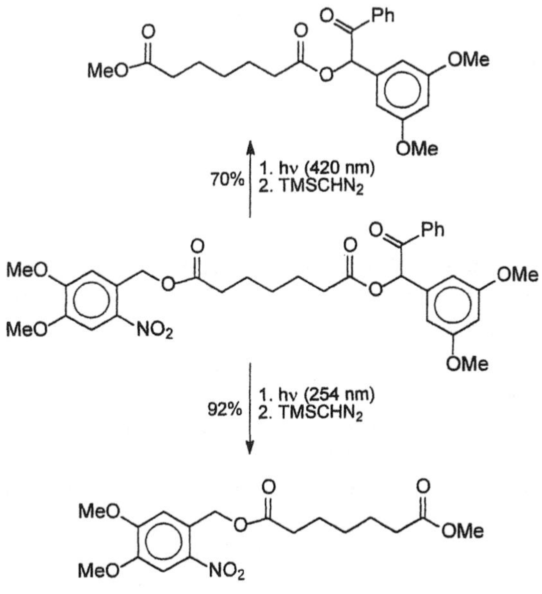

Scheme 29

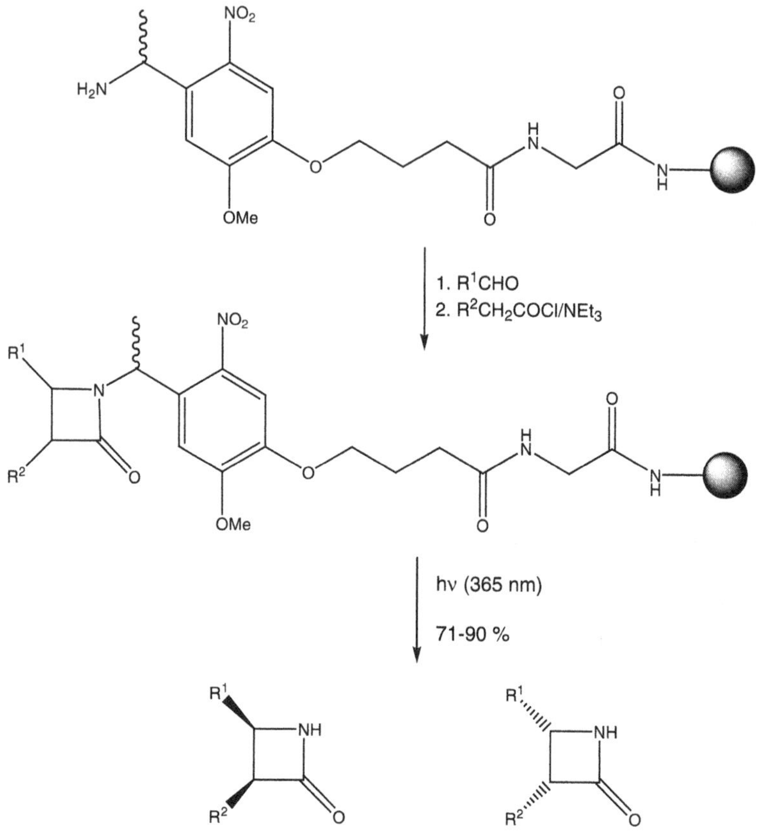

Scheme 30

convenient. Attached to a conventional solid support (polystyrene beads) through a glycine unit, it has been found stable towards 95% trifluoroacetic acid, but is smoothly cleaved by irradiation (95% in 3 h). The amine function of the benzylamine has been used for the combinatorial synthesis of N-unsubstituted β-lactams [287] (Sch. 30).

Another illustrative example is the application to spatially addressable parallel chemical synthesis. In this application, the starting compounds are linked to a glass plate through an amino linker, and irradiation through a mask leads to selective liberation of nitroveratroylcarbonyl (NVOC) amino-protected moieties. In this way, treatment with NVOC-protected activated esters of amino acids gives coupling at the illuminated positions only. The procedure can be repeated by using various masks and a large number of

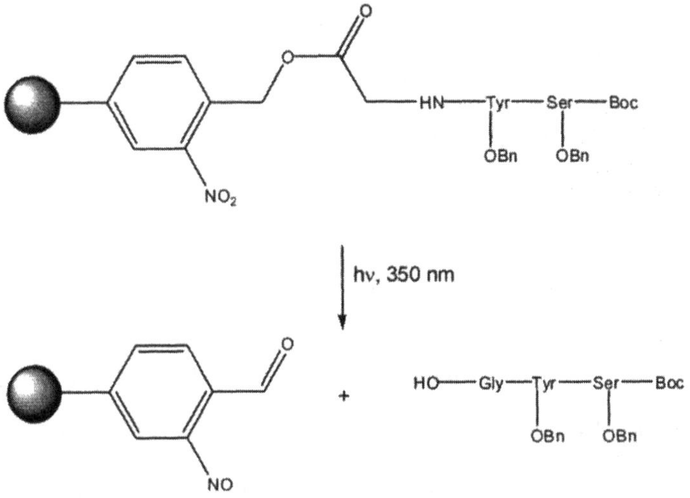

Scheme 31

different peptides can be generated on the plate and directly evaluated for their binding properties [301,302] (Sch. 31).

This principle has been extended in various directions, e.g., for carrying out a photochemical solid phase synthesis in microchip format based on the used of the dimethoxybenzoin moiety [303].

REFERENCES

1. Gomberg M. J Am Chem Soc 1913; 35:1035.
2. Lifschitz J. Ber Dtsch Chem Ges 1919; 52:1919.
3. Weyde E, Frankenburger W, Zimmermann E. Naturwiss 1930; 18:206.
4. Weyde E, Frankenburger W. Trans Far Soc 1931; 27:561.
5. Manring LE, Peters KS. J Phys Chem 1984; 88:3516.
6. Holmes EO. J Phys Chem 1957; 61:434.
7. Holmes EO. J Phys Chem 1958; 62:884.
8. Sporer AH. Trans Far Soc 1961; 57:983.
9. Herz ML. J Am Chem Soc 1975; 97:6777.
10. Harris L, Kaminsky J. J Am Chem Soc 1935; 57:1154.
11. Calvert JG, Rechen HJL. J Am Chem Soc 1952; 74:2101.
12. Fischer GL, LeBlanc JC, Johns HE. Photochem Photobiol 1967; 6:757.
13. Zimmerman HE, Sandel VR. J Am Chem Soc 1963; 85:915.
14. Cristol SJ, Bindel TH. Org Photochem 1983; 6:327.

15. Fleming S, Pincock J. In: Ramamurthy V, Schanze KS, eds. Mol Supramol Photochem. Vol. 3. 1999:211.
16. Paternò E, Chieffi G. Gazz Chim It 1909; 39 II:415.
17. Paternò E, Chieffi G. Gazz Chim It 1910; 40 II:321.
18. Cohen WD. Rec Trav Chim Pay-Bas 1920; 39:243.
19. Schönberg A, Mustafa A. J Chem Soc 1943; 276.
20. Schönberg A, Mustafa A. J Chem Soc 1944; 67.
21. Schönberg A, Mustafa A. J Chem Soc 1945; 657.
22. Sachs F, Hilpert S. Ber Dtsch Chem Ges 1904; 37:3425.
23. Libman J. J Am Chem Soc 1975; 97:4139.
24. Arnold DA, Maroulis AJ. J Am Chem Soc 1976; 98:5931.
25. Pincok JA. Acc Chem Res 1997; 30:43.
26. Larson JR, Epiotis ND, McMurchie LE, Shaik SS. J Org Chem 1980; 45:1388.
27. Michl J, Bonačić-Koutecký V. Electronic Aspects of Organic Photochemistry. New York: Wiley, 1990.
28. Geiger MV, Turro NJ, Waddel WH. Photochem Photobiol 1977; 25:15.
29. Zimmerman HE. J Am Chem Soc 1995; 117:9019.
30. DeCosta DP, Howell N, Pincock AL, Pincock JA, Rifai S. J Org Chem 2000; 65:4698.
31. Cozens FL, Pincock AL, Pincock JA, Smith R. J Org Chem 1998; 63:434.
32. Wagner PJ, Truman RJ, Puchalski HE, Wake R. J Am Chem Soc 1986; 108:7727.
33. Wagner P, Park BS. Org Photochem 1991; 11:227.
34. Wilson RM, Schnapp KA, Patterson WS. J Am Chem Soc 1992; 114:10987.
35. Wilson RM, Hannemann K, Peters K, Peters EM. J Am Chem Soc 1987; 109:4741.
36. Nakayama T, Torii Y, Nagahara T, Hamanoue K. J Photochem Photobiol A: Chem 1998; 119:1.
37. Döpp D. In: Horspool WM, Song PS, eds. CRC Handbook of Organic Photochemistry and Photobiology. Boca Raton, FL: CRC, 1995:1019.
38. Yip R, Sharma DK, Giasson R, Gravel D. J Phys Chem 1985; 89:5328.
39. Yip RW, Wen YX, Gravel D, Giasson R, Sharma DK. J Phys Chem 1991; 95:6068.
40. Schneider S, Fink M, Bug R, Schupp H. J Photochem Photobiol A: Chem 1991; 55:329.
41. Gravel D, Giasson R, Blanchet D, Yip RW, Sharma DK. Can J Chem 1991; 69:1193.
42. Wettermark G, Black E, Dogliotti L. Photochem Photobiol 1965; 4:229.
43. Albini A, Sulpizio A. In: Fox MA, Chanon M, eds. Photoinduced Electron Transfer. Vol. C. Amsterdam: Elsevier, 1988:88.
44. Albini A, Fasani E, Mella M. Top Curr Chem 1993; 168:143.
45. Popielartz R, Arnold DR. J Am Chem Soc 1990; 112:3068.
46. Maslak P. Top Curr Chem 1993; 168:1.
47. Lewis FD. Acc Chem Res 1986; 19:401.

48. Dinnocenzo JP, Farid S, Goodman JL, Gould IR, Todd WR, Mattes SL. J Am Chem Soc 1989; 111:8973.
49. Gaillard RE, Whitten DG. Acc Chem Res 1996; 29:292.
50. Nicholas AMP, Arnold DR. Can J Chem 1982; 60:2165.
51. Freccero M, Pratt A, Albini A, Long C. J Am Chem Soc 1998; 120:284.
52. Maslak P, Chapman WH, Vallombroso TM. J Am Chem Soc 1995; 117:12380.
53. Burton RD, Bartberger MD, Zhang Y, Eyler JR, Schanze KS. J Am Chem Soc 1996; 118:5655.
54. Baciocchi E, Bietti M, Putignani L, Steenken S. J Am Chem Soc 1996; 118:5962.
55. Kornblum N, Cheng L, Davies TM, Earl GW, Holy NL, Kerber RC, Kerstner WM, Manthley JW, Musser MT, Pinnick HW, Snow DH, Stuchal FW, Swiger RT. J Org Chem 1987; 52:196.
56. Lan JY, Schuster GB. J Am Chem Soc 1985; 107:6710.
57. Lan JY, Schuster GB. Tetrahedron Lett 1986; 27:4261.
58. Wan P, Shukla D. Chem Rev 1993; 93:571.
59. Wan P, Krogh E, Chak B. J Am Chem Soc 1988; 110:4073.
60. Wan P, Budac C, Krogh E. J Chem Soc, Chem Commun 1990; 255.
61. Budac D, Wan P. J Org Chem 1992; 57:887.
62. Shukla D, Wan P. J Photochem Photobiol A: Chem 1998; 113:53.
63. Russo-Caia C, Steenken S. Phys Chem Chem Phys 2002; 4:1478.
64. Cristol SG, Stull DP, Daussin RG. J Am Chem Soc 1980; 100:6674.
65. Cristol S, Schloemer GC. J Am Chem Soc 1972; 94:5916.
66. Cristol SJ, Greenwald BE. Tetrahedron Lett 1976; 2105.
67. Cristol SJ, Stull DP, McEntee TE. J Org Chem 1978; 43:1956.
68. Bartl J, Steenken S, Mayr H, McClelland RA. J Am Chem Soc 1990; 112:6918.
69. Deniz AA, Li B, Peters KS. J Phys Chem 1995; 99:12209.
70. Lipson M, Deniz AA, Peters KS. J Phys Chem 1996; 100:3580.
71. Deniz AA, Peters KS. J Am Chem Soc 1996; 118:2992.
72. Slocum GH, Schuster GB. J Org Chem 1984; 49:2177.
73. Arnold B, Donald L, Jurgens A, Pincock JA. Can J Chem 1985; 63:3140.
74. McGowan WM, Hilinski EF. J Am Chem Soc 1995; 117:9019.
75. Seiler P, Wirz J. Tetrahedron Lett 1971; 1683.
76. Seiler P, Wirz J. Helv Chim Acta 1972; 55:2693.
77. Kornblum N, Michel RE, Kerber RC. J Am Chem Soc 1966; 88:5660.
78. Russel GA, Danen WC. J Am Chem Soc 1966; 88:5663.
79. Irie M. J Am Chem Soc 1983; 105:2078.
80. Wan P, Yates K, Boyd MK. J Org Chem 1985; 50:2881.
81. Minto RE, Das PK. J Am Chem Soc 1989; 111:8858.
82. McClelland RA, Banait N, Steenken S. J Am Chem Soc 1989; 111:2929.
83. Meklemberg SL, Hilinski EF. J Am Chem Soc 1989; 111:5471.
84. Turro NJ, Wan P. J Photochem 1985; 28:93.
85. Wan P. J Org Chem 1985; 50:2583.
86. Wan P, Chak B. J Chem Soc, Perkin Trans 2 1986; 1751.
87. Hall B, Wan P. J Phototochem Photobiol A: Chem 1991; 56:35.

88. Wan P, Chack B, Krogh E. J Photochem Photobiol A: Chem 1989; 46:49.
89. Diao L, Yang C, Wan P. J Am Chem Soc 1995; 117:5369.
90. Wan P, Hennig D. J Chem Soc, Chem Commun 1987; 939.
91. Hamai S, Kukubun H. Bull Chem Soc Jpn 1974; 47:2085.
92. Hamai S, Kukubun H. Z Phys Chem (Frankfurt) 1974; 88:211.
93. Brousmiche D, Wan P. Chem Commun 1998; 491.
94. Huang CG, Wan P. J Chem Soc, Chem Commun 1988; 1193.
95. Huang CG, Wan P. J Org Chem 1991; 56:4846.
96. Huang CG, Beveridge KA, Wan P. J Am Chem Soc 1991; 113:7676.
97. Shi Y, MacKinnon A, Howard JAK, Wan P. J Photochem Photobiol A: Chem 1998; 113:271.
98. Shi Y, Wan P. Chem Commun 1995; 273.
99. Fisher M, Shi Y, Zhao B, Sniekus V, Wan P. Can J Chem 1999; 77:868.
100. Burnham KS, Schuster GB. J Am Chem Soc 1998; 120:12619.
101. Yang C, Wan P. J Photochem Photobiol A: Chem 1994; 80:227.
102. Benn R, Dreeskamp H, Schuchmann HP, von Sonntag C. Z Naturforsch 1979; 34b:1002.
103. Pitchumani K, Devanathan S, Ramamurthy V. J Photochem Photobiol 1992; 69:201.
104. Parman T, Pincock JA, Wedge PJ. Can J Chem 1994; 72:1254.
105. Zimmerman HE, Somasekara S. J Am Chem Soc 1963; 85:922.
106. McClelland RA, Kanagasabapathy VM, Banait N, Steenken S. J Am Chem Soc 1989; 111:3966.
107. Liu XG, Binkley RW. J Carbohydr Chem 1993; 12:779.
108. Liu XG, Binkley RW, Yeh P. J Carbohydr Chem 1992; 11:1053.
109. Jaeger DA. J Am Chem Soc 1975; 97:902.
110. Jaeger DA, Angelos GH. Tetrahedron Lett 1981; 22:803.
111. Liilis V, McKenna J, McKenna JM, Smith MJ, Taylor PS, Williams JH. J Chem Soc, Perkin Trans 2 1980; 83.
112. Hilborn JW, MacKnight E, Pincock JA, Wedge PJ. J Am Chem Soc 1994; 116:3337.
113. Pincock JA, Wedge PJ. J Org Chem 1994; 59:5587.
114. Hilborn JW, Pincock JA. J Am Chem Soc 1991; 113:2683.
115. DeCosta DP, Pincock JA. J Am Chem Soc 1989; 111:8948.
116. DeCosta DP, Pincock JA. J Am Chem Soc 1993; 115:2180.
117. Pincock JA. Acc Chem Res 1997; 30:43.
118. Pincock JA. In: Horspool WM, Song PS, eds. Handbook of Organic Photochemistry and Photobiology. Boca Raton: CRC Press, 1994:393.
119. Parman T, Pincock JA, Wedge PJ. Can J Chem 1994; 72:1254.
120. Barltrop JA, Schofield P. J Chem Soc 1965; 4758.
121. Cameron JF, Fréchet JMJ. J Org Chem 1990; 55:5919.
122. Cameron JF, Fréchet JMJ. J Am Chem Soc 1991; 113:4303.
123. Beecher JE, Cameron JF, Fréchet JMJ. J Mater Chem 1992; 2:811.
124. Burgess K, Jacutin SE, Lim D, Shitangkoon A. J Org Chem 1997; 62:5165.

125. Cameron JF, Wilson CG, Fréchet JMJ. J Am Chem Soc 1996; 118:12925.
126. Cameron JF, Willson GG, Fréchet JMJ. J Chem Soc, Perkin Trans 1 1997; 2429.
127. Omelanzcuk J, Sopchik AE, Lee SG, Akutagawa K, Cairns M, Bentrude WG. J Am Chem Soc 1988; 110:6908.
128. Cairns SM, Bentrude WG. Tetrahedron Lett 1989; 30:1025.
129. Bentrude WG, Lee SG, Akutagawa K, Ye W, Charbonnel Y, Omelanczuk J. Phosphorus Sulfur Relat Elem 1987; 30:105.
130. Koptyug IV, Slugget GW, Ghatlia ND, Landis MS, Turro NJ, Ganapathy S, Bentrude WG. J Phys Chem 1996; 100:14581.
131. Koptyug IV, Ghatlia ND, Sluggett GW, Turro NJ, Ganapathy S, Bentrude WG. J Am Chem Soc 1995; 117:9486.
132. Bhanthumnavin W, Bentrude WG. J Org Chem 2001; 66:980.
133. Bentrude WG, Mullah KB. J Org Chem 1991; 56:7218.
134. Mullah KB, Bentrude WG. Nucleosides Nucleotides 1984; 13:127.
135. Ganapathy S, Sekhar BBVS, Cairns SM, Akutagawa K, Bentrude WG. J Am Chem Soc 1999; 121:2085.
136. Yamaoka T, Adachi H, Matsumoto K, Watanabe H, Shirosaki T. J Chem Soc, Perkin Trans 2 1990; 663.
137. Naitoh K, Yamaoka T. J Chem Soc, Perkin Trans 2 1992; 663.
138. Shirai M, Tsunooka M. Prog Polym Sci 1996; 21:1.
139. Houlikan FM, Nalaman O, Kometani JM, Reichmanis E. J Imag Sci Technol 1997; 41:35.
140. Sugimoto A, Kimoto S, Inoue H. J Chem Res (S) 1996; 506.
141. Er Rhaimini A, Moshsinaly N, Mornet R. Tetrahedron Lett 1990; 31:5727.
142. Salerno CP, Resat M, Magde D, Kraut J. J Am Chem Soc 1997; 119:3403.
143. Ramesh D, Wieboldt R, Billington AP, Carpenter BK, Hess GP. J Org Chem 1993; 58:4599.
144. Rodenbaugh R, Fraser-Reid B, Geysen HM. Tetrahedron Lett 1997; 38:7653.
145. Nakatani K, Higashida N, Saito I. Tetrahedron Lett 1997; 38:5005.
146. Ratcliff MA, Kochi JK. J Org Chem 1971; 36:3112.
147. Larson JR, Epiotis ND, McMurchie LE, Shaik SS. J Org Chem 1980; 45:951.
148. Bremner JB, Wizenberg KN. Aust J Chem 1978; 31:313.
149. Appleton DC, Bull DC, Givens RS, Lillis V, McKenna J, McKenna JM, Thackeray S, Walley AR. J Chem Soc, Perkin Trans 2 1980; 77.
150. Foster B, Gaillard B, Pincock AL, Pincock JA, Sehmbey C. Can J Chem 1987; 65:1599.
151. Modica E, Zanaletti R, Freccero M, Mella M. J Org Chem 2001; 66:41.
152. Norris RK, Smyth-King RJ. J Chem Soc, Chem Commun 1981; 79.
153. Griffin CE, Kaufman ML. Tetrahedron Lett 1965; 773.
154. Imrie C, Modro TA, Rohwer ER, Wagener CCP. J Org Chem 1993; 58:5643.
155. Breslin DT, Saeva FD. J Org Chem 1988; 53:713.
156. Alonsono EO, Jouhnston LJ, Scaiano JC, Toscano VG. Can J Chem 1992; 70:1784.
157. Maycock AL, Berchtold GA. J Org Chem 1970; 35:2532.

158. Saeva FD, Breslin DT, Luss HR. J Am Chem Soc 1991; 113:5333.
159. Okamoto Y, Iwamoto N, Takamuku S. J Chem Soc, Chem Commun 1986; 1516.
160. Okamoto Y, Iwamoto N, Toki S, Takamuku S. Bull Chem Soc Jpn 1987; 60:277.
161. Fleming SA, Rawlins DB, Robins MJ. Tetrahedron Lett 1990; 31:4995.
162. Fleming SA, Jensen AW. J Org Chem 1993; 58:7135.
163. Flemig SA, Rawlins D, Samano V, Robins MJ. J Org Chem 1992; 57:5968.
164. Givens RS, Hrinczenko B, Liu JHS, Matuszeski B, Thole-Collison J. J Am Chem Soc 1984; 106:1779.
165. Givens RS, Matuszewski B. Tetrahedron Lett 1978; 861.
166. Amiri AS, Mellor JM. J Photochem 1978; 861.
167. Gould IR, Tung C, Turro NJ, Givens RS, Matuszewski B. J Am Chem Soc 1984; 106:1789; Guo Y, Jenks WS. J Org Chem 1995; 60:548.
168. Jenks WS, Gregory DD, Guo Y, Lee W, Tetzlaff T. In: Ramamurthy V, Schanze KS, eds. Mol Supramol Photochem 1997; 1:1.
169. Spears K, Gray TH, Huang D. J Phys Chem 1986; 90:779.
170. Cosa JJ, Gsponer HDPE. J Photochem Photobiol A: Chem 1989; 48:303.
171. McClelland RA, Kangasabapathy VM, Banait NS, Steenken S. J Am Chem Soc 1989; 111:3966.
172. Budac D, Wan P. J Photochem Photobiol A: Chem 1992; 67:135.
173. Krogh E, Wan P. J Am Chem Soc 1992; 114:705.
174. Wan P, Muralidharan S. J Am Chem Soc 1988; 110:4336.
175. Steenken S, McClelland RA. J Am Chem Soc 1989; 111:4967.
176. Popielartz R, Arnold DR. J Am Chem Soc 1992; 112:3068.
177. Maslak P, Chapman WH. J Org Chem 1996; 61:2647.
178. Maslak P, Chapman WH, Tetrahedron 1990; 46:2715.
179. Ishiguro K, Osaki T, Sawaki Y, Chem Lett 1992; 743.
180. Arnold DR, Lamont LJ. Can J Chem 1989; 67:2119.
181. Arnold DR, Lamont LJ, Perrot AL, Can J Chem 1991; 69:225.
182. Perrot AL, de Lijser HP, Arnold DR. Can J Chem 1997; 75:384.
183. Arnold DR, Du X, Chen J. Can J Chem 1995; 73:307.
184. Yamada S, Tanaka T, Akiyama S, Ohashi M. J Chem Soc, Perkin Trans 2 1992; 449.
185. Gan H, Kellet MA, Leon JW, Kloeppner LR, Leinhos U, Gould IR, Farid S, Whitten DG. J Photochem Photobiol A: Chem 1994; 82:211.
186. Perrier S, Sankararaman S, Kochi JK. J Chem Soc, Perkin Trans 2 1993; 825.
187. Albini A, Mella M. Tetrahedron 1986; 42:6219.
188. Ci X, Kellet MA, Whitten DG. J Am Chem Soc 1991; 113:3893.
189. Gan H, Leinhos U, Gould IR, Whitten DG. J Phys Chem 1995; 99:3566.
190. Lucia LA, Wang Y, Nafisi K, Netzel TL, Schanze KS. J Phs Chem 1995; 99:11801.
191. Chen L, Farahat MS, Gaillard ER, Farid S, Whitten DJ. J Photochem Photobiol A: Chem 1996; 95:21.

192. Zhang W, Yang L, Wu LM, Liu YC, Liu ZL. J Chem Soc, Perkin Trans 2 1998; 1189.
193. Chen L, Lucia L, Whitten DG. J Am Chem Soc 1998; 120:439.
194. Albini A, Fasani E, Freccero M. Adv Electron Transf Chem 1996; 5:103.
195. Lewis FD, Petisce JR. Tetrahedron 1986; 42:6207.
196. Ohashi M, Aoyashi N, Yamada S. J Chem Soc, Perkin Trans 1 1990; 1335.
197. Boggeri E, Fasani E, Mella M, Albini A. J Chem Soc, Perkin Trans 2 1991; 2097.
198. Bardi L, Fasani E, Albini A. J Chem Soc, Perkin Trans 2 1994; 545.
199. Albini A, Spreti S. Tetrahedron 1984; 40:2975.
200. Albini A, Fasani E, Oberti R. Tetrahedron 1982; 38:1034.
201. Albini A, Fasani E, Montessoro E. Z Naturf 1984; 39b:1409.
202. Albini A, Fasani E, Sulpizio A. J Am Chem Soc 1984; 106:3362.
203. Albini A, Fasani E, Mella M. J Am Chem Soc 1986; 108:4119.
204. Albini A, Sulpizio A. J Org Chem 1989; 54:2147.
205. d'Alessandro N, Fasani E, Mella M, Albini A. J Chem Soc, Perkin Trans 2 1991; 1977.
206. Sulpizio A, Albini A, d'Alessandro N, Fasani E, Pietra S. J Am Chem Soc 1989; 111:5773.
207. d'Alessandro N, Albini A, Mariano PS. J Org Chem 1993; 58:937.
208. Fasani E, d'Alessandro N, Albini A, Mariano PS. J Org Chem 1994; 59:829.
209. Mizuno K, Terasaka K, Yasueda M, Otsuji Y. Chem Lett 1988; 145.
210. Eaton DF. J Am Chem Soc 1980; 102:3280.
211. Albini A, Siviero E, Mella M, Long C, Pratt A. J Chem Soc, Perkin Trans 2 1995; 1895.
212. Maslak P, Kula J. Tetrahedron Lett 1990; 31:4969.
213. Wong JW, Natalie KJ Jr, Nwokogu GC, Pisipati JS, Flaherty PT, Greenwood TD, Wolfe JF. J Org Chem 1997; 62:6152.
214. Mella M, Fagnoni M, Albini A. Eur J Org Chem 1999; 2137.
215. Cermenati L, Albini A. J Adv Oxid Technol 2002; 5:58.
216. Cermenati L, Richter C, Albini A. J Chem Soc, Chem Commun 1998; 805.
217. Cermenati L, Mella M, Albini A. Tetrahedron 1998; 54:2575.
218. Ikeda H, Takasaki T, Takahashi Y, Konno A, Matsumoto M, Hoshi Y, Aoki T, Suzuki T, Goodman J, Miyashi T. J Org Chem 1999; 64:1640.
219. Ikeda H, Minegishi T, Abe H, Konno A, Goodman JL, Miyashi T. J Am Chem Soc 1998; 120:87.
220. Wagner PJ, Truman RJ, Puchalski HE, Wake R. J Am Chem Soc 1986; 108:7727.
221. Sulpizio A, Mella M, Albini A. Tetrahedron 1989; 45:7545.
222. Cermenati L, Freccero M, Venturello P, Albini A. J Am Chem Soc 1995; 117:7869.
223. Freccero M, Fasani E, Albini A. J Org Chem 1993; 58:1740.
224. Moore RF, Waters WA. J Chem Soc 1953; 3405.
225. Jones G, Haney WA, Phan XT. J Am Chem Soc 1988; 110:1922.

226. Borg RM, Heuckenroth RO, Lan AJY, Quillen SL, Mariano PS. J Am Chem Soc 1987; 109:2728.
227. Albini A, Spreti S. Z Naturforsch 1986; 41b:1286.
228. Juillard M, Galadi A, Chanon M. J Photochem Photobiol A: Chem 1990; 54:79.
229. Juillard M, Legris C, Chanon M. J Photochem Photobiol A: Chem 1991; 61:137.
230. Albini A, Spreti S. J Chem Soc, Perkin Trans 2 1987; 1175.
231. Baciocchi E, Rol C, Sebastiani GV, Taglieri L. J Org Chem 1994; 59:5272.
232. Baciocchi E, Bietti M, Ferrero MI, Rol C, Sebastiani GV. Acta Chem Scand 1998; 52:160.
233. Baciocchi E, Rosato GC, Rol C, Sebastiani GV. Tetrahedron Lett 1992; 33:5437.
234. Butts CP, Eberson L, Hartshorn MP, Robinson WT, Timmerman-Vaugham DJ, Young DAW. Aust J Chem 1989; 48:95.
235. Eberson L, Hartshorn MP, Timmerman-Vaugham DJ. Acta Chem Scand 1996; 50:1121.
236. Masnovi JM, Sankararaman S, Kochi JK. J Am Chem Soc 1989; 111:2263.
237. Eberson L, Calvert JL, Hartshorn MP, Robinson WT. Acta Chem Scand 1993; 47:1025.
238. Butts CP, Eberson L, Fulton KL, Hartshorn MP, Robinson WT, Timmerman-Vaugham DJ. Acta Chem Scand 1996; 50:991.
239. Baciocchi E, Rol C, Rosato GC, Sebastiani GV. J Chem Soc, Chem Commun 1992; 59.
240. Schaap AP, Siddiqui S, Pradad G, Palomino E, Lopez L. J Photochem 1984; 25:167.
241. Mizuno K, Ichinose N, Otsuji Y. J Org Chem 1992; 57:1855.
242. Mizuno K, Kamiyama N, Ichinose N, Otsuji Y. Tetrahedron 1985; 41:2207.
243. Ichinose N, Mizuno K, Tamai T, Otsuji Y, Mizuno K, Ichinose N, Tamai T, Otsuji Y. J Org Chem 1992; 57:4669.
244. Albini A, Arnold DR. Can J Chem 1978; 56:2985.
245. Schaap AP, Lopez L, Gagnon SD. J Am Chem Soc 1983; 105:663.
246. Scaap AP, Siddiqui S, Prasad G, Rahman AFMN, Oliver JP. J Am Chem Soc 1984; 106:6087.
247. Gaebert C, Mattay J. Tetrahedron 1997; 53:14297.
248. Hasegawa E, Koshii S, Horaguchi T, Shimizu T. J Org Chem 1992; 57:6342.
249. Albrecht E, Averdung J, Bischof EW, Heidbreder A, Kirschberg T, Meller F, Mattay J. J Photochem Photobiol A: Chem 1994; 82:219.
250. Shima K, Sazaki A, Nakabayashi K, Yasuda M. Bull Chem Soc Jpn 1992; 65:1472.
251. Bochet CG. J Chem Soc, Perkin Trans 1 2002; 125.
252. Patchornik A, Amit B, Woodward RB. J Am Chem Soc 1970; 92:6333.
253. Okada S, Yamashita S, Furuta T, Iwamura M. Photochem Photobiol 1995; 61:431.
254. Pirrung MC, Lee YR, Park K, Springer JB. J Org Chem 1999; 64:5042.

255. Kalbag SM, Roeske RW. J Am Chem Soc 1975; 97:440.
256. Zehavi U, Amit B, Patchornik A. J Org Chem 1972; 37:2281.
257. Zehavi U, Patchorink A. J Org Chem 1972; 37:2285.
258. Bartholomew DG, Broom AD. J Chem Soc, Chem Commun 1975; 38.
259. Rubinstein M, Amit B, Patchornik A. Tetrahedron Lett 1975; 17:1445.
260. Gravel D, Hebert J, Thoraval D. Can J Chem 1983; 61:400.
261. Blanc A, Bochet CG. J Org Chem 2003; 68:1138.
262. Chamberlin JW. J Org Chem 1966; 31:1658.
263. Birr C, Lochinger W, Stahncke G, Lang P. Liebigs Ann Chem 1972; 763:162.
264. Misteic A, Boyd MK. Tetrahedron Lett 1998; 39:1653.
265. Furuta T, Hirayama Y, Iwamura M. Org Lett 2001; 3:1809.
266. Sheenan JC, Wilson RM. J Am Chem Soc 1964; 86:5277.
267. Sheenan JC, Wilson RM, Oxford AW. J Am Chem Soc 1971; 93:7222.
268. Pirrung MC, Shuey SW. J Org Chem 1994; 59:3890.
269. Givens RS, Kuepper LW III. Chem Rev 1993; 93:55.
270. Givens RS, Park CH. Tetrahedron Lett 1996; 37:6259.
271. Bochet CG. Angew Chem, Int Ed 2001; 40:6341.
272. Blanc A, Bochet CG. J Org Chem 2002; 67:5567.
273. Rich DH, Gurwara SK. J Am Chem Soc 1975; 97:1575.
274. Rich HD, Gurwara SK. Tetrahedron Lett 1975; 301.
275. Hammer RP, Albericio F, Gear L, Barany G. Int J Pept Protein Res 1990; 36:31.
276. Lloyd-Williams P, Gairi M, Albericio F, Giralt E. Tetrahedron 1993; 49:10069.
277. Ruhland B, Bhandari E, Gordon EM, Gallop MA. J Am Chem Soc 1996; 118:253.
278. Nicolau KC, Wissinger N, Pastor J, De Roose F. J Am Chem Soc 1997; 119:449.
279. Nicolau KC, Watanabe N, Li J, Pastor J, Wissinger N. Angew Chem Int Ed 1998; 37:1559.
280. Burbaum JJ, Ohlmeyer MHJ, Reader JC, Henderso I, Dillard LW, Li G, Randle TL, Sigal NH, Chelsjy D, Baldwin JJ. Proc Natl Acad Sci USA 1995; 92:6027.
281. Ohlmeyer MHJ, Swanson RN, Dillard LW, Reader JC, Asouline G, Kobayashi R, Wigler M, Still WC. Proc Natl Acad Sci USA 1993; 90:10922.
282. Ajayaghosh A, Pillai VNR. Tetrahedron 1988; 44:6661.
283. Ajayaghosh A, Pillai VNR. J Org Chem 1987; 52:5714.
284. Ajayaghosh A, Pillai VNR. Tetrahedron Lett 1995; 36:777.
285. Holmes CP, Jones DG. J Org Chem 1995; 60:2318.
286. Holmes CP, Chinn JP, Look GC, Gordon EM, Gallop MA. J Org Chem 1995; 60:7328.
287. Ruhland B, Bhandari A, Gordon EM, Gallop MA. J Am Chem Soc 1996; 118:253.
288. Holmes CP. J Org Chem 1997; 62:2370.
289. Shi Y, Corrie JET, Wan P. J Org Chem 1997; 62:8278.

290. Peach JM, Pratt AJ, Snaith JS. Tetrahedron 1995; 51:10013.
291. Baldwin JE, McConnaughie AW, Moloney MG, Pratt AJ, Shim SB. Tetrahedron 1990; 46:6879.
292. Thirlwell H, Corrie JET, Reid GP, Trentham DR, Ferenczi MA, Biophys J 1994; 59:3890.
293. Pirrung MC, Bradley JC. J Org Chem 1995; 60:6270.
294. Rock RS, Chan SI. J Org Chem 1996; 61:1526.
295. Bellof D, Mutter M. Chimia 1985; 39:317.
296. Routledge A, Abell C, Balasubramanian S. Tetrahedron Lett 1997; 38:1227.
297. Lan AJY, Heuckenroth RO, Mariano PS. J Am Chem Soc 1987; 109:2738.
298. Ho DG, Mariano PS. J Org Chem 1988; 53:5113.
299. Cho IS, Chang SSS, Ho C, Lee CP, Ammon HL, Mariano PS. Heterocycles 1991; 32:2161.
300. Barker B, Diao L, Wan P. J Photochem Photobiol A: Chem 1997; 104:91.
301. Fodor SPA, Read JL, Pirrung MC, Stryer L, Lu AT, Solas D. Science 1991; 251:767.
302. McGall GH, Barone AD, Diggelmann M, Fodor SPA, Gentalen E, Ngo N. J Am Chem Soc 1997; 119:5081.
303. Pirrung MC, Fallon L, McGall G. J Org Chem 1998; 63:241.

16

Photoinduced Aromatic Nucleophilic Substitution Reactions

Roberto A. Rossi
Universidad Nacional de Córdoba, Ciudad Universitaria, Córdoba, Argentina

16.1. HISTORICAL REMARKS

Nucleophilic substitution is feasible through processes that involve electron transfer (ET) steps. In these reactions a compound bearing an adequate leaving group is substituted at the *ipso* position by a nucleophile.

The $S_{RN}1$ reaction, which stands for *Unimolecular Radical Nucleophilic Substitution*, is a chain process that involves radicals and radical anions as intermediates. The $S_{RN}1$ mechanism was first proposed for the substitution of alkyl derivatives bearing electron withdrawing groups (EWG) in the α position and a suitable leaving group [1,2]. In 1970 Bunnett applied this mechanism to unactivated aromatic halides [3]. While studying reactions of unactivated aryl halides, such as 1-halo-2,3,5-trimethylbenzenes, with amide ions in liquid ammonia by the benzyne mechanism, it was observed that the chloro and bromo derivatives fulfilled the expected ratio of isomers, but the iodo derivatives gave more of the straightforward *ipso* substitution product. An important observation was that solvated electrons from dissolution of alkali metals in liquid ammonia catalyzed the reactions. It was postulated that the aryl halides capture an electron to form the radical anion faster than the deprotonation by the amide ions to afford a benzyne intermediate. The radical anion fragments to afford aryl radical that to follows the $S_{RN}1$ cycle [4].

By way of studying alternate methods to generate aryl radicals, Bunnett et al. found that halobenzenes react with acetone enolate ions under irradiation in liquid ammonia to afford good yields of substitution products [5].

Several reviews have been published in relation to aromatic $S_{RN}1$ reactions [6–9], and to the synthetic applications of the process [10,11]. In this chapter an introductory short survey of the general mechanistic aspects, the aromatic substrates and nucleophiles involved in the photoinduced $S_{RN}1$ mechanism will be presented. The main emphasis being related to its synthetic capability, target applications and to the more recent advances in the field.

16.2. STATE OF THE ART MECHANISTIC MODELS

The $S_{RN}1$ mechanism is a chain process, whose main steps are presented in Sch. 1.

Overall, Eqs. (1)–(3) depict a nucleophilic substitution Eq. (4) in which radicals and radical anions are intermediates. Once the radical anion of the substrate is formed it fragments into a radical and the anion of the leaving group (Eq. (1)). The aryl radical can react with the nucleophile to furnish a radical anion (Eq. (2)), which by ET to the substrate forms the intermediates needed to continue the propagation cycle (Eq. (3)). The mechanism has termination steps that depend on the substrate, the nucleophile and experimental conditions. Not many initiation events are needed, but in this case, the propagation cycle must be fast and efficient to allow for long chains to build up.

For a given aryl moiety, there is a rough correlation between the reduction potential and its reactivity in $S_{RN}1$ reactions [12]. The order of reduction potential in liquid ammonia, $PhI > PhBr > PhNMe_3I > PhSPh > PhCl > PhF > PhOPh$ coincides with the reactivity order determined under photoinitiation. By competition experiments of pairs of halobenzenes toward pinacolone enolate ions under photoinitiation, the span in reactivity from PhF to PhI was found to be about 100,000 [13].

$$(ArX)^{\cdot -} \longrightarrow Ar^{\cdot} + X^{-} \qquad (1)$$

$$Ar^{\cdot} + Nu^{-} \longrightarrow (ArNu)^{\cdot -} \qquad (2)$$

$$\underline{(ArNu)^{\cdot -} + ArX \longrightarrow ArNu + (ArX)^{\cdot -}} \qquad (3)$$

$$ArX + Nu^{-} \longrightarrow ArNu + X^{-} \qquad (4)$$

Scheme 1

The most common mechanistic evidence used is inhibition by radical traps such as di-*t*-butylnitroxide, 2,2,6,6-tetramethyl-1-piperidinyloxy, galvinoxyl, oxygen, etc., as well as by good electron acceptors such as *p*-dinitrobenzene. To assess the formation of radicals along the propagation cycle, radical probes have also been used, which afford products derived from ring closure, ring opening, and from rearrangement of the radicals.

This chain process requires an initiation step. In a few systems a thermal (spontaneous) initiation is observed. One of the methods consists in the formation of the radical anion of the substrate by reaction with alkali metals in liquid ammonia. Electrochemical initiation is an approach that has been successful in a considerable number of cases with aromatic and heteroaromatic substrates [14]. Other possibilities include the use of Fe^{+2} [15–18], SmI_2 [19], or Na(Hg) [20]. However, the most extensively used method is the photoinitiation, and it has proven extremely suitable for synthetic purposes Even though it is a widely used initiation method, there are not many mechanistic studies of this step.

These photoinduced ET can conceivably be accomplished by one of the following mechanisms: (1) homolytic cleavage of the C–X bond; (2) ET from the excited ArX to its ground state; (3) ET from the excited nucleophile to the ArX, generating a radical anion that enters the cycle; (4) ET within an excited charge transfer complex; and (5) photoejection of an electron from the excited nucleophile.

Depending on the nature of the substrate, the nucleophile and the experimental conditions, any of these mechanisms can be a method to initiate the reaction.

In photoinitiated reactions, the reactivity of the substrate–nucleophile pair can be modified by changing the solvent and the irradiation source. For instance iodobenzene does not react under irradiation (pyrex-filtered flask) with acetophenone enolate anion in liquid ammonia [21], but it does react in DMSO [22]. However, the reaction occurs in liquid ammonia by irradiation in an immersion well [23].

Quantum yields of photoinitiated reactions have been used as a qualitative measure of the chain length. The quantum yields for substitution of haloarenes with nitrile-stabilized carbanions range from 7 to 31 in liquid ammonia [24]. Quantum yields from 20 to 50 have been determined for the substitution of iodobenzene by $(EtO)_2PO^-$ ions [25].

This mechanism offers the possibility to afford the substitution of an unreactive nucleophile through *entrainment* conditions. *Entrainment* is useful when the nucleophile is rather unreactive at initiation, but quite reactive at propagation. Under these conditions the addition of another nucleophile, more reactive at initiation, increases the generation of

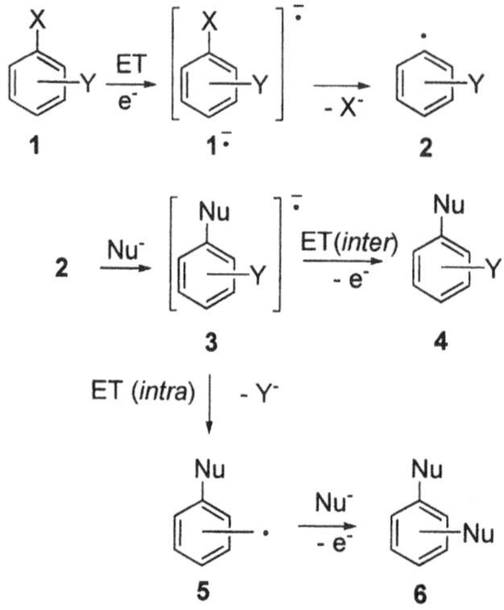

Scheme 2

intermediates and allows the less reactive nucleophile in the initiation to start its own propagation.

One of the main goals of the $S_{RN}1$ mechanism is the possibility to obtain disubstituted compounds when the reaction is performed with substrates bearing two leaving groups. Few examples are known of trisubstitutions [26,27].

When substrate **1** accepts one electron, it forms the radical anion **1⁻**, which by fragmentation at the more labile C-leaving group bond affords a radical **2** that by reaction with the nucleophile forms the radical anion of the monosubstituted compound **3**. This radical anion can transfer its extra electron to the substrate and in this case the monosubstituted compound **4** with retention of one leaving group is formed (Sch. 2).

Alternatively, the intramolecular ET to the second C–Y σ bond results in fragmentation to form a radical **5**, which by coupling with a second molecule of the nucleophile affords the disubstitution compound **6** (Sch. 2). In this case, the monosubstitution product **4** is not an intermediate in the formation of compound **6**. However, in some systems, the monosubstitution product **4** is indeed an intermediate to form **6** by two consecutive $S_{RN}1$ reactions.

The relative ratio between the intra and the intermolecular ET steps depends on the second leaving group (Y), its electron affinity in relation to the substrate that acts as another acceptor, its relative position with respect to the Nu moiety, as well as on the nature of the nucleophile.

16.3. SCOPE AND LIMITATIONS

16.3.1. Substrates

Participating substrates include unactivated aromatic and heteroaromatic halides. In addition to halides, other leaving groups are known (i.e.,: $(EtO)_2P(O)$, RS (R = Ar, alkyl), ArSO, $ArSO_2$, PhSe, Ph_2S^+, RSN_2 (R = t-Bu, Ph), N_2BF_4, R_3N^+, and N_2^+).

Many substituents are compatible with the reaction, such as alkyl groups, OR, OAr, SAr, CF_3, CO_2R, NH_2, NHCOR, NHBoc, SO_2R, CN, COAr, NR_2, and F. Even though the reaction is not inhibited by the presence of negatively-charged substituents as carboxylate ions, other charged groups such as oxyanions hinder the process.

Nitro-substituted phenyl halides produce radical anions that fragment with a rather low rate ($\cong 10^{-2}$–10^2/s) [28]. For this reason the nitro group is not a suitable substituent for most aromatic $S_{RN}1$ reactions. However, exceptions are found with o-iodonitrobenzene [29] whose radical anion has a relatively high rate of fragmentation, and nitroaryldiazo phenyl or $tert$-butyl sulfides [30].

16.3.2. Nucleophiles: Carbanions

Carbanions from hydrocarbons, nitriles, ketones, esters, N,N-dialkyl acetamides and thioamides, and mono and dianions from β-dicarbonyl compounds are some of the most common nucleophiles through which a new C–C bond can be formed. This C–C bond formation is also achieved by reaction with aromatic alkoxides. Among the nitrogen nucleophiles known to react are amide ions to form anilines; however, the anions from aromatic amines, pyrroles, diazoles and triazoles, react with aromatic substrates to afford C-arylation.

The $S_{RN}1$ mechanism has proved to be an important route to ring closure reactions. The syntheses of indoles, isocarbostyrils, binaphthyls, etc., and a number of natural products has been achieved by this process.

The main feature of carbanions derived from nitriles lies in the dependence on the aromatic substrate involved; thus, two different outcomes of the substitution reaction are possible: formation of the substitution compound by ET to the substrate from the radical anion intermediate **7**, formed by coupling of phenyl radicals and acetonitrile anion, or formation of products from elimination of the cyano group as is the case with phenyl halides [31,32] (Sch. 3). The same reactivity pattern is found with halothiophenes [33].

The α-arylated nitriles are exclusively formed by reaction of alkanenitrile anions with compounds that have a π MO of low energy to

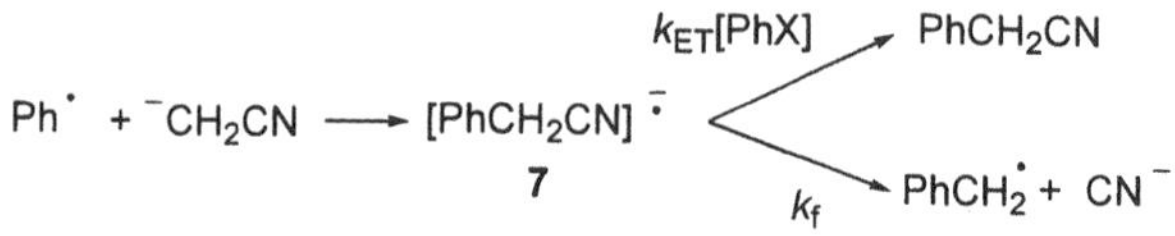

$$Ph^{\cdot} + {}^{-}CH_2CN \longrightarrow [PhCH_2CN]^{\overline{\cdot}}$$

$$\text{7}$$

with $k_{ET}[PhX]$ pathway to $PhCH_2CN$ and k_f pathway to $PhCH_2^{\cdot} + CN^{-}$

Scheme 3

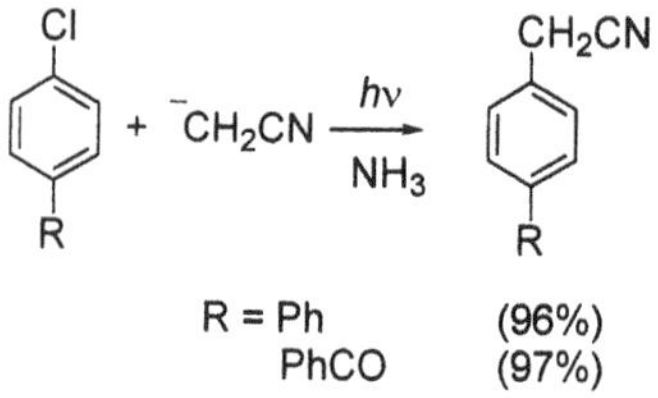

R = Ph   (96%)
PhCO   (97%)

Scheme 4

o- (90%)
m- (90%)
p- (63%)

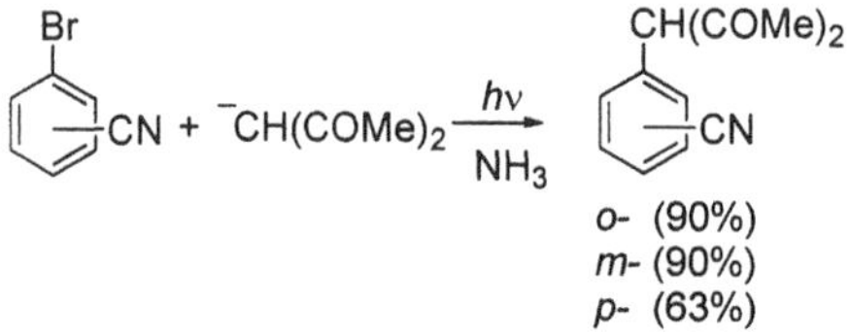

Scheme 5

stabilize the radical anion intermediate of the substitution product, as in the reaction of halogen derivatives of naphthalene, benzophenone, quinoline, pyridine, phenanthrene, and biphenyl with acetonitrile anion [34–37]. Few examples are depicted in Sch. 4 [35].

On the other hand, the reaction with $^{-}CH_2NO_2$ anions with aromatic radicals always affords products from fragmentation of the radical anion intermediates [22,38].

1,3-Dianions from β dicarbonyl compounds react quite well through the terminal carbon site, under irradiation [21,39]. Monoanions of β-dicarbonyl compounds do not react with 2-bromopyridine [40], 2-chloroquinoline [41], or 2-bromobenzamide, but they do react with more electrophilic substrates such as halobenzonitriles [42], bromocyanopyridines [42], 2-chloro-(trifluoromethyl)pyridines [43], and iodoquinolines [44]. In these reactions, excellent yields of arylation products are obtained (Sch. 5) [42].

16.3.3. Other Nucleophiles

Anions derived from tin, phosphorus, arsenic, antimony, sulfur, selenium, and tellurium are known to react at the heteroatom to form a new carbon-heteroatom bond in good yields.

A special behavior has been shown by diphenylarsenide, diphenyl-stibide, benzeneselenate, and benzenetellurate anions. With these nucleophiles scrambling of the aromatic ring has been reported [45–48]. The reaction of *p*-iodoanisole with benzeneselenate ions is illustrated in Sch. 6 [45].

According to the proposed mechanism, the radical anion formed by the coupling of *p*-anisyl radicals and benzeneselenate ions, can undergo three competitive reactions: reversion to starting materials, ET to the substrate to give the expected substitution product, and fragmentation at the Ph–Se bond to form phenyl radicals and the new nucleophile *p*-methoxybenzeneselenate ion. The nucleophile *p*-methoxybenzeneselenate ion can couple with *p*-anisyl radicals to give the symmetrical *p*-dianisylselenide, while phenyl radicals can react with benzeneselenate ions to form diphenylselenide. The fragmentation of the radical anion indicates that the C–Se σ* MO is of similar energy to the π* MO of the system [49]. However, when the π* MO is lower in energy than the C–Se σ* MO (for example with 1-naphthyl, 2-quinolyl, 4-biphenyl, and 9-phenanthryl), only the straightforward substitution products arylphenylselenides are formed in good yields (52–98% yields) [50].

This fragmentation process explains the formation of scrambled products by reaction of diphenylarsenide ions with some aromatic halides and the straightforward substitution products with others [46,47]. On the other hand, only scrambling of aryl rings is obtained with all the substrates studied and diphenylstibide anion [47].

The photoinitiated reaction of iodobenzene with ethanethiolate ion in liquid ammonia affords not only ethyl phenyl sulfide but also benzenethio-late ion, trapped as the benzyl derivative (Sch. 7) [51].

This product distribution is attributed to the fragmentation reaction at the S-ethyl bond of the radical anion intermediate, a reaction that competes with the ET step that leads to substitution. However, straightforward substitution takes place in the reaction of the ethanethiolate ion with aryl

$$\text{AnI} + \text{PhSe}^- \xrightarrow[\text{NH}_3]{h\nu} \text{AnSePh} + \text{An}_2\text{Se} + \text{Ph}_2\text{Se}$$

An = *p*-anisyl   (20%)   (25%)   (19%)

Scheme 6

PhI + EtS⁻ $\xrightarrow[\text{2. BnCl}]{\text{1. } h\nu, \text{NH}_3}$ PhSBn + PhSEt
                                            (44%)   (30%)

Scheme 7

bromides substituted by EWG that stabilize the radical anion intermediates (70–95% yield) [52]. The competition between ET and S-alkyl bond fragmentation of the radical anion intermediate depends on the stability of the aromatic moiety, and on the frangibility of the S-alkyl bond, the latter being related to the stability of the alkyl radical formed [49,53].

Aryl iodides react with benzenethiolate ions in liquid ammonia under irradiation to afford aryl phenyl sulfides in good yields [51,54–59].

16.3.4. Solvents

The main requirements of a solvent to be utilized in $S_{RN}1$ reactions are that: (1) it should dissolve both the organic substrate and the ionic alkali metal salt of the nucleophile; (2) it should not have hydrogen atoms that can be readily abstracted by aryl radicals; (3) there must not be protons that can be ionized by the bases or the basic nucleophiles and radical anions involved in the reaction; and (4) it should not undergo ET reactions with the various intermediates of the reaction. In addition to these requirements, the solvent should not absorb significantly in the wavelength range normally used in photoinitiated reactions (300–400 nm).

The effect of different solvents has been studied in photoinitiated $S_{RN}1$ reactions. However, the marked dependence of solvent effects on the nature of the aromatic substrate, the nucleophile, its counterion and the temperature at which the reaction is carried out often make comparisons cumbersome.

The photoinitiated reaction of diethylphosphite ion with iodobenzene has been examined in different solvents [60]. The reaction of acetone enolate ion with 2-chloroquinoline, under the same irradiation times, suffers a strong dependence on solvent nature. Thus, the yield of substitution product decreases going from liquid ammonia (90%) to THF (82%), DMF (74%), dimethoxyethane (28%), diethyl ether (9%), or benzene (4%) [61]. For this reason, the reactions are usually performed in liquid ammonia or DMSO under irradiation.

When the photoinduced reaction of pinacolone enolate ion with bromobenzene is carried out in DMF as solvent, it leads to the formation of benzene (28%), while in liquid ammonia this byproduct is almost precluded. The reaction in DMSO produces little benzene. On the other hand, THF

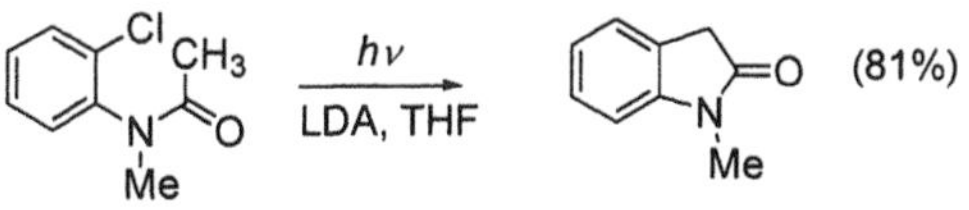

Scheme 8

almost completely inhibits the photoinitiated $S_{RN}1$ reaction, leading slowly to benzene as the major product. Coupling of the aryl radical with the solvent can also be observed in THF [23]. However, THF has proved to be an efficient solvent in intramolecular ring closure reactions. One example to this approach is the synthesis of oxindoles from N-alkyl-N-acyl-o-chloroaniline (Sch. 8) [62,63].

In the photostimulated reactions of Me_3Sn^- ions with chloroarenes, the yields of substitution products are almost similar either in liquid ammonia or diglyme; 1-chloronaphthalene affords the substitution product in 90 and 88% yields respectively [64,65]. m-Dichlorobenzene reacts with Me_3Sn^- ions in diglyme under irradiation to yield disubstitution product in 85% yield, a slightly lower yield if compared with the reaction in liquid ammonia (90% yield). The reaction of 1,3,5-trichlorobenzene with Me_3Sn^- ions in liquid ammonia affords 71% yield of the trisubstitution product [26]. However, only 28% yield is obtained in the photoinitiated reaction in diglyme [65]. These results indicate that in diglyme as solvent, the yields are lower compared with those obtained in liquid ammonia. With one or two leaving groups, the yields are slightly lower, and with three leaving groups the yields are even much lower compared with the reactions performed in liquid ammonia. These findings indicate that the reduction process is more important in diglyme than in liquid ammonia.

4,4''-Dichloro-[1,1',3',1'']terphenyl is completely insoluble in liquid ammonia, but soluble in diglyme. In the photostimulated reaction with Me_3Sn^- ions, the disubstitution product **8** is obtained in fairly good yields (Sch. 9) [65].

16.4. SYNTHETIC POTENTIAL: REACTIVITY AND SELECTIVITY PATTERN

16.4.1. Ketone Enolate Ions

The enolate ions of ketones have become the most widely studied nucleophiles in $S_{RN}1$ reactions [21–23,40,66]. The enolate ions of acyclic and cyclic ketones (from cyclobutanone to cyclooctanone) [21,40] react with aryl and heteroaryl halides to afford the α-arylation products. No

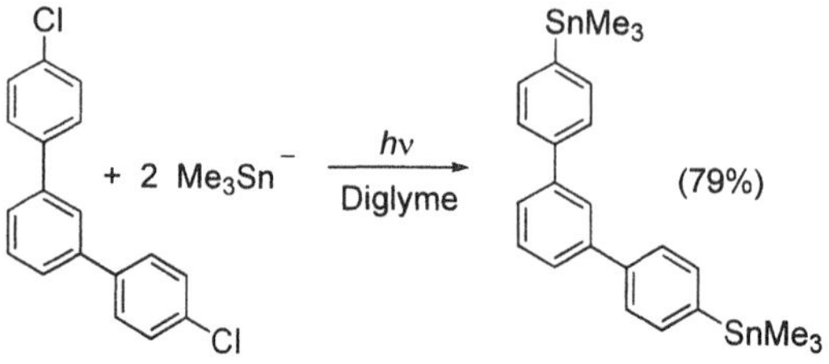

Scheme 9

R = F   (92%)
   OMe   (84%)
   NHMe   (50%)
   NHCOBu-*t* (99%)

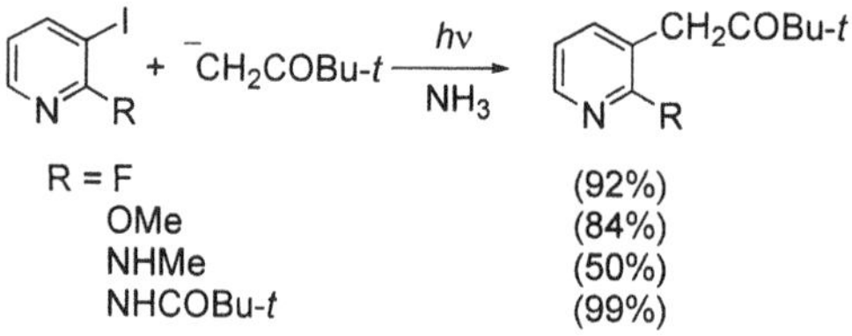

Scheme 10

substitution product is formed with the enolate ion of cyclohex-2-en-1-one [21]. Examples are shown in Sch. 10 [67].

In some reactions, α,α-diarylation can also occur, but generally in low yields (>15%). α-Diarylation is caused by ionization in the basic medium of the monoarylated product. In order to minimize the diarylation, a nucleophile/substrate ratio of 3–5 is recommended. Use of an excess of *t*-BuOK (10–20%) is also convenient to avoid the possibility of aldol condensation of the ketones.

Isomeric enolate ions can be formed from unsymmetric dialkyl ketones, and the distribution of the two possible arylation products is mainly determined by the equilibrium concentration of the various possible enolate ions. However, the selectivity also depends on the structure of the attacking radical. Reactions with the enolate ions from 2-butanone afford arylation preferentially at the more substituted α carbon to render about twice as much 3-phenyl-2-butanone as 1-phenyl-2-butanone [68,69]; however, in the reaction with the anion derived from *i*-propyl methyl ketone, the 1-phenyl derivative is predominantly formed [68]. When there is a substituent ortho- to the leaving group, the attack at the primary α carbon is enhanced [69,70].

A special behavior is shown by the enolate ions of aromatic ketones. For example, the enolate anion of acetophenone fails to react with

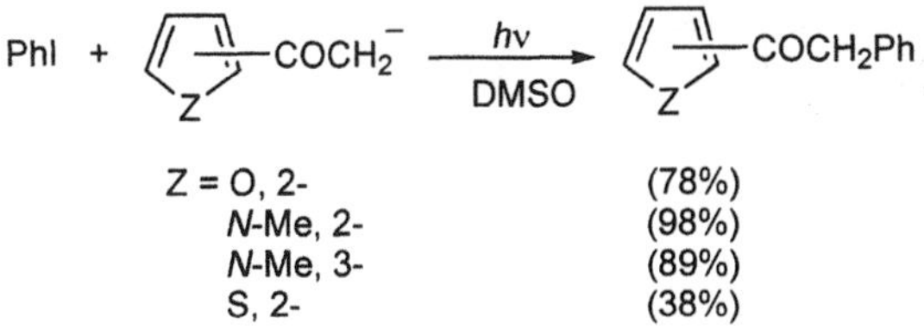

PhI + —COCH$_2^-$ $\xrightarrow[\text{DMSO}]{hv}$ —COCH$_2$Ph

Z = O, 2-   (78%)
      N-Me, 2-   (98%)
      N-Me, 3-   (89%)
      S, 2-   (38%)

Scheme 11

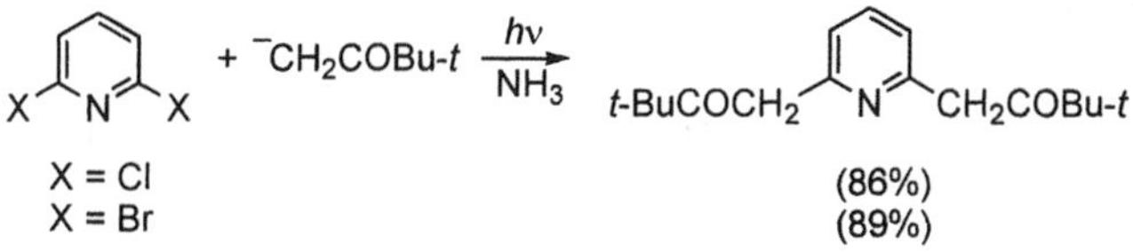

+ $^-$CH$_2$COBu-t $\xrightarrow[\text{NH}_3]{hv}$ t-BuCOCH$_2$ CH$_2$COBu-t

X = Cl   (86%)
X = Br   (89%)

Scheme 12

halobenzenes or halonaphthalenes in liquid ammonia under photoinitiation [21,34]. Heteroarylation is possible in liquid ammonia [66c,71–73] or DMSO [22,73] under photoinitiation.

The anions of other aromatic ketones such as methyl 2-naphthylketone [74], 2-acetylfuran [16,72,75,76], 2-[16,75,76], and 3-acetylthiophene [73], 2- and 3-acetyl-N-methylpyrrole [17], anthrone [22], and tetralone [72,73,77] can also be arylated (Sch. 11).

The photostimulated reactions of o- and p-dibromobenzene with the anion of pinacolone lead to the disubstitution products in 62% [78], and 64% yields [79], respectively.

Disubstitution also occurs in the reaction of 2,6- and 2,5-dibromopyridines, and 2,3-, 3,5-, and 2,6-dichloropyridines with the enolate ion of pinacolone [40,80]. Examples are show in Sch. 12.

When dihalobenzenes (Cl, Br, I) react with aliphatic ketone enolate ions, disubstitution products are furnished. On the other hand, mainly monosubstitution with retention of one halogen is obtained in the photostimulated reactions of o-iodohalobenzenes (X = I, Br, Cl) with the enolate ions of aromatic ketones, such as acetophenone, propiophenone, and 1-(2-naphthyl)ethanone in DMSO. These results are explained in terms of the energetics of the intramolecular ET from the ArCO-π system to the C–X σ bond in the monosubstituted radical anions proposed as intermediates [81]. For example see Sch. 13.

Benzo[b]furan [69], furo[3,2-h]quinolines [73], and furo[3,2-b]pyridines [73] can be synthesized through the reaction of enolate ions of ketones or

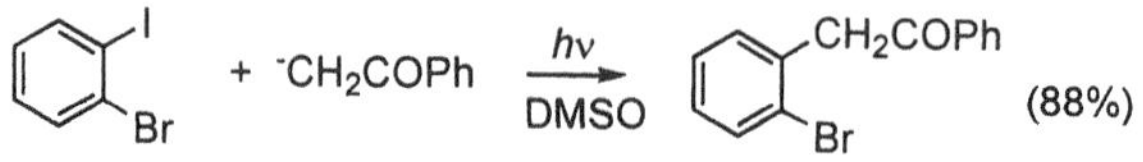

Scheme 13

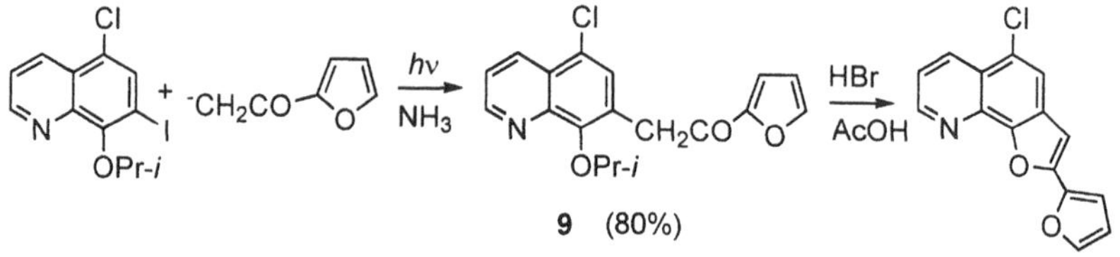

9 (80%)

Scheme 14

$R^1 = OMe$, $R^2 = H$ (72%)
$R^1 = OPr\text{-}i$, $R^2 = OMe$ (70%)

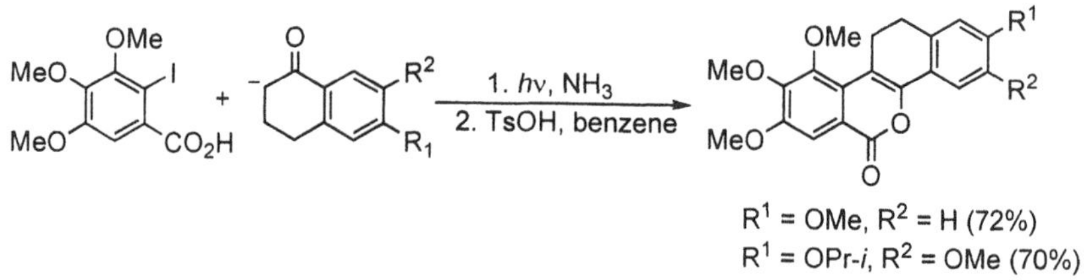

Scheme 15

aldehydes with aryl halides substituted with an *ortho*-alkoxide group. The *o*-hydroxyaryl ketones, formed after deblocking the alkoxide function of the $S_{RN}1$ product **9**, undergo spontaneous cyclodehydration to give the furan derivatives quantitatively (Sch. 14) [73].

Lactones can be obtained through the photoinitiated reaction of (*o*-iodophenyl)acetic acid derivatives with different enolate ions in good yields [70d,77]. Examples are shown in Sch. 15 [77].

16.4.2. Carbanions Derived from Esters, Acids, and *N,N*-Disubstituted Amides

α-Arylacetic esters can be obtained through the photostimulated reaction of aryl and heteroaryl halides with *t*-butyl acetate ions [23,82,83].

Other nucleophiles reported to react under photoinitiation are carbanions derived from ethyl phenylacetate [80,82], and methyl diphenylacetate

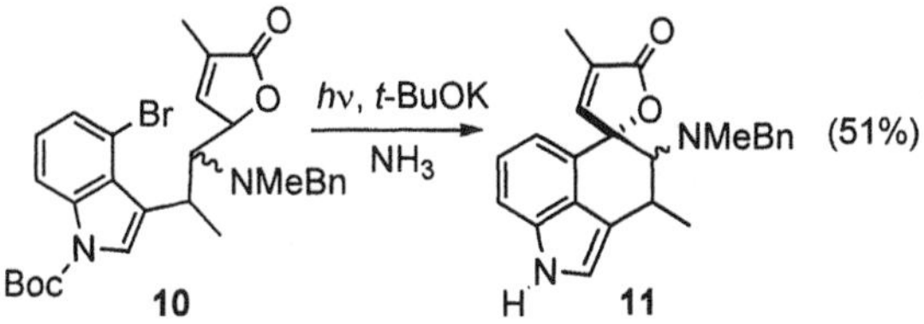

Scheme 16

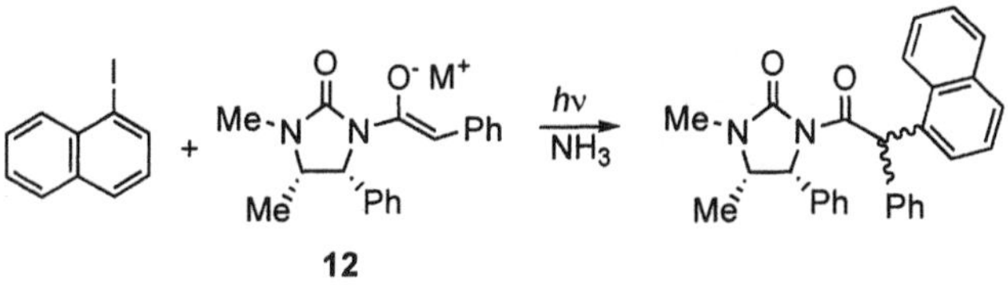

Scheme 17

[80]. With these anions, substitution of halopyridines [82], and disubstitution of dihalopyridines [80] can be achieved.

Low yields of substitution products are obtained with anions of tertiary esters that bear β-hydrogens. In these reactions, reductive dehalogenation of the substrates competes with the substitution reaction [23,83].

The $S_{RN}1$ mechanism has proved to be an excellent alternative to achieve the intramolecular cyclization of substrate **10** into the ring closure product **11** (Sch. 16), which constitutes one of the steps in the synthesis of an Ergot type alkaloid [84].

The carbanions derived from *N,N*-disubstituted amides and lactams react with certain aromatic halides in liquid ammonia under photostimulation [85,86] to form the expected α-arylated compounds in good yields. Unsymmetrical α, α-diaryl amides can be formed by reaction of aryl halides with the anion of the α-aryl-*N,N*-dimethyl acetamides [85].

An interesting example of stereoselective coupling of an aromatic radical with a nucleophile is found in the reaction of 1-iodonaphthalene with the imide anion **12**, containing a chiral auxiliary. In this reaction the diasteromeric isomers of the substitution compound are formed (Sch. 17) [87]. This reaction is highly dependent on the metal counter ion used. All the ions studied [Li, Na, K, Cs, Ti(IV)] are selective, but the highest stereoselection is reached with Li at low temperature (−78°) and with Ti(IV) (ca. 99% de) [87].

The only example known of carbanions from carboxylate ions is the reaction of dianion **13** formed when phenylacetic acid is treated with two

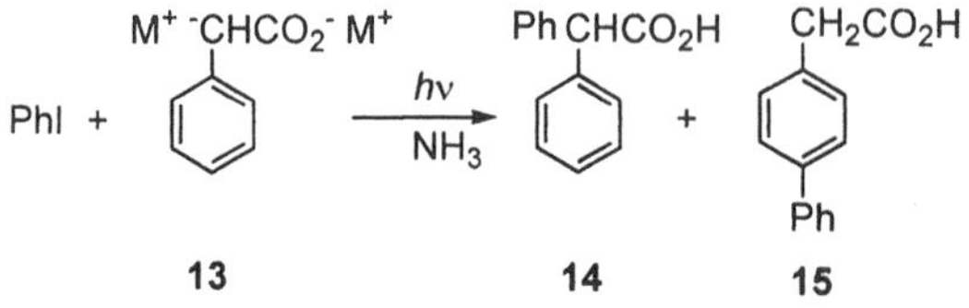

Scheme 18

equivalents of amide ions in liquid ammonia. This dianion, when irradiated in the presence of haloarenes, can afford products of α-arylation **14** and *p*-arylation **15** depending on the counterion employed. For example, with K$^+$ as counterion, only **15** is formed (73%) (Sch. 18), but with Li$^+$, only **14** is observed (77%) and with Na$^+$ both products are obtained in comparable yields [88].

16.4.3. Indole Syntheses

An important example of substitution followed by a spontaneous ring closure reaction is the synthesis of indoles by the photostimulated reactions of *o*-iodoaniline **16** with carbanions derived from aliphatic ketones in liquid ammonia to give 2-substituted indoles (Sch. 19) [89].

Unsubstituted and substituted *o*-bromo and *o*-chloroanilines are adequate substrates for obtaining indoles substituted with Me, Ph, or MeO groups in 50-90% yield [70a].

The syntheses of benzo[*e*] and benzo[*g*]indoles have been performed by reaction of 2-amino-1-bromo or 1-amino-2-bromonaphthalene with the anion of pinacolone (67% and 82% yields, respectively). The reaction with the anion $^-CH_2COCH(OMe)_2$, affords the 2-formyl indoles (52% and 65% yields, respectively), which can be oxidized to the corresponding acids [90].

The photostimulated reactions of *vic* aminohalo pyridines with acetone or pinacolone enolate ions lead to azaindoles in high yields (75–95%) [67]. When the amino group is protected as pivaloylamino derivative, (as in **17**, Sch. 20), the analogs of substitution compounds **18** obtained by the photoinduced reaction of 2-amino-3-iodo-, 3-amino-4-iodo- and 4-amino-3-iodopyridines with acetone or pinacolone enolate ions, afford 5-, 6- and 7-azaindoles in almost quantitative yields by cyclization upon deprotection of the amino group and dehydration under acidic conditions (for example, see Sch. 20) [67].

All the previously cited syntheses have been carried out in liquid ammonia. The reactions afford poor results when aromatic ketones are

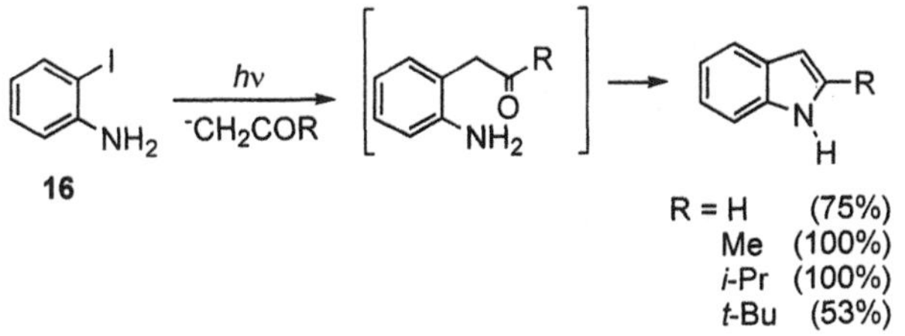

R = H   (75%)
Me (100%)
i-Pr (100%)
t-Bu (53%)

Scheme 19

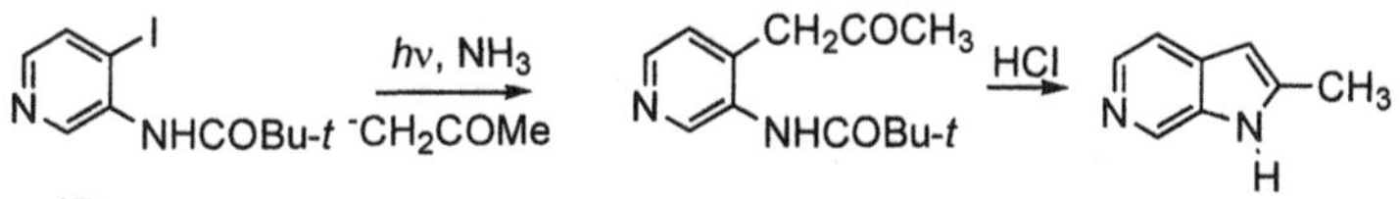

Scheme 20

n = 1 (71%)
n = 2 (58%)
n = 3 (71%)

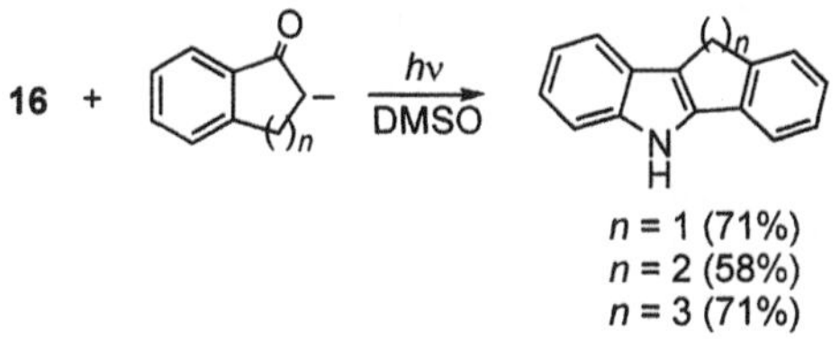

Scheme 21

used in this solvent, but these anions react readily in DMSO. The reactions of **16** with the enolate anions of aromatic ketones afford 2-substituted indoles (2-phenyl-1H-indole (88%) [91], 2-naphthalen-2-yl-1H-indole (73%) [91], 2-(1-methyl-1H-pyrrol-2-yl)-1H-indole (63%) [91], 2-pyridin-2-yl-1H-indole (64%) [92], and 2-pyridin-4-yl-1H-indole (74%) [92]. The photo-stimulated reaction of 1-iodo-2-naphthalen-2-ylamine with acetophenone enolate ion in DMSO gives 2-phenyl-3H-benzo(e)indole in 48% yield [92].

Substrate **16** reacts under irradiation in DMSO with the enolate anions of 1-indanone (n = 1), 1-tetralone (n = 2), and 6,7,8,9-tetrahydro-benzocyclohepten-5-one (n = 3) to afford fused indoles (Sch. 21) [92].

The 5,6-dihydro-indeno[2,1-b]indole **19a** is obtained by the photo-stimulated reaction of **16** with the enolate anion of 2-indanone in DMSO and in liquid ammonia (Sch. 22, n = 1) [92].

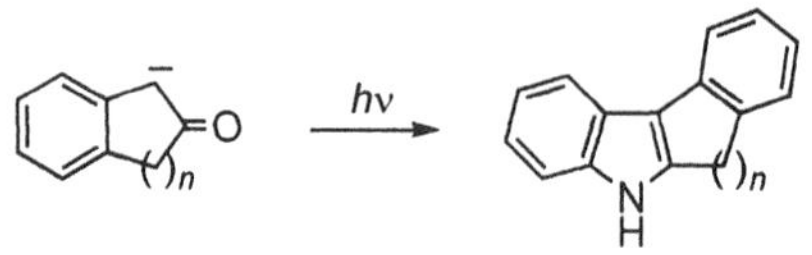

19a: $n = 1$ (50-54%)
19b: $n = 2$ (56-73%)

Scheme 22

In the photostimulated reaction of **16** with 2-tetralone, the 5,7-dihydro-6*H*-benzo[*c*]carbazole **19b** is obtained in 56% yield. However, the yield of **19b** is increased to 73% when the reaction is performed under photostimulation in liquid ammonia. (Sch. 22, $n = 2$) [92].

16.4.4. Tandem Ring Closure—S$_{RN}$1 Reactions

Ring closure reactions taking place by intramolecular addition of an aromatic radical to a double bond have been widely studied on both their regio- and stereochemical aspects [93]. Aryl halides and diazonium salts substituted at ortho- position with a *O*-allyl or *N*-allyl chain were used for the preparation of 2,3-dihydrobenzofuranes and 2,3-dihydro-1*H*-indoles under different reaction conditions. The reaction pattern involves the generation of an aryl radical **20**, which reacts with the double bond in a 5-exo trig fashion to afford the exocyclic radical **21**, plausible of reduction by a hydrogen donor to obtain the reduced-cyclized product **22** (Sch. 23) [93d,94].

Although this general scheme was employed for preparing hetero-cyclic compounds, there are only few examples where the ring closure takes place in the propagation step of a S$_{RN}$1 reaction, and in most cases the studies have been geared toward gathering kinetic information [93d,95]. The rate constants for the reactions of aryl radicals with nucleophiles are in the range 10^7–10^{10}/M/s [12b,c,96]. The rate constant for the ring closure reaction of *o*-allyloxy phenyl radical is 4.9×10^8/s [97]. The fact that ring closure is a unimolecular process, and therefore does not depend on nucleophile concentration as opposed to the bimolecular coupling reaction, makes it feasible to obtain, under controlled experimental conditions, 3-substituted 2,3-dihydrobenzofurans, and 2,3-dihydro-1*H*-indoles in one-pot reactions by a tandem cyclization-S$_{RN}$1 sequence [98].

1-Allyloxy-2-chlorobenzene **23** (Z = O) reacts with Me$_3$Sn$^-$ and Ph$_2$P$^-$ ions in liquid ammonia under photostimulation to furnish

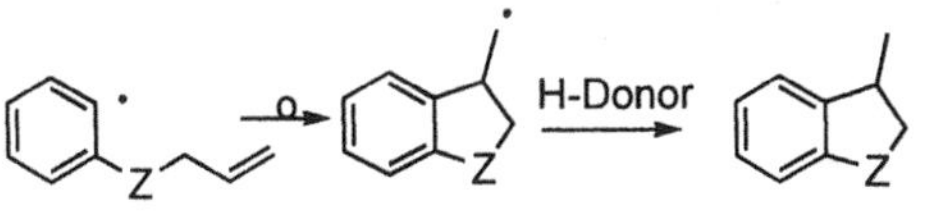

Scheme 23

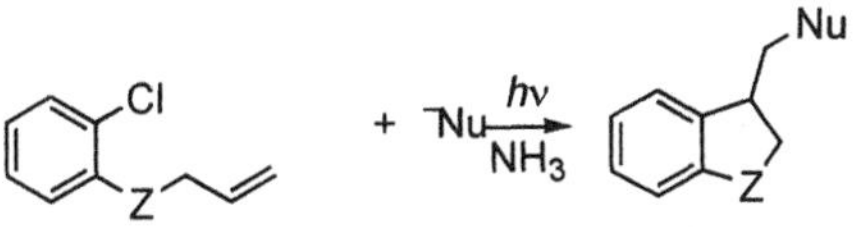

Z = O	Nu = Me₃Sn	(87%)

$Z = O$ $Nu = Me_3Sn$ (87%)
O Ph_2P (75%)
NH Me_3Sn (30%)
N-allyl Me_3Sn (97%)
N-allyl Ph_2P (89%)
N-acetyl Me_2Sn (97%)
N-acetyl Ph_2P (76%)

Scheme 24

cyclized-substituted compounds **24** (Z = O) (the phosphine was isolated as the oxide after oxidation with H_2O_2) (Sch. 24) [98].

When N-allyl-(2-chloro-phenyl)-amine **23** (Z = NH) is allowed to react with Me_3Sn^- ions in liquid ammonia, a low yield of product **24** (Z = NH) is obtained (Sch. 24). When the amino group is protected as **23** (Z = N-allyl, N-acetyl derivatives), good yields of ring closure—substitution products **24** (Z = N-allyl, N-acetyl) are obtained [98].

Similar results are found with 2-allyloxy-1-chloro-naphthalene, that affords the diphenylphosphine (98% yield) and trimethylstannane (84% yield) of (1,2-dihydro-naphtho[2,1-b]furan-1-ylmethyl) derivatives.

It is known that acetone enolate anion does not react with primary alkyl radicals, and that nitromethane anion is not capable of initiating the $S_{RN}1$ reactions even under irradiation [99]. Thus, the photostimulated reactions of **25** with nitromethane anion as nucleophile and acetone enolate anion as entrainment reagent (which enables $S_{RN}1$ initiation but cannot compete with the coupling of the methylene radical with nitromethane anion after cyclization) render the cyclized products **26** (Sch. 25) [98].

The photostimulated reaction of 2-allyloxy-1-bromo-naphthalene **27** under the same experimental conditions, affords 1-(2-nitro-ethyl)-1,2-dihydro-naphtho[2,1-b]furan **28** in good yield (Sch. 26) [98].

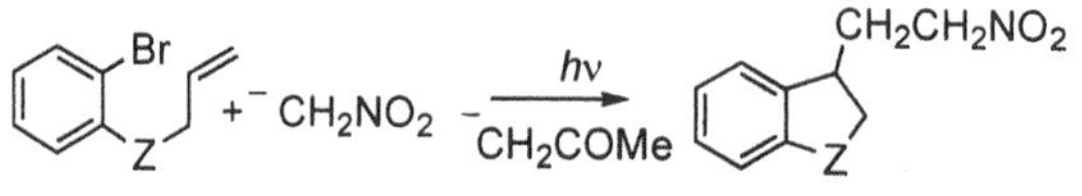

Scheme 25

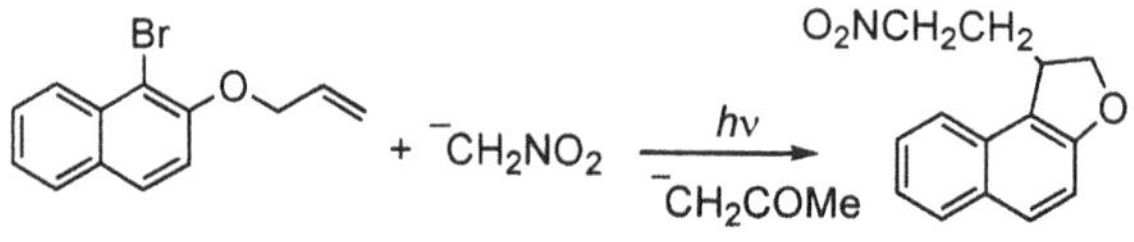

Scheme 26

16.4.5. Reactions with Me₃Sn⁻ Ions and Ensuing Processes

Aryl chlorides undergo substitution by Me_3Sn^- ions in liquid ammonia under irradiation to afford $ArSnMe_3$ in high yields (88–100%) [26,64]. For instance, *p*-chloroanisole does not react with Me_3Sn^- ions in the dark, but upon irradiation, the substitution product is obtained in ca. 100% [64]. On the other hand, aryl bromides and iodides react by an HME (halogen metal exchange) pathway [64]. Not only aryl chlorides are used as starting substrates for conversion into Me_3SnAr, but phenols can also be utilized through the intermediacy of $(EtO)_2P(O)Ar$. These syntheses proceed in high yields [100]. Other leaving groups have been employed, such as aryltrimethylammonium salts synthesized from anilines, which by photo-stimulated reaction with Me_3Sn^- ions in liquid ammonia afford $ArSnMe_3$ in good to excellent yields [101,65].

The photostimulated reactions of *o*-, *m*-, and *p*-dichlorobenzenes with Me_3Sn^- ions furnish products **29** arising from disubstitution (Sch. 27) [26,64].

The synthesis of multidentate Lewis acids is of great interest [102]. The photostimulated reactions of [*o*-(diethoxy-phosphoryl)-phenyl]-phosphonic acid diethyl ester with Me_3Sn^- ions in liquid ammonia affords the disubstitution product *o-bis*-trimethylstannanyl-benzene **29a** in 67% yield [103]. Under the same experimental conditions, 2,2′-*bis*(diethylphosphoxy)-1,1′-binaphthyl **30** affords the disubstitution product **31** in high yields (Sch. 28) [103].

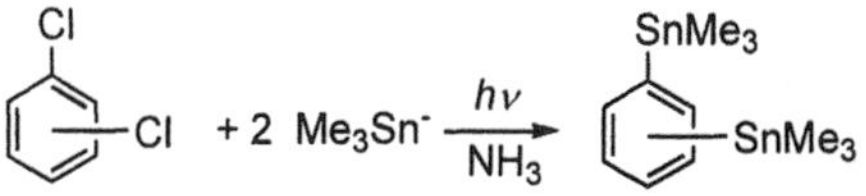

29a, *o*- (58%)
29b, *m*- (90%)
29c, *p*- (88%)

Scheme 27

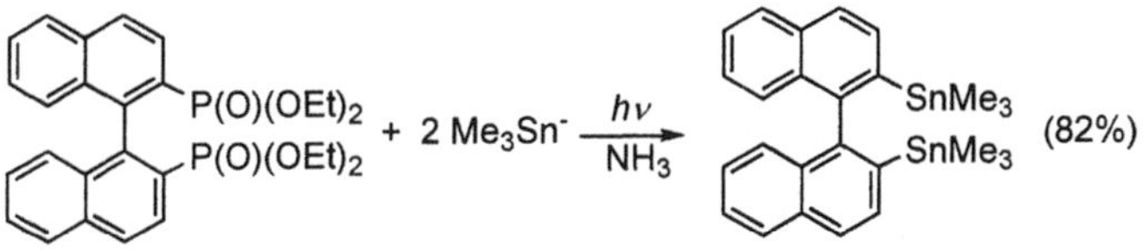

30 **31**

Scheme 28

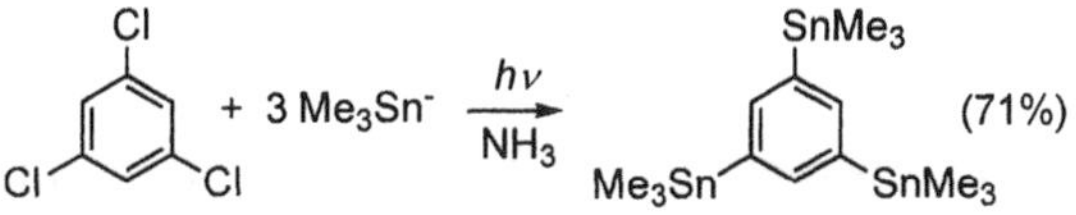

Scheme 29

Trisubstitution is achieved by the photostimulated reaction of 1,3,5-trichlorobenzene in liquid ammonia (Sch. 29) [26].

2,5-Dichloropyridine reacts in the dark (88% of disubstitution), whereas 2,6- and 3,5-dichloropyridines require photostimulation to afford ca. 80% yield of disubstitution products [26]. Aryltrialkylstannanes are valuable intermediates in organic synthesis, and the fact that they can be easily synthesized through the $S_{RN}1$ mechanism, opens up important synthetic routes to different reaction schemes. For over a decade, the Pd(0)-catalyzed cross-coupling of organotin compounds with electrophiles, known as the Stille reaction, has been a very important tool in product design [104].

The coupling of the tin-products obtained from $S_{RN}1$ reactions with a iodoarene through a Pd(0)-catalyzed process has been developed to obtain polyarylated compounds in one-pot reactions as shown in Sch. 30 [105].

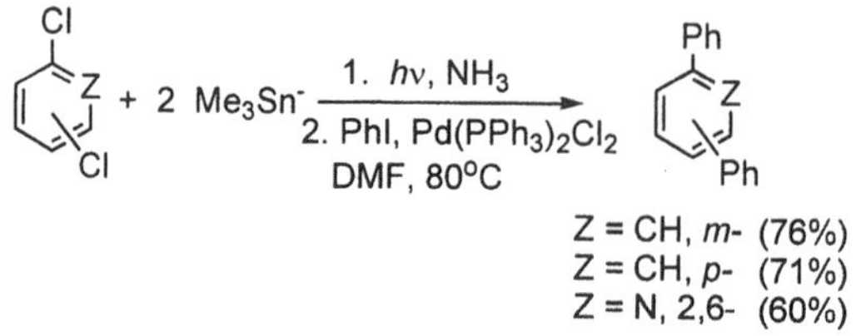

Z = CH, m- (76%)
Z = CH, p- (71%)
Z = N, 2,6- (60%)

Scheme 30

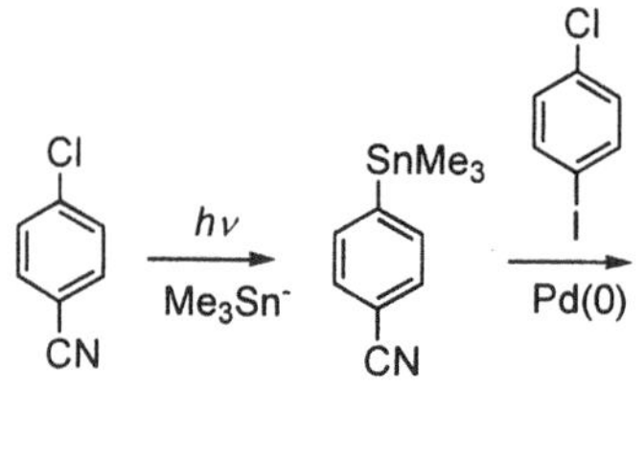

32 (85%)   33 (94%)

Scheme 31

1,3,5-*tris*(trimethylstannanyl) benzene (Sch. 29), upon treatment with iodobenzene and Pd(0), furnishes 1,3,5-triphenylbenzene in 89% yields [105]. This has been the first report of a trisubstitution by a cross coupling reaction catalyzed by Pd(0). Similar yields of 1,3,5-triphenylbenzene can be obtained from 1,3,5-trichlorobenzene in one-pot reaction (61%) [105]. The fact that chloroarenes react with Me_3Sn^- ions under irradiation to form aryltrimethyl stannanes, and that in the Pd(0)-catalyzed reaction with stannanes the reactivity of iodoarenes is much greater than that of chloroarenes, a substrate bearing both leaving groups, chlorine and iodine, will react faster by the C–I bond via a chemoselective cross-coupling reaction with a stannane under Pd(0) catalysis. This will allow the remainder leaving group, chlorine, to react later in another $S_{RN}1$-type reaction to form an organotin intermediate which can subsequently furnish product $Ar-Ar^1-Ar^2$ by a cross-coupling Pd(0)-catalyzed reaction.

The photostimulated reaction of *p*-chlorobenzonitrile with Me_3Sn^- ions affords the stannane **32**, which by a Pd(0)-catalyzed reaction with *p*-chloroiodobenzene furnishes product **33** (Sch. 31) [65].

By a photostimulated reaction of Me_3Sn^- ions in the presence of **33**, stannane **34** is obtained. A second Pd(0)-catalyzed reaction with 1-iodonaphthalene renders product **35** (Sch. 32) [65].

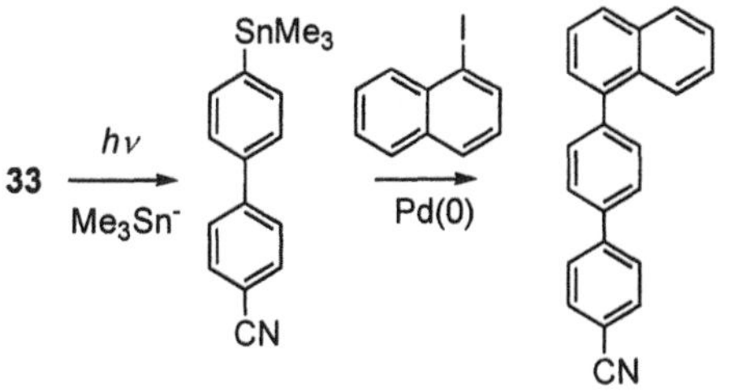

34 (94%) 35 (93%)

Scheme 32

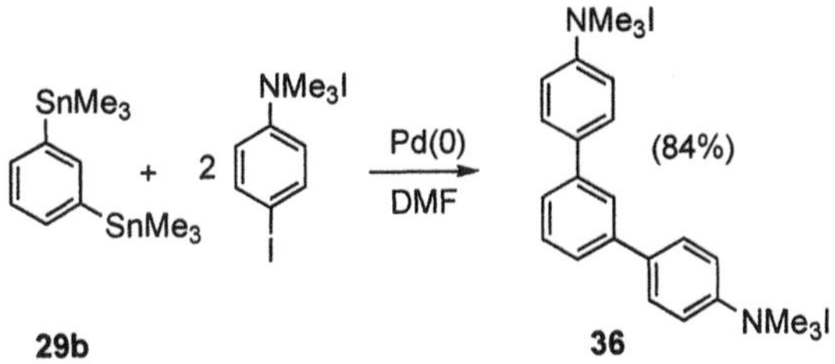

29b 36

Scheme 33

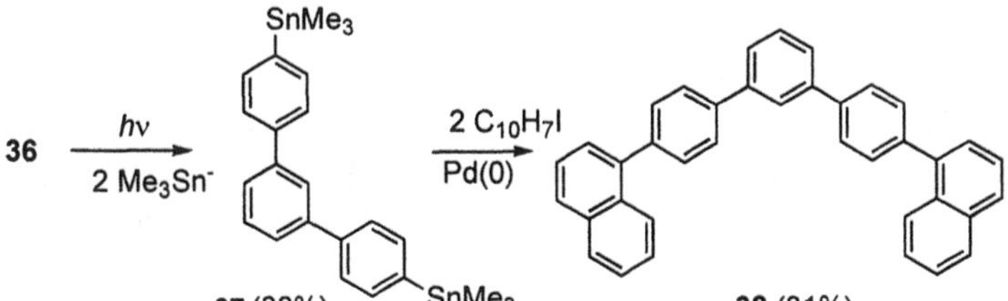

37 (83%) 38 (81%)

Scheme 34

When the distannane **29b** (synthesized according to Sch. 27) is allowed to react under Pd(0) catalysis with 4-iodophenyltrimethylammonium iodide, product **36** is obtained (Sch. 33) [65].

Compound **36** reacts under irradiation with Me_3Sn^- ions by the $S_{RN}1$ mechanism to afford the distannane **37** in good yield. In turn, **37** can undergo, under Pd(0) catalysis, coupling with 1-iodonaphthalene to afford the polyaromatic compound **38** (Sch. 34) [65].

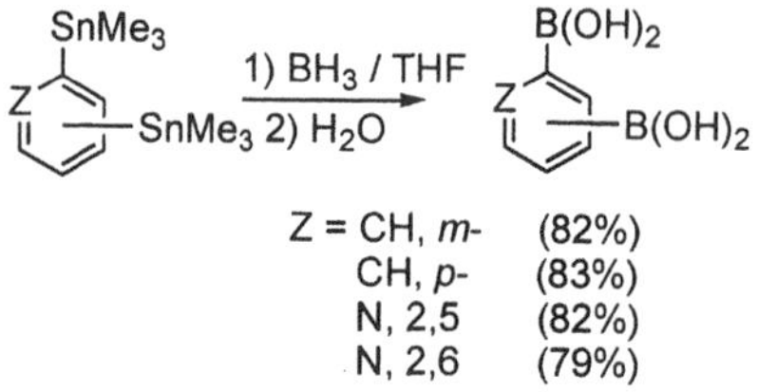

Scheme 35

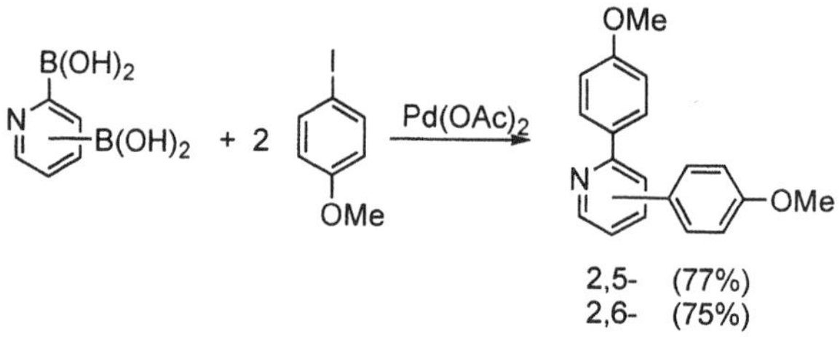

Scheme 36

Aryltrimethyl stannanes are easily transformed into boronic acids by transmetalation with borane [106]. Thus, 1,4- and 1,3-*bis*(trimethylstannanyl)benzenes as well as 2,5- and 2,6-*bis*(trimethylstannyl)pyridines synthesized by the $S_{RN}1$ mechanism, react with borane in THF to give intermediates which on hydrolysis lead to benzene- and pyridinediboronic acids (Sch. 35) [107].

While oxidation of the benzenediboronic acids with alkaline hydrogen peroxide affords the corresponding 1,3- and 1,4-dihydroxybenzenes (70–72%), the pyridinediboronic acids lead via a double Pd(0) catalyzed Suzuki reaction with 4-iodoanisole to 2,5- and 2,6-*bis*-(4-methoxyphenyl)pyridines (Sch. 36) [107].

16.5. REPRESENTATIVE EXPERIMENTAL PROCEDURES

16.5.1. Reactions in Liquid Ammonia

Liquid ammonia is supplied in cylinders from which it can be removed as a liquid or gaseous form. All the operations with this solvent must be conducted in an efficient fume-hood due to the toxicity and pungent odor of the gas. Given ammonia's low boiling point ($-33.4\,^\circ$C), an efficient condenser is required to perform these reactions. A Dewar condenser containing a slurry mixture of liquid nitrogen and EtOH or CO_2/EtOH will normally suffice. The liquid ammonia has to be distilled from a preliminary vessel in which it is dried with Na metal until a blue color persists (about 10–15 min).

Reactions are conducted in a three-necked round-bottomed flask with a gas inlet and under magnetic stirring. A positive pressure of an inert gas such as nitrogen or argon is also required. A detailed description of the experimental equipment has been published [6,108].

Photochemical reactors are commercially available or can be custom-built. They consist of an oval mirror-type wall of ca. 30 cm maximum radius equipped with two Hanovia 450-W or Philips HPT 400-W high-pressure mercury lamps inserted into a water refrigerated Pyrex flask, or equipped with a 300-W Osram sunlamp vessel (Pyrex flask).

16.5.2. Synthesis of 2,4-Dimethyl-2-(2-pyridyl)-3-pentanone [40]

2-Bromopyridine (3.16 g, 20 mmol) is added to an enolate solution prepared from 8.56 g (75 mmol) of 2,4-dimethyl-3-pentanone and 4.12 g (75 mmol) of KNH_2 in 300 mL of liquid ammonia. After the mixture has been irradiated for 1 h, diethyl ether is added and the ethereal suspension remaining after evaporation of the ammonia is decanted through a filter and the residual salts are washed with diethyl ether. This reaction affords high yield of the substitution product (Sch. 37).

17.5.3. Synthesis of 5,6,7-Trimethoxy-3-methyl-1(2H)-isoquinolone [109]

To liquid ammonia (25 mL) under argon contained in a three-necked flask fitted with a dry-ice condenser, acetone is added (4 mmol) along with one equivalent amount of t-BuOK. To the solution of the acetone enolate thus formed, 2-bromo-3,4,5-trimethoxybenzamide is added and the reaction mixture is irradiated in a photochemical reactor. The reaction is quenched by addition of NH_4Cl (0.5 g) when all the substrate has been allowed to react. The ammonia is evaporated and slightly acidified water (50 mL

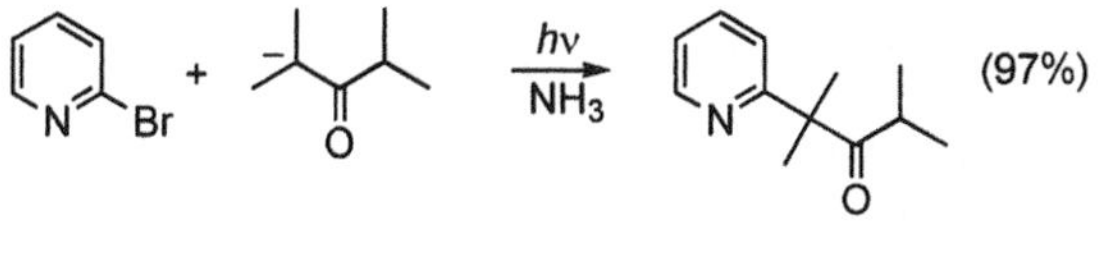

Scheme 37

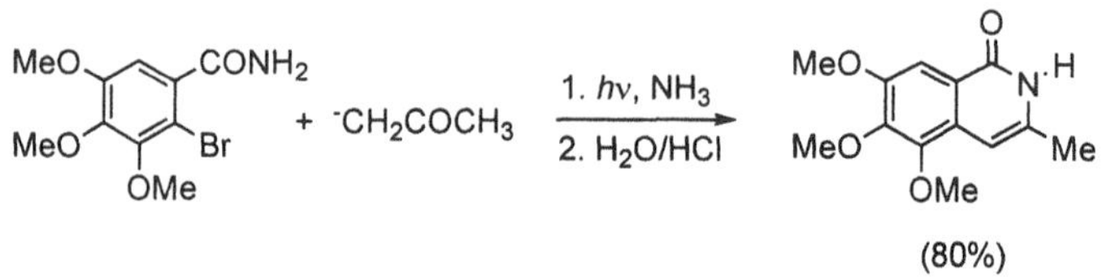

Scheme 38

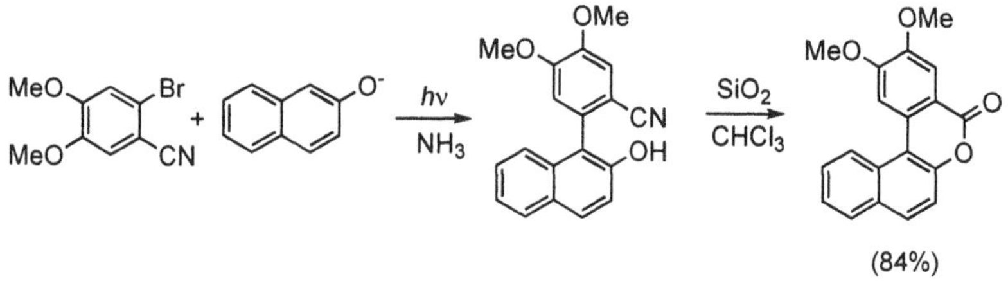

Scheme 39

containing 1 mL of 2 M HCl) is added. Extraction with CH_2Cl_2 followed by purification yields the product in good yields (Sch. 38).

16.5.4. Synthesis of 2,3-dimethoxy-6-oxa-benzo[*c*]phenanthren-5-one [110]

Aryl radicals couple with 2-naphthoxide ions at the C_1 position of the anion offering an $S_{RN}1$ route to the title compound when an appropriate aromatic substrates is used [43,111]. Into a 100 mL two-necked Pyrex flask containing *t*-BuOK (3 mmol), ammonia (50 mL) is condensed through a dry ice condenser cooled at $-78\,°C$. Under argon atmosphere, 2-naphthol (3 mmol) and the 2-bromo-4,5-dimethoxybenzonitrile (1 mmol) are successively introduced. The reaction mixture is irradiated in a photochemical reactor for 1 h and the solvent is evaporated. Acidic water (100 mL) is poured and the aqueous phase is extracted with ethyl acetate. After evaporation, the residue is dissolved in $CHCl_3$, 1 g of SiO_2 is added and boiled. The silica gel is removed by filtration, washed by acetone, and the product is isolated in good yields (Sch. 39).

16.5.5. Photostimulated Reactions with Me_3Sn^- Ions

Liquid Ammonia [26]: Into a three-necked, 500-mL, round-bottomed flask, 250 mL of ammonia previously dried with Na metal under nitrogen are condensed. Me_3SnCl (0.83 mmol) and Na metal (1.66 mmol) were added. To this solution 0.75 mmol of 2-chloropyridine is then added and irradiated for 90 min. The reaction is quenched by adding NH_4NO_3 in excess, and the ammonia is allowed to evaporate. The residue is dissolved with water, then extracted with diethyl ether. The product 2-pyridyltrimethylstannane is obtained in 88% yield.

Diglyme [65]: The Me_3Sn^- ions (6.00 mmol) are prepared in liquid ammonia by the procedure described before, and the ammonia is allowed to evaporate. Diglyme (20 mL) and 1-chloronaphthalene (1.50 mmol) are added to the residue and the is mixture irradiated for 40 min at ca. 0 °C. The reaction is quenched by adding an aqueous solution of NH_4NO_3 in excess, and extracted with diethyl ether. The ether extract is washed twice with water, dried, and 1-naphthyltrimethylstannane is obtained in 88% yield.

16.5.6. Synthesis of 5,10-Dihydro-indeno[1,2-*b*]indole [92]

The reaction is carried out in a 50 mL three-neck round-bottomed flask equipped with a nitrogen inlet and a magnetic stirrer bar. *t*-BuOK (2.55 mmol, 0.286 g) are added to 20 mL of dry and deoxygenated DMSO under nitrogen atmosphere. 1-Tetralone (2.50 mmol, 0.365 g) is then added. *o*-Iodoaniline (0.50 mmol, 0.110 g) is poured after 15 min and the reaction mixture is irradiated for 3 h. The reaction is quenched with an excess of NH_4NO_3 and water (60 mL). The mixture is extracted with methylene chloride (3 × 20 mL), the organic extract is washed with water (2 × 20 mL), dried over Na_2SO_4, and the indole was obtained in good yields (Sch. 40).

16.6. TARGET MOLECULES: NATURAL AND NONNATURAL PRODUCT STRUCTURES

16.6.1. Fluorobiprophen [86a]

Fluoride ion is not a good leaving group in $S_{RN}1$ reactions. For instance *m*-fluoroiodobenzene [66c] and 2-fluoro-3-iodopyridine [67] react with carbanions with retention of fluorine. Synthesis of anti-inflammatory drugs, such as fluorobiprophen **39** can be achieved by reaction of 4-bromo-2-fluorobiphenyl with $^-CH_2COMe$ ions followed by methylation and oxidative demethylation (Sch. 41).

16.6.2. 3-Methyl-*2H*-isoquinolin-1-one Derivatives [109]

3-Methyl-*2H*-isoquinolin-1-one derivatives **41** are accessible by the reaction of substituted *o*-bromobenzamides **40** with carbanions. One such example is

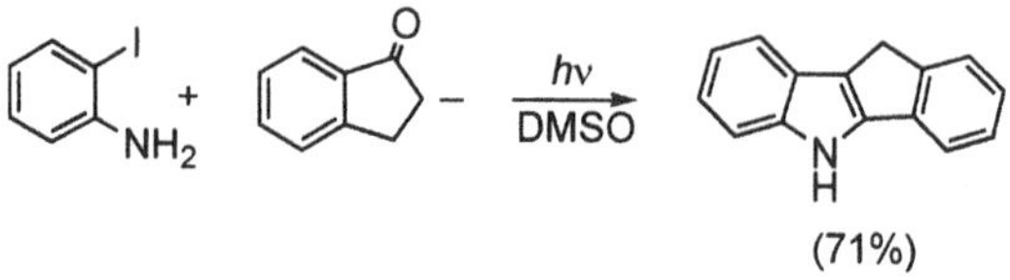

(71%)

Scheme 40

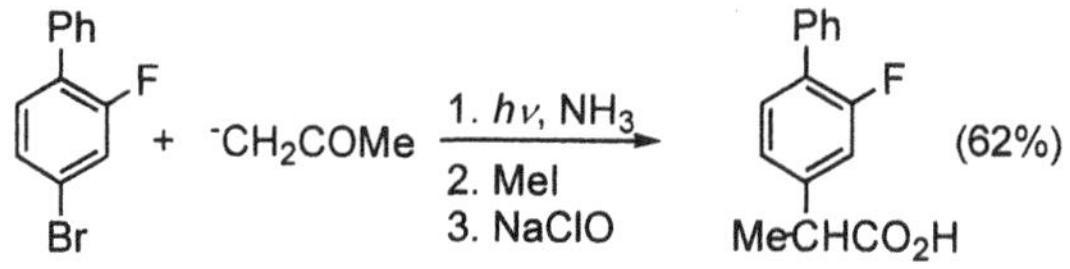

(62%)

39

Scheme 41

(70-80%)

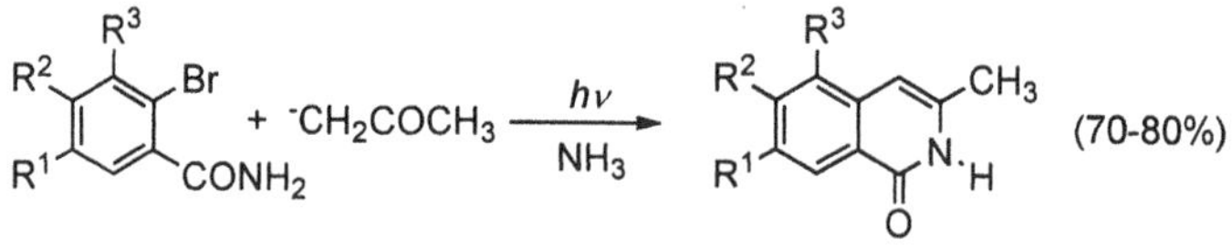

40   **41**

Scheme 42

the photostimulated reaction with acetone enolate ions (Sch. 42). The yields of the cyclization are better when the benzamides are not *N*-methylated.

16.6.3. Benzo[*c*]chromen-6-one Derivatives [110]

Another approach to ring closure reaction is the *o*-arylation of substituted phenoxide ions by *o*-bromobenzonitrile followed by SiO$_2$-catalysed lactonization. The phenoxide ions of the amino acid (*S*)-tyrosine, protected as *N,O*-diacetyl methyl ester, does not racemize under the standard S$_{RN}$1 conditions and can be used to obtain the optically active benzo[*c*]chromen-6-one (the *O*-acetyl is hydrolyzed in the reaction media to furnish the phenoxide ion) (Sch. 43). Racemic dibenzopyranones are obtained by the reaction of the anion from the *N*-acetyl methyl ester of (*R*)-hydroxyphenylglycine with *o*-bromobenzonitrile 2-cyano-4,5-dimethoxybromobenzene (65 and 79% respectively) [110].

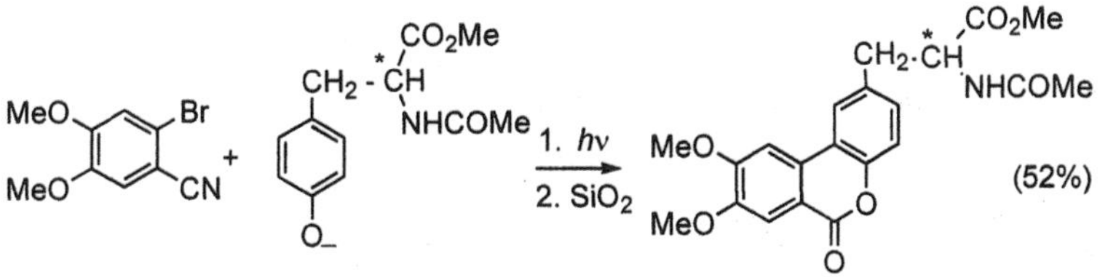

Scheme 43

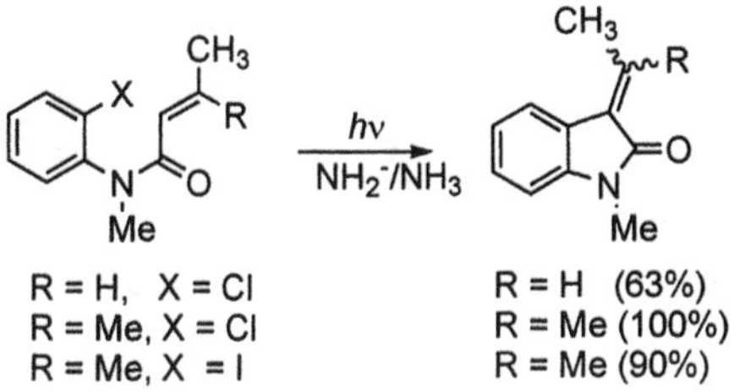

R = H, X = Cl   R = H   (63%)
R = Me, X = Cl   R = Me (100%)
R = Me, X = I   R = Me (90%)

Scheme 44

42   **43**

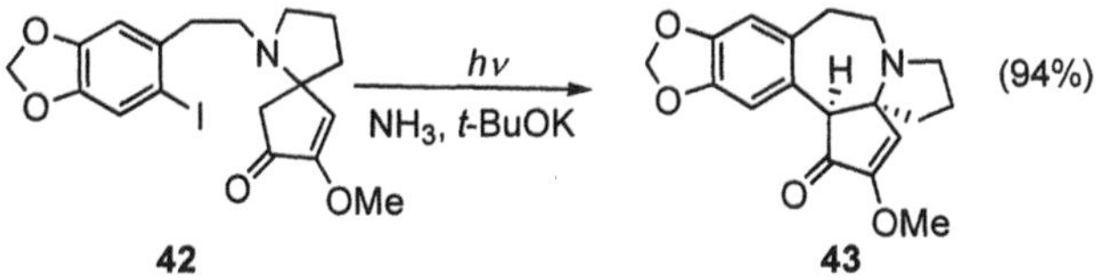

Scheme 45

16.6.4. 3-Alkylideneoxindoles [63]

N-Methyl α,β-unsaturated anilides undergo intramolecular arylation exclusively at the α-position to afford 3-alkylideneoxindoles. The best results are obtained with KNH_2/NH_3 under photoinitiation (Sch. 44).

16.6.5. Cephalotaxinone [112]

One of the first reports of an intramolecular $S_{RN}1$ reaction is the synthesis of cephalotaxinone (**43**). The iodoketone **42** cyclizes in liquid ammonia under photostimulation affording product **43** in excellent yields (Sch. 45).

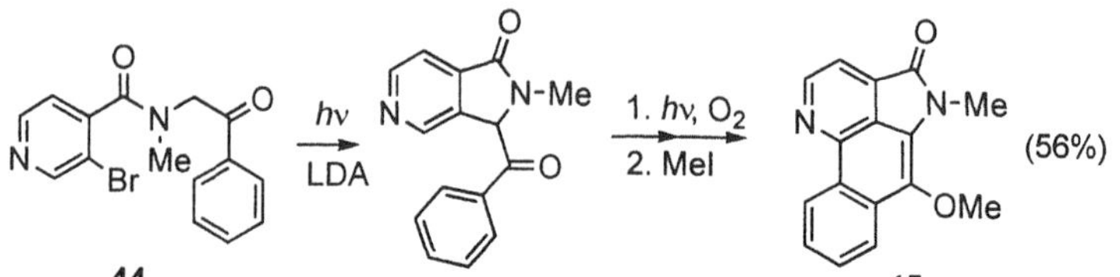

44   **45**

Scheme 46

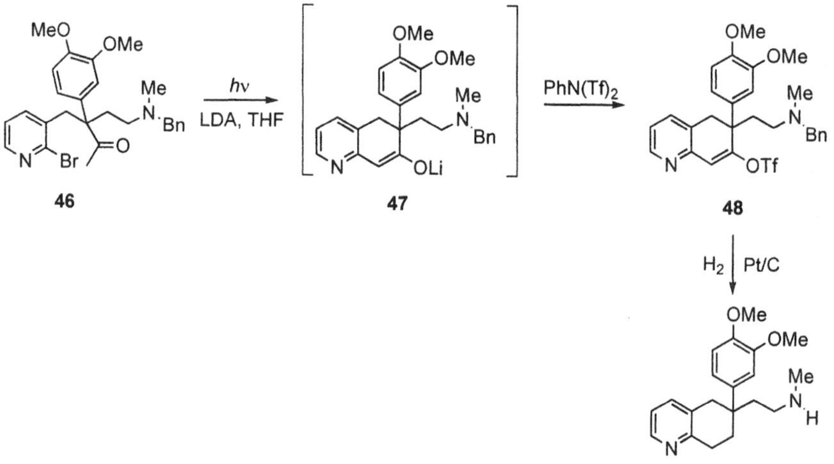

46   **47**   **48**

49

Scheme 47

16.6.6. Eupoulauramine [113]

The key step in the synthesis of the azaphenanthrene alkaloid eupoulaur-
amine **45**, is an intramolecular S$_{RN}$1 reaction of 3-bromo-*N*-methyl-*N*-
(2-oxo-2-phenyl-ethyl)-isonicotinamide **44**, followed by in situ stilbene
photocyclization, and further methylation (Sch. 46).

16.6.7. (±)Tortuosamine [114]

Treatment of substrate **46** with LDA (THF, 0 °C) under irradiation effected
cyclization to afford the enolate **47**, which is trapped by reaction with
N-phenyltriflimide to give the vinyl triflate **48** in 59% yield. Catalytic
hydrogenation reduced the vinyl triflate to the saturated hydrocarbon with

concomitant removal of the benzyl group to afford (±)tortuosamine **49** in 91% yield (Sch. 47).

REFERENCES

1. Kornblum N, Michel RE, Kerber RC. J Am Chem Soc 1966; 88:5662.
2. Russell GA, Danen WC. J Am Chem Soc 1966; 88:5663.
3. Kim JK, Bunnett JF. J Am Chem Soc 1970; 92:7463.
4. Kim JK, Bunnett JF. J Am Chem Soc 1970; 92:7464.
5. Rossi RA, Bunnett JF. J Org Chem 1973; 38:1407.
6. Rossi RA, de Rossi RH. Aromatic Substitution by the $S_{RN}1$ Mechanism. Washington, D.C.: American Chemical Society, 1983.
7. Norris RK. Nucleophilic coupling with aryl radicals. In: Trost BM, ed. Comprehensive Organic Synthesis. Vol. 4. Pergamon Press, 1991:451–482.
8. Beugelmans R. The photostimulated $S_{RN}1$ process: reactions of haloarenes with enolates. In: Horspool WM, ed. CRC Handbook of Organic Photochemistry and Photobiology. Boca Raton: CRC Press Inc., 1994:1200–1217.
9. (a) Rossi RA, Pierini AB, Peñéñory AB. Recent advances in the $S_{RN}1$ reaction of organic halides. In: Patai S, Rappoport Z, eds. The Chemistry of Functional Groups. Supl. D2, Ch 24. Chichester: Wiley, 1995:1395–1485; (b) Rossi RA, Pierini AB, Peñéñory AB, Chem Rev 2003; 103:71–167.
10. Rossi RA, Baumgartner MT. Synthesis of heterocycles by the $S_{RN}1$ mechanism. In: Attanasi OA, Spinelli D, eds. Targets in Heterocyclic Systems: Chemistry and Properties. Vol 3. Soc Chimica Italiana, 1999:215–243.
11. Rossi RA, Pierini AB, Santiago AN. Aromatic substitution by the $S_{RN}1$ reaction. In: Paquette LA, Bittman R, eds. Organic Reactions. John Wiley & Sons, Inc., 1999:1–271.
12. (a) Andrieux CP, Savéant J-M, Su KB. J Phys Chem 1986; 90:3815; (b) Amatore C, Combellas C, Pinson J, Oturan MA, Robveille S, Savéant J-M, Thiébault A. J Am Chem Soc 1985; 107:4846; (c) Amatore C, Oturan MA, Pinson J, Savéant J-M, Thiébault A. J Am Chem Soc 1985; 107:3451; (d) Andrieux CP, Savéant J-M, Zann D. New J Chem 1984; 8:107.
13. Bunnett JF. Acc Chem Res 1978; 11:413.
14. (a) Andrieux CP, Hapiot P, Savéant J-M. Chem Rev 1990; 90:723; (b) Savéant J-M. Acc Chem Res 1993; 26:455; (c) Savéant J-M. Adv Phys Org Chem 1990; 26:1–130; (d) Savéant J-M. New J Chem 1992; 16:131; (e) Savéant J-M. Adv Electron Transfer Chem 1994; 4:53; (f) Savéant J-M. Adv Phys Org Chem 2000; 35:117.
15. Van Leeuwen M, McKillop A. J Chem Soc, Perkin Trans 1 1993; 2433.
16. Baumgartner MT, Gallego MH, Pierini AB. J Org Chem 1998; 63:6394.
17. Baumgartner MT, Pierini AB, Rossi RA. J Org Chem 1999; 64:6487.
18. Murguia MC, Rossi RA. Tetrahedron Lett 1997; 38:1355.
19. (a) Nazareno MA, Rossi RA. Tetrahedron Lett 1994; 35:5185; (b) Zhang YM, Guo HY, Heteroatom Chem 2001; 12:539.

20.   (a) Austin E, Alonso RA, Rossi RA. J Org Chem 1991; 56:4486; (b) Austin E, Ferrayoli CG, Alonso RA, Rossi RA. Tetrahedron 1993; 49:4495.
21.   Bunnett JF, Sundberg JE. J Org Chem 1976; 41:1702.
22.   Borosky GL, Pierini AB, Rossi RA. J Org Chem 1992; 57:247.
23.   Semmelhack MF, Bargar TM. J Am Chem Soc 1980; 102:7765.
24.   Wu BQ, Zeng FW, Ge M, Cheng X, Wu G. Sci China 1991; 34:777.
25.   Hoz S, Bunnett JF. J Am Chem Soc 1977; 99:4690.
26.   Córsico EF, Rossi RA. Synlett 2000; 227.
27.   Kashimura T, Kudo K, Mori S, Sugita N. Chem Lett 1986; 299.
28.   (a) Compton RG, Dryfe RAW, Fisher AC. J Electroanal Chem 1993; 361:275; (b) Compton RG, Dryfe RAW, Fisher AC. J Chem Soc, Perkin Trans 2 1994; 1581; (c) Compton RG, Dryfe RAW, Eklund JC, Page SD, Hirst J, Nei LB, Fleet GWJ, Hsia KY, Bethell D, Martingale LJ. J Chem Soc, Perkin Trans 2 1995; 1673; (d) Teherani T, Bard AJ. Acta Chem Scand B 37 1983; 413.
29.   Galli C. Tetrahedron 1988; 44:5205.
30.   (a) Dell'Erba C, Novi M, Petrillo G, Tavani C, Bellandi P. Tetrahedron 1991; 47:333; (b) Petrillo G, Novi M, Garbarino G, Dell'Erba C, Tetrahedron 1987; 43:4625; (c) Novi M, Petrillo G, Dell'Erba C, Tetrahedron Lett 1987; 28:1345.
31.   Bunnett JF, Gloor BF. J Org Chem 1973; 38:4156.
32.   Rossi RA, de Rossi RH, Pierini AB. J Org Chem 1979; 44:2662.
33.   Goldfarb YL, Ikubov AP, Belen'kii LI. Zhur Geter Soedini 1979; 1044; Chem Abstr 1979; 91:193081n.
34.   Rossi RA, de Rossi RH, López AF. J Am Chem Soc 1976; 98:1252.
35.   Rossi RA, de Rossi RH, López AF. J Org Chem 1976; 41:3371.
36.   Yakubov AP, Belen'kii LI, Goldfarb YL. Iz Akad Nauk SSSR, Ser Khim 1981; 812; Chem Abstr 1982; 96:104049r.
37.   Moon MP, Komin AP, Wolfe JF, Morris GF. J Org Chem 1983; 48:2392.
38.   Amatore C, Gareil M, Oturan MA, Pinson J, Savéant JM, Thiébault A. J Org Chem 1986; 51:3757.
39.   Wolfe JF, Greene JC, Hudlicky T.J Org Chem 1972; 37:3199.
40.   Komin AP, Wolfe JF. J Org Chem 1977; 42:2481.
41.   Hay JV, Wolfe JF. J Am Chem Soc 1975; 97:3702.
42.   Beugelmans R, Bois-Choussy M, Boudet B. Tetrahedron 1982; 38:3479.
43.   Beugelmans R, Chastanet J. Tetrahedron 1993; 49:7883.
44.   Beugelmans R, Bois-Choussy M, Gayral P, Rigothier MC. Eur J Med Chem 1988; 23:539.
45.   Pierini AB, Peñéñory AB, Rossi RA. J Org Chem 1984; 49:486.
46.   Rossi RA, Alonso RA, Palacios SM. J Org Chem 1981; 46:2498.
47.   Alonso RA, Rossi RA. J Org Chem 1982; 47:77.
48.   Pierini AB, Rossi RA. J Organomet Chem 1979; 168:163.
49.   Rossi RA. Acc Chem Res 1982; 15:164.
50.   Pierini AB, Rossi RA. J Org Chem 1979; 44:4667.
51.   Bunnett JF, Creary X. J Org Chem 1975; 40:3740.
52.   Beugelmans R, Bois-Choussy M, Boudet B. Tetrahedron 1983; 39:4153.

53. Rossi RA, Palacios SM. J Org Chem 1981; 46:5300.
54. Bunnett JF, Creary X. J Org Chem 1974; 39:3173.
55. Julliard M, Chanon M. J Photochem 1986; 34:231.
56. Peñéñory AB, Rossi RA. J Phys Org Chem 1990; 3:266.
57. Zoltewicz JA, Locko GA. J Org Chem 1983; 48:4214.
58. Novi M, Garbarino G, Petrillo G, Dell'Erba C. J Org Chem 1987; 52:5382.
59. Chbani M, Bouillon J-P, Chastanet J, Souflaoui M, Beugelmans R. Bull Soc Chim Fr 1995; 132:1053.
60. Bunnett JF, Scamehorn RG, Traber RP. J Org Chem 1976; 41:3677.
61. Moon MP, Wolfe JF. J Org Chem 1979; 44:4081.
62. Wolfe JF, Sleevi MC, Goehring RR. J Am Chem Soc 1980; 102:3646.
63. Goehring RR, Sachdeva YP, Pisipatis JS, Sleevi MC, Wolfe JF. J Am Chem Soc 1985; 107:435.
64. Yammal CC, Podestá JC, Rossi RA. J Org Chem 1992; 57:5720.
65. Córsico EF, Rossi RA. J Org Chem 2002; 67:3311.
66. (a) Dillender SC, Greenwood TD, Hendi MS, Wolfe JF. J Org Chem 1986; 51:1184; (b) Hay JV, Hudlicky T, Wolfe JF. J Am Chem Soc 1975; 97:374; (c) Bunnett JF, Sundberg JE. Chem Pharm Bull 1975; 23:2620.
67. Estel L, Marsais F, Queguiner G. J Org Chem 1988; 53:2740.
68. Rossi RA, Bunnett JF. J Org Chem 1973; 38:3020.
69. Beugelmans R, Ginsburg H. J Chem Soc, Chem Commun 1980; 508.
70. (a) Bard RR, Bunnett JF. J Org Chem 1980; 45:1546; (b) Beugelmans R, Bois-Choussy M. Synthesis 1981; 730; (c) Beugelmans R, Boudet B, Quintero L. Tetrahedron Lett 1980; 21:1943; (d) Beugelmans R.; Ginsburg H, Heterocycles 1985; 23:1197.
71. Oostvee EA, van der Plas HC. Recl Trav Chim Pays-Bas 1979; 98:441.
72. Nair V, Chamberlain SD. J Am Chem Soc 1985; 107:2183.
73. Beugelmans R, Bois-Choussy M. Heterocycles 1987; 26:1863.
74. Beugelmans R, Bois-Choussy M, Tang Q. J Org Chem 1987; 52:3880.
75. Nair V, Chamberlain SD. J Org Chem 1985; 50:5069.
76. Dell'Erba C, Novi M, Petrillo G, Tavani C. Tetrahedron 1993; 49:235.
77. Beugelmans R, Chastanet J, Ginsburg H, Quinteros-Cortes L, Roussi G. J Org Chem 1985; 50:4933.
78. Bunnett JF, Singh P. J Org Chem 1981; 46:5022.
79. Alonso RA, Rossi RA. J Org Chem 1980; 45:4760.
80. Carver DR, Greenwood TD, Hubbard JS, Komin AP, Sachdeva YP, Wolfe JF. J Org Chem 1983; 48:1180.
81. Baumgartner MT, Jiménez LB, Pierini AB, Rossi RA. J Chem Soc, Perkin Trans 2 2002; 1092.
82. Wong JW, Natalie KJ, Nwokogu GC, Pisipati JS, Flaherty PT, Greenwood DT, Wolfe JF. J Org Chem 1997; 62:6152.
83. Semmelhack MF, Bargar TM. J Org Chem 1977; 42:1481.
84. Liras S, Lynch CL, Fryer AM, Vu BT, Martin SF. J Am Chem Soc 2001; 123:5918.
85. Palacios SM, Asis SE, Rossi RA. Bull Soc Chim Fr 1993; 130:111.

86. (a) Ferrayoli CG, Palacios SM, Alonso RA. J Chem Soc, Perkin Trans 1 1995; 1635; (b) Rossi RA, Alonso RA. J Org Chem 1980; 45:1239; (c) Alonso, RA, Rodriguez, CH, Rossi, RA, J Org Chem 1989; 54:5983.
87. Lotz GA, Palacios SM, Rossi RA. Tetrahedron Lett 1994; 35:7711.
88. Nwokogu GC, Wong JW, Greenwood TD, Wolfe JF. Org Lett 2000; 2:2643.
89. (a) Beugelmans R, Roussi G. J Chem Soc, Chem Commun 1979; 950; (b) Beugelmans R, Roussi G. Tetrahedron 1981; 37:393.
90. Beugelmans R, Chbani M. Bull Soc Chim Fr 1995; 132:729.
91. Baumgartner MT, Nazareno MA, Murguía MC, Pierini AB, Rossi RA. Synthesis-Stuttgart 1999; 2053.
92. Barolo SM, Lukach AE, Rossi RA. J Org Chem 2003; 68:2807.
93. (a) Beckwith ALJ, Gerba WB. J Chem Soc, Perkin Trans 2 1975; 593; (b) Beckwith ALJ. Tetrahedron 1981; 37:3073; (c) Beckwith ALJ, Gerba S. Aust J Chem 1992; 45:289; (d) Annunziata A, Galli C, Marinelli M, Pau T. Eur J Org Chem 2001; 1323.
94. (a) Beckwith ALJ, Abeywikrema AN. Tetrahedron Lett 1986; 27:109; (b) Dittami JP, Ramanathan H. Tetrahedron Lett 1988; 29:45; (c) Boisvert G, Giasson R. Tetrahedron Lett 1992; 33:6587.
95. Beckwith ALJ, Palacios SM. J Phys Org Chem 1991; 4:404.
96. (a) Amatore C, Pinson J, Savéant J-M, Thiébault A. J Am Chem Soc 1981; 103:6930; (b) Amatore C, Oturan MA, Pinson J, Savéant J-M, Thiébault A. J Am Chem Soc 1984; 106:6318.
97. Abeywickrema AN, Beckwith ALJ. J Chem Soc, Chem Commun 1986; 464.
98. Vaillard SE, Postigo A, Rossi RA. J Org Chem 2002; 67:8500.
99. (a) Borosky GL, Pierini AB, Rossi RA. J Org Chem 1990; 55:3705; (b) Rossi RA, Pierini AB, Borosky GL. J Chem Soc, Perkin Trans 2. 1994; 2577; (c) Peñéñory AB, Rossi RA. Gazz Chim Ital 1995; 125:605; (d) Lukach AE, Rossi RA. J Org Chem 1999; 64:5826.
100. (a) Chopa AB, Lockhart MT, Silbestri G. Organometallics 2000; 19:2249; (b) Chopa AB, Lockhart MT, Dorn VB. Organometallics 2002; 21:1425.
101. Chopa AB, Lockhart MT, Silbestri G. Organometallics 2001; 20:3358.
102. Altmann R, Jurkschat K, Schürman M, Dakternieks D, Duthie A. Organometallics 1998; 17:5858 and references cited therein.
103. Chopa AB, Lockhart MT, Silbestri G. Organometallics 2002; 21:5874.
104. For reviews see (a) Mitchell TN. In: Diedrich F, Stang PG, eds. Metal Catalysed Cross-Coupling Reactions. Weinheim: Verlag GmbH, Wiley VCH, 1988:197–202; (b) Farina V, Krishnamurthy V, Scott WJ. In: Paquette LA, ed. The Stille Reaction. Organic Reactions 1997, Vol. 50. New York: John Wiley & Sons, Inc., 1997:1–676.
105. Córsico EF, Rossi RA. Synlett 2000; 230.
106. Faraoni MB, Koll LC, Mandolesi SD, Zuñiga AE, Podestá JC. J Organomet Chem 2000; 613:236.
107. Mandolesi SD, Vaillard SE, Podestá JC, Rossi RA.Organometallics 2002; 21:4886.

108. Rabideau PW, Marcinow Z. Org Reactions 1992; 42:1
109. Beugelmans R, Ginsburg H, Bois-Choussy M. J Chem Soc, Perkin Trans1 1982; 1149.
110. Beugelmans R, Bois-Choussy M, Chastanet J, Legleuher M, Zhu. J Heterocycles 1993; 36:2723.
111. (a) Pierini AB, Baumgartner MT, Rossi, RA. Tetrahedron Lett 1988; 29:3451; (b) Beugelmans R, Bois-Choussy M. Tetrahedron Lett 1988; 29:1289; (c) Baumgartner MT, Pierini AB, Rossi RA. Tetrahedron Lett 1992; 33:2323.
112. (a) Semmelhack MF, Chong BP, Stauffer RD, Rogerson TD, Chong A, Jones LD. J Am Chem Soc 1975; 97:2507; (b) Semmelhack MF, Stauffer RD, Rogerson TD. Tetrahedron Lett 1973; 4519; (c) Weinreb SM, Semmelhack MF. Acc Chem Res 1975; 8:158.
113. Goehring RR. Tetrahedron Lett 1992; 33:6045.
114. Goehring RR. Tetrahedron Lett 1994; 35:8145.

17

Ortho-, *Meta-*, and *Para*-Photocycloaddition of Arenes

Norbert Hoffmann
*UMR CNRS et Université de Reims Champagne-Ardenne,
Reims, France*

17.1. HISTORICAL BACKGROUND

While thermal reactions of aromatic compounds are mostly characterized by the retention of aromaticity in the final products, chemical reactions of the electronically excited states most frequently lead to a large variety of nonaromatic products. Photochemical reactions enrich considerably the chemistry of aromatic compounds and thereby offer many unusual applications to organic synthesis. This context was immediately recognized when the first *ortho* or [2+2] photocycloaddition was described [1]. The presence of reactive functional groups as well as ring constrain of the polycyclic compounds make the products very attractive as intermediates for organic synthesis. The *meta* or [2+3] photocycloaddition was discovered almost at the same time [2]. These reaction were frequently applied to the synthesis of natural products. For instance, many polycyclic terpene derived products are easily available when an intramolecular *meta* photocycloaddition is used as key step. Several reviews dealing with photocycloaddition were published before [3–7]. The aim of this chapter is therefore to summarize the principle items of reactivity and selectivity of these reactions and to encourage to apply them to organic synthesis.

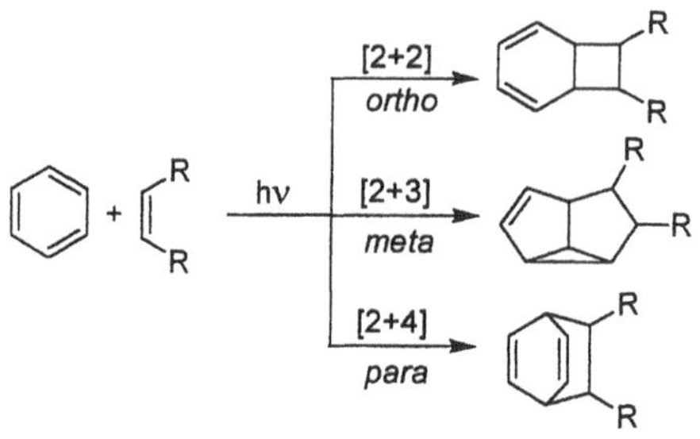

Scheme 1

17.2. STATE OF THE ART MECHANISTIC MODELS

Three types of cycloaddition products are generally obtained (Sch. 1). While [2+2] (*ortho*) and [2+3] (*meta*) cycloaddition are frequently described, the [2+4] (*para* or photo-Diels-Alder reaction) pathway is rarely observed in benzene ring systems. With naphthalene systems however, the *para* cycloaddition occurs more frequently [6,8]. The photo-Diels–Alder reaction and other photocyclization reactions are also observed with anthracene derivatives and higher condensed aromatic compounds. However, these reaction are not treated in this chapter since they are caused by the particular photophysical and photochemical properties of these compounds [6,9].

While the *meta* photocycloaddition occurs at the π,π^* singlet state of the arene [4], different cases must be distinguished for the *ortho* cyclo-addition depending on the structure of the substrates [5]: (a) excitation of a ground state charge-transfer complex; (b) excitation of the alkene (or alkyne) reaction partner; (c) excitation of the arene partner and reaction at the singlet state; and (d) excitation of the arene followed by intersystem crossing and reaction with the alkene at the triplet state.

At the singlet excited state, *ortho* and *meta* photocycloadditions are often competitive processes and physicochemical investigations were carried out to rationalize the modes of cycloaddition of arenes with alkenes. In the context of the study of photochemical electron transfer reactions, it has been proposed that the difference of the redox potentials of the reaction partners might play an important role in this competition [10]. Such a discussion involves the intervention of an exciplex as intermediate. The Rehm-Weller equation [11] was used to quantify the relationship. When an electron transfer process is strongly endergonic ($\Delta G > 1.5\,\text{eV}$), the *meta* cycloaddition should be favored. When such a process is less endergonic ($1 < \Delta G < 1.5\,\text{eV}$), the *ortho* addition dominate [12]. This means that the charge separation in the exciplex is more expressed in the case of the *ortho* photocycloaddition.

Earlier theoretical treatments based on molecular orbital interactions revealed that very polar intermediates or transition states are involved in both reactions, but these studies could not differentiate between the two cycloaddition modes [13]. However, a more recent treatment suggests that a common unpolar biradical intermediate exists [14]. The distinction between the *ortho* and the *meta* mode than occurs by dynamic effects mainly influenced by the substituents.

Recent results revealed that the *ortho* photocycloaddition frequently occurs concomitantly with the *meta* photocycloaddition even in cases where only *meta* derivatives are obtained. Especially in the case or electron-donor substituted benzene derivatives, the competitive *ortho* cycloaddition is less stereo- or regioselective and the resulting products are less stable. Aside the *meta* adducts as main products, complex mixtures of *ortho* products or products resulting from rearrangements of these primary photoproducts have been obtained [15,16]. Improved separation techniques recently enabled a better characterization of these products. Furthermore, using particular conditions like an acidic reaction medium, the intermediates resulting from *ortho* photocycloaddition could be transformed selectively in more stable final products [17].

17.3. SCOPE AND LIMITATIONS

17.3.1. *meta*-Photocycloaddition

The *meta* photocycloaddition is frequently observed in cases of alkyl or electron-donor substituted benzene derivatives. Two *exo/endo* isomers **2a,b** are obtained in high yields in the reaction of anisole **1** with cyclopentene (Sch. 2) [18]. The reaction was also efficiently carried out in an intramolecular way. Two efficient reactions of this type are depicted in Sch. 2. The azatriquinane **3** [19] and the silane derivative **4** [20] were obtained in high yields. The formation of different regioisomers can be controlled by the heteroatoms in the side chain. The longer C–Si bonds particularly favor the formation of compound **4**.

17.3.2. *ortho*-Photocycloaddition

Substituents exerting a mesomeric effects on the substrates often favor the *ortho* photocycloaddition. This reaction mode is frequently observed when the aromatic ring carries electron-withdrawing substituents and the alkene moiety carries electron donor groups as well as in the reversed case with

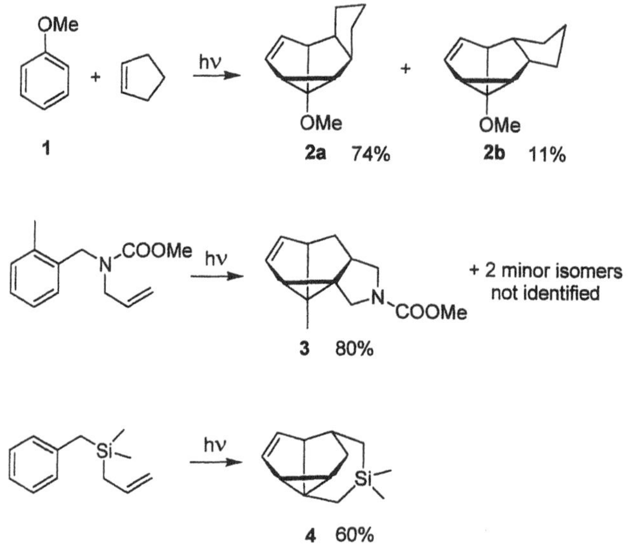

Scheme 2

electron donor substituents at the aromatic reaction partner and electron-withdrawing groups at the alkene [21]. In many cases, the reaction resembles the [2+2]-photocycloaddition of alkenes to α,β-unsaturated carbonyls, nitriles, or carboxylic acid derivatives occurring at the $^3\pi\pi^*$ state [22].

For instance, anisole **1** reacted with crotononitrile **5** to yield different stereoisomers of the *ortho* adduct **6** (Sch. 3) [23]. In many cases, the primary formed *ortho* adducts are not stable and undergo rearrangements to yield the final products. This aspect was intensively studied with intramolecular reactions. Upon irradiation, the *para*-hydroxybenzonitril derivative **7** undergo [2+2] photocycloaddition and gave the stable final product **8** by rearrangements via intermediates **A** and **B** [24,25]. The same reaction was recently applied in fluoroorganic chemistry, by the analogous transformation of partially fluorinated derivatives of **7** [26]. A corresponding meta hydroxybenzonitrile derivative was not reactive [24]. The presence of a second hydroxy substituent as in the 1,3-dihydroxybenzonitrile derivative **9** significantly enhances the reactivity [27]. In this case, the intermediate **C** resulting from the [2+2] photocycloaddition is rapidly transformed into the stable final product **10** via tautomerization. Corresponding acetophenone derivatives like **11** were also successfully transformed into cycloadducts like **12** [28,29]. It was shown, that in this case, π,π* triplet states are involved which exclude the competition of the *meta* addition.

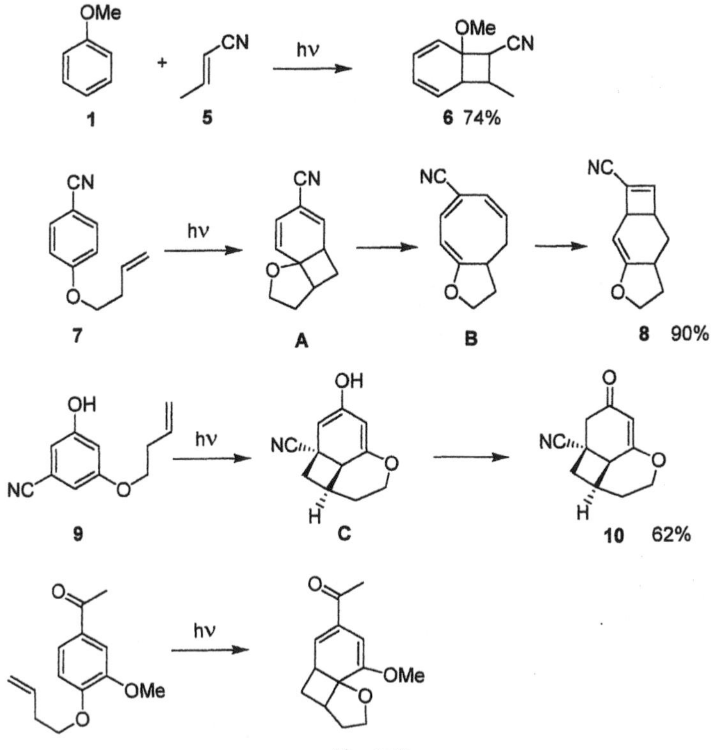

Scheme 3

The competition of *ortho* and *meta* photocycloaddition is much more expressed when the mesomeric effects of the substituents are weak [30,31]. A more precise analysis of the products revealed that *ortho* and even *para* side products are formed in minor amounts in cases were normally the *meta* cycloaddition should be observed as dominant reaction [32]. Bichromophroric substrates carrying electron donor substituents on the benzene ring and any electron active groups on the alkene moiety range in this category [15,31,33].

Due to their reduced aromatic character, naphthalene derivatives also frequently undergo *ortho* cycloaddition with alkenes (Sch. 4). 1-Cyanonaphthalene **13** reacts with cyclohexadiene to yield the regioisomeric *ortho* adducts **14** and **15** in good yields [34]. Two regioisomers **17** and **18** were also obtained from the intramolecular reaction of the 1-cyanonaphthalene derivative **16** [35]. The proportion of the two isomers depends on the solvent polarity.

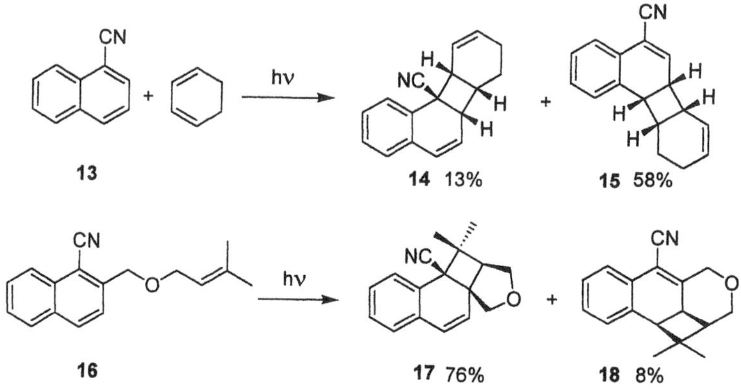

13   **14** 13%   **15** 58%

16   **17** 76%   **18** 8%

Scheme 4

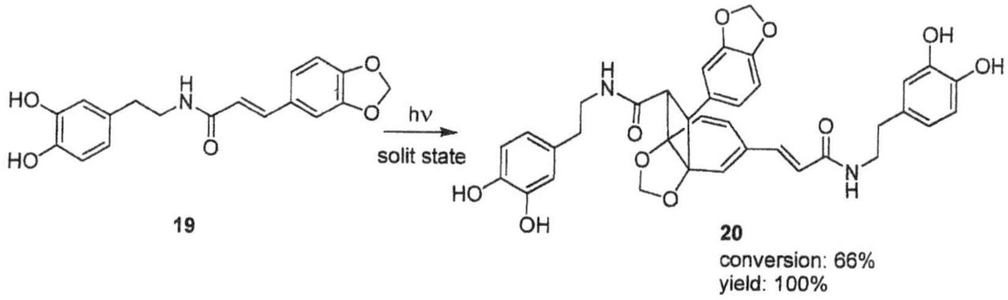

19   **20**
conversion: 66%
yield: 100%

Scheme 5

Due to the precise orientation in the solid phase, unusual photochemical reactions can take place. In this context, an *ortho* photocycloaddition is observed with a cinnamoyl derivative **19** and the photodimer **20** can be isolated with quantitative yields (Sch. 5) [36].

17.3.3. *para*-Photocycloaddition

Only few reports deal with a *para* photocycloaddition as the major reaction path. Recently, however, several cinnamide derivatives like **21** were efficiently transformed into the corresponding *para* adducts **22** (Sch. 6) [37]. Yields higher than 90% could be achieved. The *para* photocycloaddition is also observed with naphthalene derivatives like 1-acetylnaphthalene **23** and captodative enamino nitriles as **24** [38]. Other captodative substituted alkenes [39] as well as the fluorinated uracil derivative **26** [40] are transformed in the same way. Especially in the cases of **21** and **23**, the

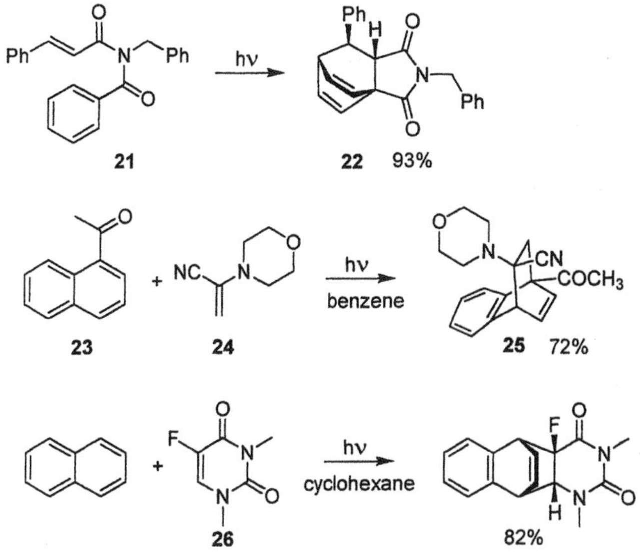

Scheme 6

reactivity of biradical intermediates possess a significant influence on the stereoselective formation of the *para* adducts.

17.4. SYNTHETIC POTENTIAL: REACTIVITY AND SELECTIVITY PATTERN

17.4.1. *Meta* Photocycloaddition

The photochemical cycloaddition of aromatic compounds leads to products with a high degree of functionality which makes them interesting for the application in organic synthesis. Furthermore and especially in the case of the *meta* addition, a large number of regio- and stereoisomers can be formed. However, electronic and steric influences of the substiutents reduce considerably the number of isomers. In intermolecular *meta* additions, the electron active substiutent on the aromatic ring has a dominant influence on the regioselectivity. After photochemical excitation of the aromatic reaction partner, exciplexes **D** and **E** are formed with the olefin (Sch. 7). Intermediates **F**, **G**, and **H** are generated by the formation of two σ-bonds in the *meta* orientation. A concerted formation of all σ-bonds or the formation of prefulvene intermediate (formation of a σ-bond inside the aromatic ring) followed by addition of the alkene are less probable [4]. All the intermediates

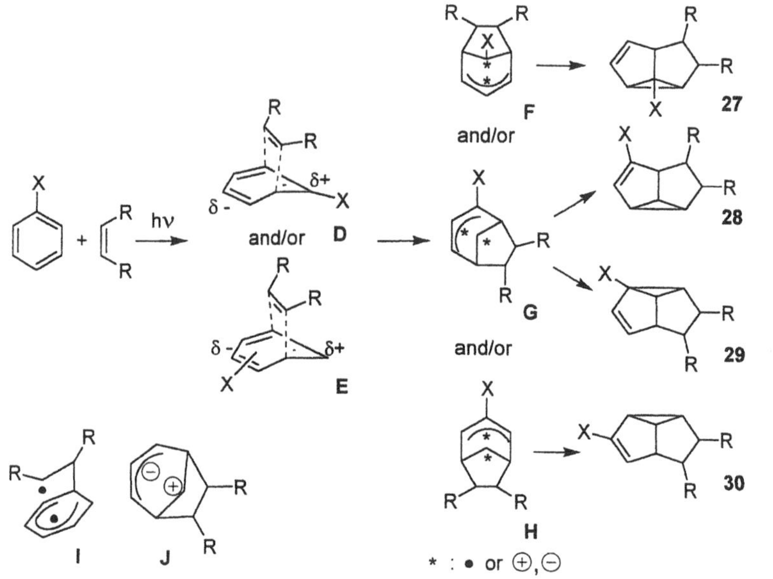

Scheme 7

depicted in the reaction pathways in Sch. 7 are highly polarized. Nevertheless, intermediates **F**, **G**, and **H** are frequently described as biradical intermediates but as the reaction occurs at the singlet state, the polar mesomeric forms contributes essentially to the nature of these species. The negative partial charge preferentially occupies a position where it is delocalized inside a π electron system, while the positive partial charge is localized on the bridgehead carbon atom of the bicyclo[3.2.1]octene system. When X is an electron donor substituent, the intermediates **D** and **F** are preferred and the final products of type **27** are obtained afterthe formation of a third σ-bond. Consequently, the regioisomers **28**, **29**, and less frequently **30** are favored in the case of electron withdrawing substituents. As mentioned above, recent theoretical studies based on valence bond calculations suggested the nonpolar biradical form **I** as intermediate for both the *ortho* and *meta* photocycloaddition [14]. However, a polar form like **J** suggested by earlier theoretical treatments [13], better accounts for the experimental observations as far as the regioselectivity is concerned. The following examples illustrate the regiodirecting effects of substituents on the aromatic ring.

When anisole **1** was irradiated in the presence of dioxole **31**, the products **32a,b** of the same regioisomer were isolated in good overall yield (Sch. 8) [41]. The high regioselectivity of the reaction can be explained by

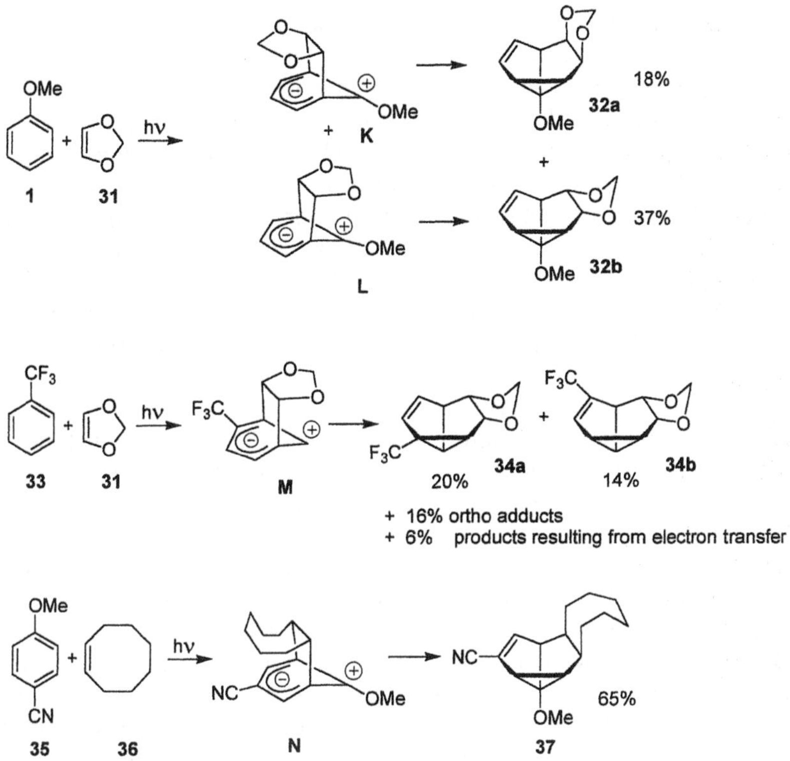

Scheme 8

the fact that the positive charge in the corresponding intermediates **K** and **L** is stabilized by a methoxy substituent. Electron withdrawing substituents stabilize the negative charge in the seven membered ring moiety of **M**. Therefore, the reaction of trifluoromethylbenzene **33** with **31** yields the adducts **34a,b** which are related to products **28** and **29** in Sch. 7 [42]. In the presence of both an electron donating and an electron withdrawing substiutent on the benzene ring (for instance in the reaction of *p*-methoxybenzonitrile **35** with cyclooctene **36**), the first one is preferentially localized at the five membered ring moiety of **N** while the latter one is positioned on the seven membered ring. The corresponding adduct **37** was obtained in high yield [43].

Aside from regioisomers, *exo/endo* diastereoisomers are formed when substituted olefinic derivatives are transformed. The *exo/endo* ratio is manly determined on the stage of the exciplex when secondary orbital interactions occur. These interactions may be compared to those which influences the *exo/endo* ratio in the Diels–Alder reaction. Coulomb forces or steric interactions between the electronically excited arene and the alkene in its ground state also become effective. Many investigations have been carried out with cyclic alkenes. In these cases the *endo* isomers generally prevail (products **2a,b** in Sch. 2, and product **37** in Sch. 8). However, when electron rich alkenes like dioxoles (**31**) are transformed the formation of *exo* isomers is favored. Due to an interaction of the oxygen atoms with the positive charge in the intermediates **L** and **M** (Sch. 8) the corresponding isomers **32b** and **34a,b** of the final products are isolated in excess.

Due to its interest for preparative organic chemistry, the intramolecular version of the *meta* photocycloaddition was also frequently studied. The regiochemistry is mainly controlled by substituents on the aromatic ring in the same way as observed for intermolecular reactions. Furtheron, the substituents on the side chain or heteroatoms which are included, have considerable influence on the ratio of diastereoisomers. Two main reaction path ways are depicted in Sch. 9. 1,3 (via intermediate **O**) and 2,4 additions (via intermediate **P**) are most frequently observed while the other attacks are seldom. The 1,3 mode is favored when X has electron donating character (e.g., $X =$ OMe) leading to the final products **38** and **39**. The formation of products **40** and **41** via 2,6 is principally observed when $X =$ H. In this case, the alkyl moiety of the side chain plays the role of an electron donating substituent. Further, the substiuents on the side chain and heteroatoms in it have a strong influence on the regioisomer composition. The product **4**

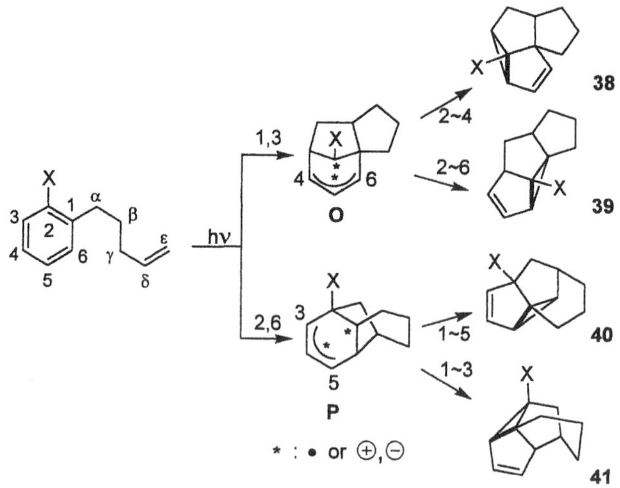

Scheme 9

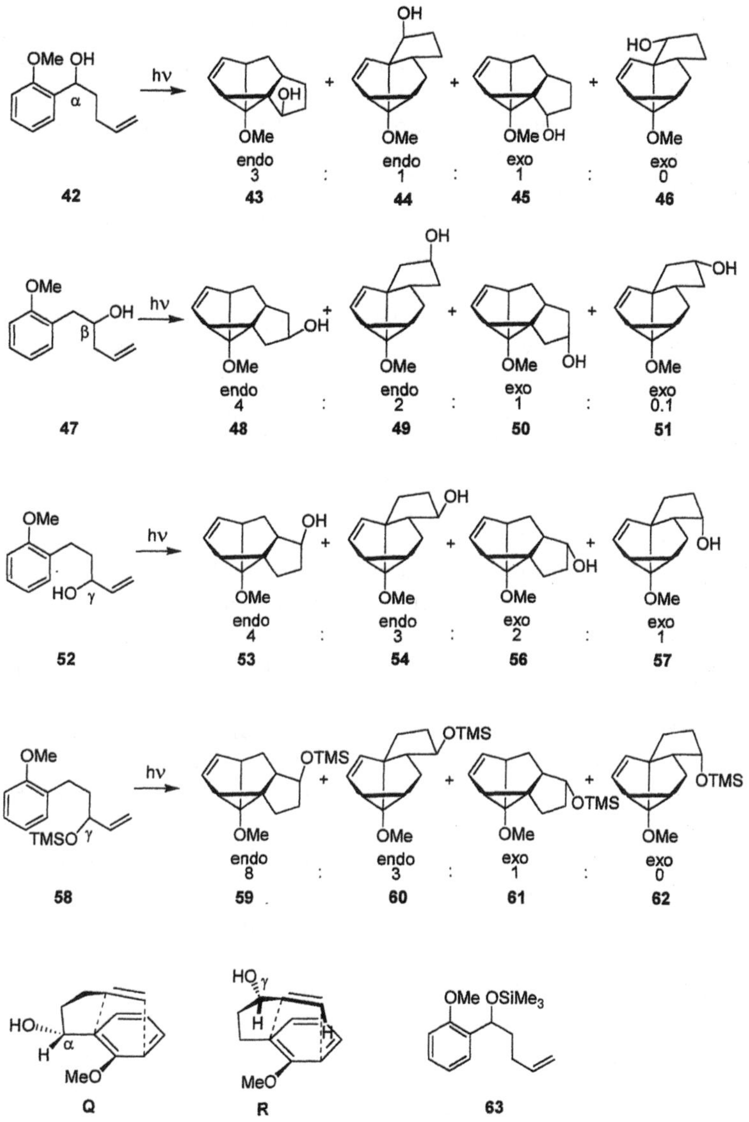

Scheme 10

(Sch. 2) results from a 2,6-attack of the side chain [20]. The orientation of the cyclopropane formation in the following step is less strictly controlled.

The examples depicted in Sch. 10 illustrate convincingly the influence of the electronally active substituent on the aromatic ring as well as the directing effect of the unsaturated side chain [44]. The regio- and dia-stereoselectivity of the photocycloaddition of the anisole derivatives **42**, **47**, **52**, and **58** carrying an unsaturated side chain in the *ortho* position was studied in detail. Furthermore, a hydroxy substituent is fixed in different positions of the side chain. All products result from a 1,3 attack of the side chain on the aromatic ring (transition states **Q** or **R**). Products **43**, **45**, **48**, **50**, **53**, **56**, **59**, and **61** result from a 2 ∼ 6 cyclopropanation step while the isomers **44**, **49**, **51**, **54**, **57**, and **60** are formed via 2 ∼ 4 cyclopropanation (compare Sch. 9). The 2 ∼ 6 cyclization dominates and factors varying between 1.5 and 4 are observed. When the hydroxy group in the γ position is replaced by the more bulky substituent OTMS in compound **58** this ratio is significantly increased from 1.5 to 4. A comparable effect cannot be observed from a substituent exchange in the α position (**63**).

Side differentiating effects like allylic 1,3-strain [45] are also observed leading to *exo*/*endo* isomers. This effect is particularly expressed in the reaction of compound **42**. In this case, a methoxy group is involved in the corresponding steric interaction (compare transition state **Q**). A lower side differentiation is observed in the reaction of compounds **52** and **58**. In the latter case, the allylic 1,3-strain effect is located on the olefinic side chain and the only one of the hydrogen atoms at the methylene carbon is involved in the corresponding interaction (transition state **R**). The 1,3-allylic strain effect was particularly important in the application of the reaction to natural product synthesis (see below) since it diminish the number of isomers.

Furthermore, from the complete study, it was concluded that hydrogen bonds effects can be ruled out as regio- and stereodirecting factors in the present examples.

The synthesis of enantiopure or enantio-enriched compounds is one of the major topics in organic chemistry. Until now, only few attempts were undertaken to obtain enantiopure products from photocycloaddition of arenes. Using optically pure starting materials provides one of the strategies followed in this context. Upon irradiation, compound **64**, which was obtained by asymmetric synthesis, yields products **65** and **66** via intramolecular *meta* cycloaddition (Sch. 11) [46]. The attack is highly regioselective in the 1,3-positions (transition state **S**). The high *endo* selectivity is induced by a 1,3-allylic strain effect as discussed above. Only the cyclopropanation step is unselective. The 2 ∼ 6 cyclization is favored and leads to the formation of **66** (compare Sch. 9). However, both isomers **65** and **66** are interconvertible via a vinylcyclopropane rearrangement. For application to organic synthesis of natural products (see below), the desired isomer **65** can thus been obtained via photoequilibration of compound **66**.

The facial diastereoselectivty of the reaction is also completely con-trolled, when a diol based tether is used as chiral auxiliary (Sch. 12) [47].

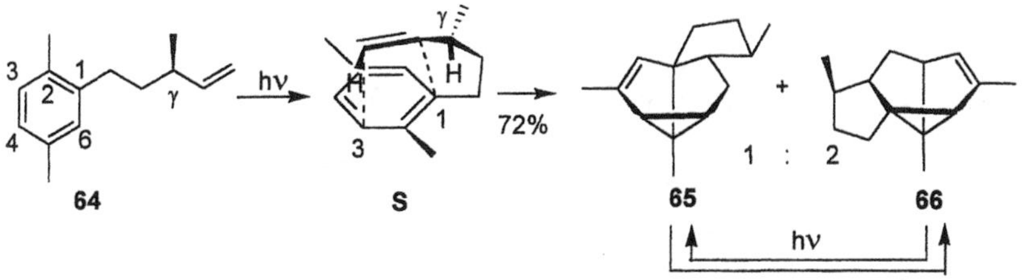

Scheme 11

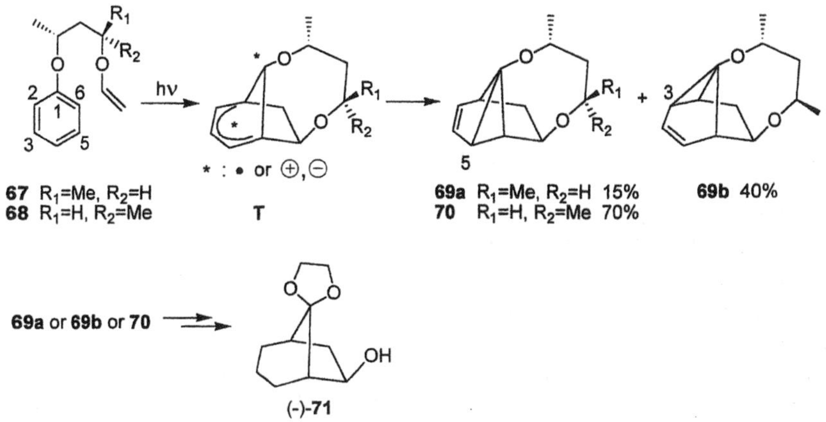

Scheme 12

Irradiation of compounds **67** and **68** yields cycloadducts exclusively via one facial 2,6 attack induced by the alkoxy substituent on the benzene ring leading to intermediate **T**. In the case of **68** possessing a side chain formally derived from the *meso* 2,4-pentanediol, the cyclopropanation step occurs also selectively. Only the $1 \sim 5$ cyclopropanation occurred leading to **70**, while in the case of **67**, the two isomers **69a** and **69b** can be isolated. Compounds **69a,b** and **70** were transformed into enantiomerically pure $(-)$-**71**.

Recently, an attempt was made to induce chirality in a *meta* cycloaddition via complexation with cyclodextrins (Sch. 13) [48]. 1/1 inclusion complexes of **72** and β-cyclodextrin can be synthesized and irradiated at $\lambda = 300$ nm. Two regioisomers **73a** and **73b** are isolated with different enantiomeric excesses. The results can be explained by interactions of different intensity with the chiral environment at the transition state **U**. In the case of pathway (a) leading to **73a** the interaction with the cyclodextrin is more expressed then for pathway (b) leading to **73b**. Thus, the formation of **73a** occurs with higher enantioselectivity.

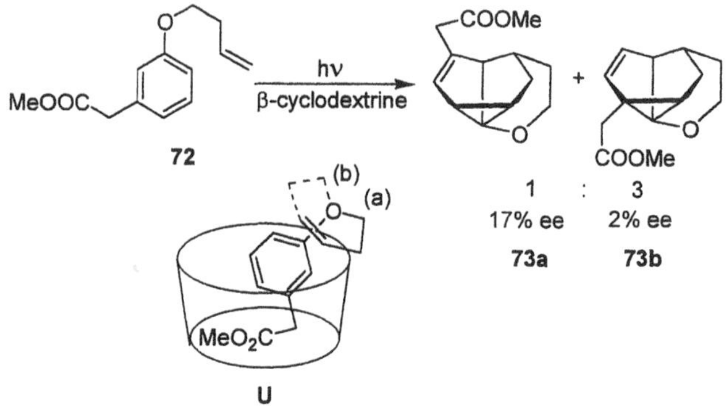

Scheme 13

17.4.2. *Ortho* Photocycloaddition

As discussed above, the *ortho* and the *meta* photocycloaddition often occur concomitantly and the chemoselectivity is mostly influenced by substituents. The low stability of the primary cycloadducts from the *ortho* pathway as well as their tendency to rearrange, decreases the efficiency of these reactions. Additives like Broensted acids are capable to catalyze selective rearrangements of intermediates which leads to stable final products. Several bichromophoric aromatic compounds like resorcinol derivatives or derivatives of salicylic acid possess a weak photochemical reactivity. When compounds **74** are irradiated in neutral reaction media, only a poor and unselective reactivity is observed. However, in the presence of acid, benzocyclobutenes **75a,b** are isolated in good yields (Sch. 14) [17,49]. In the first step, a reversible *ortho* cycloaddition takes place. The resulting intermediates **V** and **V'** are transformed via acid catalyzed rearrangements into the stable final products. The photochemical reactivity of the salicylic acid derivative **77** is considerably enhanced when the reaction is carried out in an acidic medium [50]. In this case, the equilibra between several intermediates are driven towards the final products **78** via an acid-catalyzed addition of methanol to enol ether **W**. Similar observations were made for the photochemical reactions of comparable derivatives of the 3,5-dihydroxybenzoic acid [27,51] and chromanone [52]. An intramolecular *ortho* photocycloaddition between a benzene ring and a naphthalene ring was observed for the first time when the corresponding reaction was carried out in an acidic medium [53]. The primary adducts are transformed into stable final products via acid catalysed or photochemical rearrangements depending on the wavelength [54]. The intermolecular reaction of 6-chloro-*N,N*-dimethyluracil with benzene or alkylbenzenes such as xylenes or

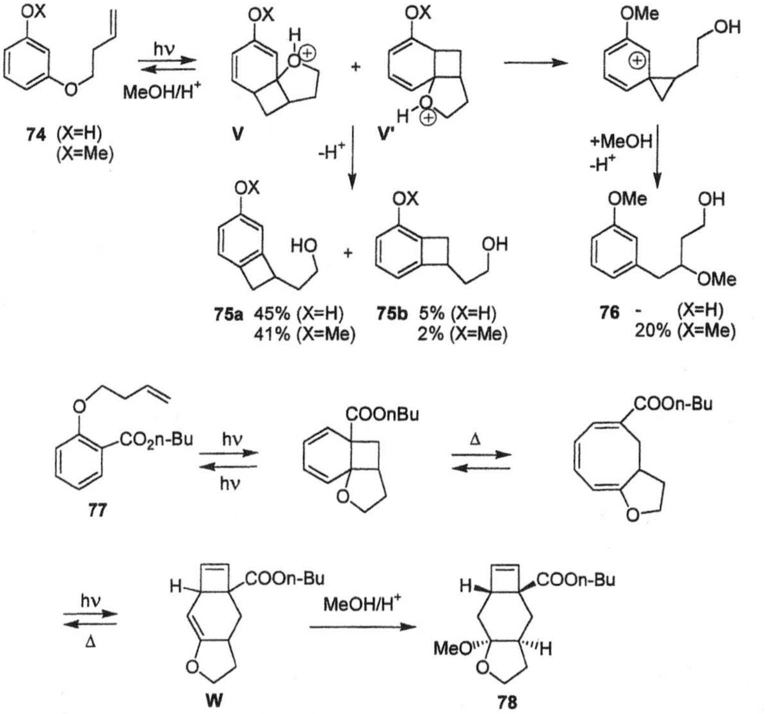

Scheme 14

mesitylene are significantly influenced in the presence of trifluoroacetic acid [55]. In these cases again, the initial *ortho* photocycloaddition is followed by acid catalyzed rearrangements leading to the final product mixture.

Until now, only few attempts were made to control the absolute configuration of the chiral polycyclic products. When salicylic esters of type **77** (Sch. 14) possessing an asymmetric alcohol moiety as chiral auxiliary instead of the *n*-butyl group are irradiated, the corresponding products can be isolated with 17% ee [50]. Better results can be obtained with the additionally acylated substrates **79a–c** (Sch. 15). Compounds **80a–c** were isolated with diastereomeric excesses up to 90% [29,56]. These compounds were formed via [2+2] photocycloaddition yielding the intermediates **X**. A thermal rearrangement leads to **80a–c**. Compounds **81a–c** are produced in the same time via a further photochemical rearrangement but due to thermal reversibility of the last step, **80a–d** could be isolated with yields up to 90% when the reaction was completed.

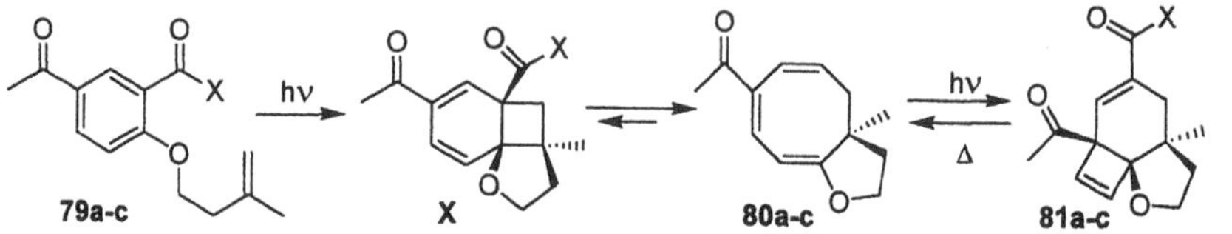

79a-c   **X**   **80a-c**   **81a-c**

80x	Auxiliary	%de
a	(S)-(-)-1-phenylethylamine	35
b	(2R,5R)-(-)-2,5-dimethylpyrrolidine	90
c	(7R)-(+)-camphorsultam	90

Scheme 15

17.4.3. *para*-Photocycloaddition

The photo Diels–Alder reaction of α-acetylnaphthalene **82** with the chiral α-enaminonitril **83** yielded the cycloadduct **84** with excellent diastereoselectivity (Sch. 16) [57]. The intermediary formed biradical **Y** is particular stable due to delocalization of the radical on the aromatic moiety and to a captodative effect on the enamine part. The chiral induction occurred in two steps [58]. In the first step, a stereogenic center is created in the α-position of the acyl group. In the second step of the diastereoselection, one of the two diasteromeric intermediates undergoes preferentially cyclization to yield the final product **84**, while the other one is more readily decomposed to form the starting material. For a more detailed discussion of the mechanism see Ref. [59].

17.5. REPRESENTATIVE EXPERIMENTAL PROCEDURES

17.5.1. *meta*-Photocycloaddition [60]

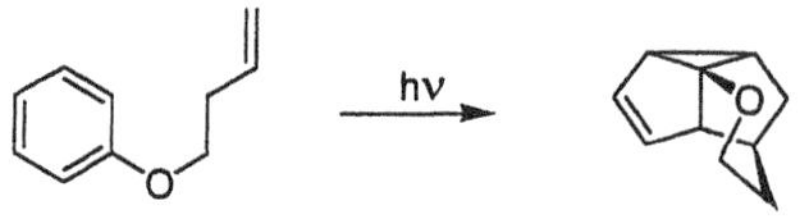

85   **86**

A solution of 96 mg (65 mmol) of **85** in 60 mL of cyclohexane in a quartz tube is deoxygenated by argon bubbling. Irradiation is carried out by means of a Rayonet photochemical reactor using 254 nm irradiation lamps during 4 h. The solution is evaporated and the residue is separated by preparative HPLC (RSiL, 10 mm, 25 × 1 cm, 5 mL/min) using hexane/ethyl acetate (30:1) as eluant. Compound **86** (30 mg) is obtained as colorless oil, while

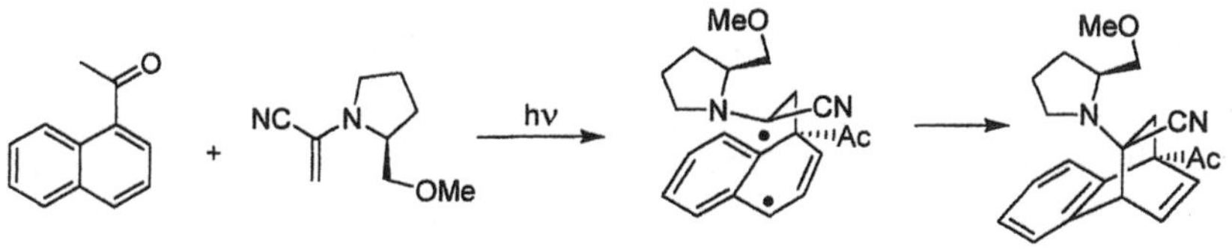

Scheme 16

the substrate is partially recovered (32 mg). Yield: 31% (46% with respect to transformed **85**). For another example see [61,62].

17.5.2. *Ortho* Photocycloaddition [63]

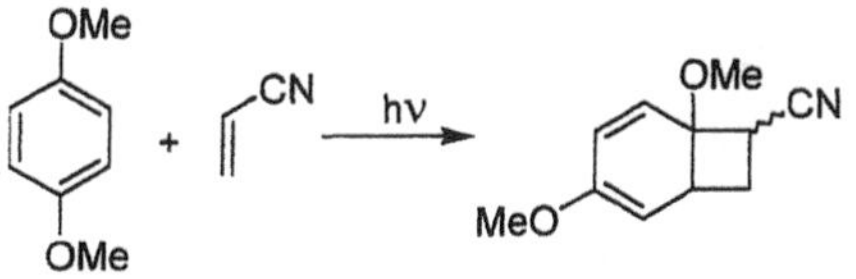

A solution of 1.38 g (0.01 mol) of **87** and 5.3 g (0.1 mol) **88** in 100 mL of cyclohexane in a fused silica tube was irradiated (six 15 W low pressure mercury arc lamps) under nitrogen until complete consumption of the arene had occurred. The cyclohexane solution was decanted from the immiscible orange oil of acrylonitrile cyclobutane dimers and rotary evaporated. The photoadduct **89** comprised 95% of the oil residue and was further purified by flash chromatography (32–62 mm silica) with petroleum ether/diethyl ether (5:3) as eluant. For an intramolecular reaction of this type see [25].

17.5.3. *Para* Photocycloaddition [64]

A solution of 1.70 g (10 mmol) of **23** (Sch. 6) and 1.38 g (10 mmol) of **24** in 100 mL of dry cyclohexane was irradiated with 150 W high-pressure mercury lamp through a water cooled immersion well made of Duran glass ($\lambda > 280$ nm) under continuous purging with dry argon (or nitrogen) for 2 h, during which time usually 26% of starting materials were converted and 70 mg crystals of **27**, m.p. 152–156 °C, precipitated (irradiation may be extended to 60% conversion but byproducts tend to accumulate). Crystallization from ethyl acetate/hexane gave 558 mg (70% based on converted starting material) of **25** of m.p. 157–159 °C. The photolysis solution on concentration gave 2.29 g of a 1:1 mixture of starting material.

17.5.4. Photocycloaddition in Acidic Media [17]

A solution of **74** ($X =$ Me) (Sch. 14) (350 mg, 2 mmol) and H_2SO_4 (0.4 mmol) in methanol (65 mL) was distributed in 4 quartz tubes (diameter: 0.5 cm) and irradiated in a Rayonet reactor ($\lambda = 254$ nm) for 4 h. Solid $NaHCO_3$ (approximately 500 mg) was added and the solvent was evaporated. The residue was purified by flash chromatography (eluant: ethyl acetate/petroleum ether: 1/5). Conversion: 86%, yield (based on the conversion of **74** ($X =$ Me)): **75a,b** ($X =$ Me) (isolated as mixture): 130 mg (43%), **76**: 72 mg (20%).

The reaction was increased to a multigram scale [65]. Solutions of **74** (7.2 g, 40 mmol) and H_2SO_4 (8 mmol) in methanol (1300 mL) were irradiated in a circulation system ($\lambda = 254$ nm) for 23 h ($T = 10$–15 °C). The solution was treated as described above. Yield: 6.1 g (57%), **76**: 1.7 g (23%).

17.6. TARGET MOLECULES: NATURAL AND NONNATURAL PRODUCT STRUCTURES

The photocycloadditions of readily accessible aromatic compounds lead to highly functionalyzed products in only one step. They can be used in a flexible manner for the synthesis of numerous target structures. The total synthesis of these products is considerably shortened and simplified by applying the photochemical cycloadditions as key steps which was particularly demonstrated with the transformation of many intramolecular *meta* cycloadducts [66].

Among the variety of terpenes which were synthesized with an intramolecular *meta* photocycloaddition as key step, triquinane derivatives were particularly well studied. Two isomers of these compounds are readily accessible via intramolecular *meta* cycloaddition in position 1,3 (Sch. 9). The following cyclopropanation reaction of intermediate **O** controls which of the angular or linear isomers is formed (Sch. 17) [67]. As this step is almost always unselective, both isomers are concomitantly formed. However, one of the isomers can be obtained predominantly via an additional photochemical equilibration step (compare Sch. 11). The triquinane frame is obtained by rupture of the distant C–C bond of the cyclopropane in connection to the 10-membered ring moiety.

The angular triquinane derivatives isocumene **90** [68], silphinene **91** [69], silphiperfol-6-ene **92** [70], 7αH-silphiperfol-5-ene **93** ($R_1 =$ H, $R_2 =$ Me) [70], 7βH-silphiperfol-5-ene **93** ($R_1 =$ H, $R_2 =$ Me) [70], subergorgic acid **94** [71], crinipellin B **95** (formal synthesis) [72], (−)-retigeranic acid **96** (formal asymmetric synthesis) [73] were synthesized via intramolecular *meta* cycloaddition as key step (Sch. 18). In the same way, linear triquinane

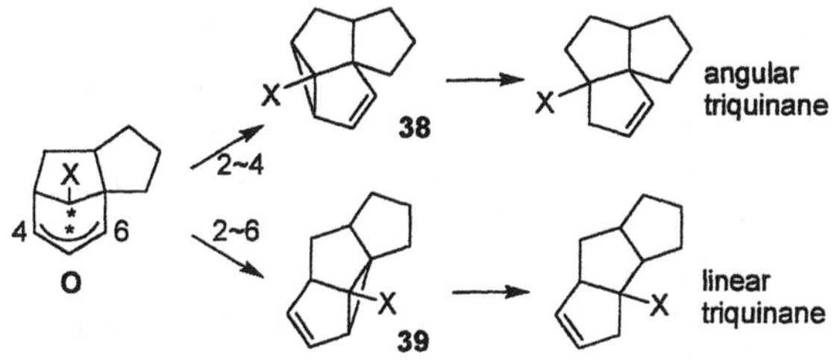

Scheme 17

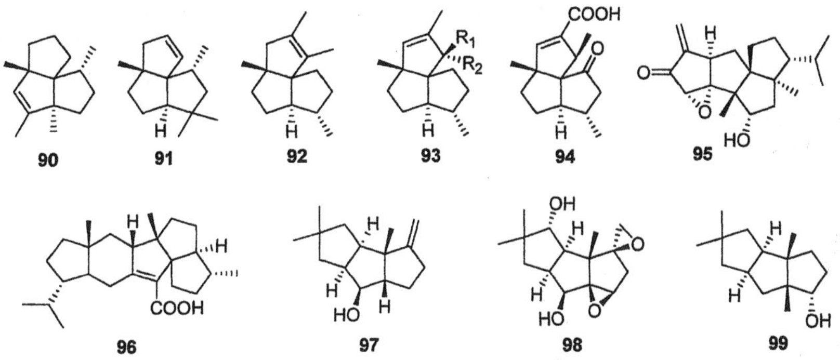

Scheme 18

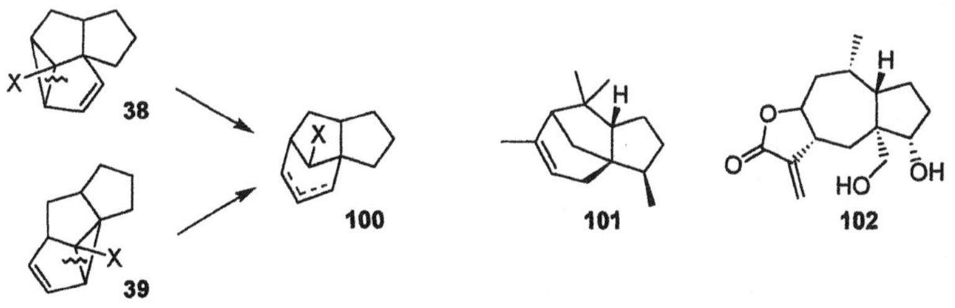

Scheme 19

derivatives were obtained: hirsutene **97** [74], coriolin **98** (formal synthesis) [75], ceratopicanol **99** [76].

When the inner C–C bonds of the cyclopropane, adjacent to the olefinic bond of the cycloadducts **38** and **39**, are cleaved, C_1 bridged tricyclic compounds **100** (compare [74]) like cedrene **101** (Sch. 19) [19,61] are readily available. If both inner bonds of the cyclopropane ring are cleaved during the synthesis, the perhydroazulen skeleton can be obtained easily like in the synthesis of rudmollin **102** [77].

Fenestranes represent particular targets. Like steroids, they are conformationally rigid, strained, and chemically robust molecules. These properties make them interesting for application in various fields [78]. Fenestranes **103** [79], **104** [80], and **105** [78] (Sch. 20) were synthesized via *meta* photocycloadditions as photochemical key step.

Propellane derivatives are available via intermolecular *meta* cycloaddition. Compounds **106** [81,82] and modhephene **107** [81] were obtained using this type of photocycloaddition (Sch. 21). Isoiridomyrmecin **108** [83] and decarboxyquadrone **109** [82] can be synthesized via the same photochemical key step.

Until now, only few attempts were made to apply the *ortho* photocycloaddition to organic synthesis. Particular interest lies in the field of compounds possessing interesting biological activity (Sch. 22). The cyclohexane-1,3-dione carboxylic acid derivative **110** can be used as core structure of new herbicides [27,51,84]. Benzocyclobutenes **75a,b** obtained

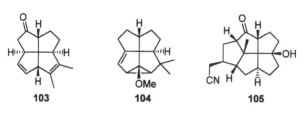

103   **104**   **105**

Scheme 20

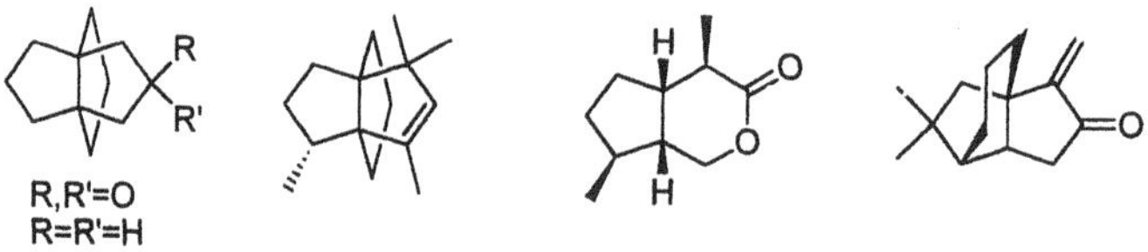

106   **107**   **108**   **109**

R,R'=O
R=R'=H

Scheme 21

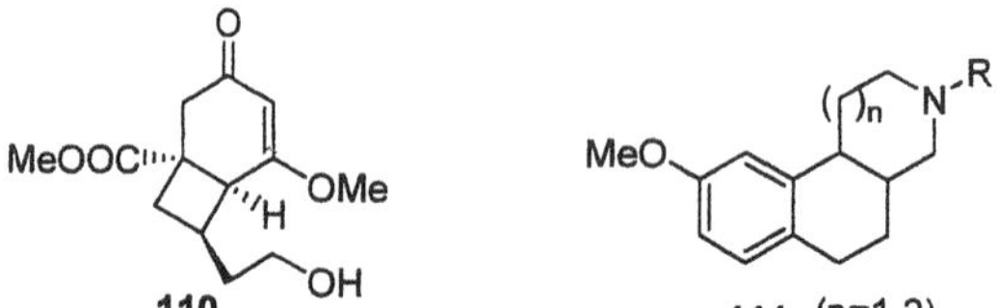

110   **111** (n=1,2)

Scheme 22

via *ortho* photocycloaddition in acidic media (Sch. 14) can be transformed into nitrogen containing heterocycles **111** possessing an affinity to dopamine receptors [65].

REFERENCES

1. (a) Ayer DE, Büchi G. US Patent 2,805,252, 1957; (b) Ayer DE, Büchi G, Robb EW. J Org Chem 1963; 85:2257–2263; (c) Kolzenburg G, Fuss PG, Mannsfeld, S-P, Schenck GO. Tetrahedron Lett 1966; 1861–1866.
2. (a) Wilzbach KE, Kaplan L. J Am Chem Soc 1966; 88:2066–2067; (b) Bryce-Smith D, Gilbert A, Orger BH. Chem Commun 1966; 512–514.
3. (a) Morrison H. Acc Chem Res 1979; 12:383–389; (b) Wender PA, von Geldern TW. Aromatic compounds: isomerisation and cycloaddition. In: Coyle JD, ed. Organic Photochemistry in Organic Synthesis. London: The Royal Society of Chemistry, 1986:226–255; (c) Wender PA, Siggel L, Nuss JM. Arene–alkene photocycloaddition reactions. In: Padwa A, ed. Organic Photochemistry. Vol. 10. New York: Marcel Dekker, Inc., 1989:357–473; (d) Wender PA, Siggel L, Nuss JM. [3+2] and [5+2] Arene–alkene photocycloadditions. In: Trost BM, Fleming I, Paquette LA, eds. Comprehensive Organic Synthesis. Vol. 5. Oxford: Pergamon Press, 1991:645–673; (e) Wender PA, Dore TM. Intra- and intramolecular cycloadditions of benzene derivatives. In: Horspool WM, Song P-S, eds. CRC Handbook of Organic Photochemistry and Photobiology. Boca Raton: CRC Press, 1995:280–290.
4. Cornelisse J. Chem Rev 1993; 93:615–669.
5. Cornelisse J, de Haan R. Ortho photocycloaddition of alkenes and alkynes to the benzene ring. In: Ramamurthy V, Schanze KS, eds. Molecular and Supramolecular Photochemistry. Vol. 8. New York: Marcel Dekker, Inc., 2001:1–126.
6. Mizuno K, Maeda H, Sugimoto A, Chiyonobu K. Photocycloaddition and photoaddition reactions of aromatic compounds. In: Ramamurthy V, Schanze KS, eds. Molecular and Supramolecular Photochemistry. Vol 8. New York: Marcel Dekker, Inc., 2001:127–316.
7. Wagner PJ. Acc Chem Res 2001; 34:1–8.
8. McCullough JJ. Chem Rev 1987; 87:811–860.
9. See: (a) Becker H-D. Chem Rev 1993; 93:145–172; (b) Bouas-Laurent H, Castellan A, Desvergne J-P, Lapouyade R. Chem Soc Rev 2000; 29:43–55; (c) Bouas-Laurent H, Castellan A, Desvergne J-P, Lapouyade R. Chem Soc Rev 2001; 30:248–263 and literature cited therein.
10. (a) Bryce-Smith D, Gilbert A, Orger B, Tyrrell H. Chem Commun 1974; 334–336; (b) Mattay J, Leismann H, Scharf H-D. Chem Ber 1979; 112:577–599; (c) Leismann H, Mattay J, Scharf H-D. J Am Chem Soc 1984; 106:3985–3991; (d) Mattay J. Tetrahedron 1985; 41:2393–2404; (e) Mattay J. Tetrahedron 1985; 41:2405–2417.
11. (a) Rehm D, Weller A. Isr J Chem 1970; 8:259–271; (b) Weller A. Z Physik Chem NF (Munich) 1982; 133:93–98.

12. Müller F, Mattay J. Chem Rev 1993; 93:99–117.
13. (a) Bryce-Smith D, Gilbert A. Tetrahedron 1976; 32:1309–1326; (b) Bryce-Smith D, Gilbert A. Tetrahedron 1977; 33:2459–2489; (c) Houk KN. Pure & Appl Chem 1982; 54:1633–1650; (d) van der Hart JA, Mulder JJC, Cornelisse J. J Mol Struct (Theochem) 1987; 151:1–10; (e) van der Hart JA, Mulder JJC, Cornelisse J. J Photochem Photobiol A: Chem 1991; 61:3–13; (f) Stehouwer AM, van der Hart JA, Mulder JJC, Cornelisse J. J Mol Struct (Theochem) 1992; 92:333–338; (g) van der Hart JA, Mulder JJC, Cornelisse J. J Photochem Photobiol A: Chem 1995; 86:141–148.
14. Clifford S, Bearpark MJ, Bernardi F, Olivucci M, Robb MA, Smith BR. J Am Chem Soc 1996; 118:7353–7360.
15. De Keukeleire D, He S-L, Blakemore D, Gilbert A. J Photochem Photobiol A: Chem 1994; 80:233–240.
16. Avent AG, Byrne PW, Penkett CS. Org Lett 1999; 1:2073–2075.
17. Hoffmann N, Pete J-P. J Org Chem 1997; 62:6952–6960.
18. (a) De Vaal P, Osselton EM, Krijnen ES, Lodder G, Cornelisse J. Recl Trav Chim Pays-Bas 1988; 107:407–411. See also: (b) Cornelisse J, Merritt VY, Srinivasan S. J Am Chem Soc 1973; 95:6197–6203; (c) Bryce-Smith D, Dadson WM, Gilbert A, Orger BH, Tyrrell HM. Tetrahedron Lett 1978, 1093–1096; (d) Ors JA, Srinivasan R. J Org Chem 1977; 42:1321–1327; (e) Jans AWH, Van Dijk-Knepper JJ, Cornelisse J. Recl Trav Chim Pays-Bas 1982; 101:275–276.
19. Blakemore DC, Gilbert A. Tetrahedron Lett 1994; 35:5267–5270.
20. Amey DM, Blakemore DC, Drew MGB, Gilbert A, Heath P. J Photochem Photobiol A: Chem 1997; 102:173–178.
21. Also compare: Nuss JM, Chinn JP, Murphy MM. J Am Chem Soc 1995; 117:6801–6802.
22. Crimmins MT, Reinhold TL. Organic Reactions 1993; 44:297–588.
23. Ohashi M, Tanaka Y, Yamada S. Tetrahedron Lett 1977; 3629–3632.
24. Al-Qaradawi SY, Cosstick KB, Gilbert A. J Chem Soc, Perkin Trans 1 1992; 1145–1148.
25. Gilbert A, Cosstick KB. 11-Cyano-4-oxatricyclo[7.2.0.0^{3,7}]undeca-2,10-diene. In: Mattay J, Griesbeck AG, eds. Photochemical Key Steps in Organic Synthesis. Weinheim: VCH, 1994:175–176.
26. Wagner PJ, Lee J-I. Tetrahedron Lett 2002; 43:3569–3571.
27. Hoffmann N, Pete J-P. Synthesis 2001; 1236–1242.
28. Wagner PJ, Sakamoto M, Madkour AE. J Am Chem Soc 1992; 114:7298–7299.
29. Wagner PJ. Acc Chem Res 2001; 34:1–8.
30. Al-Qaradawi S, Gilbert A, Jones DT. Recl Trav Chim Pays-Bas 1995; 114:485–491.
31. Vizvárdi K, Toppet S, Hoornaert GJ, De Keukeleire D, Bakó P, Van der Eycken E. J Photochem Photobiol A: Chem 2000; 133:135–146.
32. De Keukeleire D. Aldrichimica Acta 1994; 27:59–69.
33. Gilbert A, Taylor GN, Bin Samsudin MW. J Chem Soc, Perkin Trans 1 1980; 869–876.

34. (a) Noh T, Kim D, Kim Y-J. J Org Chem 1998; 63:1212–1216. See also: (b) Albini A, Giavarini F. J Org Chem 1988; 53:5601–5607.
35. Yoshimi Y, Konishi S, Maeda H, Mizuno K. Synthesis 2001; 1197–1202.
36. Ito Y, Horie S, Shindo Y. Org Lett 2001; 3:2411–2413.
37. Kishikawa K, Akimoto S, Kohmoto S, Yamamoto M, Yamada K. J Chem Soc, Perkin Trans 1 1997; 77–84.
38. Döpp D, Krüger C, Memarian HR, Tsay Y-H. Angw Chem Int Ed Engl 1985; 24:1048–1049.
39. (a) Döpp D, Memarian HR, Krüger C, Raabe E. Chem Ber 1989; 122:585–588; (b) Memarian HR, Nasr-Esfahani M, Boese R, Döpp D. Liebigs Ann/Recueil 1997; 1023–1027.
40. Ohkura K, Sugaoi T, Nishijiama K, Kuge Y, Seki K. Tetrahedron Lett 2002; 43:3113–3115.
41. Mattay J, Runsink J, Piccirilli JA, Jans AWH, Cornelisse J. J Chem Soc, Perkin Trans 1 1987; 15–20.
42. Mattay J, Runsink J, Gersdorf J, Rumbach T, Ly C. Hev Chim Acta 1986; 69:442–455.
43. (a) Al-Jalal N, Drew MGB, Gilbert A. Chem Commun 1985; 85–86; (b) Al-Jalal N, Heath P. Tetrahedron 1988; 44:1449–1459.
44. (a) Timmermans JL, Wamelink MP, Lodder G, Cornelisse J. Eur J Org Chem 1999; 463–470. See also: (b) Barentsen HM, Sieval AB, Cornelisse J. Tetrahedron 1995; 51:7495–7520.
45. (a) Hoffmann RW. Chem Rev 1989; 89:1841–1860; (b) Broeker JL, Hoffmann RW, Houk KN. J Am Chem Soc 1991; 113:5006–5017.
46. Wender PA, Singh SK. Tetrahedron Lett 1990; 31:2517–2520.
47. Sugimura T, Nishiyama N, Tai A, Hakushi T. Tetrahedron Asymm 1994; 5:1163–1166.
48. Vizvardi K, Desmet K, Luyten I, Sandra P, Hoornaert G, Van der Eycken E. Org Lett 2001; 3:1173–1175.
49. Hoffmann N, Pete J-P. Tetrahedron Lett 1996; 37:2027–2030.
50. Hoffmann N, Pete J-P. Tetrahedron Lett 1995; 36:2623–2626.
51. Hoffmann N, Pete J-P. Tetrahedron Lett 1998; 39:5027–5030.
52. (a) Kalena GP, Pradhan PP, Banerji A, Tetrahedron Lett 1992; 33:7775–7778; (b) Kalena GP, Pradhan PP, Swaranlatha Y, Singh TP, Banerji A. Tetrahedron Lett 1997; 38:5551–5554; (c) Kalena GP, Pradhan P, Banerji A. Tetrahedron 1999; 55:3209–3218.
53. Hoffmann N, Pete J-P, Inoue Y, Mori T. J Org Chem 2002; 67:2315–2322.
54. Hoffmann N. Tetrahedron 2002; 58:7933–7941.
55. (a) Ohkura K, Kanazashi N, Seki K. Chem Pharm Bull 1993; 41:239–243; (b) Ohkura K, Seki K, Hiramatsu H, Aoe K, Terashima M. Heterocycles 1997; 44:467–478; (c) Ohkura K, Noguchi Y, Seki K. Heterocycles 1998; 47:429–437; (d) Ohkura K, Nishijima K, Sakushima A, Seki K. Heterocycles 2000; 53: 1247–1250.
56. Wagner PJ, McMahon K. J Am Chem Soc 1994; 116:10827–10828.
57. Döpp D, Pies M. J Chem Soc, Chem Commun 1987; 1734–1735.

58. For the control of the diastereoselectivity in multistep reactions see: (a) Buschmann H, Scharf H-D, Hoffmann N, Esser P. Angew Chem Intern Ed Engl 1991; 30:477–515; (b) Cainelli G, Giacomini D, Galletti P. Chem Commun 1999; 567–572.

59. Döpp D. Photocycloaddition with captodative alkenes. In: Ramamurthy V, Schanze KS, eds. Molecular and Supramolecular Photochemistry. Vol. 6. New York: Marcel Dekker, Inc., 2000:101–148.

60. De Keukeleire D. *exo*-2-(2-Hydroxyethyl)bicyclo[3.2.1]oct-6-en-8-one. In: Mattay J, Griesbeck AG, eds. Photochemical Key Steps in Organic Synthesis. Weinheim: VCH, 1994:178–180.

61. Wender PA, Howbert JJ. J Am Chem Soc 1981; 103:688–690.

62. Wender PA, Howbert JJ, Dore TM. The total synthesis of (+/−)-α-cedrene. In: Mattay J, Griesbeck AG, eds. Photochemical Key Steps in Organic Synthesis. Weinheim: VCH, 1994:181–185.

63. Gilbert A, Al-Jalal N. 1-Cyano-4-methoxybenzocyclobutene. In: Mattay J, Griesbeck AG, eds. Photochemical Key Steps in Organic Synthesis. Weinheim: VCH, 1994:173–174.

64. Döpp D, Bredehorn J, Memarian H-R, Mühlbacher B, Weber J. *rel*-(1*R*,4*R*)-1-Acetyl-1,4-dihydro-1,4-ethano-naphthalene-9-one. In: Mattay J, Griesbeck AG, eds. Photochemical Key Steps in Organic Synthesis. Weinheim: VCH, 1994:186–188.

65. (a) Verrat C, Hoffmann N, Pete JP. Synlett 2000; 1166–1168; (b) Verrat C. Ph.D. dissertation, Université de Reims Champagne-Ardenne, 2000.

66. Wender PA, Ternansky R, deLong M, Singh S, Olivero A, Rice K. Pure & Appl Chem 1990; 62:1597–1602.

67. Amey DM, Gilbert A, Jones DT. J Chem Soc, Perkin Trans 2 1998; 213–218.

68. Wender PA, Dreyer GB. Tetrahedron 1981; 37:4445–4450.

69. Wender PA, Ternansky RJ. Tetrahedron Lett 1985; 26:2625–2628.

70. Wender PA, Singh SK, Tetrahedron Lett 1985; 26:5987–5990.

71. Wender PA, deLong M, Tetrahedron Lett 1990; 31:5429–5432.

72. Wender PA, Dore TM, Tetrahedron Lett 1998; 39:8589–8592.

73. Wender PA, Singh SK, Tetrahedron Lett 1990; 31:2517–2520.

74. Wender PA, Howbert JJ, Tetrahedron Lett 1982; 23:3983–3986.

75. Wender PA, Howbert JJ, Tetrahedron Lett 1983; 24:5325–5328.

76. Baralotto C, Chanon M, Julliard M. J Org Chem 1996; 61:3576–3577.

77. Wender PA, Fisher K, Tetrahedron Lett 1986; 27:1857–1860.

78. Wender PA, Dore TM, deLong MA. Tetrahedron Lett 1996; 37:7687–7690.

79. Mani J, Keese B. Tetrahedron 1985; 41:5697–5701.

80. Mani J, Schüttel S, Zhang C, Bigler P, Müller C, Keese R. Helv Chim Acta 1998; 72:487–495; See also: Zhang C, Bourgin D, Keese R. Tetrahedron 1991; 47:3059–3074.

81. Wender PA, Dreyer GB. J Am Chem Soc 1982; 104:5805–5807.

82. Metha G, Pramod K, Subrahmanyam D. J Chem Soc, Chem Commun 1986; 247–248.

83. Wender PA, Dreyer GB. Tetrahedron Lett 1983; 24:4543–4546.

18

Medium Effects on Photochemical Processes: Organized and Confined Media

Lakshmi S. Kaanumalle, Arunkumar Natarajan, and V. Ramamurthy
Tulane University, New Orleans, Louisiana, U.S.A.

18.1 INTRODUCTION

Chemists have long recognized the important role the reaction media (solvents with varying characteristics) play in controlling the rates, product distributions and stereochemistry of organic reactions. Recently, much effort has been directed toward the use of organized media to modify the photochemical reactivities achieved with isotropic liquids [1,2]. A principal goal of such studies is to utilize the order of the medium to increase the rate and selectivity of a chemical process in much the same way that enzymes modify the reactivity of the substrates to which they are bound. Notable among the many ordered or constrained systems are micelles, microemulsions, liquid crystals, inclusion complexes, monolayers and solid phases such as porous solids (silica, alumina, clay, and zeolites) and crystals [3–77]. The differences in chemical reactivities occurring in ordered media as compared to isotropic solvent phases are largely due to the physical restraints imposed by the environment. These include preventing large amplitude conformational, configurational and translational changes along the reaction coordinate, and modifying diffusional characteristics of reactants and facilitating alternate reaction pathways. The intrinsic reactivity of a molecule in an ordered medium is frequently of secondary importance in comparison to features such as site symmetry, nearest neighbor separation, and other geometric considerations.

This chapter deals with several organized media, each with its own characteristics. To appreciate the effect of organized and constraining media on photoreactions, it is essential to have some knowledge of their physical characteristics [71–73,78–88]. For a detailed understanding of the structures and characteristics of various media one should consult the large number of monographs and reviews available on this topic.

18.2. PHOTOCHEMISTRY OF VISION–PROTEIN AS A CONFINING MEDIUM

The effect of an organized medium on photoreactions is nicely illustrated by the photoprocess that occurs in our eye [89–98]. The key element in the vision process is rhodopsin, a membrane protein present in the outer segments of rod cells in the retina. Rhodopsin is made up of the protein opsin and the co-factor 11-*cis* retinal. The latter is held in place by formation of a protonated Schiff base with the lysine residue of opsin. Rhodopsin is an excellent switch that converts a light signal into the electrical response of the photoreceptor cells. The rhodopsin in the retina captures light to trigger a biochemical cascade that the brain interprets as sight. The overall process is shown in Fig. 1. Absorption of a photon by protonated 11-*cis* retinyl Schiff base causes its isomerization to the *all-trans* form leading to a conformational change of the protein moiety and deprotonation of the iminium ion to the *all-trans* Schiff base. This process causes activation of transducin which through a series of steps transmits a visual signal in the brain.

Because the event that is of interest to us is the selective photoisomerization of the protonated 11-*cis* retinyl Schiff base to the *all trans* isomer we will concentrate only on that. Clearly the three dimensional structure of the protein and the shape of the binding cavity are crucial to understanding the vision process. The X-ray crystal structure of dark-adapted rhodopsin reveals a highly organized heptahelical transmembrane bundle with 11-*cis* retinal as a key co-factor (Fig. 2). The 11-*cis* chromophore is maintained in the 6*s-cis*, 12*s-trans*, *anti* C=N conformation. The selectivity of the binding site of opsin is reflected in the relative rates of pigment formation, which favor the native 11-*cis* retinal and those isomers and analogs closest in shape. This aside, it is clear that specific protein-retinal interactions are present as revealed by the strong circular dichroism activity of the retinyl chromophore when bound to opsin, in spite of the absence of a single chiral center in the chromophore itself. Thus the protein offers a unique "reaction environment" and maintains the retinyl chromophore in the specific

Figure 1 Crystal structure of rhodopsin at 2.8 Å resolution (Palczewski K, Kumasaka T, Hori T, Behnke CA, Motoshima H, Fox BA, Trong IL, Teller DC, Okada T, Stenkamp RE, Yamamoto M, Miyano M. Science 2000; 289:739–745).

conformation required to trigger a complex sequence of events after the absorption of a photon.

There are a number of unique aspects of the photobehavior of the retinyl Schiff base bound to a protein. First, the isomerization is specific; although there are several isomerizable C=C bonds, 11-*cis* retinal is isomerized specifically to the *all trans* form (only the C_{11}–C_{12} double bond isomerizes from *cis* to *trans* geometry). Second, the isomerization quantum yield is high (0.67) compared to that in isotropic solution (~0.15). Third, the rotation around the C_{11}–C_{12} double bond is initiated within 200 femtoseconds (fs) of light absorption making this one of the fastest known photoreactions. The slow relaxation of the protein surrounding the retinyl chromophore allows low temperature characterization of various intermediates during the isomerization of 11-*cis* retinal to the *all trans* geometry.

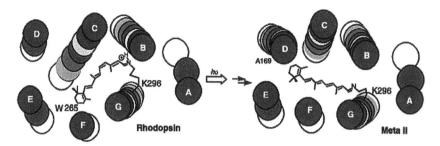

Figure 2 Photoconversion of 11-*cis* to *all trans* rhodopsin and associated protein movement. (Borhan B, Souto Ml, Imai H, Shichida Y, Nakanishi K. Science 2000; 288:2209–2212).

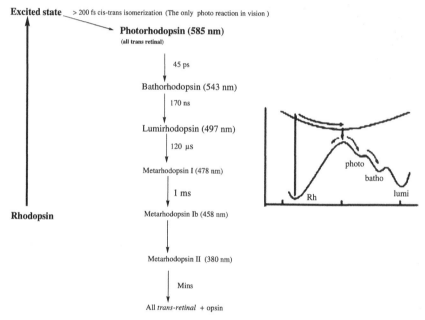

Figure 3 Intermediates during the photoconversion of 11-*cis* to *all trans* rhodopsin.

The sequence of events involved in the vision process are listed in Fig. 3. The overall process triggered by the 11-*cis* retinyl chromophore is very specific to the protein medium. Through *cis–trans* isomerization of the chromophore, light energy is transduced into chemical free energy, which in turn is utilized to cause conformational changes in the protein and ultimately activate the retinal G-protein. The important role of the protein environment, both in terms of specific electronic interactions and certain spatial constraints, is revealed from the above discussion.

18.3. A COMPARISON OF PROTEIN MATRIX AND ISOTROPIC SOLUTION AS REACTION MEDIA

The example of vision demonstrates the profound influence of a protein matrix on the photochemistry of its constituent cofactor (guest molecule). This occurs by stabilization of unstable conformers and strained geometries and by fixation of the relative arrangements of systems of co-factors and generation of contacts between co-factors. Although the complexity of the structure of the protein precludes their use in "everyday laboratory control" of photoreactions, the lessons learned from the example of vision (and photosynthesis) are useful in designing media that provide better control of photoreactions than that obtained in isotropic solution. Let us compare the site (termed the reaction cavity) at which the reaction occurs in a protein and an isotropic solution medium.

The space available to a reactant molecule in an isotropic solution is restricted only by the size of the flask in which the reaction is carried out and by the diffusional characteristics and excited state lifetime of the reactant. On the other hand, in a protein the space available for a reactant is highly restricted and may in fact be only as large as the reacting molecule. Secondly, unlike the situation in isotropic solution, in a protein the reactant molecules are preorganized, thus minimizing entropic requirements for reaction. Thirdly in a protein the relatively small free space surrounding the reactant and the slow relaxation of the immediate surroundings allow only limited shape changes of the reactant compared to solution media.

From the descriptions above one may conclude that selectivity in photochemical reactions could be achieved if a "reaction cavity" resembling that of a protein could be created. Although the reaction cavity provided by a protein has not been duplicated in the laboratory, the reaction cavities that are much more confined, ordered and organized than the one offered by an isotropic solution have been exploited to achieve selectivity in photochemical reactions. A few examples are highlighted in the following section.

18.4. ORGANIZED MEDIA MIMIC A PROTEIN: SELECTED EXAMPLES

Trans-cinnamic acid upon irradiation in solution yields the *cis* isomer with a quantum yield of 0.62. On the other hand, irradiation of the crystalline

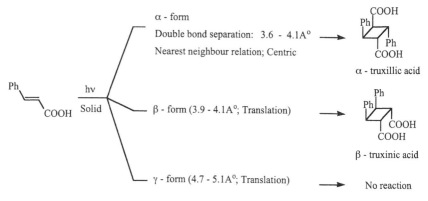

Scheme 1

material leads to [2 + 2] dimerization rather than isomerization [99,100]. In fact, substituted *trans*-cinnamic acids can be crystallized in three polymorphs and each one shows a different behavior (Sch. 1). Extensive studies by Schmidt have resulted in a set of "topochemical rules" which relate the configuration of the product and the crystal structure of the reactant [101].

1. The product formed in the crystalline state is governed by the environment rather than by the intrinsic reactivity of the individual molecules.
2. The proximity and the degree of parallelism of the reacting double bonds are crucial for dimerization. Generally, the reactive double bonds are expected to be within 4.2 Å and be parallel to each other.
3. There is a one-to-one relationship between the configuration and symmetry of the product with the symmetry between the reactants in the crystal.

Figure 4 provides molecular packing of α and β forms of cinnamic acids. The packing arrangement nicely accounts for the observed dimers from α and β-forms of the crystal. The above topochemical postulates rationalize the absence of reactions in the γ-modification (distance of separation between C=C bonds is over 4.7 Å).

The solid state photochemistry of *trans*-cinnamic acid illustrates the similarity in the nature of the control exerted by the crystal and the protein. In the crystal as in the protein the reactant molecules are preorganized. The photobehavior of *trans*-1,2-*bis*(4-pyridyl)ethylene further highlights this analogy [102]. *Trans*-1,2-*bis*(4-pyridyl)ethylene upon irradiation in solution,

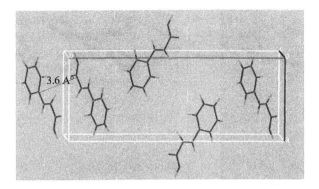

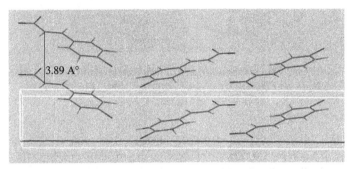

Figure 4 Packing arrangment of *trans*-cinnamic acid (top) and *para*-chloro-*trans*-cinnamic acid (bottom). The top packing represents the α-form and the bottom one the β-form.

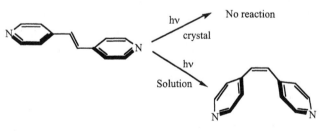

Scheme 2

not surprisingly, undergoes *cis–trans* isomerization. Irradiation of crystals of *trans*-1,2-*bis*(4-pyridyl)ethylene does not give any products (Sch. 2). The absence of dimers from *trans*-1,2-*bis*(4-pyridyl)ethylene is surprising given the fact that *trans*-cinnamic acid dimerizes in the crystalline state.

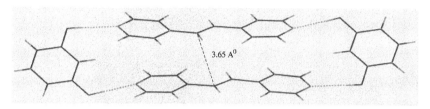

Template   Reactants   Template   Product

Scheme 3

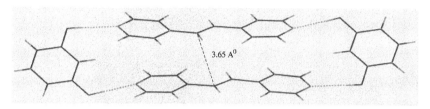

3.65 A⁰

Figure 5   Packing arrangment of resorcinol-*trans*-1,2-*bis*(4-pyridyl)ethylene complex. (Macgillivray LR, Reid JL, Ripmeester JA. J Am Chem Soc 2000; 122:7817–7818.)

However, scrutiny of the crystal structure reveals that *trans*-1,2-*bis*(4-pyridyl)ethylene molecules crystallize in a layered structure in which olefins of neighboring molecules are separated by more than 6.52 Å. Such a large distance does not allow dimerization. This molecule can be engineered to dimerize in the crystalline state if it is co-crystallized in the presence of 1,3-dihydroxybenzene (Sch. 3). Irradiation of a mixture of 1,3-dihydroxybenzene and *trans*-1,2-*bis*(4-pyridyl)ethylene in solution resulted only in geometric isomerization. On the other hand, in the crystalline state a single photodimer was obtained in quantitative yield. In this case hydrogen bonding between 1,3-dihydroxybenzene and *trans*-1,2-*bis*(4-pyridyl)ethylene keeps the olefins parallel to each other and within 4.2 Å, thus facilitating the dimerization process (Fig. 5).

Another example of the effect of confined medium is found during photo-Fries rearrangement of naphthyl esters in zeolites [103,104]. Upon photolysis in isotropic solution 1-naphthyl benzoate undergoes the photo-Fries rearrangement to yield both *ortho* (2-) and *para* (4-) phenolic ketones (Sch. 4). When this ester is included in NaY zeolite and irradiated the main product (96%) is the ortho isomer. This remarkable "ortho-selectivity" within zeolites has been rationalized on the basis of interactions of the reactant 1-naphthyl benzoate and intermediate radicals with the sodium ion. Due to restrictions imposed by the medium the benzoyl radical, once formed, is compelled to react only with the accessible ortho position.

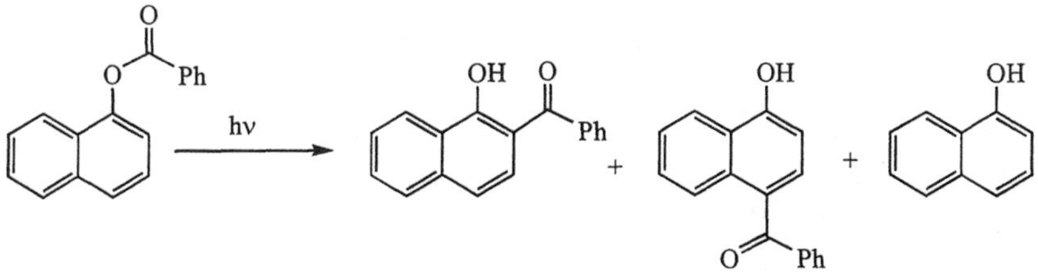

Hexane	70	26	4
NaY	96	1	10

Scheme 4

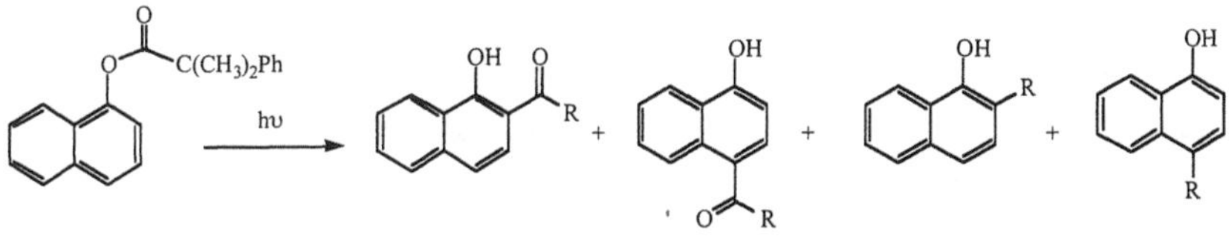

Hexane	28%	15%	9%	14%
NaY zeolite	99%	---	---	---

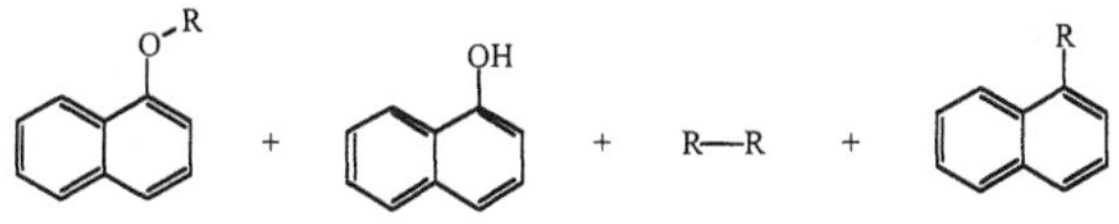

Hexane	--	13%	15%	6%
NaY zeolite	---	---	---	---

R = -C(CH₃)₂Ph

Scheme 5

An impressive illustration of the influence of the medium on the photo-Fries reaction product distribution can be found in Sch. 5 where the products of photolysis of 1-naphthyl 2-methyl 2-phenyl propanoate in hexane solution and within NaY are compared. Remarkably, whereas in solution eight products are formed, within NaY zeolite a single product dominates the product mixture. Cations present within zeolites help to anchor the reactants, intermediates and products to the surfaces of the zeolite and

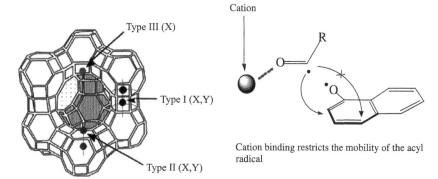

Figure 6 Left: Structure of the supercage and location of cations in X and Y zeolites. Right: A cartoon representation of the proposed interaction of cation with the radical intermediates.

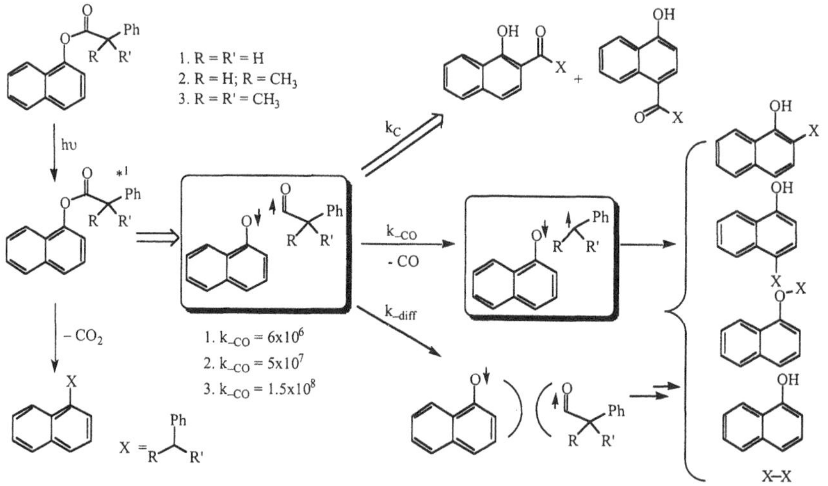

Scheme 6

control the reactivity of the intermediates (Fig. 6). In this case, the cation binding not only forced the phenyl acyl radical to react at the ortho position preferentially but also prevented it from decarbonylating (Sch. 6).

Upon irradiation in acetonitrile/aqueous solution, of 1-cyanonaptha-lene in presence of *all trans*-farnesyl acetate, the latter undergoes only *cis–trans* isomerization via an electron transfer process. No products from any other reactions were isolated. However, when the same combination of

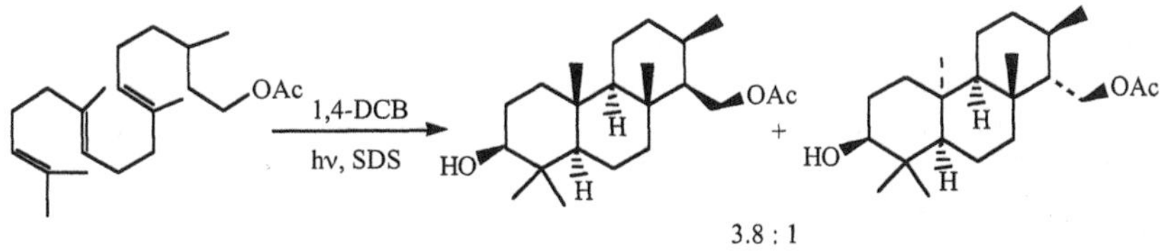
OAc
1,4-DCB
hν, SDS
HO
H
OAc
+
3.8 : 1

Scheme 7

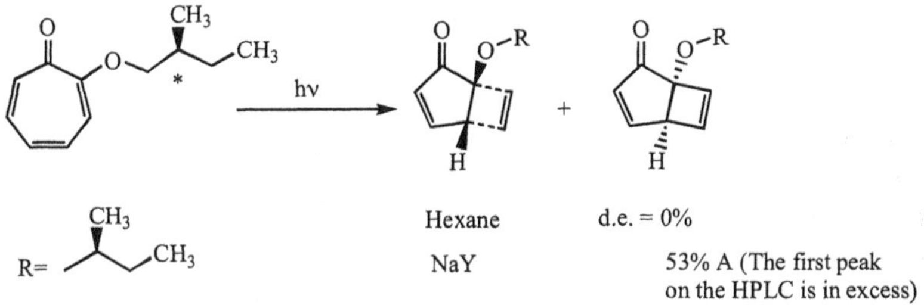
CH3
O
CH3
hν
O-R
H
+
R=
Hexane
d.e. = 0%
NaY
53% A (The first peak
on the HPLC is in excess)

Scheme 8

molecules were irradiated in sodium dodecyl sulfate micellar solution the main product was the result of intramolecular cyclization (Sch. 7) [105]. The cyclization is stereo- and regioselective to yield a trans fused product in which the water molecule has added stereoselectively. Such a striking selective cyclization occurs only in a micellar medium.

Irradiation of (*S*)-tropolone 2-methyl butyl ether in solution yields a 4-electron electrocyclization product as a 1:1 diastereomeric mixture (Sch. 8) [106]. In solution the presence of the chiral auxiliary in proximity to the reactive center has no influence on the product stereochemistry. When irradiated within NaY zeolite, however, the same molecule affords the cyclized product in ~53% diastereomeric excess. The restricted space of the zeolite supercage apparently forces communication between the chiral center and the reaction site.

From the above examples it is clear that reaction cavity provided by an organized or confining medium has unique features that mimics some of the features of proteins. While crystals and zeolites provide reaction cavities that are inflexible, there is a whole spectrum of organized and confined media (e.g., micelles, host-guest complexes, monolayers and bilayers, liquid crystals etc.,) that allow different degrees of freedom to the reactant molecules. These systems demonstrate clever usage of favorable entropy that is so important in natural systems. One should keep in mind that with a chemist-chosen organized medium although one may not reach the remarkable selectivity observed with a protein, it is often possible to achieve product selectivity far greater than that observed in an isotropic solution. The methods developed by this approach are simple, cost-effective and general. An understanding of the unique features of each organized medium should enable one to choose the appropriate medium to achieve the desired selectivity.

18.5. REACTION CAVITY CONCEPT

The term "reaction cavity" was originally used by Cohen to describe reactions in crystals [107]. He identified the reaction cavity as the space occupied by the reacting partners in crystals and used this model to provide a deeper understanding of the topochemical control of their reactions. Selectivity seen for reactions in crystals, according to this model, arises due to lattice restraints on the motions of the atoms in reactant molecules within the reaction cavity. In other words, severe distortion of the reaction cavity will not be tolerated and only reactions that proceed without much distortion of the cavity are allowed in a crystal (Fig. 7). Crystals possess time independent structures; the atoms that form the walls of the reaction cavity are fairly rigid and exhibit only limited motions (e.g., lattice vibrational modes) during the time periods necessary to convert excited state molecules to their photoproducts. Therefore, in the Cohen model, the space required to accommodate the displacement of reactant atoms from their original positions during a chemical reaction must be built largely into the reaction cavity. In this section the characteristics of a reaction cavity with some modifications as applied to various confined media are discussed [108,109].

A chemical reaction can be viewed as a phenomenon dealing with molecular shape (topology) changes. Whether a particular reaction will take place will depend upon whether the product can fit within the space occupied by the reactant. The space occupied by the reactant is the reaction cavity. Since the boundaries of a reaction cavity are undefined in an isotropic solution, size matching of the reactant, products and the reaction cavity is not important in this medium. On the other hand, when the reaction cavity has a well-defined boundary, as in most organized assemblies (especially in solid state), size matching can become important and occasionally may even become the sole factor controlling the feasibility of a reaction (Fig. 8).

The reaction cavity concept, which emphasizes the shape changes that occur as the reactant guest is transformed into the product, is generally

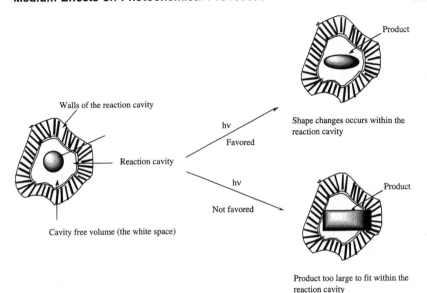

Figure 7 Reaction cavity concept illustrated: cavity free volume controls the product formation.

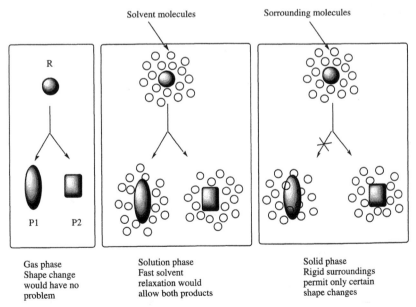

Figure 8 A comparison of reaction cavity in gas phase, solution and solid state.

useful for understanding and predicting the photobehavior of guest molecules included in organized media. The reaction cavity as applied to most organized media possesses the following features: (a) A reaction cavity is an enclosure that reduces the mobility of the reactant molecules and provides a boundary which the reactant molecules may not cross without overcoming an energy barrier. (b) The size and shape of reaction cavities among organized media may vary. (c) The free volume within a reaction cavity is an important parameter, whose shape, size, location, directionality, and dynamics control in large part the extent to which the medium influences a photoreaction. (d) When the atoms/molecules constituting the walls of the reaction cavity are stationary, the free volume necessary to convert a guest molecule to its photoproducts must be built into the reaction cavity. On the other hand, in systems where the walls are flexible, the free volume may become available during the course of a reaction. For these media, the free volume in a reaction cavity, is modified by structural fluctuations of the medium and cannot be readily represented by static molecular models (Fig. 9). (e) The reaction cavity may contain specific

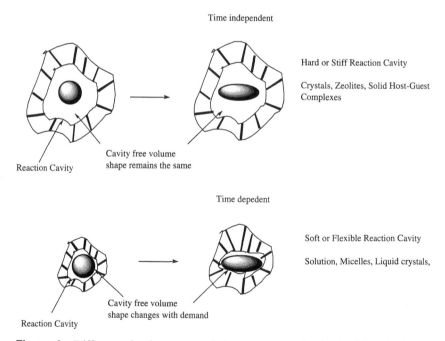

Figure 9 Difference in the nature of the reaction cavity in "soft" and "hard" reaction media. Top: Hard reaction cavity, cavity free volume fixed. Bottom: Soft reaction cavity, flexible cavity free volume.

Passive Reaction Cavity
Guest-cavity interactions very weak and non-directional

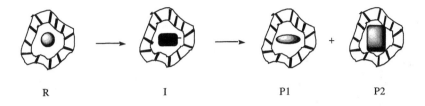

Active Reaction Cavity
Guest-cavity interactions strong and directional

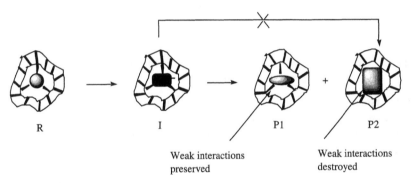

Figure 10 Representation of passive and active reaction cavities.

functional groups or atoms that may interact strongly (attractively or repulsively) with guest molecules, the transition state or the intermediates as the guest proceeds to products. Such specific interactions may lead to unique product selectivity and enhance or decrease the quantum yields for reactions (Fig. 10). (f) Unlike isotropic fluid solutions, organized media may contain more than one type of reaction cavity in which guest molecules reside. If interchange of molecules among different types of reaction cavities is slow on the time scale of excited state processes, prediction of quantum yields and product distributions requires detailed structural and dynamic knowledge of the system, especially the nature and relative abundance of the different reaction cavities. The above features are common to most organized media. In the following sections each of the above properties of an organized medium is discussed with examples.

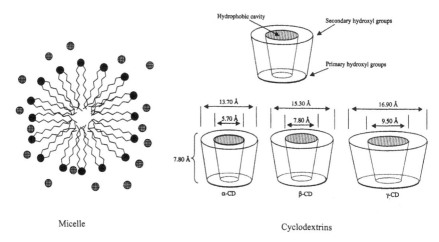

Micelle   Cyclodextrins

Figure 11   Structure of micelles and cyclodextrins.

18.5.1.   Characteristics of a Reaction Cavity Illustrated with Examples: Boundary

Unlike isotropic media, where molecules have equal mobility and con-formational flexibility in all dimensions, in an organized medium their mobility and flexibility are restricted or constrained in at least one dimension. For example, the reaction cavities of a micelle and cyclodextrins are made up of a hydrophobic core and a hydrophilic exterior (Fig. 11). A highly polar boundary separates the hydrophobic core from the aqueous exterior. Such a boundary provides unique features to these media that are absent in isotropic solution. Translational motion of a guest present within the reaction cavity is hindered by the well-defined boundary.

The photochemistry of dibenzylketones (DBK) within micelles nicely illustrates how the characteristics of a reaction cavity defined by a boundary (and an interface) can be exploited [11,110–121]. Photolysis of 3-(4-methylphenyl)-1-phenylacetone (MeDBK) results in a 1:2:1 mixture of three products (AA, BB, AB) as shown in Sch. 9. Photolysis of the same molecule solubilized in a micelle (hexadecyltrimethylammonium chloride) gives a single product (AB). This remarkable change in product distribution is due to the cage effect provided by the micellar structure. The plot of the percentage cage effect $[AB - AA - BB]/AA + BB + AB]$ vs. surfactant concentration clearly indicates the importance of micelles in altering the course of the photoreaction (Fig. 12). The change in product distribution occurs at or above the critical micelle concentration. The selective formation of AB in a micelle is consistent with the entrapment of the radical

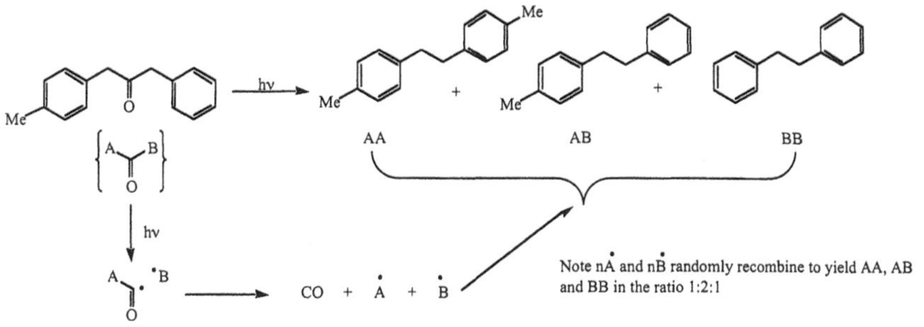

Note nA and nB randomly recombine to yield AA, AB and BB in the ratio 1:2:1

Scheme 9

intermediates in a small hydrophobic reaction cavity (Sch. 10). One of the differences between an isotropic solution and the micelle is the size of the reaction cavity. In principle, when a reaction is carried out in an isotropic medium, the excited molecule, has access to the entire volume of the solution in the flask. On the other hand, when the reaction is carried out in a micellar medium, the molecule may be confined to a single micelle during the time required for a reaction. The space available for the reacting molecule has thus shrunk from a large beaker to a small nano-beaker (a single micelle). In isotropic solution the A° and B° radicals generated from DBK have unrestricted translational mobility (depending on the viscosity of the solution), enabling them to rapidly diffuse apart and find partners based only on a statistical probability (AA, AB, BB). On the other hand, in a micellar medium, at low occupancy levels, only one pair of A° and B° radicals is likely to be generated within a single micelle. The movement of radicals generated within a single micelle is restricted by the hydrophobic-hydrophilic boundary and they are forced to find a partner within the same micelle (AB), thus resulting in coupling of the A and B radicals.

The increase in the cage effect with the increase in the number of methyl groups is consistent with the hydrophobicity of the secondary radical pair (A° and B°). For example, the cage effect for DBK is 31% whereas that for 4,4'-di-*tert*-butyl DBK is 95%. The results of photolysis of MeDBK in detergents of varying chain length between 6 and 14 shown in Fig. 13 suggest that the rate of radical escape from a micellar cage is inversely proportional to the size of the micelle. The larger the micelle, the greater is the retention of the radical pair. The estimated exit rates of the benzyl radical from micelles of different sizes (sodium decyl sulfate: $2.7 \times 10^6 \, s^{-1}$; sodium dodecyl sulfate: $1.8 \times 10^6 \, s^{-1}$ and sodium tetradecyl sulfate: $1.2 \times 10^6 \, s^{-1}$) correlate well with the observed trends in the cage effect. Thus, the extent to which the radicals (A and B) can escape the micellar reaction cavity as

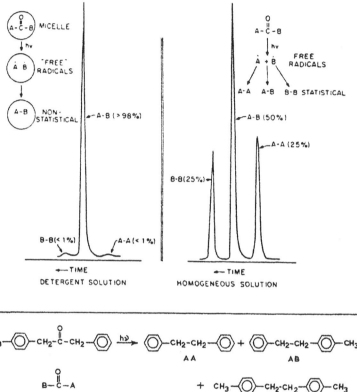

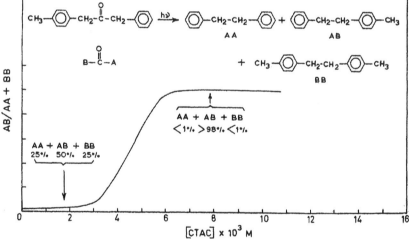

Figure 12 Cage effect during the photolysis of dibenzyl ketone. Top: GC traces of products in micelles and solution. Bottom: Cage effect with respect to the detergent concentration. Note the sudden change in the cage effect at the cmc.

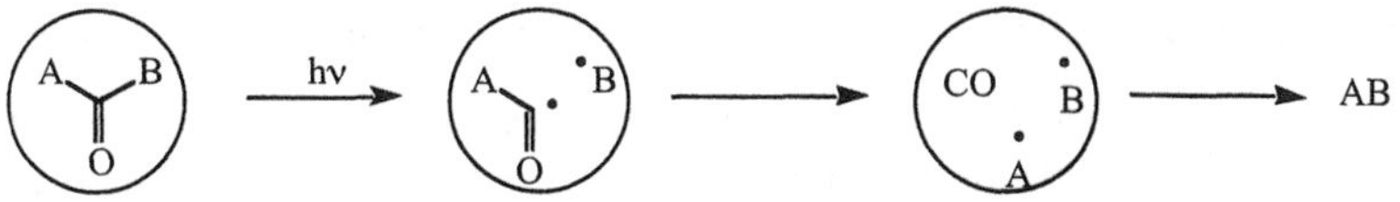

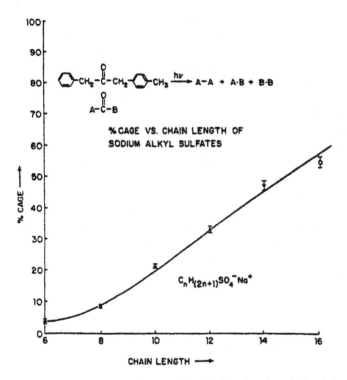

$$\text{Cage effect} = \frac{AB - (AA + BB)}{(AA + AB + BB)}$$

Scheme 10

Figure 13 Variation of cage effect during the photolysis of dibenzyl ketone with respect to detergent chain length.

estimated experimentally by measuring the cage effect depends upon both the size of the reaction cavity (micellar size) and the hydrophobicity of the radical pair. The more hydrophobic the radical, less it would venture outside the micellar reaction cavity. The larger the micelle, the less the radical pair escapes the reaction cavity.

18.5.2. Characteristics of a Reaction Cavity: Interface

In some organized media the boundary of the reaction cavity may separate two phases of very different properties. For example the boundary of a micelle and cyclodextrins separates the nonpolar hydrocarbon interior from the polar aqueous exterior (Fig. 11). These interfacial boundaries allow one to expose a reactant molecule simultaneously to two phases with different properties. This works best when the reactant molecule has both hydrophobic and hydrophilic components. For example, consider a molecule with nonpolar hydrocarbon part and polar functional groups such as OH, COOH, and CO. The hydrocarbon part of the molecule is hydrophobic and the OH, COOH, or CO portion is hydrophilic. When solubilized in water with the help of a micelle or cyclodextrin, such molecules will adopt a specific arrangement or conformation with the hydrophobic part in the hydrocarbon like phase and the hydrophilic part exposed to the aqueous phase. The alignment of the reactants at the interface might restrict the randomness in their orientation and hence might lead to regioselectivity during bimolecular reactions. The conformation of the molecules at the interface may also be different from that in the bulk homogeneous phase. Under such conditions it may be possible to initiate reactions that wouldn't otherwise be possible. Such examples are discussed in this section. Micelles, vesicles, cyclodextrins, and monolayers are a few media that allow exploitation of the interface.

Photocycloaddition of 3-n-butylcyclopentanone in the presence of 1-octene in a potassium decanoate micelle yields the the head-to-head photoadduct as the major product, consistent with the notion that a micellar interface helps to orient the reactant molecules (Fig. 14). When the orientation of the reactive olefin is further controlled by adding a hydrophilic group,

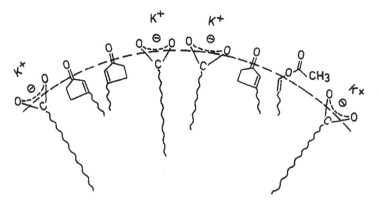

Figure 14 A cartoon representation of alignment of enones and enes with long chain substituents on a micellar-water interface.

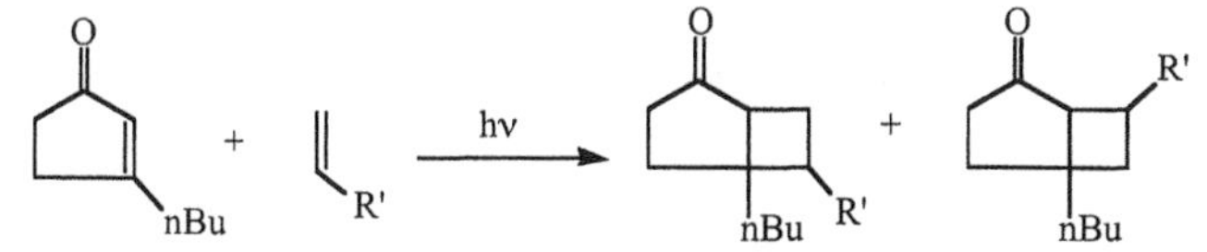

$R' = C_6H_{13}$	cyclohexane	53	47
	diethyl ether	47	53
	Potassium dodecanoate micelle	88	12

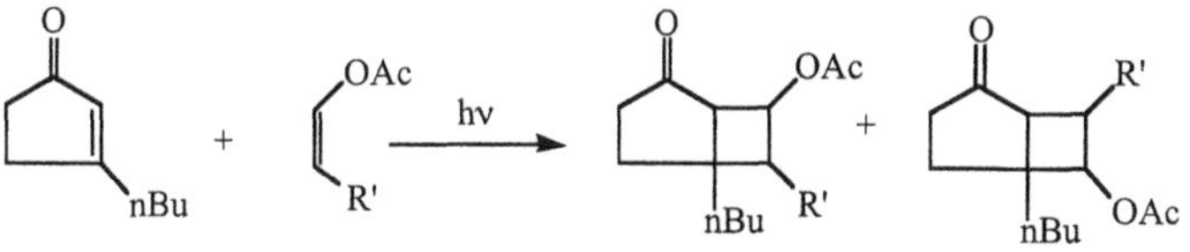

$R' = C_5H_{11}$	Methanol	0	100
	cyclohexane	0	100
	Potassium dodecanoate micelle	70	30

Scheme 11

as in the case of heptenyl acetate instead of 1-heptene, a dramatic switch in the regiochemistry of the product is observed (Sch. 11). One might question the value of selectivity observed in micellar media when compared to that in crystals. It is important to keep in mind that not all molecules can be crystallized and not all molecules react in the crystalline state. Therefore, identifying a medium where all molecules can be oriented and made to selectively react is important.

Photoproduct selectivity in unimolecular reactions can be achieved through conformational control of the reactant molecule at the interface. For example, benzoin alkyl ethers undergo Norrish type I reaction as the only photoreaction in organic solvents (Sch. 12) [125]. The Norrish-Yang reaction, although feasible, is not observed probably due to the over-whelmingly faster type I process. Incorporation of such systems in a micellar medium suppresses the type I pathway via the cage effect. Under such circumstances, the normally slow Norrish-Yang reaction becomes detectable in certain benzoin alkyl ethers and is the major pathway for benzoin methyl ether in cetyl trimethyl ammonium bromide and cetyl trimethyl ammonium chloride micelles (Sch. 13). The cage effect alone is not

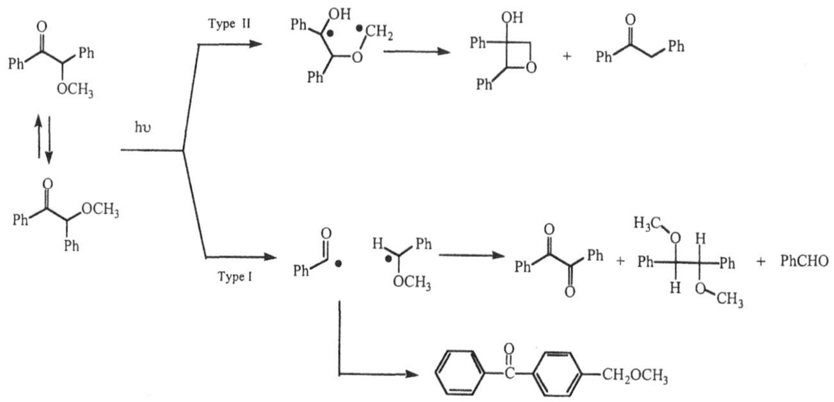

Scheme 12

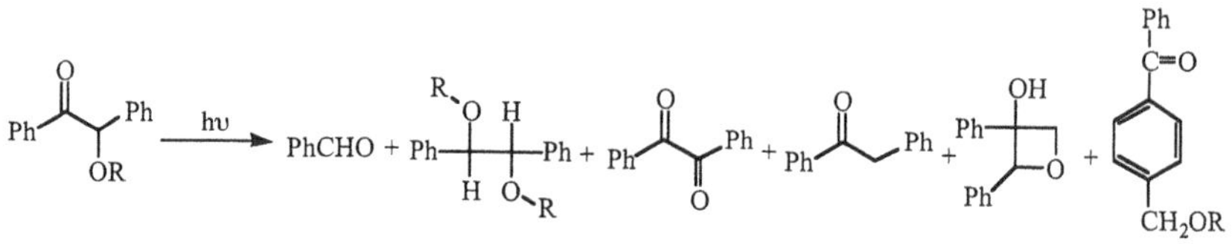

R = CH₃

Benzene	27%	39%	23%	--	--	--
Cetyl trimethyl ammonium chloride micelle	14%	8%	7%	4%	52%	--

R = CH₂-(CH₂)₆-CH₃

Benzene	12%	49%	--	3%	--	--
Cetyl trimethyl ammonium chloride micelle	3%	36%	--	2%	--	45%

Scheme 13

sufficient to account for the Norrish-Yang products. It is the conformational control at the interface that allows formation of Norrish-Yang
products. Of the two conformations A and B shown in Sch. 14 for benzoin
methyl ether, the specific conformer B at the micellar interface (probably
due to the preference for the polar carbonyl and methoxy groups to be in the
aqueous phase) leads to the Norrish-Yang products. Consistent with this
rationale, benzoin octyl ether failed to give the Norrish-Yang products and

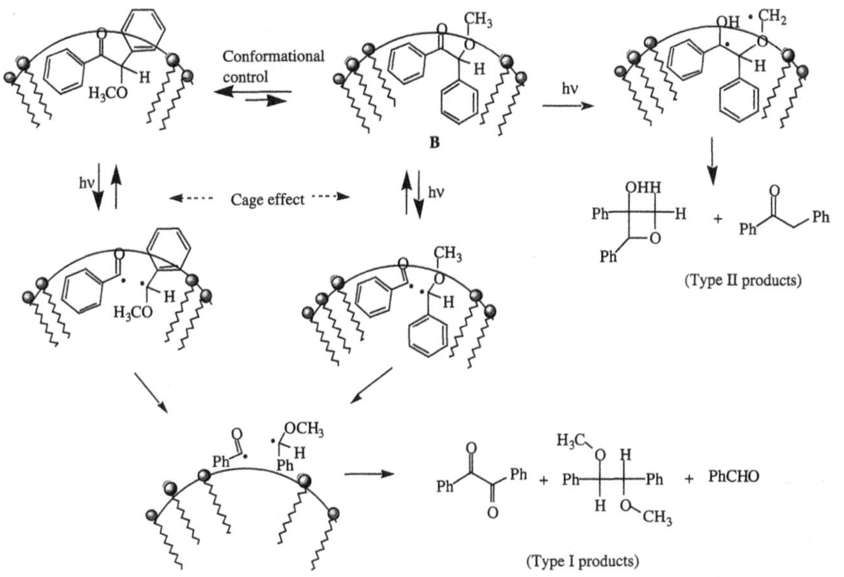

Scheme 14

afforded only the *para*-substituted benzophenone in the same micellar media. In benzoin octyl ether, the long alkyl chain would prefer to remain in the hydrocarbon interior leading to predominantly conformer **A** at the micellar interface. Under such conditions the cage effect is not sufficient to favor the Norrish-Yang reaction.

18.5.3. Characteristics of a Reaction Cavity: Free Volume

In order to accommodate the shape changes that occur as the reactants transform to products, the reaction cavity must contain some amount of free volume. In isotropic liquids, the free volume is highly mobile by virtue of the motions (translation, rotation, vibration, internal rotation, etc.,) of the constituent molecules. In organized media, the free volume may be essentially immobile, as in crystals, or have mobility ranging from that found in crystals to that of isotropic liquids. Since the surfaces of silica, the interplanar regions of clay, and the interiors of zeolites, possess time-independent structures similarly to those of crystals (whose relaxation times are much longer than the period necessary to transform a reactant molecule to its products), the free volume needed to accommodate the shape changes occurring during the course of a reaction must be an intrinsic part of the structure. The reaction cavities of such media possess "stiff" walls and may be characterized as being "hard" (Fig. 9).

Distinct differences exist between the rigidly organized structures discussed above and media such as micelles, microemulsions, molecular aggregates, and liquid crystals. In the latter group, the guest reaction cavities may contain minimal intrinsic free volume at the time of photoexcitation. However, since the molecules constituting these organized assemblies are mobile, the reaction cavity can respond to shape changes as the reaction proceeds. The extent and ease with which each medium can accommodate shape changes will determine the selectivity that can be obtained. These media are said to possess "flexible" reaction cavity walls and the cavities may be characterized as being "soft" (Fig. 9). The ease of a medium's response to shape changes occurring during the course of a reaction depends on the microviscosity of the medium and the extent of co-operative motions involving guest and host molecules. For a guest molecule to react in the environment provided by a restrictive host, necessary adjustments must be made by the medium within the time frame determined by the rate limiting paths leading to the transition state(s). No reaction will occur if the host medium fails to respond in a "timely" fashion.

The importance of free volume is nowhere more evident than in photo-reactions of crystalline molecules. Irradiation of crystalline 7-chlorocoumarin yields a single dimer (syn head-to-head) (Sch. 15) [126]. The packing arrangement shown in Fig. 15 reveals that there are two potentially reactive pairs of 7-chlorocoumarin molecules in a unit cell. One pair, being translationally related, has a center-to-center distance of 4.45 Å (favored to yield the syn head-to-head dimer). The other pair, being centrosymmetrically related, has a center-to-center distance of 4.12 Å (favored to yield the anti head-to-tail dimer). Despite the favorable arrangement of the centrosymmetric pair the dimer is obtained from the translationally related pair only. Lattice energy calculations reveal that the relative increase in lattice energy would be much higher if the centrosymmetrically related pair were to react. In other words, the free volume around the translationally related pair is much larger than that near the centrosymmetrically related pair whose double bonds are initially closer. Lack of free volume in the most topochemically favored pair leads to no reaction while presence of sufficient free volume allows dimerization of the less favored pair.

A comparison of the solid state photochemistry of two crystals, 7-methoxycoumarin and methyl *m*-bromo-cinnamate, further exemplifies the importance of the existence of free volume near the reaction site (Sch. 15) [127–129]. In spite of the fact that the reactive double bonds are rotated by 65° with respect to each other and a center-to-center double-bond distance of 3.83 Å (Fig. 16), photodimerization occurs in crystals of 7-methoxy-coumarin to give the syn head-to-tail isomer (Sch. 15). On the other hand, methyl *m*-bromo-cinnamate, which also has a nonideal arrangement of the

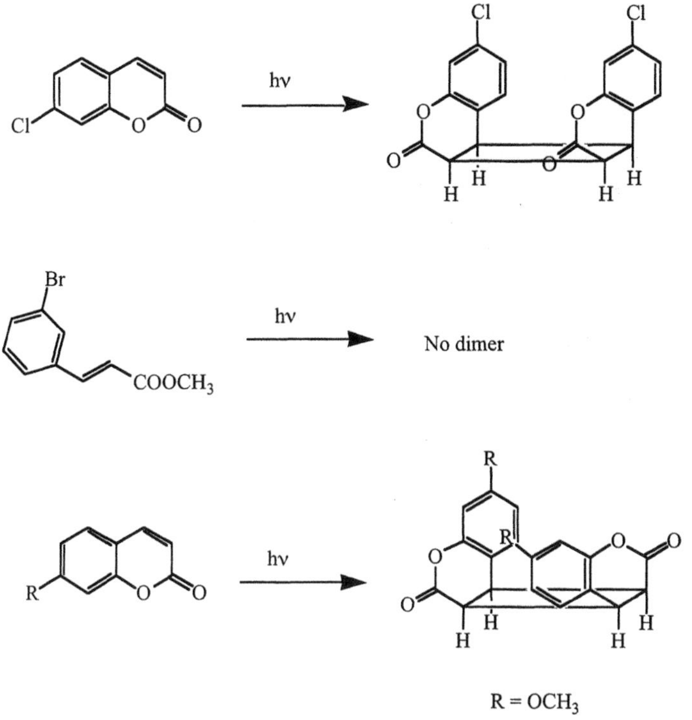

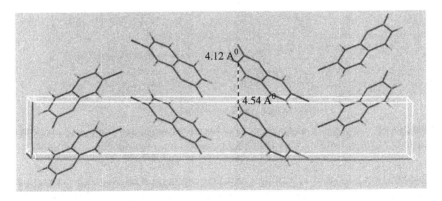

Scheme 15

Figure 15 Packing arrangement of 7-chlorocoumarin molecules in crystals. Note that in principle dimerization can occur between centrosymmetric or mirror symmetric pairs of molecules. Both are within the stipulated topochemical distance.

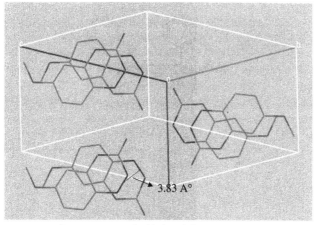

Center to center double bond distance 3.83 A°

Figure 16 Packing arrangement of 7-methoxycoumarin molecules in crystals. In spite of being within 3.83 Å the two molecules are not parallel to one another.

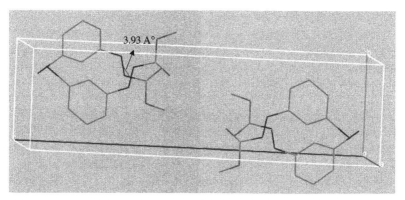

Center to center double bond distance 3.93 A°

Figure 17 Packing arrangement of *meta*-bromoethylcinnamate molecules in crystals. In spite of being within 3.93 Å the two molecules are not parallel to one another.

double bonds in the crystalline state (the distance between the centers of adjacent double bonds is 3.93 Å; the double bonds are rotated and make an angle of 28° when projected down the line joining the centers of the bonds; Fig. 17) does not undergo photochemical dimerization in the crystalline state. While both molecules are poorly oriented, one yields a dimer while the other does not. The increase in lattice energy upon reorientation of the reactive pair of 7-methoxycoumarin in the crystal lattice was estimated to be roughly the same order of magnitude as for many photoreactive crystals with favorably oriented pairs [127–129]. On the other hand, the lattice energy increase necessary to align the molecules parallel to each other in a geometry suitable for dimerization is enormous for methyl *m*-bromo-cinnamate. Such a large increase in the lattice energy indicates a resistance from the cavity wall for the required reorganization of methyl *m*-bromo-cinnamate. To rephrase the above conclusion in terms of free volume within a reaction cavity, the reaction cavity occupied by a pair of 7-methoxycoumarin molecules contains a greater amount of free volume than that present in crystals of methyl *m*-bromo-cinnamate. Because of the greater free volume, the reaction cavities in 7-methoxycoumarin crystals are tolerant towards large motions executed by a pair of 7-methoxycoumarin molecules during dimerization. The lesson to be learnt is to identify the location of free volume. It is important that the free volume is present where it is needed.

18.5.4. Characteristics of a Reaction Cavity: Presence of Weak Forces

An important feature of reaction cavities in many organized media is that they possess functionalities that can selectively interact with the solute molecules. The reaction cavity wall is termed *"active"* when there is an attractive or repulsive interaction between a guest molecule and a wall of the cavity and the interaction may serve as a template for the guest as it proceeds to products. The presence of active walls requires that one consider possible wall-guest interactions during the prediction (and rationalization) of product selectivity. If sufficiently attractive or repulsive, the interactions will influence the location or conformation of guest molecules in the cavity. Additionally, if the interactions are to have a discernable influence on the course of the transformations, they must persist for times at least comparable to those required for reaction to occur. While it is easy to recognize the presence of specific binding or repulsion between the ground state guest molecules and host framework, it is important to note that new potentially stronger or weaker interactions may develop between the walls and functional groups created in intermediates during the course of a reaction, especially with electronically excited molecules. Interactions may vary from weak van der Waals forces, to hydrogen bonds to strong electrostatic forces between charged centers. For example, a number of hosts capable of forming inclusion complexes (such as cyclodextrin, urea and deoxycholic acid) possess functional groups that can form hydrogen bonds with guest molecules. Silica surfaces possess silanol groups, which

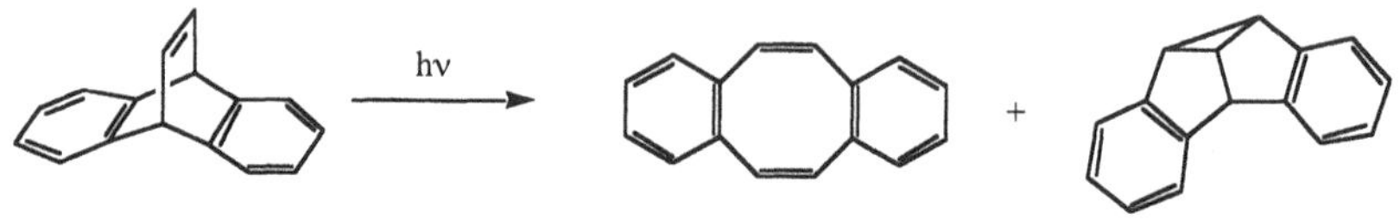

Acetonitrile	77	23
Acetone	--	100
KY	75	25
RbY	35	65
CsY	28	72
TlY	1	99

Scheme 16

may orient the adsorbed molecules through hydrogen bonding. Surfaces of clays and zeolites often carry a large number of cations, which can interact electrostatically with guests. Most reaction cavity walls may, in fact, be "*active*" to some extent. Understanding this feature generally allows one to exploit the medium to the maximum extent. We provide below several examples chosen from studies in zeolites that illustrate the importance of cavity-guest interactions.

Supercages of X and Y zeolites contain a large number of exchangeable cations (Fig. 6). Of the various cations, those of type II and III, present within the supercages, can interact with the included guest molecules. By careful choice of the cations one can control the chemistry that takes place inside the supercage.

The external heavy cation effect has also been used to fine-tune product distributions within a zeolite. For example, irradiation of dibenzobarrelene included in zeolite KY gives dibenzocyclooctatetraene as the major product, whereas photolysis in TlY yields dibenzosemibulvalene as the only product (Sch. 16) [130,131]. The former is the S_1-derived product whereas the latter comes from T_1. The unusually strong external heavy-atom effect has been observed due to the close approach between the guest molecule and the heavy atom (cation), which is re-inforced by the zeolite supercage, and by the presence of more than one heavy atom (cation) per supercage. It is the weak interaction between the reaction cavity and the guest that forces

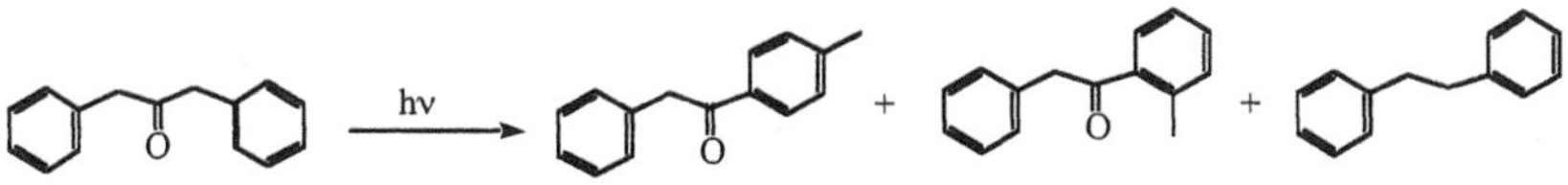

Hexane	-	-	> 99
KY (1.44 A°)	20	-	80
RbY (1.58 A°)	20	10	70
CsY (1.84 A°)	52	11	37
TlY (1.40 A°)	16	60	23

Scheme 17

the guest molecule to show unique behavior that is different from that in isotropic media.

An example of how the cations in zeolites can control the product distribution by interacting with reactive intermediates is found in the photochemistry of zeolite-included dibenzylketone [132,133]. Irradiation of dibenzylketone included in KY yields 1,2-diphenylethane as the major product, whereas photolysis in TlY the rearranged products, 1-phenyl-4-methylacetophenone and 1-phenyl-2-methylacetophenone (Sch. 17), are formed. As illustrated in Sch. 18, the heavy cation is able to enhance the intersystem crossing rate of the primary triplet radical pair ($C_6H_5CH_2 °$ and $C_6H_5CH_2CO °$) that is formed from the reactive dibenzylketone triplet. The combined cage and heavy cation effects influence the chemistry of dibenzylketone within zeolites.

Interaction with cations may also result in "state switching" in the case of enones. Enones possess two reactive triplets of $n\pi^*$ and $\pi\pi^*$ configuration. Depending on the nature of the lowest triplet, different products are generally obtained. One of the often-used approaches to control the nature of the lowest triplet state is to use solvents of different polarity. With Li^+ and Na^+ exchanged zeolites it has been possible to make the $\pi\pi^*$ state the lowest triplet [134,135]. Enones which even in polar solvents give products from the $n\pi^*$ state give products resulting from $\pi\pi^*$ state when irradiated within LiY or NaY. For example, irradiation of testosterone acetate in 2-propanol results in the slow reduction of the enone double bond to yield 5α-isomer where the hydrogen abstraction has occurred from the α-face (Sch. 19). In contrast, the major product from irradiation of cholesteryl acetate included in NaY is the 5β-isomer (note that the hydrogen source within zeolite is hexane in which the zeolite is irradiated as slurry). It is known that α-addition occurs from the $n\pi^*$ triplet and β-addition from the $\pi\pi^*$ triplet state. Based on the products one may

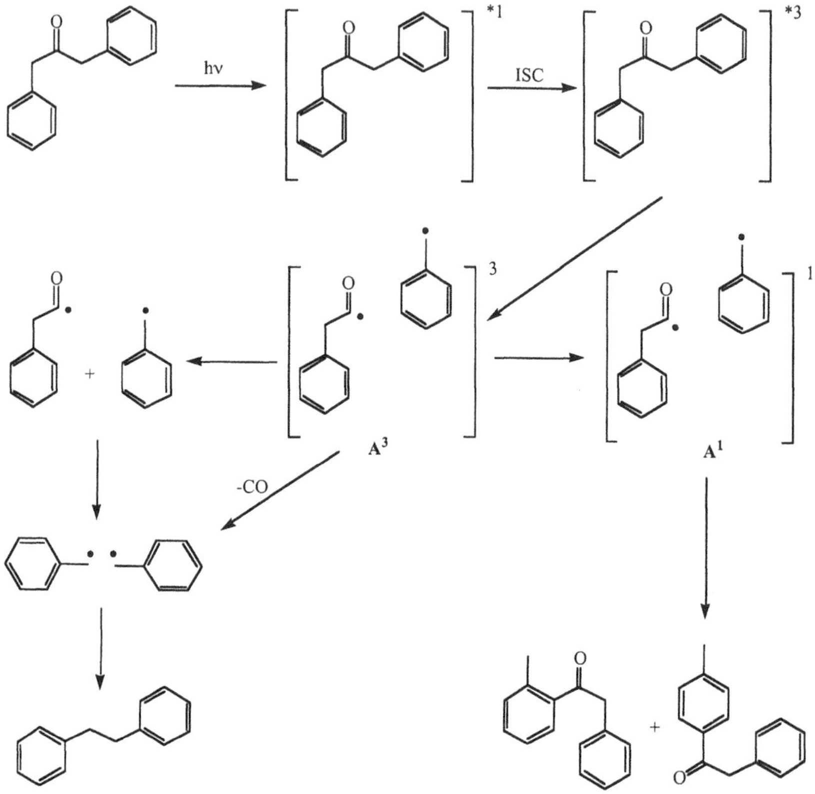

Scheme 18

conclude that the reduction occurs within NaY from the $\pi\pi^*$ triplet whereas in 2-propanol solution, the $n\pi^*$ triplet is the reactive state. Lack of reaction when the irradiation is conducted in a Y zeolite with no cation demonstrates the importance of the cation in this excited state switching (Sch. 19). Recognition of the importance of weak interactions within a reaction cavity has led to achievement of the above selectivity.

Differential interactions between cations in zeolites and the products of a photoreaction may result in selectivity. One such example is the selective photoisomerization of *trans*-1,2-diphenylcyclopropane to the *cis* isomer [136]. Triplet sensitization of 1,2-diphenylcyclopropane in solution results in a photostationary state mixture consisting of ~55% *cis* and 45% *trans* isomers. When the same process is carried out within NaY zeolite the *cis* isomer is formed in excess of 95%. The preference for the *cis* isomer within NaY is attributed to the preferential binding of the cation to the *cis*

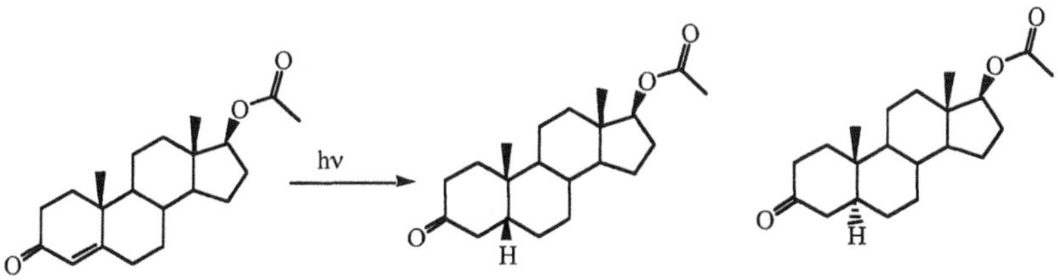

hv

5-β

ππ*

5-α

nπ*

	ππ*	nπ*
Hexane	No reaction	
2-propanol	27	73
NaY (Si/Al: 2.5)	85	15
NaY (Si/Al: 300)	No reaction	

Scheme 19

isomer. As illustrated in Fig. 18 cation binding is proposed to result in a barrier for the rotation of the *cis* to the *trans* isomer on the excited state surface. Consistent with this proposal, the presence of water, which would be expected to bind to the cation, eliminates the effect of the sodium ion on the photochemistry of 1,2-diphenylcyclopropane.

18.6. CAUTION IN EMPLOYING ORGANIZED MEDIA

18.6.1. Microheterogeneity in Organized Media

Fluid solutions allow the solute molecules to experience an average microenvironment by virtue of the fast relaxation times of the solvent molecules and/or the high mobility of the solute molecules. However, in organized and highly viscous media, solvent relaxation and rates of guest diffusion may be slower than the time period of a photoreaction, leading to reaction in a variety of reaction cavities ("sites"). On the other hand, even in the absence of migration of molecules between sites, one could imagine a situation in which all reacting molecules experience identical environments because all sites are equivalent. Such is expected to be the case in the reaction cavities present in proteins. If all reaction cavities in biological systems were not identical one would not observe the remarkable selectivity for which enzymatic reactions are well known.

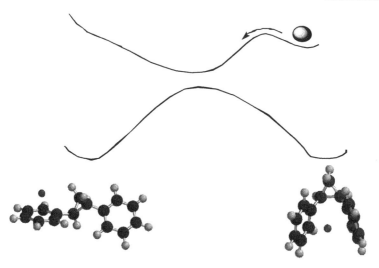

Cation binding restricts the cis to trans conversion

Figure 18 Proposed energy surface diagram for the isomerization of 1,2-diphenyl-cyclopropane. Cation binding to the cis isomer imposes barrier for rotation in the excited state.

However, this is not the case in most nonbiological organized media. Even perfect crystals have at least two types of reaction cavities corresponding to molecules in the bulk and those on the surface. Normally, crystals have defects that may also lead to differences in reactivity. Diffusion of molecules among these site types is not expected to occur as rapidly as single molecules are transformed to photoproducts. Thus from a mechanistic point of view at least three different reaction cavity types (surface, bulk, and defect) should be expected in crystalline media. Generally, the relationship between the rates of hopping of molecules among various sites and the rates of photoreactions will determine the importance of multiple sites on reactions in organized media, and therefore, the selectivity of the product distribution.

Scheme 20 summarizes the above situation using a minimum number of sites and competing processes. In this scheme, two sites—square well type (X) and spherical well type (Y)—are available for reactant molecules (A). For the sake of convenience, molecules residing at sites X and Y are labeled A_X and A_Y. Excitation of these molecules gives rise to A_X^* and A_Y^*. The photoreactivity of excited molecules in each site will be identical if they equilibrate between sites X and Y before becoming photoproducts. In media with time-independent structures, such as crystals, equilibration requires

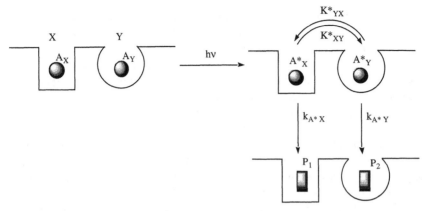

Scheme 20

(I) Translational of guest from site X to site Y

(II) Fluctuations of the media (Time dependent structures-shapes of sites change with time)

(III) Energy transfer between molecules present at sites X and Y

Scheme 21

diffusion of molecules of A; in media with time-dependent structures, such as micelles and liquid crystals, equilibration can be accomplished via fluctuations in the microstructure of the reaction cavities as well as translational motion of A (Sch. 21). An additional mechanism for site selective reactions or equilibration of A_X^* and A_Y^* molecules can be achieved via energy migration (e.g., energy hopping, exciton migration or Foster energy transfer).

Three limiting dynamic situations can be envisioned for Sch. 20: (I) the rate of excited state reaction is slower than the rate of interconversion between A_X^* and A_Y^* (equilibrium is established between A_X^* and A_Y^* before decay; k_{XY}^* and $k_{YX}^* >> k_{A_X}^*$ and $k_{A_Y}^*$); (II) the rate of excited state reaction is faster than the rate of interconversion between A_X^* and A_Y^*; (III) the rate of interconversion between A_X^* and A_Y^* and the rates of reaction (decay) of excited guest molecules at these sites are comparable (assume for instance, $k_{A_Y}^* >> k_{Y_X}^*$ and $k_{X_Y}^* >> k_{A_X}^*$ and $k_{A_Y}^*$).

Situation (I) corresponds to a fluid isotropic solution where a uniform time averaged environment should exist. Under such conditions single exponential decay would be expected for the guest excited states and the photoreactivity should be predictable on the basis of a single effective reaction cavity. In situation (II) there should be two kinetically distinct excited states in two noninterconverting sites resulting in nonexponential decay of the excited state of A. The quantum efficiency of product formation and the product distribution may depend upon the percent conversion. An example of mechanism (II) is provided in Sch. 22 [137]. The ratio of products **A**, **B**, and **C** has been shown to depend on the crystal size. With the size of the crystal the ratio of molecules present on the surface and in the interior changes which results in different extents of reactions from two the distinct sites namely, surface and interior.

According to mechanism (III) even if the concentration of A in one of the two sites (Y) is much smaller than the amount in site X, appreciable photoreaction from A_Y may occur through $A_X^* \rightarrow A_Y^*$ energy transfer as outlined in Sch. 21 or if the quantum yield of reaction from A_Y^* is much larger than that from A_X^*. An example in which situation (III) operates is the photodimerization of 9-cyanoanthracene in the crystalline state presented above (Sch. 23) [138,139]. On a statistical basis, many more molecules within the crystalline bulk phase are expected to be excited than those at defect sites. However, the reaction cavities capable of supporting reaction are specific to defect sites. Efficient photodimerization is believed to occur from exciton migration from the inert bulk sites to the defect sites.

Terms such as microheterogeneous, anisotropic, and nonhomogeneous that are used to describe a number of organized media derive from the unique features (multiple sites) described above.

18.6.2. Failure of Rule of Homology

Organic chemistry is built on the principle of "homology." For example, understanding the chemical behavior of one functional group in a single molecule allows for the chemistry to be extended to other homologues of the series with slight modifications due to steric and electronic features of

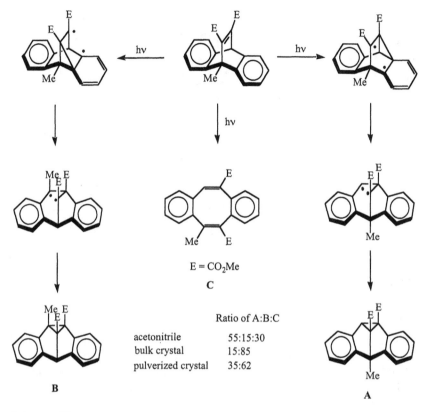

Scheme 22

the substituent(s). But this may not hold with photochemistry carried out in organized media. This should not be a surprise to those who are aware of the high specificity with which protein binds to molecules. For example, of the possible eight isomers of retinal, only the 11-*cis* isomer binds to opsin and functions as a trigger. Thus one cannot generalize the known behavior of 11-*cis* retinal with opsin to other isomers.

A small variation in the structure of the reactant can result in large changes, such as molecular packing of the reactant as a crystal, orientation at the interfaces of micelles and monolayers, and alignment within liquid crystalline media. These differences can have dramatic effects on the product distributions. As a result, a high product selectivity shown by one member of a homologous series in a particular organized medium may not be representative of all its members. Therefore, the study of a number of homologues is critical in understanding the basic principles governing

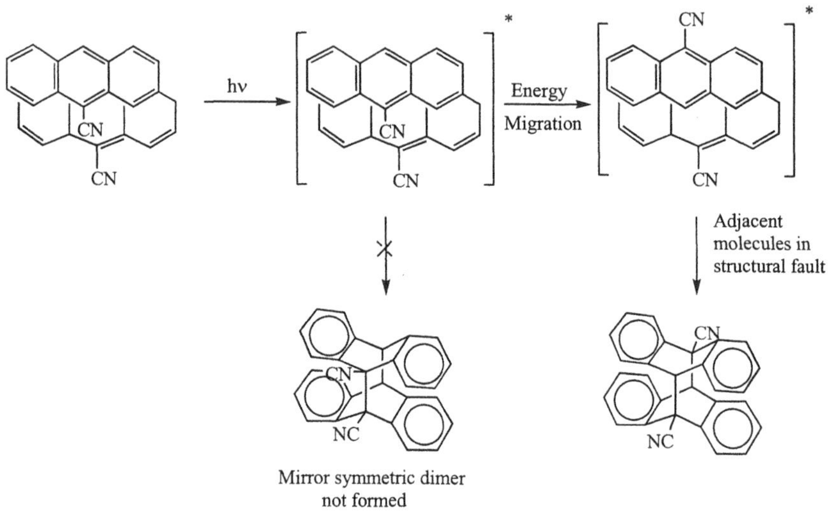

Scheme 23

selectivity in a particular medium. We illustrate this concept with the photochemistry of dialkyl ketones in a liquid crystalline medium [140–142].

An ordered medium chosen for illustration is n-butyl stearate. Depending on the temperature *n*-butyl stearate can exist as an isotropic liquid, liquid crystalline smectic B, or solid phase. *n*-Butyl stearate is isotropic above 26.1° and liquid crystalline (smectic B) between 15 and 26.1° (Fig. 19). The influence of medium on the Norrish-Yang reaction of ketones was monitored by measuring the ratio of fragmentation (E) to cyclization (C) products from the 1,4-diradical intermediate (Sch. 24). In the scheme the E/C ratios for three homologous 2-alkanones (Sch. 24) varying only in chain length are provided. Consistent with the principle of "homology," the E/C ratio is nearly the same for all ketones in the isotropic liquid *n*-butyl stearate medium. The E/C ratios, however, are not the same for all ketones in crystalline or liquid crystalline phases. Why not all ketones behave in the same manner in *n*-butyl stearate liquid crystalline smectic B? In the liquid crystalline (smectic B) phase of *n*-butyl stearate, the individual *n*-butyl stearate molecules remain in extended conformations and are orthogonal to the layers which their hexagonal close packing define. The selectivity is greatest when the extended length of the ketone is similar to that of the stearoyl portion of *n*-butyl stearate which least disrupts the host order. Only those ketones that are well-ordered initially will be subjected to the influence of the ordered medium when they attempt to change their shape (react to give a new product). Ketones, which are significantly shorter or longer than

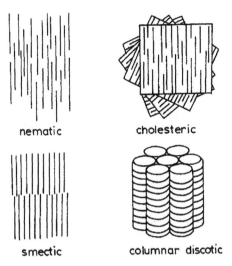

Figure 19 A cartoon representation of molecular organization in various liquid crystalline media.

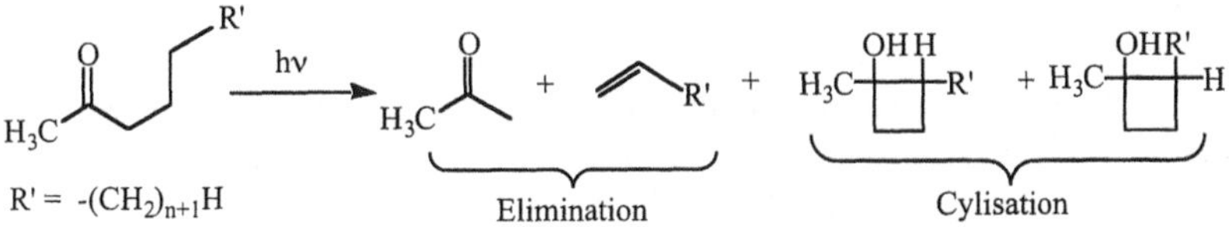

R' = -(CH₂)ₙ₊₁H

$$R' = -(CH_2)_{n+1}H$$

Elimination   Cylisation

E/C ratio in n-butylstearate at various temperatures

n = 6		n = 13		n = 15	
Temp	E/C	Temp	E/C	Temp	E/C
30°C	2.9	30°C	3.7	30°C	3.9
20	3.1	20	14.8	20	10.4
10	2.6	10	31.3	10	18.5
0	3.6	0	42.3	0	--

Scheme 24

n-butyl stearate should disturb their local environment to a greater extent than the ones, which are, near the length of n-butyl stearate. These ketones create their own unique environment and the medium has a little role to play in controlling their shape changes. Such ketones will not be subjected to environmental effect to the same degree as the ones, which fit nicely in the

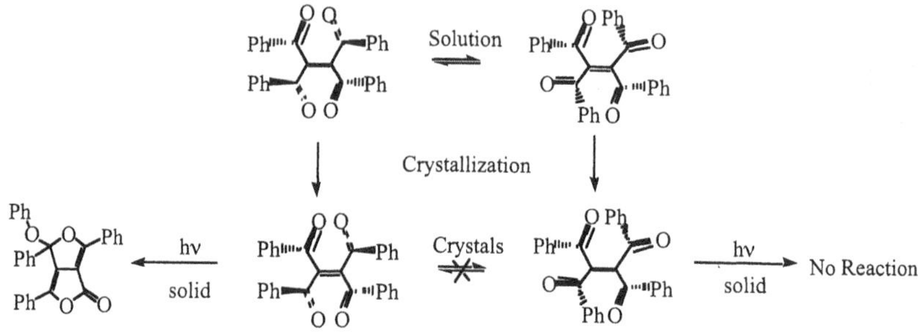

Scheme 25

ordered medium. Thus not all ketones with different lengths are expected to behave similarly in a liquid crystalline matrix and, as presented in Sch. 24, they do not.

Even more important point to note is that often times one cannot even extend the chemistry of the same molecule in the same medium under different conditions. The best examples are found in the crystalline state where molecules tend to crystallize as different polymorphs. It has been very well established that the different polymorphic forms of a compound show significant differences in photochemical behavior. A classic example that shows polymorphic form dependent reactivity is cinnamic acid. Substituted cinnamic acids crystallize in three polymorphic forms namely α, β, and γ and provide photodimers characteristic of the crystal packing (Sch. 1). One of the interesting crystal structure dependent cyclizations is that of tetrabenzoyl-ethylene [143]. Photolysis of only one of the two dimorphic modifications gives rise to a product while the other is inert to UV radiation (Sch. 25).

18.6.3. New Mechanisms may Emerge in a Confined/Organized Media

This chapter thus far has revealed that molecules when confined may behave differently from that in solution where they are "free." In a confined medium, reactant molecules, of the various reaction paths available to them most often choose one, not necessarily the one with lowest activation energy. The photoreactions become selective essentially because the confined medium poses new barriers along certain reaction co-ordinates which cannot be overcome by thermal energy and/or favor certain pathways by lowering entropic barriers. In a confined medium entropy plays an important role. These characteristics might favor new pathways (and mechanisms) at the expense of pathways well-established in solution chemistry. That is, a new

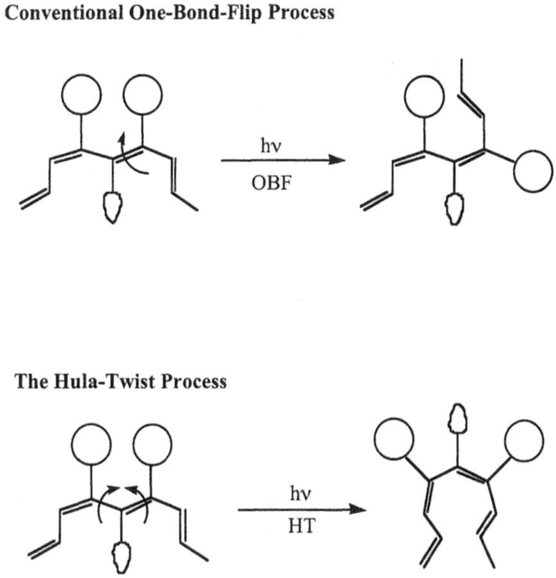

Conventional One-Bond-Flip Process

$\xrightarrow[\text{OBF}]{h\nu}$

The Hula-Twist Process

$\xrightarrow[\text{HT}]{h\nu}$

Scheme 26

reaction mechanism that is not important in solution might become available to a molecule in a confined medium. We illustrate this phenomenon with one example below.

One of the most fundamental reactions of alkenes, geometric isomerization, as presented in the beginning of this chapter, occurs within the highly confined space of our eye. The question is how does this process occur in the confined space provided by the protein opsin? Does it follow the mechanism that has been developed based on solution studies or does it follow a pathway, which is unlikely under laboratory conditions? The conventional model of photochemical geometric isomerization is torsional relaxation of the double bond which involves a one bond flip process (OBF), i.e., turning over one half of the molecule (Sch. 26) [144–149]. This is a three dimensional process requiring the presence of a large free volume within a reaction cavity. In the absence of a large globular (three dimensioanl) free volume this process would be prevented by the walls of the reaction cavity. The fact that only one (11-*cis*) of the many isomers (sixteen in total) of retinal is favored by the protein opsin suggests that the reaction cavity has definite shape and there is unlikely to be large free volume around the retinyl Schiff base. While soft reaction cavities of proteins (unlike crystals and zeolites, which possess rigid and hard reaction cavities) are likely to accommodate shape changes in the reactant (retinal), it will do so at its own pace (relaxation time which is likely

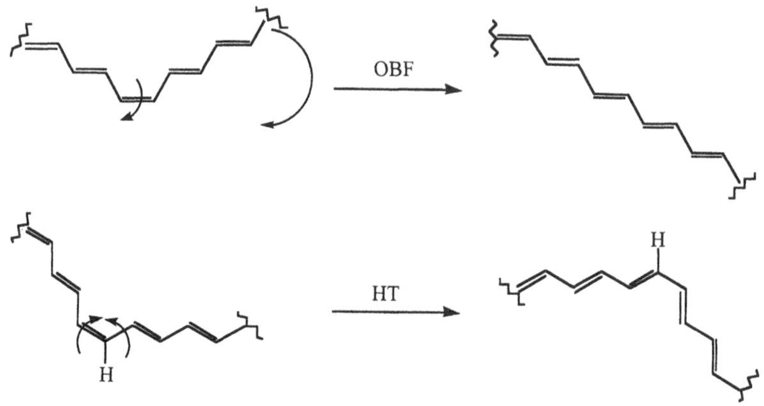

Scheme 27

to be in the milli-to-microsecond range). Therefore, any reactions occurring in less than this time scale have to be the ones that do not require large shape changes of the reactant and consequently the surroundings. The photo-isomerization of protein bound 11-*cis* retinyl Schiff base (to the all *trans*) occurs within a picosecond. As indicated above the volume demanding one bond isomerization cannot occur within the reaction cavities of a protein in this short time scale. A new mechanism known as "hula-twist" accommodates this ultra-fast isomerization process.

The "hula-twist"'s process (HT), unlike conventional one bond process, requires less volume change during *cis* to *trans* conversion (Sch. 26) [144–149]. The hula-twist is more of a two dimensional rather than a three dimensional process. The difference between conventional one bond flip and hula-twist is illustrated in Sch. 27 with 5-*cis*-decapentaene as an example. During the *cis* to *trans* conversion via one bond flip mechanism one half of the molecule undergoes a 180° flip, i.e., half of the molecule rises from the molecular plane and sweeps a 180° motion before it rests on the same plane in a different geometry. On the other hand, a hula-twist process which involves a simultaneous rotation of two adjacent bonds (a single and a double bond) or a 180° translocation of one C–H unit results in *cis* to *trans* isomerization. This process as shown in Sch. 27 results in simultaneous configurational (*cis* to *trans*) and conformational changes (transoid to cisoid). An important difference between hula-twist and one bond flip is that the former is essentially a two dimensional process. Only one C–H unit of the molecules rise above the molecular plane during *cis* to *trans* conversion. Therefore, the volume demand on this process is much less than that during one bond flip. In the hula-twist process only the central atom moves in a sweeping semicircular manner (in and out of the plane of the molecule) while the two terminal atoms translate sideways in the original plane of the molecule. Remember that a mechanism cannot alter the shape and size of the reactant and product molecules, but what it can do is to make sure that the intermediate structures that are involved during the shape change is not too large. The point to be noted is that between one bond flip and hula-twist, the latter would be preferred under conditions where the reaction cavity has only a limited free volume. Therefore, with media constraints, the conventional one bond flip process could be completely suppressed, revealing otherwise less probable hula-twist process.

The photoisomerization of *all-s-trans-all-trans* 1,3,5,7-octatetraene at 4.3 K illustrates the need for a new mechanism to explain the observed behavior [150]. Upon irradiation, *all-s-trans-all-trans* 1,3,5,7-octatetraene at 4.3 K undergoes conformational change from *all-s-trans* to *2-s-cis*. Based on NEER principle (NonEquilibrium of Excited state Rotamers), that holds good in solution, the above transformation is not expected. NEER postulate and one bond flip mechanism allow only *trans* to *cis* conversion; rotations of single bonds are prevented as the bond order between the original C–C bonds increases in the excited state. However, the above simple photochemical reaction is explainable based on a hula-twist process. The free volume available for the *all-s-trans-all-trans* 1,3,5,7-octatetraene in the *n*-octane matrix at 4.3 K is very small and under such conditions, the only volume conserving process that this molecule can undergo is hula-twist at carbon-2.

The hula-twist is most likely responsible for a variety of biological events (rhodopsin, bacteriorhodopsin, photoactive yellow protein, phyto-chrome etc.,) where the reaction cavities offered by proteins are restricted, confined and lack free volume. The examples discussed in this section caution us that one should be prepared to view the events in a confined medium with a fresh mind, and the knowledge gained from solution studies should be used only as a starting point. Photobehavior of organic molecules in a confined medium as a whole argues that we are yet to fully develop a paradigm that would allow us to rationalize and predict the photochemical behavior of even simple organic molecules in a more complex surrounding.

18.7. POWER OF ORGANIZED MEDIA ILLUSTRATED WITH THEIR INFLUENCE ON ASYMMETRIC PHOTOREACTIONS

From the above presentation it is evident that one could control the distribution of products in a photochemical reaction through an intelligent use of an organized/confined medium. We provided examples in which by careful choice of the organized/confined medium, one could restrict the number of different reactions an excited molecule undergo, control the number of products a reaction yields, and influence the distribution of regio and sterochemistry of photoproducts. The unique nature of organized media is nowhere more evident than in their ability to control the formation of optical isomers of photoproducts. In this section we discuss a number of examples in which chiral induction has been achieved with the use of organized/confined medium. These serve to bring out the power of the medium in influencing the excited state behavior of an organic molecule.

Asymmetric photochemical reactions have not been studied as intensively as their ground state counterparts. In the past, chiral solvents, chiral auxiliaries, circularly polarized light, and chiral sensitizers have been utilized to accomplish enantioselective photoreactions [151–157]. The highest chiral induction achieved by any of these approaches in solution at ambient temperature and pressure has been approximately 30% enantiomeric excess (e.e). Enatiomeric excesses of 2–10% are more common in photochemical reactions under the above conditions. The crystalline state and solid host-guest assemblies have, on the other hand, provided the most encouraging results, and enantiomeric excesses of >90% have been obtained. In this section, the use of crystals, organic host-guest assemblies and zeolite-guest systems is illustrated in the context of stereochemical control of photoreactions. It should be recognized that although a solid medium can provide extraordinary control on the enantioselectivity of a photoreaction, in many cases the factors that control the selectivity and the generality of the method are yet to be established. That is, no one has devised a set of rules that allows a crystal to be engineered so that its molecules upon irradiation will produce a product of desired stereochemistry.

18.7.1. Crystalline Media

To appreciate the strategies employed to accomplish chiral induction in the crystalline state an understanding of the relationship between crystal symmetry and the stereochemistry of the products is essential. In addition to their individual symmetry, the molecules present in a crystal are related by various other symmetry elements. The molecular and crystal symmetries need not be the same. The 230 space groups representing the different packing arrangements in which molecules may crystallize can be divided into two classes. One of which is comprised of the 65 chiral space groups containing only the symmetry elements of translations, rotations, and combinations thereof, i.e., the corresponding screw operations. The most common chiral space groups for organic molecules are monoclinic $P2_1$, P1, and orthorhombic $P2_12_12_1$. The remaining space groups are achiral. Resolved chiral molecules must crystallize in chiral space groups, but a racemic compound may aggregate to form either an achiral racemic crystal or undergo a spontaneous resolution in which the two enantiomers segregate as a conglomerate of chiral crystals. Achiral molecules may crystallize in either achiral or chiral space groups. The relationship between an achiral compound and a chiral crystal is similar to that of a square brick and a spiral staircase. Achiral bricks, when properly placed can give a chiral staircase. When they crystallize in a chiral space group, achiral molecules reside in a chiral environment imposed by the lattice. Among the most common examples of achiral molecules that crystallize in chiral space groups are quartz, sodium chlorate, urea, maleic anhydride and the γ-form of glycine.

Chiral phototransformations of achiral molecules have been accomplished by two approaches, both aimed at influencing the achiral molecule to crystallize in a chiral space group. The first approach relies on the fact that achiral molecules occasionally, but unpredictably, crystallize in chiral space groups [158–160]. In the second "ionic chiral auxiliary approach" a salt is formed between an acid and a base, one of which is chiral 161–163]. The chiral salt, which is required to crystallize in a chiral space group, favors asymmetric induction during a photoreaction. After reaction the salt is hydrolyzed to yield an enantiomerically enriched product. These approaches are briefly outlined below. The message from these studies is that high enantioselectivity can be achieved in an organized assembly even though is not possible in an isotropic solution, at least with the methods currently available.

18.7.1.1. Achiral Molecules as Chiral Crystals

In this approach the chirality of the product is determined at the initial crystallization stage. The photochemical reaction then transforms the chirality of the crystal to that of the product. An achiral molecule crystallizing in a chiral space group has equal probability of crystallizing into each of two enantiomorphous forms. To ensure the homochirality of the crystalline phase, large single crystals must be grown. In well-designed chiral matrices composed of nonchiral molecules, the chiral environment can exert an asymmetric influence on a lattice-controlled solid-state transformation. Stereospecific solid state reactions of chiral crystals formed from achiral reactants are defined as absolute asymmetric syntheses. It is important to note that absolute asymmetric synthesis cannot be planned a priori. Although a certain amount of good fortune may favor the formation of chiral crystals from achiral reactants, difficulties in predicting the factors controlling the crystallization process have impeded the development of this approach.

When chiral crystals are formed from achiral molecules, one question always arises: whether all crystals in one batch have the same absolute configuration or a mixture of both enatiomorphic crystals are present. If it is the latter, the crystals will not be useful for conducting asymmetric photoreactions. In general, crystallization will result in nonenantiomeric crystallization, meaning one enantiomeric form of the crystal will be present in slight excess over the other (the ratio will not be exactly 50:50); i.e., one enantiomer will predominate over the other. When crystallization is carried out slowly and carefully there is a higher probability of obtaining crystals where the enantiomeric ratio is much different from 50:50. From this batch of crystals one might be able to isolate crystals all having the same absolute configuration. Such crystals serve as a seed to prepare large chiral single crystals from asymmetric compounds. Thus once enantiomorphic crystals are isolated, a large amount of the desired chiral crystals can be selectively prepared by seeding during the crystallization process.

The diisopropyl ester upon irradiation in solution undergoes the Zimmerman reaction to yield the dibenzosemibullvalene (Sch. 28) [164]. The product, which contains four chiral centers, is formed as a racemic mixture. The above crystals are dimorphic, and one of the forms grown from the melt is chiral (space group $P2_12_12_1$). Irradiation of large single crystals (20–85 mg) give the expected dibenzosemibullvalene but the optical activity of the product, depends on the space group of the parent crystal. Crystals

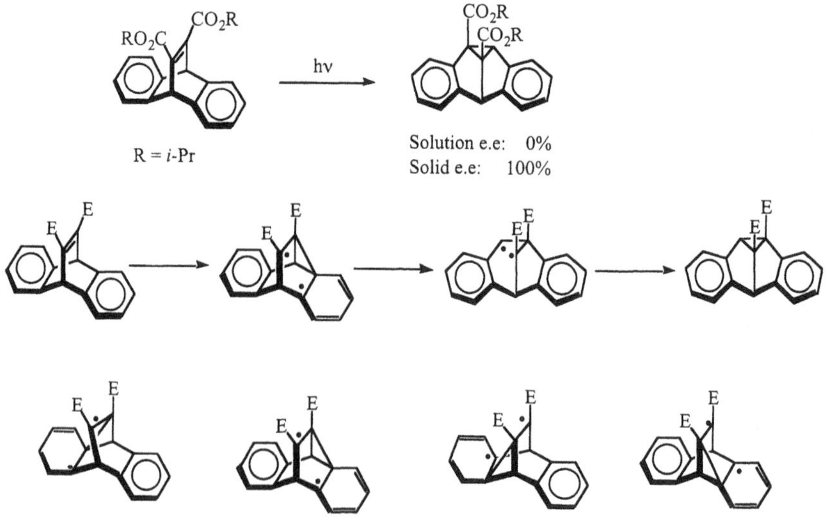

Scheme 28

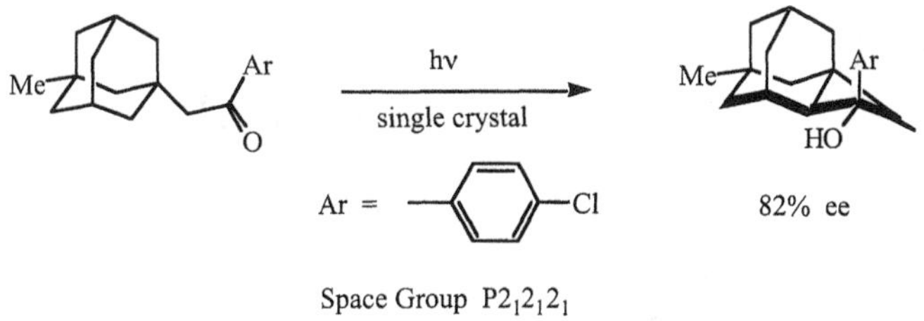

Scheme 29

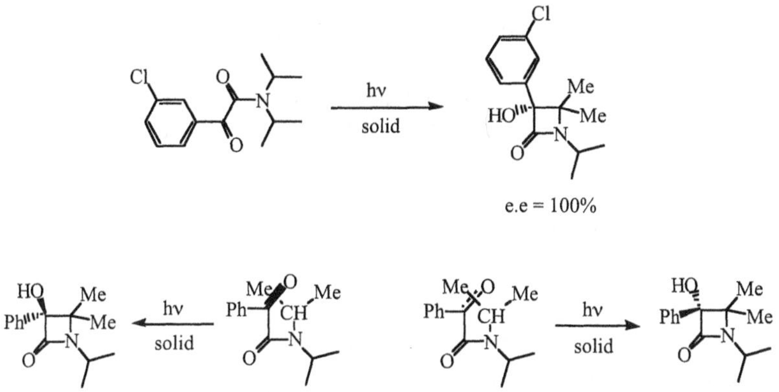

Scheme 30

belonging to achiral space group (Pbca) gave product with no optical activity and those belonging to chiral space group $P2_12_12_1$ gave the product with average specific rotation of $24.2 \pm 2.9°$. The enantiomeric excess of the product from the chiral crystal was shown to be 100%! (Sch. 28). A point to note is that since the packing of the crystal can change with conversion (as the product is formed within the parent reactant crystal) the % e.e. is likely to depend on the extent of conversion.

The achiral adamantyl ketone in Sch. 29 crystallizes in the chiral space group $P2_12_12_1$ [164]. Upon irradiation of single crystals of this ketone, a cyclobutanol is obtained via the Norrish-Yang reaction, while in solution one obtains a mixture of four cyclobutanol isomers. In the solid state a single cyclobutanol is obtained with e.e. >80%.

Upon irradiation in solution N,N-diisopropylphenylgloxylamide yields oxazolidin-4-one and 3-hydroxyazetidin-2-one products of an electron transfer process (Sch. 30) [165–167]. The products, as expected, are racemic.

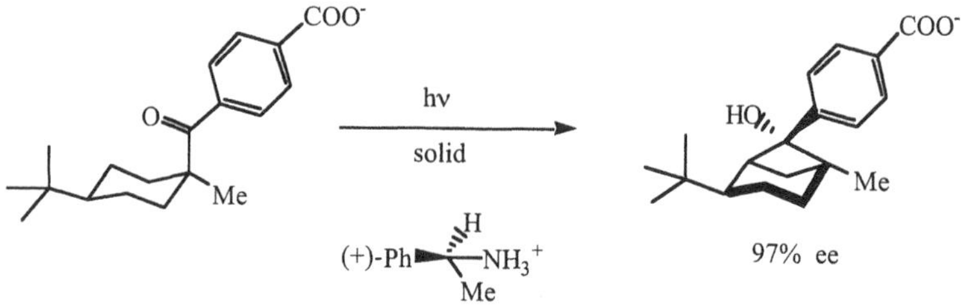

Scheme 31

When a polycrystalline sample of *N,N*-diisopropylphenylgloxylamide was irradiated, 3-hydroxyazetidin-2-one was obtained as the major product but still with no enantioselectivity. However, irradiation of a 5–10 mg single crystal of *N,N*-diisopropylphenylgloxylamide gave optically enriched 3-hydroxyazetidin-2-one (e.e. 93%). Single crystals of *N,N*-diisopropylphenylgloxylamide of (+) or (−) rotation could be separated and individual irradiation of the two gave opposite enantiomers of 3-hydroxyazetidin-2-one.

18.7.1.2. The Ionic Chiral Auxiliary Approach

This approach involves formation of salts most commonly between prochiral carboxylic acid-containing photoreactants and optically pure nonabsorbing amines. While the carboxylic acids are achiral the salts are chiral and these salts, which necessarily crystallize in chiral space groups, provide the asymmetric media in which to carry out the reactions. Clearly the opposite approach, namely forming a salt using a reactive achiral amine and a chiral acid, is also valid.

Irradiation of achiral cyclohexyl phenyl ketone (Sch. 31) in solution yields a racemic cyclobutanol [168]. The acid derivative of this ketone was used as the substrate and pseudoephedrine was used as the chiral amine. Irradiation of the crystalline salt gave the cyclobutanol in 99.5% e.e! This ketone contains two prochiral hydrogen atoms (Sch. 32). While in solution both hydrogen atoms are abstracted with equal efficiency, in the crystal there is a clear preference for one. The two main effects of the medium on this reaction are: (a) ordering of the reactants in a single homochiral conformation that permits abstraction of only one of the two possible enantiotopic hydrogens (Sch. 32) and (b) prevention of the rotation about the C_1–C_2 bond by the medium even when the parent crystal lattice disintegrates as the reaction progresses. The extent of e.e. obtained varies with the chiral auxiliary and conversion. The ionic chiral auxiliary approach

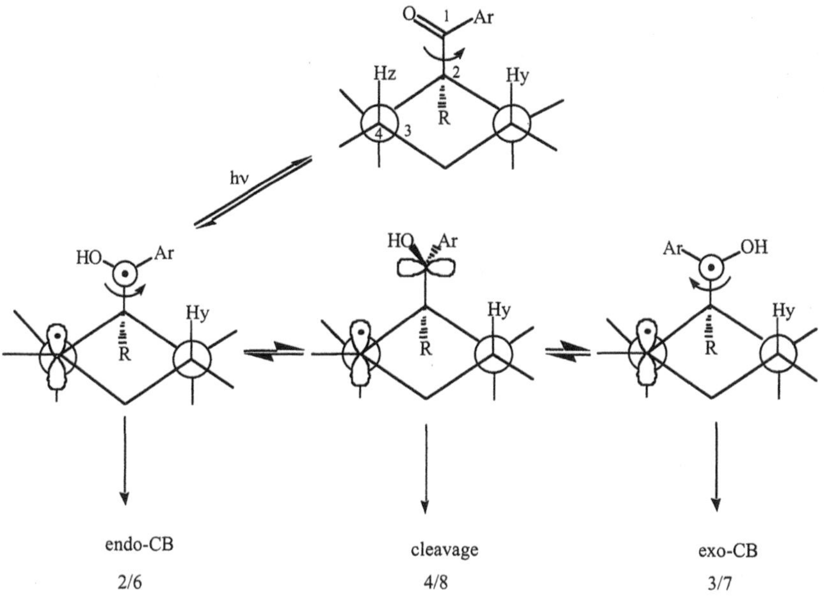

Scheme 32

has been shown to be general by examining several systems that undergo the Norrish Yang and other reactions (Schs. 33 and 34) [161–163,169–174].

One of the limitations of the ionic chiral auxiliary approach is that the reactants must contain acidic or basic functional groups that allow salt formation. Also the approach is still empirical since it is not yet obvious which chiral amine or acid will give the best result.

18.7.2. Achiral Molecules within Chiral Hosts

There are many host systems that are inherently chiral or attain chirality upon crystallization. When achiral molecules are included in these chiral hosts induced chirality may result. If this occurs the induced chirality of the guest reactant can be preserved when it is converted to a product [175–178]. This has been shown to be possible with several systems. Once again, the reactants are nonchiral and give racemic products in isotropic solution. Structures of a few chiral molecules that serve as hosts are listed in Sch. 35. The following examples in which 1,6-*bis*(*o*-chlorophenyl)-1,6-diphenyl-2,4-dyine-1,6-diol and cyclodextrin serves as the host illustrate the point.

Consider the intramolecular γ-hydrogen abstraction in α-oxoamides that we discussed in Sch. 30 [179–181]. One approach to obtain an enantiomerically enriched lactam is to irradiate chiral crystals of

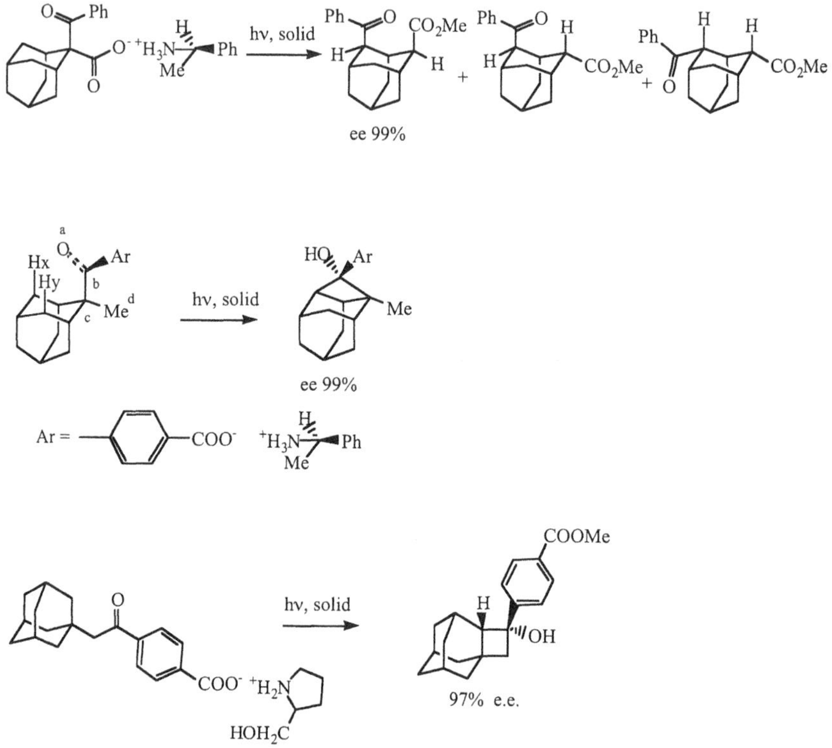

Scheme 33

α-oxoamides. However, not all α-oxoamides give chiral crystals. A more general approach for formation of chiral lactams involves use of optically pure 1,6-*bis*(*o*-chlorophenyl)-1,6-diphenyl-2,4-dyine-1,6-diol or some other resolved clathrate forming compound as a chiral host. For example, when a host-guest complex of 1,6-*bis*(*o*-chlorophenyl)-1,6-diphenyl-2,4-dyne-1,6-diol and *N*,*N*-dimethyl α-oxobenzenacetamide was irradiated the corresponding lactam was formed in 100% e.e. This result is remarkable when one recognizes that irradiation in solution leads to a racemic mixture. The X-ray crystal structure of the complex of *N*,*N*-dimethyl α-oxobenzenacetamide with 1,6-*bis*(*o*-chlorophenyl)-1,6-diphenyl-2,4-dyne-1,6-diol indicates the inclusion of a single conformer of the guest and the enantioselectivity is controlled by the conformation about the O=C–C=O single bond (Sch. 36). Hydrogen bonds between the OH group of the host and C=O groups of the guest hold the guest in a fixed conformation and prevent free rotation of the O=C–C=O single bond. In solution, free rotation of this bond results in formation of a racemic mixture.

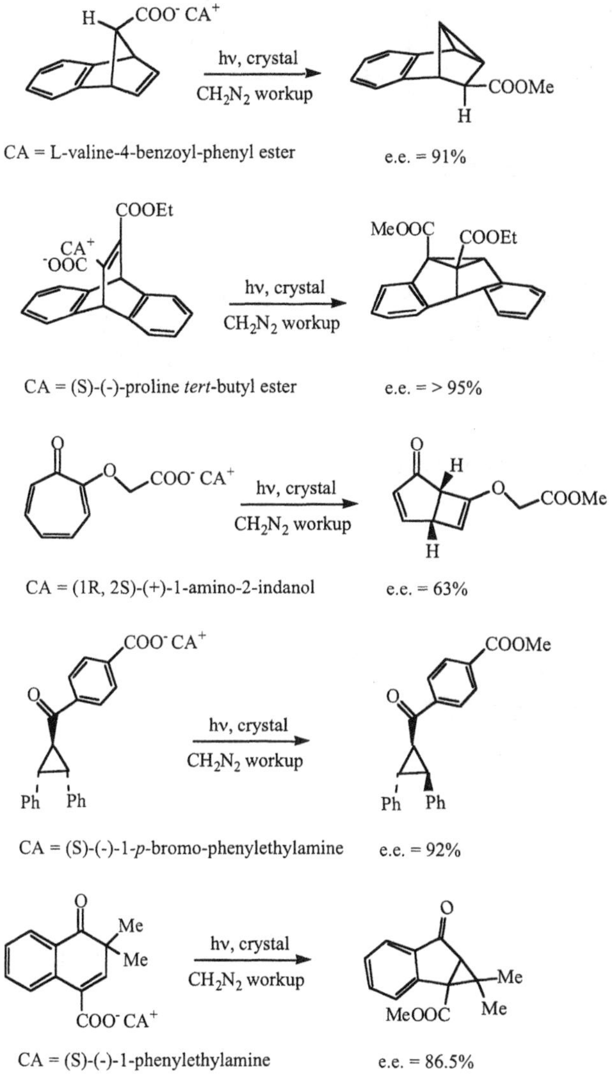

CA = L-valine-4-benzoyl-phenyl ester   e.e. = 91%

CA = (S)-(-)-proline *tert*-butyl ester   e.e. = > 95%

CA = (1R, 2S)-(+)-1-amino-2-indanol   e.e. = 63%

CA = (S)-(-)-1-*p*-bromo-phenylethylamine   e.e. = 92%

CA = (S)-(-)-1-phenylethylamine   e.e. = 86.5%

Scheme 34

The next example deals with the enantioselective photocyclization of α-tropolone alkyl ethers [182]. Cyclization occurs through an allowed 4e⁻-disrotation that can result in opposite optical isomers depending on the direction of rotation (Sch. 37). Racemic products are obtained in solution, but irradiation of crystalline inclusion complexes of α-tropolone alkyl

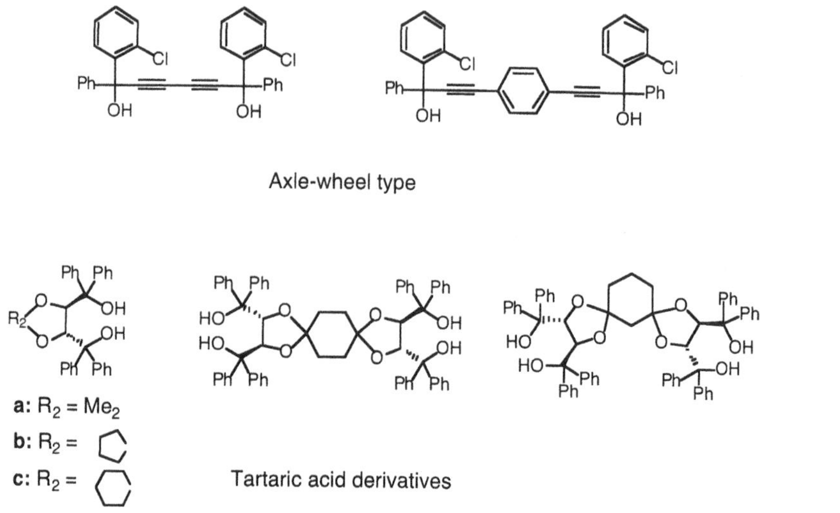

Axle-wheel type

a: $R_2 = Me_2$

b: $R_2 = $

c: $R_2 = $

Tartaric acid derivatives

Scheme 35

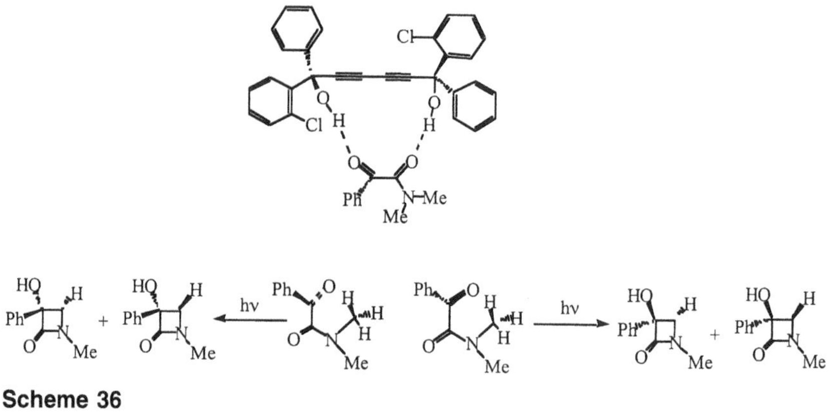

Scheme 36

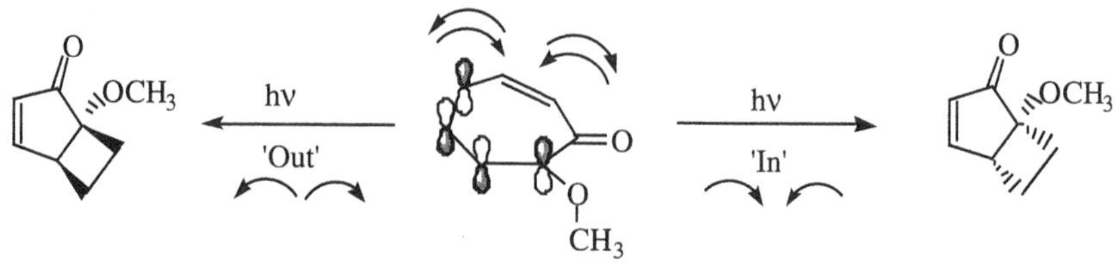

Scheme 37

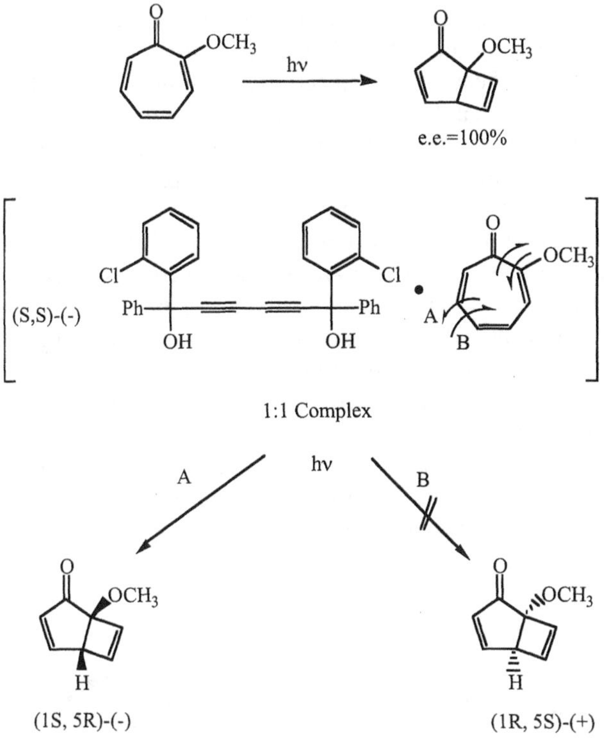

Scheme 38

ethers with 1,6-*bis*(*o*-chlorophenyl)-1,6-diphenyl-2,4-dyne-1,6-diol gave cyclic products of 100% optical purity. The X-ray crystal structure of the 1:1 α-tropolone ethyl ether complex with 1,6-*bis*(*o*-chlorophenyl)-1,6-diphenyl-2,4-dyne-1,6-diol shows that the guest molecule is hydrogen bonded to two host molecules. The enantiomeric control results from the chiral environment provided by the host and from the differences in the space available for rotation of the planar molecule to the product at both sides (Sch. 38).

The success of the above host-guest approach is limited to guests that can form crystalline complexes with the host without disturbing the host's macro-structure. Another disadvantage is the necessity of separating the host from the reaction mixture once photolysis is complete. The reactivity of molecules in the crystalline state and in solid host-guest assemblies is controlled by the details of molecular packing. Currently, molecular packing and consequently the chemical reactivity in the crystalline state, cannot be reliably predicted. Therefore even after successfully crystallizing a molecule in a chiral space group or complexing a molecule with a chiral host or a chiral auxiliary, there is no guarantee that the guest will react in the crystalline state. Hence even though crystalline and host-guest assemblies have been very useful in conducting enantioselective photoreactions, their general applicability thus far has been limited. In spite of this limitation the above examples have shown that very high enantioselectivity can be obtained in a photoreaction once the reactant is preorganized.

Cyclodextrins, one of the most commonly used host systems both in thermal and photochemical studies are inexpensive and possess hydrophobic chiral cavities which in aqueous solution can include a variety of organic molecules. Cyclodextrin-guest solid state complexes can be prepared by precipitating from saturated cyclodextrin-water solutions containing organic molecule. Irradiation of tropolone methyl ether-α-cyclodextrin solid complexes give the photocyclization product (Sch. 37) in 28% e.e [183]. Value of e.e. upto 17% is obtained during the intramolecular *meta*-photocycloaddition of phenoxyalkenes and e.e upto 60% is obtained during photocyclization of *N*-methyl pyridone (Sch. 39) [184,185]. Although, inclusion complexes of cyclodextrin and achiral guest molecules yield optically active products with

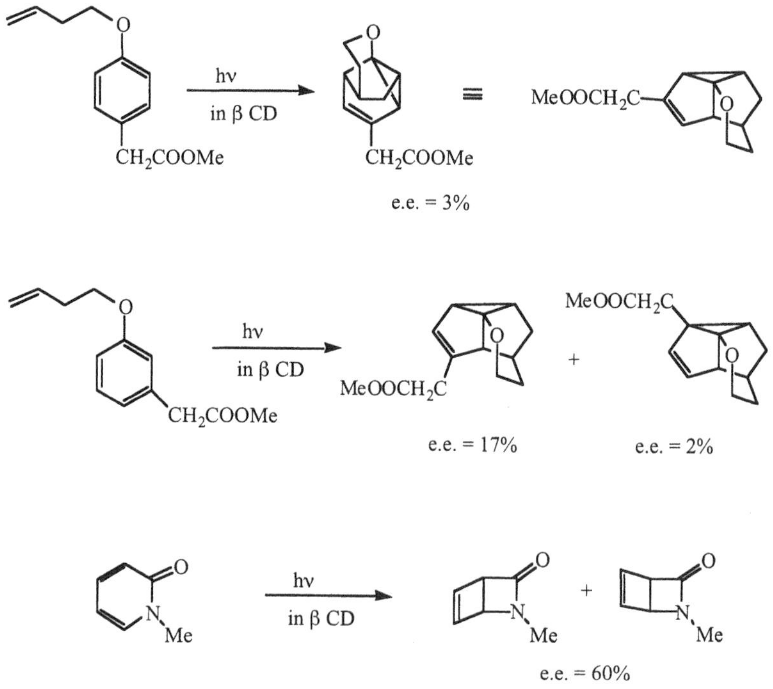

Scheme 39

low to moderate e.e. it is important to recognize that the same molecules yield only racemic products in isotropic solution.

18.7.3. Enantioselective Photoreactions within Zeolites

An ideal approach to achieving chiral induction in a constrained medium such as zeolite would be to make use of a chiral medium. No zeolite that can accommodate organic molecules, currently exists in a stable chiral form. Though zeolite beta and titanosilicate ETS-10 have unstable chiral polymorphs, no pure enantiomorphous forms have been isolated. Although many other zeolites can, theoretically, exist in chiral forms (e.g., ZSM-5 and ZSM-11) none has been isolated in such a state. In the absence of readily available chiral zeolites, one is left with the choice of creating an asymmetric environment within zeolites by the adsorption of chiral organic molecules.

In order to provide the asymmetric environment lacking in zeolites during the reaction a chiral source had to be employed. For this purpose, in the approach known as the chiral inductor method (CIM), where optically pure chiral inductors such as ephedrine were used, the nonchiral surface of the zeolite becomes "locally chiral" in the presence of a chiral inductor [186–188]. This simple method affords easy isolation of the product as the chiral inductor and the reactant is not connected through either a covalent or an ionic bond.

The chiral inductor that is used to modify the zeolite interior will determine the magnitude of the enantioselectivity of the photoproduct. The suitability of a chiral inductor for a particular study depends on its inertness under the given photochemical condition, its shape, size (in relation to that of the reactant molecule and the free volume of the zeolite cavity) and the nature of the interaction(s) that will develop between the chiral agent and the reactant molecule/transition state/reactive intermediate. One should recognize that no single chiral agent might be ideal for two different reactions or at times structurally differing substrates undergoing the same reaction. These are inherent problems of chiral chemistry.

To examine the viability of CIM a number of photoreactions (electrocyclic reactions, Zimmerman (di-π) reaction, oxa-di-π-methane rearrangement, Yang cyclization, geometric isomerization of 1,2-diphenyl-cyclopropane derivatives, and Schenk-ene reaction) which yield racemic products even in presence of chiral inductors in solution have been explored (Sch. 40) [187,189–200]. Highly encouraging enantiomeric excesses (ee) on two photoreactions within NaY have been obtained: photocyclization of tropolone ethylphenyl ether (Eq. (1), Sch. 40) and Yang cyclization of phenyl benzonorbornyl ketone (Eq. (3), Sch. 40). The ability of zeolites to drive a photoreaction that gives racemic products in solution to ee >60% provides

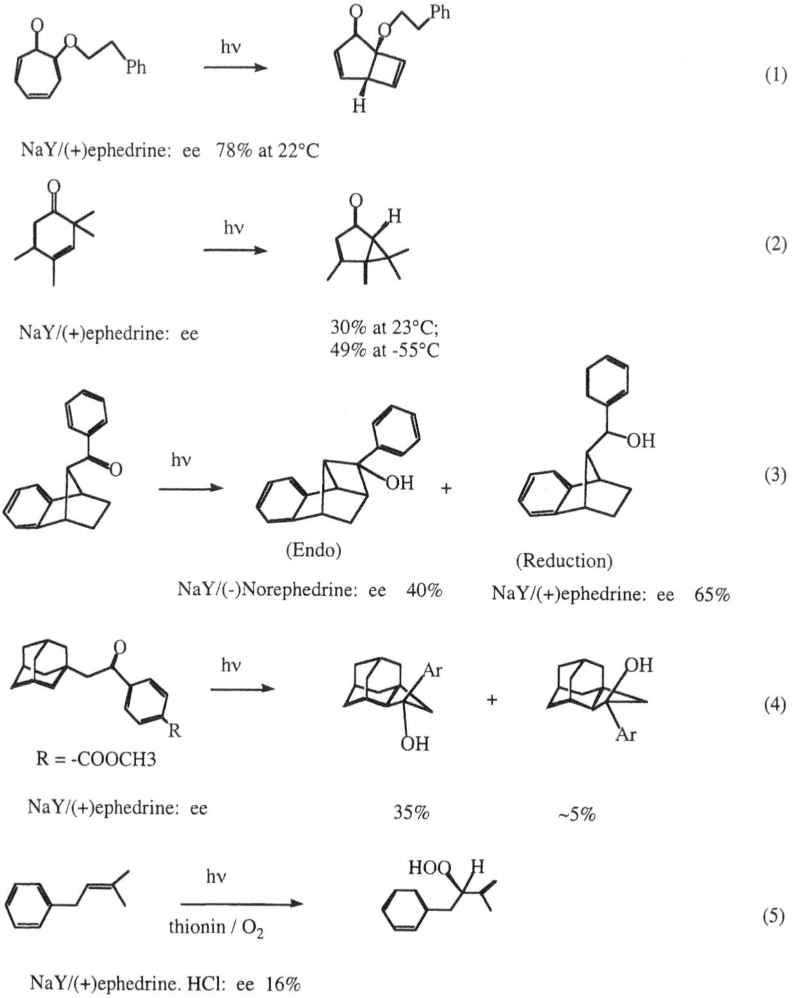
hν
Ph
(1)
NaY/(+)ephedrine: ee 78% at 22°C
hν
(2)
NaY/(+)ephedrine: ee
30% at 23°C;
49% at -55°C
hν
OH
+
(3)
(Endo)
(Reduction)
NaY/(-)Norephedrine: ee 40%
NaY/(+)ephedrine: ee 65%
hν
R
R = -COOCH3
Ar
OH
+
Ar
(4)
NaY/(+)ephedrine: ee
35%
~5%
hν
thionin / O_2
HOO H
(5)
NaY/(+)ephedrine. HCl: ee 16%

Scheme 40

hope of identifying conditions necessary to achieve high ee for a number of photoreactions with zeolite as a reaction medium. The following generalizations have resulted from the above studies: (a) moderate but encouraging ee (15–70%) can be obtained in zeolites for systems that only result in racemic products in solution; (b) not all chiral inductors work well within a zeolite. Best results are obtained with ephedrine, norephedrine and pseudoephedrine; (c) the extent of ee obtained is inversely related to the water content of the zeolite; and (d) the ee depends on the nature of the

(I) Chiral Inductor Method (CIM)

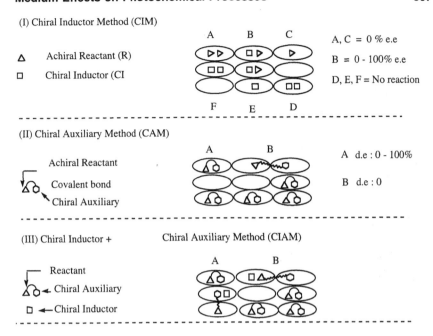

Scheme 41

alkali cation present in a zeolite. For example, the ee on photocyclization of tropolone ethylphenyl ether, within (+)-ephedrine adsorbed, various cation exchanged zeolites are as follows: LiY: 22%; NaY: 68%; KY: 11%; and RbY: 2%.

The strategy of employing chirally modified zeolites as a reaction medium requiring the inclusion of two different molecules, a chiral inductor (CI) and a reactant (R), within the interior space of an achiral zeolite, by its very nature does not allow quantitative asymmetric induction. The expected six possible statistical distribution of the two different molecules CI and R when included within zeolites X and Y shown in Sch. 41-I are: cages containing two R molecules (type A), one R and one CI (type B), single R (type C), two CI (type D), a single CI (type E), and no CI and R molecules (type F). The products obtained from the photoreaction of R represent the sum of reactions that occur in cages of types A, B, and C, of which, B alone leads to asymmetric induction.

Obtaining high asymmetric induction therefore requires the placement of every reactant molecule next to a chiral inductor molecule (type B situation); i.e., enhancement of the ratio of type B cages to the sum of types A and C. In the chiral auxiliary method (CAM) the chiral perturber is connected to the reactant via a covalent bond [186,187,197–205]. In this

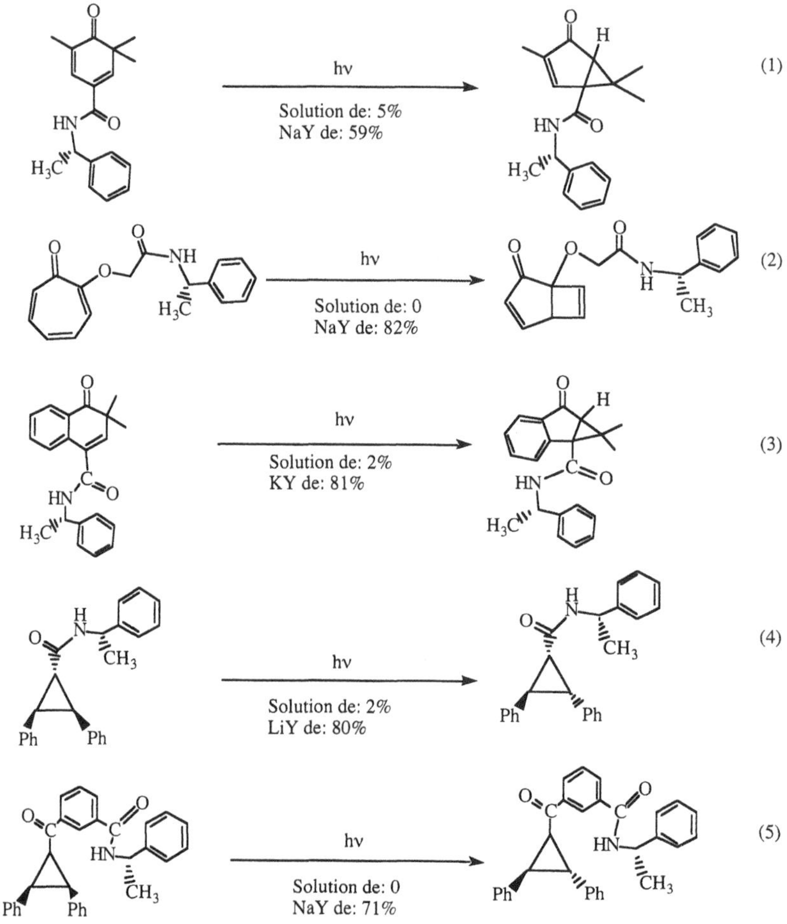

Scheme 42

approach, most cages are expected to contain both the reactant as well as the chiral inductor components within the same cage. The CAM has been tested with several reactions (electrocyclic reactions, oxa-di-π-methane rearrangement, Yang cyclization, and geometric isomerization of 1,2-diphenylcyclopropanes; for selected examples see Sch. 42) and have found that the diastereomeric excesses (de) obtained within zeolites are far superior to that in solution; de >75% have been obtained within MY zeolites for several systems which yield photoproducts in 1:1 diastereomeric ratio in solution. The observed generality suggests the phenomenon responsible for the enhanced asymmetric induction within zeolites to be independent of the reaction.

It is possible that the reactant and covalently linked chiral inductor still remain in different cages (type B in Sch. 41-II) by adopting an extended conformation that could result in <100% de. An asymmetric environment has been provided to such molecules by using a chirally modified Y zeolite as the reaction medium (Sch. 2-III; CIAM). Within (−)-ephedrine modified NaY the de with TMBE increased from 53 to 90% while it decreased from 59 to 3% in the case of 1-phenylethylamide of 2,6,6-trimethylcyclohexa-2,4-diene-1-one-4-carboxylic acid (Eq. (1) Sch. 42). Thus the combination of the chiral inductor and the chiral auxiliary has led to a limited success [195,198]. However, the 90% de obtained with TMBE within (−)-ephedrine modified NaY is the highest thus far reported for any photochemical reaction in a noncrystalline medium. As shown in Fig. 29 the 20% decrease in the maximum de obtained with (+)-ephedrine from that to its antipode (90% in (−)-ephedrine and 70% in (+)-ephedrine) suggests the reactions to occur in two types of cages, one that contains TMBE alone and the second that contains TMBE and a chiral inductor (type A and type B respectively in Sch. 41-III.

One of the drawbacks of the use of zeolite as a reaction medium is the difficulty in controlling the distribution of reactants and chiral inductors as illustrated in Sch. 41-I. This problem could be overcome by localizing the photoreaction to those cages in which the reactant is next to a chiral inductor (type B in Sch. 41-I). This concept has been explored with the photoreduction of ketones by amines as a probe reaction [206]. The ketone that has been examined is phenyl cyclohexyl ketone (Sch. 43). This ketone upon excitation in solution gives an intramolecular γ-hydrogen abstraction Norrish type II product. However, when included within a chirally (ephedrine, pseudoephedrine or norephedrine) modified zeolite, it gave the intermolecular reduction product, α-cyclohexyl benzyl alcohol. The ratio of the intermolecular reduction to Norrish type II product was dependent on the nature (primary, secondary or tertiary) and amount of the chiral amine. These observations are indicative of the reduction occurring only in cages that contain a chiral

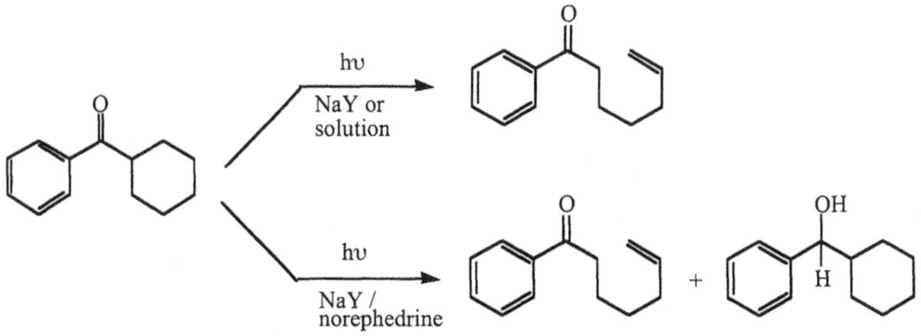

Scheme 43

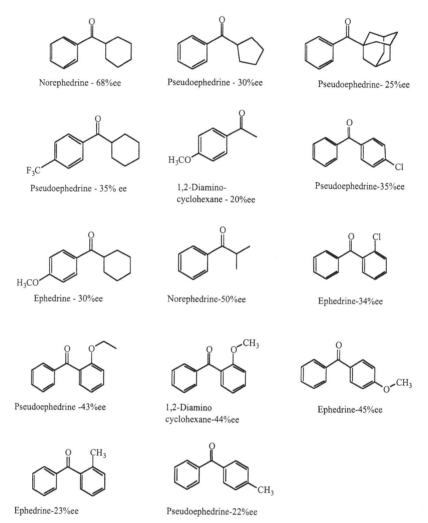

Norephedrine - 68%ee   Pseudoephedrine - 30%ee   Pseudoephedrine- 25%ee

Pseudoephedrine - 35% ee   1,2-Diamino-   Pseudoephedrine-35%ee
                              cyclohexane - 20%ee

Ephedrine - 30%ee   Norephedrine-50%ee   Ephedrine-34%ee

Pseudoephedrine -43%ee   1,2-Diamino   Ephedrine-45%ee
                              cyclohexane-44%ee

Ephedrine-23%ee   Pseudoephedrine-22%ee

Scheme 44

inductor. Using norephedrine as the chiral inductor the ee obtained on the
reduction product is 68%. As expected, the enhanced isomer is reversed with
the antipode of the chiral inductor. It is important to note that under similar
conditions in solution no ee was obtained in the reduction product. The
strategy presented above with phenyl cyclohexyl ketone has been established
to be general by investigating a number of aryl alkyl and diaryl ketones
(Sch. 44). Despite the high ee obtained by this approach where entire reaction
occurs within chirally modified cages, the %ee is not quantitative.

18.8. SUMMARY

When planning to employ an organized medium to control a photochemical reaction one should consider the "medium" in the same manner as any other reagent. Just as no single reagent can perform all transformations–reduction, oxidation, cycloaddition etc., no single organized/confined medium will be useful under all conditions. The choice of an appropriate medium to achieve a desired selectivity must be made on the basis of the reaction and the reactants. Since unifying principles to aid this choice are not yet available, one must rely on intuition, logic and the literature as the guiding factors in achieving selectivity with constrained and organized media.

ACKNOWLEDGMENTS

VR thanks the National Science Foundation for support of the research (CHE-9904187 and CHE-0212042). Reaction cavity concept presented in this review was originally developed by M. D. Cohen for reactions of organic crystals and fruitful collaboration between R. G. Weiss and G. S. Hammond and VR resulted in the generalization of the concept to organized media. VR thanks R. G. Weiss, N. J. Turro, R. S. H. Liu, and J. R. Scheffer for sharing their ideas on photoreactions in organized media.

REFERENCES

1. Kalyanasumdaram K, ed. Photochemistry in Microheterogeneous Systems. New York: Academic Press, 1987.
2. Ramamurthy V, ed. Photochemistry in Organized & Constrained Media. New York: VCH, 1991.
3. Weiss RG. Tetrahedron 1988; 44:3413–3475.
4. Weiss RG, Treanor RL, Nunez A. Pure & Appl Chem 1988; 60:999–1008.
5. Turro NJ, Barton KJ, Tomalia DA. Acc Chem Res 1991; 24:332–340.
6. Gould IR, Turro NJ, Zimmt MB. In: Bethell D, ed. Advances in Physical Organic Chemistry. Vol. 20. London: Academic Press, 1984:1–53.
7. Thomas JK. Acc Chem Res 1977; 10:133–138.
8. Whitten DG, Hopf FR, Quina FH, Sprintschnik G, Sprintschnik HW. Pure & Appl Chem 1977; 49:379–388.
9. Whitten DG, Eaker DW, Horsey BE, Schmehl RH, Worsham PR. Ber Bunsenges Phys Chem 1978; 82:858–867.
10. Whitten DG, Russell JC, Schmehl RH. Tetrahedron 1982; 18:2455–2487.
11. Turro NJ, Gratzel M, Braun AM. Angew Chem Int Ed Engl 1980; 19:675–696.

12. Thomas JK Chem Rev. 1980; 80:283–299.
13. Whitten DG. Acc Chem Res 1993; 26:502–509.
14. Whitten DG. Angew Chem Int Ed Engl 1979; 18:440–450.
15. Devanathan S, Syamala MS, Ramamurthy V. Proc Ind Acad Sci 1987; 98: 391–407.
16. Syamala MS, Reddy GD, Rao BN, Ramamurthy V. Curr Sci 1986; 55:875–886.
17. Balzani V. Tetrahedron 1992; 48:10443–10514.
18. Takahashi K. Chem Rev 1998; 98:2013–2033.
19. Bortolus P, Monti S. In: von Bunau G, ed. Advances in Photochemistry. Vol. 21. New York: John Wiley, 1996:1–133.
20. Turro NJ. Chem Commun 2002; 2279–2292.
21. Turro NJ, Garcia M-G, In: Ramamurthy V, ed. Photochemistry in Organized & Constrained Media. New York: VCH, 1991:1–38.
22. Ramamurthy V. Tetrahedron 1986; 42:5753–5839.
23. Turro NJ. Pure & Appl Chem 1995; 67:199–208.
24. Mayo PD. Pure & Appl Chem 1982; 54:1623–1632.
25. Turro NJ. Pure & Appl Chem 1986; 58:1219–1228.
26. Kumar CV, Raju BB. In: Ramamurthy V, Schanze KS, eds. Molecular and Supramolecular Photochemistry. Vol. 8. New York: Marcel Dekker, Inc., 2001: 505–576.
27. Takagi K, Shichi, T. In: Ramamurthy V, Schanze KS, eds. Molecular and Supramolecular Photochemistry. Vol. 5. New York: Marcel Dekker, Inc., 2000: 31–110.
28. Turro NJ, Buchachenko AL, Tarasov VF. Acc Chem Res 1995; 28:69–80.
29. Turro NJ. Acc Chem Res 2000; 33:637–646.
30. Ramamurthy V, Robbins RJ, Thomas KJ, Lakshminarasimhan PH. In: Whitesell JK, ed. Organised Molecular Assemblies in the Solid State. New York: John Wiley, 1999:63–140.
31. Joy A, Warrier M, Ramamurthy V. Spectrum 1999; 12:1–8.
32. Ramamurthy V. Chimia 1992; 46:359–376.
33. Vasenkov S, Frei H. In: Ramamurthy V, Schanze K, eds. Molecular and Supramolecular Photochemistry. Vol. 5. New York: Marcel Dekker, Inc., 2000: 299–327.
34. Yoon KB. Chem Rev 1993; 93:321–339.
35. Thomas JK. Chem Rev 1993; 93:301–320.
36. Schmidt GMJ. In Reactivity of the Photoexcited Organic Molecule Proc Conf Chem 13th. New York: Wiley Interscience, 1967:227–284.
37. Irie, M. In: Schanze KS, ed. Molecular and Supramolecular Photochemistry. Vol. 5. New York: Marcel Dekker, Inc., 2000:111–141.
38. Shklover VE, Timofeeva TV. Russ Chem Rev 1985; 54:619–644.
39. Hollingsworth MD, McBride JM. Adv Photochem 1990; 15:279–379.
40. McBride JM. Acc Chem Res 1983; New York: John Wiley, 16:304–312.
41. Green BS, Yellin AR, Cohen MD. In Topics in Stereochemistry. Vol. 16. 1986:131–218.
42. Cohen MD. Mol Cryst Liq Cryst 1979; 50:1–10.

43. Theocharis CR. In: Patai S, Rappoport Z, eds. The Chemistry of Enones. New York: John Wiley, 1989:1133–1176.
44. Thomas JM, Morsi SE, Desvergne JP. In: Bethell H, ed. Advances in Physical Organic Chemistry. London: Academic press, 1977:63–151.
45. Ramamurthy V, Venkatesan K. Chem Rev 1987; 87:433–481.
46. Toda F. Crys Eng Commun 2002; 4:215–222.
47. MacGillivray LR. Crys Eng Commun 2002, 4:37.
48. Keating AE, Garcia-Garibay MA. In: Ramamurthy V, Schanze KS, eds. Molecular and Supramolecular Photochemistry. New York: Marcel Dekker, Inc., 1998:195–248.
49. Garcia-Garibay MA, Constable AE, Jernelius J, Choi T, Cimeciyan D, Shin SH. In Physical Supramolecular Chemistry. Vol. 485. NATO ASI Series, 1996:289–312.
50. Ihmels H, Scheffer JR. Tetrahedron 1999; 55:885–907.
51. Scheffer JR, Garcia-Garibay M, Nalamasu O. Org Photochem 1987; 8:249–347.
52. Scheffer JR. Acc Chem Res 1980; 13:283–290.
53. Ito Y. In: Matsuura T, ed. Photochemistry on Solid Surfaces. Vol. 47. Amsterdam: Elsevier, 1989:469–480.
54. Ito Y. In: Ramamurthy V, Schanze K, eds. Molecular and Supramolecular Photochemistry. Vol. 3. New York: Marcel Dekker, Inc., 1999:1–70.
55. Ito Y. Synthesis 1998; 1–32.
56. Ogawa M, Kuroda K. Chem Rev 1995; 95:399–438.
57. Yoon KB. In: Ramamurthy V, Schanze K, eds. Molecular and Supramolecular Photochemistry. Vol. 5. New York: Marcel Dekker, Inc., 2000:147–255.
58. Frei H, Blatter F, Sun H. Chem Tech 1996; 26:24–30.
59. Ramamurthy V, Lakshminarasimhan PH, Grey CP, Johnston LJ. Chem Commun 1998; 2411–2424.
60. Ramamurthy VJ. Photochem Photobiol C 2000; 1:145–166.
61. Scaiano JC, Garcia H. Acc Chem Res 1999; 32:783–793.
62. Hashimoto S. In: Ramamurthy V, Schanze K, eds. Molecular and Supramolecular Photochemistry. Vol. 5. New York: Marcel Dekker, Inc., 2000: 253–294.
63. Tung C-H, Wu L-Z, Zhang L-P, Chen B. Acc Chem Res 2003; 36:39–47.
64. Calzaferri G, Maas H, Pauchard M, Pfeenniger M, Megelski S, Devaux, A. In: Neckers DC, Bunau G. v, Jenks WS, eds. Advances in Photochemistry. Vol. 27. New York: John Wiley, 2002.
65. Hashimoto SJ. Photochem Photobiol C 2003; 4:19–49.
66. Clennan EL. In: Ramamurthy V, Schanze K, eds. Photochemistry of Organic Molecules in Isotropic and Anisotropic Media. Vol. 9. New York: Marcel Dekker, Inc., 2003:275–308.
67. Sivaguru J, Shailaja J, Ramamurthy V. In: Dutta PK, ed. Handbook of Zeolite Science and Technology. New York: Marcel Dekker, Inc., 2003:515–589.
68. Yoon KB. In: Auerbach SM, Carrado KA, Dutta PK, eds. Handbook of Zeolite Science and Technology. New York: Marcel Dekker, Inc., 2003: 591–720.

69. Maas H, Huber S, Khatyr A, Pfenniger M, Meyer M, Calzaferri, G. In: Ramamurthy V, Schanze K, eds. Photochemistry of Organic Molecules in Isotropic and Anisotropic Media. Vol. 9. New York: Marcel Dekker, Inc., 2003: 309–351.
70. Garcia H, Roth HD. Chem Rev 2002; 102:3947–4008.
71. Gutsche DC, ed. Calixarenes. Royal Society of Chemistry, 1989.
72. Cram DJ, Cram JM, eds. Container Molecules and Their Guests. Royal Society of Chemistry, 1993.
73. Szejtli J, Osa T, eds. Cyclodextrins. Vol. 3. New York: Pergamon, 1996.
74. Balzani V, Scandola F, eds. Supramolecular Photochemistry. Ellis Horwood, 1991.
75. Vogtle F, ed. Supramolecular Chemistry. New York: John Wiley & Sons, 1991.
76. Lehn J-M. Supramolecular Chemistry. New York: VCH, 1995.
77. Schneider, H-J, Durr H, eds. Frontiers in Supramolecular Organic Chemistry and Photochemistry. New York: VCH, 1991.
78. Auerbach SM, Carrado KA, Dutta PK, eds. Handbook of Zeolite Science and Technology. New York: Marcel Dekker, 2003.
79. Bekkum HV, Flanigen EM, Jansen JC, eds. Introduction to Zeolite Science and Practice. Vol. 58. New York: Elsevier, 1991.
80. Breck DW. Zeolite Molecular Sieves. Malabar: Robert E. Krieger, 1974.
81. Anpo M, ed. Surface Photochemistry. New York: John Wiley, 1996.
82. Anpo M, Matsuura T, eds. Photochemistry on Solid Surfaces. Amsterdam: Elsevier, Vol. 47. 1989.
83. Desiraju GR, ed. Organic Solid State Chemistry. Elsevier, 1987.
84. Toda F, ed. Organic Solid-State Reactions. Kluwer Academic, 2002.
85. Iler RK. The Chemistry of Silica Solubility. Polymerization, Colloid and Surface Properties, and Biochemistry. New York: John Wiley, 1979.
86. Alberti G, Bein T, eds. Solid-state Supramolecular Chemistry: Two- and Three-dimensional Inorganic Networks. Vol. 7. New York: Pergamon, 1996.
87. Fendler JH, ed. Membrane Mimetic Chemistry. New York: John Wiley, 1982.
88. Fendler JH, Fendler EJ, eds. Catalysis in Micellar and Macromolecular Systems. New York: Academic Press, 1975.
89. Gartner W. Angew Chem Int Ed Engl 2001; 40:2977–2981.
90. Pepe IM. J Photochem Photobiol B 1999; 48:1–10.
91. Shichida Y, Imai H. Cell Mol Life Sci 1998; 54:1299–1315.
92. Borhan B, Souto Ml, Imai H, Shichida Y, Nakanishi K. Science 2000; 288:2209–2212.
93. Palczewski K, Kumasaka T, Hori T, Behnke CA, Motoshima H, Fox BA, Trong IL, Teller DC, Okada T, Stenkamp RE, Yamamoto M, Miyano M. Science 2000; 289:739–745.
94. Chosrowjan H, Mataga N, Shibata Y, Tachibanaki S, Kandori H, Shichida Y, Okada T, Kouyama TJ. Am Chem Soc 1998; 120:9706–9707.
95. Kandori H, Katsuta Y, Ito M, Sasabe HJ. Am Chem Soc 1995; 117: 2669–2670.
96. Kandori H, Sasabe H, Nakanishi K, Yoshizawa T, Mizukami T, Shichida YJ. Am Chem Soc 1996; 118:1002–1005.

97. Nakanishi K, Chen, A-H, Derguini F, Franklin P, Hu S, Wang J. Pure & Appl Chem 1994; 66:981–988.
98. Shichi, H. From Biochemistry of Vision. New York: Academic Press, 1983:73–90.
99. Cohen MD, Schmidt GMJ. J Chem Soc 1964; 383:1996–2000.
100. Cohen MD, Schmidt GMJ, Sonntag FI. J Chem Soc 1964; 384:2000–2013.
101. Schmidt GMJ, et al. In: Ginsburg D, ed. Solid State Photochemistry. New York: Verlag Chemie, 1976.
102. Macgillivray LR, Reid JL, Ripmeester JA. J Am Chem Soc 2000; 122:7817–7818.
103. Gu W, Warrier M, Ramamurthy V, Weiss RG. J Am Chem Soc 1999; 121:9467–9468.
104. Pitchumani K, Warrier M, Cui C, Weiss RG, Ramamurthy V. Tetrahedron Lett 1996; 37:6251–6254.
105. Hoffmann U, Gao Y, Pandey B, Klinge S, Warzecha, K-D, Kruger C, Roth HD, Demuth MJ. Am Chem Soc 1993; 115:10358–10359.
106. Joy A, Uppili S, Netherton MR, Scheffer JR, Ramamurthy V. J Am Chem Soc 2000; 122:728–729.
107. Cohen MD. Angew Chem Int Ed Engl 1975; 14:386–393.
108. Weiss RG, Ramamurthy V, Hammond GS. Acc Chem Res 1993; 26: 530–536.
109. Ramamurthy V, Weiss RG, Hammond GS. In: Neckers DC, ed. Advances in Photochemistry. Vol. 18. John Wiley, 1993:67–234.
110. Turro NJ, Cherry WR. J Am Chem Soc 1978; 100:7431–7432.
111. Robbins WK, Eastman RH. J Am Chem Soc 1970; 92:6077–6079.
112. Turro NJ, Weed GC. J Am Chem Soc 1983; 105:1861–1868.
113. Turro NJ, Zimmt MB, Lei XG, Gould IR, Nitsche KS, Cha Y. J Phys Chem 1987; 91:4544–4548.
114. Turro NJ, Chow, M-F, Chung C-J, Kraeutler B. J Am Chem Soc 1981; 103:3886–3891.
115. Turro NJ, Anderson DR, Chow M-F, Chung C-J, Kraeutler B. J Am Chem Soc 1981; 103:3892–3896.
116. Gould IR, Zimmt MB, Turro NJ, Baretz BH, Lehr GF. J Am Chem Soc 1985; 107:4607–4612.
117. Baretz BH, Turro NJ. J Am Chem Soc 1983; 105:1309–1316.
118. Turro NJ, Gould IR, Baretz HB. J Phys Chem 1983; 87:531–532.
119. Lunazzi L, Ingold KU, Scaiano JC. J Phys Chem 1983; 87:529–530.
120. Tsentalovich YP, Fischer H. J Chem Soc, Perkin Trans 2 1994; 729–733.
121. Turro NJ. Proc Natl Acad Sci USA 1983; 80:609–621.
122. de Mayo P, Sydnes LK. Chem Commun 1980; 994–995.
123. Lee, K-H, de Mayo P. Chem Commun 1979; 493–495.
124. Berenjian N, de Mayo P, Sturgeon M-E, Sydnes LK, Weedon AC. Can J Chem 1982; 60:425–436.
125. Devanathan S, Ramamurthy V. J Phy Org Chem 1988; 1:91–102.
126. Gnanaguru K, Ramasubbu N, Venkatesan K, Ramamurthy V. J Photochem 1984; 27:355–362.

127. Bhadbade MM, Murthy GS, Venkatesan K, Ramamurthy V. Chem Phys Lett 1984; 109:259–263.

128. Gnanaguru K, Ramasubbu N, Venkatesan K, Ramamurthy V. J Org Chem 1985; 50:2337–2346.

129. Murthy GS, Arjunan P, Venkatesan K, Ramamurthy V. Tetrahedron 1987; 43:1225–1240.

130. Pitchumani K, Warrier M, Scheffer JR, Ramamurthy, V. Chem Commun 1998; 1197–1198.

131. Pitchumani K, Warrier M, Kaanumalle LS, Ramamurthy V. Tetrahedron 2003; 59:5763.

132. Warrier M, Turro NJ, Ramamurthy V. Tetrahedron Lett 2000; 41:7163–7167.

133. Warrier M, Kaanumalle LS, Ramamurthy V. Can J Chem 2003. *In press.*

134. Rao VJ, Uppili S, Corbin DR, Schwarz S, Lustig SR, Ramamurthy V. J Am Chem Soc 1998; 120:2480–2481.

135. Shailaja J, Lakshminarasimhan PH, Pradhan AR, Sunoj RB, Jockusch S, Karthikeyan S, Uppili S, Chandrasekhar J, Turro NJ, Ramamurthy V. J Phys Chem A 2003; 107:3187–3198.

136. Lakshminarasimhan P, Sunoj RB, Chandrasekhar J, Ramamurthy V. J Am Chem Soc 2000; 122:4815–4816.

137. Pokkuluri PR, Scheffer JR, Trotter J. Tetrahedron Lett 1989; 30:1601–1604.

138. Craig DP, Sarti-Fantoni, P. Chem Commun 1969; 20:742–743.

139. Cohen MD, Ludmer Z, Thomas JM, Williams JO. Proc R Soc London, Ser A. 1971; 324:459–468.

140. He Z, Weiss RG. J Am Chem Soc 1990; 112:5535–5541.

141. Treanor RL, Weiss RG. Tetrahedron 1987; 43:1371–1391.

142. Nunez A, Hammond GS, Weiss RG. J Am Chem Soc 1992; 114:10258–10271.

143. Canon JR, Patrick VA, Raston CL, White AH. Aust J Chem 1978, 31, 1265.

144. Liu RSH. Hammond GS. Chem Eur J 2001; 7:2–10.

145. Liu RSH. Acc Chem Res 2001; 34:555–562.

146. Liu RSH, Browne DT. Acc Chem Res 1986; 19:42–48.

147. Liu RSH, Hammond GS. Proc Natl Acad Sci USA 2000; 97:11153–11158.

148. Muller AM, Lochbrunner S, Schmid WE, Fuβ, W. Angew Chem Int Ed Engl 1998; 37:505–507.

149. Liu RSH, Asato AE. Proc Natl Acad Sci USA 1985; 82:259–263.

150. Kohler BE. Chem Rev 1993; 93:41–54.

151. Rau H. Chem Rev 1983; 83:535–547.

152. Demuth M, Mikhail G. Synthesis 1989; 145–162.

153. Buschman H, Scharf H-D, Hoffmann N, Esser P. Angew Chem Int Ed Engl 1991; 30:477–515.

154. Inoue Y. Chem Rev 1992; 92:741–770.

155. Pete J-P. In: Neckers DC, Volman DH, Von Bunan G, eds. Adv Photochem. Vol. 21. John Wiley, 1996:135–216.

156. Everitt SRL, Inoue Y. In: Ramamurthy V, Schanze K, eds. Molecular and Supramolecular Photochemistry. Vol. 3. New York: Marcel Dekker, Inc., 1999: 71–130.

157. Griesbeck AG, Meierhenrich UJ. Angew Chem Int Ed Engl 2002; 41: 3147–3154.
158. Sakamoto M. Chem Eur J 1997; 3:684–689.
159. Green BS, Lahav M, Rabinovich D. Acc Chem Res 1979; 12:191–197.
160. Addadi L, Lahav M. Pure & Appl Chem 1979; 51:1269–1284.
161. Scheffer JR. Can J Chem 2001; 79:349–357.
162. Leibovitch M, Olovsson G, Scheffer JR, Trotter J. Pure & Appl. Chem 1997; 69:815–823.
163. Gamlin JN, Jones R, Leibovitch M, Patrick B, Scheffer JR, Trotter J. Acc Chem Res 1996; 29:203–209.
164. Evans SV, Garcia-Garibay M, Omkaram N, Scheffer JR, Trotter J, Wireko F. J Am Chem Soc 1986; 108:5648–5650.
165. Toda F, Miyamoto H. J Chem Soc, Perkins Trans 1 1993; 1:1129–1132.
166. Hashizume D, Kogo H, Sekine A, Ohashi Y, Miyamoto H, Toda F. J Chem Soc, Perkin Trans 2 1996; 61–66.
167. Sekine A, Hori K, Ohashi Y, Yagi M, Toda F. J Am Chem Soc 1989; 111: 697–699.
168. Leibovitch M, Olovsson G, Scheffer JR, Trotter J. J Am Chem Soc 1998; 120:12755–12769.
169. Cheung E, Kang T, Netherton MR, Scheffer JR, Trotter J. J Am Chem Soc 2000; 122:11753–11754.
170. Cheung E, Netherton MR, Scheffer JR, Trotter J. J Am Chem Soc 1999; 121:2919–2920.
171. Cheung E, Rademacher K, Scheffer JR, Trotter J. Tetrahedron Lett 1999; 40:8733–8736.
172. Janz KM, Scheffer JR. Tetrahedron Lett 1999; 40:8725–8728.
173. Cheung E, Kang T, Raymond JR, Scheffer JR, Trotter J. Tetrahedron Lett 1999; 40:8729–8732.
174. Cheung E, Netherton MR, Scheffer JR, Trotter J. Tetrahedron Lett 1999; 40:8737–8740.
175. Tanaka K, Toda F. In: Toda F, ed. Organic Photoreactions in the Solid State. New York: Kluwer, 2002:109–157.
176. Toda, F. Acc Chem Res 1995; 28:480–486.
177. Toda F, Tanaka K, Miyamoto, H. In: Ramamurthy V, Schanze KS, eds. Understanding and Manipulating Excited-State Processes. Vol. 8. New York: Marcel Dekker, Inc., 2001:385–425.
178. Toda F. In: Weber E, ed. Mol. Inclusion Mol Recognit-Clathrates 2. Vol. 149. Springer-Verlag, 1988:211–238.
179. Toda F, Tanaka K, Yagi M. Tetrahedron 1987; 43:1495–1502.
180. Toda F, Miyamoto H, Matsukawa R. J Chem Soc, Perkin Trans 1 1992; 1461–1462.
181. Kaftory M, Toda F, Tanaka K, Yagi M. Mol Cryst Liq Cryst 1990; 186: 167–176.
182. Toda F, Tanaka K. J Chem Soc Chem Commun 1986; 1429–1430.
183. Koodanjeri S, Joy A, Ramamurthy V. Tetrahedron 2000; 56:7003–7009.
184. Vizvardi K, Desmet K, Luyten I, Sandra P, Hoornaert G, Eycken EVD. Org Lett 2001; 1173.

185. Shailaja J, Karthikeyan S, Ramamurthy V. Tetrahedron Lett 2002; 43.
186. Sivaguru J, Natarajan A, Kaanumalle LS, Shailaja J, Uppili S, Joy A, Ramamurthy V. Acc Chem Res 2003; 36:509–521.
187. Sivaguru J, Shailaja J, Uppili S, Ponchot K, Joy A, Arunkumar N, Ramamurthy, V. In: Toda F, ed. Organic Solid-State Reactions. Kluwer Academic, 2002:159–188.
188. Joy A, Ramamurthy V. Chem Eur J 2000; 6:1287–1293.
189. Shailaja J, Sivaguru J, Uppili S, Joy A, Ramamurthy V. Microporous and Mesoporous Mater 2001; 48:319.
190. Leibovitch M, Olovsson G, Sundarababu G, Ramamurthy V, Scheffer JR, Trotter J. J Am Chem Soc 1996; 118:1219–1220.
191. Sundarababu G, Leibovitch M, Corbin DR, Scheffer JR, Ramamurthy V. Chem Commun 1996; 2159–2160.
192. Joy A, Robbins R, Pitchumani K, Ramamurthy V. Tetrahedron Lett 1997, 8825.
193. Joy A, Ramamurthy V, Scheffer JR, Corbin DR. Chem Commun 1998; 1379–1380.
194. Joy A, Corbin DR, Ramamurthy V. In: Tracey MMJ, Marcus BK, Bisher BE, Higgings JB, eds. Proceedings of 12th International Zeolite Conference. Warrendale, 1999:2095.
195. Joy A, Uppili S, Netherton MR, Scheffer JR, Ramamurthy V. J Am Chem Soc 2000; 122:728–729.
196. Joy A, Scheffer JR, Ramamurthy V. Org Lett 2000; 2:119–121.
197. Cheung E, Chong KCW, Jayaraman S, Ramamurthy V, Scheffer JR, Trotter J. Org Lett 2000; 2:2801–2804.
198. Uppili S, Ramamurthy V. Org Lett 2002; 4:87–90.
199. Natarajan A, Wang K, Ramamurthy V, Scheffer JR, Patrick B. Org Lett 2002; 4:1443–1446.
200. Natarajan A, Joy A, Kaanumalle LS, Scheffer JR, Ramamurthy V. J Org Chem 2002; 67:8339–8350.
201. Jayaraman S, Uppili S, Natarajan A, Joy A, Chong KCW, Netherton MR, Zenova A, Scheffer JR, Ramamurthy V. Tetrahedron Lett 2000; 41: 8231–8235.
202. Chong KCW, Sivaguru J, Shichi T, Yoshimi Y, Ramamurthy V, Scheffer JR. J Am Chem Soc 2002; 124:2858–2859.
203. Sivaguru J, Scheffer JR, Chandrasekhar J, Ramamurthy V. Chem Commun 2002; 830–831.
204. Kaanumalle LS, Sivaguru J, Sunoj RB, Lakshminarasimhan PH, Chandrasekhar J, Ramamurthy V. J Org Chem 2002; 67:8711–8720.
205. Kaanumalle LS, Sivaguru J, Arunkumar N, Karthikeyan S, Ramamurthy V. Chem Commun 2003; 116–117.
206. Shailaja J, Ponchot KJ, Ramamurthy V. Org Lett 2000; 2:937–940.

Index